Organic Molecular Photophysics

Volume 2

WILEY MONOGRAPHS IN CHEMICAL PHYSICS

Editors

John B. Birks, *Reader in Physics, University of Manchester*

Sean P. McGlynn, *Professor of Chemistry, Louisiana State University*

Photophysics of Aromatic Molecules:
J. B. Birks, Department of Physics, University of Manchester

Atomic & Molecular Radiation Physics: *L. G. Christophorou, Oak Ridge National Laboratory, Tennessee*

The Jahn–Teller Effect in Molecules and Crystals:
R. Englman, Soreq Nuclear Research Centre, Yavne

Organic Molecular Photophysics: Volumes 1 and 2: *edited by J. B. Birks, Department of Physics, University of Manchester*

Internal Rotation in Molecules:
edited by W. J. Orville-Thomas, Department of Chemistry and Applied Chemistry, University of Salford

Organic Molecular Photophysics

Volume 2

Editor

John B. Birks

Reader in Physics,
University of Manchester

A Wiley–Interscience Publication

JOHN WILEY & SONS

London · New York · Toronto · Sydney

Library of Congress Cataloging in Publication Data: (Revised)

Birks, John Betteley.
Organic molecular photophysics.

(Wiley monographs in chemical physics)
Includes bibliographical references.
1. Molecular spectra. 2. Chemistry, Organic. I. Title.
DNLM: 1. Chemistry, Organic. 2. Spectrum analysis.
QC454.M6 B6190.
QC454.M6B55 547′.308′5 72–8594

ISBN 0 471 07421 7 (v.2)

Printed in Great Britain by J. W. Arrowsmith Ltd., Bristol, England

'With a little help from my friends[1] . . .
my friends pictured within[2].'

1. The Beatles
2. Sir Edward Elgar, Enigma Variations

Preface

Organic molecular photophysics, to use a musical analogy, is an intricate set of variations on a common theme, the fate of the excited organic molecule. The variations may be played radiatively or radiationlessly on the singlet or the triplet manifolds, which are linked by intersystem crossing. The scales may be ascending, as in absorption, or descending, as in emission.

Each instrument introduced into the orchestra produces its own variations. The nanosecond light flash and its sharper monochromatic companion, the pulsed laser, enable single notes to be struck instantaneously. Time-resolved spectrometry can then record the transient music which is not heard under normal concert conditions. Reduction of the cacophony has allowed some quiet, delicate tunes to be played. The pure resonances of single vibronic level fluorescence have been plucked like the strings of a harp. The gentle overtones of the higher excited state fluorescences have emerged from their previous obscurity. The feeble strains of the alkane fluorescences have been recorded. There are also many ways to modulate the music. The molecular pitch can be tuned by compression or by deuteration, it can be modified by alkyl or heavy-atom substitution, and its intensity can be changed by varying the excitation pitch. The music can be influenced by the environment: in the vapour phase by the pressure or the carrier gas; in solution by the viscosity, polarity and nature of the solvent; and in the crystal phase by the lattice-structure or by defects.

The theme is orchestrated on the monophotonic, monomolecular scale, but it has many variations on the biphotonic and bimolecular or biexcitonic scales. The biphotonic and multiphotonic scales were originally explored with monochromatic lasers. The recent concert debut of the tunable dye laser, which spans nearly an octave, has added a new biphotonic instrument to the orchestra. The bimolecular or biexcitonic scale is an intricate fugue in which none, one or two of the molecules may be initially excited, and which has homopolar and heteropolar variations. The homopolar fugue includes dimers, excimers, energy migration, self quenching, homofusion and homofission. The corresponding heteropolar fugue includes donor–acceptor complexes, exciplexes, energy transfer, impurity quenching, heterofusion and heterofission. Most of these intermolecular tunes also occur intramolecularly, so that the bimolecular fugues have their counterpoints on the monomolecular scale.

Some of these variations were familiar to the editor when he embarked on the manuscript of *Photophysics of Aromatic Molecules* (*PAM*) in 1967, and he learnt of many others during its composition. The two volumes of *Organic Molecular Photophysics* (*OMP*) grew from the realization that the scores of many variations were incomplete, and that others had still to be recorded. A group of experienced instrumentalists and composers were invited to join the orchestra for this purpose. It seems appropriate that the dedication of *OMP* is musical in origin, and that it includes Elgar's dedication of his *Enigma Variations*, an apt pseudonym for *Organic Molecular Photophysics*.

PAM and the two volumes of *OMP* provide a contemporary rendition of the unfinished symphony, so far as it is known to the present orchestra and conductor. Some passages in the score are open to alternative interpretations. There are many deafening silences, revealing gaps in our current knowledge of the music. The challenge is now to the reader, to the experimentalist to play his instruments, to the theorist to compose or make new transcriptions, in order to hear and interpret more music. Whether the reader contemplates an isolated molecule, or gazes into a crystal, or seeks the solution within a solution, it is hoped that this three-part work will provide an adequate introduction to the theme and its many enigmatic variations.

March 1974

JOHN B. BIRKS
Manchester

Contributing Authors

BEENS, H.	*Chemisch Laboratorium, Free University, Amsterdam, Netherlands (now at Koninklijke/ Shell-Laboratorium, Amsterdam, Netherlands)*
BIRKS, J. B.	*The Schuster Laboratory, University of Manchester, Manchester, U.K.*
CUNDALL, R. B.	*Department of Chemistry, University of Nottingham, Nottingham, U.K.*
DRUGER, S. D.	*Department of Physics, State University of New York, Binghampton, New York, U.S.A. (now at Department of Physics and Astronomy, University of Rochester, Rochester, New York, U.S.A.)*
GEACINTOV, N. E.	*Department of Chemistry and Radiation and Solid State Laboratory, New York University, New York, U.S.A.*
HENRY, B. R.	*Department of Chemistry, University of Manitoba, Winnipeg, Canada*
LIPPERT, E.	*Iwan N. Stranski-Institut für Physikalische und Theoretische Chemie, Berlin West 12, Germany*
OGILVIE, S. MCD.	*Department of Chemistry, University of Nottingham, Nottingham, U.K.*
SIEBRAND, W.	*Division of Chemistry, National Research Council of Canada, Ottawa, Canada*
SWENBERG, C. E.	*Department of Physics and Radiation and Solid State Laboratory, New York University, New York, U.S.A.*

VOLTZ, R. — *Centre de Recherches Nucléaires et Université Louis Pasteur, Laboratoire de Physique des Rayonnements et d'Electronique Nucléaire, Strasbourg, France*

WELLER, A. — *Max-Planck-Institut für biophysikalische Chemie, Göttingen, Germany*

WILKINSON, F. — *School of Chemical Sciences, University of East Anglia, Norwich, U.K.*

Contents

3. Triplet quantum yields and singlet–triplet intersystem crossing

FRANK WILKINSON

4. Excited molecular π-complexes in solution

HENK BEENS and ALBERT WELLER

5. Theory of molecular decay processes

RENÉ VOLTZ

6. Radiationless transitions—a postscript

BRYAN R. HENRY and WILLEM SIEBRAND

7. Theory of charge transport processes in organic molecular solids

STEPHEN D. DRUGER

1 *Laser-spectroscopic studies of reorientation and other relaxation processes in solution*

Ernst Lippert

1.1 Fluorescence and relaxation in solution

1.1.1 *Very fast photochemical reactions*

Until recently it has been assumed[1] that 'the rotational motion of molecules can only be studied by some method which yields the time correlation function of some orientation-dependent molecular parameter'. It turns out that the direct measurement of reorientation and relaxation processes in solution is possible by laser techniques. Recent results, current methods and future possibilities will be described in this chapter. Reorientation and other relaxation processes of molecules in liquids correspond to changes in the distances and angles of the molecular axes of symmetry and inertia due to changes in the forces and energies of inter- and intra-molecular interactions.

Let us consider a dilute solution at temperature T of a polar solute with molecular electric dipole moments μ_g and μ_e in its electronic ground state S_0 and its first excited electronic singlet state S_1, respectively, in a polar solvent of dielectric constant D and refractive index n. In order to investigate directly an orientational relaxation process in such a system, we need an experimental method that produces a rapid change of the effective electrostatic forces and permits the observation of the kinetics of the relaxation process produced thereby. Compared with the relaxation rate, the rate of the initiating process must be very fast; it should approximate to a δ-function. This may be achieved (i) by rapidly switching on and off an external electric field, or (ii) by irradiation with pulses of photons of energy $h\nu = hc_0\tilde{\nu} = \Delta E = E_e - E_g$, where suffixes e and g refer to the energy of the excited and the ground state, respectively, c_0 is the speed of light, h is Planck's constant, ν is the frequency and $\tilde{\nu}$ is the wavenumber of a photon.

The change (i) of an external electric field initiates an orientation relaxation process of the solute that can be followed, for instance, by the time dependence of the absorption of polarized light. The irradiation (ii) with photons $hc_0\tilde{\nu}$ causes $S_1 \leftarrow S_0$ absorption transitions which, for the case $\mu_e \neq \mu_g$, start an orientation relaxation process of the surrounding solvent molecules in the new internal electric field of the solute. This process can be followed, for instance, by the time dependence of the Stokes' red shift of the fluorescence spectrum.

Method (i) can only be usefully applied to solutes possessing a very large molecular dipole moment. In this chapter we shall deal with method (ii) which is applicable to almost every solute, since a molecule changes all its physical and chemical properties in undergoing an electronic transition.

The system 2-amino-7-nitro-fluorene/*ortho*-dichlorobenzene, ANF/ODB, for instance, has a fluorescence decay time (S_1 excitation lifetime) of about[2]

$$\tau_F^s \approx 2 \text{ ns} \tag{1.1}$$

The superscript s indicates that τ_F is that determined by radiationless transitions and by the Einstein coefficient for spontaneous emission, A_{eg} (see e.g. reference 3, p. 48 ff, p. 69, p. 85 ff, etc.), but not by induced emission, $B_{eg} = B_{ge}$.

The fluorescence emission and the corresponding absorption transition of this system are situated in the spectral region of $\lambda \approx 300$ nm ($\tilde{\nu} \approx$ 33,333 cm^{-1}). The duration $\delta\tau$ of a radiative electronic transition is about the transit time of one wavelength

$$\delta\tau_A = \frac{1}{\nu_A} \approx \delta\tau_F = \frac{1}{\nu_F} = \frac{\lambda_F}{c_0} = \frac{1}{\tilde{\nu}_F c_0} \approx 1 \text{ fs} \equiv 10^{-15} \text{ s} \tag{1.2}$$

where suffixes A and F refer to absorption and emission, respectively.

The macroscopic relaxation time of the dipole–dipole relaxation of the pure solvent ODB has been measured[4,5] as

$$\tau_R \approx 10 \text{ ps} \tag{1.3}$$

at room temperature.

Hence the absorption and emission processes $S_1 \leftarrow S_0$ and $S_1 \rightarrow S_0$, respectively, of the system are rapid 'jumps' compared with the relaxation phenomena to be observed, and fluorescence parameters are adequate for their observation, provided they are affected by the relaxation process and their time-dependence can be measured.

It should be noted that even the fastest photochemical reactions can in principle be observed by laser techniques, since they cannot be faster than the duration of one vibrational mode, which is of the order of a fraction of a picosecond. The reaction is the faster, the higher the force constant, the smaller the mass and the shorter the distance along the reaction coordinate. Hence proton transfer reactions along hydrogen bonds are among the fastest chemical reactions. Even here the vibrational $\tilde{\nu} \approx 3333\ \text{cm}^{-1}$ wavenumber corresponds to $\tau_c \approx 10$ fs and in comparison the absorption process of, for example, ANF still behaves as a 'jump'. For an example see §1.3.3(*bab*).

1.1.2 *Stokes' red shift and intermolecular interaction energy*

Compared with vibration and relaxation times the duration of a radiative electronic transition is so short that during such a transition the position and the motion of all atoms remain almost unchanged (equations (1.1)–(1.3)) ('Franck–Condon principle').

After an electronic absorption process from the equilibrium ground state S_0^{eq} has occurred the solute/solvent system immediately begins a relaxation process from its Franck–Condon state S_1^{FC} to reach its thermodynamical equilibrium state S_1^{eq} in which all atoms vibrate around their new equilibrium positions. The excess vibrational energy Q_1 is completely transferred into heat by collisions provided

$$\tau_R \ll \tau_F \tag{1.4}$$

Eventually the fluorescence photon is emitted and the system at first reaches a Franck–Condon ground state S_0^{FC}. Then the inverse relaxation process starts to bring the system into its original ground state S_0^{eq}. According to Pauling[6] almost the same amount of energy, $Q_0 \approx Q_1$, is transferred into heat in this process. Even in inert solvents only the mean amount of energy

$$h\nu_F = h\nu_A - (Q_1 + Q_0)/N_L \tag{1.5}$$

where N_L is Loschmidt's constant (Avogadro's number), is available for fluorescence ('Stokes' red shift'). Where strong interactions contribute to the red shift (for dipole–dipole interactions, shifts of up to 50% of $h\nu_A$ have been observed[7]) the effect is called an 'anomalous Stokes' red shift'. Even if the fluorescence quantum efficiency equals unity, a fraction of the excitation energy is transformed and dissipated as heat.

Each absorption process is the origin of a photophysical relaxation cycle (Figure 1.1). In systems of the polar solute/polar solvent type the anomalous Stokes' red shift increases with $(\mu_e - \mu_g)$ and with the polarity of the solvent.

From a second-order perturbation treatment by Ooshika,[9] Lippert,[10] and Mataga,[11] the red shift depends on the dielectric properties of the

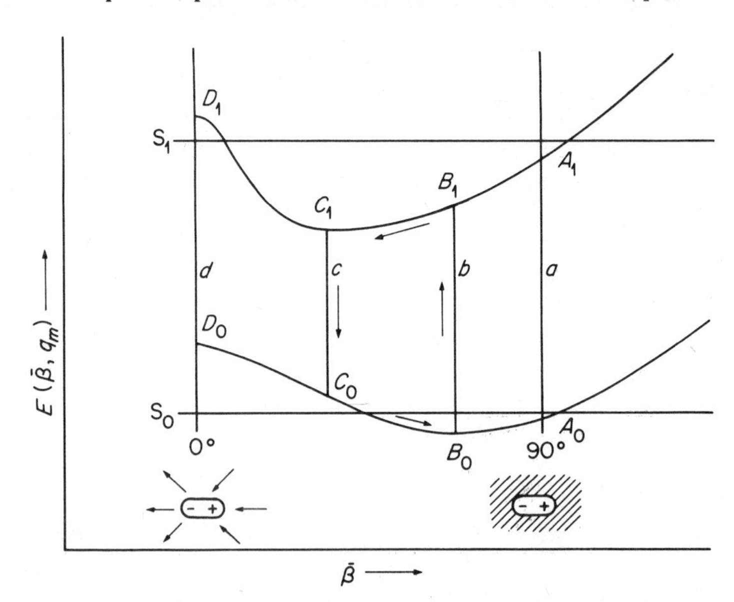

Figure 1.1 Activation/deactivation cycle. The energy E of the solute molecule/solvent cage system is plotted as a function of the reaction coordinate, for instance $\bar{\beta}$, the mean degree of orientation of the dipole moments of the nearest-neighbour polar solvent molecules relative to the electric field produced by a solute molecule, the dipole moment of which is higher in the first excited singlet state, S_1, than in the electronic ground state, S_0. The excess energy $h(\nu_A - \nu_F)$ is assumed to be distributed rapidly to the environment of the system. Straight lines refer to the free molecule[8]

solvent and is given by the equation

$$\tilde{\nu}_A - \tilde{\nu}_F = \frac{2(\mu_e - \mu_g)^2}{hc_0a^3}\left(\frac{D-1}{2D+1} - \frac{n^2-1}{2n^2+1}\right) \tag{1.6}$$

for an ideal spherical polar molecule with radius a. Equation (1.6), which has been modified for molecules with an ellipsoidal shape,[7] allows the estimation of the electric dipole moment μ_e of a molecule in its fluorescent state (ANF, $\mu_e = 25$ Debye units) if μ_g is known.

1.1.3 *Molecular adsorption in mixed solvents*

In mixed solvents constituting a homogeneous binary mixture of components of different polarity the anomalous Stokes' red shift of the fluorescence of a polar solute as a function of the percentage of the admixture shows the characteristics of a Langmuir's adsorption isotherm (Figure 1.2).[18]

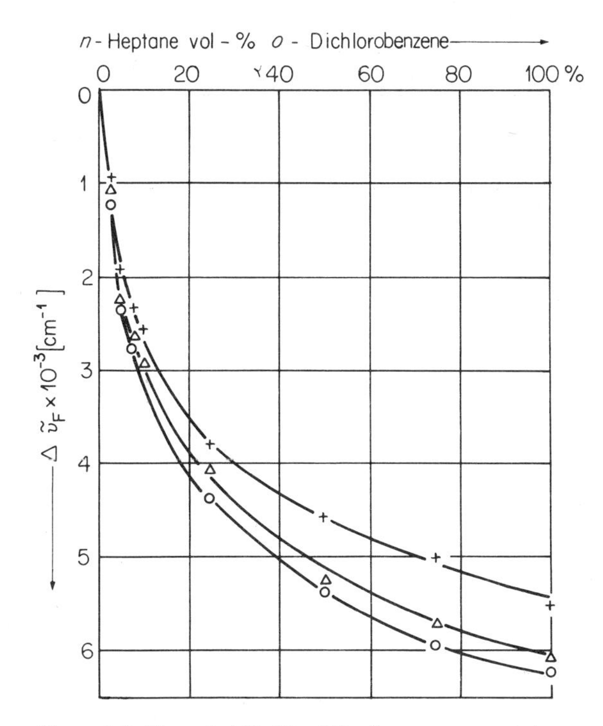

Figure 1.2 The red shift $\Delta\tilde{\nu}_F$ of the fluorescence maximum of 4-dimethylamino-4′-nitro-stilbene in mixtures of *n*-heptane and ODB at 80°C (+), 20°C (△) and −5°C (○)[17]

Small amounts of a polar solvent added to a solution of a polar solute in an inert solvent yield a large red shift. The maximum of the spontaneous fluorescence varies continuously with the composition of the mixed solvent. Starting with an inert solvent the red shift at first increases linearly with the increasing percentage of the polar solvent concentration, but eventually reaches its final value asymptotically.

For a quantitative treatment it has been assumed[18] that there is only a certain number n_{max} of sites on the surface of a solute molecule that can be occupied by nearest-neighbour polar solvent molecules. For most systems investigated so far, $n_{max} = 10$ is a reasonable value. It was further assumed that each polar solvent molecule M_A that is adsorbed on the surface of an excited solute molecule has the same interaction energy W_i and produces the same red shift of the solute fluorescence, irrespective of whether it takes the first or the last site of the surface. The last sites occupied will, of course, give a smaller total interaction energy because of the repulsive forces between the polar solvent molecules already adsorbed and a smaller red shift; but the last sites will only be occupied if the concentration c of polar solvent molecules is sufficiently high that molecules at larger distances from the solute molecule also contribute to the red shift.

The total interaction energy W of a solute with its neighbourhood is then given by the number n of adsorbed polar solvent molecules

$$W = n \,.\, W_i \tag{1.7}$$

and the red shift will be proportional to it. (For the definition of the symbol y for the red shift see equation (1.17).)

$$y = \frac{\tilde{\nu}_0 - \tilde{\nu}_F}{\tilde{\nu}_0 - \tilde{\nu}_{max}} = \frac{W}{W_{max}} = \frac{nW_i}{n_{max} W_i} = \frac{n}{n_{max}} \tag{1.8}$$

The value of W_i of a given system can be determined using the mass action law. The polar solvent molecules M may either be free in solution or adsorbed on a site S of the surface of the solute molecule, i.e. M_L or M_A, respectively:

$$M_L + S \rightleftarrows M_A \tag{1.9}$$

The total concentration c of polar solvent molecules is the sum of the concentrations c_L and c_A of those molecules that are free in the solution or adsorbed, respectively,

$$c = c_L + c_A \tag{1.10}$$

The equilibrium is determined by the concentration c_s of sites available:

$$K = \frac{c_A}{c_L \,.\, c_s} \approx \frac{c_A}{c \,.\, c_s} \tag{1.11}$$

$$Kc_L = \frac{c_A}{c_s} = \frac{n_A}{n_{max} - n_A} = \frac{n_A/n_{max}}{1 - n_A/n_{max}} \approx Kc \tag{1.12}$$

$$1/y = 1 + 1/Kc \tag{1.13}$$

Adsorption enthalpies obey expressions of the type

$$K = \text{const } T^q \, e^{-\Delta H_i/RT} \tag{1.14}$$

and hence it can be derived that

$$W_i = RT\left[\ln\left(\frac{1}{y} - 1\right) + q \ln T + \ln c + \text{const.}\right] \tag{1.15}$$

For the systems discussed here W_i had been determined from $y = y(T, c)$ and values of 1·5 to 1·9 kcal/mol had been obtained. $W = n \,.\, W_i \approx 12$ to 15 kcal/mol is in accordance with the value for W obtained from solute–dipole/solvent–dielectric interactions using the solute electric dipole moment differences ($\mu_e - \mu_g$) as calculated from equation (1.6).[19] Even for dye lasers the admixture of small amounts of polar solvents to the solution in an inert solvent allows one to control the spectral position of the induced emission.[20]

From a kinetic point of view the model has successfully been used to give a discrete as well as a continuous description of the frequency/time behaviour of the anomalous Stokes' red shift in binary polar mixtures by assuming the polar solvent molecules either to be oriented or non-oriented in the reaction field of the solute.[21,22]

1.2 Dye laser and relaxation phenomena

1.2.1 *Intensity effects*

It has already been pointed out that a fluorescing system reaches thermal equilibrium in its electronic excited state after a radiative transition only if $\tau_R \ll \tau_F$. Only then will the maximum Stokes' red shift be achieved. If τ_F is shortened, it will become less than τ_R and approach zero. Hence the red shift will decrease, or from the point of view of the observer, the fluorescence will be shifted to the blue. For the system ANF/ODB such a 'blue shift' has been observed with laser excitation due to the effect of induced emission (Figure 1.3).

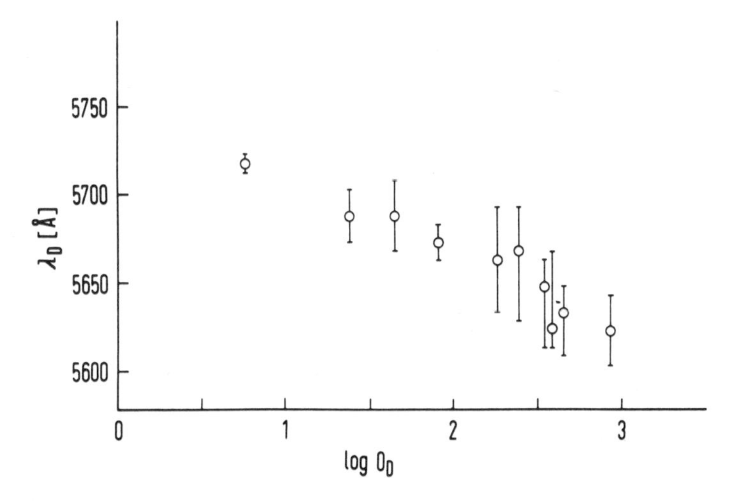

Figure 1.3 Plot of the wavelength λ_D of the maximum of the dye laser photon flux of the system ANF/ODB against dye laser output power O_D in relative units. Bars indicate the accuracy with which the maximum of the spectrum could be fixed[2]

The effect of induced emission is to shorten the lifetime τ_F of the excited state and thereby decrease the time that is available for better orientation of the ODB solvent cage around the strong polar excited ANF molecule.[12,13,14] The effect increases with the density of fluorescence photons in the sample, and it can be enhanced by increasing the excitation intensity, the solute concentration, or the Q-factor of the resonator.[2] During excitation by a frequency-doubled ruby laser pulse of a duration of some nanoseconds the pumped dye laser spectrum is shifted more and more to the blue since soon after passing the threshold, only solvent cages which are more or less non-relaxed are available for induced emission.

There are now mode-locked neodymium laser techniques available, which allow the selection of single excitation pulses which are about 2 ps broad. These allow the direct step-by-step measurement of relaxation processes by polarization and delay techniques in absorption or fluorescence. The ratio of fluorescence lifetimes due to induced and spontaneous emission is given by

$$\frac{\tau_F^i}{\tau_F^s} = \frac{8\pi c_0 n^2}{3\Phi_{FM}\lambda_D^4} \cdot \frac{\int_0^\infty f^s(\lambda)\,d\lambda}{P_{\lambda_D} \cdot f(\lambda_D)} \tag{1.16}$$

where Φ_{FM} is the fluorescence quantum yield (see reference 3, equation (4.31)). $f^s(\lambda)$ is the spectral quanta distribution of spontaneous emission with its maximum normalized to unity, and P_{λ_D} is the photon flux density of the dye laser radiation.[2]

The red shift $\tilde{\nu}_0 - \tilde{\nu}_F$ increases with time t, presumably by an exponential law, and it reaches its final value $\tilde{\nu}_0 - \tilde{\nu}_\infty$ asymptotically (the subscripts 0 and ∞ refer to the time t of emission after excitation)

$$y \equiv \frac{\tilde{\nu}_0 - \tilde{\nu}_F}{\tilde{\nu}_0 - \tilde{\nu}_\infty} = 1 - e^{-kt} \tag{1.17}$$

Let us assume that the initial slope of the red shift is steeper the faster the relaxation process is, the smaller τ_R is, or the larger the ratio τ_F/τ_R is. Let us consider that the wavenumbers $\tilde{\nu}$ are mean values of many processes so that we can substitute τ_F for t; then for a given relaxation time τ_R we may make the initial assumption that

$$y \equiv \frac{\tilde{\nu}_0 - \tilde{\nu}_F}{\tilde{\nu}_0 - \tilde{\nu}_\infty} = (1 - e^{-k\tau_F^2/\tau_R})^p \tag{1.18}$$

Even if the relaxation time becomes very large ($\tau_R \rightarrow \infty$), the Stokes' red shift still reaches its maximum, $\tilde{\nu}_F = \tilde{\nu}_\infty$, if the observation time is long enough ($\tau_F \rightarrow \infty$). It should be noted that $y(\tau_F)$ has an initial finite slope when starting from $\tilde{\nu}_F = \tilde{\nu}_0$ at $\tau_F = 0$, only if the right-hand side of equation (1.18) has the exponent $p = \frac{1}{2}$ and that only then the slope is

$$\left(\frac{\partial y}{\partial \tau_F}\right)_{\tau_F \to 0} = \left(\frac{k}{\tau_R}\right)^{\frac{1}{2}} \tag{1.19}$$

as it should be, as well as

$$\left(\frac{\partial y}{\partial \tau_F}\right)_{\tau_F \to \infty} = 0 \tag{1.20}$$

as observed.

1.2.2 *Temperature effects*

For a given finite value of τ_F equation (1.18) states that the red shift tends asymptotically to zero, $\tilde{\nu}_F \to \tilde{\nu}_0$, if $\tau_R \to \infty$, that is if the temperature $T \to 0$ or $(1/T) \to \infty$. The dependence $\tau_R(1/T)$ might be written[15] as

$$\tau_R = \frac{a}{T} e^{b/T} \tag{1.21}$$

$$\frac{d\tau_R}{dT} = -\frac{\tau_R}{T}\left(1 + \frac{b}{T}\right) \tag{1.21a}$$

We assume that τ_F does not strongly depend on temperature, compared with τ_R. The observed red shift approaches asymptotically to zero as $T \to 0$[16,17] because with decreasing temperature the viscosity of the solution increases and during the short lifetime of the fluorescent state the polar solvent molecules like ODB have insufficient time to orient themselves in the reaction field of the solute:

$$\left(\frac{\partial y}{\partial T}\right)_{T \to 0} = 0 \tag{1.22}$$

$$\left(\frac{\partial y}{\partial \tau_R}\right)_{\tau_R \to \infty} = 0 \tag{1.23}$$

On the other hand if T increases, $(1/T) \to 0$, $\tau_R \to 0$, the system reaches its thermal equilibrium state immediately with the absorption process, but there will be almost no anomalous Stokes' red shift since thermal motion acts against better orientation. The derivative of the red shift with respect to temperature should be zero ($\tilde{\nu}_F = \tilde{\nu}_0$) when cooling down from high temperatures, but the derivative with respect to τ_R should be infinite:

$$\left(\frac{\partial y}{\partial T}\right)_{T \to \infty} = 0 \tag{1.24}$$

$$\left(\frac{\partial y}{\partial \tau_R}\right)_{\tau_R \to 0} = \infty \tag{1.25}$$

If the dependence of the red shift on the relaxation time is also exponential in nature and with the same coefficient k for a given relaxing system let us assume that

$$y = (1 - e^{-k\tau_R})^p \tag{1.26}$$

Conditions (1.24) and (1.25) are fulfilled if the right-hand side of equation (1.26) has an exponent p with $0 < p < 1$. Hence $p = \frac{1}{2}$ would be a satisfactory value.

The function $y(\tau_R(T))$ must have a maximum between $\infty > T > 0$ or $0 < \tau_R < \infty$, namely at $\tau_R \approx \tau_F$, as has been found experimentally in 1959 (Figure 1.4) for several dyes and several solvents,

$$\left(\frac{\partial y}{\partial \tau_R}\right)_{\tau_R = \tau_F} = 0 \tag{1.27}$$

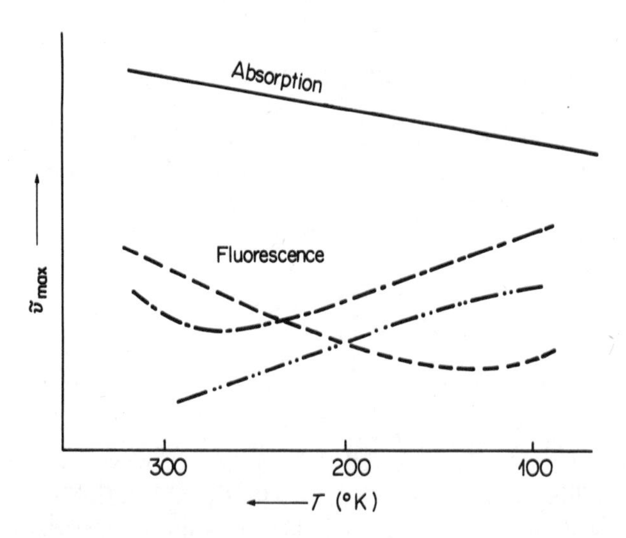

Figure 1.4 The effect of temperature on the spectral position of electronic transitions: schematic representation of experimental results for merocyanine dyes.[17] ——— absorption in and relative to any solvent; –·–·–·– fluorescence in ethyl alcohol; –··–·· fluorescence in *iso*-butyl alcohol; ----- fluorescence in butylchloride, butyronitrile, ethylacetate, triethylamine

This maximum is a property of the system and occurs at a certain temperature of inversion according to equation (1.21) which we designate T_u ('Umkehrtemperatur'). For comparison also see reference 23.

As is shown in the schematic diagram (Figure 1.4), even the spectral position of the absorption peak depends on temperature, since the dielectric polarizability of the solvent increases with decreasing temperature.

In the following we shall consider the Stokes' red shift $(\tilde{\nu}_0 - \tilde{\nu}_F)$ relative to $\tilde{\nu}_0 \equiv \tilde{\nu}_{00}$, the 00-transition of absorption, where $\tilde{\nu}_F$ and $\tilde{\nu}_\infty$ are the 00-transitions of fluorescence after the times t and ∞, respectively.

1.2.3 *A general expression*

A general equation that governs the behaviour of the anomalous Stokes' red shift in solution as a function of temperature T and fluorescence photon output O_D by means of the corresponding relaxation time $\tau_R(T)$ and the lifetime $\tau_F(O_D)$ of the fluorescing state is obtained by multiplying the positive square roots of equations (1.18) and (1.26)

$$y \equiv \frac{\tilde{\nu}_0 - \tilde{\nu}_F}{\tilde{\nu}_0 - \tilde{\nu}_\infty} = (1 - e^{-k\tau_R})^{\frac{1}{2}}(1 - e^{-k\tau_F/\tau_R})^{\frac{1}{2}} \tag{1.28}$$

Equation (1.28) fulfils all the conditions of (1.20), (1.22), (1.23), (1.24), (1.25) and (1.27). In addition $y = 0$ if $\tau_F = 0$, or if $\tau_R = 0$, or $\tau_R = \infty$, i.e. $T = \infty$ or $T = 0$, respectively, for given τ_F, and

$$y_{\tau_F \to \infty} = (1 - e^{-k\tau_R})^{\frac{1}{2}} = \begin{cases} 0 \text{ if } T \to \infty \\ 1 \text{ if } T \to 0 \end{cases} \tag{1.29}$$

$$\left(\frac{\partial y}{\partial \tau_F}\right)_{\tau_F \to 0} = \left[\frac{k}{\tau_R}(1 - e^{-k\tau_R})\right]^{\frac{1}{2}} = \begin{cases} k \text{ if } T \to \infty \\ 0 \text{ if } T \to 0 \end{cases} \tag{1.30}$$

$$y_{\tau_F = \tau_R = \tau} = 1 - e^{-k\tau} = \begin{cases} 0 \text{ if } T \to \infty \\ 1 \text{ if } T \to 0 \end{cases} \tag{1.31}$$

$$\frac{\partial y}{\partial \tau} = k\, e^{-k\tau} = \begin{cases} k \text{ if } T \to \infty \\ 0 \text{ if } T \to 0 \end{cases} \tag{1.32}$$

From equation (1.31) it is possible to determine the order of magnitude of the parameter k of the system ANF/ODB. For example, if

$$y_r = \frac{1 - \tilde{\nu}_F/\tilde{\nu}_0}{1 - \tilde{\nu}_\infty/\tilde{\nu}_0} = \frac{1 - \frac{3}{4}}{1 - \frac{2}{4}} = \frac{1}{2} \tag{1.33}$$

at $\tau_F = \tau_R = 15$ ps $= \tau_r$ at room temperature then

$$k = \frac{1}{\tau_r} \ln 2 = 0{\cdot}5 \,.\, 10^{11}\ \mathrm{s}^{-1} \tag{1.34}$$

This is of the order of magnitude of the corresponding normal relaxation rate.

Figure 1.5 may be used to illustrate the region where observations have been undertaken. For the system ANF/ODB, for instance,[2] τ_F has been shortened from values of $\tau_F \gg \tau_R$ down to a value not far from τ_R, and the expected blue shift has been observed. For a certain value of $\tau_F \gg \tau_R$ also an increase of temperature from $-25°$C to $+45°$C gave the expected blue shift of the spontaneous fluorescence as well as of the induced emission.

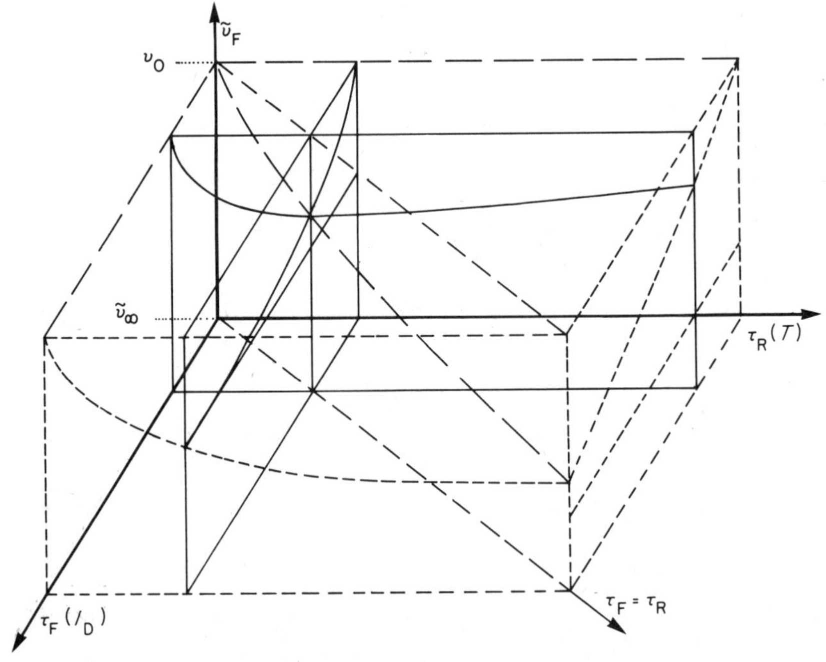

Figure 1.5 Schematic diagram of the anomalous Stokes' red shift $\tilde{\nu}_0 - \tilde{\nu}_F$ as a function of the lifetime τ_F of a fluorescing electronic state (i.e. of the photon flux of a dye laser) and of the relaxation time τ_R (i.e. of the change of viscosity η by temperature T). $\tilde{\nu}_0 \equiv \tilde{\nu}_A^{00}$ is the (temperature-dependent) wavenumber of the $0 - 0$ transition in absorption and $\tilde{\nu}_F \equiv \tilde{\nu}_F^{00}$ the corresponding $0 - 0$ transition of fluorescence. $\tilde{\nu}_0 - \tilde{\nu}_F$ is presented relative to $\tilde{\nu}_0 - \tilde{\nu}_\infty$, the maximum red shift, which would be achieved asymptotically for $\tau_F \to \infty$ and $\tau_R \to \infty$

For the system 3-dimethylamino-6-monomethylamino-*N*-methylphthalimide ($2{\cdot}58 \,.\, 10^{-3}$ M) in glycerin a blue shift of the laser peak has been observed by decreasing temperature, as well as by increasing the excitation intensity[14] in the region of $\tau_F \approx \tau_R$. It has been pointed out that more than one relaxation process and relaxation time can contribute to the Stokes' red shift.

1.3 Orientational relaxation and multiple fluorescence

1.3.1 *Multiple fluorescence*

The molecular fluorescence of solutions mostly consists of only a single electronic transition, the $S_1 \rightarrow S_0$ emission from the lowest excited state that has the same multiplicity as the ground state. In such cases the vibrational fine structure of the fluorescence band is approximately the mirror image of the vibrational pattern of the corresponding absorption band, $S_1 \leftarrow S_0$ (Kasha's rule[24]). Crystallized anthracene, however, containing small impurities of tetracene, shows the yellow-green fluorescence of tetracene instead of the blue fluorescence of anthracene.[25] For solutions of aromatic compounds exceptions from Kasha's rule are also known. Such violations in a solute/solvent system are important in the application of dye laser techniques to photochemistry.

If an adiabatic[26] photoreaction competes with the fluorescence of an electronically excited state of the solute, i.e. if the photoreaction yields an electronically excited reaction product, then in principle both the initial solute and the reaction product fluoresce. In such cases not one but two fluorescence bands will be observed. Birks,[3] therefore, introduced the term 'dual fluorescence'. But even in systems showing dual fluorescence the emissions of both the initial and the final species may obey Kasha's rule with respect to the corresponding absorption bands.

There are also systems known in which more than one adiabatic photoreaction occurs and for which more than two fluorescence bands have been observed. An example will be given in §3.3(*bab*). For the description of such observations the term 'triple fluorescence' or more generally 'multiple fluorescence' should be used.[27]

The spectral position of the reaction product fluorescence band is always situated at the long-wavelength side of the initial species fluorescence band as may be seen from Figure 1.6.

The terms dual, triple and multiple fluorescence will be limited to effects which are inherent to the system as described above, i.e. which are unavoidable due to adiabatic photoreactions which compete with the emission of fluorescence light. Effects of recombination delayed fluorescence, therefore, are excluded.

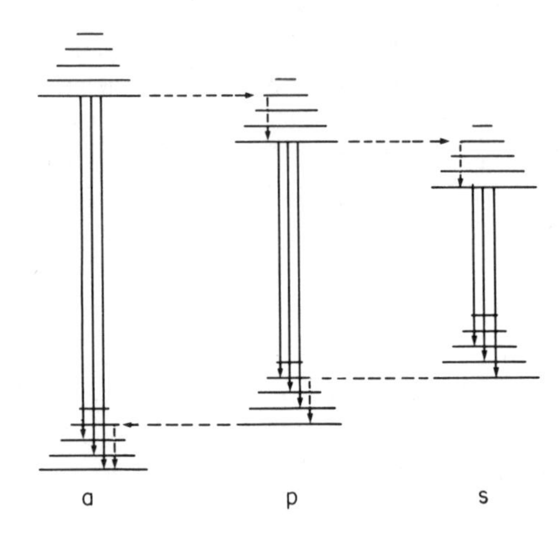

Figure 1.6 Energies of electronic states of a solution exhibiting triple fluorescence: (a) fluorescence of the absorbing species, (p) fluorescence of the primary adiabatic photoreaction product, and (s) fluorescence of a secondary adiabatic photoreaction product.

The rate constant of an adiabatic photoreaction can be determined by two different laser techniques, namely

(a) by excitation with very short pulses and observation of the time-resolved fluorescence spectrum, i.e. the increase and decrease of the various fluorescence band intensities;[28a] and

(b) by excitation with very intense pulses and observation of the ratio of the various fluorescence band intensities as a function of the excitation intensity, i.e. as a function of the lifetimes of the fluorescent excited states.[28b]

A synopsis of the types of adiabatic photoreactions allowing for dual fluorescence is given in Table 1.1. For each type only one or two of the first and most prominent examples are given. A special sequence of these reaction types has been chosen in order to give a better survey over the possible types as far as they have been studied experimentally.

The spectral positions and the intensity ratio of the two fluorescence bands depend on the polarity of the solvent if the dipole moments of the fluorescing states have different values. This is the case for most systems showing dual fluorescence. In the following sections some remarks will be made to the different sources of origin for dual fluorescence given in the synoptical Table 1.1.

1.3.2 *Adiabatic photophysical reactions*

(The letters at the top of the following sections refer to Table 1.1.)

(*aaa*) Azulene is the most carefully studied example of dual fluorescence.

Table 1.1 Synoptical table of dual fluorescence in solution

(*a*) *Photophysical primary processes*
(*aa*) Reduced internal conversion between S_2 and S_1 if these states
  (*aaa*) are separated by a large energy interval:
    Cryptocyanine in methyl alcohol,[29] azulene in several solvents;[30] the fluorescence transitions are $S_1 \rightarrow S_0$ and $S_2 \rightarrow S_0$.
  (*aab*) are separated by a small energy interval:
    p-Dimethylaminobenzonitrile in polar solvents;[31] it fluoresces from ^1L_a and ^1L_b in solvated states.
(*ab*) Intramolecular electron donor/acceptor interactions
  (*aba*) between *n*- and π-electronic systems:
    9,10-Diazaphenanthrene and derivatives in inert solvents;[32] $\pi^* \rightarrow \pi$ and $\pi^* \rightarrow n$ fluorescences are possible,
  (*abb*) between two π-electronic systems which are strongly sterically hindered:
    9,9′-Dianthryl in polar solvents;[33]
    the absorbing half fluoresces as well as the entire molecule,
  (*abc*) between two unsaturated groups separated by a saturated chain:
    1,3-Diphenylpropane in deoxygenated cyclohexane;[34] it fluoresces from the phenyl-group and the intramolecular excimer.
(*ac*) Energy transfer between two non-equal unsaturated groups
  (*aca*) intramolecular with separation by a saturated chain $(CH_2)_n$:
    1-Naphthyl-9-anthryl alkanes, if $n \neq 3$;[35]
    even if the naphthyl group absorbs, sometimes the anthryl group fluoresces,
  (*acb*) intermolecular with molecular separations up to about 100 Å:
    Pyrene and Sevron yellow in rigid solutions;[36]
    even if pyrene absorbs sometimes Sevron yellow fluoresces.

(*b*) *Photochemical primary processes*
(*ba*) Proton transfer processes along hydrogen bonds
  (*baa*) intramolecular between basic and acidic sites:
    Salicylic acid in non-polar solvents,[37]
    1,3,6-trimethyl-7-hydroxy-lumacine in 0·1 N H_2SO_4,[38]
    keto/enol- and lactim/lactam-tautomers, both of which fluoresce,
  (*bab*) intermolecular between basic and acidic molecules:
    β-Naphthol in aqueous solutions;[39]
    it fluoresces from the neutral molecule as well as the anion.
(*bb*) Intermolecular electron donor/acceptor interactions
  (*bba*) between equal molecules:
    Pyrene in several solvents;[40]
    it fluoresces from the monomer as well as the excimer,
  (*bbb*) between non-equal molecules:
    Perylene dimethylaniline in non-polar solvents;[41]
    it fluoresces from the monomer as well as the hetero-excimer (exciplex).

This is due to a large energy gap between the states S_2 and S_1 of a solute.[42,43,44] The larger the energy difference the smaller is the rate of $S_2 \rightarrow S_1$ internal conversion and the higher, therefore, is the probability for $S_2 \rightarrow S_0$ fluorescence. $S_2 \rightarrow S_0$ fluorescence have also been observed for pyrene[45] and naphthalene[46] in the vapour phase. The subject is more fully discussed by Birks (Chapter 9, Vol. 2).

(*aab*) Earlier work on the dual fluorescence of solutions of *p*-dimethyl-amino-benzonitrile (DMAB) and *p*-diethylamino-benzonitrile (DEAB) has been summarized by Birks (pp. 164–166 of reference 3). In non-polar solvents the highly polar state ^1L_a is energetically situated near but above the state ^1L_b. In polar solvents, however, the intermolecular interaction energy of the solvent cage and the solute shifts the ^1L_a state below the energy of the ^1L_b state. The corresponding *a*-fluorescence is situated, therefore, at the long-wavelength side of the *b*-fluorescence and exhibits an anomalous strong red shift with solvent polarity.[17,31,52] With butyl-chloride as polar solvent the red shift increases when cooling from 293°K to 148°K, but the ratio of the fluorescence intensities I_a/I_b decreases because with increasing viscosity the relative number of cages that approach the equilibrium state decreases (Figure 1.7). Transitions from

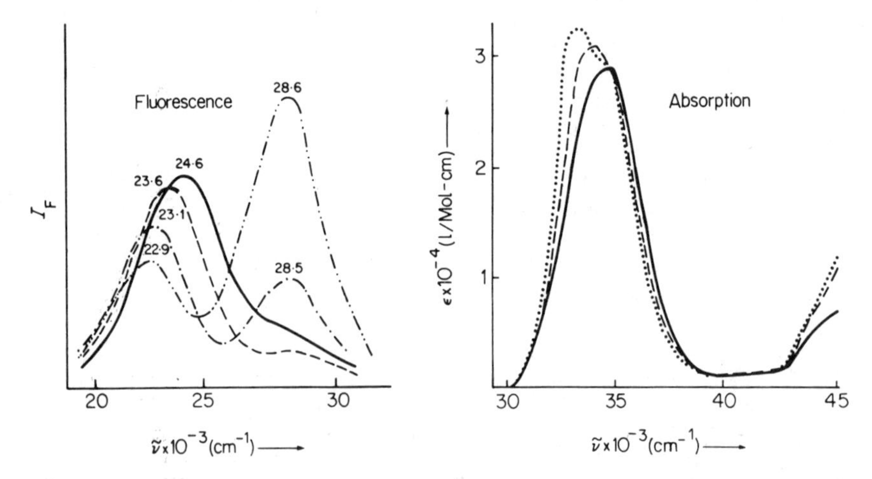

Figure 1.7 Effect of temperature on the fluorescence and absorption spectra of DEAB in butylchloride/methylcyclohexane/*iso*pentane 12:3:1. Fluorescence (not corrected) of 5×10^{-5} M solution at: ——— 293°K; ---- 234°K; -·-·-· 173°K; -··-··-··- 148°K. Absorption of $1{\cdot}6 \times 10^{-5}$ M solution at: ——— 293°K; ----- 193°K; ········ 113°K[17]

states ^1L_a and ^1L_b to the ground state are considered to have transition moments in the plane of the aromatic system but perpendicular to each other (Figure 1.8). Grabowski *et al.*[49] consider the *a*- and *b*-fluorescences to originate from a molecule with the nitrogen lone pair orbital axis of the $-N(CH_3)_2$ group oriented either perpendicular or parallel to the plane of the aromatic ring. These opinions are not necessarily in contradiction with each other.

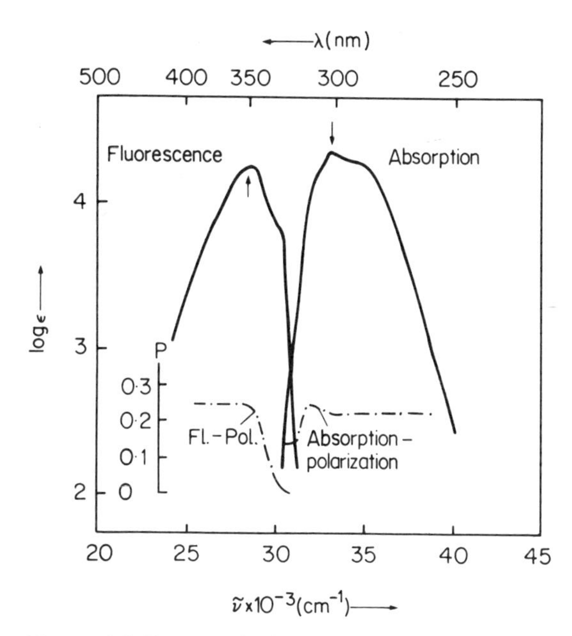

Figure 1.8 Degree of absorption and fluorescence polarization $p = (I_{\parallel} - I_{\perp})/(I_{\parallel} + I_{\perp})$ of 10^{-4} M solution of DMAB in EtOH at 90°K[17]

McGlynn *et al.*[47] claimed to have identified a dimer absorption band at about 29,000 cm^{-1}, but Mataga *et al.*[48] showed that this became negligible after recrystallization five times from *n*-hexane. They thus confirmed that the spectra ascribed to the dimer of DMAB are due to an impurity. Mataga *et al.*[50] measured the time-resolved fluorescence spectra and the fluorescence decay curves of the *a*- and *b*-fluorescence and demonstrated clearly the relaxation process forming the ^1L_a-solvated state. The *b*-fluorescence approaches its intensity maximum after 4 ns and the *a*-fluorescence after about 7 ns after the end of a short excitation pulse.

(*aba*) No nitrogen heterocyclics other than 9,10-diazaphenanthrene (benzo[c]quinoline) and derivatives are known to exhibit dual fluorescence of the type $\pi^*\pi$ and π^*n, but this effect is well known for aldehydes and ketones in the gas phase.[51]

(*abb*) A molecule, such as 9,9′-dianthryl, which consists of two (or more) π-electronic systems which are oriented perpendicular to each other because of strong steric hindrance (Figure 1.9) exhibits a strange fluorescence behaviour in polar solvents, even though the absorption

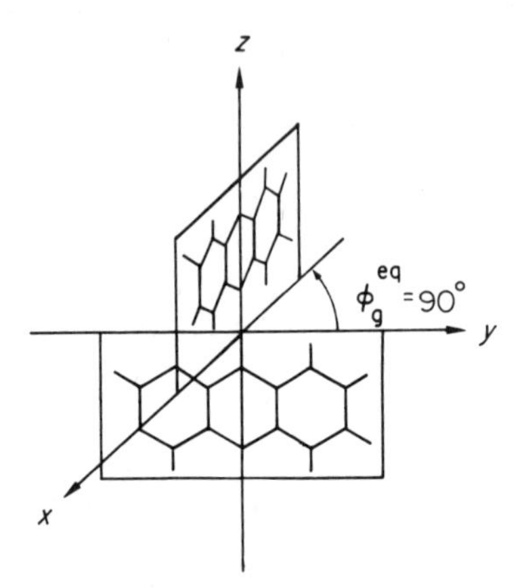

Figure 1.9 Conformation of 9,9′-dianthryl in its electronic ground state

spectrum is merely the superposition of the absorption spectra of the isolated π-electronic systems.[33,53,54] There occurs a threshold solvent polarity below which the Stokes' red shift is almost constant as it is normally the case for non-polar solutes. But if a certain solvent polarity parameter

$$\Delta f' = \frac{D-1}{2D+1} - \frac{1}{2}\frac{n^2-1}{2n^2+1} \tag{1.35}$$

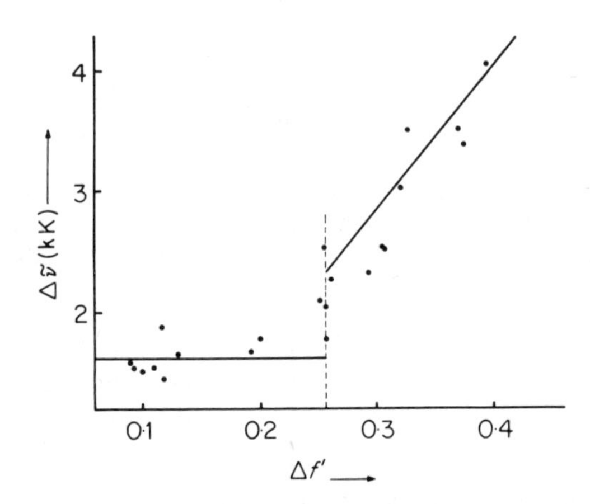

Figure 1.10 Stokes' red shift of 9,9′-dianthryl in different solvents. Each point refers to a certain solvent. The values are taken from Table 1 in reference 33a

exceeds the threshold value a new long wavelength structureless fluorescence band appears at the expense of the intensity of the initial normal fluorescence band. The new fluorescence band is shifted to the red proportionally to the polarity $\Delta f'$ of different solvents (Figure 1.10). An increase of the solvent polarity and the induced non-equivalence of the two anthryl-halves M_1 and M_2 of the 9,9'-dianthryl molecule can also be achieved by cooling a polar solution.[12,33] The above-mentioned effect of dual fluorescence has first been observed with solutions of 9,9'-dianthryl. The first excited singlet state of the free dianthryl molecule achieved by absorption is of the type ψ ($M_1^* . M_2$) and possesses two minima of the potential energy at twisting angles of about $\Theta = (90 \pm 12)^\circ$ around the central C—C bond (Figure 1.11) whereas the ground state has merely one minimum at $\Theta = 90^\circ$. The M_i represent the different anthryl groups.

The appearance of two potential minima in S_1 is the result of better interaction between the two π-electronic systems in case one of them is excited. The better overlapping in S_1 renders possible a charge-transfer interaction between the two anthryl groups. Increasing solvent polarity eventually induces an intramolecular univalent redox process. The molecule will achieve an excited state to which polar structures of the type $\psi(M_1^+ . M_2^-)$ contribute. This structure alone describes the electronic structure if $\Theta = 90^\circ$ and represents a doublet/doublet anion/cation biradical.

The factor $\frac{1}{2}$ in equation (1.35) as compared with the factor in brackets (Δf) of equation (1.6) comes from the fact that the dianthryl molecule absorbs from a non-polar to a non-polar state, but fluoresces from a polar to a non-polar state.[33b] The inclination of the straight line above the threshold value of $\Delta f'$ in Figure 1.10 is determined by a dipole moment of about $\mu_e = 20$ Debye units in the polar excited state, corresponding to a charge separation of one electronic charge along a distance of about 4 Å. Above that threshold value the stabilization energy for the polar CT structure is supplied by the solute/solvent intermolecular interaction energy.

The effect of dual fluorescence due to an induced intramolecular charge transfer interaction produced by a splitting of the potential minimum along a twisting coordinate has also been observed for weak polar compounds such as *N*-naphthylcarbazoles[53] and for the strong polar solute *p*-(9'-anthryl)-*N*,*N*-dimethylaniline. The latter compound is laser-active and shows the expected blue shift of the dye laser output with increasing concentration,[55] i.e. with increasing fluorescence photon density which shortens τ_F as demonstrated in Figure 1.3. The kinetics of the cooperative reorientation of polar solvent molecule and its mechanism have not yet been investigated.

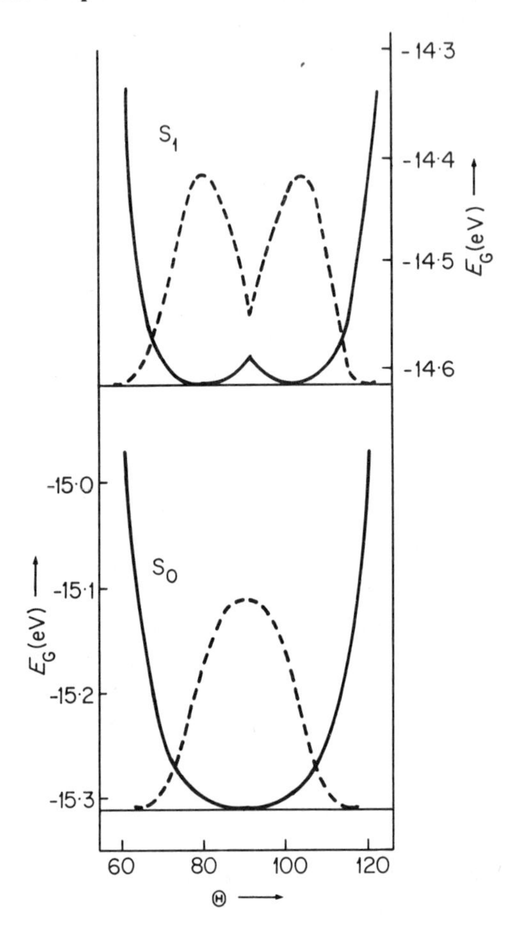

Figure 1.11 The torsional distribution (relative abundance) ----- and the potential energy ——— of the free 9,9′-dianthryl molecule as a function of torsional angle around the central C—C band, calculated[33b] for S_0 and S_1

Molecules like tetraphenylethylene[56] which possess only medium steric hindrance and mean twisting angles $0° < \Theta < 90°$ of the π-electronic systems in S_0 and no splitting but an angular shift of the potential minimum in S_1 do not exhibit dual fluorescence but one broad fluorescence band. The frequency/time behaviour of that fluorescence band displays the pattern shown in Figure 1.12; the shorter the lifetime τ_F^s of spontaneous emission the smaller the reorientation and the smaller is the red shift.

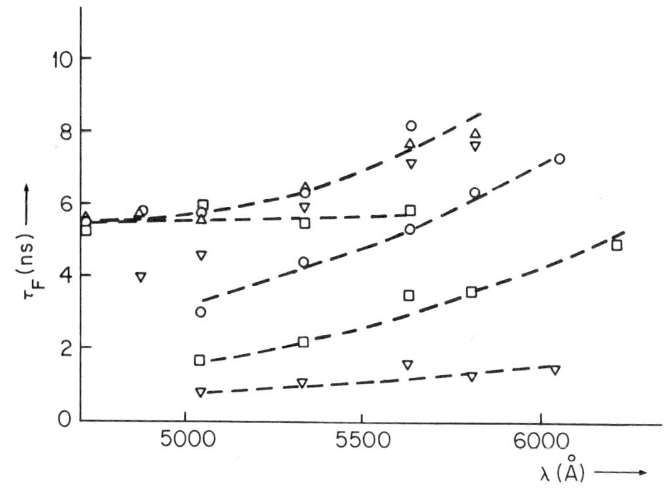

Figure 1.12 Fluorescence decay time τ_F versus wavelength λ for solutions of tetraphenylethylene ($c = 10^{-4}$–10^{-3} M) in methylcyclohexane/*iso*pentane (1:1) at various temperatures:[56] ▽ 128, ■ 117, ● 113, ▼104, ○ 99, ▲ 93 and □ 86°K

The effect is highest at temperatures for which $\tau_F \approx \tau_R$, where τ_R refers to the relaxation processes that allow for the reorientation to the equilibrium angles Θ in S_1. At higher as well as at lower temperatures almost no relaxation process will occur, as has been outlined in §1.2. The lower the temperature the smaller is the quenching rate, the higher the fluorescence quantum efficiency, and the higher, therefore, the mean lifetime τ_F^s.

(*abc*) If the connecting chain of methylene-groups between an electron donor D and an electron acceptor A in a molecule D—$(CH_2)_n$—A consists of $n = 3$ groups, then sandwich excimers of types (*bb*) can be formed. But even if $n = 1$ or $n = 2$ a charge transfer interaction between an electron donor D = dimethylamine and an electron acceptor A = 9-anthryl has been observed[57] with a threshold value $\Delta f' \approx 2{\cdot}0$ above which the compounds show dual fluorescence of the dianthryl type[57a] (*abb*).

(*ac*) For further examples and rccent results of energy transfer see references 3 and 58.

1.3.3 *Adiabatic photochemical reactions*

(The letters at the top of the following sections refer to Table 1.1.)

(*baa*) Quite often tautomerization is not merely a one-proton transfer process along an intramolecular hydrogen bond as is the case for salicylic acid in inert solvents, but several protons change places along molecular chains of hydrogen bonds, i.e. along at least one water molecule in aqueous solutions of 1,3,6-trimethyl-7-hydroxy-lumacine

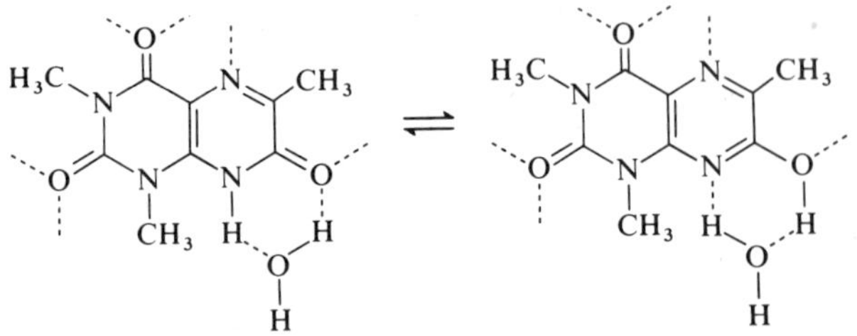

The complex (lumacine compound) : (water molecule) can be discussed as one molecule in a physical sense. Both tautomers have their own fluorescence band and the dissolved system exhibits, therefore, a dual fluorescence, whereas the absorption spectrum is almost only due to the 7-OH isomer.[38] Recently a dual fluorescence of alloxazine induced by acetic acid has been observed.[71]

The reason for this behaviour is the fact that in the first excited singlet state S_1 of aromatic compounds the basic groups behave more basic, and the acidic groups behave more acidic, compared with the ground state S_0. The most striking example known is 7-azaindole in inert solvents.[59] The doubly hydrogen bonded dimer exists in S_0 and S_1 and the thermal equilibrium is shifted from left to right by excitation:

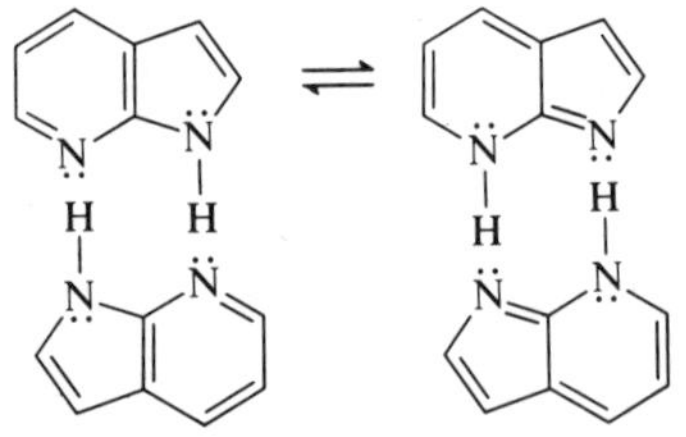

A molecular exciton effect facilitates the cooperative two-proton reversible transfer.[60]

(*bab*) Ground-state (S_0) ionization constants of solutions can be determined spectroscopically by means of absorption spectroscopy. If the neutral molecule is a base **B** then let us consider the reaction

$$BH^+ + H_2O \rightleftarrows B + H_3^+O \tag{1.36}$$

$$K_{BH^+} = [B].[H_3^+O]/[BH^+] \tag{1.37}$$

$$pK_+ = -\log_{10} K_{BH^+} \tag{1.38}$$

$$= pH - \log [B]/[BH^+] \tag{1.39}$$

If the total concentration of the base

$$c = [B] + [BH^+] \tag{1.40}$$

then according to Beer's law at a given wavelength

$$\frac{1}{d}\log\frac{I_0}{I} = \varepsilon c = \varepsilon([B] + [BH^+]) = \varepsilon_B[B] + \varepsilon_+[BH^+] \tag{1.41}$$

$$[B](\varepsilon - \varepsilon_B) = [BH^+](\varepsilon_+ - \varepsilon) \tag{1.42}$$

where ε is the apparent extinction coefficient and ε_+ and ε_B those of BH^+ and B, respectively. From equations (1.39) and (1.42) it follows that

$$pK_+ = pH - \log\frac{\varepsilon_+ - \varepsilon}{\varepsilon - \varepsilon_B} \tag{1.43}$$

If the neutral molecule acts as an acid A, i.e. as a proton donor, then

$$pK = pH - \log\frac{\varepsilon_A - \varepsilon}{\varepsilon - \varepsilon_-} \tag{1.44}$$

Constants of tautomerization are calculated in the same way. All these methods had until recently been applicable only to solutes in their ground state, but experimental methods have now been developed which enable one for the first time to conduct absorption spectroscopic measurements from S_1 in the picosecond scale by applying short but broad spectral pulses.[61]

pK* values for S_1 states can also be determined by fluorescence spectroscopic measurements (for a recent critical summary see reference 62).

$$pK^* = -\log_{10}\frac{[B^*][H_3^+O]}{[B^*H^+]} \tag{1.45}$$

$$= -0{\cdot}434 \ln_e \frac{[B^*][H_3^+O]}{[B^*H^+]} \tag{1.46}$$

$$= 0{\cdot}434\frac{\Delta G^*}{RT} \tag{1.47}$$

$$= 0{\cdot}434\frac{\Delta H^* - T\,\Delta S^*}{RT} \tag{1.48}$$

Often the factor $0{\cdot}434 = \log_{10} e$ is missing in the literature, and the published pK* values are then in error by a factor $\approx \frac{1}{2}$.

If we assume that the entropy change contributions cancel each other, then

$$pK^* - pK = \frac{\Delta H^* - \Delta H}{RT \ln 10} \tag{1.49}$$

Let us define the mean energy difference $\Delta h \equiv \Delta H/N_L$ of a solute/solvent system by the reaction enthalpy ΔH; then

$$\Delta h^* - \Delta h = E(S_1^{eq} - S_0^{eq})_B - E(S_1^{eq} - S_0^{eq})_{BH^+} \tag{1.50}$$

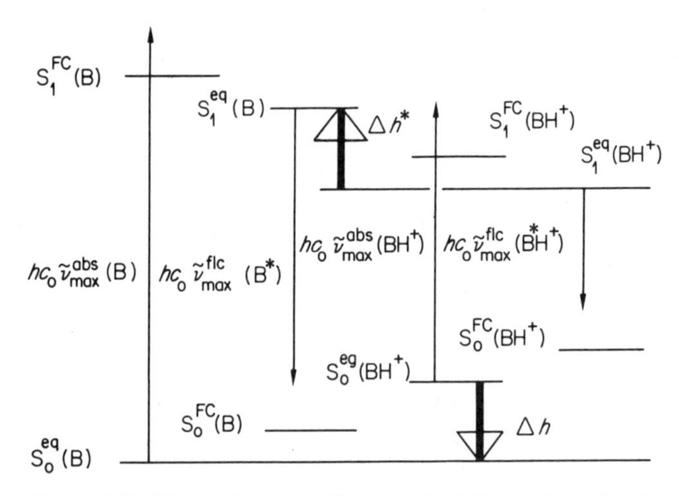

Figure 1.13 Schematic energy diagram of adiabatic photochemical reactions

(Figure 1.13). If we follow Pauling's[6] assumption

$$E(S_1^{FC} - S_1^{eq}) = E(S_0^{FC} - S_0^{eq}) \tag{1.51}$$

$$2E(S_1^{eq} - S_0^{eq}) = E(S_1^{FC} - S_0^{eq}) + E(S_1^{eq} - S_0^{FC}) \tag{1.52}$$

$$E(S_1^{eq} - S_0^{eq}) = \frac{hc_0}{2}(\tilde{\nu}_{00}^{abs} + \tilde{\nu}_{00}^{flc}) \tag{1.53}$$

$$\Delta h^* - \Delta h = \frac{hc_0}{2}[\tilde{\nu}_{00}^{abs}(\mathrm{B}) + \tilde{\nu}_{00}^{flc}(\mathrm{B^*}) - \tilde{\nu}_{00}^{abs}(\mathrm{BH^+}) - \tilde{\nu}_{00}^{flc}(\mathrm{B^*H^+})] \tag{1.54}$$

$$\Delta h^* - \Delta h = \frac{hc_0}{2}(\Delta\tilde{\nu}_{00}^{abs} + \Delta\tilde{\nu}_{00}^{flc}) \tag{1.55}$$

where Δ always refers to reaction (1.36). From equations (1.49) and (1.55) one obtains the final equation

$$\text{pK}^* = \text{pK} + \frac{hc_0}{kT} \cdot \frac{\Delta\tilde{\nu}_{00}^{\text{abs}} + \Delta\tilde{\nu}_{00}^{\text{flc}}}{2 . \ln 10} \tag{1.56}$$

for the determination of pK*-value of solutes which undergo adiabatic photo-ionization reactions. In most cases the 00-transitions are not observable and it is necessary, therefore, to make two further assumptions concerning equal band shape

$$(\tilde{\nu}_{\text{max}}^{\text{abs}} - \tilde{\nu}_{00}^{\text{abs}})_{\text{B}} = (\tilde{\nu}_{\text{max}}^{\text{abs}} - \tilde{\nu}_{00}^{\text{abs}})_{\text{BH}^+} \tag{1.57}$$

$$(\tilde{\nu}_{00}^{\text{flc}} - \tilde{\nu}_{\text{max}}^{\text{flc}})_{\text{B}} = (\tilde{\nu}_{00}^{\text{flc}} - \tilde{\nu}_{\text{max}}^{\text{flc}})_{\text{B*H}^+} \tag{1.58}$$

In practice one determines the difference between two mean values which can be obtained graphically from equations (1.56), (1.57) and (1.58)

$$\text{pK}^* = \text{pK} + \frac{hc_0}{\ln 10 . kT}\left(\frac{(\tilde{\nu}_{\text{max}}^{\text{abs}} + \tilde{\nu}_{\text{max}}^{\text{flc}})_{\text{B}}}{2} - \frac{(\tilde{\nu}_{\text{max}}^{\text{abs}} + \tilde{\nu}_{\text{max}}^{\text{flc}})_{\text{BH}^+}}{2}\right) \tag{1.59}$$

Values thus obtained for lumacine derivates are given in reference 63. It is found that lumacine, 3-methyl-lumacine, and 2-methoxy-4-hydroxy-

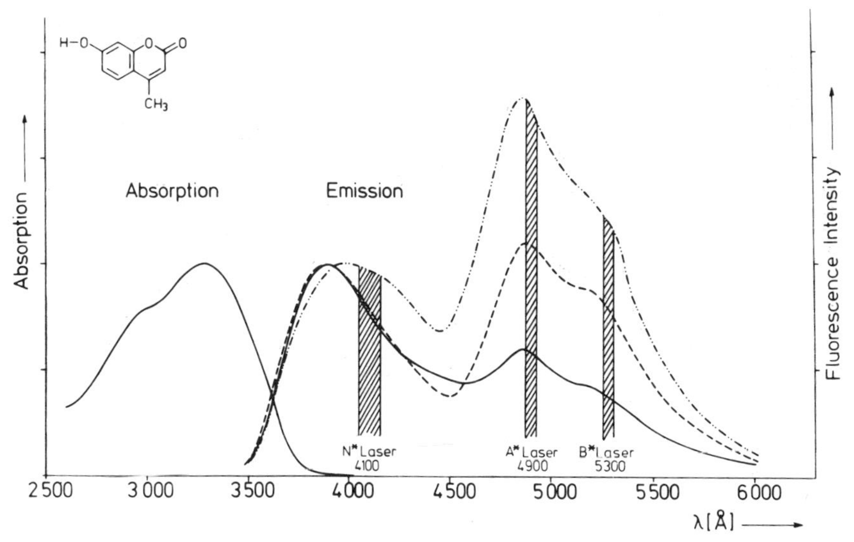

Figure 1.14 Absorption and emission spectra of HCl-acidified ethanolic 4-MU solutions. 10^{-4} M, 4-MU, with ——— 0·01 M HCl, ----- 0·1 M HCl, –··–··– 1 M/1 HCl. The emission spectra were obtained with an excitation wavelength of 3472 Å. The height of the first peak of the emission spectrum has in each case been normalized to the absorption peak[65]

pteridine do not exist as neutral molecules, but either as cations or as anions in their first excited singlet state for every pH-value of aqueous solutions.

The kinetics of photo-ionization can be studied by dye laser techniques if the originally excited compound, N*, as well as the adiabatic photo-

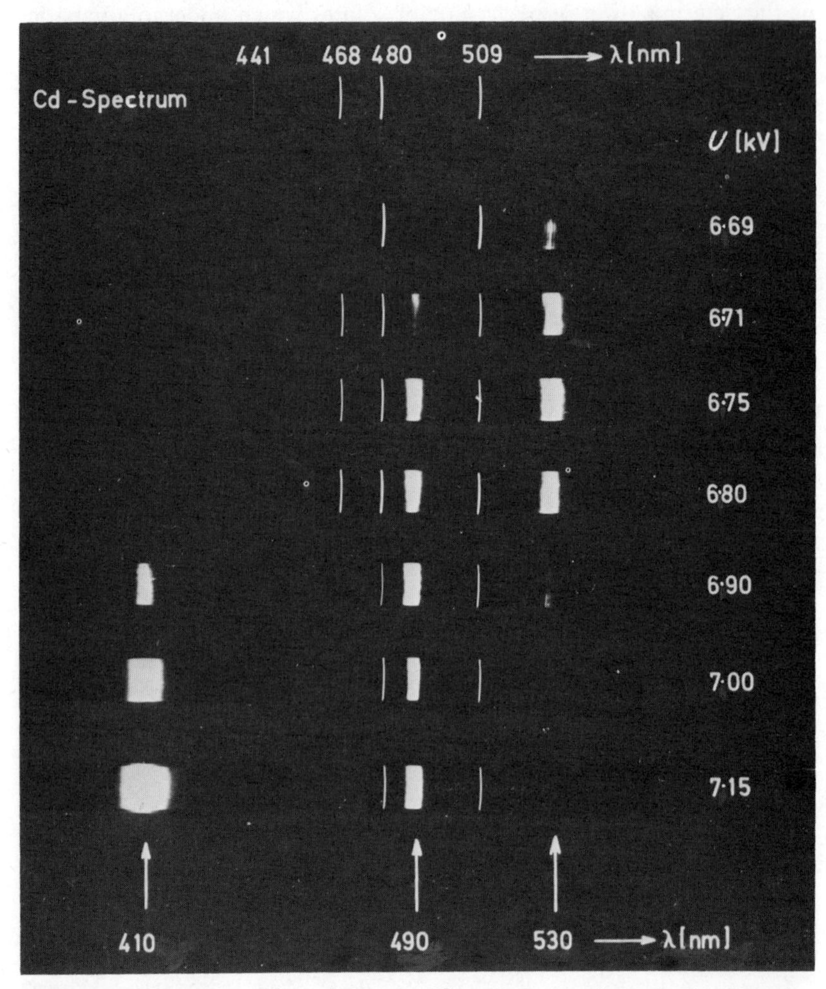

Figure 1.15 Dependence of the 4-MU dye laser spectra on the excitation intensity. $c_{4\text{-MU}} = 5 \times 10^{-2}$ M, $c_{HCl} = 0{\cdot}1$ M, U = ruby laser flashlamp voltage in kV, as U is proportional to the 6943 Å ruby laser output. The dye solutions were pumped using a transversal arrangement. The second-harmonic-generated output of a Q-switched ruby laser is focused into a 1 cm square quartz cell containing the acidified 4-MU solution. The maximum output of the ruby laser is 300 MW at 6943 Å, the full half-width of the laser pulses being 15 ns. The dye laser resonator is merely the dye cell in order to obtain identical resonator conditions for all three dye laser emissions[65]

reaction products, A* and B* etc., are laser-active, i.e. if the excitation photon density is so high that the laser thresholds of fluorescence photon densities from N*, A*, B*, . . . , all are exceeded. An example is 4-methylumbelliferone (4-MU) in an ethanolic solution acidified by hydrochloric acid.[64,68] At about 0·1 M/HCl the system shows the effect of multiple fluorescence and in a certain range of excitation intensities all three fluorescence transitions are laser-active simultaneously (Figure 1.14). Those three laser emissions occur at 4100 Å, 4900 Å and 5300 Å, respectively. The third emission only occurs within a HCl-concentration range of 10^{-2} M < c(HCl) < 2·5 M. This emission is quenched if the laser threshold for the emission of the neutral form of 4-methylumbelliferone at 4100 Å is exceeded (Figure 1.15).

These excited molecules deactivate by spontaneous or induced emission, or by radiationless processes, to unstable ground state cations which rapidly dissociate to give a neutral molecule N and a proton. The latter can be seen from the fact that the absorption spectra are independent of the acid concentration of the dye solution. It has been found that the long-wavelength laser emission of B* is quenched if

(i) the B* formation rate is too small due to insufficient proton concentration,
(ii) a quenching process originating from the presence of Cl^--ions becomes relevant in strongly acidified solutions,[70]
(iii) the threshold of the neutral 4-MU laser is exceeded, using high excitation intensities,
(iv) the proton concentration is too high. This may inhibit a ring opening reaction.

The process (iii) for the quenching of B* might be due either to the high concentration of N* and A* if one of these acts as a quencher for B* or due to special dye laser kinetics in the non-stationary starting period.

It has been shown[64] that the formation rate of B* is probably slower than that of A*. This fact would explain the sensitivity of B* laser emission to various processes which compete with the population of the B* state. As to the population of the states A* and B*, two different formation mechanisms seem possible:

(i) A parallel reaction

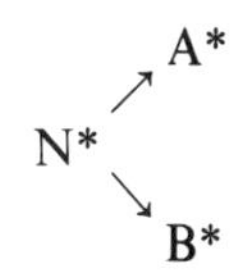

corresponding for instance to a direct protonation either on the ring-oxygen atom or on the carbonyl group.

(ii) A stepwise population N* → A* → B* corresponding to a protonation followed by a further reaction step which leads to the energetically favourable species B*. This step may be a tautomerization or a ring opening reaction.

The adiabatic photoreaction products A* and B* of the reaction

$$N + H^{+} \xrightarrow{h\nu_a} N^{*} + H^{+} \rightarrow \tag{1.60}$$

of 4-MU

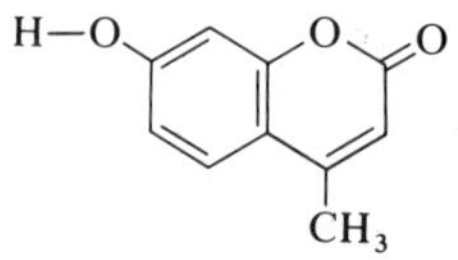

might either be

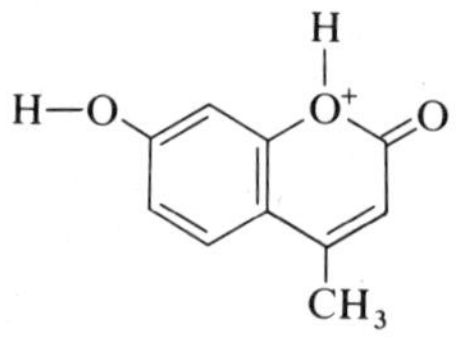

or

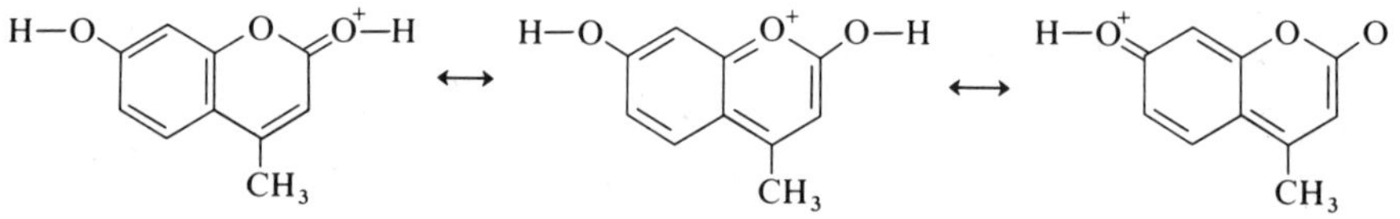

or

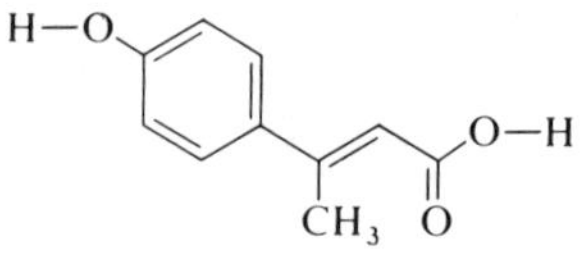

or

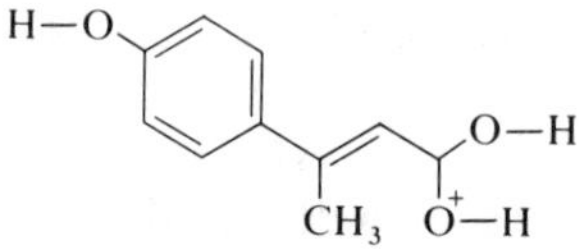

The fluorescence of the photoproducts A* and B* is missing, however, if the 7-OH group is absent or ethylated to the 7-OC_2H_5 group. The possible product of the ring-opening reaction of 4-MU, *o*-hydroxy-β-methyl-cinnamic acid, fluoresces green like B*. In ethanolic solutions with 3 M $HClO_4$ or with 5 M H_2O laser peaks have been observed at 4250 Å and 4400 Å which are attributed to the doubly protonated 4-MU* $(H^+)_2$ and to the de-protonated anion, respectively.[69]

(*bb*) Detailed studies have been made for the exciplex formation process in the case of aromatic hydrocarbon-aliphatic amine systems. From the time-resolved fluorescence studies at various temperatures, it has been made clear that, in a very viscous supercooled liquid state, a fraction of the exciplexes are formed immediately after the excitation and for these exciplexes the solvent reorientation is not yet accomplished, and the solvent relaxation occurs subsequently leading to the time-dependent red shift of the emission. Exciplexes are formed to some extent even in the glassy state at lower temperatures where, however, the solvent relaxation does not occur during the excited state lifetime. The mechanism of the relaxed exciplex formation from the Franck–Condon excited state has been discussed in comparison with the case of the tetracyanobenzene-toluene EDA complex.[66]

Fluorescence rise and decay processes were measured and rate parameters were determined for the pyrene-tri-*n*-butylamine and pyrene-*N*,*N*-diethylaniline exciplex systems in various solvents. It has been concluded that the additional activation energy over that for the diffusional motion is necessary for either of the exciplex formation and deactivating quenching processes in the case of the pyrene-tri-*n*-butylamine system and the rate determining step for these processes is the electron transfer in the encounter collision leading to the non-relaxed electron transfer state.[67]

1.4 References

1. F. Kohler, *The Liquid State*, p. 203, Verlag Chemie, Weinheim, 1972.
2. B. Gronau, E. Lippert and W. Rapp, *Ber. Bunsenges*, **76**, 432 (1972).
3. J. B. Birks, *Photophysics of Aromatic Molecules*, p. 162ff., Wiley, London, 1970.
4. E. Fischer, *Z. Naturforsch.*, **49**, 707 (1949).
5. R. J. Meakins, *Trans. Faraday Soc.*, **54**, 1160 (1958).
6. L. Pauling, *Proc. Nat. Acad. Sci. USA*, **25**, 577 (1939).
7. E. Lippert, *Z. Elektrochem.*, **61**, 962 (1957).
8. E. Lippert, *2. Intern. Farbensymposium Elmau 1964*, p. 342 (Ed. W. Foerst), Kongressband, Verlag Chemie, Heidelberg, 1965.
9. Y. Ooshika, *J. Phys. Soc. Japan*, **9**, 594 (1954).
10. E. Lippert, *Z. Naturforsch.*, **10a**, 541 (1955).
11. N. Mataga, Y. Kaifu and M. Koizumi, *Bull. Chem. Soc. Japan*, **29**, 465 (1956).

12. E. Lippert, *Accts. Chem. Res.*, **3**, 74 (1970).
13. H. E. Lessing, E. Lippert and W. Rapp, *Chem. Phys. Letters*, **7**, 247 (1970).
14. A. V. Aristov, N. G. Bakhshiev, V. A. Kuzin and I. V. Piterskaya, *Opt. and Spectrosc.*, **30**, 75 (1971).
15. C. P. Smyth, *Dielectric behaviour and structure*, pp. 63, 103 and 119, McGraw-Hill, New York, 1965.
16. E. Lippert, W. Lüder and F. Moll, *Spectrochimica Acta*, **10**, 858 (1959).
17. E. Lippert, W. Lüder, F. Moll, W. Nägele, W. Voss, H. Prigge and I. Seibold-Blankenstein, *Angew. Chemie*, **73**, 695 (1961).
18. E. Lippert and F. Moll, *Z. Elektrochem.*, **58**, 718 (1954).
19. E. Lippert, *Z. Phys. Chem. N.F.*, **6**, 125 (1956).
20. A. V. Aristov and V. A. Kuzin, *Opt. and Spectrosc.*, **30**, 77 (1971).
21. W. Rapp, H.-H. Klingenberg and H.-E. Lessing, *Ber. Bunsenges.*, **75**, 883 (1971).
22. N. G. Bakhshiev, Y. T. Mazurenko and I. V. Piterskaya, *Opt. and Spectrosc.*, **21**, 307 (1966).
23. H. Baba and Chie Mugiya, *Bull. Chem. Soc. Japan*, **43**, 13 (1970).
24. M. Kasha, *Disc. Faraday Soc.*, **9**, 14 (1950).
25. A. Winterstein and K. Schoen, *Naturwiss.*, **22**, 237 (1934).
26. Th. Förster, *Ber. Bunsenges.*, **73**, 737 (1969); *Pure Appl. Chem.*, **34**, 225 (1973).
27. E. Lippert, *Physik des flüssigen Zustands, St. Georgen*, (Ed. F. Kohler), **IV**, 31 (1972).
28a. M. R. Topp and P. M. Rentzepis, *Phys. Rev.*, **A3**, 358 (1971).
28b. Th. Kindt, E. Lippert, and W. Rapp, to be published.
29. A. Müller and E. Pflüger, *Chem. Phys. Letters*, **2**, 155 (1968).
30. P. M. Rentzepis, *Chem. Phys. Letters*, **3**, 717 (1969).
31. H. Boos, E. Lippert and W. Lüder, *Advances in Molecular Spectroscopy*, p. 443, (Ed. A. Mangini), Pergamon Press, London, 1962.
32. E. M. Kreidler, *Dissertation*, Stuttgart, 1966.
33. (a) E. Lippert and F. Schneider, *Ber. Bunsenges.*, **72**, 1155 (1968); (b) E. Lippert and F. Schneider, *Ber. Bunsenges.*, **74**, 624 (1970).
34. F. Hirayama, *J. Chem. Phys.*, **42**, 3163 (1965).
35. O. Schnepp and M. Levy, *J. Amer. Chem. Soc.*, **84**, 172 (1962).
36. R. G. Bennett, *J. Chem. Phys.*, **41**, 3037 (1964).
37. A. Weller, *Z. Phys. Chem. N.F.*, **17**, 224 (1958).
38. (a) E. Lippert and H. Prigge, *Ber. Bunsenges.*, **69**, 458 (1965); (b) E. Lippert and H. Prigge, *Ber. Bunsenges.*, **64**, 662 (1960).
39. Th. Förster, *Z. Elektrochem.*, **54**, 531 (1950).
40. Th. Förster and K. Kasper, *Z. Phys. Chem. N.F.*, **1**, 19 (1954).
41. H. Leonhardt and A. Weller, *Ber. Bunsenges.*, **67**, 791 (1963).
42. P. Kröning and M. Glandien, *Z. Phys. Chem. N.F.*, **72**, 76 (1970).
43. S. Murata, C. Iwanaga, T. Toda and H. Kokubun, *Ber. Bunsenges.*, **76**, 1176 (1972).
44. D. Huppert, J. Jortner and P. Rentzepis, *J. Chem. Phys.*, **56**, 4826 (1972).
45. H. Baba, A. Nakajima, M. Aoi and K. Chihara, *J. Chem. Phys.*, **55**, 2433 (1971).
46. P. Wannier, P. Rentzepis and J. Jortner, *Chem. Phys. Letters*, **10**, 193 (1971).
47. O. S. Khalil, R. H. Hofeldt and S. P. McGlynn, *Chem. Phys. Letters*, **17**, 479 (1972); *Spectroscopy Letters*, **6(3)**, 147 (1973).
48. N. Mataga and N. Nakashima, Private communication, 1973.
49. K. Rotkiewicz, K. H. Grellman and Z. Grabowski, *Chem. Phys. Letters*, in press.
50. N. Nakashima, H. Inoue and N. Mataga, *Bull. Chem. Soc. Chim. Japan*, in press.

51. G. Herzberg, *Molecular Spectra and Molecular Structure, Vol. III*, D. Van Nostrand, Princeton, 1966.
52. E. Lippert, *Luminescence of Organic and Inorganic Materials*, p. 271 (Eds. H. P. Kallmann and G. M. Spruch), Wiley, New York, 1962.
53. (a) M. Zander, *Z. Naturforsch.*, **24a**, 254 (1969); (b) M. Zander, *Z. Naturforsch.*, **25a**, 445 (1970).
54. F. Schneider and M. Zander, *Ber. Bunsenges.*, **75**, 887 (1971).
55. N. Nakashima, N. Mataga, C. Yamanaka, R. Ide and S. Misumi, *Chem. Phys. Letters*, in press (1973).
56. E. Lippert, H. H. Klingenberg and W. Rapp, *Chem. Phys. Letters*, **18**, 417 (1973).
57. N. Mataga, *New Frontier*, 839 (1973).
57a. E. A. Chandross and H. T. Thomas, *Chem. Phys. Letters*, **9**, 393 (1971).
58. Th. Förster, *Comprehensive Biochemistry* (Eds. M. Florkin and E. Stotz), Elsevier, Amsterdam, 1967.
59. K. C. Ingham, M. Abu-Elgheit and M. A. El-Bayoumi, *J. Amer. Chem. Soc.*, **93**, 5023 (1971).
60. C. A. Taylor, M. A. El-Bayoumi and M. Kasha, *Proc. Nat. Acad. Sci.*, **63**, 253 (1969).
61. G. E. Busch, R. P. Jones and P. M. Rentzepis, *Chem. Phys. Letters*, **18**, 178 (1973).
62. N. Mataga and T. Kubota, *Molecular Interactions and Electronic Spectra*, Marcel Dekker, Inc., New York, 1970.
63. E. Lippert and H. Prigge, *Ber. Bunsenges.*, **44**, 662 (1960).
64. (a) C. V. Shank, A. Dienes, A. M. Trozzolo and J. A. Myer, *App. Phys. Letters*, **16**, 405 (1970); (b) A. Dienes, C. V. Shank and A. M. Trozzolo, *App. Phys. Letters*, **17**, 189 (1970); (c) C. V. Shank, A. Dienes and W. T. Silfvast, *App. Phys. Letters*, **17**, 307 (1970).
65. Th. Kindt, E. Lippert and W. Rapp, *Z. Naturforsch.*, **27a**, 1371 (1972).
66. F. Ushio and N. Mataga, *Z. Phys. Chem. N.F.*, **79**, 150 (1972).
67. N. Nashima, N. Mataga and C. Yamanaka, *Intern. J. Chem. Kinetics*, in press (1973).
68. M. Nakashima, J. A. Sousa and R. C. Clapp, *Nature Phys. Sci.*, **235**, 16 (1972).
69. A. Bergman and J. Jortner, *J. Luminescence*, **6**, 390 (1973).
70. A. Bergman, R. David and J. Jortner, *Opt. Comm.*, **4**, 431 (1972).
71. P. Emmerich and J. Koziel, private communication (1973).

2 *The photophysics of benzene in fluid media*

Robert B. Cundall and Stephen McD. Ogilvie

2.1 Introduction

An extensive literature already exists on the behaviour of excited states of benzene in all phases. In spite of the considerable amount of accumulated information, the character and properties of these states are still very far from being clearly defined. This is true even for the lowest singlet and triplet states, upon which most research effort has been directed. The special interest in benzene arises from the fact that it is the simplest aromatic molecule in terms of structure, and there is the hope that understanding of its photophysical and photochemical properties will provide a pattern for a full appreciation of the behaviour of excited $\pi\pi^*$ states in general. The achievements of molecular spectroscopy provide, for benzene, a wealth of knowledge about vibronic and rotational level structure which is unique for a polyatomic molecule. It is therefore surprising to find that the first and second singlet excited states have only recently been conclusively shown to have orbital arrangements which are ${}^1B_{2u}$ and ${}^1B_{1u}$, respectively. The lowest triplet state, which has been much studied in low-temperature phosphorescence, is almost universally assigned to ${}^3B_{1u}$. Also, some states for which there is no convincing spectroscopic evidence have been proposed to explain some of the recent and puzzling observations on singlet excited state relaxation.

The high symmetry (D_{6h}) of the benzene ring, which allows the remarkably detailed analysis of the gas-phase spectrum, prevents it being the general model which we would like for the whole field of aromatic photochemistry. Nevertheless, this simplicity makes this molecule the most suitable which we have available for accomplishing the amalgamation of spectroscopic data with photophysical and photochemical information for elucidation and quantitative assignment of the various radiative and non-radiative transitions. This objective is still far from being achieved and in this article only experimental evidence will be discussed. The emphasis and selection of material reflects our own interests in the measurement of intersystem crossing yields over a variety of conditions which, together with fluorescence data, makes it possible to measure the efficiency of the most important photophysical processes. The results show the fluorescence and triplet formation from the first excited singlet state of benzene and its substituted derivatives does not usually account for all the excitation energy, except at low temperature. The possible internal conversion processes and chemical reactions which must be invoked to explain this discrepancy remain obscure, and this is the major problem in benzene photophysics at the present time. It is hoped that this article gives an overall picture of the state of experimental investigation and indicates directions for further research.

For convenience the subject matter is divided into essentially separate accounts of the gas and liquid phases.

2.2 Benzene in the gas phase

The spectroscopy of benzene in the vapour phase is a classic study extending over several decades, but the aspects of benzene photophysics which will be described have developed during the last twelve years from three researches published in the early 1960s. These were (i) the estimation of the fluorescence lifetime of benzene vapour by Donovan and Duncan,[1] (ii) the determination of the fluorescence and triplet yields of benzene using the biacetyl technique by Ishikawa and Noyes,[2,3] and (iii) development of the butene-2 isomerization technique for triplet estimation by Cundall and Palmer.[4] These, and similar techniques, have been developed and extensively improved, and at the same time detailed comparisons have been made with further analysis of the high-resolution ultraviolet spectrum.[5]

The unusually discrete nature of the vibronic levels seen in the benzene spectrum allows pumping of particular selected levels with suitably tuned monochromatic light. At sufficiently low pressures electronic relaxation of these vibronic states occurs before collision. The study of single vibronic level fluorescence is probably the most interesting and active area for study of benzene and its simple derivatives at the moment (1973). A masterly review of this subject and relevant aspects of benzene spectroscopy has been provided by Parmenter[6] and accordingly this review will deal with the subject from a more general standpoint (see also Stockburger, Chapter 2, Vol. 1).

Many of the electronic relaxation studies on benzene vapour have used the practically convenient mercury 253·7 nm resonance line for excitation. It is unfortunate that this line excites five or six vibrational levels about 2000 cm^{-1} above the zero-point level,[7] but analysis of the emission spectrum shows that at pressures above 10 torr fluorescence comes from a Boltzmann distribution of states. Whether all electronic processes in such systems occur after equilibration of vibrational energy is an important question which requires an answer. It is certainly not the case if excitation is brought about by light of wavelengths less than 252 nm.

Most of the experimental data for benzene vapour has been obtained at room temperature (*c.* 293–300°K), so unless otherwise stated it is implied that all data quoted are for these conditions.

2.2.1 *Absorption spectrum*

The electron molecular orbitals of benzene in order of increasing energy are a_{2u}, e_{1g}, e_{2u} and b_{2g}. Single electron excitation out of the e_{1g} orbital of the ground state configuration $(a_{2u})^2(e_{1g})^4$ can give the configurations $(a_{2u})^2(e_{1g})^3(e_{2u})^1$ and $(a_{2u})^2(e_{1g})^3(b_{2g})^1$. The former corresponds to B_{2u}, B_{1u} and E_{1u} excited states and the latter gives an E_{2g} state. Calculation[8] and experiment are consistent for the singlet and triplet state energies, and values for the zeroth levels are shown in Figure 2.1. The

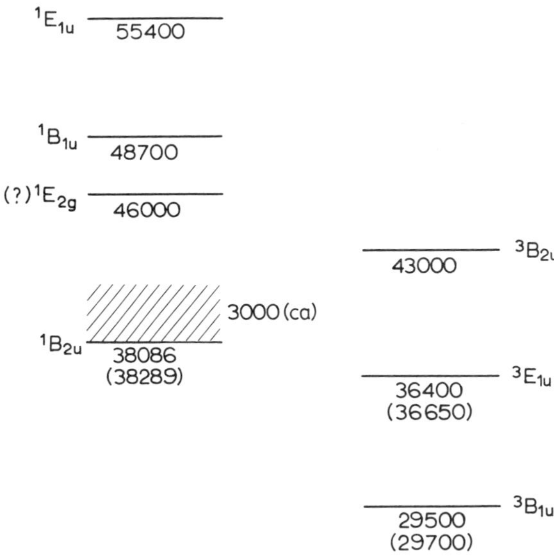

Figure 2.1 The zeroth energy levels (cm^{-1}) of the lower π, π^* states of benzene. The values, except for the $^1E_{2g}$ state, are derived from gas-phase spectra. The shaded region is the region in which fluorescence and intersystem crossing occurs in the vapour phase. The figures in brackets are for benzene-d$_6$

transition $^1E_{2g} \leftarrow {}^1A_{1g}$ is symmetry forbidden and would be expected on theoretical grounds to be at around 210 nm. Recent studies of benzene in rare gas and other matrices show evidence of a hitherto unreported transition, which is largely obscured by the more intense $^1B_{1u} \leftarrow {}^1A_{1g}$ transition, near 46,000 cm^{-1} which has been attributed to excitation of the $^1E_{2g}$ state.[9]

In spite of detailed work, there is still much to learn about the excitation of benzene. The transition in the region of 260 nm is certainly established as the $^1B_{2u} \leftarrow {}^1A_{1g}$ ($f = 0{\cdot}001$, $\varepsilon \approx 200$, $\lambda_{max} = 264$ nm) transition.[10] The second transition has broadened vibrational structure due to fast non-radiative decay, but it is without doubt mainly due to the $^1B_{1u} \leftarrow {}^1A_{1g}$ transition ($f \approx 0{\cdot}1$, $\varepsilon \approx 6000$, $\lambda_{max} = 210$ nm). Both of these are symmetry forbidden. The third transition, ascribed mainly to $^1E_{1u} \leftarrow {}^1A_{1g}$, is fully allowed and it has a maximum at 180 nm, with $\varepsilon \approx 10^5$ and $f = 1$. Due to the high symmetry of the ground $^1A_{1g}$ state antisymmetric e_{2g} vibrations are required to contribute to the intensity of the two lower transitions through some degree of mixing of the lower $^1B_{2u}$ and $^1B_{1u}$ states with the higher, but allowed, $^1E_{1u}$ state. The degenerate mode providing this intensity has frequencies of 606 cm^{-1} and 521 cm^{-1} in the ground and excited states. As a consequence of the symmetry forbidden character of the $^1B_{2u} \leftarrow {}^1A_{1g}$ transition, the electronic origin (0, 0) in the 260 nm band does not appear in the gas phase spectrum in either absorption or emission, nor do vibronic transitions other than those allowed for a vibrationally stimulated $^1B_{2u} \leftarrow {}^1A_{1g}$ transition. The differences in vibrational spectral structure observed in condensed phases arise from environmental perturbations.

Theoretically, singlet–triplet transitions are possible with very low probability, but only the $T_1 \leftarrow S_0$ transition has been detected in the vapour and liquid states in the presence of oxygen at about 310 nm.[11] The characters of the T_1 state[12] amd T_2 state[13] have been identified, but the energy and nature of T_3 is based on tenuous evidence, for which support has been obtained from electron scattering measurements. The T_2 and T_3 states are of interest in understanding non-radiative processes which occur from the $^1B_{2u}$ and higher singlet states.

2.2.2 *Processes involving excited states*

The absorption of light can raise a molecule to a high vibrational level of the excited singlet state. As a rule this initially formed vibronic state rapidly equilibrates itself with its surroundings by loss of vibrational energy and then more slowly relaxes its electronic excitation by fluorescence, internal conversion, intersystem crossing, or chemical reaction. The vibrational deactivation of the triplet state formed by intersystem crossing is also fast, otherwise the reverse $S \leftarrow T$ transition would occur. By working with gas-phase systems at different pressures, the relative efficiencies of the different processes can be assessed.

2.2.2.1 *Fluorescence*

The fluorescence yield of benzene vapour depends on the pressure, temperature and exciting wavelength. The results can be conveniently discussed for two different pressure conditions.

(a) *The high-pressure region.* At total pressures above 10 torr and an excitation wavelength λ_{ex} greater than 250 nm, the fluorescence quantum yield ϕ_{FM} has an essentially constant value of 0·18.[3,14,15] At λ_{ex} less than 250 nm both Poole[14] and Noyes and Harter[16] observed that ϕ_{FM} decreases with λ_{ex} and it becomes zero at $\lambda_{ex} < 240$ nm. Poole[14] also determined the fluorescence quantum yields of benzene and benzene-d_6 in the range 5–15 torr at 30, 60 and 90°C. The yields decrease with increasing temperature both in the presence and absence of 80 torr of cyclohexane. Under all conditions the fluorescence quantum yield of C_6D_6 exceeds that of C_6H_6.

By measuring the quenching of fluorescence, Lee and co-workers[17] determined the rates of intermolecular electronic energy transfer from the $^1B_{2u}$ state of benzene in the high-pressure region to a variety of π-bonded molecules. The most efficient quenchers were carbonyl compounds and carbon disulphide, while conjugated dienes had quenching rate constants which were about ten times smaller. The importance of overlap between the energy loss spectrum of the donor and the excitation spectrum of the acceptor for efficient transfer of energy was also demonstrated. Steric hindrance influences the overall quenching rate constant for singlet energy transfer between $^1B_{2u}$ benzene and alphatic ketones.[18] The efficiency of energy transfer is found to be reduced by increased 2-methyl substitution around the carbonyl group, showing that close interaction of the excited state of benzene and chromophore of the acceptor is necessary for energy exchange. The transfer efficiency is close to the encounter rate calculated by hard sphere collision theory.

Morikawa and Cvetanovic[19] have also determined the rates of collisional deactivation of the $^1B_{2u}$ state by a series of mono- and di-olefins. Mono- and non-conjugated dienes showed only a small quenching effect, but conjugated dienes were more efficient and introduction of both internal and terminal methyl groups increased the quenching effectiveness of 1,3-butadiene. The methyl substituents evidently exert an electronic effect, and consequently it was concluded that an electron–donor–acceptor complex is formed during the quenching process:

$$C_6H_6(^1B_{2u}) + Q \underset{k_2}{\overset{k_1}{\rightleftharpoons}} C_6H_6Q \xrightarrow{k_d} \text{quenching products}$$

The stability of the postulated intermediate increases with the electron donating ability of the olefin, and the quenching rate constant k_q is given by $k_1 k_d (k_d + k_2)^{-1}$.

Photoaddition of benzene excited in the 260 nm region to *cis*-butene-2 has been found in both the gas[20,21] and liquid phases.[22] Morikawa, Brownstein and Cvetanovic[21] found four different adducts were formed from each butene-2 isomer in the gas phase, and they concluded that fluorescence quenching and stable adduct formation are related processes.

Morikawa and Cvetanovic[23] showed that oxygen ($^3\Sigma_g$) quenches the fluorescence of benzene at about every collision without noticeably decreasing the concentration of benzene triplet. It appears therefore that the $^1B_{2u}$ state is converted to the $^3B_{1u}$ triplet as follows[24]

$$C_6H_6(^1B_{2u}) + O_2(^3\Sigma_g) \rightarrow C_6H_6(^3B_{1u}) + O_2(^1\Delta_g)$$

Chlorinated compounds are also very effective quenchers[25] of the $^1B_{2u}$ state.

The experimental data, taken as a whole, show that the $^1B_{2u}$ state of benzene can be quenched by four mechanisms, (i) electronic energy transfer (carbonyl compounds), (ii) enhancement of $S_1 - T_1$ intersystem crossing (O_2, heavy-atom-containing molecules), (iii) charge transfer interaction (chlorinated molecules and olefins) and (iv) chemical addition (olefins).

The fluorescence decay time of the $^1B_{2u}$ state under the appropriate conditions is necessary for the calculation of the absolute quenching rate constants. A number of determinations of this parameter have been made and are discussed later.

(b) *The intermediate- and low-pressure region.* Kistiakowsky and Parmenter[26] found that the fluorescence quantum yield of benzene excited at 253·7 nm increased from 0·18 to about 0·3 when the total pressure of the system (C_6H_6 together with cyclohexane, or CO_2) was reduced from 10 to 0·46 torr. In an extension of this work Anderson and Kistiakowsky[27,28] showed that the fluorescence quantum yields of benzene-h_6, benzene-d_6 and benzene 1,4–d_2 excited at 253·7 nm have constant values of 0·39, 0·53 and 0·38 in the pressure range 0·4 to 10^{-2} torr. The range of benzene pressure was extended down to 7×10^{-5} torr by Parmenter and White[29] and to 7×10^{-3} torr by Douglas and Mathews.[30] In each case the low-pressure values of the fluorescence yield remained constant and did not show any tendency to approach unity.

Kistiakowsky and Parmenter[26] discussed their results in terms of a postulated mechanism in which three vibrational levels of the $^1B_{2u}$ state were involved, each with the same probability of fluorescence. The best-fit curve gave the relative probabilities for intersystem crossing to be 1, 1·7 and 0·5 from the lowest, intermediate, and highest levels. Alternative mechanisms can, of course, be made to fit the data. Strickler and Watts[31] suggested an alternative two-vibronic-level mechanism. They assumed a level S′ (radiative rate k'_{FM}) with one quantum of e_{2g} vibration and another level S (radiative rate k_{FM}) not vibrationally excited in this way, which is populated by deactivation of S′. If $k'_{FM} = 3k_{FM}$ the data are reproduced with identical intersystem crossing rates from both states. A multistep cascade process is a realistic alternative scheme, but the value of a curve-fitting analysis is dubious, especially in view of the fact that several levels are excited by the 253·7 nm light. Direct spectroscopic observations of individual vibronic states are possible for benzene, and these can show the detailed nature of the relaxation of different vibronic states.

Parmenter and Schuyler[32,33] studied the fluorescence spectra originating from individual vibronic levels of benzene, excited by light sufficiently monochromatic (*c.* 40 cm^{-1} band spread) to preclude excitation of multiple levels at pressures less than 0·2 torr to prevent collisional redistribution of energy prior to fluorescence. The spectra are similar in appearance and consist of bands whose displacement from the exciting line are the same for all levels. When excitation is into one of the lower vibrational levels of $^1B_{2u}$ the spectra are sharp and structured, but excitation into higher levels tends to give more diffuse and complex spectra due to intramolecular redistribution of the vibrational energy. This is exemplified in Figure 2.2. The fluorescence yield drops sharply to zero

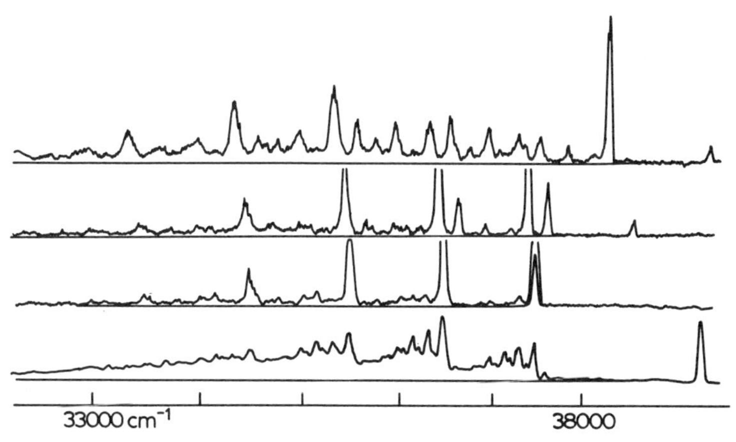

Figure 2.2 Single vibronic level fluorescence spectra from the $^1B_{2u}$ state of benzene. The first three spectra are from the vibrational levels, $(\nu'_6 - \nu'_1)$ (at 1446 cm^{-1} above the zero-point energy), ν'_6 (523 cm^{-1}) and zero-point energy respectively from the top. They were obtained by tuning a 40 cm^{-1} band from a Xe continuum at an appropriate absorption band in 0·1 torr of benzene. Analysis of these spectra show they originate from the levels excited by absorption. The bottom spectrum is fluorescence for a Boltzmann distribution of vibronic levels of $^1B_{2u}$ benzene obtained by adding 40 torr of *iso*pentane to 0·1 torr of benzene excited at 253·6 nm (after Parmenter and Schuyler[36])

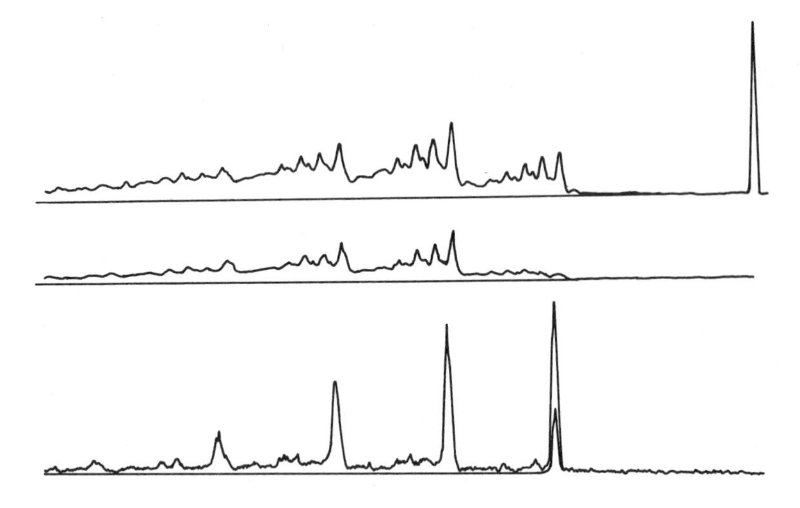

Figure 2.3 A comparison of the spectrum of fluorescence from the zero-point level with those from a 300°K Boltzmann distribution of vibrational levels in the $^1B_{2u}$ state of benzene vapour. The bottom spectrum is from the zero-point level. The upper spectra are from Boltzmann distributions of vibrational levels obtained by addition of *iso*pentane. The top spectrum is from 0·1 torr of C_6H_6 in the presence of 100 torr of *iso*pentane using the 253·7 nm Hg line for excitation. The centre spectrum is obtained from 8 torr of benzene with 92 torr of added *iso*pentane also excited by light which produces the zero-point energy level. The centre spectrum resembles the upper with exception of the short wave group of bands due to reabsorption of fluorescence by benzene at high pressures (after Parmenter and Schuyler[32])

when vibrational levels of energy greater than 2500 cm^{-1} above the zeroth energy level are excited, due to the sudden onset of an efficient competing intramolecular radiationless process. Figure 2.3 shows the simple fluorescence emission seen from the zeroth vibrational level of the $^1B_{2u}$ state, in comparison with that obtained after equilibration to the Boltzmann distribution. The single vibronic level excitation technique demonstrates that the lowest vibrational level of the $^1B_{2u}$ state emits to comparatively few excited vibrational levels of the ground state to give a simple fluorescence spectrum, whose structure is governed by vibrational symmetry conditions. When benzene vapour is excited to higher vibrational levels and subsequently relaxes to the Boltzmann distribution, the emission spectrum is very complex and consists of many bands with an underlying continuum. Emission occurs from many levels to an appropriately wider range of vibrational levels of the ground state. The collision cross-section for energy redistribution or relaxation is large and about five times the gas kinetic hard sphere collision cross-section. Redistribution of vibrational energy in excited benzene is more efficient than energy transfer,[34] and detailed studies have been made on the vibrational relaxation of the $^1B_{2u}$ state following 253·7 nm irradiation.[35] Parmenter and Schuyler[36] have also determined the fluorescence quantum yields of three vibronic levels in isolated benzene molecules excited to the first singlet state. Values were obtained at pressures less than 0·1 torr for the zero-point, v_0^1 (523 cm^{-1}) and $v_6^1 + v_1^1$ (1446 cm^{-1}) levels. The respective fluorescence quantum yields of 0·22, 0·27 and 0·25 were obtained by comparison with the high-pressure value of $\phi_{FM} = 0{\cdot}18$ due to Noyes[3,15] and Poole[14] with the same pressure of benzene together with 40 torr of *iso*pentane.

Measurements of single vibronic level fluorescence are not usually possible for substituted benzenes since the density of vibrational levels is very high and the rigorous symmetry rules, which apply to benzene, are relaxed by substitution and rapid redistribution of vibrational energy occurs.

2.2.2.2 *Lifetime of the excited singlet state*

Donovan and Duncan[1] first measured the lifetime of the fluorescence of benzene vapour excited by a high-frequency electrodeless discharge in the low-pressure range. Excitation conditions were ill-defined, and the data have been superseded by values obtained by the modern techniques of pulse sampling and photon counting. Accurate determinations of rate constants for the various primary processes over a very wide range of conditions have been achieved.

Breuer and Lee,[37,38] using 4–7 ns duration pulses of light between 230 and 280 nm, determined the fluorescence decay time τ_M to be 60 ns by a pulse sampling method for pressures of benzene between 5 and 40 torr at room temperature. The corresponding value for benzene-d$_6$ is $\tau_M = 95 \pm 5$ ns. Nishikawa and Ludwig[39] have also measured τ_M as a function of excitation wavelength λ_{ex}. The benzene pressure p was varied between 4 and 75 torr and values of λ_{ex} of 260, 253 and 247 nm with a band width of 2 nm were used. At $p > 20$ torr, the decay is exponential and independent of λ_{ex}. At lower pressures the decays are non-exponential at $\lambda_{ex} = 253$ and 247 nm. At $\lambda_{ex} = 260$ nm the lifetime extrapolated to zero pressure is $\tau_M = 77 \pm 3$ ns. Chen and Schlag[40] measured τ_M for benzene in the pressure range 5–85 torr by a phase-shift method. Excitation was at $\lambda_{ex} = 259$ nm with a bandwidth of 1 nm, which was considered to populate only the first vibronic state. From the data they calculated the zero-pressure lifetime of the 6_1^0 state to be 80 ns, in contrast to 1·3 ns for the 6_0^1 state. Burton and Hunziker[41] also made phase shift measurements on benzene vapour at $\lambda_{ex} = 253{\cdot}7$ nm, between 3 and 10 torr in the presence of 1 atm of either N_2 or Ar, and they measured a lifetime of 80 ± 10 ns for the decay of the $6_1^0\ 1_1^0$ state at 274 nm. Selinger and Ware[42] consider the phase shift method to be unsuitable for low-pressure lifetime measurements, and they used a single photon counting technique. The fluorescence decay times were measured for benzene-h_6 and benzene-d_6 excited at 266·2, 259, 253 and 247·4 nm with 2 nm band pass, corresponding to the $B_0^0(6_1^0\ 1_0^0)$, $A_0^0(6_0^1\ 1_0^0)$, $A_1^0(6_0^1\ 1_1^1)$ and $A_0^2(6_0^1\ 1_0^2)$ transitions. In some experiments pressures of 0·1 torr were used under conditions comparable with the low pressure fluorescence yield measurements of Schuyler and Parmenter.[43] At $p = 0{\cdot}1$ torr τ_M decreases as the excitation energy increases, in the order 90, 80, 68 and 50 ns for benzene-h_6, and in the order 164, 125, 84 and 69 ns for benzene-d_6. At $p = 0{\cdot}5$ torr the corresponding τ_M values decrease to 84, 70, 55 and 41 ns for benzene-h_6, and to 184, 115, 78 and 61 ns for benzene-d_6. At a pressure of 50 torr of benzene-h_6, or of 0·5 torr of benzene-h_6 with 50 torr of *iso*pentane, $\tau_M = 72$ ns independent of λ_{ex} from 266·2 to 247·4 nm; for C_6D_6 under similar conditions $\tau_M = 92$ ns. The lifetimes of 0·5 torr of benzene-h_6 and benzene-d_6 decreased to 57 ns and 62 ns in the presence of 500 and 400 torr of *iso*pentane, respectively.

Luria[44] has investigated the fluorescence self-quenching of benzene vapour by lifetime measurements. In agreement with Selinger and Ware,[42] the lifetime is found to be independent of λ_{ex} between 245 and 265 nm at high pressures and, in accord with Nishikawa and Ludwig[39] and Selinger and Ware,[42] τ_M depends on benzene pressure. In contrast others have found the lifetime to be independent of benzene pressure.[37,38,41] The lifetimes (in ns) at 15 torr and self-quenching rate constants (in $10^8\ M^{-1}\ s^{-1}$) at different temperatures found by Luria[44] are 69 ± 2, 4·0, (25°C), 67 ± 2, 4·3 (30°C), 62 ± 2, 4·9 (40°C), 59 ± 2, 5·5 (50°C), 55 ± 2, 6·3 (60°C). The activation energy for self-quenching is 950 cm^{-1}; a value higher than that of 460 cm^{-1} found by Poole,[14] and roughly corresponding to 923 cm^{-1}, the ν_2 mode frequency. The bimolecular rate constant is $8{\cdot}4 \times 10^{10}\ M^{-1}\ s^{-1}$ for quenching by oxygen and is $3{\cdot}6 \times 10^8\ M^{-1}\ s^{-1}$ for quenching by xenon.

Although there have been a number of investigations of the effects of excitation to single well-defined vibronic levels in the collision-free region, the definitive work is certainly that of Spears and Rice.[45] Their fluorescence quantum yield and decay time data are shown in Table 2.1. It is evident that particular combinations of vibrations critically determine the magnitude of the radiative and non-radiative transition probabilities, and large variations are possible even for states with the same excess energy. There appears to be a monotonic dependence of the non-radiative lifetime on the number of quanta in a vibrational progression, and the

(1971).)

Vibronic state	Exciting transition		% of other transitions overlapped	Energy above origin (cm^{-1})	Observed[a] lifetime (ns)	Quantum[b] yield	Radiative lifetime (ns)	Non-radiative lifetime (ns)	Estimated[c] range of the pure vibronic lifetime (ns)
0	6^0_1	B^0_0	27	0	100	0·22	455	128	100–107
6^1	6^1_0	A^0_0	8	521	79	0·27[d]	290	108	79–80
$6^1 16^1$	$6^1_0 16^1_1$	A^1_0	8	764	77	0·26[d]	295	104	77–78
1^1	$6^0_1 1^1_0$	B^0_1	28 (A^1_0)	923	83	0·21	395	105	83–85
$6^1 16^2$	$6^1_0 16^2_2$	A^2_0	25	1007	66	0·23	285	86	61–68
6^2	6^2_1	C^0_0	13	1042	72	0·29[d]	250	101	72–78
10^2	$6^0_1 10^2_0$	M^0_1	52	1170	52	0·16	325	62	25–52
$6^1 1^1$	$6^1_0 1^1_0$	A^0_1	9	1444	71	0·21[d]	340	90	71–72
?	?	E^0_0	21	1470	65	0·19	340	80	63–66
$6^1 11^2$	$6^1_0 11^2_0$	K^0_0	7	1547	51	0·18	285	62	50–51
$6^1 1^1 16^1$	$6^1_0 1^1_0 16^1_1$	A^1_1	24	1687	61	0·16[d]	380	73	58–61
$6^1 10^1$	$6^1_0 10^2_0$	0^0_0	43	1691	52	0·24	215	68	45–52
1^2	$6^0_1 1^2_0$	B^2_0	48 (A^1_1)	1846	62	0·16	390	74	61–62
$6^1 1^1 16^2$	$6^1_0 1^1_0 16^2_2$	A^2_1	26	1930	49[c]	0·20[e]	245	61	41–49
$6^2 1^1$	$6^2_1 1^1_0$	C^0_1	9 (A^0_0) 10 (E^1_0)	1965	61	0·22[d]	280	78	59–61
$6^1 5^2$	$6^1_1 5^2_0$	W^0_0	49 (M^0_2)	2070	55	0·06	920	59	55–60
$6^1 1^2$	$6^1_0 1^2_0$	A^0_2	15	2367	55	0·17[d]	320	66	55–57
?	?	E^0_1	14	2393	39[e]	0·26[e,f]	150	53	35–39
$6^1 1^2 16^1$	$6^1_0 1^2_0 16^1_1$	A^1_2	31	2610	42	0·08	525	46	36–42
$6^1 10^2 1^1$	$6^1_0 10^2_0 1^1_0$	0^0_1	18	2614	42	0·06	700	45	41–42
1^3	$6^0_1 1^3_0$	B^0_3	31 (A^1_2)	2769	49	0·07	700	53	49–55
$6^2 1^2$	$6^2_1 1^2_0$	C^0_2	19 (E^1_1)	2888	47	0·03	1470[g]	48	44–47

[a] Entries subject to 1·2% uncertainty.

[b] All quantum yields stated relative to $\phi_F(A^0_0) = 0{\cdot}27$, measured by C. S. Parmenter and M. W. Schuyler. Quantum yields accurate to 15% (see notes d and f).

[c] The maximum range of lifetime for the pure vibronic state (see text).

[d] Quantum yields probably accurate to 5% (relative to A^0_0).

[e] Only one measurement for this state.

[f] Quantum yield is only accurate to 30%

[g] The large radiative lifetime implies that the observed lifetime is too large, probably due to band overlap (see text).

rate of change of lifetime within one progression is quasi-linear in the energy range 0–3300 cm^{-1}.

The results have been discussed by Spears and Rice[45] in terms of the current theories of non-radiative transitions. It should be remarked in this respect that it is not certain that intersystem crossing is the only process competing with fluorescence, even in the lower vibrational levels of the $^1B_{2u}$ state. The sudden increase in non-radiative decay found at 3000 cm^{-1} is due to a process, commonly referred to as Channel 3, which is not intersystem crossing and which may involve a chemical rearrangement, as will be discussed below.

The very large increase in the non-radiative processes at above 3000 cm^{-1} in benzene is not general behaviour for the lowest excited singlet states of all benzenoid molecules. For example, Rockley and Phillips[46] find no indication of such a decay channel in *p*-fluorotoluene even for vibrational energies in excess of 5000 cm^{-1}.

Abramson, Spears and Rice[47] have also carried out comprehensive single vibronic level studies on benzene-d_6 and fluorobenzene. The data for C_6D_6 are shown in Table 2.2. The same trends as those noted in C_6H_6 are again apparent. In this paper the data are discussed in terms of the theories of radiationless processes developed by Freed, Jortner and Rice.

2.2.2.3 *Triplet state formation*

Identification of the non-radiative processes mentioned previously requires measurement of triplet yields. Since the benzene triplet does not detectably phosphoresce in the vapour phase, photosensitization methods have been used for the measurement of the yield of this and related triplet states. The biacetyl and olefin isomerization techniques, which have been used most extensively for this purpose, arose from experiments which provided the first established examples of triplet excitation energy transfer from polyatomic molecules in the gas phase.

Ishikawa and Noyes[3] found that irradiation of benzene vapour in the 260 nm band sensitized the phosphorescence from small pressures of added biacetyl without any significant indications of fluorescence from excited biacetyl. Analysis of the kinetics showed that the excited species of benzene involved in the process was longer lived by several orders of magnitude than the $^1B_{2u}$ state of benzene, and it was assumed to be the triplet ($^3B_{1u}$) state. The fact that energy transfer to biacetyl can also occur from excited singlet states and the need to know the phosphorescence efficiency of biacetyl accurately are disadvantages of this particular technique for estimation of triplet state parameters. The states of biacetyl formed by singlet–singlet energy transfer are high in energy and undergo rapid decomposition under gas-phase conditions.

Table 2.2 Lifetimes and quantum yields: C_6D_6 [a]. (A. S. Abramson, K. G. Spears and S. A. Rice, *J. Chem. Phys.*, **56**, 2291 (1972).)

Exciting transition	Energy above[b] origin (cm^{-1})	Relative quantum yields[e]	τ_{obs} (ns)	ϕ_F [d]	τ_r (ns)	τ_{nr} (ns)
$B^0_0(6^0_1)$	0	0·84 ± 0·13	200 ± 14	0·35 ± 0·05	565 ± 105	309 ± 30
$B^1_0(6^0_1 16^1_1)$	205	0·86 ± 0·12	163 ± 28	0·36 ± 0·06	450 ± 100	255 ± 50
$A^0_0(6^1_0)$	498	1·00 ± 0·05	144 ± 2	0·42 ± 0·03	345 ± 25	249 ± 13
$A^1_0(6^1_0 16^1_1)$	703	1·05 ± 0·05	132 ± 2	0·44 ± 0·03	300 ± 25	234 ± 12
$B^0_1(6^0_1 1^1_0)$	879	0·88 ± 0·06	125 ± 6	0·37 ± 0·03	340 ± 30	199 ± 14
'K^0_0'	880?	1·12 ± 0·06	116 ± 2	0·47 ± 0·03	250 ± 20	220 ± 12
$A^2_0(6^1_0 16^2_2)$	914	0·89 ± 0·09	110 ± 4	0·37 ± 0·04	300 ± 40	176 ± 12
$C^0_0(5^2_1)$	996	1·14 ± 0·06	118 ± 2	0·48 ± 0·03	250 ± 20	227 ± 13
$A^0_1(6^1_0 1^1_0)$	1377	0·76 ± 0·04	110 ± 2	0·32 ± 0·02	345 ± 25	161 ± 5
$O^0_0(6^1_0 10^2_0)$	1408	0·83 ± 0·05	90 ± 1	0·35 ± 0·02	260 ± 15	139 ± 4
$A^1_1(6^1_0 1^1_0 16^1_1)$	1582	0·61 ± 0·03	83 ± 1	0·26 ± 0·02	325 ± 25	111 ± 3
$B^0_2(6^0_1 1^2_0)$	1758	0·66 ± 0·03	83 ± 3	0·28 ± 0·02	300 ± 20	115 ± 5
$A^2_1(6^1_0 1^1_0 16^2_2)$	1787	0·60 ± 0·03	76 ± 1	0·26 ± 0·02	300 ± 20	101 ± 3
$W^0_0(6^1_0 5^2_0)$	1825	0·28 ± 0·03	78 ± 4	0·12 ± 0·01	705 ± 85	88 ± 4
$M^0_1(6^1_0 1^1_0 10^1_1)$	1832	0·64 ± 0·03	87 ± 2	0·27 ± 0·02	325 ± 25	119 ± 4
$C^0_1(6^2_1 1^1_0)$	1875	0·70 ± 0·03	76 ± 1	0·29 ± 0·02	260 ± 20	107 ± 3
$A^0_2(6^1_0 1^2_0)$	2256	0·45 ± 0·02	69 ± 1	0·19 ± 0·01	360 ± 25	85 ± 2
'E^0_1'	2282?	0·53 ± 0·02	65 ± 1	0·22 ± 0·01	290 ± 20	83 ± 2
$Q^0_0(7^1_0)$	2320	0·40 ± 0·04	59 ± 1	0·17 ± 0·02	350 ± 40	71 ± 2
$A^1_2(6^1_0 1^2_0 16^1_1)$	2461	0·35 ± 0·02	55 ± 1	0·14 ± 0·01	380 ± 25	64 ± 1
$B^0_3(6^0_1 1^3_0)$	2637	0·19 ± 0·22	78 ± 1	0·08 ± 0·01	990 ± 110	85 ± 7
$C^0_2(6^2_1 1^2_0)$	2754	0·22 ± 0·02	54 ± 1	0·09 ± 0·01	605 ± 70	60 ± 6

[a] Assignments of F. M. Garforth and C. K. Ingold, *J. Chem. Soc.* 1948–433, Table I, as modified by J. H. Callomon, I. M. Dunn, and I. M. Mills in *Phil. Trans. Roy. Soc. London*, **A259**, 1104 (1966), Table VI.

[b] Based upon J. H. Callomon, T. M. Dunn, and I. M. Mills, *Phil. Trans. Roy. Soc. London*, **A259**, 1104 (1966).

[c] Relative to 6^1.

[d] Based upon value of 0·27 for 6^1 of benzene given by C. S. Parmenter and M. W. Schuyler, *Chem. Phys. Letters*, **6**, 339 (1970).

The olefin isomerization method arose from the observation of Cundall and Palmer[4] that the gas phase *cis–trans* isomerization of butene-2 was sensitized by an excited state of benzene. The suggestion that this was due to triplet energy transfer was confirmed by the fact that the fluorescence of benzene was not markedly affected by the presence of olefin.[48] The reaction has proved to be a simple and accurate monitor of triplet state yields for molecules like benzene in both vapour and liquid phases. Nothing is known about the triplet state of olefins, like butene-2, formed by triplet transfer, except that they are very short lived ($<10^{-10}$ s). No energy transfer effects from such a state have been established, probably due to rapid $T - S_0$ intersystem crossing in the olefins arising from intersection of the potential surfaces of the triplet and ground singlet states. The oxygen perturbation method[49] has given 78 kcal mole^{-1} for the triplet energy of butene-2, and electron scattering techniques provide similar values, but in view of the dependence of $T_1 - S_0$ energy separation on the degree of twisting around the C–C double bond such measurements must be accepted with caution in predicting possibilities for energy transfer. The butene-2 method is restricted to molecules with relatively high triplet energies. Di-olefins like the *cis*- and *trans*-piperylene isomers with lower energy triplet states may be used in the same way, but interactions with excited singlet states can be more significant.

The rate constant for transfer of triplet excitation to *cis*-butene-2 has been estimated by Haniger and Lee[51] to be $9 \pm 2 \times 10^9$ M^{-1} s^{-1}. A direct determination by Hunziker and Wendt[52] using the modulation technique gave $8{\cdot}95 \pm 0{\cdot}33 \times 10^9$ M^{-1} s^{-1} for C_6D_6. In spite of the good agreement, the reliability of the latter value could be affected by the dominance of second-order excitation state interactions which apply under the experimental conditions required by this technique.

Although olefins like butene-2 interact with the fluorescent excited singlet state, the effect with benzene is small and a correction can be made when necessary if high olefin concentrations are used for triplet scavenging. This correction is required in the case of some fluorinated benzene derivatives in which the triplet lifetimes are very short.

From consideration of all the available evidence, it seems satisfactory to assume that a common triplet state B* is formed for butene-2 by excitation of both the *cis* and *trans* isomer and that relaxation is adequately characterized by the simple kinetic steps

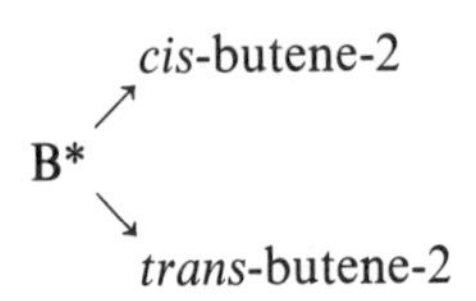

When *cis*-butene-2 is present at a concentration sufficiently high to scavenge effectively all the sensitizer triplet states the yield $\phi_{c\to t}$ for conversion of *cis*- to *trans*-butene-2 at low conversions is given by

$$\phi_{c\to t} = \phi_{\mathrm{TM}} b_t \quad (2.1)$$

where ϕ_{TM} is the yield of triplet sensitizer and b_t is the fraction of olefin triplets which relax to the *trans*-isomer. For a similar experiment with *trans*-butene-2 as scavenger

$$\phi_{t\to c} = \phi_{\mathrm{TM}} b_c \quad (2.2)$$

Since $b_c + b_t = 1$,

$$\phi_{\mathrm{TM}} = \phi_{c\to t} + \phi_{t\to c} \quad (2.3)$$

The measurement of initial isomerization yields in the presence of sufficient olefin therefore gives ϕ_{TM} with the minimum of mechanistic assumptions. The ratio b_t/b_c is unity[53] within the limits of error when triplet benzene is involved although it should be noted that this is not the case for all sensitizers and measurements of both $\phi_{c\to t}$ and $\phi_{t\to c}$ are always desirable.

If the amount of butene-2 is insufficient for complete scavenging of the aromatic triplet the *cis–trans* rearrangement yield is given by

$$\phi_{c\to t} = \phi_{\mathrm{TM}} b_t \left(\frac{k_{\mathrm{tr}}[cis\text{-}\mathrm{C_4H_8}]}{k_{\mathrm{T}} + k_{\mathrm{tr}}[cis\text{-}\mathrm{C_4H_8}]} \right) \quad (2.4)$$

where k_{tr} is the rate constant for triplet energy transfer and $k_{\mathrm{T}}(= \tau_{\mathrm{T}}^{-1})$ is the first- or pseudo-first-order decay constant for the aromatic triplet. Plots of $[cis\text{-}\mathrm{C_4H_8}]^{-1}$ *vs.* $\phi_{c\to t}^{-1}$ may be used to determine ϕ_{TM} by assuming a value for k_{tr}.[51–53]

Cundall and Davies[54] showed that the triplet quantum yield of benzene, as determined by the isomerization technique, is $\phi_{\mathrm{TM}} = 0{\cdot}72$ for excitation at 253·7 nm over the range 2–19 torr (branching ratio = 1). A value of $\phi_{\mathrm{TM}} = 0{\cdot}63$ was obtained for benzene-d$_6$. Noyes and Harter[16] also used the isomerization method to determine quantum yields of triplet formation for benzene at several excitation wavelengths. Between 252 and 266·8 nm they observed $\phi_{\mathrm{TM}} = 0{\cdot}72$, apart from an unacceptable value at 254 nm. The benzene pressure was varied between 6·5 and 26 torr with a constant mole fraction (0·38) of *cis*-butene-2. At $\lambda_{\mathrm{ex}} = 242$ nm ϕ_{TM} was considerably reduced, in confirmation of values obtained by the biacetyl method.

The $^3B_{1u}$ triplet state of benzene undergoes energy transfer to a number of other molecules and sensitizes various reactions. Only some examples which give indications of benzene triplet yields will be cited. Ho and

Noyes[55] found that dissociation of ketene is sensitized by singlet and triplet benzene. The fraction of triplet benzene formed after 252 nm excitation was 0·71, as estimated from the reaction of the resultant methylenes when the pressure was about 10 torr. Denschlag and Lee[56] have also investigated the benzene-sensitized photolysis of cyclobutanone and cyclobutanone-2-t in the gas phase with a benzene pressure of about 2 torr. A value of 0·71 for the triplet yield of benzene at infinite cyclobutanone pressure was deduced. In view of the complexity of these systems the yields are in better agreement with other methods than expected. Use of pyrazine[57] as photosensitizer gives $\phi_{TM} = 0{\cdot}79$ for benzene.

All the data, which have been reported in the literature for the gas-phase triplet yields, are for room temperature. Dunnicliff,[58] in our laboratory, has found that as the temperature is increased the quantum yields of *both* fluorescence and triplet formation decrease.

It is necessary for fruitful development of theory that the relaxation processes from single vibronic levels should be elucidated by measurements of intersystem crossing yields. This is a difficult task and has not yet been accomplished. Attempts to use the butene-2 technique at low pressures have only met with limited success. Sigal[59,60] studied the behaviour of irradiated C_6D_6/*cis*-C_4H_8-2 mixtures below 1 torr, the butene-2 isomer composition being determined by long-path infra-red spectrophotometry. The triplet yields appeared to decrease with both benzene and butene-2 pressures. Two alternative explanations were offered; (i) intersystem crossing, which may involve unimolecular and bimolecular components, and (ii) intersystem crossing which depends on the vibrational energy and increases with vibrational level. The latter suggestion is not supported by the observed decrease in triplet yield with wavelength, and it seems likely that surface effects could have seriously influenced the results of Sigal's experiments.

Kistiakowsky and Parmenter[26] studied the benzene-sensitized isomerization of *cis*-butene-2 at a pressure of 0·047 torr. The formation of the *trans* isomer shows that apparently triplet states were formed by intramolecular intersystem crossing, even at this low pressure. The system was further investigated by Anderson and Kistiakowsky[28] in the range 0·5 to 0·1 torr. A mechanism involving first-order rates for formation and destruction of triplet benzene, the latter competing with energy transfer to butene-2 molecules, was used to describe the results. At lower pressures isomerization of butene-2 proceeded with quantum yields much greater than unity. This could have been due to a surface effect or have arisen from an impurity induced chain reaction. This type of effect could explain some of Sigal's[59,60] data also.

Properties of the $^3B_{1u}$ state have been obtained by competitive quenching methods. Haninger and Lee have made a study of the efficiencies of various π-bonded molecules in quenching the benzene triplet using both the isomerization[61] and biacetyl techniques.[62] The increased effectiveness of olefin with alkyl substitution demonstrates the electrophilic character of the benzene triplet. Similar findings have been reported by Morikawa and Cvetanovic,[63] who made a comparison of the effect of cadmium (^3P) atoms and triplet benzene on the rate of isomerization of butene-2 in competition with a series of olefins. The triplet state of cadmium (5^3P_1) was quenched at approximately the same rate by all the olefins used, whereas the benzene triplet showed a variation in the relative quenching rate constant between 0·22 for ethylene and 2·67 for tetramethylene (*cis*-butene-2 = 1). Oxygen was only 30% as efficient as *cis*-butene-2, whilst the inert gases, SF_6, Xe, CO_2 and N_2O were all found to be inefficient triplet quenchers. Schmidt and Lee[64] also measured the relative rates of triplet energy transfer from benzene to mono- and di-olefins, and also CS_2, by competition with the sensitized isomerization of *cis*-butene-2. A deuterium effect on energy transfer was observed and an exchange interaction process for the energy transfer mechanism was proposed. However the observed correlations with olefin structure appear to require a contribution from some form of chemical interaction with an electrophilic $^3B_{1u}$ state of benzene during energy transfer. Exchange within the relatively long duration of a sticky collision or encounter may overcome any restricting Franck–Condon conditions, by allowing some measure of rotation around the olefin bond before energy transfer. This would be impossible if simple exchange transfer occurred at normal encounter separation distances without any strong interaction.

Cis- and *trans*-dideuteroethylene molecules which have been electronically excited by benzene photosensitization undergo some hydrogen atom scrambling to form the 1,1-dideuteroethylene isomer as well as *cis–trans* isomerization.[65] The extent to which each of these alternative processes occurs depends on the electronic energy transferred from the sensitizer as was shown by the use of sixteen aromatic sensitizers other than benzene. The relative energies of the lowest triplet states were estimated from the varying proportions of isotopic scrambling.

2.2.2.4 *Lifetime of the triplet state*

Several estimates of the lifetime τ_T of the benzene triplet in the gas phase have been based on competitive kinetics. Ishikawa and Noyes,[3] and Cundall and Davies[54] estimated $\tau_T \sim 10\mu s$ from the dependence of yields on pressure. In a re-examination of the butene-2 system. Lee[53] estimated $\tau_T \sim 100\ \mu s$. An investigation of the benzene/cyclopentanone system[66] allowed estimation of $\tau_T > 0{\cdot}3\ \mu s$. It is possible that impurities and the quenching effects of photoproducts could affect results at low partial pressures in such competitive experiments and direct measurements are required. Parmenter and Ring[67] used a flash sensitization technique, which allowed direct observation of triplet–triplet transfer. A mixture of 20 torr of benzene and 0·01 torr of biacetyl was submitted to a 20 J 3 μs flash of 250–260 nm light. From the time dependence of the biacetyl phosphorescence emission a lifetime of $\tau_T = 26 \pm 5\ \mu s$ was deduced. Lee[53] refers to a private communication from Parmenter in which a $^3B_{1u}$ lifetime of $70 \pm 20\ \mu s$ is given for a 2 torr benzene system. The determination of $3{\cdot}42 \times 10^{10}\ \mathrm{M}^{-1}\ \mathrm{s}^{-1}$ for energy transfer of the benzene triplet to biacetyl by the same method allows values for the rate constants for $^3B_{1u}$ triplet quenching to be calculated from the relative values.[62]

A modulated transient absorption technique has been developed by Burton and Hunziker[68] which uses Hg(6^3P_0) sensitization to populate the benzene triplet state. The decay is considered to be pseudo-first order and mainly due, because of the experimental conditions, to bimolecular triplet–triplet interaction. For benzene-h_6 and benzene-d_6 triplet decay rates of $3 \times 10^3\ \mathrm{s}^{-1}$ and $2 \times 10^3\ \mathrm{s}^{-1}$ were calculated for pressures of less than 1 torr in the presence of 300–800 torr of nitrogen. The absorption band at 231·5 nm, seen in this system, is assigned to an allowed $^3E_{2g} \leftarrow {}^3B_{1u}$ transition[69] with oscillator strengths of 0·07 and 0·06 for C_6H_6 and C_6D_6. Because of the high exciting light intensities used, the results obtained by this technique have proved controversial and further developments must be awaited.

Using a technique that has been invaluable in other areas, observations of an increase in the concentration of naphthalene triplet in the pulse radiolysis of benzene/naphthalene vapour mixtures at 390°K allow calculation of $1{\cdot}3 \times 10^5\ \mathrm{s}^{-1}$ ($\tau_T = 7{\cdot}7\ \mu s$) for the decay constant of the $^3B_{1u}$ state of C_6H_6 at a partial pressure of 18 torr.[70]

Hunter and Stock[71] have recently used a new technique which involves measurement of the frequency dependence of the acoustic signal resulting from transformation of electronic energy through vibrational to translational energy, following initial intensity-modulated excitation of the first benzene singlet. This was used to measure the triplet lifetime over a range of benzene pressures (0 to 6 torr). As suggested by some of the earlier work, the decay of the benzene triplet depends upon benzene pressure. A pressure-independent radiationless process has a rate constant of $(1{\cdot}7 \pm 0{\cdot}1) \times 10^3\ \mathrm{s}^{-1}$, and a collision efficiency of $1{\cdot}6 \times 10^{-4}$ is found for a second-order self-quenching process. The authors claim that their technique effectively discounts the possibilities of deactivation by any trace oxygen impurity. Clearly this appears to be a useful method and its full exploitation could be very fruitful. A strong perturbation of the triplet state during collisions with ground state molecules seems the most likely mechanism for the observed dependence of τ_T on benzene pressure. A triplet excimer involvement is in accord with the following scheme

$$\begin{array}{ccc} {}^3M^* + {}^1M & \underset{k_{-1}}{\overset{k_1[{}^1M]}{\rightleftharpoons}} & {}^3M_2^* \\ \downarrow k_2 & & \downarrow k_3 \\ {}^1M & & 2\,{}'M \end{array}$$

The reciprocal of the triplet lifetime is given by

$$\tau_T^{-1} = k_2 + \{k_3(k_{-1} + k_3)^{-1}\}k_1[{}^1M] \qquad (2.5)$$

Analysis of the results shows that $k_3(k_{-1} + k_3)^{-1}$ is $1{\cdot}6 \times 10^{-4}$ and $k_{-1} \gg k_3$, a situation opposite to that used to explain the shortening of triplet state lifetimes of some aromatic hydrocarbons in solution ($k_3 \gg k_{-1}$) where interaction between triplet and ground state molecules also explains the shortening of the triplet lifetime.[72] Flash photolysis experiments with anthracene and naphthalene in the vapour phase indicate that ϕ_{TM} tends to zero at low pressures, although at higher pressure ϕ_{TM} becomes constant.[73] Ashpole, Formosinho and Porter[74] ascribe these results to reversible intersystem crossing between singlet and triplet states when the pressure is sufficiently low for vibrational relaxation within the triplet manifold to be improbable. Such behaviour does *not* seem to apply to the benzene vapour system. The self-quenching mechanism proposed to explain the short lifetime of the benzene triplet state applies to the liquid phase also.

2.2.2.5 *Benzene ring isomerization*

A chemical reaction may explain the ill-defined internal conversion process (Channel 3) and an extensive literature deals with the organic photochemistry reactions of benzene and its derivatives. Foote, Mallon and Pitts,[75] and Shindo and Lipsky[76] have investigated the photochemistry of benzene vapour excited to its S_3 state. Both sets of data extrapolated to zero pressure give a photochemical quantum yield ϕ_c = 1·0, and the isomer formed was subsequently shown to be fulvene and a polymer.[77] Wilzbach, Kaplan and Harkness[78] showed that benzene-1,3,5d_3 rearranged to the 1,2,4 isomer when excited at 253·7 nm in the vapour phase, and they deduced that this occurs via a benzvalene intermediate with $\phi_c \approx 0{\cdot}03$. Subsequently Kaplan and Wilzbach[79] succeeded in isolating benzvalene from benzene excited within the first absorption band ($\phi_c \approx 0{\cdot}02$). The quantum yield of benzvalene formation was increased by the addition of neopentane and *cis*-butene-2 and also by excitation at shorter wavelengths. The interpretation seems to be that added gases stabilize 'hot' benzvalene and do so most effectively for those with least energy. The interesting finding that *cis*-butene-2 increased the steady state concentration of benzvalene from $2 \times 10^{-3}\%$ to $1{\cdot}3\%$ at 253·7 nm was attributed to a reduction in the sensitized decomposition of benzvalene by triplet benzene.

Bryce-Smith and Longuet-Higgins[22] have suggested a scheme which seems to provide, by and large, a satisfactory description of the facts. The pre-fulvene and Dewar-like diradical intermediates are consistent with the symmetry of the $^1B_{2u}$ and $^3B_{1u}$ excited states which give rise to them. Jano and Mori[80] have discussed the formation of fulvene from gaseous benzene at 184·9 nm, and they agree with the proposal of Bryce-Smith that it is formed by internal conversion of the S_3 to S_1 state which eventually either forms pre-fulvene, which can lead to fulvene, benzvalene, or revert to the ground state S_0. Non-dissociative reaction may occur from an upper excited singlet state of benzene as demonstrated by formation of the Dewar isomer directly from the S_2, and possibly S_3, states,[81] after 206 nm irradiation of benzene in solution.

For heavily substituted fluorobenzenes Dewar isomeric forms may arise from the excited singlet states in apparent contradiction of the symmetry relations.[82] The rigour of such orbital symmetry rules must be further established.

All the work reviewed so far shows that chemical relaxation channels leading to non-benzenoid isomers exist from thermally accessible vibrational levels in S_1. Their role in the non-radiative relaxation processes is not yet established, however. An extensive speculation on the radiationless relaxation of the S_1 state in the gas phase and the significance of isomerization by Phillips, Lemaire, Burton and Noyes[83] requires that electronic relaxation other than by fluorescence and intersystem crossing occurs by transitions to an isomer through favourable overlap of potential surfaces. This seemingly attractive theory, which could remove many theoretical difficulties in interpreting the behaviour of the $^1B_{2u}$ state, is not in apparent agreement with the low yields of products which are generally found after irradiation of benzene and its derivatives, even under conditions when $\phi_{FM} + \phi_{TM} \ll 1$.

The data on the effect of exciting wavelength show that relaxation from the higher levels of the S_1 state is accompanied by the disappearance of all indications of triplet formation together with *perhaps* some increase in photochemical reaction. The work of Kaplan and Wilzbach[79] provides the most secure evidence for an increase in the benzvalene yield in going from 253·7 nm to 237 nm. The relative rate of non-radiative decay, which increases sharply when the vibrational energy exceeds 2500 cm^{-1}, is not satisfactorily explained. An $S_1 - S_0$ transition is regarded as unlikely by Parmenter.[6] There is no apparent reason why a thermally induced $S_1 - T_2$ transition should set in with such suddenness (the increase in rate is about 10^3) and $T_3 \leftarrow S_1$ is too endothermic to be involved. The failure to detect sensitized phosphorescence[84] or butene-2 isomerization at shorter wavelengths discounts the formation of triplet states with $\tau > 10^{-11}$ s. The situation is clearly unsatisfactory and it must be stressed that the yields of products detected and measured are far below the values of unity required for a satisfactory explanation of the nature of Channel 3.

Another argument against benzvalene formation providing a complete explanation of the Channel 3 non-radiative processes in benzene has been advanced by Spears and Rice.[45] The principal axes of rotation of benzvalene and benzene are in different directions and the law of conservation of angular momentum must inhibit conversion of one to the other when the angular momentum of either species is large. The rotational structure of the absorption bands shows that there are states in which angular momenta are high and low, and the formation of benzvalene should change with consequent variation in the fluorescence yield. Parmenter and Schuh[85] have not found any such changes in the radiative or non-radiative rate constants when the rotational envelopes of the 6^0_1, 6^1_0, $6^1_1\,1^1$ and $6^1_0\,1^2_0$ vibrational levels, in which angular momenta change, are scanned.

A stimulating suggestion has been provided by Callomon, Parkin and Lopez-Delgado[86] which may prove to be crucial to understanding the nature of the non-radiative process. Many, but not all, bands of the 260 nm absorption system of benzene above 3000 cm^{-1} are diffuse when photographed under high resolution. The threshold for the onset of this diffuseness is almost exactly at the highest level above the zero-point level from which fluorescence can be excited, and it indicates the sharply defined onset of the non-radiative process shown by the other methods. The level broadening indicates that the lifetime of states produced by absorption is about $5{\cdot}3 \times 10^{-12}$ s. Predissociation is excluded by failure to detect hydrogen, and so another excited state is suggested. $^{1,3}E_{2g}$ states are mentioned as possibilities, but the $^3E_{2g}$ $(\sigma - \pi^*)$ state is tentatively proposed by Callomon *et al.*[96] The $^1E_{2g}$ state is however rather more favourable in view of the failure to detect any triplet states. The subsequent fate of the E_{2g}, or any other state for that matter, still leaves another problem that requires further explanation (see Birks, Chapter 1, Vol. 1).

2.2.2.6 *The behaviour of higher excited states*

As already mentioned, early[84] work by Lipsky showed that in the vapour phase internal conversion to S_1, and T_1 formation by intersystem crossing from higher singlet states, were not efficient processes. Gregory,

Hirayama and Lipsky[87] have recently observed $S_3 \rightarrow S_0$ and $S_2 \rightarrow S_0$ emissions from *p*-xylene vapour (4 torr, 25°C) following excitation at 184·9 nm. The emission at $\lambda_{max} \approx 195$ nm is the $S_3 \rightarrow S_0$ fluorescence $[\phi_F(S_3) \approx 10^{-5}]$. From the calculated oscillator strength it appears that $k_{nr}(S_3) \approx 10^{14}\ s^{-1}$. The *p*-xylene $S_2 \rightarrow S_0$ emission is at $\lambda_{max} = 243$ nm $(\phi_F(S_2) = 1{\cdot}8 \times 10^{-5})$. Similar observations were made with toluene (λ_{max} at 192 and 238 nm) and mesitylene (λ_{max} at 205 and 246 nm). The $S_3 \rightarrow S_0$ transition in benzene is blue shifted into the excitation band, but the dectable $S_2 \rightarrow S_0$ transition has λ_{max} at 233 nm. Increase in *p*-xylene pressure from 5 to 115 torr affects neither the intensity nor the spectrum of S_2 and S_3 emissions, whereas the S_1 emission intensifies, blue-shifts and narrows approaching that found for 253·7 nm excitation, although the yield is 10^{-4} times less. Addition of 1 atm of neohexane to 5 torr of *p*-xylene produces no change in intensity or spectrum of even the S_1 fluorescence. Oxygen does not quench the S_2 and S_3 fluorescence and it quenches the S_1 fluorescence only on the blue side. A much more efficient and uniform quenching of S_1 emission by oxygen is found for excitation at 253·7 nm. This is explained by a variation in the lifetimes of the vibrational levels from which emission arises; longer wavelength transitions originate from very highly excited vibrational levels, which are very short lived, and populated by direct non-radiative transitions. Further experiments on benzene and its derivatives open up new areas of study on internal conversion processes in aromatic hydrocarbon vapours.

2.2.2.7 *Concluding remarks*

An as yet incomplete picture of the photophysical behaviour of benzene vapour at room temperature is emerging. The main features of processes

Table 2.3 Relaxation processes for the thermal vibrational distribution of states of $^1B_{2u}$ benzene in 10–20 torr vapour near 300°K for excitation at or near 254 nm

		C_6H_6	C_6D_6
ϕ_{FM}	fluorescence quantum yield	0·18 ± 0·04	0·32 ± ?
τ_M	lifetime	75 ± 10 ns	92 ± 10 ns
$k_r(= k_{FM})$	radiative relaxation rate constant	$2{\cdot}5 \times 10^6\ s^{-1}$	$3{\cdot}5 \times 10^6\ s^{-1}$
$k_{nr}(= k_{IM})$	non-radiative rate constant	$11{\cdot}2 \times 10^6\ s^{-1}$	$7{\cdot}4 \times 10^6\ s^{-1}$
ϕ_{TM}	quantum yield for intersystem crossing	0·73 ± 0·03	0·63 ± ?
k_2	rate constant for collision induced relaxation to gas	10^{-2} to 10^{-3} $torr^{-1}\ s^{-1}$	as C_6H_6
ϕ_{isom}	isomerization yield for benzvalene	0·03	—

occurring when excitation is in region of the commonly used 254 nm excitation region at pressures above 10 torr are now fairly well established (Table 2.3). One of the most surprising omissions from the available data is the almost complete lack, apart from some results of Poole,[14] of any information about the effect of temperature on the photophysical behaviour in the gas phase.

2.3 Benzene in the liquid phase

In the vapour state molecules are, on average, widely separated and interactions between them are slight, except within the duration of a loosely defined collision. In the liquid phase molecules still retain some translational freedom, but because of their close proximity to one another they are in a continuous, if individually varying, state of interaction. This influences the vibrational and electronic energy levels, and all the photophysical processes are affected in some degree. Further, higher concentrations than those usually achieved in the gas phase make the role of excimers and exciplexes much more important.

A 'hot' molecule, in Boltzmann equilibrium with its environment, can be achieved by absorption of a photon, but it is usually assumed that the excess vibrational energy is rapidly dispersed ($k \geqslant 10^{12}\,s^{-1}$). Radiationless relaxation of the $^1B_{2u}$ state when the excess energy is greater than 3000 cm^{-1}, and of higher singlet excited states, occurs with a rate of this magnitude in the gas phase. Channel 3 or similar rapid processes may also occur in other phases. The commonly identified photophysical processes such as fluorescence, phosphorescence and intersystem crossing usually occur over a longer time scale ($k \approx 10^8\,s^{-1}$). In the liquid and solid phases emissions have only recently been observed from excited states which are not thermalized (see Chapter 9, Vol. 2). Prior to this, evidence for the participation of excited states of benzene other than the $^1B_{2u}$, $^1B_{1g}$ excimer, and $^3B_{1u}$ states has been deduced from an examination of effects of varying excitation wavelength, and so forth, rather than from direct observation.

2.3.1 *Absorption spectrum*

The absorption spectrum of dilute solutions of benzene in transparent solvents at wavelength longer than 200 nm is well known. In hexane, for example, the $S_1 \leftarrow S_0$ peak is at 39,300 cm^{-1} ($\varepsilon_{max} = 250$), that for $S_2 \leftarrow S_0$ at 49,000 cm^{-1} ($\varepsilon_{max} = 8800$) and for $S_3 \leftarrow S_0$ is at 54,300 cm^{-1} ($\varepsilon_{max} = 68{,}000$). These are essentially the same electronic transitions in practically the same wavelength regions as those already described for the gas phase. In solution solvent molecules affect the solute in both its ground and excited states and the spectra in absorption and emission change to some extent in response to this influence. It has been known for some time[88] that the spectrum of benzene changes slightly with increase in dielectric constant and refractive index of the solvent, but there is no discontinuity between the gas, liquid and solid phases so that the effect cannot be strong.

In addition to the production of spectral shifts, the vibrational structure of the spectra are influenced by environmental interaction. The loss of vibrational structure is due to specific solvent–solute interactions, whilst the changes of the spectral profile and energy are due to modification of the shape and position of the potential energy surfaces. Lawson, Hirayama and Lipsky[89] observed both the fluorescence and absorption spectra of benzene in the vapour phase and various solvents. In perfluoro-*n*-hexane the spectra retain most of the fine structure of the vapour phase, showing that solvent–solute interactions are minimal, and the spectral shifts are also negligible. The vibrational fine structure is absent in *n*-hexane due to an increase in specific solvent–solute interaction. The spectrum broadens further in going from saturated solvents through aromatic to strongly polar media like ethanol, where the interaction is greatest. Deuterium

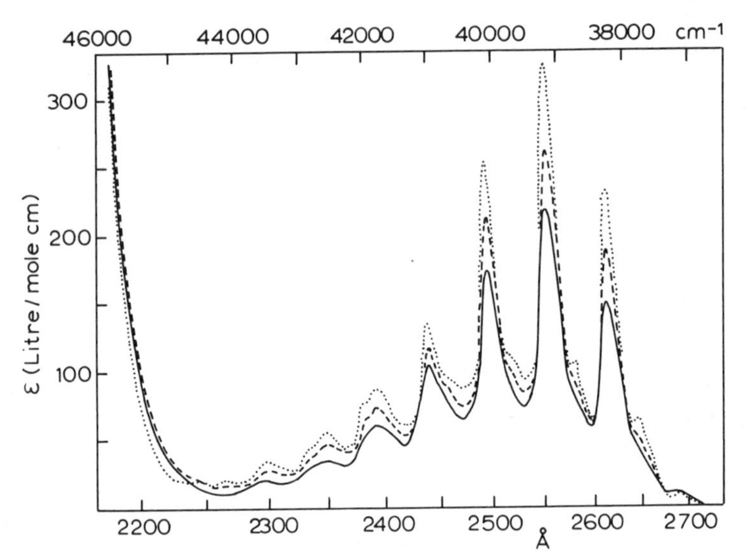

Figure 2.4 The absorption spectra of benzene in the pure liquid state for the $^1B_{2u} - {}^1A_{1g}$ transition. ——— liquid phase at 299°K, ---- liquid phase at just above the freezing point (5·45°C) and ······ solid phase just below the freezing point. The molar extinction coefficient is exact only for the liquid phase data at 299°K (after T. Inagaki, *J. Chem. Phys.*, **57**, 2526 (1972))

substitution in benzene leads to a blue shift of the spectra due to a reduction of the zero-point vibrational energy.

Studies of liquid benzene absorption have been comparatively neglected until recently because of experimental difficulties. Very weak transitions in the region above 280 nm are due to the spin-forbidden $T_1 \leftarrow S_0$ transition with a peak near 340 nm which is induced by the presence of O_2[11]. Yguerabide[90] observed the absorption spectrum of liquid benzene down to 233 nm, but he was unable to determine the absorption coefficient. As suspected, the spectrum was not very different from that in dilute solutions in cyclohexane. Inagaki[91] has more recently measured the absorption spectra of pure benzene at several temperatures from 26°C to just below the melting point at 5·45°C. The three absorption bands of quite different intensities, which were observed, correspond to the π–π^* singlet transitions seen in the vapour phase (Figures 2.4 and 2.5). The $^1B_{2u} \leftarrow {}^1A_{1g}$ transition still has some vibrational structure and no significant shifts in the vibronic bands with temperature are seen. An oscillator strength of 2×10^{-3} calculated for this transition compares with a value of $1{\cdot}6 \times 10^{-3}$ in the vapour phase and in hydrocarbon solution. The small liquid–solid shifts are less than the vapour–solid shifts (*c.* 250 cm^{-1}), and they are all towards the red. Oscillator strengths for the $^1B_{1u} \leftarrow {}^1A_{1g}$

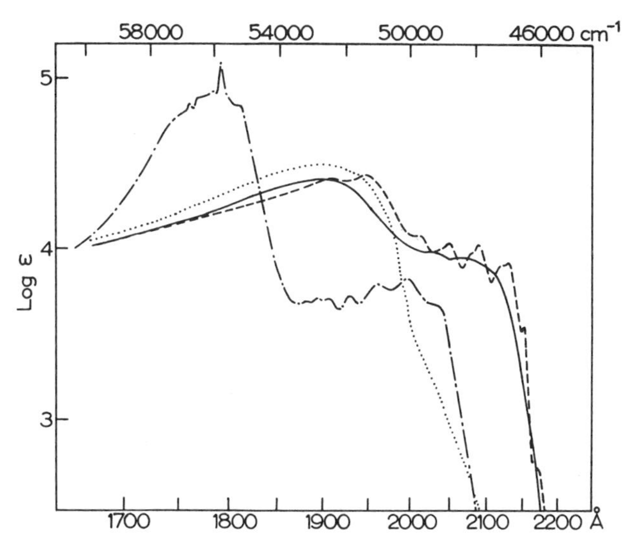

Figure 2.5 A comparison of the absorption spectra of benzene in the vapour, solid and liquid phases. ——— liquid (film) at room temperature, ---- solid at near 4·2°K, –·–·–·– vapour and ······ liquid (reflectance measurement) (after T. Inagaki, *J. Chem. Phys.*, **57**, 2526 (1972))

and $^1E_{1u} \leftarrow {}^1A_{1g}$ transitions in the solid state are 0·095 and 0·77, in good agreement with the values in the other phases.

The usual interpretation of solvent effects on spectra is based on the Onsager theory of dielectrics.[92] Instead of this macroscopic treatment, a microscopic model has been based on the interaction energy between solute and solvent molecules and applied with apparent success to the $S_1 \leftarrow S_0$ and $S_2 \leftarrow S_0$ transitions for benzene in the solvents, cyclohexane, benzene, dioxan, acetonitrile and water.[93]

The absorption spectrum of benzene has been investigated in both the vapour[94] and solid[95] state over a wide range of wavelengths (6 to 35 eV) and temperatures, but the results do not affect the interpretation of any photophysical data we shall present. Such data are of course very relevant to the effects of high-energy ionizing radiation on benzene-containing systems.

$T_1 \leftarrow S_0$ and $T_q \leftarrow S_0$ transitions are spin forbidden and this accounts for the very low $T_1 \leftarrow S_0$ radiative transition probabilities of unsubstituted aromatic hydrocarbons. Oxygen perturbation and heavy atom effects both enhance the intensity of the $T_1 \leftarrow S_0$ absorption.[11] In the O_2-perturbed system a vibronic structure corresponding to the vibrational levels of T_1 is seen in the absorption spectrum along with an underlying structureless contact charge transfer band of the molecular complex formed by an interaction between the aromatic hydrocarbon and perturbing O_2 molecule.[96] The oxygen perturbation technique has been successfully used to observe $T_1 \leftarrow S_0$ and $T_2 \leftarrow S_0$ transitions in benzene-h_6 and benzene-d_6 at 4·2°K.[13] The $T_2 \leftarrow S_0$ transition is 100 times weaker than the $S_1 \leftarrow S_0$ transition, and T_2 is only 900 cm^{-1} below S_1. The low temperature used in these experiments eliminates hot bands from the $S_1 \leftarrow S_0$ transitions and the $T_2 \leftarrow S_0$ band is observed directly in absorption. The $T_2 \leftarrow S_0$ transition has also been observed in absorption with perturbation by nitric oxide as well as by O_2.[97]

Transitions from T_1 to higher excited triplet states are spin-allowed and either (i) photostationary excitation techniques at high intensities or (ii) flash photolysis or pulse radiolysis methods, can be used, in principle at least, for their observation. The triplet–triplet absorption bands in benzene are still incompletely known and assigned.

Godfrey and Porter[98] first observed a broad structureless band, with a possible maximum at 41,600 cm^{-1}, in a low-temperature glass. The same band has been observed for C_6H_6 and C_6D_6 in the gas phase by Burton and Hunziker[68] (λ_{max} = 231·5 nm). Early assignments to a $^3E_{2g} \leftarrow {}^3B_{1u}$ transition were in accord with the calculations of Pariser.[99] Astier and co-workers[100] observed another triplet–triplet band in alkane solutions at 93–113°K which they attributed to the $^3E_{2g}^- \leftarrow {}^3B_{1u}^+$ transition, an assignment never universally accepted and now ascribed to the $^3E_{2g}^+ - {}^3B_{1u}^+$ transition.[69]

The development of nanosecond laser flash photolysis and pulse radiolysis provides methods for study of the absorption spectroscopy of short-lived intermediates but no absorption of the $^1B_{2u}$ state has been observed in the range 270–1000 nm in spite of an earlier assignment.[101] The prediction by Pariser of a $^1E_{2g} \leftarrow {}^1B_{2u}$ transition at about 350 nm has not been confirmed.[102–104]

The spectra of the excimers of benzene and its simple derivatives have been observed by the nanosecond excitation techniques.[102–104] The maxima of the excimer absorptions of benzene derivatives lie in the 500 nm region and for benzene in particular have a temperature dependence identical to that for excimer fluorescence. Interaction of the $^1B_{2u}$, and $^1E_{1u}$ excited molecular states with a ground state ^1A_g molecule gives excimer states which split into pairs, $^1B_{1g}$ and $^1B_{2u}$, and $^1E_{1g}$ and $^1E_{1u}$.[105] The excimer absorption band observed at about 500 nm is assigned to the $^1E_{1u} \leftarrow {}^1B_{1g}$ transition.

Multiphotonic excitation provides a means of populating higher states. Liquid benzene has been excited by pulses of 347·1 nm light from a Q-switched ruby laser,[106] and the absorption and fluorescence intensities depend on the square of the light intensity. Therefore, biphotonic excitation appears to occur and it is followed by internal conversion from a higher state together with $^1B_{1g}$ excimer formation. The authors' reasons for postulating participation of the $^1E_{2g}$ state as the initially formal higher state are not clear, nor is the mechanism which gives rise to the fluorescing excimer.

2.3.2 *Processes involving excited states*

In solution the excited state, or molecule which absorbs light, is exposed to the influence of the surrounding molecules of solvent which may produce significant changes in its photophysical properties. These are characteristic of the molecule together with its environment, rather than of the isolated molecule as is the case for benzene vapour at low pressure. As with the absorption spectra, the fluorescence spectra are shifted to longer wavelengths with increase in dielectric constant and refractive index of the solvent. Dispersion forces between solvent and solute molecules broaden the vibrational structure of the spectra. The observations of Lawson, Hirayama and Lipsky[89] on the spectra of benzene in the vapour and different solvents show this clearly and also the remarkable retention of fine structure in fully fluorinated solvents like perfluoro-*n*-hexane. This latter effect is due to the small specific solvent–solute interactions in this particular type of solvent.

2.3.2.1 *Effect of excitation at shorter wavelengths*

Braun, Kato and Lipsky[84] have shown that the quantum yield of fluorescence depends on the energy of the exciting photon for benzene, and its derivatives, in both the liquid state and in transparent solvents, as well as in the gas phase. These results were confirmed by Birks, Conte and Walker[107] for benzene and a number of other aromatic compounds. It seems clear from these data that $S_n \rightarrow S_{n-1}$ internal conversion where $n \geqslant 2$ does not occur with unit quantum efficiency. The quantum yield for benzene fluorescence in the liquid state, unlike the vapour phase, is constant over the first absorption band, but it decreases at shorter wavelengths. The data of Braun *et al.*[84] were expressed in terms of a parameter β_F which is the quotient, $(\phi_F)_{ex}/(\phi_F)_1$, where $(\phi_F)_1$ is the fluorescence yield for excitation into S_1 and $(\phi_F)_{ex}$ that for excitation into a higher state S_n. Initially it was assumed that the values of β_F provided a direct measure of the overall efficiency of conversion of the higher state S_n to the fluorescent state S_1. Lawson, Hirayama and Lipsky[108] modified this simple view for the following reasons. A quencher, Q, like O_2 or CCl_4, when added to a system in which benzene is excited into the S_3 (184·9 nm) or S_1 (253·7 nm) states reduces both the fluorescence yields, but the ratio β_F increases from 0·22 at $[Q] = 0$ to a limiting value of 0·38 at high values of [Q]. This is explained by formation of a small yield of a photoproduct, probably fulvene, which acts as an excited state quencher. The effect of this product, is least when the light penetration is greatest as is the case at 253·7 nm. At sufficiently high concentrations of Q the value of β_F is considered to be a true measure of the efficiency of internal conversion, q_{MH}, for $S_3 \rightarrow S_1$. When the experiments were carried out in the presence of 2,5-diphenyloxazole (PPO) instead of O_2 or CCl_4 the same limiting value for β_F was obtained, and transfer from higher excited states to the added quenchers was apparently unimportant. Laor and Weinreb[109] also observed an increase in β_F for benzene with increasing solute concentration but, unlike Lawson *et al.*, they considered that kinetic analysis required transfer of energy to the solute from higher singlet states of the solvent. Dilution of the pure liquid with an inert solvent such as cyclohexane increases β_F for all aromatic liquids except benzene, for which β_F decreases with dilution (see later). Lawson *et al.*[89] considered this to arise from two opposing effects, (i) a reduction in concentration of quenching photoproduct and (ii) a decrease in q_{MH} in going from an aromatic to an aliphatic solvent environment (the opposite appears true for benzene). Birks[107] proposed another explanation which involves formation of excimers of higher states

$$^1M^{**} + {}^1M \rightleftharpoons {}^1D^{**}$$

and the efficiencies of internal conversion, q_{DH}, of these states to $^1D^*$ being less than q_{MH}, except in benzene where $q_{DH} > q_{MH}$. On the basis of the available data it is not possible to distinguish between these alternative models.

More recent experimental results are those of Fuchs, Heisel and Voltz[110] in which attention was directed towards an understanding of the fate of the higher vibronic levels which are formed after the interaction of benzene with high-energy radiations in condensed media. Since the fluorescence yields for the neat aromatic liquids are inconveniently low, the scintillator 2-(1-naphthyl),5-phenyl-oxazole (α-NPO) was added in small concentrations to accept electronic excitation energy from the aromatic host material. Results were obtained with benzene, benzene-d_6, toluene, toluene-d_8, *p*-xylene, and mesitylene. The data show that for exciting wavelengths above about 175 nm the behaviour is similar to that described in the earlier reports, but at shorter wavelengths (< 175 nm) a drastic change in behaviour occurs (Figures 2.6 and 2.7). A critical energy, W_0, at which

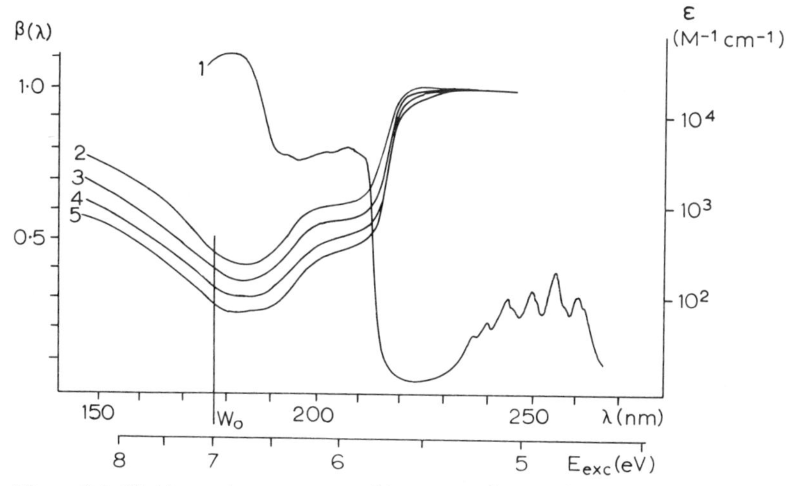

Figure 2.6 (1) Absorption spectrum of benzene solvent. Fluorescence excitation spectra of benzene solutions of α-NPO at various concentrations, 10^{-1} M (2); 3×10^{-2} M (3); $1{\cdot}5 \times 10^{-2}$ M (4); 5×10^{-3} M (5) (after Fuchs, Heisel and Voltz[110])

β_F has a minimum value, is interpreted as being the threshold for molecular autoionization and a distinction between (i) excited states ($E_{ex} < W_0$) and (ii) superexcited states ($E_{ex} > W_0$) of molecules is required. For (i) the characteristic deviations already reported are explained by the combination of photophysical effects which arise from the ^1L_a and ^1B_b states and especially a low internal conversion efficiency to S_1. The relative fluorescence

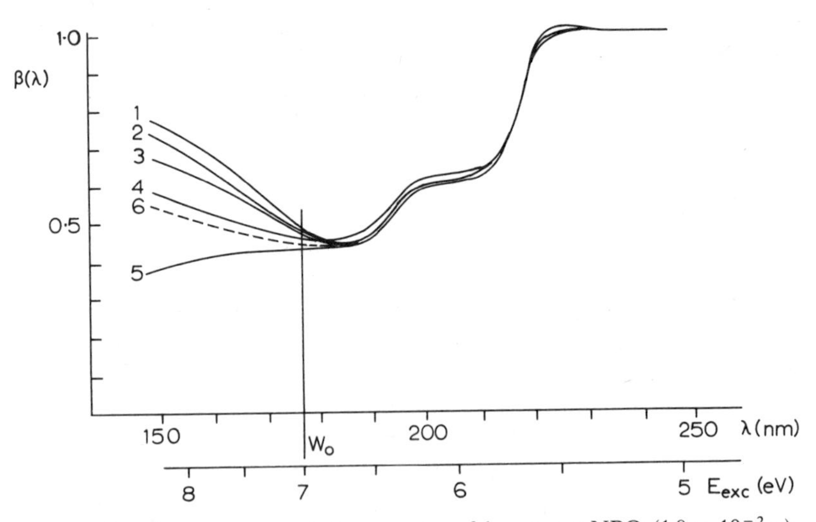

Figure 2.7 Fluorescence excitation spectra of benzene α-NPO ($1{\cdot}8 \times 10^{-2}$ M) solutions containing chloroform or oxygen: $[CHCl_3] = 0$ (1); 5×10^{-2} M (2); 10^{-1} M (3); 2×10^{-1} M (4); 5×10^{-1} M (5); solution under 1 atm of O_2 (6) (after Fuchs, Heisel and Voltz[110])

efficiencies, β_F, are constant, but different, over each of the main absorption bands ($^1B_{2u}$, $^1B_{1u}$ and $^1E_{1u}$) and so the effect depends on some form of electronic transformation and it is not noticeably affected by vibrational energy. Stern–Volmer analyses of the data for excitation into the S_2 and S_3 states show different quenching efficiencies for these states and values of the Stern–Volmer constants, $k_{t2}\tau_2$ and $k_{t3}\tau_3$ were determined (where k_{t2} and k_{t3} are the bimolecular rate constants for energy transfer and τ_2 and τ_3 the S_2 and S_3 lifetimes). Both parameters are about three orders of magnitude lower than those for the corresponding value for the lowest excited singlet state S_1, $k_{t1}\tau_1$. For benzene, $k_{t2}\tau_2$ is $5{\cdot}2$ M^{-1} and $k_{t3}\tau_3$ is 1 M^{-1}, and for C_6D_6 the corresponding values are 4·5 and 1 M^{-1}. This type of result is similar to the evidence of Laor and Weinreb,[109] but at variance with the conclusion of Lawson *et al.*,[89] and indicates that energy transfer to the quencher from upper levels of an aromatic solvent cannot, apparently, be ignored. The influence of O_2 and $CHCl_3$ on S_2 and S_3 states is found to be negligible, in agreement also with Laor and Weinreb.[109] Voltz also takes exception to the possibility of a direct $S_3 \rightarrow S_1$ transition and considers that a succession of transitions $S_3 \rightarrow S_2 \rightarrow S_1$ is required by the 'energy gap' law.

In region (ii) where $E_{ex} > W_0$, superexcited states are formed and β_F values increase indicating that different precursors from the ^1L_a and

^1B_b levels are involved. These precursors, unlike the S_2 and S_3 states, are strongly quenched by $CHCl_3$ and O_2 both of which are known to be efficient electron scavengers, and the necessary conclusion seems to be that ionization of the higher states is extensively involved at the shorter excitation wavelengths. The energy W_0 is found, for the aromatic liquids examined, to be approximately equal to [I.P.(gas) − 2] eV, the energy required for autoionization into a broad conduction band, a process competing with solvent-assisted vibronic relaxation by internal conversion or intersystem crossing. The sum of P_+ and P_-, the polarization energies of the holes and electrons formed by ionization is therefore about 2 eV. The energy levels and transitions of the benzene solvent and solute F envisaged are summarized in Figure 2.8.

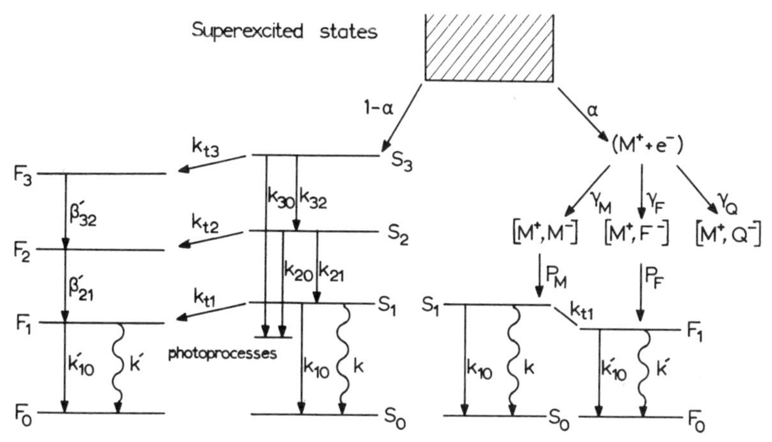

Figure 2.8 Energy levels and transitions in benzene solutions containing a solute F, irradiated in the vacuum ultraviolet (after Fuchs, Heisel and Voltz[110])

This type of experiment considerably improves our knowledge of ionizing energy effects as well as providing information on the nature of internal conversion processes from higher states, and more data of this kind are required. It should be pointed out that the reported inefficiency of $CHCl_3$ in deactivating the S_3 state represents a complete change in the view previously held by a number of workers, about the action of this solute in irradiated systems.[111]

Another important advance has very recently been achieved by Hirayama, Gregory and Lipsky[112] using techniques which permit the measurement of fluorescence yields as low as 10^{-6}. They recorded the emission spectrum from oxygenated solutions of neat benzene and other aromatic liquids excited at 184·9 nm, and they recorded a new emission

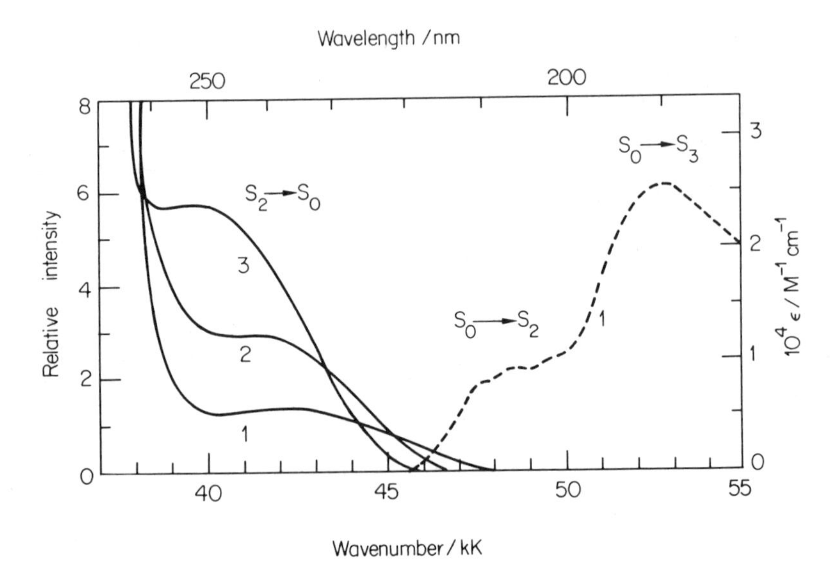

Figure 2.9 Fluorescence from higher states of benzene and other aromatic hydrocarbon liquids, with excitation at 184·9 nm. Absorption and 'corrected' emission spectra of oxygenated neat benzene (curve 1), toluene (curve 2) and *p*-xylene (curve 3). Only the neat benzene absorption spectrum is shown (after Hirayama, Gregory and Lipsky[112])

at shorter wavelengths than the well established $S_1 \to S_0$ fluorescence (Figure 2.9). Subtraction of the tail of the residual S_1 emission gives $\lambda_{max} \approx 235$ nm and $\phi_F(S_2) = 8 \times 10^{-6}$ for benzene. The presence of oxygen is needed to reduce the stronger $S_1 \to S_0$ emission and reveal the new fluorescence assigned to a radiative $S_2 \to S_0$ transition. Similar results were obtained with liquid toluene and *p*-xylene. No shorter wavelength emissions were seen from the neat liquids, but it is interesting to note the observation of weak emissions, possibly due to the $S_3 \to S_0$ transition, from *p*-xylene ($\lambda_{max} \simeq 205$ nm) and mesitylene ($\lambda_{max} \simeq 210$ nm) at concentrations below 0·1 M in *iso*octane. It is not suggested that these emissions account for any appreciable proportion of the relaxation of S_2 or S_3 states (see Chapter 9, Vol. 2).

Other experimental methods have given information on internal conversion of excited singlet states of benzene. Lumb, Braga and Pereira[113] measured both the fluorescence and phosphorescence excitation spectra of rigid solutions of benzene in EOA (3:3:1, ether:*iso*octane:alcohol) at 77°K. The two excitation spectra are identical showing that ϕ_{FM} and ϕ_{TM} vary in the same manner with λ_{ex}. For a 0·112 M solution β-values for fluorescence or phosphorescence decrease below unity at $\lambda_{ex} \approx 243$ nm. Some of the apparent change in β-values observed arises from absorption of exciting light by the solvent, for which a correction could not be made, but a definite decrease in β occurs at about 220 nm and for excitation of pure benzene into the S_2 state β is *c.* 0·4 at 77°K and room temperature. The values of β at the shorter wavelengths increase with decreasing concentration. The reasons are not clear although a surface exciton quenching mechanism provides a plausible explanation.

As already mentioned, Lawson *et al.*[89] measured the quantum efficiency, q_{MH}, for internal conversion from S_3 (184·9 nm) to S_1 for benzene in several solvents by measurements in the presence of 0·15 M CCl_4 or with solutions saturated with oxygen at 1 atm and they found that the values vary markedly. $q_{MH} = 0{\cdot}45$ in benzene, 0·37 to 0·28 in alcohols and other polar solvents, 0·25–0·22 in aliphatic hydrocarbons, 0·04 in n-C_6F_{14}, and ≈ 0 in the vapour phase. A correlation between q_{MH} and the intensity of the Ham band, which is a weak low-energy absorption band observed in pure benzene and benzene in solution in tetrahydrofuran, ethyl ether, acetonitrile and very weakly in solution in cyclohexane, was noted. The Ham band is due to the symmetry-forbidden $0 - 0$ absorption transition to the $^1B_{2u}$ state, which is absent from the low-pressure vapour spectrum. The enhancement of the $0 - 0$ band in solution arises from a solvent perturbation which is effective in mixing of the $S_3(^1E_{1u})$ and $S_1(^1B_{2u})$ states.[113] In the absence of such a solvent induced coupling between these states the direct transition from $^1E_{1u}$ to $^1A_{1g}$ is apparently a major mode of relaxation for S_3.

In toluene and *p*-xylene the substituent methyl groups enhance the $0 - 0$ transition in the $S_1 \leftarrow S_0$ absorption, which is observable in the vapour phase, and the fluorescence intensities and values of q_{MH} in these molecules are greater than the values for benzene under the same conditions.

2.3.2.2 *Photophysical processes involving the lowest excited states*

For solutions of benzene the general decay pattern of the lowest excited states may be represented as follows:

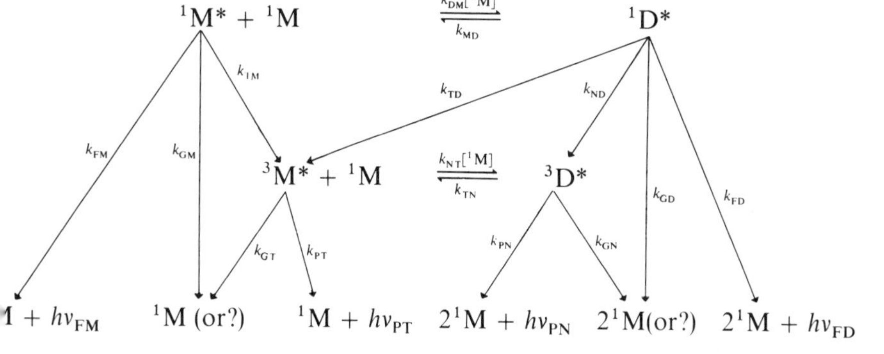

There are quantitative data for most of these processes in the case of benzene and the behaviour of the lower excited states will be treated according to the above scheme. The solvent provides very rapid vibrational relaxation of the excited states and also affects the efficiency of the various primary processes.

After Birks[92] the following parameters are defined:

$$q_{FM} = k_{FM}/(k_{FM} + k_{TM} + k_{GM}) = k_{FM}\tau_M \tag{2.6}$$

$$q_{FD} = k_{FD}/(k_{FD} + k_{ND} + k_{TD} + k_{GD}) = k_{FD}\tau_D \tag{2.7}$$

q_{FM} and q_{FD} are the fluorescence efficiencies of the $^1B_{2u}$ state of benzene and its excimer. τ_M and τ_D are the intrinsic lifetimes. The observed quantum yields for excited monomer and excimer emission of a solution of concentration c are

$$\phi_{FM} = q_{FM}/(1 + c/c_h) \tag{2.8}$$

and

$$\phi_{FD} = q_{FD}/(1 + c_h/c) \tag{2.9}$$

where c_h is the 'half-value' concentration at which $\phi_{FM} = \frac{1}{2}q_{FM}$ and $\phi_{FD} = \frac{1}{2}q_{FD}$.

The relationships are considerably simplified in the 'high-temperature' region where

$$k_{DM}[^1M], k_{MD} \gg \tau_M^{-1}, \tau_D^{-1} \tag{2.10}$$

The equilibrium constant for excimer formation is

$$K_e = k_{DM}/k_{MD} = \tau_D/\tau_M c_h \tag{2.11}$$

and

$$\Delta G^0 = -RT \ln K_e = \Delta H^0 - T\,\Delta S^0 \tag{2.12}$$

where ΔG^0, ΔH^0 and ΔS^0 are the free energy, enthalpy and entropy of excimer formation, respectively.

Without making some assumptions, a scheme of this complexity cannot be applied to the analysis of data. One of the least satisfactory is the supposition that individual rate constants do not vary with composition for any particular binary solution system.

2.3.2.3 *Influence of environment on fluorescence decay of the $^1B_{2u}$ state*

The fluorescence radiative lifetime, τ_{FM} of an aromatic molecule in solution is related to the absorption intensity by the Strickler and Berg expression[115]

$$k_{FM} = 1/\tau_{FM} = 2{\cdot}88 \times 10^{-9} n^2 \langle \bar{\nu}^{-3} \rangle_{Av}^{-1} (g_l/g_u) \int \varepsilon \, d \ln \bar{\nu} \tag{2.13}$$

where $\langle \bar{\nu}^{-3} \rangle_{\text{Av}}^{-1}$ is the reciprocal of the mean value of $\bar{\nu}^{-3}$ over the fluorescence spectrum, n is the refractive index of the solution, g_l and g_u are the multiplicities of the lower and upper electronic states, and ε is the extinction coefficient. The absorption band integral is for thc transition, which is the reverse of that observed in emission. The derivation assumes a mirror symmetry relation between the emission and absorption spectra and also that there is a negligible change in configuration between the S_1 and S_0 states. Birks and Dyson[116] have used a modified expression which allows for changes in dispersion of the medium. There have been few systematic studies of the effect of medium on the photophysical processes, but some data of this type are available for benzene.

Eastman[117] has shown that the fluorescence of benzene undergoes thermal quenching in solution by an effect not due to O_2 or other impurities. At 77°K the fluorescence quantum yields of benzene in n-hexane, methanol, 30/70 *iso*butanol/*iso*pentane, and EPA are within $\pm 20\%$ of the yield at 298°K in the vapour. At the highest temperature used (330°K) the yield approached zero in all four solvents at concentrations of 10^{-4} to 10^{-2} M, and it was suggested that the thermal quenching is not due to intersystem crossing, which should not require any appreciable activation energy, but to another process with an activation energy of about 2000 cm^{-1}, induced by solute–solvent interaction. Eastman and Rehfeld[118] have examined the fluorescence quantum yields and UV absorption of benzene in forty solvents at 25°C (Table 2.4). The oscillator strength of the transition giving rise to the 260 nm band is independent of the solvent,[119] except in $CHCl_3$ and CCl_4, which provides evidence for the formation of charge-transfer complexes in the ground state with these two solvents.[120] The absorption frequency decreases and the line width of the vibrational fine structure increases with increase of refractive index of the solvent. The Strickler–Berg equation gives, without use of the n^2 correction term, $k_{FM} = 2{\cdot}5 \times 10^6 \text{ s}^{-1}$ as the average for benzene in twenty-three solvents as well as in the vapour phase at 25°C. Combination of $\phi_{FM} = 0{\cdot}061$ for benzene in cyclohexane with Berlman's[121] value of $\tau_{FM} = 29$ ns, also in cyclohexane solution, gives $k_{FM} = \phi_{FM}/\tau_M = 2{\cdot}1 \times 10^6 \text{ s}^{-1}$. The agreement, within the limits of error, is taken as convincing evidence that all the radiationless processes competing with fluorescence occur from the same state as that which fluoresces and that which is also susceptible to solvent quenching. Eastman's results confirm earlier observations[114,122] that the $0 - 0$ band is induced by the solvent. This occurs more strongly the more polar the solvent. The only clear empirical observation which correlates with the fluorescence quenching power of the solvent is the $(0 - 0)$ absorption intensity, suggesting that the induced quenching depends on the same intermolecular forces as those responsible for the

Table 2.4 Some photophysical properties of benzene in different solvents (J. W. Eastman and S. J. Rehfeld, *J. Phys. Chem.*, **74**, 1438 (1970))

No.[a]	Solvent	Refractive index n^b	Wavelength of maximum $S_0 - S_1$ absorption (Å)	$\bar{\nu}_{00}$ + 522 cm^{-1} (10^3 cm^{-1})	Maximum $S_0 - S_1$ molar extinction coefficient ε_{max} (M cm)$^{-1}$ $\pm$ 4	Fluorescence yield, at 25°C ϕ_{FM}	$10^{-8} \sum k_{nr}$[c] (s^{-1})
1	Vapour	1·000	2529	38·61	—	0·18	0·114
13	Perfluoro-*n*-hexane	1·2514	2530	38·61	325	—	—
27	Methanol	1·331	2543	38·42	268	0·027	0·90
40	D_2O	1·338	2537	38·51	179	0·0052	5·0
39	Water	1·333	2539	38·49	188	0·0058	4·3
36	Acetonitrile	1·346	2543	38·43	227	0·029	0·84
2	2-Methylbutane	1·351	2542	38·42	267	0·044	0·54
23	Diethyl ether	1·350	2543	38·42	262	0·047	0·51
4	*n*-Pentane	1·357 (15°C)	2542	38·43	258	0·054	0·44
14	Tetramethylsilane	1·359 (19°C)	2543	38·40	267	0·054	0·44
22	Ethanol	1·359	2544	38·40	265	0·033	0·73
5	*n*-Hexane	1·375	2543	38·40	259	0·058	0·41
20	2-Propanol	1·375	2543	38·42	263	0·037	0·65
21	1-Propanol	1·384	2545	38·39	245	—	—
16	2-Methyl-2-propanol	1·384	2544	38·40	262	0·031	0·78
35	Dimethyl sulphate	1·387	2540	38·45	221	—	—
3	2,2,4-Trimethylpentane	1·392	2543	38·42	251	0·057	0·41
18	2-Methyl-1-propanol	1·398	2544	38·39	258	0·042	0·57
6	*n*-Octane	1·398	2544	38·40	248	0·057	0·41
10	Cyclopentane	1·404	2544	38·40	261	0·056	0·42
19	1-Butanol	1·399	2545	38·37	256	0·045	0·53

7	*n*-Nonane	1·405	2546	38·37	232	0·059	0·40
15	3-Methyl-1-butanol	1·408	2446	38·37	238	0·045	0·53
8	Methylcyclohexane	1·424	2544	38·39	258	0·046	0·52
11	Cyclohexane	1·424	2545	38·39	261	0·061	0·38
33	2-Methoxyethanol	1·402	2546	38·36	232	—	—
31	2-Ethoxyethanol	1·408	2546	38·37	240	—	—
32	1,4-Dioxane	1·423	2544	38·39	241	0·044	0·54
28	2-(2-Ethoxyethoxy)-ethanol		2546	38·36	233	—	—
17	1-Octanol	1·429	2546	38·36	260	—	—
25	1,2-Propanediol	1·433	2547	38·34	206	0·029	0·84
9	*n*-Hexadecane	1·432	2547	38·36	217	0·063	0·37
12	1-Dodecanol		2546	38·36	230	—	—
30	Dichloromethane	1·425	2548	38·34	205	0·031	0·78
26	1,2-Ethanediol	1·432	2547	38·33	192	0·029	0·84
29	1,2-Dichloroethane	1·448	2548	38·33	206	0·031	0·78
34	1,2,3-Propanetriol	1·473	2549	38·30	197	0·025	0·98
37	Chloroform	1·448	2551	38·26	225	0·0049	5·1
24	Benzene	1·498	2552	38·24	202	—	—
38	Carbon tetrachloride	1·604 (15°C)	2557	38·20	261	0·271	—

[a] Solvents listed in order of increasing refractive index. Both the maximum wavelength of the $S_0 - S_1$ absorption and the $(\bar{\nu}_{00} + 522)$ cm^{-1} transition generally shift to lower energies as refractive index decreases. The numbers (1–40) give the order of solvent-induced (0 – 0) absorption and radiationless conversion.

[b] Values for Na D-line light at 20 °C, unless otherwise stated.

[c] Spontaneous emission rate assumed to be $k_{FM} = 2{\cdot}5 \times 10^6\ s^{-1}$, an average value calculated by means of the Strickler–Berg equation for twenty-three solvents (n^2 correction ignored).

induced absorption. There is little temperature dependence of the (0 – 0) absorption band intensity, and so it is inferred that the thermal quenching arises from thermal population of solvent-*promoted* vibrational levels above the zero-point level of the $^1B_{2u}$ state.

The work of Gregory and Helman,[123] amongst others, also shows that the fluorescence rate constants in dilute solution and in the vapour phase are identical, and that the phase effects are almost solely on the rate of internal conversion. At room temperature the value of k_{FM} in methylcyclohexane solution is $2{\cdot}5 \times 10^6\ s^{-1}$ in exact agreement with the high-pressure values in the gas phase[39,42] and close to the value of $2{\cdot}4 \times 10^6\ s^{-1}$ for the radiative decay rate of the $^1B_0^0$ state.[42]

Cundall and Pereira[124] measured fluorescence quantum yields and lifetimes under the same conditions for a number of aromatic hydrocarbons including benzene, and the effects of temperature and environment on k_{FM} and $\sum k_{nr}$ were determined without use of other workers' data. The radiative lifetimes calculated from $\tau_{FM} = \tau_M/\phi_{FM}$ show some temperature dependence. In the case of aromatic hydrocarbons other than benzene, correction of the lifetime τ_{FM} by the Strickler–Berg-based relationship $\tau_{FM}^0 = \tau_{FM} n^2$, where n is the refractive index of the solvent applicable to the conditions of the experiment, yields temperature-independent τ_{FM}^0 values, which are practically independent of the solvent. Some of the slight temperature dependence of τ_{FM} or τ_{FM}^0 observed for benzene in cyclohexane, methylcyclohexane and ethanol was explained as follows. Strickler and Berg[115] arrived at their expression for k_{FM} by expanding the electronic transition moment integral for the transition from the lower to the upper electronic state in a power series with the normal coordinates Q of the molecule as follows

$$M_{lu}(Q) = M_{lu}(0) + \sum_r \left(\frac{\partial M_{lu}}{\partial Q_r}\right) Q_r + \cdots$$

The zeroth-order term is considered dominant as a rule but Strickler and Berg pointed out that for forbidden or weakly allowed transitions higher-order terms may not be negligible, and this would make the expression for k_{FM} dependent on temperature.

An important result of the work of Eastman,[118] Gregory and Helman,[123] and Cundall and Pereira[124] is the doubt which must be attached to the validity of the n^2 factor in the equation of Strickler and Berg, and others in the case of benzene. The calculated values of k_{FM} for benzene in solution and in the vapour are certainly not in serious discord if the n^2 corrections, supposedly required by theory, are not applied. The discrepancy between the corrected solution and gas-phase (where $n \simeq 1$) values for k_{FM} is unacceptably large.[124]

The only useful effect of applying the n^2 correction seems to be some improvement in the constancy of k_{FM} with temperature. Whilst accepting the physical basis for initially introducing the n^2 correction it must be concluded from examination of the experimental data that the electrostatic environment for the transition dipole is effectively provided by the molecule being excited, and that the effect of solvent on the probability of the transition is of secondary importance. This problem of interpretation has only become apparent through the recent availability of accurate yield and lifetime data. More and better experiments should clarify the situation.

2.3.2.4 *Effect of excimer formation on decay of the* $^1B_{2u}$ *state*

Since the early reports of Dammers-de-Klerk[125] and Ivanova, Mokeeva and Sveshnikov[126] on the concentration dependence of benzene fluorescence in solution several studies have been published with interpretations based on the excimer formation model. There have been criticisms of the possible simplifications, but the scheme of Birks, Braga and Lumb[127] seems sufficiently well established for an adequate explanation of the available results.

The variations of monomer and excimer fluorescence yields and observed lifetimes for benzene in cyclohexane as a function of temperature are shown in Figures 2.10 and 2.11. The derived values of q_{FM}, q_{FD}, k_{FM} and τ_D are shown in Figures 2.12, 2.13 and 2.14.

By using equations (2.8) and (2.9) it is possible to evaluate intrinsic fluorescence yields q_{FM} and q_{FD} from the observations of fluorescence spectra at different concentrations. Studies by Cundall and Robinson[128] show q_{FM} to decrease and c_h to increase with increasing temperature. q_{FD} for benzene increases with temperature.

For a solution of concentration c, τ_D is evaluated from measurements of τ_M, c_h, and the observed fluorescence lifetime τ using the equation, at different temperatures,

$$\tau(c + c_h) = \tau_M c_h + \tau_D c \tag{2.14}$$

The values of τ_D found decrease with temperature. The binding energy of the excimer, E_b, is obtained from an Arrhenius plot of $\log K$ *vs.* T^{-1}, taking account of the observed temperature dependence of

$$k_{FD} = k'_{FD} \exp(-E_{FD}/\mathbf{k}T) \quad \text{and} \quad k_{FM} = k'_{FM} \exp(-E_{FM}/\mathbf{k}T).$$

The equations

$$K = \frac{\phi_{FD}}{\phi_{FM} c} = \frac{q_{FD}}{q_{FM} c_h} = \frac{k_{FD} k_{DM}}{k_{FM} k_{MD}} \tag{2.15}$$

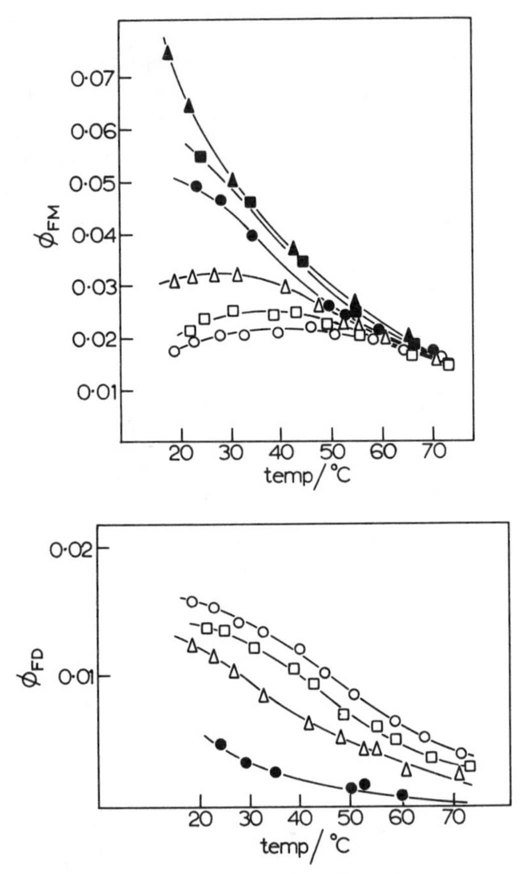

Figure 2.10 Benzene. (a) Variation of monomer fluorescence quantum yield ϕ_{FM} with concentration and temperature. ▲ 0·12 M, ■ 0·51 M, ● 1·12 M, △ 4·5 M, □ 7·9 M, ○ 11·2 M. (b) Variation of excimer fluorescence quantum yield ϕ_{FD} with concentration and temperature. ○ 11·2 M, □ 7·9 M, △ 4·5 M, ● 1·12 M

hold if the high temperature approximation is valid and the slope of such a plot yields ($E_b - E_{FD} + E_{FM}$) where E_b is the excimer binding energy. Published values obtained for this sum are 0·22 eV in cyclohexane (Birks *et al.*[127]), 0.27 eV in methylcyclohexane (Hirayama and Lipsky[129]) and 0·21 eV in cyclohexane (Cundall and Robinson[128]). Within the limits of experimental error $E_{FM} = 0$, the data of Cundall and Robinson show that $E_{FD} = 0{\cdot}15$ eV, so that $E_b = 0{\cdot}35$ eV. K_e is also given by $\tau_D/\tau_M c_h$ and

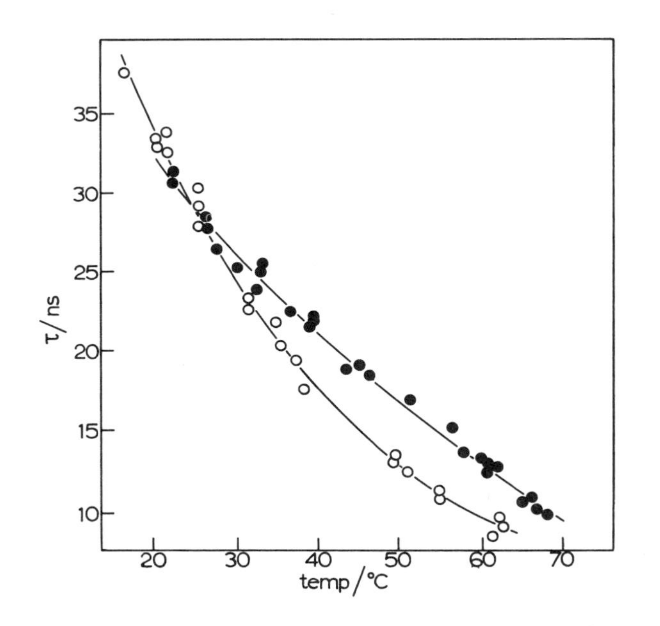

Figure 2.11 Temperature dependence of the observed fluorescence lifetime τ of (a) 10^{-2} M benzene in cyclohexane ○, and (b) liquid benzene ●

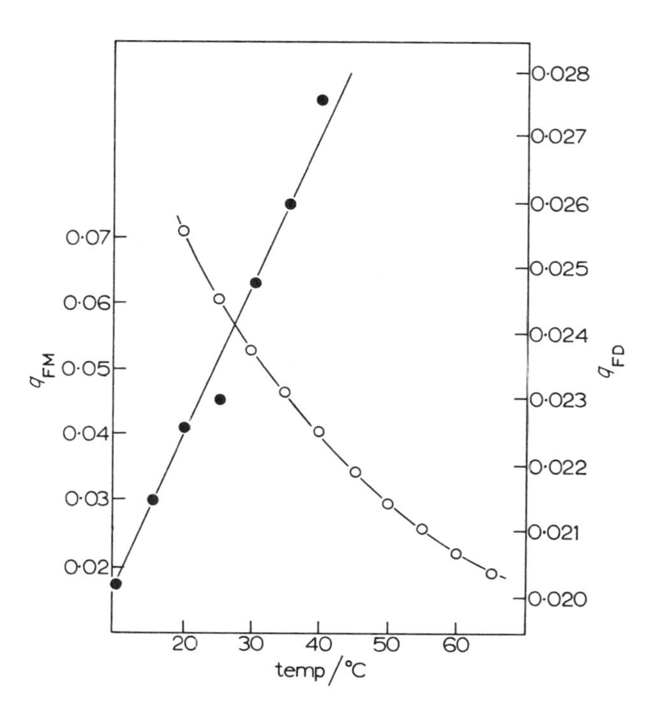

Figure 2.12 Benzene. Temperature dependence of (a) monomer fluorescence quantum efficiency q_{FM} ○, and (b) excimer fluorescence quantum efficiency q_{FD} ●

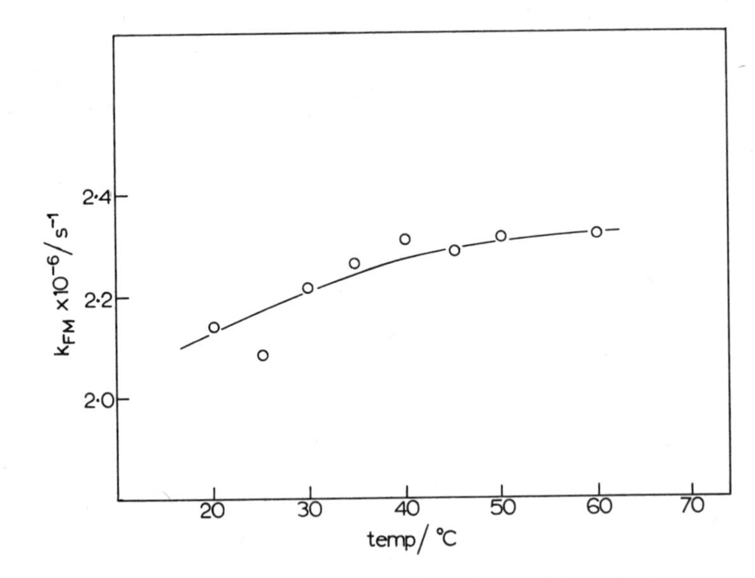

Figure 2.13 Benzene. Temperature dependence of the radiative rate constant k_{FM} for monomer decay (obtained from q_{FM}/τ_M)

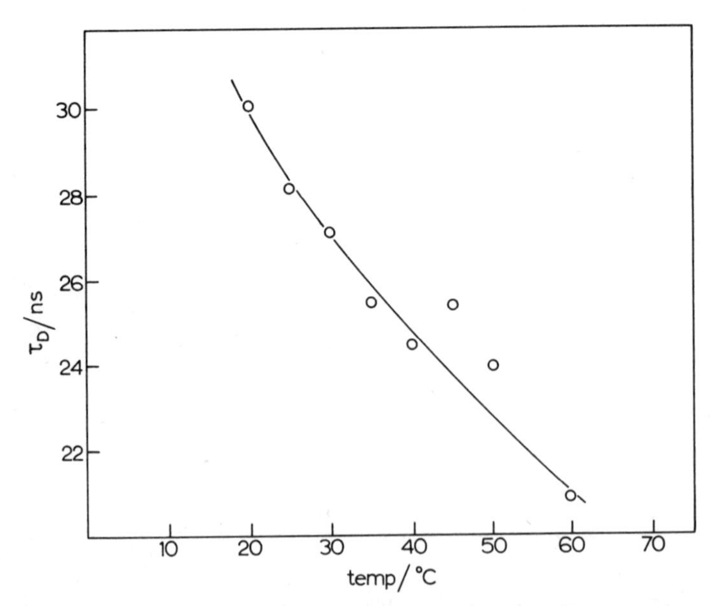

Figure 2.14 Benzene. Temperature dependence of the excimer fluorescence lifetime τ_D (obtained from $\tau(c + c_h) = \tau_M c_h + \tau_D c$)

from the variation of this quantity with temperature it can also be deduced that $E_b = 0{\cdot}34 \pm 0{\cdot}3$ eV in good agreement with the above calculation and comparable with 0·36 eV due to Hirayama and Lipsky who only estimated a value for E_{FD}. The value of 0·34 eV for the excimer binding energy is within the range predicted by theory.[130] The entropy change, ΔS^0, associated with excimer formation from the Cundall and Robinson[128] data is $-29{\cdot}6$ cal mol^{-1} °K^{-1}, which is to be compared with values of $-20{\cdot}3$ and -28 cal mol^{-1} °K^{-1} due to Birks,[127] and Hirayama and Lipsky.[129] The ground state repulsion energy R_m may be calculated from the expression

$$M_0 - D_m = V_m = E_b + R_m \tag{2.16}$$

where M_0 is the $0 - 0$ energy of the $S_1 - S_0$ transition, D_m is the energy of the peak excimer transition, and V_m is the interaction potential of the excimer.

The zero-point energy of the $^1B_{2u}$ state of benzene, M_0, is 4·71 eV and from the position of the peak in the excimer emission spectrum at 31,500 cm^{-1} $D_m = 3{\cdot}91$ eV. (Both these quantities are independent of composition.) R_m is determined to be 0·46 eV, in close agreement with value obtained by Birks *et al.*[127]

In summary, the Birks excimer model is consistent with data for benzene in a solvent like cyclohexane, but the temperature dependence of the excimer fluorescence rate affects earlier estimates of the excimer binding energy. The activation energy for excimer emission may arise from the dipole forbidden nature of excimer emission[131] and it is likely that a large interplanar vibration is necessary for emission, rather than twisting with respect to an axis perpendicular to the plane of the benzene molecules.[130]

Helman[132] measured the fluorescence decay times of neat C_6H_6 and C_6D_6 excited by 250 nm light to be 27 and 32 ns, respectively, and he found that dilution with a solvent (cyclo-C_6H_{12} or cyclo-C_6D_{12}) had no effect. In later experiments Gregory and Helman[123] measured the decay of benzene in methylcyclohexane in the range 169 to 298°K. At and below 195°K the rate for excimer formation was directly measurable, and excimer formation was shown under all conditions to be diffusion controlled, however it was observed that excimer dissociation occurred more rapidly in methylcyclohexane than in benzene. The radiative rate constants for fluorescence of the benzene monomer and excimer at 298°K are $k_{FM} = 2{\cdot}52\ (\pm 0{\cdot}12) \times 10^6\ s^{-1}$ and $k_{FD} = 0{\cdot}65\ (\pm 0{\cdot}03) \times 10^6\ s^{-1}$, and at 195°K are $k_{FM} = 1{\cdot}66\ (\pm 0{\cdot}11) \times 10^6\ s^{-1}$ and $k_{FD} = 0{\cdot}42\ (\pm 0{\cdot}02) \times 10^6\ s^{-1}$. The ratios of the rate constants at 298 and 195°K are $1{\cdot}52 \pm 0{\cdot}12$ for the monomer and $1{\cdot}55 \pm 0{\cdot}10$ for the excimer, so that the marked dependence of k_{FD} at above ambient temperatures found by Cundall and Robinson[128] is not apparent in the lower temperature range data of Gregory and Helman, although there are, as already indicated, some theoretical grounds for expecting k_{FD} to require some degree of vibrational excitation.

Aliphatic amines and *s*-tetracyanobenzene form fluorescent electron donor–acceptor complexes or exciplexes with singlet excited aromatic molecules including benzene.[133] Such complexes, although not necessarily fluorescent, occur in quenching of benzene fluorescence by molecules like CCl_4, $CHCl_3$, CH_2Cl_2 and also possibly O_2.[134,135]

We shall not review energy-transfer mechanisms except where relevant to benzene photophysics. Quenching of fluorescence can occur by (i) long-range transfer due to dipole–dipole or multipole interaction, (ii) diffusion of excited state and quenching solute and (iii) migration of solvent excitation energy between solvent molecules. Birks and Conte[136] in 1968 proposed that energy migration in liquids like benzene, toluene, *p*-xylene and mesitylene can be due to excimer formation and dissociation and that singlet transfer can occur by a sequence of these processes. Dilution of the solvent decreases the extent of excitation migration, but it increases the transfer distance and in benzene and similar liquids the transfer efficiency remains practically constant.

Since that time a number of papers on the subject have appeared which support this mechanism, for example, those of Yguerabide,[90,134] and Ludwig and Amata.[137] One of the more recent is that of Klein, Voltz and Laustriat[138] in whose argument the singlet excitation migration process is regarded as being restricted by the formation of excimers which act as transient energy traps. The model is tested by a detailed comparison with data on the effects of CCl_4 as a fluorescence quencher.

A summary of fluorescence data for benzene in the liquid state is given in Table 2.5.

2.3.2.5 *Triplet yields and decay*

The quantum yields of benzene triplet states in solution have been determined by sensitization methods, since direct spectroscopic measurements on benzene have not so far proved possible. The olefin isomerization and biacetyl techniques applied successfully to the vapour have been found suitable for this purpose, and the rationale of application is essentially the same as discussed for the gas phase.

Several determinations of the triplet quantum yield ϕ_T of benzene in the liquid phase at room temperature have been made. For liquid benzene at 25°C, excited at 253·7 nm, a value of $\phi_T = 0{\cdot}57$ was obtained by Cundall and Tippett[139] using the butene-2 isomerization method. Some confirmation of this value was obtained by the addition of xenon to such a

Table 2.5 Data for the first excited singlet state of benzene in the liquid state (hydrocarbon solvents)

Parameter	Value	Temp./°C	Solvent	Referencc
C_6H_6				
q_{FM}	0·050	20	Hexane	Lumb and Weyl[a]
	0·061	25	Cyclohexane	Eastman and Rehfeld[b]
	0·07	Ambient	Cyclohexane	Berlman[c]
	0·071	25	Methylcyclohexane	Hirayama and Lipsky[d]
	0·060	25	Hexane	Braga, Pereira, Lumb and Barradas[e]
	0·058	25	Hexane	Eastman[f]
	0·061	25	Cyclohexane	Evans[g]
	0·060	25	Cyclohexane	Cundall and Robinson[h]
	see Table 2.6			Cundall and Pereira[i]
q_{FD}	0·012	20	Hexane	Lumb and Weyl[a]
	0·016	25	Methylcyclohexane	Hirayama and Lipsky[d]
	0·023	25	Cyclohexane	Cundall and Robinson[h]
c_h(M)	4·8	24	Hexane	Lumb[j]
	2·1	20	Hexane	Birks, Braga and Lumb[k]
	5·0	25	Monomer Methyl cyclohexane	Hirayama and Lipsky[d]
	2·9	25	Excimer Methyl cyclohexane	Hirayama and Lipsky[d]
	5·0	20	Monomer Cyclo-hexane	Yguerabide[l]
	16·1	20	Excimer Cyclo-hexane	Yguerabide[l]
	5·0	25	Monomer Cyclo-hexane	Cundall and Robinson[h]
	4·65	25	Excimer Cyclo-hexane	Cundall and Robinson[h]
E_b(eV)	0·22	—	Cyclohexane	Birks, Braga and Lumb[k]
	0·36	—	Methylcyclohexane	Hirayama and Lipsky[d]
	0·35	—	Cyclohexane	Cundall and Robinson[h]
τ_M(ns)	32·0	20	*n*-Nonane	Ludwig and Amata[m]
	29·0	Ambient	Cyclohexane	Berlman[o]
	24	20	Hexane	Birks, Braga and Lumb[j]
	27·5	24	Hexane	
	29·5	25	Cyclohexane	Cundall and Robinson[h]
	26	Room temp.	Cyclohexane	Ivanova, Kudryashov and Sveshnikov[t]
	26			
	27·9	25	Methylcyclohexane	Gregory and Helman[n]
	see Table 2.6			Cundall and Pereira[i]

Table 2.5 (*continued*)

Parameter	Value	Temp./°C	Solvent	Reference
C_6H_6				
τ_D(ns)	12·3	Ambient	Ionizing radiation studies	Christophorou, Carter and Abu-Zeid[o]
	12	20	Hexane	Birks, Braga and Lumb[k]
	28	25	Cyclohexane	Cundall and Robinson[h]
τ(ns)	27·1	20	Pure liquid	Ludwig and Amata[m]
	27	20	Pure liquid	Helman[p]
	27·1	25	Pure liquid	Amata, Burton, Helman, Ludwig and Rodemayer[q]
	32·0		llM with *n*-Nonane	Amata, Burton, Helman, Ludwig and Rodemayer[q]
	26		Pure liquid	Birks and Munro[r]
	23–30	25	Pure liquid, depends upon degassing procedure	Wagner, Christophorou and Carter[s]
	29·5	25	Pure liquid	Cundall and Robinson[h]
	31·5	25	Pure liquid	Greenleaf and King[v]
C_6D_6				
q_{FM}	0·066	25	Cyclohexane	Berlman[c]
	0·042	24	Methylcyclohexane	Cundall, Ogilvie and Robinson[w]
τ_M	26·6 ns	25	Cyclohexane	Berlman[c]
	32	25	Cyclohexane	Helman[p]
	30	25	Methylcyclohexane	Cundall, Ogilivie and Robinson[w]
τ	32	25	Pure liquid	Helman[p]

[a] M. D. Lumb and D. A. Weyl, *J. Molec. Spectrosc.*, **23**, 365 (1967).
[b] J. W. Eastman and S. J. Rehfeld, *J. Phys. Chem.*, **75**, 991 (1971).
[c] I. B. Berlman, *Handbook of Fluorescence Spectra of Aromatic Molecules*, Academic Press, New York, 1965.
[d] F. Hirayama and S. Lipsky, *J. Chem. Phys.*, **51**, 1939 (1969).
[e] C. L. Braga, L. C. Pereira, M. D. Lumb and M. I. Barradas, *Rev. fiz. Quin. Eng. Lourenco Marques*, **1**, 1 (1969).
[f] J. W. Eastman, *J. Chem. Phys.*, **49**, 4617 (1968).
[g] G. B. Evans, Unpublished data, Nottingham.
[h] R. B. Cundall and D. A. Robinson, *J. Chem. Soc. Faraday Transactions II*, **68**, 1133 (1972).
[i] R. B. Cundall and L. C. Pereira, *Chem. Phys. Letters*, **18**, 371 (1973).
[j] M. D. Lumb, Private communication.
[k] J. B. Birks, C. L. Braga and M. D. Lumb, *Proc. Roy. Soc.*, **A283**, 83 (1965).
[l] J. Yguerabide, *J. Chem. Phys.*, **49**, 326 (1968).
[m] P. K. Ludwig and C. D. Amata, *J. Chem. Phys.*, **49**, 326 (1968).
[n] T. A. Gregory and W. P. Helman, *J. Chem. Phys.*, **56**, 377 (1972).
[o] L. G. Christophorou, J. G. Carter and M.-E. M. Abu-Zeid, *J. Chem. Phys.*, **49**, 3775 (1968).
[p] W. P. Helman, *J. Chem. Phys.*, **51**, 354 (1969).

concentration (*c.* 0·26 M) that the value of ϕ_T became practically unity, due to spin–orbit catalysis of the $S_1 - T_1$ conversion. The rate constants for the xenon-catalysed intersystem crossing transitions, $S_1 \rightarrow T_1$ and $T_1 \rightarrow S_0$, were 2×10^8 M^{-1} s^{-1} and 3×10^8 M^{-1} s^{-1}, respectively. Since this was for the neat liquid, these values apply to the simultaneous quenching of the excimer and monomer excited singlet states.

Hentz and Perkey[140] also obtained $\phi_T = 0·58$ for liquid benzene under practically the same conditions as those used by Cundall and Tippett.[139] Impurity and intensity effects do not appear significant in affecting either the measured triplet yield or lifetime. Sandros[141] reported $\phi_T = 0·25$ for 0·029 M benzene in cyclohexane, a value in agreement with $\phi_T = 0·24$ (for which no details are available) attributed to Hammond and Lamola[142] from a measurement of the *cis–trans* isomerization of piperylene. Cundall, Pereira and Robinson[143] found that ϕ_T values obtained by the butene-2-technique were concentration dependent, e.g. 0·56 for 11·2 M (neat benzene), 0·47 for 5·1 M benzene in cyclohexane, 0·24 for 0·11 M benzene in cyclohexane at 24°C. At 6×10^{-2} M in methylcyclohexane $\phi_T = 0·25$ for C_6H_6 and 0·23(5) for C_6D_6 at 26°C, in agreement with the values of 0·252 and 0·240 in cyclohexane at 20°C, and also with those obtained by Sandros,[141] using the triplet-sensitized phosphorescence of biacetyl. The triplet quantum yield shows a temperature dependence; for a 0·112 M solution in cyclohexane the ϕ_T values were 0·208, 0·141 and 0·104 at 28, 45 and 60°C.[144] Hentz and Thibault[145] using *trans* → *cis* isomerization of octene-2 obtained $\phi_T = 0·71$ for neat benzene decreasing to 0·24 for 0·05 M benzene in methylcyclohexane at 22°C, and $\phi_T = 0·94$ for 0·56 M benzene and 0·85 for 0·056 M benzene at −78°C. Similar results have been obtained at low temperatures for benzene by Cundall and Pereira.[146] It seems established that values of $\phi_F + \phi_T$ tend to 1 at low temperature for benzene.

The lifetime of the benzene triplet in the liquid state has been a subject for some disagreement which now seems to be resolved. Cundall and Griffiths[147,148] obtained indications from the benzene/butene-2 system that the lifetime of the benzene triplet was short *c.* 10^{-8} s^{-1} at room temperature and that this value decreased as the temperature was raised. The value of $k_{ET}\tau_T = 21·2$ M^{-1} at 25°C, where k_{ET} is the overall rate constant for excitation energy transfer from triplet benzene to butene-2 and τ_T is the benzene triplet lifetime.

---

[q] C. D. Amata, M. Burton, W. P. Helman, P. K. Ludwig and S. A. Rodemayer, *J. Chem. Phys.*, **48**, 2374 (1968).
[r] J. B. Birks and I. H. Munro, *Prog. Reaction Kinetics*, **4**, 237 (1967).
[s] O. E. Wagner, L. G. Christophorou and J. G. Carter, *Chem. Phys. Letters*, **4**, 224 (1969).
[t] T. V. Ivanova, P. I. Kudryashov and B. Y. Sveshnikov, *Soviet Phys. Doklady*, **6**, 407 (1961).
[u] K. Sandros, *Acta Chem. Scand.*, **25**, 3651 (1971).
[v] J. R. Greenleaf and T. A. King, *Proc. Intern. Conference Luminescence*, Budapest, 212 (1966).
[w] R. B. Cundall, S. McD. Ogilvie and D. A. Robinson, *J. Photochem.*, **1**, 417 (1972–73).

The short triplet lifetime was confirmed by Dubois[149] by experiments with biacetyl in neat benzene which failed to provide evidence for triplet energy transfer. However, values of 1–2 μs for the benzene triplet lifetime in dilute solutions in cyclohexane (10^{-2} M) were obtained by Lipsky,[150] Sandros[151] and Russian workers.[152,153] Experiments with *cis*-butene-2 solutions of *o*-xylene in methylcyclohexane showed that τ_T values for the aromatic triplet decreased with increasing concentration of *o*-xylene from *c.* 1 μs to a few ns in going from 10^{-2} M to the neat liquid.[154] This is the case with benzene also,[160] and it provides an explanation for the apparent discrepancies. All the estimates of the triplet lifetimes by sensitization techniques require assumption of a value for the rate constant for energy transfer, usually about 5×10^9 M^{-1} s^{-1} (e.g. see reference 144) at room temperature. Thomas and Mani[103] in their studies of the pulse radiolysis of benzene and benzene/cyclohexane mixtures found a transitory species with an absorption maximum at 320 nm and $t_{\frac{1}{2}}$ for decay of 112 ns. The addition of naphthalene, biacetyl, O_2 and piperylene suggests that the species responsible for the 320 nm absorption is a free radical formed by reaction of triplet benzene with a ground state benzene molecule. The build-in times of this species are 3 ns in pure benzene and 20 ns in 10% benzene and they are unaffected by temperature. Laser flash photolysis[104] evidence is consistent with this interpretation of the mechanism for triplet decay and a value of $\phi_T = 0{\cdot}39$ has been estimated for the triplet quantum yield in 0·3 M solution. These experiments represent the closest yet achieved to the direct observation of the benzene triplet state and its decay in illuminated solutions at room temperature.

Table 2.6 Values of fluorescence yields (ϕ_{FM}), fluorescence decay times (τ_M) and triplet yields (ϕ_T) for dilute solutions of benzene at room temperature (25°C). (R. B. Cundall and L. C. Pereira, *Chem. Phys. Letters*, **18**, 371 (1973))

Solvent	ϕ_{FM}	τ_M (ns)	ϕ_T
perfluoro-*n*-hexane	0·065	39	0·24
cyclohexane	0·060[a]	34[a]	0·23
methylcyclohexane	0·052[a]	30[a]	0·24
ethanol	0·040[a]	26[a]	0·15
methanol	0·029[a]	20[a]	0·13
acetonitrile	0·025[a]	15[a]	0·09
water	0·006[b]	2·2[b]	0·07

[a] R. B. Cundall and L. C. Pereira, *J. Chem. Soc. Faraday Transactions II*, **68**, 1152 (1972).
[b] M. Luria and G. Stein, *J. Phys. Chem.*, **76**, 165 (1972).

Values of ϕ_T as well as ϕ_{FM} and τ_M were measured in a series of solvents for low concentrations at 25°C by Cundall and Pereira.[155] As seen from the data in Table 2.6, the rate constants for intersystem crossing, as well as the corresponding values for the fluorescence, except for solutions in water, show little solvent effect. In water solutions the yields are very low, and experimental errors could be significant. In the case of methylcyclohexane solutions where data were available for evaluation of $\phi_T/\tau_M = k_{ISC}$ ($= k_{TM}$ in dilute solution), a slight temperature dependence was found.[144] The early interpretation in terms of a simple Arrhenius expression for k_{ISC} is probably incorrect.[156] The concentration used (0·112 M) may have been sufficiently high for some temperature-dependent intersystem crossing from the excimer.

Cundall, Ogilvie and Robinson[157] find C_6D_6 to be similar in its behaviour to C_6H_6. Table 2.7 shows ϕ_T is slightly higher (*c.* 10%) for

Table 2.7 Triplet yield measurements on benzene-d_6 (10^{-2} M) in methylcyclohexane at different temperatures. (R. B. Cundall, S. McD. Ogilvie and D. A. Robinson, *J. Photochem.*, **1**, 417 (1972/73))

Temp. (°C)	*cis*-butene-2 (M)	ϕ_{ISOM}	ϕ_T	ϕ_{IC}[a]
27	0·170	0·124 (7)	0·249	0·713
28	0·171	0·122 (3)	0·244	0·717
33	0·164	0·108 (4)	0·217	0·749
42	0·160	0·088 (4)	0·177	0·794
52	0·168	0·075 (4)	0·151	0·830
63	0·193	0·060 (5)	0·121	0·862

[a] $\phi_{IC} = 1 - \phi_F - \phi_T$.

C_6D_6 than for C_6H_6 in methylcyclohexane (10^{-2} M). In cyclohexane the C_6D_6 triplet yield (0·24) is fractionally smaller than for C_6H_6 (0·25) at 20°C. There is no isotope effect apparent like that manifested in the gas phase, and it is a process other than fluorescence or intersystem crossing which is most markedly affected. The slight temperature dependence of k_{ISC} (Table 2.8) for C_6D_6, similar to that detected for C_6H_6, is explained by the two alternative intersystem crossing processes (i) $S_1 \rightarrow T_1$ and (ii) $S_1 \rightarrow T_2$. The slighter temperature dependence for C_6D_6 probably arises from closer spacing of vibrational levels. There is no marked effect on the $S_1 \rightarrow T_1$ transition equivalent to that proposed for explaining the effect of deuteration[158] on the $T_1 \rightarrow S_0$ transition, but it is in accord with the recent proposition of Johnson and Ziegler[159] that the radiationless $S_1 \rightarrow T_1$ transition rate is insensitive to deuteration.

Table 2.8 S_1 decay rate parameters for benzene-d_6 (10^{-2} M) in methylcyclohexane. (R. B. Cundall, S. McD. Ogilvie and D. A. Robinson, *J. Photochem.*, **1**, 417 (1972/73))

Temp. (°C)	k_{FM} ($\times 10^{-6}$) (s^{-1})	k_{ISC} ($\times 10^{-6}$)[a] (s^{-1})	k_{IC} ($\times 10^{-6}$)[b] (s^{-1})
25	1·37	8·67	23·3
30	1·40	8·75	27·7
35	1·42	8·80	32·5
40	1·43	8·97	37·9
45	1·43	9·18	44·3
50	1·45	9·56	51·5
55	1·43	9·72	58·8
60	1·45	9·85	65·6
65	1·45	9·83	73·5

[a] $k_{ISC} = k_{TM}$ in dilute solution.
[b] $k_{IC} = k_{GM}$ in dilute solution.

Hentz and Thibault[145] obtained their value of $\phi_T = 0{\cdot}71$ for pure benzene at 22°C by extrapolation of results in solution for $\phi_{trans \to cis}$ for octene-2 according to the equation

$$\phi_{t \to c}^{-1} = (0{\cdot}5\phi_T)\{1 + 1/k_{ET}\tau_T[trans\text{-}C_8H_{16}\text{-}2]\} \tag{2.17}$$

where k_{ET} is the rate constant for the process $C_8H_{16} + 3M^* \rightarrow {}^3C_8H_{16} + {}^1M$. A plot of $\phi_{t \to c}^{-1}$ *vs.* $[trans\text{-}C_8H_{16}\text{-}2]^{-1}$ yields ϕ_T and $k_{ET}\tau_T$ from the slope and intercept. Hentz and Thibault maintained that the value of Cundall and Tippett for ϕ_T in liquid benzene was low, because the concentration of triplet scavenger used was insufficient. Cundall and Robinson[160] regard the system as more complex than the suggested scheme of Hentz and Thibault, who assumed that both monomeric triplet and triplet excimer could be scavenged by olefin under all conditions. This may not always be the case. Plots of $\phi_{c \to t}^{-1}$ *vs.* $[\text{Bu-2}]^{-1}$ show some evidence of curvature at high olefin concentrations, and it has been suggested that if a term for butene-2 quenching of triplet excimer is included the rate of isomerization is

$$b[\text{Bu}]\{k_{ETM}[{}^3M^*] + k_{ETD}[{}^3M_2^*]\} \tag{2.18}$$

On evaluation of steady state conditions the rate of isomerization is of the form

$$\frac{bQ[\text{Bu}] + bR[\text{Bu}]^2}{N + O[\text{Bu}] + P[\text{Bu}]^2} \tag{2.19}$$

where N, O, P, Q and R are complex functions of rate constants and excited state concentrations. Plots of ϕ^{-1} *vs.* $[Bu]^{-1}$ are complex, but at high concentrations equation (2.19) reduces to

$$bR[Bu]^2/\{N + P[Bu]^2\} \tag{2.20}$$

and reciprocal plots should be paraboloid. Plots of ϕ_{ISC}^{-1} *vs.* $[Bu]^{-2}$ are more nearly linear and from such plots the limiting intercepts for isomerization yields in neat benzene were calculated to be 0·295, which is within 3% of the directly measured value at 25°C for higher olefin concentrations. If at some temperatures the added olefin does not quench all the triplets no matter how high the concentration, the plots would give $\phi_{T(meas)} < \phi_{T(total)}$. This could occur if the triplet excimer decayed by a fast process in competition with $3B_2 \rightarrow {}^3B^* + B$ and if olefins like butene-2 or octene-2 were not efficient in accepting triplet transfer from the triplet excimer. This is very probable since the very short lifetime of the triplet state of benzene is explained by rapid radiationless decay of the triplet excimer and there is evidence that this decay is fast at higher temperatures.

The total triplet yield

$$\phi_T = \phi_{TM} + \phi_{TD} = f_M q_{TM} + f_D q_{TD} \tag{2.21}$$

where q_{TM} and q_{TD} are the quantum efficiencies of intersystem crossing from the singlet excited monomer and excimer, and f_M and f_D are the fractions of excited singlet states in the monomeric and excimer forms. Since $f_M + f_D = 1$, a simple evaluation of both q_{TM} and q_{TD} can be carried out from a knowledge of f_M and ϕ_T. A temperature study showed that q_{TD} did not appear to vary simply with temperature, an anomaly attributed to the failure of butene-2 to quench all the triplet excimer formed. At 25°C it was calculated from data for neat benzene and methylcyclohexane/benzene solutions that the maximum quantum yield which can be attributed to internal conversion of the singlet excimer is $\leqslant 19\%$, a value much less than for the monomeric ${}^1B_{2u}$ state (70%).

The total specific decay rate k_D for the benzene excimer has been found to be independent of temperature between 298 and 169°K by Gregory and Helman.[123] $\tau_D (= 1/k_D)$ varies between $35{\cdot}9 \pm 0{\cdot}3$ and $36{\cdot}1 \pm 0{\cdot}4\ s^{-1}$ over this range. This result is consistent with the excimer only undergoing fluorescence or intersystem crossing. Over the same temperature range the total rate constant (τ_M^{-1}) for monomer ${}^1B_{2u}$ decay changes from $35{\cdot}9 \pm 0{\cdot}2$ to $9{\cdot}7 \pm 0{\cdot}1 \times 10^6\ s^{-1}$.

The lifetime of the monomeric triplet is not known with any certainty, but there is no reason to think that the values in both the liquid and gas phases are inconsistent with the low-temperature behaviour in glasses, where it has been considered that the thermally activated $T_1 \rightarrow S_0$

radiationless transition arises from distortion of the regular hexagonal forms of the benzene ring.[161] The concentration quenching mechanism[72,103,154] appears to involve some form of chemical or charge-transfer interaction, reducing the forbidden character for relaxation to the ground state. There have been many investigations of the effect of environment on benzene phosphorescence at low temperature. These results will not be discussed here (see Chapter 9, Vol. 2), but it is perhaps appropriate to mention the recent work of Haaland and Nieman.[162] At temperatures below 65°K it was clear from a study of solvent effects on the lifetime of benzene phosphorescence that hydrocarbons are not inert solvents, and they can cause changes of about 50% in τ_T values at low temperatures. Perfluorocarbons were the least perturbing class of solvents. This is consistent with the previous discussions on the use of these solvents for investigations of intersystem crossing and internal conversion of the excited singlet state in fluid media.

2.3.2.6 *Radiationless processes, other than intersystem crossing*

The results of the effect of temperature on the fluorescence quantum yield ϕ_{FM} of benzene have been interpreted as involving a temperature-dependent radiationless process. ϕ_{FM} can be expressed in the form

$$\phi_{FM} = \frac{k_{FM}}{k_{FM} + k^0_{IM} + k'_{IM}\exp(-W_{IM}/\mathbf{k}T)} \tag{2.22}$$

where the subscript IM represents internal quenching transitions.

Benzene has been examined in dilute hexane and ethanol solutions.[163] By a curve-fitting process values of $k^0_{IM} = 13{\cdot}3 \times 10^6\ s^{-1}$ and $8{\cdot}6 \times 10^6\ s^{-1}$, $1{\cdot}2 \times 10^{12}\ s^{-1}$ and $1{\cdot}2 \times 10^{11}\ s^{-1}$, and $W_{IM} = 0{\cdot}35$ eV and $0{\cdot}22$ eV were derived for the hexane and ethanol solutions, respectively. In liquid benzene $k^0_I = 0{\cdot}45 \times 10^6\ s^{-1}$, $k'_I = 8{\cdot}2 \times 10^{11}\ s^{-1}$ and $W_I = 0{\cdot}30$ eV.

Without any knowledge of triplet yields the nature of the thermally induced process is not known. The increased formation of phosphorescing triplet states at low temperatures indicates that it could be either internal conversion or chemical reaction. Cundall and Robinson[144] measured ϕ_{FM}, τ_M, and ϕ_T (Figure 2.15) for benzene in 0·112 M solution in cyclohexane and they found that $\phi_{IC}(=1 - \phi_{FM} - \phi_T)$ increases with temperature and that one or more processes, other than fluorescence or triplet formation, are dominant in causing the disappearance of the $^1B_{2u}$ state at room temperature and above (Table 2.9). $k'_{IC} = 7{\cdot}9 \times 10^{12}\ s^{-1}$ and $W_{IC} = 0{\cdot}325$ eV for benzene in cyclohexane. The internal conversion process may perhaps be better described by a more complex expression of the type $k_{IC} = k^0_{IC} + \sum k'_{IC}\exp(-W_{IC}/\mathbf{k}T)$ where k^0_{IC} can have a value

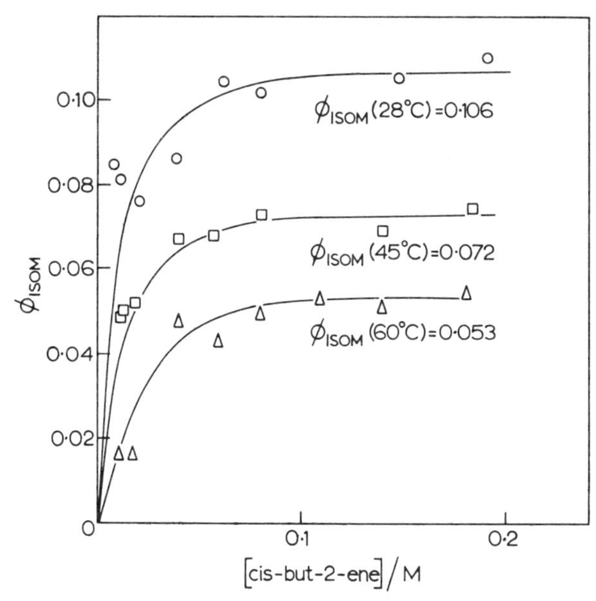

Figure 2.15 Variation in ϕ_{ISOM} with *cis*-but-2-ene concentration at three temperatures (28, 45 and 60°C) for 0·112 M benzene in cyclohexane

close to $0{\cdot}5 \times 10^6\ s^{-1}$ estimated theoretically for the $S_1 \rightarrow S_0$ internal conversion in benzene by Siebrand and Williams.[164] Similar analysis of data for C_6D_6 in methylcyclohexane gives $k'_{IC} = 4{\cdot}6 \times 10^{11}\ s^{-1}$ and $W_{IC} = 0{\cdot}25$ eV.[157] The temperature dependence of intersystem crossing is slight, and somewhat greater than the small dependence found for fluorescence. Fluorescence and intersystem crossing should therefore be dominant at low temperatures, as is found to be the case. Helman[132]

Table 2.9 Fluorescent lifetimes and quantum yields for fluorescence, intersystem crossing, and internal conversion after excitation by 254 nm light at different temperatures. (R. B. Cundall and D. A. Robinson, *J. Chem. Soc. Faraday Transactions II*, **68**, 1145 (1972))

Temp. (°C)	ϕ_{FM}	ϕ_T	τ_M/s ($\times 10^9$)	ϕ_{IC}[a]
28	0·055	0·208	25·7	0·737
45	0·035	0·141	14·8	0·824
60	0·022	0·104	9·4	0·874

[a] Total yield of radiationless $S_1 \rightarrow S_0$ decay i.e., $[1 - (\phi_{FM} + \phi_T)]$.

calculated an activation energy of 0·28 eV for a process assigned to internal conversion in both C_6D_6 and C_6H_6 which he obtained from the effect of temperature on the fluorescence decay times between 5 and 35°C. Using Helman's lifetime data, Sandros[151] obtained W_{IC} values of 0·288 eV for C_6H_6 and 0·292 eV for C_6D_6 in cyclohexane, and high k'_{IC} values of about 10^{12} s^{-1}.

From their measurements of ϕ_{FM} of benzene in a wide variety of solvents at room temperature, Eastman and Rehfeld[118] established a relationship between ϕ_{FM} and the enhancement of the 0 − 0 band in absorption, and they concluded that solvent-induced internal quenching depends on the same intermolecular forces as those responsible for induced absorption. The process(es) involved are more clearly defined by the simultaneous measurement of triplet quantum yields, fluorescence lifetimes and fluorescence quantum yields as carried out by Cundall and Pereira.[155] From their results it is apparent that internal conversion is the process by far the most affected in all the solvents studied except water (Table 2.10). The

Table 2.10 Rate constants for fluorescence (k_{FM}), intersystem crossing (k_{ISC}) and internal conversion (k_{IC}) for dilute solutions and gaseous benzene, at room temperature (λ_{exc} = 253·7 nm). (R. B. Cundall and L. C. Pereira, *Chem. Phys. Letters*, **18**, 371 (1973))

Solvent	k_{FM} (10^{-6}) (s^{-1})	k_{ISC} (10^{-6}) (s^{-1})	k_{IC} (10^{-6}) (s^{-1})	$k_{FM}/k_{ISC} \times 10^{-1}$
Perfluoro-*n*-hexane	1·66	6·10	17·8	2·72
Cyclohexane	1·76	6·75	19·9	2·61
Methylcyclohexane	1·73	8·00	23·6	2·16
Ethanol	1·54	5·80	31·1	2·66
Methanol	1·46	6·50	42·0	2·25
Acetonitrile	1·67	6·00	59·0	2·78
Water	2·7	32	460	0·84
Gas phase	2·5	9·8	1·17	2·55

correlation between the Ham band enhancement and the promotion of internal conversion is very similar to that found by Lawson *et al.* for solvent promotion of the $S_3 \rightarrow S_1$ internal conversion in benzene. The temperature dependence allows some analysis assuming a simple Arrhenius relationship to be valid (Table 2.11). The change of the magnitude of the frequency factor k'_{IC} is the most significant effect of the environment on the $^1B_{2u}$ state. A tentative suggestion[155] for the nature of the internal conversion from the $^1B_{2u}$ state is a vibrationally activated transition to a (σ, π^*) state as proposed by Callomon *et al.*[86] σ electrons are partially

Table 2.11 Arrhenius parameters for the temperature dependent non-radiative process (internal conversion) of the $^1B_{2u}$ state of benzene. (R. B. Cundall and L. C. Pereira, *Chem. Phys. Letters*, **18**, 371 (1973))

Solvent	k'_{IC} (s^{-1})	W_{IC} (eV)
Cyclohexane[a]	$2{\cdot}8 \times 10^{11}$	0·25
Methylcyclohexane[a]	$4{\cdot}6 \times 10^{11}$	0·25 (3)
Ethanol[a]	$2{\cdot}5 \times 10^{12}$	0·29 (4)
Methanol[b]	1·1 × 10·14	0·37 (5)

[a] R. B. Cundall and L. C. Pereira, *J. Chem. Soc. Faraday II*, **68**, 1152 (1972).
[b] J. W. Eastman, *J. Chem. Phys.*, **49**, 4617 (1968).

shielded and less affected by the solvent than the π and π^* orbitals. On the basis of this model the effect of solvent polarity is likely to be a change of the energy levels of (π, π^*) and (σ, π^*) states in such a way as to increase the value of W_{IC}. The differences in internal conversion between the gas and liquid phases, for excitation to low vibrational levels of the $^1B_{2u}$ state may arise from the marked decrease in W_{IC} associated with the change in energy-level density in going from the vapour to a solvent, even for one of such low polarity as perfluoro-*n*-hexane. In the gas phase there are no solvent perturbations to affect the k'_{IC} factor, with the result that internal conversion is less efficient in the gas phase than in solution for excitation at longer wavelengths.

The temperature dependence for internal conversion in cyclohexane solution agrees with Eastman's[117] activation energy of 0·26 eV and frequency factor of 10^{13} s^{-1}. Lumb and Pereira[165] also obtained an activation energy of 0·26 ± 0·02 eV and a frequency factor of 10^{11} s^{-1} in solid benzene for a process they attributed to isomerization of the first excited singlet state.

It appears from the similarity of the results of experiments in all phases that basically the same non-radiative transition is involved in the decay of the $^1B_{2u}$ state of benzene in all phases.

The nature of the 'internal conversion' process is obscure. The decomposition or isomerization quantum yields for benzene in solution and rigid glasses are small, usually about 0·01, and in general very much less than in the gas phase due to collisional stabilization. If an intermediate of the prefulvene type is involved several quanta of ring deforming vibrations would be required and this could explain the activation energy for 'internal conversion'. However the lack of appreciable amounts of products does not favour this as the major process. More work on this controversial problem is required.

Chemical reactions of excited benzene with the solvent cannot be ignored in accounting for the disappearance of excited states of benzene in solution and it is unfortunate that experimental difficulties have not allowed clear results to be obtained. This is illustrated by the following examples. In liquid benzene the products

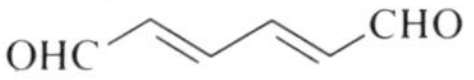

OHC CHO and $OHC(CH{=}CH)_5CHO$

have been reported as oxygen photoaddition products.[166] In 1968 it was reported that benzene, irradiated in aqueous solution at pH 5–7, added one molecule of O_2 per benzene molecule and the product was[167]

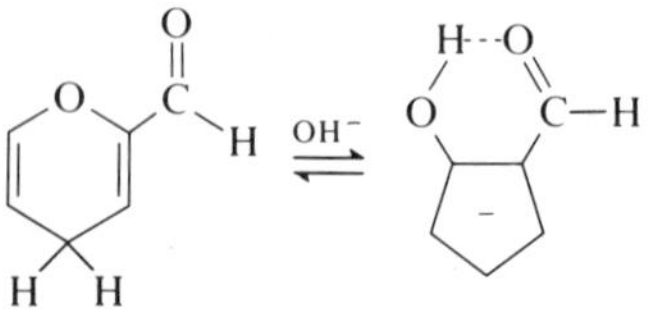

O C H $\underset{}{\overset{OH^-}{\rightleftharpoons}}$ H--O O C—H H H

with a quantum yield of 0·1. The ${}^1B_{2u}$ state is initially produced but the species reacting with O_2 was thought to be either triplet benzene or a labile isomer.[168] It has now been reported that the structure originally proposed is incorrect and that the compounds involved are

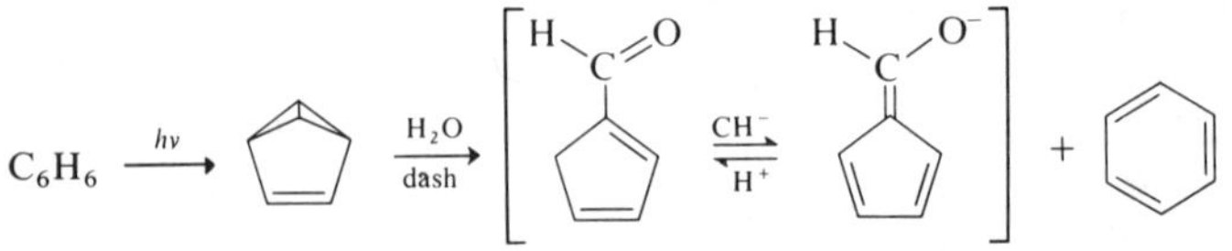

$C_6H_6 \xrightarrow{h\nu}$ $\xrightarrow[\text{dash}]{H_2O}$ [H C O $\underset{H^+}{\overset{CH^-}{\rightleftharpoons}}$ H C O^-] +

and oxygen is not essential in this scheme.[169] Luria and Stein[170] still uphold their explanation but this is difficult to reconcile with the fact that Kaplan *et al.* maintain their finding is true even with O_2 absent. It is relevant to the use of olefins as triplet scavengers to point out that aromatic–olefin systems yield addition products, although usually in small yields.[22,171]

2.3.2.7 *Concluding remarks*

Data for benzene in the liquid phase are slowly accumulating and the importance of a solvent dependent radiationless process, other than intersystem crossing, similar to that found in the vapour phase, is established for the ${}^1B_{2u}$ state. An excitation wavelength dependence of radiationless decay from different vibrational levels of the S_1 state has not been observed: rapid energy transfer and relaxation due to solvent interaction with the vibronic states is probably responsible for this. There is no evidence that there is an important internal conversion process for the excimer.

2.4 References

1. J. W. Donovan and A. B. F. Duncan, *J. Chem. Phys.*, **35**, 1389 (1961).
2. H. Ishikawa and W. A. Noyes, Jr., *J. Amer. Chem. Soc.*, **84**, 1502 (1962).
3. H. Ishikawa and W. A. Noyes, Jr., *J. Chem. Phys.*, **37**, 583 (1962).
4. R. B. Cundall and T. F. Palmer, *Trans. Faraday Soc.*, **56**, 1211 (1960).
5. T. M. Dunn, *Studies on Chemical Structure and Reactivity* (Ed. J. H. Ridd), Methuen, London, 1966, and references quoted therein.
6. C. S. Parmenter, *Advances in Chemical Physics*, **22**, 365 (1972).
7. G. H. Atkinson and C. S. Parmenter, *J. Phys. Chem.*, **75**, 1564 (1971); G. H. Atkinson, C. S. Parmenter and M. W. Schuyler, *J. Phys. Chem.*, **75**, 1572 (1971).
8. S. P. McGlynn, T. Azumi and M. Kinoshita, *Molecular Spectroscopy of the Triplet State*, Prentice-Hall, Englewood Cliffs, N.J., 1969.
9. G. C. Morris and J. G. Angus, *J. Molec. Spec.*, **45**, 271 (1973); A. M. Taleb, I. H. Munro and J. B. Birks, *Chem. Phys. Letters*, **21**, 454 (1973); see also J. Karowski, *Chem. Phys. Letters*, **18**, 47 (1973).
10. J. H. Callomon, T. M. Dunn and I. M. Mills, *Phil. Trans. Roy. Soc.*, **259**, 499 (1966).
11. D. F. Evans, *J. Chem. Soc.*, 1351, 3885 (1957).
12. A. C. Albrecht, *J. Chem. Phys.*, **38**, 354 (1963).
13. S. D. Colson and E. R. Bernstein, *J. Chem. Phys.*, **43**, 2661 (1961).
14. J. A. Poole, *J. Phys. Chem.*, **69**, 1343 (1965).
15. W. A. Noyes, Jr., W. A. Mulac and D. A. Harter, *J. Chem. Phys.*, **44**, 2100 (1966).
16. W. A. Noyes, Jr. and D. A. Harter, *J. Chem. Phys.*, **46**, 674 (1967). In a more recent reference, *J. Phys. Chem.*, **75**, 2741 (1971), a critical review of the methods of measurement of triplet yields is presented by these authors.
17. E. K. C. Lee, M. W. Schmidt, R. G. Shortridge, Jr. and G. A. Haninger, Jr., *J. Phys. Chem.*, **73**, 1805 (1969).
18. K. Janda and F. S. Wettack, *J. Amer. Chem. Soc.*, **94**, 305 (1972).
19. A. Morikawa and R. J. Cvetanovic, *J. Chem. Phys.*, **49**, 1214 (1968).
20. L. Kaplan and K. E. Wilzbach, *J. Amer. Chem. Soc.*, **88**, 2066 (1966) and **90**, 3291 (1968).
21. A. Morikawa, S. Brownstein and R. J. Cvetanovic, *J. Amer. Chem. Soc.*, **92**, 1471 (1970); R. Srinivasan, *Tetrahedron Letters*, 4551 (1971).
22. D. Bryce-Smith and H. C. Longuet-Higgins, *Chem. Comm.*, 593 (1963).
23. A. Morikawa and R. J. Cvetanovic, *J. Chem. Phys.*, **52**, 3237 (1970).
24. D. R. Snelling, *Chem. Phys. Letters*, **2**, 346 (1968).
25. G. Das Gupta and D. Phillips, *J. Chem. Soc. Faraday Transactions II*, **68**, 2003 (1972).
26. G. B. Kistiakowsky and C. S. Parmenter, *J. Chem. Phys.*, **42**, 2942 (1965).
27. E. M. Anderson and G. B. Kistiakowsky, *J. Chem. Phys.*, **48**, 4787 (1968).
28. E. M. Anderson and G. B. Kistiakowsky, *J. Chem. Phys.*, **51**, 182 (1969).
29. C. S. Parmenter and A. H. White, *J. Chem. Phys.*, **50**, 1631 (1969).
30. A. E. Douglas and C. W. Mathews, *J. Chem. Phys.*, **44**, 4788 (1968).
31. S. J. Strickler and R. J. Watts, *J. Chem. Phys.*, **44**, 426 (1966).

32. C. S. Parmenter and M. W. Schuyler, *J. Chem. Phys.*, **52**, 5366 (1970).
33. C. S. Parmenter and M. W. Schuyler, *Transitions Non-Radiatives dans les Molécules. J. Chim. Phys.*, **67**, 92 (1970).
34. M. Stockburger and H. J. Kemper, *Ber. Bunsenges. Phys. Chem.*, **72**, 1044 (1968); L. M. Logan, I. Buduls and I. G. Ross, *Molecular Luminescence*, (Ed. E. C. Lim), W. A. Benjamin, Inc., New York, p. 53, 1969.
35. H. J. Kemper and M. Stockburger, *J. Chem. Phys.*, **53**, 268 (1970); J. Blondeau and M. Stockburger, *Ber. Bunsenges. Phys. Chem.*, **75**, 450 (1971).
36. C. S. Parmenter and M. W. Schuyler, *Chem. Phys. Letters*, **6**, 339 (1970)
37. G. M. Breuer and E. K. C. Lee, *J. Chem. Phys.*, **51**, 3130 (1969).
38. G. M. Breuer and E. K. C. Lee, *J. Chem. Phys.*, **51**, 3615 (1969).
39. M. Nishikawa and P. K. Ludwig, *J. Chem. Phys.*, **52**, 107 (1970).
40. T. L. Chen and E. W. Schlag, *Molecular Luminescence*, (Ed. E. C. Lim), W. A. Benjamin, Inc., New York, p. 381, 1969.
41. C. S. Burton and H. E. Hunziker, *J. Chem. Phys.*, **52**, 3302 (1970).
42. B. K. Selinger and W. R. Ware, *J. Chem. Phys.*, **52**, 5482 (1970); **53**, 3160 (1970).
43. W. R. Ware, B. K. Selinger, C. S. Parmenter and M. W. Schuyler, *Chem. Phys. Letters*, **6**, 342 (1970).
44. M. Luria, *Israel J. Chem.*, **10**, 721 (1972).
45. K. G. Spears and S. A. Rice, *J. Chem. Phys.*, **55**, 5561 (1971).
46. M. G. Rockley and D. Phillips, *Chem. Phys. Letters*, **21**, 181 (1973).
47. A. S. Abramson, K. G. Spears and S. A. Rice, *J. Chem. Phys.*, **56**, 2291 (1972).
48. R. B. Cundall, F. J. Fletcher and D. G. Milne, *Trans. Faraday Soc.*, **60**, 1146 (1964).
49. G. P. Semeluk and R. D. Stevens, *Canad. J. Chem.*, **49**, 2452 (1971).
50. H. H. Brongessma, J. A. van der Hart and L. J. Oosterhoff, *Fast Reactions and Primary Processes in Chemical Kinetics* (Ed. S. Claesson), Interscience, 1967; J. H. Moore, *J. Phys. Chem.*, **76**, 1130 (1972).
51. G. A. Haninger, Jr. and E. K. C. Lee, *J. Phys. Chem.*, **73**, 1815 (1969).
52. H. E. Hunziker and H. R. Wendt, *Chem. Phys. Letters*, **12**, 181 (1971).
53. E. K. C. Lee, H. O. Denschlag and G. A. Haninger, Jr., *J. Chem. Phys.*, **48**, 4547 (1968).
54. R. B. Cundall and A. S. Davies, *Trans. Faraday Soc.*, **62**, 1151 (1966).
55. S. Y. Ho and W. A. Noyes, Jr., *J. Amer. Chem. Soc.*, **89**, 5091 (1967).
56. H. O. Denschlag and E. K. C. Lee, *J. Amer. Chem. Soc.*, **90**, 3628 (1968).
57. K. Nakamura, *J. Amer. Chem. Soc.*, **93**, 3138 (1971).
58. K. Dunnicliff, Unpublished results.
59. P. Sigal, *J. Chem. Phys.*, **42**, 1953 (1965).
60. P. Sigal, *J. Chem. Phys.*, **46**, 1043 (1967).
61. G. A. Haninger, Jr. and E. K. C. Lee, *J. Phys. Chem.*, **71**, 3104 (1967).
62. G. A. Haninger, Jr. and E. K. C. Lee, *J. Phys. Chem.*, **73**, 1815 (1969).
63. A. Morikawa and R. J. Cvetanovic, *Canad. J. Chem.*, **46**, 1813 (1968).
64. M. W. Schmidt and E. K. C. Lee, *J. Amer. Chem. Soc.*, **92**, 3579 (1970).
65. S. Hirokami and S. Sato, *Canad. J. Chem.*, **45**, 3181 (1967); T. Terao, S. Hirokami, S. Sato and R. J. Cvetanovic, *Canad. J. Chem.*, **44**, 2173 (1966).
66. E. K. C. Lee, *J. Phys. Chem.*, **71**, 2804 (1967).
67. C. S. Parmenter and B. L. Ring, *J. Chem. Phys.*, **46**, 1998 (1967).
68. C. S. Burton and H. E. Hunziker, *Chem. Phys. Letters*, **6**, 352 (1970).
69. J. Leclercq and J. M. Leclercq, *Chem. Phys. Letters*, **18**, 411 (1973).
70. M. Nishikawa and M. C. Sauer, *J. Chem. Phys.*, **51**, 1 (1969).

71. T. F. Hunter and M. G. Stock, *Chem. Phys. Letters*, **22**, 368 (1973).
72. J. Langelaar, G. Jansen, R. P. H. Rettschnick and G. Hoytink, *Chem. Phys. Letters*, **12**, 86 (1971).
73. G. Porter and P. West, *Proc. Roy. Soc.*, **A279**, 302 (1964).
74. C. W. Ashpole, G. Porter and S. J. Formosinho, *Proc. Roy. Soc.*, **A323**, 11 (1971).
75. J. K. Foote, M. H. Mallon and J. N. Pitts, Jr., *J. Amer. Chem. Soc.*, **88**, 3698 (1966).
76. K. Shindo and S. Lipsky, *J. Chem. Phys.*, **45**, 2292 (1966).
77. H. R. Ward, J. S. Wishnock and P. D. Sherman, Jr., *J. Amer. Chem. Soc.*, **89**, 162 (1967).
78. K. E. Wilzbach, L. Kaplan and A. L. Harkness, *J. Amer. Chem. Soc.*, **90**, 1116 (1968).
79. L. Kaplan and K. E. Wilzbach, *J. Amer. Chem. Soc.*, **90**, 3291 (1968).
80. I. Jano and Y. Mori, *Chem. Phys. Letters*, **2**, 185 (1968).
81. D. Bryce-Smith, A. Gilbert and D. A. Robinson, *Angew-Chem.*, **83**, 803 (1971).
82. G. Camaggi, F. Gozzo and C. Cevidalli, *Chem. Comm.*, 313 (1966); G. P. Semeluk and R. D. S. Stevens, *Chem. Comm.*, 1720 (1970).
83. D. Phillips, J. Lemaire, C. Burton and W. A. Noyes, Jr., *Advances in Photochemistry, Vol. 5*, Wiley, New York, 1968.
84. C. L. Braun, S. Kato and S. Lipsky, *J. Chem. Phys.*, **39**, 1645 (1963).
85. C. S. Parmenter and M. D. Schuh, *Chem. Phys. Letters*, **13**, 120 (1972).
86. J. H. Callomon, J. E. Parkin and R. Lopez-Delgado, *Chem. Phys. Letters*, **13**, 125 (1972); V. Lawetz, W. Siebrand and G. Orlandi, *Chem. Phys. Letters*, **16**, 448 (1972).
87. T. A. Gregory, F. Hirayama and S. Lipsky, *J. Chem. Phys.*, **58**, 4697 (1973).
88. J. R. Platt and H. B. Klevens, *Chem. Rev.*, **41**, 301 (1947); W. Robertson and S. Babb, *J. Chem. Phys.*, **28**, 953 (1958).
89. C. W. Lawson, F. Hirayama and S. Lipsky, *J. Chem. Phys.*, **51**, 1590 (1969).
90. J. Yguerabide, *J. Chem. Phys.*, **49**, 1018 (1968).
91. T. Inagaki, *J. Chem. Phys.*, **57**, 2526 (1972).
92. J. B. Birks, *Photophysics of Aromatic Molecules*, p. 113, Wiley, London, 1970.
93. M. J. Mantione and J. P. Daudey, *Chem. Phys. Letters*, **6**, 93 (1970).
94. E. E. Koch and A. Otto, *Chem. Phys. Letters*, **12**, 476 (1972).
95. M. Brith, R. Lubart and I. T. Steinberger, *J. Chem. Phys.*, **54**, 5104 (1971).
96. H. Tsubomura and R. S. Mulliken, *J. Amer. Chem. Soc.*, **82**, 5966 (1960).
97. E. R. Bernstein and S. D. Colson, *J. Chem. Phys.*, **45**, 3873 (1966).
98. T. S. Godfrey and G. Porter, *Trans. Faraday Soc.*, **62**, 7 (1966).
99. R. Pariser, *J. Chem. Phys.*, **24**, 250 (1956).
100. R. Astier, A. Bokobza and Y. Meyer, *J. Chem. Phys.*, **67**, 137 (1970); R. Astier and Y. Meyer, *Chem. Phys. Letters*, **3**, 399 (1969).
101. R. Bonneau, J. Faure and J. Joussot-Dubien, *Chem. Phys. Letters*, **2**, 65 (1968); J. B. Birks, *Chem. Phys. Letters*, **3**, 567 (1969); R. Bonneau, R. Joussot-Dubien and R. Bensasson, *Chem. Phys. Letters*, **3**, 353 (1969).
102. R. Cooper and J. K. Thomas, *J. Chem. Phys.*, **48**, 5103 (1968).
103. J. K. Thomas and I. Mani, *J. Chem. Phys.*, **51**, 1834 (1969).
104. R. Bensasson, J. T. Richards and J. K. Thomas, *Chem. Phys. Letters*, **9**, 13 (1971).
105. M. T. Vala, I. H. Hillier, S. A. Rice and J. Jortner, *J. Chem. Phys.*, **44**, 23 (1967).
106. J. T. Richards and J. K. Thomas, *Chem. Phys. Letters*, **5**, 527 (1970).

107. J. B. Birks, J. C. Conte and G. Walker, *J. Phys. B.* (*Atom. Molec. Phys.*), **1**, 934 (1968).
108. C. W. Lawson, F. Hirayama and S. Lipsky, *Molecular Luminescence*, p. 837 (Ed. E. C. Lim), W. A. Benjamin, Inc., New York, 1969.
109. U. Laor and A. Weinreb, *J. Chem. Phys.*, **43**, 1565 (1965).
110. C. Fuchs, F. Heisel and R. Voltz, *J. Phys. Chem.*, **76**, 3867 (1972).
111. R. Voltz, *Radiat. Res. Rev.*, **1**, 301 (1968).
112. F. Hirayama, T. A. Gregory and S. Lipsky, *J. Chem. Phys.*, **58**, 4696 (1973).
113. M. D. Lumb, C. L. Braga and L. C. Pereira, *Trans. Faraday Soc.*, **65**, 1992 (1969).
114. M. Koyanagi, *J. Molec. Spectrosc.*, **25**, 273 (1968).
115. S. J. Strickler and R. A. Berg, *J. Chem. Phys.*, **37**, 814 (1962).
116. J. B. Birks and D. J. Dysen, *Proc. Roy. Soc.*, **A275**, 135 (1963).
117. J. W. Eastman, *J. Chem. Phys.*, **49**, 4617 (1968); *Spectrochim. Acta*, **26A**, 1545 (1970).
118. J. W. Eastman and S. J. Rehfeld, *J. Phys. Chem.*, **74**, 1438 (1970).
119. N. S. Bayliss and L. Hulme, *Aust. J. Chem.*, **6**, 257 (1953).
120. R. F. Weimer and J. M. Prausnitz, *J. Chem. Phys.*, **42**, 3643 (1965).
121. I. B. Berlman, *Handbook of Fluorescence Spectra of Aromatic Molecules*, Academic Press, New York, N.Y., 1971.
122. N. S. Bayliss and N. W. Grant, *Spectrochim. Acta*, **18**, 1287 (1962); M. Koyanagi and Y. Kanda, *Spectrochim. Acta*, **20**, 993 (1964).
123. T. A. Gregory and W. P. Helman, *J. Chem. Phys.*, **56**, 377 (1972).
124. R. B. Cundall and L. C. Pereira, *J. Chem. Soc. Faraday Transactions II*, **68**, 1152 (1972).
125. A. Dammers-de-Klerk, *Molec. Phys.*, **1**, 141 (1958).
126. T. V. Ivanova, G. A. Mokeeva and B. Y. Sveshnikov, *Opt. and Spectrosc.*, **12**, 325 (1962).
127. J. B. Birks, C. L. Braga and M. D. Lumb, *Proc. Roy. Soc.*, **A 283**, 83 (1965).
128. R. B. Cundall and D. A. Robinson, *J. Chem. Soc. Faraday Transactions II*, **68**, 113 (1972).
129. F. Hirayama and S. Lipsky, *J. Chem. Phys.*, **51**, 1939 (1969).
130. D. B. Chestnut, C. J. Fritchie and H. E. Simmons, *J. Chem. Phys.*, **42**, 1127 (1965); M. T. Vala, I. H. Hillier, S. A. Rice and J. Jortner, *J. Chem. Phys.*, **49**, 362 (1966).
131. T. Azumi and H. Azumi, *Bull. Chem. Soc. Japan*, **39**, 2317 (1966). A. K. Chandra and E. C. Lim, *J. Chem. Phys.*, **48**, 2589 (1968); **49**, 5066 (1968).
132. W. P. Helman, *J. Chem. Phys.*, **51**, 354 (1969).
133. M. G. Kuzmin and L. N. Guseva, *Chem. Phys. Letters*, **6**, 525 (1970); K. Egawa, N. Nakashima, N. Mataga and C. Yomanaka, *Chem. Phys. Letters*, **8**, 108 (1971); **3**, 71 (1969); R. Potashnik and M. Ottolenghi, *Chem. Phys. Letters*, **6**, 525 (1970); H. Masuhara, M. Shimada, N. Tsujino and N. Mataga, *Bull. Chem. Soc. Japan*, **44**, 3310 (1971); H. Masuhura and N. Mataga, *Bull. Chem. Soc. Japan*, **45**, 43 (1972).
134. J. Yguerabide, *J. Chem. Phys.*, **49**, 1026 (1968).
135. J. Klein, V. Pinaud-Plazanet and G. Laustriat, *J. Chim. Phys.*, **67**, 302 (1970).
136. J. B. Birks and J. C. Conte, *Proc. Roy. Soc.*, **A 303**, 85 (1968).
137. P. K. Ludwig and C. D. Amata, *J. Chem. Phys.*, **49**, 326, 333 (1968).
138. J. Klein, R. Voltz and G. Laustriat, *J. Chim. Phys.*, **67**, 704 (1970).
139. R. B. Cundall and W. Tippett, *Trans. Faraday Soc.*, **66**, 350 (1970).
140. R. R. Hentz and L. M. Perkey, *J. Phys. Chem.*, **74**, 3047 (1970).

141. K. Sandros, *Acta Chem. Scand.*, **23**, 2815 (1969).
142. A. A. Lamola and G. S. Hammond, quoted by J. G. Calvert and J. N. Pitts, Jr., *Photochemistry*, Wiley, New York, 1966.
143. R. B. Cundall, L. C. Pereira and D. A. Robinson, *Chem. Phys. Letters*, **13**, 253 (1972).
144. R. B. Cundall and D. A. Robinson, *J. Chem. Soc. Faraday Transactions II*, **68**, 1145 (1972).
145. R. R. Hentz and R. M. Thibault, *J. Phys. Chem.*, **77**, 1105 (1973).
146. R. B. Cundall and L. C. Pereira, unpublished results.
147. R. B. Cundall and P. A. Griffiths, *Disc. Faraday Soc.*, **36**, 11 (1963); R. G. Kaufmann, *Disc. Faraday Soc.*, **36**, 262 (1963).
148. R. B. Cundall and P. A. Griffiths, *Trans. Faraday Soc.*, **61**, 1968 (1965).
149. J. T. Dubois and F. Wilkinson, *J. Chem. Phys.*, **38**, 2541 (1963); **39** 377 (1963); J. W. van Loben Sels and J. T. Dubois, *J. Chem. Phys.*, **45**, 1522 (1966).
150. S. Lipsky, *J. Chem. Phys.*, **38**, 2786 (1963).
151. K. Sandros, *Acta Chem. Scand.*, **25**, 3651 (1971).
152. I. A. Lisovskaya, M. D. Zakharova and I. G. Kaplan, *Optics and Spectroscopy*, **23**, 348 (1967).
153. V. A. Krongaus, *Doklady Akad. Nauk. SSR*, **155**, 658 (1964).
154. R. B. Cundall and A. J. R. Voss, *Chem. Comm.*, **1969**, 116 (1969).
155. R. B. Cundall and L. C. Pereira, *Chem. Phys. Letters.*, **18**, 371 (1973).
156. R. B. Cundall and D. A. Robinson, *Chem. Phys. Letters*, **14**, 438 (1972).
157. R. B. Cundall, S. McD. Ogilvie and D. A. Robinson, *J. Photochem.*, **1**, 417 (1972/73).
158. E.g. J. W. Rabalais, H. J. Maria and S. P. McGlynn, *J. Chem. Phys.*, **51**, 2259, (1969); T. E. Martin and A. H. Kalantar, *Molecular Luminescence*, p. 437 (Ed. E. C. Lim), W. A. Benjamin, Inc., New York, 1969; G. W. Robinson and R. P. Frosch, *J. Chem. Phys.*, **38**, 1187 (1963).
159. P. M. Johnson and L. Zeigler, *J. Chem. Phys.*, **56**, 2169 (1972).
160. R. B. Cundall and D. A. Robinson, *J. Chem. Soc., Faraday Transactions II*, **68**, 1691 (1972).
161. M. S. de Groot and J. H. van der Waals, *Molec. Phys.*, **6**, 545 (1963); G. F. Hatch, M. D. Erlitz and G. C. Nieman, *Molecular Luminescence*, p. 21 (Ed. E. C. Lim), W. A. Benjamin, Inc., New York, 1969.
162. D. M. Haaland and G. C. Nieman, *J. Chem. Phys.*, **59**, 1013 (1973).
163. J. R. Greenleaf and T. A. King, *Proc. Intern. Conf. Luminescence*, Budapest, p. 212 (1966); J. R. Greenleaf, M. D. Lumb and J. B. Birks, *J. Phys.*, **B1**, 1157 (1968).
164. W. Siebrand and D. F. Williams, *J. Chem. Phys.*, **49**, 1860 (1968).
165. M. D. Lumb and L. C. Pereira, *Organic Scintillators and Liquid Scintillation Counting*, p. 239 (Eds. D. L. Horrocks and C. Peng), Academic Press, New York, 1971.
166. K. Wei, J. C. Mani and J. N. Pitts, *J. Amer. Chem. Soc.*, **89**, 4225 (1967).
167. G. Farenhorst, *Tetrahedron Letters*, 4835 (1968).
168. M. Luria and G. Stein, *Chem. Comm.*, 1650 (1970); J. Irina and K. C. Kurien, *Chem. Ind.*, 763 (1972); *Chem. Comm.*, 738 (1973).
169. L. Kaplan, L. A. Wendling and K. E. Wilzbach, *J. Amer. Chem. Soc.*, **93**, 3819, 3821 (1971).
170. M. Luria and G. Stein, *J. Phys. Chem.*, **76**, 165 (1972).
171. K. E. Wilzbach and L. Kaplan, *J. Amer. Chem. Soc.*, **88**, 2066 (1966); **93**, 2073 (1971); J. Cornelisse, V. Y. Merritt and R. Srinivasan, *J. Amer. Chem. Soc.*, **95**, 6197 (1973).

3 *Triplet quantum yields and singlet–triplet intersystem crossing*

Frank Wilkinson

3.1 Introduction

Since the ground states of most organic molecules are singlet states, direct production of triplet states by singlet–triplet absorption is usually extremely inefficient due to the very low extinction coefficients of these radiative transitions which are spin forbidden. Thus singlet–triplet absorptions due to $n \rightarrow \pi^*$ and $\pi \rightarrow \pi^*$ transitions in molecules composed of atoms no heavier than fluorine have extinction coefficients of $\sim 10^{-2}$ and $\sim 10^{-5}$ l mol^{-1} cm^{-1} respectively at their band maxima. Despite the fact that these transitions are very weak, there is ample evidence that triplet states of organic molecules are produced in high yields following absorption of light. This can be explained if one postulates that absorption results in excited singlet states which undergo non-radiative transitions to other electronic states which finally produce triplet states.

Radiationless transitions are discussed by Henry and Siebrand (Chapter 4, Vol. 1), by Voltz (Chapter 5, Vol. 2) and in references 1–6. As far as this chapter is concerned we may define a non-radiative electronic transition as an isoenergetic change in which a vibronic level in one electronic state is converted into a vibronic level in another electronic state. Thus in a radiationless transition from an upper to a lower electronic state the gain in internal energy (mainly vibrational energy) within the lower-energy electronic state is equal to the electronic energy difference. Under the conditions discussed here, i.e., mainly in solution, but also in medium- and high-pressure gases, non-radiative transitions in which electronic energy is converted into vibrational energy are rapidly followed by vibrational–vibrational energy transfer from the excited molecules to the surrounding medium thereby raising its temperature slightly.

A non-radiative transition between states of like multiplicity is called internal conversion, while one which involves states of different multiplicity, i.e., one resulting in a change of spin, is referred to as intersystem crossing.[7] Under most conditions radiationless transitions have time-independent transition probabilities, resulting in simple unimolecular decay which may be written as

$$\text{M}^*(\text{State 1}) \xrightarrow{k_\text{D}} \text{M}^*(\text{State 2})$$

where

$$-\frac{\text{d}[\text{M}^*\text{State 1})]}{\text{d}t} = k_\text{D}[\text{M}^*(\text{State 1})]$$

Robinson and Frosch[1] have developed a theory of non-radiative transitions, based on time-dependent perturbation theory, in which it is assumed that absorption of a photon leads to excitation of a Born–Oppenheimer (B–O) state, which then undergoes a non-radiative transition to another (B–O) state. Since this theory is somewhat analogous to the theory usually adopted for radiative transitions, it will be used here in preference to a more complete and more complicated theory based on a stationary states model in which the non-radiative transition depends on the time evolution of components within an exact eigenfunction of the entire molecular system.[5,6,8]

According to the time-dependent theory the rate constant for a radiationless transition in a polyatomic molecule from an upper B–O state, which is usually the lowest vibronic level, $\phi_u\chi_u(0)$, to those isoenergetic final states of the lower electronic state with vibrational energy E, $\phi_l\chi_l(E)$, which form a continuum of density ρ, is given by

$$k_\text{D} = \frac{4\pi^2}{h}\rho J^2 F \tag{3.1}$$

where J is the electronic matrix element $\langle\phi_u|H'|\phi_l\rangle$. For intersystem crossing, the perturbation operator $H' = T_\text{N} + H_\text{SO}$ where T_N is the nuclear kinetic energy operator and H_SO is the spin–orbit operator. In equation (3.1) F is called the Franck–Condon factor by analogy with radiative transitions since F depends on the overlap of the initial and final vibrational wavefunctions, i.e., $F = \sum\langle\chi_u(0)|\chi_l(E)\rangle^2$, where the sum extends over all final states with energies in the range $(E \pm \rho/2)$.

Although radiationless transitions between states of different multiplicity are subject to spin selection rules, spin-forbidden non-radiative transitions are nevertheless efficient enough to compete successfully with spin-allowed radiative transitions, thereby giving rise to efficient population of triplet states as a result of singlet–triplet intersystem crossing following singlet–singlet absorption. The fraction of molecules which are initially excited into singlet states following the absorption of light and which undergo singlet–triplet intersystem crossing is variously referred

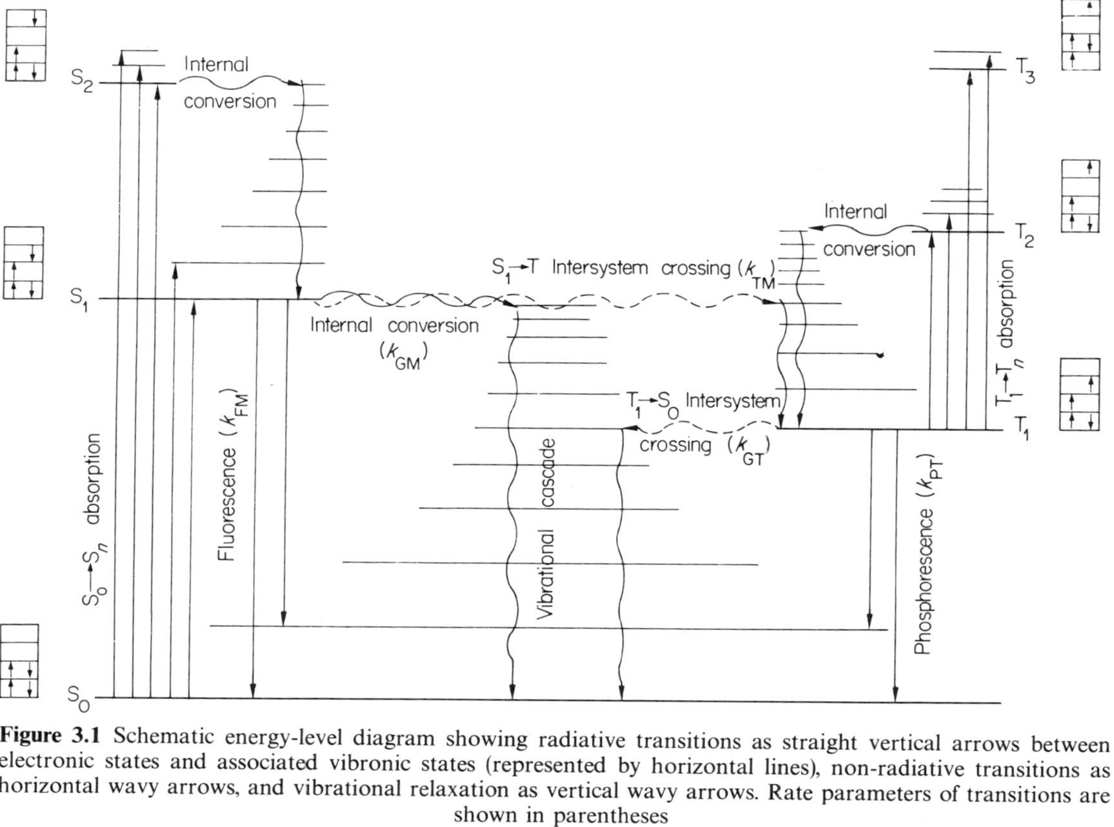

Figure 3.1 Schematic energy-level diagram showing radiative transitions as straight vertical arrows between electronic states and associated vibronic states (represented by horizontal lines), non-radiative transitions as horizontal wavy arrows, and vibrational relaxation as vertical wavy arrows. Rate parameters of transitions are shown in parentheses

to as the quantum yield of intersystem crossing, the quantum yield of triplet state production or, more simply, as the triplet quantum yield, ϕ_{TM}.

The competing intramolecular processes which affect the triplet quantum yield of a typical organic molecule in solution or in the gas phase at medium or high pressures can conveniently be discussed with the aid of a simplified energy-level diagram of the type shown in Figure 3.1. At all but very low pressures in the gas phase, internal conversion among excited singlet states, followed by vibrational energy transfer to the surrounding medium, is so fast that both fluorescence and singlet–triplet intersystem crossing usually arise from the thermally relaxed lowest excited singlet state, S_1, and are independent of exciting wavelength. In the absence of photodecomposition it follows (see Figure 3.1) that

$$\phi_{FM} + \phi_{TM} + \phi_{GM} = 1 \tag{3.2}$$

where ϕ_{FM} and ϕ_{GM} are respectively the quantum yields of fluorescence and of internal conversion of the S_1 state direct to the ground state and also

$$\phi_{TM} = \frac{k_{TM}}{k_{FM} + k_{TM} + k_{GM}} = k_{TM}\tau_M \tag{3.3}$$

where $\tau_M = (k_{FM} + k_{TM} + k_{GM})^{-1}$ is the lifetime of the lowest excited singlet state, S_1.

Non-radiative processes which result in relaxation of electronically excited molecules are of great theoretical and practical interest in photochemistry (e.g. see references 9 and 10). Therefore accurate methods for determining ϕ_{TM}, which can be combined with fluorescence lifetime values, τ_M, using equation (3.3) to yield values of radiationless transition probabilities, k_{TM}, are of considerable importance for testing theories of radiationless transitions and for establishing the relative importance of excited singlet and triplet states in photochemical systems. Several methods have been evolved in recent years for measuring ϕ_{TM} in dilute solutions and these are outlined in §3.2. Similar methods are often used for the determination of ϕ_{TM} for molecules in the gas phase at medium and high pressures and some reference to such work is included here. However, no attempt is made to discuss intersystem crossing originating from single vibronic levels in low-pressure gases (see Stockburger, Chapter 2, Vol. 1). Other aspects of singlet–triplet intersystem crossing include the use of laser photolysis to monitor vibrational relaxation of triplet states formed following intersystem crossing,[11,12] phosphorescence-microwave double resonance techniques[13] and spin polarization studies of triplet states, etc.

3.2 Determination of triplet quantum yields

3.2.1 *Direct determination from triplet–triplet absorption measurements (A)*

Triplet–triplet absorption (see Figure 3.1) is a spin-allowed process and these transitions have high extinction coefficients. This fact is exploited in flash and laser photolysis studies of triplet states, where monitoring of the characteristic absorption of triplet states gives a great deal of useful information concerning the properties of triplet states under a variety of conditions. In rigid solutions τ_T, the lifetime of the lowest triplet state of a typical organic molecule, is often so long (it can be as long as a few seconds[14]) that the photostationary state concentration of triplet states obtained during continuous irradiation with an arc lamp is sufficient for the observation of triplet–triplet absorption. In fluid media τ_T is usually much less, but triplet–triplet absorption spectra can generally be observed without difficulty immediately following excitation with a short intense flash of light[15] (e.g. see Figure 3.2). Although triplet–triplet

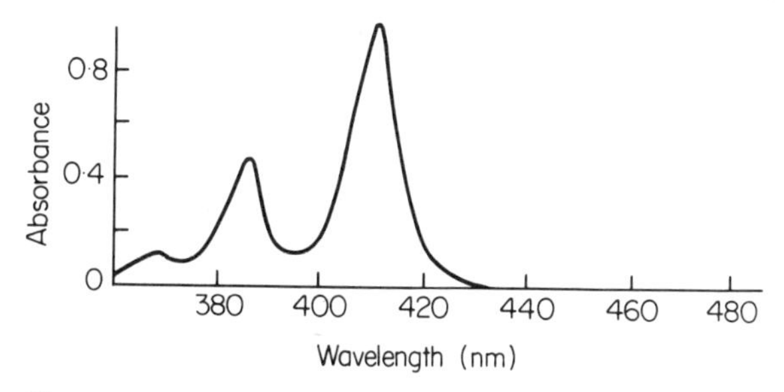

Figure 3.2 Triplet–triplet absorption spectrum of naphthalene in *n*-hexane taken 20 μs after excitation

absorption spectra are readily available and agreement between different research workers concerning the band positions and relative intensities is good, the same cannot be said for the published values of ε_T, the triplet–triplet extinction coefficients (see Labhart and Heinzelmann, Chapter 6, Vol. 1).

Inaccuracies arise for two main reasons: (a) the triplet absorption spectrum overlies the singlet (ground state) absorption spectrum and (b) the difficulty of estimating [T], the concentration of triplets, because the exciting light does not produce a uniform distribution of triplets over the analysing path length *d*. Dawson[16] has pointed out that when the analysis and excitation beams follow the same path through the reaction vessel, as is possible using beam-splitting techniques, it follows that, provided all the molecules pumped out of the ground state are in the triplet state, the observed change in absorbance ΔA at any wavelength is given by

$$\Delta A = (\varepsilon_T - \varepsilon_G) \int_0^d [T]\, dx \tag{3.4}$$

where ε_G is the extinction coefficient of the singlet ground state which can be determined by usual absorption measurements. ΔA can be determined at different analysing wavelengths in photostationary state experiments by taking the spectrum whilst irradiating, or in flash experiments by extrapolating back to zero-time changes in absorption measured as a function of time following the pulse excitation. Thus at each wavelength equation (3.4) may be rewritten as

$$\varepsilon_T = \varepsilon_G + \Delta A/K \tag{3.5}$$

since

$$K = \int_0^d [T]\, dx$$

is a constant. ε_T can be evaluated from ε_G and ΔA by adjusting K until the triplet spectrum contains no contribution recognizably correlating with the singlet spectrum, e.g. if a value of K is chosen which is either too large or too small the triplet spectrum derived from equation (3.5) will contain either troughs or peaks respectively coinciding with peaks in the singlet spectrum. Another approach made by Hadley and Keller[17] which saves the trial and error procedure is to assume that ε_T values vary in a linear manner with wavelength across a sharp band in the singlet spectrum.

Other methods for obtaining ε_T have involved extrapolation of plots of triplet absorbance against excitation intensity up to a saturation level assumed to correspond to complete conversion of all molecules into triplet states.[18] Values relative to the extinction coefficient of the ketyl radical of benzophenone have been obtained by Land[19] on the basis of a saturation energy transfer method (see Chapter 6, Vol. 1). ε_T has also been obtained from ESR measurements of the concentration of triplets.[20]

Once ε_T is known, measurements of ΔA allow the determination of the concentration of triplet states produced. In order to obtain ϕ_{TM} it is also necessary to know I_a, the intensity of the absorbed light. The incident light is usually easily obtained by actinometry; however, when excitation is with a broad band of exciting light (e.g. from a flash lamp) which excites a molecule within a broad absorption band, the overall fraction of light absorbed has to be found by suitable integrations (e.g. see Bowers and

Porter[21]). Because of the high spectral purity it follows that laser excitation allows measurements of the fraction of light absorbed to be much more accurately determined, but the spectral range available for excitation is reduced.[22]

The direct method which appears to give the most accurate values for ϕ_{TM} is that developed by Heinzelmann and Labhart.[23] This method involves measuring the very small changes in absorbance caused by excitation with a chopped continuous light source using a lock-in amplifier. Furthermore by measuring the effect on ΔA of varying the chopper frequency, the triplet lifetime is also obtained. The values of ϕ_{TM} obtained from this method are in good agreement with the most reliable values obtained from other methods (see Table 3.4).

3.2.2 *Determinations based on triplet–triplet energy transfer* (*B*)

Sensitized phosphorescencc occurring in rigid media due to triplet–triplet energy transfer, i.e., to the process

$$M^*(T_1) + A(S_0) \rightarrow M(S_0) + A^*(T_1)$$

was first reported by Terenin and Ermolaev[24] in 1952 for several donor–acceptor combinations with energy levels as shown in Figure 3.3. The radius R of the measured sphere of action within which transfer is essentially complete, and outside of which virtually no transfer occurs, indicates that this kind of transfer results from a short-range interaction, consistent with an exchange mechanism (see Table 3.1). Ermolaev[25] has defined the relative yield of sensitized phosphorescence as

$$\phi_S = \frac{\phi_{PA}^S}{\phi_{PM}^0 - \phi_{PM}}$$

where ϕ_{PA}^S is the measured sensitized phosphorescence yield of the acceptor A in a rigid solution in which the phosphorescence yield of the donor M drops to ϕ_{PM} from a value of ϕ_{PM}^0 in the absence of the acceptor. In terms of the rate parameters given in Figure 3.1 it follows that

$$\phi_S = \frac{\phi_{PA}^S}{(\phi_{PM}^0 - \phi_{PM})} = \left(\frac{k_{PA}}{k_{GT}^A + k_{PA}}\right)\left(\frac{k_{GT}^M + k_{PM}}{k_{PM}}\right) \tag{3.6}$$

where $k_{PM}(k_{PA})$ and $k_{GT}^M(k_{GT}^A)$ represent the rate constants for decay of the triplet state of molecule M(A) due to phosphorescence and intersystem crossing to the ground state, respectively. Ermolaev[25] has shown that within experimental error for several molecules at 77°K in ether–ethanol glasses

$$\phi_S = \left(\frac{\phi_{PA}}{1 - \phi_{FA}}\right)\left(\frac{1 - \phi_{FM}}{\phi_{PM}}\right)$$

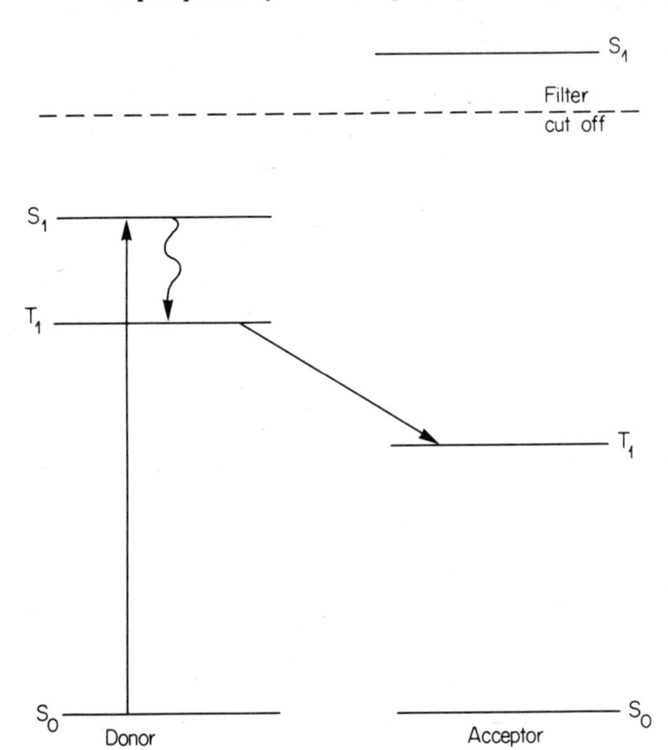

Figure 3.3 Energy levels of donor and acceptor molecules

which follows from equation (3.6) provided $\phi_{FM} + \phi_{TM} = 1$ and $\phi_{FA} + \phi_{TA} = 1$. However, the experimental error in some cases was quite substantial (see Table 3.1).

In dilute solution it is well established[26,27] that triplet–triplet electronic energy transfer between organic molecules occurs as a diffusion-controlled reaction provided the donor triplet $M^*(T_1)$ has an energy which is at least $1000\ cm^{-1}$ higher than that of the acceptor triplet $A^*(T_1)$. This is understandable since this transfer process obeys Wigner's spin rule and results from exchange interaction which requires good overlap between donor and acceptor orbitals. The transfer results in sensitized emission and/or sensitized reactions of $A^*(T_1)$ and many examples are known (e.g. see reference 28). In principle, measurements of the photosensitized yields of any well characterized triplet reaction of an acceptor molecule can be used to determine relative triplet quantum yields of a series of sensitizers. Thus if two donor molecules, M_1 and M_2, are excited in the presence of an acceptor molecule at a sufficiently high concentration so as to ensure

Table 3.1 Volumes and radii of spheres of action for triplet–triplet energy transfer and yields of sensitized phosphorescence ϕ_S in a rigid alcohol ether glass at 90 or 77°K (Ermolaev[25])

Donor	Acceptor	Volume (10^{-21} cm^3)	R (nm)	ϕ_S	
				$\frac{\phi_{PA}^{S}}{\phi_{PM}^{0} - \phi_{PM}}$	$\frac{\phi_{PA}(1 - \phi_{FM})}{\phi_{PM}(1 - \phi_{FA})}$
Benzaldehyde	Naphthalene	6·8	1·2	0·13	0·04
Benzaldehyde	1-Chloronaphthalene	7·0	1·2	0·22	0·34
Benzaldehyde	1-Bromonaphthalene	7·2	1·2	0·27	0·29
Benzophenone	Naphthalene	8·6	1·3	0·07	0·06
Benzophenone	1-Methylnaphthalene	9·5	1·3	0·07	0·05
Benzophenone	1-Chloronaphthalene	9·5	1·3	0·12	0·22
Benzophenone	1-Iodonaphthalene	8·6	1·3	0·35	0·27
Benzophenone	Quinoline	7·2	1·2	0·14	0·14
Acetophenone	Naphthalene	6·0	1·1	0·10	0·07
p-Chlorobenzaldehyde	Naphthalene	6·7	1·2	0·14	—
p-Chlorobenzaldehyde	1-Bromonaphthalene	6·2	1·1	0·49	—
o-Chlorobenzaldehyde	Naphthalene	5·4	1·1	0·11	—
m-Iodobenzaldehyde	Naphthalene	5·8	1·1	0·11	0·07
m-Iodobenzaldehyde	1-Bromonaphthalene	5·7	1·1	0·30	0·22
Xanthone	Naphthalene	9·2	1·3	0·11	—
Anthraquinone	Naphthalene	5·9	1·1	0·10	—
Anthraquinone	1-Bromonaphthalene	7·6	1·2	0·27	—
Triphenylamine	Naphthalene	9·3	1·3	0·07	—
Carbazole	Naphthalene	14	1·5	0·08	—
Phenanthrene	Naphthalene	10	1·3	0·30	0·28
Phenanthrene	1-Chloronaphthalene	11	1·4	0·73	1·10
Phenanthrene	1-Bromonaphthalene	11	1·4	0·99	0·94

virtually complete triplet–triplet transfer to give $A^*(T_1)$ and, if the measured yields of photosensitized reaction or emission of $A^*(T_1)$ are ϕ_1^S and ϕ_2^S respectively, then

$$\frac{(\phi_{TM})_1}{(\phi_{TM})_2} = \frac{\phi_1^S}{\phi_2^S} \tag{3.7}$$

However, equation (3.7) is only applicable provided sensitization occurs as a result of the physical process of triplet–triplet energy transfer (i.e. if there is no other interaction of excited states of M with the acceptor A) and provided also that the presence of the donor molecules does not interfere in any way with the reactions of $A^*(T_1)$.

3.2.2.1 *Photosensitized* cis-trans *isomerization of olefins* (*B*(*i*))

Lamola and Hammond[29] developed an elegant, yet simple, method for determining triplet quantum yields in solution by measuring the yields of photosensitized *cis* → *trans* isomerization of piperylene and other olefins in which analytical measurements were made with high accuracy using vapour-phase chromatography. The mechanism may be written as follows:

1. $M(S_0) + h\nu \rightarrow M^*(S_X) \rightarrow M^*(S_1)$	Absorption and internal conversion to $M^*(S_1)$
2. $M^*(S_1) \rightarrow M(S_0)$	All unimolecular decay other than intersystem crossing
3. $M^*(S_1) \rightarrow M^*(T_1)$	Intersystem crossing
4. $M^*(T_1) \rightarrow M(S_0)$	Triplet decay
5. $M^*(T_1)$ + *cis*-olefin → T + $M(S_0)$	Triplet–triplet energy transfer to the *cis*-olefin
6. $M^*(T_1)$ + *trans*-olefin → T + $M(S_0)$	Triplet–triplet energy transfer to the *trans*-olefin
7. T → *cis*-olefin	Decay of triplet olefin to give the *cis*-olefin
8. T → *trans*-olefin	Decay of triplet olefin to give the *trans*-olefin

In the presence of only the *cis*-olefin (i.e. in the absence of the *trans*-olefin) at a concentration such that $k_5[cis] \gg k_4$, it follows that the initial quantum yield of photosensitized *cis* → *trans* isomerization is given by

$$\phi^{S}_{c \to t} = \frac{\phi_{TM} k_8}{k_7 + k_8} \tag{3.8}$$

Similarly in the presence of only the *trans*-olefin at a concentration such that $k_6[trans] \gg k_4$, it follows that

$$\phi^{S}_{t \to c} = \frac{\phi_{TM} k_7}{k_7 + k_8} \tag{3.9}$$

Note that from equations (3.8) and (3.9)

$$\phi^{S}_{c \to t} + \phi^{S}_{t \to c} = \phi_{TM} \tag{3.10}$$

and

$$\frac{\phi^{S}_{c \to t}}{\phi^{S}_{t \to c}} = \frac{k_8}{k_7} \tag{3.11}$$

Thus for a particular acceptor $\phi^{S}_{c \to t}/\phi^{S}_{t \to c}$ should be independent of donor. In agreement with this, Lamola and Hammond[29] found that with piperylene as acceptor $\phi^{S}_{c \to t}/\phi^{S}_{t \to c} = 1{\cdot}24 \pm 0{\cdot}02$ for four different sensitizers. When a piperylene solution containing a triplet energy donor is continuously irradiated, a photostationary state is eventually reached. Lamola and Hammond showed that piperylene with several high-energy donors gave a photostationary state with

$$\frac{[trans]_{pss}}{[cis]_{pss}} = 1{\cdot}22 \pm 0{\cdot}05$$

which is exactly what one would expect for diffusion controlled energy transfer to both *cis*- and *trans*-piperylene, i.e. with $k_5 = k_6$, in which case the photostationary state would be determined only by the ratio k_8/k_7.

Lamola and Hammond[29] found that for benzophenone as sensitizer with three different acceptors, using equation (3.10)

$$\phi_{TM} = \phi^{S}_{c \to t} + \phi^{S}_{t \to c} = 1{\cdot}00 \pm 0{\cdot}02$$

Thus Lamola and Hammond[29] were able to measure triplet quantum yields relative to that of benzophenone for which a value of unity had been obtained. Unfortunately certain of these ϕ_{TM} values, e.g. those obtained for naphthalene and several of its derivatives, have been shown to be in error due to the fact that significant quenching of the sensitizer singlet states occurs at high diene concentrations.[30] This singlet-state quenching does not lead to *cis* → *trans* isomerization or to the catalysed production of triplet sensitizer which explains why low values for ϕ_{TM} were sometimes obtained.

Measurements of photosensitized isomerization are also used to determine triplet yields in the gas phase using an analogous method introduced by Cundall.[31] This method has been applied mainly to measure ϕ_{TM} for benzene and its derivatives which are used as sensitizers with *cis*-but-2-ene as the isomerizing acceptor. More recently Cundall *et al.*[32] have extended such measurements from the gas phase into solution (see Chapter 2, Vol. 2).

3.2.2.2 *Photosensitized phosphorescence of biacetyl* (*B*(*ii*))

Biacetyl is an exceptional molecule since it phosphoresces with high yield both in dilute fluid solution and in the gas phase. The use of photosensitized phosphorescence yield measurements for determining triplet quantum yields was first developed by Ishikawa and Noyes[33] to obtain the triplet quantum yield of benzene in the gas phase. The method has also been applied to dilute solutions.[34]

As an example of the way in which sensitized phosphorescence measurements can be used to determine ϕ_{TM} values we shall examine one of the kinetic treatments outlined by Sandros;[35] other approaches are given elsewhere (e.g. reference 36). Consider the mechanism

1. $M(S_0) + h\nu \xrightarrow{I_a} M^*(S_1)$	Absorption
2. $M^*(S_1) \rightarrow M(S_0)$	All unimolecular decay of S_1 except intersystem crossing
3. $M^*(S_1) \rightarrow M^*(T_1)$	Intersystem crossing
4. $M^*(T_1) \rightarrow M(S_0)$	Triplet decay
5. $M^*(S_1) + B(S_0) \rightarrow B^*(S_1) + M(S_0)$	Singlet–singlet energy transfer
6. $M^*(T_1) + B(S_0) \rightarrow B^*(T_1) + M(S_0)$	Triplet–triplet energy transfer
7. $B^*(S_1) \rightarrow B(S_0) + h\nu_F$	Fluorescence of biacetyl
8. $B^*(S_1) \rightarrow B^*(T_1)$	Intersystem crossing of biacetyl
9. $B^*(T_1) \rightarrow B(S_0) + h\nu_P$	Phosphorescence of biacetyl
10. $B^*(T_1) \rightarrow B(S_0)$	Radiationless decay of biacetyl triplet

where B represents biacetyl. Applying the steady state treatment to

$M^*(S_1)$, $M^*(T_1)$ and $B^*(T_1)$ gives

$$\frac{d[M^*(S_1)]}{dt} = I_a - [M^*(S_1)](k_2 + k_3 + k_5[B]) = 0$$

$$\frac{d[M^*(T_1)]}{dt} = k_3[M^*(S_1)] - [M^*(T_1)](k_4 + k_6[B]) = 0$$

and, since ϕ_{TM} for biacetyl is close to unity we can ignore step 7, so that

$$\frac{d[B^*(T_1)]}{dt} = [B](k_5[M^*(S_1)] + k_6[M^*(T_1)]) - [B^*(T_1)](k_9 + k_{10}) = 0$$

Combining these equations gives

$$[B^*(T_1)] = \frac{I_a[B]}{(k_2 + k_3 + k_5[B])(k_9 + k_{10})}\left(k_5 + \frac{k_3 k_6}{k_4 + k_6[B]}\right) \qquad (3.12)$$

The measured phosphorescence lifetime of biacetyl, $\tau_{TB} = (k_9 + k_{10})^{-1}$ and the steady state photomultiplier signal, I^S_{PB}, due to sensitized biacetyl phosphorescence is equal to $\alpha k_9[B^*(T_1)]$, where α is an instrumental constant. Substitution of these experimental parameters into equation (3.12) gives

$$\frac{I^S_{PB}}{\tau_{TB}} = K\frac{[B]}{(k_2 + k_3 + k_5[B])}\left(k_5 + \frac{k_3 k_6}{(k_4 + k_6[B])}\right) \qquad (3.13)$$

where $K = \alpha I_a k_9$. At constant I_a, extrapolation of (I^S_{PB}/τ_{TB}) values to infinite biacetyl concentration gives a value for K.

At low biacetyl concentrations all the transfer will be due to triplet–triplet transfer if the lifetime of the first excited triplet state of the sensitizer is much larger than that of its first excited singlet state and, provided also that $k_3 \gg k_4$, in which case equation (3.13) simplifies to give

$$\left(\frac{I^S_{PB}}{\tau_{TB}}\right)_T = \frac{K\phi_{TM}k_6[B]}{k_4 + k_6[B]} \qquad (3.14)$$

where subscript T indicates that the biacetyl phosphorescence is sensitized by the triplet state of the donor. Inversion of equation (3.14) gives

$$\left(\frac{\tau_{TB}}{I^S_{PB}}\right)_T = \frac{1}{K\phi_{TM}}\left(1 + \frac{k_4}{k_6[B]}\right) \qquad (3.15)$$

Thus a plot of $(\tau_{TB}/I^S_{PB})_T$ versus $[B]^{-1}$ allows the value of ϕ_{TM} to be calculated from the intercept and from the value of K obtained at infinitely high biacetyl concentrations. If, on the other hand, the biacetyl concentration is increased to such a high value that triplet–triplet transfer is complete, then $k_6[B]/(k_4 + k_6[B]) \approx 1$ and substitution into equation (3.13) gives

$$\frac{I^S_{PB}}{\tau_{TB}} = \frac{K}{1 + k_5\tau_M[B]}(k_5\tau_M[B] + \phi_{TM}) \tag{3.16}$$

where $\tau_M = (k_2 + k_3)^{-1}$ is the lifetime of the singlet state of the donor. Since K and ϕ_{TM} are known, this allows values of $k_5\tau_M$ to be determined. Other methods of analysis[35] are employed when singlet and triplet lifetimes of the donor are not sufficiently different to allow this kind of treatment, as is the case for benzene, toluene, etc.

Usually in the gas phase the reciprocal of the absolute sensitized phosphorescence yield of biacetyl is plotted against $[B]^{-1}$ since from equation (3.12) it follows that at low pressures, i.e., when only triplet–triplet transfer occurs,

$$\frac{1}{\phi^S_{PB}} = \frac{I_a}{k_9[B^*(T_1)]} = \frac{k_9 + k_{10}}{k_9}\left(\frac{1}{\phi_{TM}} + \frac{k_4}{\phi_{TM}k_6[B]}\right) \tag{3.17}$$

Thus ϕ_{TM} can be obtained from the intercept of such plots when taken together with an absolute value for $k_9/(k_9 + k_{10})$ of 0·145 as given in the literature.[36] Recently Jones and Brewer[36] have shown that by measuring relative sensitized and unsensitized yields of biacetyl phosphorescence in the same experiment, absolute values of ϕ_{TM} can be obtained. These authors determined ϕ_{TM} in the vapour phase by both the Cundall and the biacetyl techniques, and found very good agreement between the two methods except in the case of pyrazine vapour where certain complications arise.

3.2.2.3 *Photosensitized delayed fluorescence* (*B*(*iii*))

As mentioned above, biacetyl is exceptional since most organic molecules show very little phosphorescence in fluid solution. However, dilute fluid solutions of aromatic hydrocarbons often show delayed fluorescence which arises as a result of triplet–triplet annihilation,[37] i.e. the process

$$A^*(T_1) + A^*(T_1) \xrightarrow{pk_{TT}} A^*(S_1) + A(S_0) \text{ (triplet–triplet annihilation)}$$

where p represents the fraction of the triplet–triplet reactions which produce ground and excited singlet states. When the $A^*(S_1)$ excited states produced in this way fluoresce, i.e. undergo the following reaction

$$A^*(S_1) \rightarrow A(S_0) + h\nu_{FA} \text{ (Fluorescence)}$$

the result is delayed fluorescence. In dilute fluid solutions irradiated with a continuous light source, only a minute proportion of triplet states decay by this process, i.e. triplet decay is given by

$$-\frac{d[A^*(T_1)]}{dt} = k_{TA}[A^*(T_1)] + k_{TT}[A^*(T_1)]^2$$

where $k_{TA} \gg k_{TT}[A^*(T_1)]$. Therefore during a dark period following the production of a photostationary state concentration of triplet states, $[A^*(T_1)]_{pss}$, formed by direct excitation

$$[A^*(T_1)] = [A^*(T_1)]_{pss} \exp(-k_{TA}t)$$

and the delayed fluorescence, the intensity of which is proportional to $[A^*(T_1)]^2$ has a lifetime equal to $(2k_{TA})^{-1}$ $(= \frac{1}{2}\tau_{TA})$ under these conditions. The photostationary state concentration $[A^*(T_1)]_{pss}$ is that obtained when the rate of production of the triplet state equals its rate of decay. Thus for direct excitation of A

$$k_{TA}[A^*(T_1)]_{pss} = I_a\phi_{TA} \tag{3.18}$$

where ϕ_{TA} is the triplet quantum yield of A and I_a is the rate of absorption. The yield of delayed fluorescence for direct excitation, θ^d_{FA}, is given by the rate of production of $A^*(S_1)$ due to triplet–triplet annihilation, which is $\frac{1}{2}pk_{TT}[A^*(T_1)]^2_{pss}$ multiplied by the fraction of molecules produced in the lowest excited singlet state which emit, ϕ_{FA}, divided by I_a. Thus

$$\theta^d_{FA} = \tfrac{1}{2}pk_{TT}[A^*(T_1)]^2_{pss}\phi_{FA}/I_a \tag{3.19}$$

which from equation (3.18) gives

$$\frac{\theta^d_{FA}}{\phi_{FA}} = \tfrac{1}{2}pk_{TT}I_a(\phi_{TA}\tau_{TA})^2 \tag{3.20}$$

where τ_{TA} $(= 1/k_{TA})$ is the triplet lifetime of A.

When sensitized delayed fluorescence is observed following complete energy transfer from a donor molecule M to an acceptor A, the steady state concentration of acceptor triplet states is given by

$$k_{TA}[A^*(T_1)]_{pss} = I_a\phi_{TM} \tag{3.21}$$

(cf. equation (3.18)). Thus if the quantum yield of sensitized delayed fluorescence is θ^{sd}_{FA}, it follows from equations (3.19) and (3.21) that

$$\frac{\theta^{sd}_{FA}}{\phi_{FA}} = \tfrac{1}{2}pk_{TT}I_a(\phi_{TM}\tau_{TA})^2 \tag{3.22}$$

If the intensity of sensitized delayed fluorescence I^{sd}_{FA} emitted by two solutions containing the same acceptor but different donors and measured at the same rate of light absorption are indicated by subscripts 1 and 2, it follows from equation (3.22) that

$$\frac{(\phi_{TM})_1}{(\phi_{TM})_2} = \left(\frac{(I_{FA}^{sd})_1}{(I_{FA}^{sd})_2}\right)^{1/2} \tag{3.23}$$

Parker and Joyce[38] used this relationship to determine relative triplet quantum yields of several aromatic hydrocarbons using perylene as acceptor. As a standard they assumed ϕ_{TM} for anthracene in ethanol at 20°C to be $0{\cdot}70 = 1 - \phi_{FM}$ on the basis of reference 39 (see also §3.2.3 and Table 3.4).

3.2.2.4 *Photosensitized emission from rare-earth chelates* (*B*(*iv*))

Chelates of europium show intense luminescence in liquid solutions arising from essentially f → f transitions of the central rare-earth ion. Ermolaev and co-workers[40] have developed a method for determining triplet quantum yields using measurements of sensitized emission of the europium line at 613 nm from evacuated toluene solutions containing europium *tris*-thenoyltrifluoracetonate-1,10-phenanthroline ($Eu(TTA)_3$-phen.) as an energy acceptor. The lowest triplet ligand, as determined from the phosphorescence spectrum of $Gd(TTA)_3$phen., is at 20,400 cm^{-1} and donor molecules with higher triplet energy levels than this are considered to transfer into this level, which undergoes a radiationless decay to the emitting $^5D_0$ level (see Figure 3.4). The conditions are chosen so that there is complete transfer of energy from the triplet state of the donor molecules but no transfer from singlet states. Under these conditions in degassed solutions the luminescence intensity

$$I_{evac} = I_{nat} + I_T$$

where I_{nat} is the natural or unsensitized luminescence intensity of the chelate due to direct excitation of the chelate in the mixture and I_T is the intensity observed as a result of triplet–triplet transfer. The luminescence of $Eu(TTA)_3$phen. is not quenched by oxygen, but in undegassed solutions almost all donor triplets are quenched by oxygen and virtually none survive to undergo triplet–triplet transfer. Thus $I_{nat} = I_{un}$, where I_{un} is the intensity of luminescence observed in the undegassed mixture. It follows that

$$\frac{I_T}{I_{nat}} = \frac{I_{evac} - I_{un}}{I_{un}} = \frac{\varepsilon_M[M]}{\varepsilon_{Ch}[Ch]}\phi_{TM}$$

where ε_M and ε_{Ch} are the extinction coefficients at the exciting wavelength for the donor M and the chelate Ch. Since the limiting concentration of $Eu(TTA)_3$phen. in toluene is 2×10^{-4} mol l^{-1}, this method cannot be

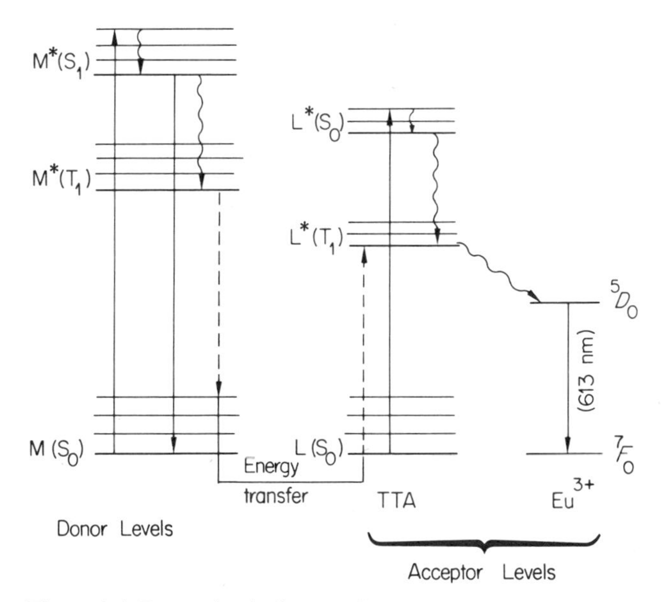

Figure 3.4 Energy-level diagram illustrating electronic energy transfer leading to the emission of the europium line at 613 nm

used to measure ϕ_{TM} for molecules which have triplet lifetimes of $< 10^{-6}$ s. However Ermolaev and Sveshnikova[41] have overcome this problem by using high concentrations of naphthalene as an intermediary energy acceptor which is able to quench completely even short-lived triplet states and then pass on the energy to the chelate. In this way ϕ_{TM} values for short-lived molecules such as benzophenone and triphenylamine were determined.

3.2.2.5 *Sensitized photoreduction* (*B*(*v*))

Koizumi and co-workers[42] have made extensive studies of the photoreduction of xanthene and thiazine dyes in the presence of certain reducing agents such as allylthiourea (ATU) and have confirmed that the mechanism for the photoreduction involves reaction of the triplet states of the dyes. Recently they have made use of measurements of the sensitized and unsensitized yields of photoreduction to obtain relative triplet yields.[43] If ϕ_R and ϕ_R^S are the quantum yields of direct and sensitized photoreduction respectively due to triplet reaction, then it follows that

$$\phi_R = \phi_{TA}\beta$$

and

$$\phi_R^S = \phi_{TM}\beta\gamma$$

where γ is the efficiency of triplet–triplet energy transfer from the donor M to the acceptor A and β is the fraction of triplet dye molecules produced which are reduced. Using molecules of well-established triplet quantum yields as sensitizers under conditions such that $\gamma \rightarrow 1$ the following relationship holds

$$\frac{\phi_{R}}{\phi_{R}^{S}} = \frac{\phi_{TA}}{\phi_{TM}}$$

This equation has been used to obtain triplet quantum yields for several dyes (see Table 3.12) and also for some donors (see Table 3.4).

3.2.3 *Intersystem crossing catalysed by external heavy atoms* (*C*)

Wilkinson and co-workers[39,44] have shown that heavy-atom quenching of fluorescence is accompanied by a corresponding increase in the triplet state production which arises because of the increase in spin–orbit coupling introduced by the heavy atoms. Thus the fluorescence quenching and the catalysed intersystem crossing are a result of the process

$$M^{*}(S_1) + Q_H \xrightarrow{k_{QM}} M^{*}(T_1) + Q_H$$

where Q_H represents a quencher which contains at least one atom of high atomic number. In the presence of a heavy-atom quencher the quantum yield of triplet state production is given by

$$\phi_{TM} = \frac{k_{TM} + k_{QM}[Q]}{k_{FM} + k_{TM} + k_{GM} + k_{QM}[Q]} \tag{3.24}$$

The fluorescence yield in the presence of quencher is given by

$$\phi_{FM} = \frac{k_{FM}}{k_{FM} + k_{TM} + k_{GM} + k_{QM}[Q]} \tag{3.25}$$

The fluorescence and triplet quantum yields in the absence of quencher, ϕ_{FM}^{0} and ϕ_{TM}^{0}, are given by

$$\phi_{FM}^{0} = \frac{k_{FM}}{k_{FM} + k_{TM} + k_{GM}} \quad \text{and} \quad \phi_{TM}^{0} = \frac{k_{TM}}{k_{FM} + k_{TM} + k_{GM}}$$

and therefore

$$\frac{\phi_{FM}^{0}}{\phi_{FM}} = 1 + k_{QM}\tau_{M}[Q] \tag{3.26}$$

where $\tau_M = (k_{FM} + k_{TM} + k_{GM})^{-1}$ is the lifetime of the excited singlet state. Combining equations (3.24) and (3.26) to eliminate [Q] gives

$$\frac{\phi^0_{FM}}{\phi_{FM}} - 1 = \phi^0_{TM}\left(\frac{\phi_{TM}\phi^0_{FM}}{\phi^0_{TM}\phi_{FM}} - 1\right) \tag{3.27}$$

The advantage of this method is that only relative measurements of the decreasing fluorescence yields and relative measurements of the increasing triplet yields need to be obtained. The former is easily obtained from fluorescence intensity measurements since

$$\frac{\phi^0_{FM}}{\phi_{FM}} = \frac{F^0}{F}$$

where F^0 and F are the fluorescence intensities in the absence and presence of a quencher, and the latter is obtained by measuring the triplet absorbance in the absence and presence of quencher under conditions where the quencher does not absorb any exciting light. Thus

$$\frac{\phi_{TM}}{\phi^0_{TM}} = \frac{A_T}{A^0_T}$$

where A_T and A^0_T are the initial triplet absorbances at the triplet absorption maximum observed with and without added quencher in flash photolysis experiments. Thus equation (3.27) may be written as

$$\frac{F^0}{F} = \phi^0_{TM}\left(\frac{A_T F^0}{A^0_T F} - 1\right) + 1 \tag{3.28}$$

Plots of F^0/F versus $[(A_T F^0/A^0_T F) - 1]$ give straight lines of slope ϕ^0_{TM} which are independent of the quencher used (see Figure 3.5). Several heavy atom quenchers have been used for this type of measurement, but xenon is an ideal quencher[45] (see Figure 3.6) since it is chemically inert, it dissolves to the extent of $\sim$0·1 mol l^{-1} in ethanol, it does not absorb and it does not affect either the absorption or the emission spectra of the molecules under study. Solvent[46] and concentration[47] dependences of triplet quantum yields of aromatic hydrocarbons have been measured accurately using this method. Quenchers used include xenon, bromobenzene, methyl and ethyl iodides, potassium iodide, tetramethyl lead, dimethyl mercury and several others. Accurate triplet quantum yields of several porphyrins have been obtained using this method (see Table 3.10).

An interesting variation on this method has been described by Owen and co-workers[48] who have studied the photochemical decomposition

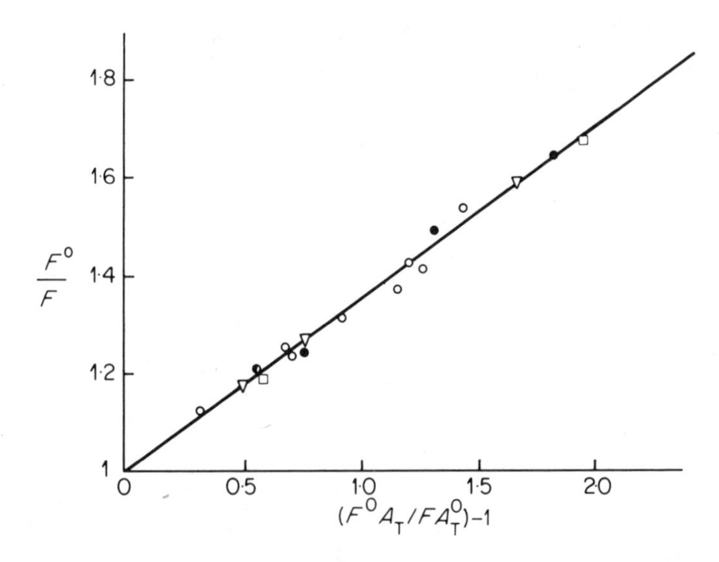

Figure 3.5 Plot showing relationship between relative fluorescence intensities and relative initial amounts of triplet-state production for 9-phenylanthracene in liquid paraffin with various amounts of quencher. ○ bromobenzene, ● bromocyclohexane, ◐ *p*-dibromobenzene, □ ethyl iodide and ▽ *n*-propyl iodide

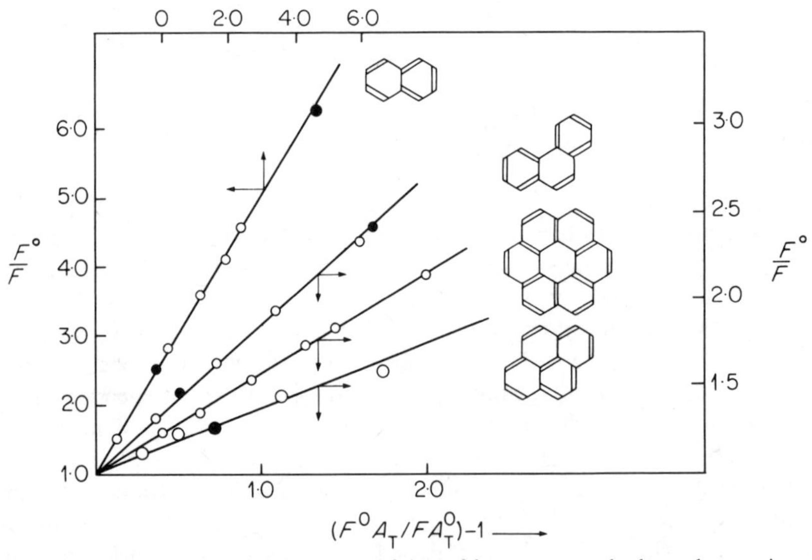

Figure 3.6 Plots for the determination of ϕ_{TM} of four aromatic hydrocarbons using xenon as a heavy-atom quencher. Naphthalene ○ $6{\cdot}2 \times 10^{-5}$ M, ● $4{\cdot}9 \times 10^{-5}$ M; Phenanthrene ○ 5×10^{-5} M; ● 3×10^{-5} M; Coronene ○ $5{\cdot}9 \times 10^{-6}$ M; Pyrene ○ 2×10^{-5} M, ● $1{\cdot}45 \times 10^{-5}$ M

of the triphenylmethyl halides. The proposed mechanism is

$Ph_3C{-}X(S_0) + h\nu \rightarrow Ph_3C{-}X^*(S_1)$ Absorption

$Ph_3C{-}X^*(S_1) \rightarrow Ph_3C{-}X(S_0) + h\nu_F$ Fluorescence

$Ph_3C{-}X^*(S_1) \rightarrow Ph_3C{-}X(S_0)$ Internal conversion

$Ph_3C{-}X^*(S_1) \rightarrow Ph_3C{-}X^*(T_1)$ Intersystem crossing

$Ph_3C{-}X^*(T_1) \rightarrow Ph_3C{-}X(S_0)$ Triplet decay

$Ph_3C{-}X^*(T_1) \rightarrow Ph_3C\cdot + X\cdot$ Decomposition

In the presence of a heavy atom quencher (e.g. 1-iodonaphthalene) the fluorescence of the triphenylmethyl halide is quenched, but there is a corresponding increase in the photochemical decomposition due to the occurrence of the process

$$Ph_3C{-}X^*(S_1) + Q_H \rightarrow Ph_3C{-}X^*(T_1) + Q_H$$

The consequent increase in the triplet concentration is thus reflected in the increased initial rate of reaction R. By analogy with equation (3.28) plots of F^0/F versus $(RF^0/R^0F - 1)$ should be linear with slope ϕ^0_{TM} (see Figure 3.7). Heavy atom fluorescence quenchers also enhance photolysis of dilute solutions of diphenylacetylene and a similar treatment has been used to obtain ϕ^0_{TM} for diphenylacetylene in different solvents (see Table 3.4).

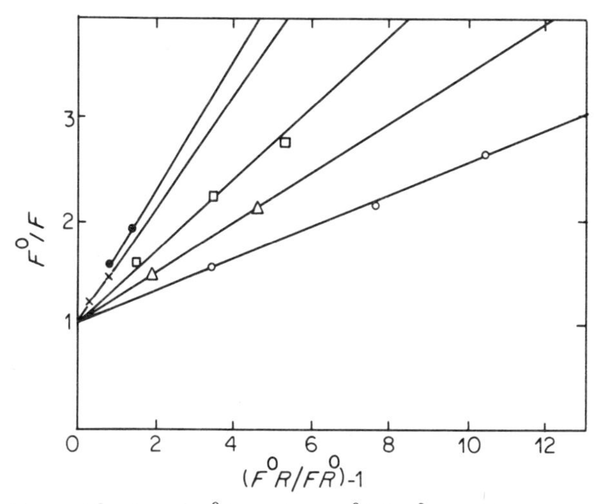

Figure 3.7 Plot of F^0/F against $(F^0R/FR^0) - 1$ for fluorescence quenching of $Ph_3C{-}X$ by 1-iodonaphthalene: ● X = I; ╳ X = Br; □ X = 1; △ X = F; ○ X = H (after Lewis and Owen[48])

3.2.4 *Photoperoxidation studies* (*D*)

Recent evidence has firmly established the role played by $O_2^*(^1\Delta_g)$ as the oxidizing intermediate in many direct and sensitized photoperoxidations (e.g. see reference 49). Stevens and Algar[50,51] have shown that oxygen quenching of the sensitizer singlet state simply catalyses intersystem crossing and that the sensitizer triplet state is the sole precursor of $O_2^*(^1\Delta_g)$ as shown in steps (5) and (7) respectively in the mechanism given below

1. $M(S_0) + h\nu \rightarrow M^*(S_1)$	Absorption
2. $M^*(S_1) \xrightarrow{k_{FM}} M(S_0) + h\nu_F$	Fluorescence
3. $M^*(S_1) \xrightarrow{k_{GM}} M(S_0)$	Internal conversion
4. $M^*(S_1) \xrightarrow{k_{TM}} M^*(T_1)$	Intersystem crossing
5. $M^*(S_1) + O_2(^3\Sigma_g^-) \rightarrow M^*(T_1) + O_2(^3\Sigma_g^-)$	Catalysed intersystem crossing
6. $M^*(T_1) \rightarrow M(S_0)$	Triplet decay
7. $M^*(T_1) + O_2(^3\Sigma_g^-) \rightarrow M(S_0) + O_2^*(^1\Delta_g)$	Electronic energy transfer
8. $A + O_2^*(^1\Delta_g) \rightarrow AO_2$	Oxidation of A
9. $O_2^*(^1\Delta_g) \rightarrow O_2(^3\Sigma_g^-)$	Singlet oxygen decay

Recent laser photolysis studies[52,53] have also given strong support for this mechanism according to which, at high oxygen concentrations when $k_5[O_2] \gg k_6$,

$$\phi_{AO_2} = \phi_{TM}\left(\frac{k_8[A]}{k_9 + k_8[A]}\right) = \phi_{TM}\theta_{AO_2} \tag{3.29}$$

where θ_{AO_2} is the fraction of $O_2^*(^1\Delta_g)$ formed which reacts with the oxidizable acceptor A and ϕ_{TM} is the triplet quantum yield of M in the presence of oxygen, i.e.

$$\phi_{TM} = \frac{k_{TM} + k_5[O_2]}{k_{FM} + k_{TM} + k_{GM} + k_5[O_2]} \tag{3.30}$$

The presence of oxygen also reduces the lifetime of the excited singlet state S_1 and the fluorescence intensity, i.e.

$$\frac{F^0}{F} = \frac{\tau_M^0}{\tau_M} = \frac{k_{FM} + k_{TM} + k_{GM} + k_5[O_2]}{k_{FM} + k_{TM} + k_{GM}} \tag{3.31}$$

Remembering that $k_{TM}\tau_M^0 = \phi_{TM}^0$ it follows from equations (3.29), (3.30) and (3.31) that

$$\phi_{AO_2}\frac{F^0}{F} = \left(\phi_{TM}^0 + \frac{F^0}{F} - 1\right)\theta_{AO_2} \tag{3.32}$$

Thus plots of $\phi_{AO_2}F^0/F$ versus $(F^0/F - 1)$ or $\phi_{AO_2}\tau_M^0/\tau_M$ versus $(\tau_M^0/\tau_M) - 1$ should be linear (see Figure 3.8) and from equation (3.32) it follows that

$$\frac{\text{Intercept}}{\text{Slope}} = \phi_{TM}^0 \tag{3.33}$$

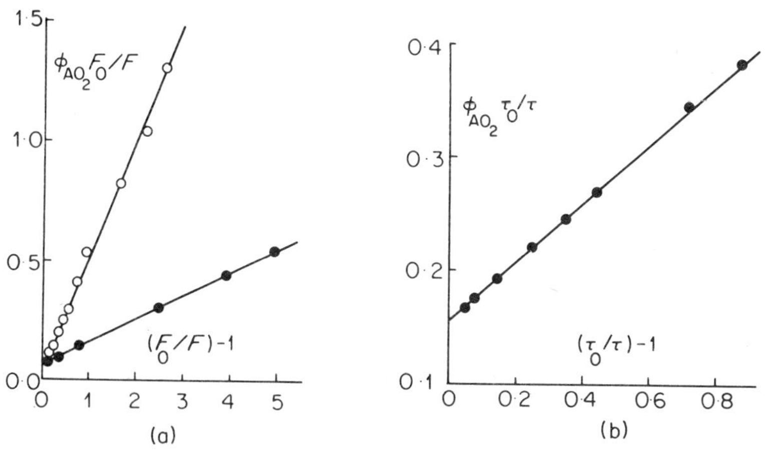

Figure 3.8 Triplet quantum yield plots from photoperoxidation studies for (a) 9,10-dimethylanthracene ○, and 9,10-dimethyl-1:2-benzanthracene ●, and (b) tetracene, in benzene solutions at 25°C (after Stevens and Algar[50,51])

Stevens and Algar[50,51] have determined the values of ϕ_{TM}^0 for several compounds using equation (3.32) and the values obtained give good agreement with other reliable methods. Previous to this Gollnick and Schenck[54] had obtained values of ϕ_{TM}^0 directly from measurements of photoperoxidation yields ϕ_{AO_2} by working under conditions and with compounds for which $k_8[A] \gg k_9$ and $k_{TM} \gg k_5[O_2]$; in which case equations (3.29) and (3.30) combine to give

$$\phi_{AO_2} = \phi_{TM}^0$$

The values obtained by Gollnick and Schenck[54] are shown in Table 3.12.

More recently Potashnik *et al.*[55] using laser photolysis have determined the initial triplet absorbance in the presence and absence of oxygen, A_T/A_T^0, for several aromatic hydrocarbons. The fraction of singlet states

Table 3.2 Intersystem crossing associated with the quenching of aromatic hydrocarbon fluorescence by molecular oxygen[55]

Hydrocarbon	Concentration (mol/l^{-1})	Solvent	Quenching fraction $X = \frac{F^0}{F} - 1$	ϕ_{TM}	$\phi_{TM}^{O_2}$ [h]
Anthracene	10^{-4}	Toluene	0·48	0·72[a]	0·95 ± 0·1
		Acetonitrile	0·45		0·9 ± 0·1
Pyrene	2×10^{-5}	Toluene	1·0	0·34[b]	1·0 ± 0·1
		Acetonitrile	1·0		0·6 ± 0·1
Coronene	10^{-4}	Toluene	1·0	0·64[e]	1·0 ± 0·2
		Acetonitrile	1·0		0·9 ± 0·2
1:2-Benzanthracene	2×10^{-4}	Toluene	0·9	0·73[c]	1·25 ± 0·1
		Acetonitrile	0·9		0·80 ± 0·1
Phenanthrene[i]	4×10^{-3}	Toluene	0·5	0·85[a]	0·90 ± 0·1
		Acetonitrile	0·5		0·55 ± 0·1
3:4-Benzopyrene	10^{-4}	Toluene	0·9	0·58[f]	0·9 ± 0·1
Chrysene	4×10^{-3}	Toluene	0·9	0·81[c]	1·1 ± 0·1
1:2:5:6-Dibenzanthracene	7×10^{-5}	Toluene	0·9	0·89[g]	1·0 ± 0·1

[a] A. R. Horrocks and F. Wilkinson, *Proc. Roy. Soc.*, **A306**, 257 (1968).
[b] The value of 0·34 was chosen as a compromise between that of 0·38 reported in a and c, and that of 0·27 reported in d.
[c] W. Heinzelmann and H. Labhart, *Chem. Phys. Letters*, **4**, 20 (1969).
[d] C. A. Parker and T. A. Joyce, *Trans. Faraday Soc.*, **62**, 2785 (1966).
[e] W. R. Dawson and J. L. Kropp, *J. Phys. Chem.*, **73**, 693 (1969). (A lower value of 0·56 is reported in a.)
[f] C. A. Parker, *Photoluminescence of Solutions*, p. 315, Elsevier, Amsterdam, 1968.
[g] A. A. Lamola and G. S. Hammond, *J. Chem. Phys.*, **43**, 2129 (1965).
[h] Error estimates are based on range of experimental accuracy of laser experiments, not on the uncertainty in ϕ_{TM}. In the calculation of $\phi_{TM}^{O_2}$, ϕ_{TM} was assumed to be independent of solvent.
[i] The phenanthrene triplet–triplet absorption was characterized by a maximum at 420 nm (see text).

which are quenched by oxygen may be expressed as X where

$$X = \frac{F^0}{F} - 1 = \frac{k_5[O_2]}{k_{FM} + k_{TM} + k_{GM} + k_5[O_2]} \tag{3.34}$$

If the yield of triplet states formed as a result of quenching the singlet state is $\phi_{TM}^{O_2}$, i.e. if we consider $M^*(T_1)$ and $O_2(^3\Sigma_g^-)$ as only one of the possible products arising from reaction of $M^*(S_1)$ and $O_2(^3\Sigma_g^-)$, then

$$\frac{\phi_{TM}}{\phi_{TM}^0} = \frac{A_T}{A_T^0} = \frac{\phi_{TM}^{O_2} k_5[O_2] + k_{TM}}{\phi_{TM}^0(k_{FM} + k_{TM} + k_{GM} + k_5[O_2])} \tag{3.35}$$

and substitution of equation (3.34) into (3.35) upon rearrangement gives

$$\phi_{TM}^{O_2} = \frac{\phi_{TM}^0}{X}\left(\frac{A_T}{A_T^0} - (1 - X)\right) \tag{3.36}$$

By taking values of ϕ_{TM}^0 which were obtained by other workers, who unfortunately used different solvents, and, assuming ϕ_{TM}^0 is independent of solvent, Potashnik *et al.*[55] showed that $\phi_{TM}^{O_2}$ calculated from equation (3.36) was equal to unity within experimental error in toluene (see Table 3.2) but in the case of pyrene and phenanthrene in acetonitrile values of $\phi_{TM}^{O_2}$ much less than unity were obtained. Two points are worth noting in connection with this work. Firstly, in the case of phenanthrene these workers report a triplet–triplet absorption band at 420 nm which is characteristic of triplet anthracene, a persistent impurity often found in phenanthrene which has not been specially purified, and, secondly, the assumption that ϕ_{TM}^0 is independent of solvent is not fully justified (see Table 3.4). Further studies are therefore necessary to establish whether oxygen quenching of excited singlet states of aromatic hydrocarbons in acetonitrile occurs by another process which competes with catalysed intersystem crossing.

3.2.5 *Determinations from flash calorimetric measurements* (*E*)

This method is based on the fact that radiationless transitions are followed by a vibrational cascade which results in a temperature rise within the surrounding medium.[56] The time dependence of this heating has been determined by measuring the resultant volume changes with a capacitor microphone transducer. A flash lamp with a half-life of ~10 μs was used for excitation and a rapid heat rise due to the decay processes arising from excited singlet states was observed, followed by a much slower heat rise corresponding to the dissipation of triplet energy. Partitioning the total heating we have

$$Q_{tot} = Q_{fast} + Q_{slow} \tag{3.37}$$

Assuming the triplet state is non-phosphorescent

$$Q_{\text{slow}} = \phi_{\text{TM}} E_{\text{T}} N_{\text{exc}} \tag{3.38}$$

where ϕ_{TM} is the triplet quantum yield, E_{T} is the triplet energy and N_{exc} is the number of molecules initially excited, which equals the number of photons absorbed from the exciting flash. The fast heat rise should be equal to the energy absorbed less the energy emitted and that still available within the triplet states which are yet to decay. Thus

$$Q_{\text{fast}} = [h\nu_{\text{a}} - \phi_{\text{FM}} h\nu_{\text{F}} - \phi_{\text{TM}} E_{\text{T}}] N_{\text{exc}} \tag{3.39}$$

Combining equations (3.37), (3.38) and (3.39) and rearranging gives

$$\phi_{\text{TM}} = \frac{Q_{\text{slow}}(h\nu_{\text{a}} - \phi_{\text{FM}} h\nu_{\text{F}})}{Q_{\text{tot}} E_{\text{T}}}$$

The ratio $Q_{\text{slow}}/Q_{\text{tot}}$ is obtained from calorimetric measurements which are averaged over 100–500 flashes while spectral measurements can be used to obtain appropriate values for ν_{a}, ν_{F} and E_{T}. ϕ_{FM} has to be determined separately. Goutermann *et al.*[57] have also developed a calorimetric method for doing this. The sensitivity and time response of the flash calorimetric apparatus is such that at present only triplets with lifetimes between 100 μs and 1 s can be measured. The value of ϕ_{TM} obtained for anthracene in ethanol compares very well with that obtained by method C (see Table 3.4).

3.2.6 *Determination from ESR intensity measurements* (*F*)

The triplet state is paramagnetic and ESR intensity measurements can be used to determine triplet concentrations. This can be done by comparing I_{R}, the integrated absorption spectrum of the allowed transition ($\Delta m = 1$) of a free radical of known concentration [R], with I_{T}, the integrated absorption of the forbidden $\Delta m = 2$ spectrum of a photostationary state concentration of triplet states, $[\text{T}]_{\text{pss}}$, using the following equation given by Aleksandrov and Pukhov[58]

$$\frac{[\text{R}]}{[\text{T}]_{\text{pss}}} = \frac{8}{15}\left(\frac{D^2 + 3E^2}{(h\nu_0)^2}\right)\frac{I_{\text{R}}}{I_{\text{T}}}$$

where $h\nu_0$ is the microwave energy and D and E are zero field splitting parameters which can be obtained from the ESR spectrum. The technique is not very sensitive and a measurable photostationary triplet state concentration is only possible for triplet states of relatively long lifetime. The steady state triplet concentration and therefore the ESR signal is proportional to $\phi_{\text{TM}}\tau_{\text{T}}$ and only values of $\phi_{\text{TM}}\tau_{\text{T}} \gg 10^{-2}$ s can be studied.

Guéron and co-workers[59] have measured triplet quantum yields of several nucleotides relative to standard compounds of known triplet yield and lifetime (e.g. naphthalene and triphenylene). By using rigid solutions of equal absorbance, identical distribution in the standard and the unknown solutions is ensured and the measured ESR intensities are related by the following equation

$$\frac{(\phi_{TM})_1}{(\phi_{TM})_2} = \frac{(I_T\tau_T)_1 B_1 i_1}{(I_T\tau_T)_2 B_2 i_2}$$

where B is the fractional Boltzmann population difference which in the high-temperature approximation is $h\nu_0/3\mathbf{k}T$. Thus B_1 is usually equal to B_2 and both can be derived; i_1 and i_2 depend on the different values of the zero field splittings and on the line widths (for further details see reference 59). At present ϕ_{TM} values obtained by this method appear to be subject to an uncertainty of about $\pm 40\%$ but the method does not interfere with the system and it does not require optically clear samples. It is likely to find many applications in biological systems.

3.2.7 *Other methods and comparisons*

Our knowledge concerning singlet–triplet intersystem crossing is by no means restricted to information arising from the application of methods A–F outlined in §§3.2.1–3.2.6 above. Considerable information is available from other photophysical and photochemical studies. Fluorescence yields and fluorescence lifetimes, for example, are both related to k_{TM}, the rate constant for singlet–triplet intersystem crossing. Thus the lack of any change in ϕ_{FM} or τ_M of a particular hydrocarbon upon deuteration would clearly indicate that k_{TM} in this compound was either negligible or unaffected by deuteration. Often methods A–F have established the relative importance of singlet–triplet intersystem crossing and this, combined with a knowledge of ϕ_{FM} and τ_M and their dependence on various parameters, e.g. temperature and pressure, gives much valuable additional information.

The phosphorescence quantum yield ϕ_{PT} depends directly on ϕ_{TM} since

$$\phi_{PT} = \phi_{TM} q_{PT} = \frac{\phi_{TM} k_{PT}}{k_{PT} + k_{GT}}$$

where q_{PT} is the phosphorescence quantum efficiency, the fraction of molecules formed in the triplet state which phosphoresce, and k_{PT} and k_{GT} are the rate constants for phosphorescence and intersystem crossing from the triplet state to the ground state, respectively. Thus it follows that $\phi_{PT} \leqslant \phi_{TM}$ and any changes in ϕ_{TM} will lead to corresponding changes in ϕ_{PT}, although it is not always easy to determine the separate contributions which arise from variations in ϕ_{TM} and/or q_{PT}.

Often photochemical studies establish the state responsible for a particular photochemical reaction and the reaction mechanism. Measurements of quantum yields of reaction give information concerning the quantum yield of production of the reacting state. If it is established that reaction is due to a triplet state, then values of ϕ_{TM} can be obtained. Since such a method of determining ϕ_{TM} requires a detailed knowledge of the mechanism of the particular photochemical reaction, which may vary from compound to compound and with solvent, temperature, exciting wavelength, etc., it has not been included among the general methods outlined above. However, enough is known about the mechanisms of several reactions to establish, for example, that the triplet quantum yields for benzophenone,[60] anthraquinone[61] and duroquinone[62] in *iso*propanol are all close to unity.

The most appropriate method for determining the triplet quantum yield of a particular compound is not always immediately apparent. Method A (§3.2.1) can be applied to all organic molecules which have accessible triplet–triplet absorption spectra. However, when continuous irradiation is used to produce a significant population of triplet states this can cause photochemical degradation which often results from double photon absorption, i.e. due to the absorption of a second quantum by an excited triplet state. Significant photodecomposition can also arise during flash excitations. Certainly the precision of the direct methods does not appear to be very good as witnessed by the large scatter in values of triplet–triplet extinction coefficients (Chapter 6, Vol. 1). The use of low-intensity, monochromatic, modulated excitation sources with phase-sensitive detection as developed by Labhart[23] seems to give much higher accuracy (see Table 3.4).

Methods B(i)–B(v) (§3.2.2) require the molecule under investigation to act as a triplet energy sensitizer and they can therefore only be applied to molecules with triplet levels higher than that of the acceptor. However, usually one of the several low-energy acceptors which are available can be employed. It is important also when using this type of method to ensure that sensitization is due to triplet–triplet energy transfer and that only the triplet state of the sensitizer transfers energy to the acceptor. Often these methods give only relative triplet yields, in which case some known standard is required.

Method C (§3.2.3) is capable of high accuracy since it involves relative measurements combined in such a way as to give absolute quantum yields of triplet production. However, it requires the molecule under investigation to be fluorescent and to be in a medium which allows one to add heavy atom fluorescence quenchers. Although this does not usually present a great difficulty, it is as well to use a very inert heavy-atom quencher such as xenon to ensure it acts only by catalysing singlet–triplet intersystem crossing. It is also advisable to use more than one heavy-atom-containing quencher to establish that the values of ϕ_{TM}^0 obtained are not dependent on the quencher.

It has been demonstrated that method D (§3.2.4) is capable of giving accurate values in good agreement with other methods. However, as with most of the other methods, it is essential to be careful in its application. It is important to ensure that the energy levels and the concentrations of donor molecules and of oxidizable acceptors used are such that the kinetic scheme which has been shown to apply for one donor is equally applicable in the case of another. Although the flash calorimetric method (method E (§3.2.5)) and the method based on ESR measurements (method F (§3.2.6)) have not so far been applied to many molecules, they appear to have high potential, especially if they can be made more sensitive.

To summarize, triplet quantum yields can be obtained by a variety of methods which can give values with an experimental accuracy of better than $\pm 5\%$ in favourable cases, which compares well with the precision available for absolute quantum yields of fluorescence. The various methods complement each other and often the values obtained by more than one method agree to within experimental error. When this is not the case, it may be that the method has been applied without due care, or that some unexpected additional process is responsible for the spurious result.

3.3 Radiationless transitions in aromatic hydrocarbons

3.3.1 *Triplet–singlet intersystem crossing*

Before discussing singlet–triplet intersystem crossing in aromatic hydrocarbons it is instructive to review briefly our knowledge concerning triplet–singlet intersystem crossing ($T_1 \rightarrow S_0$). In rigid solutions the triplet lifetime, $\tau_T = (k_{PT} + k_{GT})^{-1}$, is very much longer than τ_M and therefore can be obtained with great accuracy. Furthermore often $k_{GT} \gg k_{PT}$, so that very reliable values of k_{GT} can be obtained. There are also no complications arising from other competing radiationless transitions during the decay of T_1 and there are no electronic states lying between T_1 and S_0. It is perhaps not surprising therefore that certain generalizations can be made concerning triplet–singlet intersystem crossing in aromatic hydrocarbons. These are that heavy-atom substituents increase the value of k_{GT}; that k_{GT} is always reduced by deuteration;

and that the larger the electronic energy gap, E_T, the lower the value of k_{GT}. These last two effects are attributable to changes in the Franck–Condon factor for this radiationless transition, as can be seen by considering the schematic diagram shown in Figure 3.9. The overlap of the vibrational wavefunctions shown in Figure 3.9 is not very large since

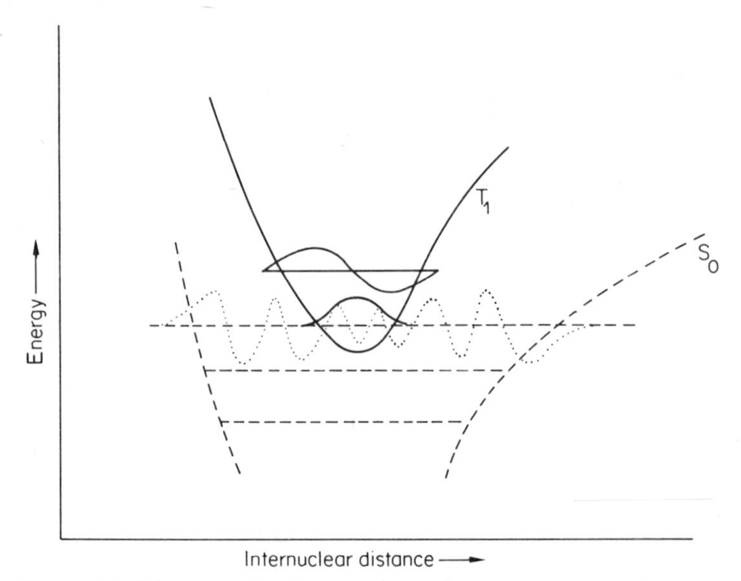

Figure 3.9 Diagram showing overlap of vibrational wavefunctions involved in intersystem crossing from $T_1 \rightarrow S_0$

the vibrational wavefunction in the S_0 state has several nodes; however, an increase in the $T_1 - S_0$ energy gap would increase the number of nodes in the isoenergetic S_0 vibronic level, as would replacing C–H vibrations which are the highest frequency vibrations in aromatic hydrocarbons by lower frequency C–D vibrations.

Siebrand[63] has shown that plots of $\log k_{GT}$ versus $(E_T - E_0)/\eta$ are linear for perprotonated and perdeuterated hydrocarbons (see Figure 3.10). η is the fraction of atoms within the hydrocarbon which are hydrogen or deuterium atoms, i.e. for the hydrocarbon C_xH_y, $\eta = y/(x+y)$, and E_0 is a constant. Siebrand[63] has also shown that this exponential dependence is to be expected from theory since according to equation (3.1)

$$k_{GT} = \frac{4\pi^2}{h}\rho J_{GT}^2 F$$

and assuming that $(4\pi^2/h)\rho J_{GT}^2$ is a constant for aromatic hydrocarbons,

k_{GT} will be directly proportional to the Franck–Condon factor F which falls off exponentially for energies greater than a certain minimum energy E_0. The dependence on the factor η demonstrates the importance of C–H (C–D) vibrations in determining the values of F in aromatic hydrocarbons for large energy gaps in excess of $E_0^{\mathrm{H}}(E_0^{\mathrm{D}})$. Siebrand[63] also showed that the value of the slopes of the plots for the perdeuterated and perprotonated hydrocarbons would be expected to be equal to $(\mu_{\mathrm{C-D}})/\mu_{\mathrm{C-H}})^{\frac{1}{2}} = 1{\cdot}36$ in reasonable agreement with the experimental value of 1·32.

The effect of heavy atoms on triplet–singlet intersystem crossing probabilities k_{GT} (see Table 3.3) is understandable, because of the increase in spin–orbit coupling which these heavy atoms introduce into the molecules

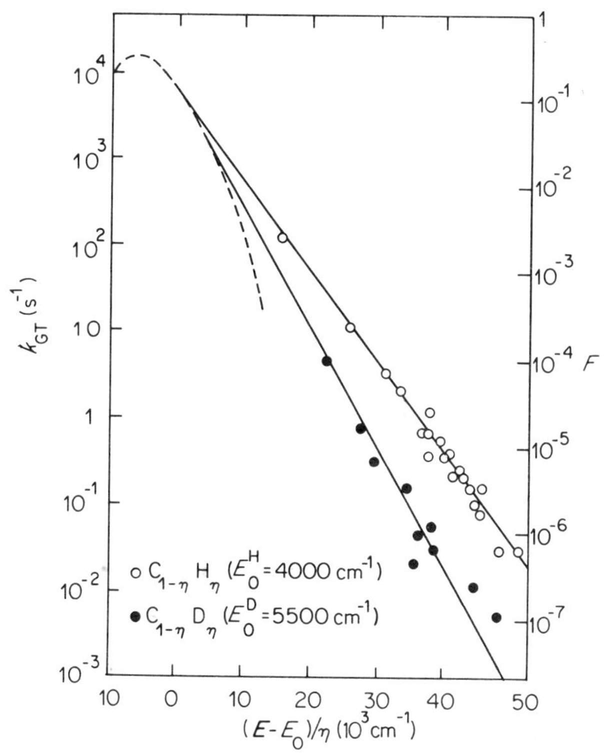

Figure 3.10 $T_1 \rightarrow S_0$ intersystem crossing rate constant k_{GT} and Franck–Condon factor F as a function of triplet-state energy E_T for perprotonated (○) and perdeuterated (●) hydrocarbons. The broken line represents F as derived from phosphorescence spectra. The two solid lines are drawn as tangents to this function thereby determining the F-scale shown (after Siebrand[63])

Table 3.3 Internal heavy atom effects in naphthalene derivatives in alcohol–ether glasses at −196°C[a]

Compound	ϕ_{FM}	ϕ_{PT}	τ_T (s)	ϕ_{TM}[b]	k_{PT} (s^{-1})[c]	k_{GT} (s^{-1})[d]	k_{TM} ($10^6 s^{-1}$)[e]
Naphthalene	0·55	0·051	2·3	0·45	0·05	0·39	0·82
1-Fluoronaphthalene	0·84	0·056	1·5	0·16	0·23	0·42	0·57
1-Chloronaphthalene	0·058	0·30	0·29	0·94	1·10	2·35	49
1-Bromonaphthalene	0·0016	0·27	0·02	1·00	13·5	36·5	1850
1-Iodonaphthalene	<0·0005	0·38	0·002	1·00	190	310	>6000

[a] V. L. Ermolaev and K. K. Svitashev, *Opt. Spectrosc.*, **7**, 399 (1965).
[b] $\phi_{TM} = 1 - \phi_{FM}$ (see Table 3.4).
[c] $k_{PT} = \phi_{PT}/\phi_{TM}\tau_T$.
[d] $k_{GT} = (\phi_{TM} - \phi_{PT})/\phi_{TM}\tau_T$.
[e] $k_{TM} = k_{FM}\phi_{TM}/\phi_{FM}$ where $k_{FM} \sim 10^6 s^{-1}$ for naphthalene and $\sim 3 \times 10^6 s^{-1}$ for the 1-halonaphthalenes.
[f] D. S. McClure, *J. Chem. Phys.*, **17**, 905 (1949).

concerned, making both radiative and non-radiative spin-forbidden processes from T_1 to the ground state more likely. Note that k_{GT} is sensitive to both internal and external heavy-atom effects.

3.3.2 $S_1 \rightarrow S_0$ *internal conversion*

Examination of the sums $\phi_{FM} + \phi_{TM}$ shown in Table 3.4 confirms that $S_1 \rightarrow S_0$ internal conversion is usually negligibly small in aromatic hydrocarbons. From equation (3.2) $\phi_{GM} = 1 - (\phi_{FM} + \phi_{TM})$ and, because experimentally determined values of both ϕ_{FM} and ϕ_{TM} are subject to considerable uncertainties, it is only if results of independent workers using different methods confirm that $\phi_{FM} + \phi_{TM} < 1$ that the presence of internal conversion within an aromatic hydrocarbon can be held to be established. Apart from benzene[64] and its derivatives (see Chapter 2, Vol. 2) and azulene, the occurrence of $S_1 \rightarrow S_0$ internal conversion in other aromatic hydrocarbons is only fully established in the case of tetracene, since three different research groups[50,65,66] using two different methods for evaluating ϕ_{TM} have shown $\phi_{GM} = 0{\cdot}18 \pm 0{\cdot}03$ at room temperature. It is probable that $\phi_{GM} \neq 0$ in the case of coronene, since Horrocks and Wilkinson[44] using method C find $\phi_{FM} + \phi_{TM} = 0{\cdot}79 \pm 0{\cdot}03$ at room temperature. Furthermore, Dawson and Kropp[67] using method A found that at room temperature for coronene-h_{12}, coronene-d_{12} and benzocoronene in solutions of poly(methylmethacrylate) ϕ_{GM} was small but probably present, although the values they obtained for ϕ_{GM} are close to their experimental uncertainties (see Table 3.4). Although only one estimate of ϕ_{TM} for pentacene is available (see Table 3.4) it does appear that $\phi_{FM} + \phi_{TM} < 1$ in this case[22] also.

Siebrand and Williams[68] have made semi-empirical calculations of values of k_{GM} based on energy-gap dependences of Franck–Condon factors found for triplet–singlet intersystem crossing in aromatic hydrocarbons which give a good fit with the experimentally determined values for k_{GM}, if the spin-allowed non-radiative process $S_1 \rightarrow S_0$ is taken to be $\sim 10^8$ times more probable than the spin-forbidden process $T_1 \rightarrow S_0$ (see also reference 69, p. 160). Of the thirteen aromatic hydrocarbons which Siebrand and Williams considered, only in the case of benzene, azulene, tetracene and coronene do their estimates for k_{GM} indicate that the quantum yield of internal conversion is likely to be greater than 0·03. Although Siebrand and Williams[68] did not consider internal conversion in pentacene, a similar treatment predicts an appreciable value for ϕ_{GM} in the case of pentacene, since it has a very low $S_1 - S_0$ energy gap.

Although aromatic hydrocarbons are not alone in showing very little $S_1 \rightarrow S_0$ internal conversion, it is important that the relationship $\phi_{FM} + \phi_{TM} = 1$ should not be applied unquestioningly. Certainly present

Table 3.4 Quantum yields of triplet state formation (ϕ_{TM}), fluorescence yields and lifetimes (ϕ_{FM} and τ_M) and rate constants for singlet–triplet intersystem crossing (k_{TM}) in dilute solutions of aromatic hydrocarbons at room temperature (except where otherwise stated)

Compound	Solvent	Method	ϕ_{TM}	Ref.	ϕ_{FM}	Ref.	τ_M (ns)	Ref.	k_{TM} ($10^6\ s^{-1}$)	$\phi_{FM} + \phi_{TM}$
Naphthalene	Ethanol	C	0·80	a	0·21	a	52	b	15·4	1·01
	Ethanol	B(iii)	0·71	c	0·21	c	52	b	13·6	0·92
	Cyclohexane	B(ii)	0·68	d	0·23	e	96	e	7·1	0·91
	Toluene	B(iv)	0·73	f	0·28	f	110	g	6·6	1·01
	Benzene	C	0·82	a	0·28	a				1·10
	Benzene	B(i)	>0·40	h	0·28	a				>0·68
	BIP (77°K)	A	0·25	i						
Naphthalene-d_8	Benzene	B(i)	>0·38	h	0·28	a				>0·66
	BIP (77°K)	A	0·25	i						
1-Methyl-naphthalene	Benzene	B(i)	>0·48	h						
2-Methyl-naphthalene	Benzene	B(i)	>0·51	h						
1-Fluoro-naphthalene	Benzene	B(i)	>0·63	h						
1-Methoxy-naphthalene	Ethanol	C	0·50	a	0·53	a				1·03
	Ethanol	B(iii)	0·46	c	0·53	c				0·99
	Benzene	B(i)	>0·26	h						
	Toluene	B(iv)	0·55	f	0·47	f	15	j	37	1·02
1-Naphthol	Benzene	B(i)	>0·27	h						
1-Naphthyl acetate	Benzene	B(i)	>0·29	h						
1-Naphthoic acid	Benzene	B(i)	>0·20	h						
1-Naphthonitrile	Benzene	B(i)	>0·17	h						
1-Naphthylamine	Benzene	B(i)	>0·15	h						

2-Naphthylamine	Cyclohexane	C	0·58	q	0·33	S				0·91
	Ethanol	C	0·32	q	0·59	T				1·01
1,5-Naphthalene disulphonic acid	Water	F	0·88	k						
Acenaphthene	Ethanol	C	0·58	a	0·39	a				0·97
	Ethanol	B(iii)	0·45	c	0·39	c				0·84
	Cyclohexane	B(ii)	0·46	d	0·60	e	46	e	10	1·06
	Benzene	B(i)	0·47	h						
Anthracene	Ethanol	C	0·72	a	0·30	a	5·2	l	138	1·02
	Ethanol	E	0·66	m	0·30	a	5·5	n	120	0·96
	Liquid paraffin	A	0·58	o	0·33	o				0·91
	Liquid paraffin	C	0·75	p	0·31	p				1·06
	Liquid paraffin	C	0·72	q	0·31	p				1·03
Anthracene-d_{10}	EPA (77°K)	A	0·53	r			5·9	j	90	
9-Methyl-anthracene	Liquid paraffin	C	0·48	s	0·51	s				0·99
	Ethanol	B(iii)	0·67	c	0·33	c	5·2	t	129	1·00
9,10-Dimethylanthracene	Ethanol	B(iii)	0·03	c	0·89	c	11	e	2·7	0·92
	Benzene	D	0·02	u	0·90	v	13·9	v	1·5	0·92
9-Ethylanthracene	Liquid paraffin	C	0·43	s	0·52	s				0·95
9-Phenylanthracene	Liquid paraffin	C	0·37	p	0·57	p				0·94
	Liquid paraffin	C	0·35	s	0·57	s				0·92
	Isopropanol	C	0·51	p	0·46	p				0·97
	Ethanol	B(iii)	0·47	c	0·49	c	5·1	t	92	0·96
	Ethanol	C	0·51	w	0·45	w	5·1	t	100	0·96
	Ethylene glycol	C	0·24	w	0·68	w				0·92
9,10-Diphenylanthracene	Liquid paraffin	C	0·12	p	0·81	p				0·93
	Ethanol	B(ii)	0·02	c	0·89	c	6·8	t	2·9	0·91
9,10-Dichloroanthracene	95% Ethanol	C	0·45	x	0·56	x				1·01

Table 3.4 (*continued*)

Compound	Solvent	Method	ϕ_{TM}	Ref.	ϕ_{FM}	Ref.	τ_M (ns)	Ref.	k_{TM} (10^6 s^{-1})	$\phi_{FM} + \phi_{TM}$
1-Anthracene sulphonic acid	Water	F	0·82	k						
Phenanthrene	Ethanol	C	0·85	a	0·13	a				0·98
	Ethanol	B(iii)	0·80	c	0·13	c				0·93
	Cyclohexane	B(ii)	0·82	d	0·24	y	61	y	13	1·06
	Benzene	B(i)	0·76	h	0·15	h				0·91
	3-Methyl-pentane	A	0·70	o	0·14	o				0·84
	BIP (77°K)	A	0·35	i						
Phenanthrene-d_{10}	BIP (77°K)	A	0·45	i						
Pyrene	Ethanol	C	0·38	z	0·65	z	530	A	0·71	1·03
	Ethanol	B(iii)	0.27	c	0·72	c	530	A	0·51	0·99
	n-Hexane	A	0·38	B						
	PMM	A	0·18	C	0·61	C	400	C	0·45	0·89
	PMM (77°K)	A	0·22	C	0·78	C	515	C	0·43	1·00
Pyrene-d_{10}	PMM	A	0·25	C	0·53	C	354	C	0·71	0·78
	PMM (197°K)	A	0·16	C	0·69	C	452	C	0·34	0·85
	PMM (77°K)	A	0·15	C	0·79	C	535	C	0·30	0·94
Tetracene	Benzene	C	0·68	D	0·15	D	5·2	v	130	0·83
	Benzene	D	0·61	u	0·19	v	5·2	v	119	0·80
	Benzene	C	0·61	E	0·19	v	5·2	v	117	0·80
	Benzene	D	0·63	F	0·16	F	5·2	F	121	0·79
	95% Ethanol	C	0·66	D	0·16	D				0·82
1:2-Benzanthracene	Hexane	C	0·77	a	0·20	a				0·97
	Ethanol	C	0·82	a	0·22	a				1·04

	Ethanol	C	0·79	E	0·20	G				0·99
	Liquid paraffin	C	0·79	E						
	n-Hexane	A	0·79	B	0·18	B				0·97
	PMM	A	0·68	G	0·23	G				0·91
	EPA (77°K)	A	0·67	G			60	j	11	
9,10-Dimethyl-1:2-benzanthracene	Benzene	D	0·66	u	0·36	v	24	v	27	1·02
Chrysene	Ethanol	C	0·85	a	0·17	a				1·02
	Ethanol	B(iii)	0·82	c	0·17	c				0·99
	Benzene	B(i)	0·67	h	0·23	h				0·90
	n-Hexane	A	0·81	B						
	EPA (77°K)	A	0·70	r			50	j	14	
Triphenylene	Ethanol	C	0·89	c	0·09	c				0·98
	Cyclohexane	B(ii)	0·86	d	0·08	e	37	e	23	0·94
	Benzene	B(i)	0·95	h						
	BIP (77°K)	A	0·54	i						
Triphenylene-d_{12}	BIP (77°K)	A	0·88	i						
Pentacene	1-Chloro-naphthalene	A	0·16	H	0·08	H				0·24
Perylene	Benzene	D	0·06	I	0·89	v	4·8	J	12·5	0·95
	Ethanol	B(iii)	0·01	K	0·87	L	6·0	M	1·7	0·88
	n-Hexane	B(iii)	0·015	K	0·98	K				1·00
	Cyclohexane	B(iii)	0·014	K	0·98	K	6·4	e	2·2	0·99
1:2:5:6-Dibenzanthracene	Benzene	C	0·90	E						
	EPA (77°K)	A	0·98	r						>0·98
	3-Methyl-pentane	A	1·03	O						1·03
Picene	EPA (77°K)	A	0·36	r						

Table 3.4 *(continued)*

Compound	Solvent	Method	ϕ_{TM}	Ref.	ϕ_{FM}	Ref.	τ_M (ns)	Ref.	k_{TM} ($10^6\ s^{-1}$)	$\phi_{FM} + \phi_{TM}$
Anthanthrene	Benzene	D	0·21	I	0·73	v	3·8	v	55	1·04
Rubrene	Benzene	D	0·04	u	0·98	v	16·4	v	2·4	1·02
1:12-Benzoperylene	PMM	A	0·53	N	0·36	N	191	N	2·8	0·88
	PMM (77°K)	A	0·72	N	0·35	N	240	N	3·0	1·07
	EPA (77°K)	A	0·59	r						
Coronene	Ethanol	C	0·56	a	0·23	a				0·79
	PMM	A	0·64	O	0·27	O	320	O	2·0	0·91
	PMM (77°K)	A	0·68	O	0·27	O	320	O	2·1	0·95
Coronene-d_{12}	PMM	A	0·66	O	0·28	O	355	O	1·9	0·94
	PMM (77°K)	A	0·67	O	0·28	O	355	O	1·9	0·95
1:2-Benzocoronene	PMM	A	0·55	P	0·27	P				0·82
	PMM	A	0·58	O	0·32	O	340	O	1·7	0·90
	PMM (77°K)	A	0·66	O	0·39	O	440	O	1·5	1·05
Biphenyl	Cyclohexane	B(ii)	0·81	d	0·18	e	16·0	e	51	0·99
	Cyclohexane	B(ii)	0·73	U	0·18	e	16·0	e	46	0·91
Fluorene	Cyclohexane	B(ii)	0·32	d	0·80	e	10	e	32	1·12
	Cyclohexane	B(ii)	0·22	U	0·80	e	10	e	22	1·02
	Ethanol	C	0·32	a	0·68	a	7·5	a	43	1·00
p-Terphenyl	*n*-Hexane	A	0·11	B						
Diphenylacetylene	Hexane	C	0·033	Q	0·0028	Q	0·0027†	Q	$1{\cdot}2 \times 10^4$	0·036
	Isopropanol	C	0·042	Q	0·0018	Q	0·0018†	Q	$2{\cdot}3 \times 10^4$	0·044
	Ethanol	C	0·056	Q	0·0016	Q	0·0015†	Q	$3{\cdot}6 \times 10^4$	0·058
	Methanol	C	0·110	Q	0·0013	Q	0·0023†	Q	$4{\cdot}7 \times 10^4$	0·111

Triphenylmethane	Cyclohexane	C	0·16	R	0·028	R	2·3†	R	69	0·19
Triphenylmethyl fluoride	Cyclohexane	C	0·25	R	0·025	R	1·5†	R	166	0·28
Triphenylmethyl chloride	Cyclohexane	C	0·35	R	0·019	R	0·7†	R	500	0·37
Triphenylmethyl bromide	Cyclohexane	C	0·57	R	0·011	R	0·3†	R	1900	0·58
Triphenylmethyl iodide	Cyclohexane	C	0·65	R	0·010	R	0·23†	R	2800	0·66

a. A. R. Horrocks and F. Wilkinson, *Proc. Roy. Soc.*, **A 306**, 257 (1968).
b. B. K. Selinger, *Austral. J. Chem.*, **19**, 825 (1966).
c. C. A. Parker, *Photoluminescence of Solutions*, Elsevier Publishing Co., Amsterdam, 1968.
d. K. Sandros, *Acta Chem. Scand.*, **22**, 2815 (1969).
e. I. B. Berlman, *Handbook of Fluorescence Spectra of Aromatic Molecules*, Academic Press, New York and London, 1965.
f. V. L. Ermolaev, E. B. Sveshnikova and E. A. Saenko, *Optics and Spectrosc.*, **22**, 86 (1967).
g. N. Mataga, M. Tomura and H. Nishimura, *Molec. Phys.*, **9**, 367 (1965).
h. A. A. Lamola and G. S. Hammond, *J. Chem. Phys.*, **43**, 2129 (1965).
i. S. G. Hadley and R. A. Keller, *J. Phys. Chem.*, **73**, 4356 (1969).
j. J. D. Laposa, E. C. Lim and R. E. Kellogg, *J. Chem. Phys.*, **42**, 3025 (1965).
k. M. Nemoto, H. Kokubun and M. Koizumi, *Bull. Chem. Soc. Japan*, **42**, 2464 (1969).
l. J. B. Birks and D. J. Dyson, *Proc. Roy. Soc.*, **A 275**, 135 (1963).
m. J. B. Callis, M. Goutermann and J. D. S. Danielson, *Rev. Sci. Inst.*, **40**, 1599 (1969).
n. W. S. Metcalf, D. R. S. Natusch, S. G. Page, E. D. Shipley and P. M. Wiggins, *J. Sci. Instr.*, **42**, 603 (1965).
o. P. G. Bowers and G. Porter, *Proc. Roy. Soc.*, **A 299**, 348 (1967).
p. T. Medinger and F. Wilkinson, *Trans. Faraday Soc.*, **61**, 620 (1965).
q. E. Vander Donckt and J. P. Van Bellingheim, *Chem. Phys. Letters*, 7, 630 (1970).
r. M. W. Windsor and W. R. Dawson, *Molec. Cryst.*, **4**, 253 (1968).
s. A. R. Horrocks, T. Medinger and F. Wilkinson, *International Symposium on Luminescence*, *The Physics and Chemistry of Scintillators*, p. 16 (Eds. N. Riehl and H. Kallmann), Verlag Karl Thiemig, Munich, 1966.
t. A. S. Cherkasov, V. A. Molchanov, T. M. Vember and K. G. Voldaikina, *Soviet Phys. Doklady*, **1**, 427 (1956).
u. B. Stevens and B. E. Algar, *J. Phys. Chem.*, **73**, 3468 (1969).
v. B. Stevens and B. E. Algar, *J. Phys. Chem.*, **72**, 2582 (1968).
w. A. R. Horrocks, T. Medinger and F. Wilkinson, *Photochem. Photobiol.*, **6**, 21 (1967).
x. A. Kearvell and F. Wilkinson, *J. Chim. Phys.*, **1970**, 125 (1970).
y. J. B. Birks and S. Georghiou, *J. Phys. B* (*At. Molec. Phys.*), **1**, 958 (1968).
z. T. Medinger and F. Wilkinson, *Trans. Faraday Soc.*, **62**, 1785 (1966).
A. B. Stevens and M. Thomaz, *Chem. Phys. Letters*, **1**, 535 (1968).
B. W. Heinzelmann and H. Labhart, *Chem. Phys. Letters*, **4**, 20 (1969).
C. J. L. Kropp, W. R. Dawson and M. W. Windsor, *J. Phys. Chem.*, **73**, 1747 (1969).
D. A. Kearvell and F. Wilkinson, *Chem. Phys. Letters*, **11**, 472 (1971).
E. E. Vander Donckt and D. Lietaer, *J. Chem. Soc. Faraday I*, **68**, 112 (1972).
F. B. Stevens and B. E. Algar, *Chem. Phys. Letters*, **1**, 58 (1967).
G. C. A. Parker, C. G. Hatchard and T. A. Joyce, *J. Molec. Spectrosc.*, **14**, 311, (1964).
H. B. Soep, A. Kellman, M. Martin and L. Lindqvist, *Chem. Phys. Letters*, **13**, 241 (1972).
I. B. Stevens and B. E. Algar, *J. Phys. Chem.*, **73**, 1711 (1969).
J. S. J. Strickler and R. A. Berg, *J. Chem. Phys.*, **37**, 814 (1962).
K. C. A. Parker and T. A. Joyce, *Chem. Comm.*, **1966**, 108 (1966).
L. W. H. Melhuish, *J. Phys. Chem.*, **65**, 229 (1961).
M. W. R. Ware, *J. Phys. Chem.*, **66**, 445 (1962).
N. W. R. Dawson and J. L. Kropp, *J. Phys. Chem.*, **73**, 175 (1969).
O. W. R. Dawson and J. L. Kropp, *J. Phys. Chem.*, **73**, 693 (1969).
P. W. R. Dawson, *J. Opt. Soc. Am.*, **58**, 222 (1968).
Q. K. C. Henson, J. L. W. Jones and E. D. Owen, *J. Chem. Soc.* (*A*), **1967**, 116 (1967).
R. H. G. Lewis and E. D. Owen, *J. Chem. Soc.* (*B*), **1967**, 422 (1967).
S. M. A. El-Bayoumi, J. P. Dalle and M. F. O'Dwyer, *J. Amer. Chem. Soc.*, **92**, 3494 (1970).
T. C. J. Seliskar, D. C. Turner, J. R. Gohlke and L. Brand, *Molecular Luminescence* (Ed. E. C. Lim), W. A. Benjamin, New York, 1969.
U. J. M. Bonnier and P. Jardon, *J. Chim. Phys.*, **67**, 571 (1970).

† Values estimated from absorption spectra and fluorescence yields.

BIP (30% 1-butanol, 70% isopentane), PMM poly(methylmethacrylate), EPA (ether-isopentane-ethanol 5:5:2 by volume).

evidence suggests that $\phi_{FM} + \phi_{TM} \neq 1$ for tetracene, coronene and pentacene as mentioned above, and also in the case of diphenylacetylene, triphenylmethyl halides (see Table 3.4), azulene,[11] several heterocyclic compounds (see Table 3.9) and some fluorescein dyes (see Table 3.12).

3.3.3 *Effect of temperature on singlet–triplet intersystem crossing*

Apart from the exceptions mentioned above, most aromatic hydrocarbons and their derivatives have only two significant modes of decay from the lowest excited singlet state S_1, namely fluorescence and singlet–triplet intersystem crossing. Usually the probability of the fluorescence is independent of temperature if allowance is made for refractive index changes (e.g. see reference 70). Very occasionally, however, when there is a state S_2 close in energy to S_1, thermally activated fluorescence from S_2 can be observed.[71] However this is not usually the case and therefore for aromatic hydrocarbons which have $\phi_{FM} + \phi_{TM} = 1$, temperature dependences of ϕ_{FM}, ϕ_{TM} and τ_M are usually all due to the temperature dependence of the singlet–triplet intersystem crossing process. Experimental results indicate that k_{TM} is composed of a temperature-independent term k_{TM}^0 and a temperature-dependent term which has an Arrhenius form, i.e.

$$k_{TM} = k_{TM}^0 + A_{TM} \exp(-E_{IS}/RT) \tag{3.40}$$

where E_{IS} represents an activation energy for intersystem crossing. Several pieces of evidence demonstrate that often temperature-dependent intersystem crossing is due to thermally activated intersystem crossing into triplet levels, T_q, lying above the S_1 state. Thus the activation energy obtained for increases in triplet state production has been shown to be equal to that obtained from the decrease in the fluorescence lifetime with increasing temperature.[72] In several compounds triplet–triplet absorption studies indicate the presence of a triplet state T_q lying slightly higher in energy than S_1.[72,73] Theoretical estimates also indicate that many such states exist. In general for intersystem crossing from S_1 we would expect[74]

$$k_{TM} = \sum_i k_{T_iM}^0 + \sum_j A_{T_jM} \exp(-E_{IS}^j/RT) \tag{3.41}$$

where the first and second terms in equation (3.41) are summed over all triplet states i lying below S_1 and all triplet states j lying above S_1 respectively. Since the temperature-independent terms $k_{T_iM}^0$ due to intersystem crossing to lower triplet levels will have Franck–Condon factors which probably decrease exponentially with increasing energy gap, it is likely that k_{TM}^{0} will be equal to the transition probability to the state T_q lying immediately below S_1. The exponential dependence on E_{IS} also makes it likely that only one higher triplet state j will contribute to k_{TM}, explaining the experimentally observed form of equation (3.40).

Since deuteration has very little effect on k_{TM} or ϕ_{TM} (see Table 3.4) it follows that direct singlet–triplet intersystem crossing to T_1, which often corresponds to a large energy gap, does not usually account for much of the observed intersystem crossing. In fact in some cases, e.g. anthracene, an inverse deuterium isotope effect is observed in the singlet–triplet intersystem crossing rate constant.[75] Jones and Calloway[70] have obtained the following values for the parameters given in equation (3.40) by measuring fluorescence lifetimes and relative triplet quantum yields as a function of temperature for perprotonated and perdeuterated naphthalene, respectively, in poly(methylmethacrylate): $k_{TM}^{0}(H)$ and $k_{TM}^{0}(D) = 2{\cdot}7 \times 10^6$ and $2{\cdot}0 \times 10^6\ s^{-1}$; $A_{TM}(H)$ and $A_{TM}(D) = 23 \times 10^6$ and $23 \times 10^6\ s^{-1}$; and $E_{IS}(H)$ and $E_{IS}(D) = 352$ and $345\ cm^{-1}$. The positive isotope effect observed in k_{TM}^{0} was interpreted as evidence for some intersystem crossing directly from S_1 to T_1. Jones and Siegel[76] find pyrene behaves in a similar manner with A_T/k_{TM}^{0} values of 44 and 29 for pyrene·d_{10} and pyrene·h_{10}, respectively.

Although the rate constant for singlet–triplet intersystem crossing of anthracene in ethanol is given by equation (3.40), in many of its *meso*-derivatives only temperature-dependent intersystem crossing is observed and ϕ_{TM} approaches zero at low temperatures.[77] Kearvell and Wilkinson[74] have proposed that intersystem crossing in anthracene goes mainly via the lower $^3B_{1g}$ state and to a lesser degree via a temperature-dependent process to produce the $^3B_{3u}$ state, while the energy levels in *meso*-derivatives are such that intersystem crossing to the $^3B_{3u}$ state predominates. The variation of E_{IS} with solvent shift of S_1 strongly suggests that solvent effects on *meso*-derivatives of anthracene are due to changes in the energy difference between S_1 and higher T_q levels.[74] Solvent dependences of triplet quantum yields of *meso*-substituted anthracenes (see Table 3.4) arise for the same reason.[46]

Widman and Huber[78] have studied fluorescence yields and lifetimes of anthracene in poly(methylmethacrylate) as a function of temperature and derived temperature-dependent rate constants for intersystem crossing. In the light of the recent theory of Henry and Siebrand,[79] they consider that the proposals made by Kearvell and Wilkinson[74] are probably correct, although Widman and Huber[78] found evidence for a second temperature-dependent process below 140°K which they think might involve a specific interaction between the solute molecule and the solvent cage. Hunter and Wyatt[80] have reported that for anthracene in methyl tetrahydrofuran a temperature-dependent intersystem crossing is observed below 170°K, which cannot adequately be explained by equation (3.41). They interpret this as being due to slow lattice relaxation in the excited singlet state. Such effects are certainly not yet fully understood, neither is the analogous temperature dependence of the $T_1 \rightarrow S_0$ triplet–singlet intersystem crossing process (for further discussion see references 81–84).

A temperature-dependent process which is, however, fairly well understood is the reverse of singlet–triplet intersystem crossing which gives rise to thermally activated delayed fluorescence (see Figure 3.11). Such

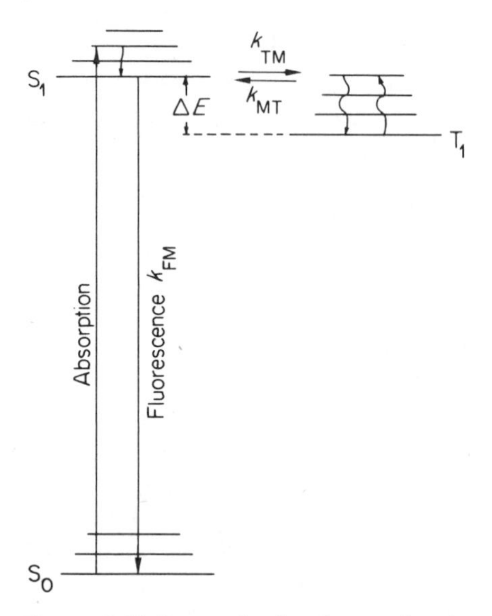

Figure 3.11 Energy-level scheme showing how thermally activated delayed fluorescence can arise

a process is likely to occur whenever the energy gap between S_1 and T_1 is comparable to the thermal energy $\mathbf{k}T$. This thermally activated triplet–singlet intersystem crossing, which gives rise to what is often known as E-type delayed fluorescence (after eosin which is a classic example), has an activation energy equal to the $S_1 - T_1$ energy gap. In eosin[85] and coronene[86] the pre-exponential factors for the reverse process are 5×10^6 and $3 \times 10^8\ s^{-1}$ respectively, which are comparable with the values of k_{TM}, the rate constant for singlet–triplet intersystem crossing in the forward direction in each of these molecules.[85,86]

3.3.4 *The internal heavy-atom effect on singlet–triplet intersystem crossing*

Ermolaev and Svitashev[87] measured the fluorescence yields of naphthalene and several of its halogen derivatives in an ethanol–ether glass at 77°K (see Table 3.3). Since there are good grounds for believing that in these compounds $\phi_{FM} + \phi_{TM} = 1$, it follows that ϕ_{TM} increases as the atomic number of the substituent increases and, since the fluorescence probabilities of these naphthalene derivatives vary only slightly, it follows that the singlet–triplet intersystem crossing probabilities k_{TM} show a marked internal heavy-atom effect (see Table 3.3). Stevens and Thomaz[88] have shown that the temperature dependences of k_{TM} for naphthalene and 1-chloronaphthalene, respectively, can be represented by equation (3.40) with $k^0_{TM} = 1{\cdot}5 \times 10^6$ and $73 \times 10^6\ s^{-1}$; $A_{TM} = 1{\cdot}1 \times 10^8$ and $36 \times 10^8\ s^{-1}$; and $E_{IS} = 630$ and $570\ cm^{-1}$, i.e. both k^0_{TM} and A_{TM} show similar relative heavy-atom effects.

At room temperature in ethanol the fluorescence yields of anthracene, 9,10-dichloroanthracene, 9-bromoanthracene and 9,10-dibromoanthracene are 0·27, 0·54, 0·02 and 0·11 respectively. In other words, no clear-cut heavy-atom effect is apparent. However, it must be remembered that the *meso*-substituted anthracenes[74,77] show marked temperature-dependent fluorescence. When the values of A_{TM} and E_{IS} are determined (see Table 3.5)

Table 3.5 Temperature-dependent intersystem crossing parameters in 95% ethanol[a]

Compound	ϕ_{FM} at 25°C	$A_{TM}(10^{10}\ s^{-1})$	$E_{IS}(cm^{-1})$	$E(S_1)(cm^{-1})$
Anthracene	0·270	0·34 ± 0·2[b]	900 ± 500[b]	26525
9-Methylanthracene	0·370	0·49 ± 0·11	790 ± 50	25797
9-Phenylanthracene	0·450	0·22 ± 0·03	620 ± 20	25780
9,10-Dichloroanthracene	0·54	7·2 ± 2·0	1570 ± 70	24850
9-Bromoanthracene	0·020	58 ± 7·0	1110 ± 20	25560
9,10-Dibromoanthracene	0·110	91 ± 35	1590 ± 50	24650

[a] A. Kearvell and F. Wilkinson, *Transitions Non-Radiatives dans les Molecules*, Paris, 1969; *J. Chim. Phys.*, **1970**, 125 (1970).

[b] These values show large error since temperature-independent intersystem crossing dominates in the case of anthracene.

it can be seen that there are two effects which are important in determining k_{TM} and therefore $\phi_{FM} = k_{FM}/(k_{FM} + k_{TM})$. Firstly and most important, *meso*-substituents lower the S_1 level relative to T_q, thereby increasing the values of E_{IS}, e.g. cf. the behaviour of 9-phenylanthracene. However, the values of A_{TM} show typical internal heavy-atom effects (see Table 3.5). Values of k^0_{TM}, A_{TM} and E_{IS} have also been determined for pyrene and its 3-chloro- and 3-bromo- derivatives in ethanol and k^0_{TM} and A_{TM} show

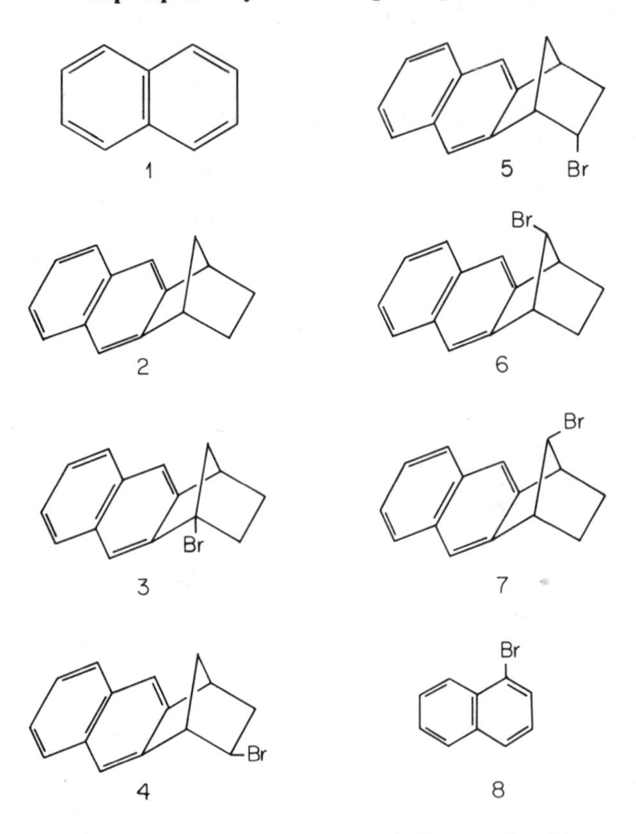

Figure 3.12 Structures of naphthonorboranes listed in Table 3.6

Table 3.6 Photophysical parameters for naphthonorboranes (see Figure 3.12) in EPA at 77°K[a]

Molecule	ϕ_{PT}/ϕ_{FM}	ϕ_{PT}	ϕ_{FM}	$k_{TM}(10^6\ s^{-1})$	$k_{GT}(s^{-1})$
1	0·09	0·03	0·33	2	0·41
2	0·14	0·057	0·41	1·5	0·24
3	32	0·32	0·010	99	2·0
4	14	0·25	0·018	55	2·4
5	46	0·47	0·010	97	2·1
6	6·1	0·45	0·07	13	0·67
7	140	0·56	0·004	250	4
8	275	0·16	0·001	520	58

[a] G. Karamos, T. Cole, P. Scribe, J. C. Dalton and N. J. Turro, *J. Amer. Chem. Soc.*, **93**, 1032 (1971).

typical internal heavy-atom effects[89] as do the k_{TM} values of the triphenylmethyl halides shown in Table 3.4.

An interesting study of how the heavy-atom effect varies as a function of geometry and distance from an aromatic hydrocarbon has been made by Karamos *et al.*,[90] who studied the series of compounds shown in Figure 3.12. The results obtained are summarized in Table 3.6. The heavy-atom effects on k_{TM} and k_{GT} do not parallel each other suggesting that they may have different mechanisms.

3.3.5 *Concentration dependence of triplet quantum yields*

Medinger and Wilkinson[47] have shown that for pyrene in ethanol the triplet quantum yield ϕ_T obtained by method C decreases with increasing pyrene concentration, although the fractional decrease is less than that of the monomer fluorescence quantum yield, ϕ_{FM}, which, as is well known, decreases at high concentrations, and is replaced by excimer fluorescence of quantum yield ϕ_{FD}. At low pyrene concentration $\phi_{FM} + \phi_{TM} = 1$. At higher concentrations $\phi_{FM} + \phi_{FD} + \phi_T < 1$, where $\phi_T = \phi_{TM} + \phi_{TD}$, and ϕ_{TD} is the triplet quantum yield of the excimer (see Table 3.7). These results were explained in terms of the following reaction scheme.

1. $M(S_0) + h\nu \rightarrow M^*(S_1)$	Absorption
2. $M^*(S_1) \rightarrow M(S_0) + h\nu_{FM}$	Monomer fluorescence
3. $M^*(S_1) \rightarrow M^*(T_1)$	Monomer intersystem crossing
4. $M^*(S_1) + M(S_0) \rightarrow M_2^*(S_1)$	Excimer formation
5. $M_2^*(S_1) \rightarrow M^*(S_1) + M(S_0)$	Excimer dissociation
6. $M_2^*(S_1) \rightarrow 2M(S_0) + h\nu_{FD}$	Excimer fluorescence
7. $M_2^*(S_1) \rightarrow M^*(T_1) + M(S_0)$	Dissociative excimer intersystem crossing
8. $M_2^*(S_1) \rightarrow 2M(S_0)$	Dissociative excimer internal conversion
9. $M^*(T_1) \rightarrow M(S_0)$	Monomer triplet decay

Equally well (7) could be the overall reaction for the two steps,

7a. $M_2^*(S_1) \rightarrow M_2^*(T_1)$	Excimer intersystem crossing
7b. $M_2^*(T_1) \rightarrow M^*(T_1) + M(S_0)$	Dissociation of triplet excimer

provided that the excimer triplet $M_2^*(T_1)$ undergoes very rapid dissociation which would explain why no triplet excimer was observed.

Table 3.7 Total triplet state quantum yields ϕ_T as a function of concentration in hexane at 300°K. ϕ_{FM} and ϕ_{FD} represent quantum yields of monomer and excimer fluorescence, respectively

Compound	Concentration mol l^{-1}	ϕ_T	Ref.	ϕ_{FM}	Ref.	ϕ_{FD}	Ref.
1:2-Benzanthracene	$0{\cdot}9 \times 10^{-5}$	0·79	a	0·18	a		
	$1{\cdot}25 \times 10^{-4}$	0·73	a	0·18	a		
	$1{\cdot}0 \times 10^{-3}$	0·54	a				
Pyrene	$0{\cdot}8 \times 10^{-5}$	0·38	a				
	$1{\cdot}0 \times 10^{-4}$	0·26	a				
	$1{\cdot}0 \times 10^{-3}$	0·12	a				
	$1{\cdot}0 \times 10^{-2}$	0·08	a				
	$5{\cdot}0 \times 10^{-2}$	0·08	a				
Pyrene in ethanol (293°K)	At infinite dilution	0·38	b	0·65	c	0	c
	$2{\cdot}0 \times 10^{-5}$	0·38	b	0·63	c	0·02	c
	$1{\cdot}0 \times 10^{-4}$	0·33	b	0·52	c	0·08	c
	$3{\cdot}0 \times 10^{-4}$	0·26	b	0·41	c	0·21	c
	$1{\cdot}0 \times 10^{-3}$	0·18	b	0·21	c	0·38	c
Fluorene[d]	$1{\cdot}3 \times 10^{-3}$	0·10	a	0·52	a		
Biphenyl[d]	$1{\cdot}0 \times 10^{-3}$	0·51	a	0·12	a		
9,10-Dihydrophenanthrene	$1{\cdot}0 \times 10^{-3}$	0·13	a	0·48	a		

[a] W. Heinzelmann and H. Labhart, *Chem. Phys. Letters*, **4**, 20 (1969).
[b] T. Medinger and F. Wilkinson, *Trans. Faraday Soc.*, **62**, 1785 (1966).
[c] C. A. Parker and C. G. Hatchard, *Trans. Faraday Soc.*, **59**, 284 (1963).
[d] Contrast low concentration values given in Table 3.4.

Heinzelmann and Labhart[23] have also studied ϕ_T as a function of concentration in *n*-hexane using method A and have found that at higher concentrations $\phi_F + \phi_T \neq 1$ (see Table 3.7 and contrast with Table 3.4). So far no evidence for absorption by triplet excimers has been obtained in this type of experiment, although part of the temperature dependence of ϕ_{FD} can be explained by intersystem crossing in excimers.[91]

3.3.6 *Singlet–triplet intersystem crossing transition probabilities*

Strictly speaking, when one considers the transition probabilities for singlet–triplet intersystem crossings in aromatic hydrocarbons one should compare values of k^0_{TM} and A_{TM} in the various compounds. Unfortunately not many accurate determinations are available. However, from the values quoted in §3.3.3 and from estimates which can be made assuming that ϕ_{FM} at 77°K represents $k_{FM}/(k_{FM} + k^0_{TM})$ and that $\phi_{FM} + \phi_{TM} = 1$, etc., it can be shown that the general features which emerge may be discussed equally well in terms of the room-temperature values given, for example, in Table 3.4. These may be summarized as follows. (i) The range of rate constants for singlet–triplet intersystem crossing in aromatic hydrocarbons of 10^6–10^8 s^{-1} confirms that this process is a spin-forbidden, radiationless transition, since it is $\sim 10^{-6}$ times less likely than internal conversion between excited singlet states. Singlet–triplet intersystem crossing also shows substantial heavy-atom effects, which depend on $\sum \xi_i^2$ where ξ_i is the spin–orbit coupling parameter of the *i*th atom in the molecule.[92] (ii) Although no energy-gap law is established for singlet–triplet intersystem crossing, and perdeuteration of aromatic hydrocarbons has little effect on k_{TM}, this is interpreted as being consistent with the proposition that Franck–Condon factors are important in these transitions and are responsible for the fact that intersystem crossing from S_1 to close-lying triplet states T_q is so much more likely than direct intersystem crossing from S_1 to T_1. (iii) A_{TM} values in several aromatic hydrocarbons are 10–100 times higher than those of k^0_{TM}. This difference could be due to an increased electronic matrix element reflecting some selection rules for intersystem crossing, or it could be due to transitions to lower triplet states usually having Franck–Condon factors 10–100 times lower than those pertaining for thermally activated $S_1 \rightarrow T_q$ intersystem crossing. (iv) Aromatic molecules which have S_1 states which are ^1L_a states, according to the Platt notation, are usually $\sim 10^2$ times more likely to undergo intersystem crossing than those with ^1L_b states as their lowest excited singlet states. ^1L_a states have π-electron density maxima at the carbon atoms, while ^1L_b states have nodal planes passing through the atoms, and Horrocks and Wilkinson[44] have suggested that this could lead to greater spin–orbit coupling in the ^1L_a states. It is interesting to note that radiative transitions from ${}^1L_a$ states to their ground states are $\sim 10^2$ higher than those from ${}^1L_b$ states, although in this case the differences are due to symmetry selection rules.[93]

3.4 Singlet–triplet intersystem crossing in molecules with (n, π^*) states

Molecules which have a ${}^1(n, \pi^*)$ state as their lowest excited singlet state usually show very weak fluorescence (often immeasurable) and strong phosphorescence. One reason for this is the relatively low radiative transition probability of ${}^1(n, \pi^*)$ states. However, this is not the complete answer, since the oscillator strengths of singlet $n \rightarrow \pi^*$ transitions are often comparable with weak singlet $\pi \rightarrow \pi^*$ transitions in aromatic hydrocarbons such as benzene and pyrene, which show appreciable fluorescence. It is apparent, therefore, that singlet–triplet intersystem crossing in molecules in which the S_1 state is a ${}^1(n, \pi^*)$ state must be much more efficient than in molecules in which the S_1 state is a ${}^1(\pi, \pi^*)$ state. El-Sayed[94] has explained this by showing that to a first-order approximation spin–orbit coupling is forbidden between states of the same configuration, whereas spin–orbit coupling between (n, π^*) states and certain (π, π^*) states is allowed to a first-order approximation. Thus he suggests the following selection rules for intersystem crossings:

$$\text{(a)} \quad {}^{1 \text{ or } 3}(n, \pi^*) \leftrightarrow {}^{3 \text{ or } 1}(\pi, \pi^*)$$

but

$$\text{(b)} \quad {}^1(n, \pi^*) \not\leftrightarrow {}^3(n, \pi^*) \quad \text{and} \quad \text{(c)} \quad {}^1(\pi, \pi^*) \not\leftrightarrow {}^3(\pi, \pi^*)$$

where process (a) is likely to be a factor of 10^3 larger than either processes (b) or (c). In addition to this increase, which is due to an increase in the matrix element for singlet–triplet intersystem crossing from a ${}^1(n, \pi^*)$ to a ${}^3(\pi, \pi^*)$ state, an increase in the Franck–Condon factor over that observed between states of the same type is also expected, because of the relatively large differences in internuclear equilibrium distances which are likely to exist for two states of different character.[95] Franck–Condon factors have been estimated for ${}^{1 \text{ or } 3}(n, \pi^*) \leftrightarrow {}^{3 \text{ or } 1}(\pi, \pi^*)$ transitions in N-heterocyclics as being ~ 100 times greater than those between states of the same type (i.e. ${}^1(n, \pi^*) \leftrightarrow {}^3(n, \pi^*)$ or ${}^1(\pi, \pi^*) \leftrightarrow {}^3(\pi, \pi^*)$) and separated by the same electronic energy gap. It follows that the efficiency of singlet–triplet intersystem crossing depends markedly on the type and relative energies of the various electronically excited states. The various possible orders of lower ${}^{1 \text{ or } 3}(n, \pi^*)$ and ${}^{1 \text{ or } 3}(\pi, \pi^*)$ states are illustrated in Figure 3.13.

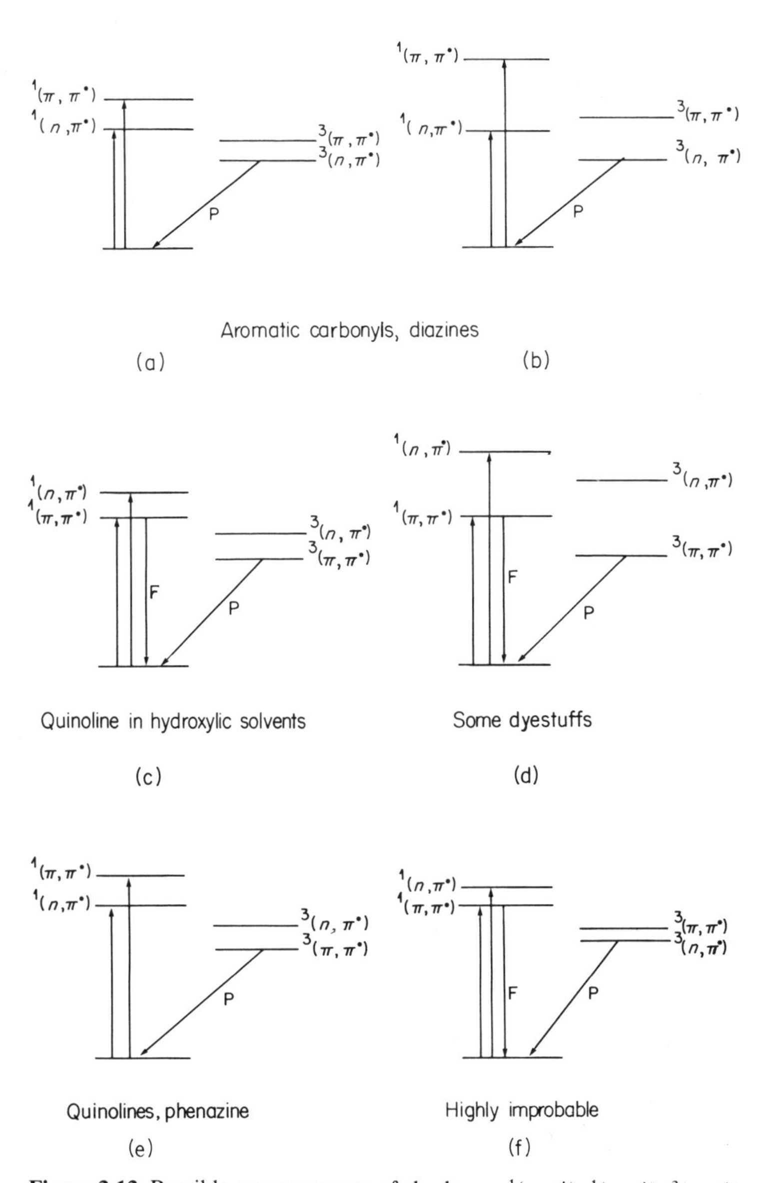

Figure 3.13 Possible arrangements of the lower $^1(\pi, \pi^*)$, $^1(n, \pi^*)$, $^3(\pi, \pi^*)$ and $^3(n, \pi^*)$ levels

Table 3.8 Triplet quantum yields (ϕ_{TM}), fluorescence yields and lifetimes (ϕ_{FM} and τ_M) and singlet–triplet intersystem crossing rate constants (k_{TM}) for carbonyl-containing compounds in dilute solution at room temperature

Compound	Solvent	Method	ϕ_{TM}	Ref.	ϕ_{FM}	Ref.	τ_M (ns)	Ref.	k_{TM} ($10^6\ s^{-1}$)	$\phi_{FM} + \phi_{TM}$
Acetone	Hexane	B(i)	0·90	a	0·001	b	2·0	c	450	0·90
Pentan-2-one	Hexane	B(i)	0·63	d			1·8	c	350	
Hexan-2-one	Hexane	B(i)	0·27	d			0·73[j]	d	370	
Di-*tert*-butyl-ketone	Hexane	B(i)	0·69	a	0·004	b	5·6	a	123	0·69
5-Methyl-hexan-2-one	Hexane	B(i)	0·11	d			0·41[j]	d	270	
Biacetyl	Benzene	B(ii)	1·0	e						1·0
Benzophenone	Benzene	B(i)	1·0	f	0					1·0
Benzophenone	Toluene	B(iv)	1·0	g	0					1·0
Benzophenone	Cyclohexane	B(ii)	1·0	h	0					1·0
p,p′-bis(Dimethylamino-benzophenone)	Benzene	B(i)	1·0	f	0					1·0
p-Phenyl-benzophenone	Cyclohexane	B(ii)	1·02	h	0					1·02
Acetophenone	Benzene	B(i)	1·0	e	0					1·0
2-Acetonaphthone	Toluene	B(iv)	1·0	i	0					1·0
2-Acetonaphthone	Benzene	B(iv)	0·84	f	0					0·84
Fluorenone	Benzene	B(i)	0·93	f	0					0·93
Benzil	Benzene	B(i)	0·92	f	0					0·92
Anthraquinone	Benzene	B(ii)	0·90	f	0					0·90

[a] N. C. Yang, E. D. Feit, M. H. Hui, N. J. Turro and J. C. Dalton, *J. Amer. Chem. Soc.*, **92**, 6974 (1970).
[b] M. O'Sullivan and A. C. Testa, *J. Amer. Chem. Soc.*, **92**, 5842 (1970).
[c] F. S. Wettack, G. D. Renkes, M. G. Rockley, N. J. Turro and J. C. Dalton, *J. Amer. Chem. Soc.*, **92**, 1794 (1970).
[d] N. C. Yang, S. P. Elliott and B. Kim, *J. Amer. Chem. Soc.*, **91**, 7551 (1969).
[e] H. L. J. Backstrom and K. Sandros, *Acta Chem. Scand.*, **14**, 48 (1960).
[f] A. A. Lamola and G. S. Hammond, *J. Chem. Phys.*, **43**, 2129 (1965).
[g] V. L. Ermolaev and E. B. Sveshnikova, *Optics and Spectrosc.*, **24**, 153 (1968).
[h] K. Sandros, *Acta Chem. Scand.*, **22**, 2815 (1969).
[i] V. L. Ermolaev, E. B. Sveshnikova and E. A. Saenko, *Optics and Spectrosc.*, **22**, 86 (1967).
[j] Estimated values (not direct measurements).

3.4.1 *Carbonyl compounds*

Carbonyl compounds usually show little fluorescence and high phosphorescence yields. Triplet quantum yields of several aliphatic and aromatic ketones have been measured, mainly using methods based on triplet–triplet energy transfer (see Table 3.8). Invariably singlet–triplet intersystem crossing in these compounds is so efficient that ϕ_{TM} values are close to unity. For this reason aromatic ketones are often used as photochemical sensitizers for triplet state reactions. Aliphatic ketones show measurable fluorescence, and recently fluorescence yields and lifetimes have been measured. This allows k_{TM} to be evaluated, and for aliphatic ketones values of $\sim 3 \times 10^8$ s^{-1} are obtained. Estimates of rate constants for intersystem crossing in aromatic ketones, such as acetophenone and benzophenone,[11,96] are $\sim 10^{11}$ s^{-1}. Such high values are usually explained by suggesting that in the aromatic ketones $S_1(n, \pi^*) \rightarrow T_q(\pi, \pi^*)$ intersystem crossing occurs and that it is very efficient, as discussed earlier. In the aliphatic ketones, however, no low-lying $^3(\pi, \pi^*)$ states are available. Thus one can rationalize the large differences in singlet–triplet intersystem crossing efficiencies for aliphatic and aromatic ketones.

Recently thermally activated delayed fluorescence has been observed for several aromatic ketones, indicating that the reverse process of $T_1 \rightarrow S_1$ intersystem crossing is also very efficient.[97]

3.4.2 *Heterocyclic compounds*

As can be seen from Table 3.9, with the exception of acridine, carbazole and dibenzofuran, most heterocyclic compounds for which triplet yields have been obtained show very low fluorescence yields. In the larger heterocyclic compounds $\phi_{FM} + \phi_{TM}$ approaches unity, and the rate constants for singlet–triplet intersystem crossing of N- and O-heterocyclics are comparable with the higher values obtained for aromatic hydrocarbons, but for heterocyclics containing the heavier atoms S and Se the rate constants are considerably larger (see Table 3.9).

Intersystem crossing in the monocyclic diazines has been discussed by Cohen and Goodman,[98] who quote values of ϕ_{TM} obtained by Stephenson and Hammond using method B(i) (see Table 3.9). It is important that these values should be checked in independent experiments, since they indicate that $\phi_{TM} + \phi_{FM} < 1$ and that internal conversion is an important process in the monocyclic diazines. The occurrence of internal conversion to the ground state is supported by the fact that the quantum yield of fluorescence of pyridazine is increased upon deuteration by a factor of 3 at 77°K; however, conversely, no change in ϕ_{FM} is observed upon deuteration of pyrazine or pyrimidine.[98]

Table 3.9 Triplet quantum yields $(\phi_T)_M$, fluorescence yields and lifetimes (ϕ_{FM} and τ_M), and singlet–triplet intersystem crossing rate constants $(k_T)_M$, for heterocyclic compounds at room temperature (except where otherwise stated)

Compound	Solvent	Method	ϕ_{TM}	Ref.	ϕ_{FM}	Ref.	τ_M (ns)	Ref.	k_{TM} (10^6 s^{-1})	$\phi_{FM} + \phi_{TM}$
Pyrazine	*n*-Hexane	B(i)	0·33	a	0·0004	a				0·33
Pyrimidine	*n*-Hexane	B(i)	0·12	a	0·0029	a				0·12
Pyridazine	*n*-Hexane	B(i)	0·20	a	0·0002	a				0·20
3,6-Dichloropyridazine	Benzene	B(i)	0·24	a	0·002	a				0·24
Quinoline	Wet benzene	B(i)	0·16	b						
	Dry benzene	B(i)	0·31	b						
	BIP (77°K)[o]	A	0·50	c						
	MIP (77°K)[o]	A	0·43	c	0	c				0·43
Isoquinoline	BIP (77°K)[o]	A	0·24	c						
	MIP (77°K)[o]	A	0·19	c						
Quinoxaline	BIP (77°K)[o]	A	0·18	c	0	c				0·18
	MIP (77°K)[o]	A	0·27	c	0	c				0·27
Phthalazine	BIP (77°K)[o]	A	0·99	d	0	d				0·99
Carbazole	Benzene	B(i)	0·36	b						
Acridine	Glycerol	E	0·23	e	0·44	e				0·67
	Ethanol	C	0·76	l	0·13	l	0·7[n]	m	1100	0·89
Benzothiophene	Cyclohexane	B(ii)	0·91	h	0·013	h	14·1[n]	j	65	0·92
Dibenzofuran	Cyclohexane	B(ii)	0·39	f	0·31	g	13·7[n]	i	28	0·70
Dibenzothiophene	Cyclohexane	B(ii)	0·96	f	0·027	g	0·79[n]	i	1200	0·96
Dibenzoselenophene	Cyclohexane	B(ii)	0·55	f	0·0003	g	$<$0·05	i	$>$11000	0·55
Tetramethyl-thiophene	Cyclohexane	B(ii)	1·0	h						1·0
Tetraphenyl-thiophene	Cyclohexane	B(ii)	0·96	h	0·016	h	0·21[n]	j	4571	0·98
Thioxanthene	Cyclohexane	B(ii)	0·78	h	0·008	h	6·6[n]	j	118	0·79
Thianthrene	Cyclohexane	B(ii)	0·94	h	0·036	h	4·7[n]	j	200	0·98
Phenoxathine	Cyclohexane	B(ii)	0·69	k	0	k				0·69
Phenoxaselenium	Cyclohexane	B(ii)	0·63	k	0	k				0·63

[a] B. J. Cohen and L. Goodman, *J. Chem. Phys.*, **46**, 713 (1967).
[b] A. A. Lamola and G. S. Hammond, *J. Chem. Phys.*, **43**, 2129 (1965).
[c] S. G. Hadley, *J. Phys. Chem.*, **75**, 2083 (1971).
[d] V. L. Alvarez and S. G. Hadley, *J. Phys. Chem.*, **76**, 3937 (1972).
[e] J. B. Callis, M. Goutermann and J. D. S. Danielson, *Rev. Sci. Inst.*, **40**, 1599 (1969).
[f] J. M. Bonnier and P. Jardon, *J. Chim. Phys.*, **67**, 571 (1970).
[g] J. M. Bonnier, P. Jardon and J. P. Blanchi, *Bull. Soc. Chim. (France)*, **12**, 4797 (1968).
[h] J. M. Bonnier and P. Jardon, *J. Chim. Phys.*, **68**, 428 (1971).
[i] J. M. Bonnier and P. Jardon, *J. Chim. Phys.*, **67**, 1385 (1970).
[j] J. M. Bonnier and P. Jardon, *J. Chim. Phys.*, **69**, 245 (1972).
[k] J. M. Bonnier and P. Jardon, *J. Chim. Phys.*, **68**, 432 (1971).
[l] F. Wilkinson and J. T. Dubois, *J. Chem. Phys.*, **48**, 2651 (1968).
[m] S. Kato, S. Minagawa and M. Koizumi, *Bull. Chem. Soc. Japan*, **37**, 1026 (1962).
[n] Values estimated from fluorescence quenching constants.
[o] BIP (30% 1-butanol, 70% *iso*pentane by volume), MIP (20% methyl-cyclohexane, 80% *iso*pentane by volume).

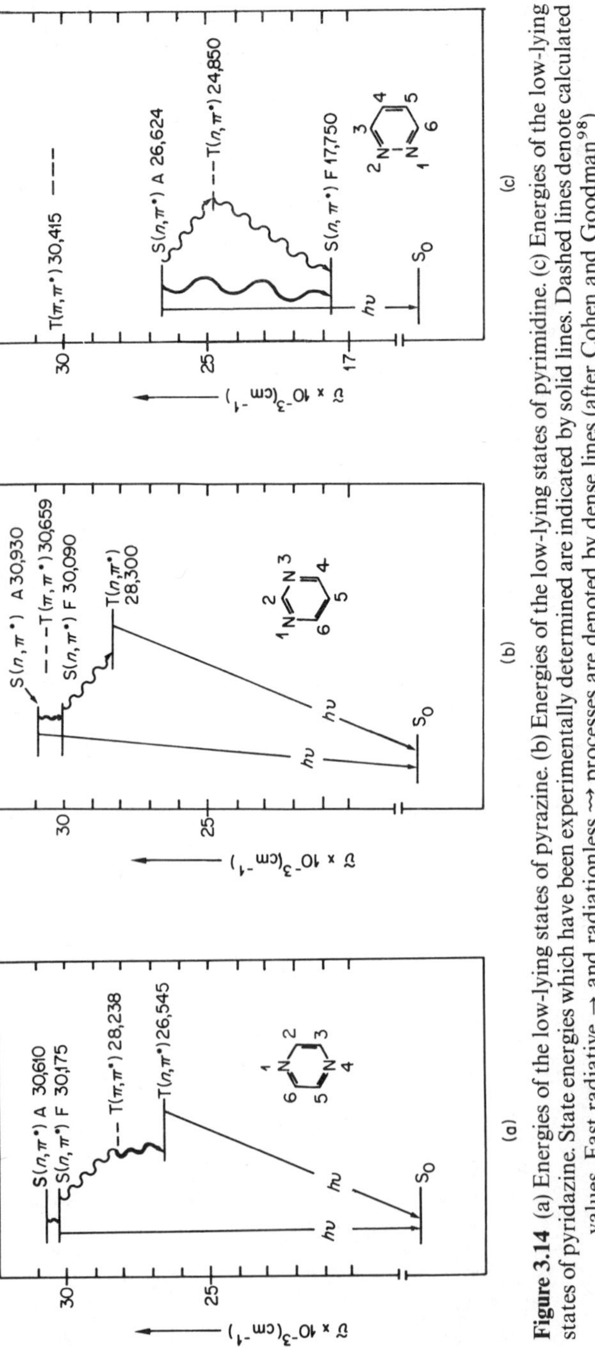

Figure 3.14 (a) Energies of the low-lying states of pyrazine. (b) Energies of the low-lying states of pyrimidine. (c) Energies of the low-lying states of pyridazine. State energies which have been experimentally determined are indicated by solid lines. Dashed lines denote calculated values. Fast radiative → and radiationless ⇝ processes are denoted by dense lines (after Cohen and Goodman[98])

The electronic states of the diazines have been extensively investigated, and Cohen and Goodman[98] discuss their results in terms of the energy-level diagrams given in Figure 3.14. The two (n, π^*) states arise because of the presence of two filled non-binding orbitals, which combine to give bonding and antibonding *n*-orbitals, n_- and n_+ respectively. The $^1(n_-, \pi^*)$ state is lower in energy than the $^1(n_+, \pi^*)$ state, and radiative transitions from the ground state to these two excited states are forbidden and allowed, respectively. Cohen and Goodman explain the low fluorescence yields in these compounds in terms of rapid internal conversion from the allowed $^1(n_+, \pi^*)$A state to the forbidden $^1(n_-, \pi^*)$F state rather than in terms of rapid singlet–triplet intersystem crossing. There are many interesting questions raised in this area, and the subject is still being actively pursued (e.g. see reference 99).

Triplet quantum yields of quinoline, *iso*quinoline and quinoxaline have been measured by Hadley[100] using method A, and the values obtained are similar to that obtained for naphthalene. Since for quinoline and quinoxaline the fluorescence yields are virtually zero, these measurements imply that there is considerable internal conversion in these molecules. The energy-level diagram given in Figure 3.15 is taken from Hadley,[100]

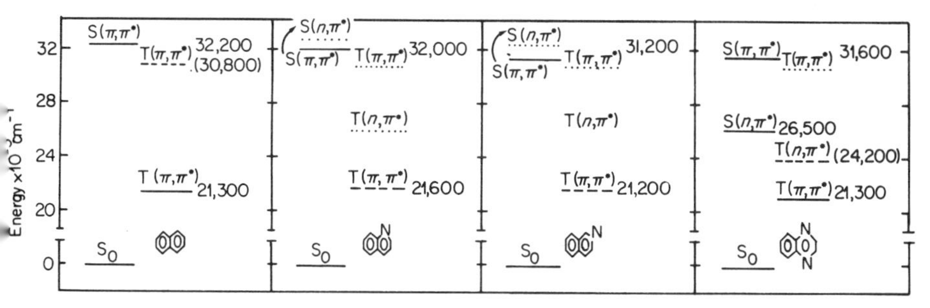

Figure 3.15 Energies of the lower singlet and triplet states of naphthalene, quinoline, *iso*quinoline and quinoxaline. ——— States observed by Hadley.[100] Energy in cm^{-1}. ----- Other states observed in pure materials (energy in cm^{-1} in parentheses). ······· States anticipated from either studies on the parent hydrocarbon, naphthalene, or from theory (energy is uncertain) (after Hadley[100])

and it is surprising in the light of these diagrams that the ϕ_{TM} values are not particularly sensitive to the order or the separation of the $^1(n, \pi^*)$, $^3(n, \pi^*)$ and $^3(\pi, \pi^*)$ states. These results are also surprising, since Lim and Yu[101] have attributed large changes in ϕ_{PT}/ϕ_{FM} as a function of solvent in the case of *iso*quinoline to enhanced singlet–triplet intersystem crossing in a hydroxylic medium as expected from the energy-level shifts. However, Hadley[100] has demonstrated that the triplet quantum yield of *iso*quinoline

Table 3.10 Quantum yields of triplet state formation (ϕ_{TM}), fluorescence yields and lifetimes (ϕ_{FM} and τ_M), and rate constants for singlet–triplet intersystem crossing (k_{TM}) in dilute solutions of porphyrins and chlorophylls at room temperature

Compound	Solvent	Method	ϕ_{TM}	Ref.	ϕ_{FM}	Ref.	τ_M (ns)	Ref.	k_{TM} (10^6 s^{-1})	$\phi_{FM} + \phi_{TM}$
Porphyrin	Toluene	C	0·90	a	0·06	a	12·5	a	72	0·96
Chlorin	Propanol	C	0·78	a	0·23	a	8·1	a	96	1·01
Tetraphenylporphyrin	Propanol	C	0·84	a	0·09	a	10·1	a	83	0·93
Tetrabenzporphyrin	Dimethyl-formamide	C	0·60	a	0·46	a	12·4	a	48	1·06
Mg-phthalocyanine	Propanol	C	0·23	a	0·76	a	7·2	a	32	0·99
Etioporphyrin I	Toluene	C	0·83	d	0·17	d	21·2	d	39	1·0
Mg-etioporphyrin I	Toluene	C	0·75	d	0·25	d	12·4	d	60	1·0
Zn-etioporphyrin I	Toluene	C	0·94	d	0·04	d	2·3	d	408	0·98
Chlorophyll a	Ether	A	0·64	b	0·32	b	5·0	e	128	0·96
	3-Methyl pentane	A	<0·1	b	<0·01	b				<0·1
	Toluene	C	0·57	d	0·39	a	4·4	e	129	0·96
	Propylene glycol	B(ii)	0·24	c						
Chlorophyll b	Ether	A	0·88	b	0·12	b	3·9	f	225	1·0
	3-Methyl pentane	A	0·17	b	<0·01	b				<0·2
	Propylene glycol	B(ii)	0·50	c						

[a] A. T. Gradyushko, A. N. Sevchenko, K. N. Solovyov and M. P. Tsvirko, *Photochem. Photobiol.*, **11**, 387 (1970).
[b] P. G. Bowers and G. Porter, *Proc. Roy. Soc.*, **A 296**, 435 (1967).
[c] C. A. Parker and T. A. Joyce, *Photochem. Photobiol.*, **6**, 395 (1967).
[d] A. T. Gradyushko and M. P. Tsvirko, *Optics and Spectrosc.*, **27**, 291 (1971).
[e] O. D. Dimitrievskii, V. L. Ermolaev and A. N. Terenin, *Dokl. Akad. Nauk. SSSR*, **114**, 1957 (1957).
[f] E. I. Rabinowitch, *J. Phys. Chem.*, **61**, 870 (1957).

is only very slightly higher in the hydroxylic solvent (BIP) than in the hydrocarbon solvent (MIP) at 77°K (see Table 3.9). Once again independent confirmation of the importance of internal conversion, this time in the case of these heterocyclic compounds (see Table 3.9), would be welcome.

Li and Lim[102] report that the quantum yield of phosphorescence of the diazine phthalazine changes by a factor of three with exciting wavelength. However, Alvarez and Hadley[103] find that phthalazine has a triplet quantum yield of unity and that this value is independent of exciting wavelength. They point out that until τ_M values for heteroaromatics are known, a direct comparison of their radiative and radiationless processes is difficult, and that it appears that there are no simple generalizations that can be made concerning heterocyclic compounds.

3.4.3 *Porphyrins and chlorophylls*

In recent years much has been learnt concerning the energy levels of porphyrins and chlorophylls from electronic absorption and emission spectroscopy, and there have been several theoretical attempts to explain these spectra (e.g. see reference 104). The fundamental role played by chlorophylls in photosynthesis makes it important to gain knowledge concerning the various modes of electronic energy relaxation in compounds of this type. Gradyushko *et al.*[105] have applied the heavy-atom-quenching method of Medinger and Wilkinson[47] to determine the triplet yields of several porphyrins and chlorophyll a (see Table 3.10). This work establishes that for these compounds $\phi_{FM} + \phi_{TM} = 1$ and that k_{TM} values of $\sim 10^8\ s^{-1}$ are obtained, i.e. comparable to those obtained for many aromatic hydrocarbons. It is interesting to note the lack of any $S_1 \rightarrow S_0$ internal conversion in these compounds, even for those which have S_1 levels as low as 15,000 cm^{-1}.

Examination of the k_{TM} values for etioporphyrin I, Mg etioporphyrin I and Zn etioporphyrin I shows a heavy-atom effect, which contrasts with the values of k_{GT} for these three compounds of 50, 7·8 and 16·2 s^{-1} respectively. The drop in k_{GT} on going from etioporphyrin I to Mg etioporphyrin I can be explained as being due to the increase in the Franck–Condon factor in the complex due to the raising of the T_1 level from 12,900 to 13,950 cm^{-1} upon complex formation.

The most interesting feature of the results on chlorophyll a and b is shown in the work of Bowers and Porter,[106] who demonstrated that the drastic decrease in fluorescence of the chlorophylls in going from a 'wet' (ether) to a 'dry' (3-methylpentane) solvent is accompanied by a decrease in the triplet quantum yield. In 'wet' solvents $\phi_{FM} + \phi_{TM} = 1$, but in 'dry' solvents the chlorophylls exist almost completely as dimers. Bowers and Porter[106] suggest that probably the excited dimer decays mainly by

Table 3.11 Triplet quantum yields (ϕ_{TM}), fluorescence yields and lifetimes (ϕ_{FM} and τ_M), and singlet–triplet intersystem crossing rate constants (k_{TM}), for amines in dilute solution at room temperature (except where otherwise stated)

Compound	Solvent	Method	ϕ_{TM}	Ref.	ϕ_{FM}	Ref.	τ_M (ns)	Ref.	k_{TM} (10^6 s^{-1})	$\phi_{FM} + \phi_{TM}$
N,*N*-Dimethylaniline	Pure liquid	A	0·95	g						>0·95
N,*N*-Diethylaniline	Toluene	B(iv)	0·90	a	0·10	a	1·5[h]	f	600	1·0
Diphenylamine	Benzene	B(i)	0·38	b						
	Toluene	B(iv)	0·32	a	0·05	a				0·37
	n-Butanol	B(iv)	0·47	a	0·09	a				0·55
	Ether–ethanol (90°K)	B(iv)	0·87	a	0·13	a	1·4[h]	f	510	1·00
	Glycerine (218°K)	B(iv)	0·10	a	0·10	a				0·20
N-Methyldiphenylamine	Toluene	B(iv)	0·62	a	0·06	a				0·68
Triphenylamine	Toluene	B(iv)	1·05	a	0·05	a				1·1
	Benzene	B(i)	0·88	b						0·88

[a] E. B. Sveshnikova and M. I. Snegov, *Optics and Spectrosc.*, **27**, 265 (1970).
[b] A. A. Lamola and G. S. Hammond, *J. Chem. Phys.*, **43**, 2129 (1965).
[c] E. Vander Donckt and D. Lietaer, *J. Chem. Soc. Faraday I*, **68**, 112 (1971).
[d] M. A. El-Bayoumi, J. P. Dalle and M. F. O'Dwyer, *J. Amer. Chem. Soc.*, **92**, 3494 (1970).
[e] C. J. Seliskar, D. C. Turner, J. R. Gohlke and L. Brand, *Molecular Luminescence*, p. 677 (Ed. E. C. Lim), W. A. Benjamin, New York, 1969.
[f] V. L. Ermolaev, *Optics and Spectrosc.*, **11**, 266 (1961).
[g] E. J. Land, J. T. Richards and J. K. Thomas, *J. Phys. Chem.*, **76**, 3805 (1972).
[h] Estimated values.

Table 3.12 Triplet quantum yields (ϕ_{TM}), fluorescence yields and lifetimes (ϕ_{FM} and τ_M), and singlet–triplet intersystem crossing rate constants (k_{TM}), for xanthene, thiazine and acridine dyes in dilute solution at room temperature

Compound	Solvent	Method	ϕ_{TM}	Ref.	ϕ_{FM}	Ref.	τ_M (ns)	Ref.	k_{TM} ($10^6\ s^{-1}$)	$\phi_{FM} + \phi_{TM}$
Fluorescein	Aqueous pH 9	A	0·05	a	0·92	a				0·97
	0·01 M NaOH	A	0·032	b	0·92	b	4·6	k	7	0·95
	0·05 M NaOH	B(v)	0·02	e	0·93	j				0·95
	Ethanol 0·01 M KOH	A	0·035	b	0·97	b	3·6	j	10	1·00
	Ethanol 0·02 M KOH	B(v)	0·007	d	0·97	b	3·6	j	2	0·98
	Methanol	D	0·03	c	0·93	c	3·8[o]	c	8	0·96
6-Hydroxy-9-phenyl-fluoron	0·01 M NaOH	A	0·02	i	0·21	i	1·2[o]	i	17	0·23
Dibromo fluorescein	Aqueous pH 9	A	0·49	a	0·65	j	2·7	j	181	1·14
Tetrabromo fluorescein (eosin)	Aqueous pH 9	A	0·71	a	0·19	a				0·90
	Aqueous pH 7·2	B(v)	0·64	e						
	0·01 M NaOH	A	0·76	b	0·22	g	1·7[o]	l		0·98
	Methanol	D	0·3	c	0·63	c	3·4[o]	c	88	0·93
	Ethanol	B(v)	0·42	d	0·48	g				0·90
	Ethanol	B(iii)	0·44	h	0·48	g				0·88
Tetraiodofluorescein (erythrosin)	Aqueous pH 9	A	1·07	a	0·02	a				1·09
	Methanol	D	0·6	c	0·08	c	0·42[o]	c	1410	0·68
	Ethanol	B(v)	1·05	d						1·05
Tetraiodo-tetrachloro fluorescein (Rose-Bengal)	Methanol	D	0·76	c	0·08	c	0·44[o]	c	1730	0·84

Table 3.12 (*continued*)

Compound	Solvent	Method	ϕ_{TM}	Ref.	ϕ_{FM}	Ref.	τ_M (ns)	Ref.	k_{TM} (10^6 s^{-1})	$\phi_{FM} + \phi_{TM}$
Acridine orange	Ethanol 0·01 M KOH	A	0·37	b	0·20	b	2·0[o]	l	185	0·57
	Ethanol 0·01 M HCl	A	0·10	b	0·46	b				0·56
	Ethanol	B(iii)	0·30	h	0·46	g	4·4	n	68	0·76
Proflavine	Ethanol 0·01 M HCl	A	0·22	b	0·40	g				0·62
	Ethanol	B(iii)	0·47	h	0·40	g	4·5[o]	l	104	0·87
Acriflavine	Ethanol	B(iii)	0·47	h	0·54	m				1·01
Methylene blue	Aqueous pH 7.2	B(v)	0·52	e						
	Ethanol	B(v)	0·52	d						
Thionine	Aqueous pH 7·2	B(v)	0·55	e						
	Ethanol	B(v)	0·62	d						

[a] P. G. Bowers and G. Porter, *Proc. Roy. Soc.*, **A 299**, 348 (1967).
[b] B. Soep, A. Kellman, M. Martin and L. Lindqvist, *Chem. Phys. Letters*, **13**, 241 (1972).
[c] K. Gollnick and G. O. Schenck, *Pure App. Chem.*, **9**, 507 (1964).
[d] M. Nemoto, H. Kokubun and M. Koizumi, *Bull. Chem. Soc. Japan*, **42**, 1223 (1969).
[e] M. Nemoto, H. Kokubun and M. Koizumi, *Bull. Chem. Soc. Japan*, **42**, 2464 (1969).
[f] J. B. Callis, M. Goutermann and J. D. S. Danielson, *Rev. Sci. Inst.*, **40**, 1599 (1969).
[g] C. A. Parker, *Photoluminescence of Solutions*, Elsevier, Amsterdam, 1968.
[h] M. Nemoto, H. Kokubun and M. Koizumi, *Chem. Comm.*, **1969**, 1095 (1969).
[i] L. Lindqvist and G. W. Lundeen, *J. Chem. Phys.*, **44**, 1711 (1966).
[j] P. G. Seybold, M. Goutermann and J. Callis, *Photochem. Photobiol.*, **9**, 229 (1969).
[k] W. R. Ware and B. A. Baldwin, *J. Chem. Phys.*, **40**, 1703 (1964).
[l] R. F. Chen, G. G. Vurek and N. Alexander, *Science*, **156**, 949 (1967).
[m] G. Weber and F. W. Teale, *Trans. Faraday Soc.*, **53**, 646 (1957).
[n] H. Rammensee and V. Zanker, *Z. Angew. Phys.*, **12**, 237 (1960).
[o] Estimated values.

undergoing a dissociative internal conversion to give two ground state monomer molecules.

3.4.4 *Amines*

Triplet quantum yields for various amines are collected together in Table 3.11. Although ϕ_{TM} is close to unity in some cases, few generalizations can be made on the basis of the meagre data available at present. It should be noted that several of the amines listed in Table 3.11 undergo photochemical reactions,[107] which may account in part for $\phi_{FM} + \phi_{TM}$ being less than unity in some of these derivatives, e.g. for diphenylamine $\phi_{FM} + \phi_{TM}$ approaches unity at low temperatures.

3.4.5 *Xanthene, thiazine and acridine dyes*

The primary photophysical processes of various dyes have attracted considerable attention, and triplet quantum yields have been determined by a variety of methods (see Table 3.12). Often commercial dyes contain considerable amounts of impurities, so that it is essential to work with purified samples. The photophysical properties of water-soluble dyes are usually pH dependent because of the different ionic forms which can exist both in the ground state and in electronically excited states.[57] Dyes often form aggregates even at low concentrations. Unfortunately in this chapter we cannot discuss any of these effects in detail.

Although $\phi_{FM} + \phi_{TM}$ for several fluorescein dyes is close to unity, according to Martin and Lindqvist[108] in the temperature range 0–50°C both fluorescein itself and 6-hydroxy-9-phenyl fluorone (HPF) exhibit temperature-dependent $S_1 \rightarrow S_0$ internal conversion in H_2O as well as in D_2O with an activation energy of $\sim 1900\ cm^{-1}$. Internal conversion is the predominant decay pathway for HPF, and the value of k_{GM} is 1·7 times higher in H_2O than in D_2O at 25°C. The value of k_{TM} is independent of temperature for both compounds, and the ratio of its value in D_2O to that in H_2O is 0·80 for HPF and 0·94 for fluorescein.

3.5 References

1. G. W. Robinson and R. P. Frosch, *J. Chem. Phys.*, **37**, 1962 (1962); *J. Chem. Phys.*, **38**, 1187 (1963).
2. S. H. Lin, *J. Chem. Phys.*, **44**, 3759 (1966).
3. W. Siebrand, *J. Chem. Phys.*, **44**, 4055 (1966).
4. J. Jortner and M. Bixon, *J. Chem. Phys.*, **48**, 715 (1968).
5. B. R. Henry and M. Kasha, *Ann. Rev. Phys. Chem.*, **19**, 161 (1968).
6. J. Jortner, S. A. Rice and R. M. Hochstrasser, *Adv. Photochem.*, **7**, 149 (1969).
7. M. Kasha, *Disc. Faraday Soc.*, **9**, 14 (1950).
8. J. Jortner, *Pure Appl. Chem.*, **24**, 165 (1970).

9. J. G. Calvert and J. N. Pitts, Jnr., *Photochemistry*, Wiley, New York, 1966.
10. R. S. Becker, *Theory and Interpretation of Fluorescence and Phosphorescence*, Wiley, New York, 1969.
11. P. M. Rentzepis, *Science*, **169**, 239 (1970).
12. S. J. Formosinho, G. Porter and M. A. West, *Chem. Phys. Letters*, **6**, 7 (1970).
13. D. S. Tinti and M. A. El-Sayed, *J. Chem. Phys.*, **54**, 2529 (1971).
14. G. N. Lewis and M. Kasha, *J. Amer. Chem. Soc.*, **66**, 2100 (1944).
15. G. Porter and M. W. Windsor, *Proc. Roy. Soc.*, **A 245**, 238 (1958).
16. W. R. Dawson, *J. Opt. Soc. Am.*, **58**, 222 (1968).
17. S. G. Hadley and R. A. Keller, *J. Phys. Chem.*, **73**, 4351 (1969).
18. L. Lindqvist, *Arkiv. Kemi.*, **16**, 79 (1960).
19. E. J. Land, *Proc. Roy. Soc.*, **A 305**, 457 (1968).
20. J. S. Brinen, *J. Chem. Phys.*, **49**, 586 (1968).
21. P. G. Bowers and G. Porter, *Proc. Roy. Soc.*, **A 299**, 348 (1967).
22. B. Soep, A. Kellman, M. Martin and L. Lindqvist, *Chem. Phys. Letters*, **13**, 241 (1972).
23. W. Heinzelmann and H. Labhart, *Chem. Phys. Letters*, **4**, 20 (1969).
24. A. N. Terenin and V. L. Ermolaev, *Dokl. Akad. Nauk. SSSR*, **85**, 547 (1952); *Trans. Faraday Soc.*, **52**, 1042 (1956).
25. V. L. Ermolaev, *Dokl. Akad. Nauk. SSSR*, **139**, 348 (1961).
26. G. Porter and F. Wilkinson, *Proc. Roy. Soc.*, **A 264**, 1 (1961).
27. K. Sandros, *Acta Chem. Scand.*, **18**, 2355 (1964).
28. A. A. Lamola and N. J. Turro, *Energy Transfer and Organic Photochemistry*, Wiley-Interscience, New York, 1969.
29. A. A. Lamola and G. S. Hammond, *J. Chem. Phys.*, **43**, 2129 (1965).
30. L. M. Stephenson, D. G. Whitten, G. F. Vesley and G. S. Hammond, *J. Amer. Chem. Soc.*, **88**, 3665, 3893 (1967).
31. R. B. Cundall, F. J. Fletcher and D. G. Milne, *J. Chem. Phys.*, **39**, 3536 (1963); *Trans. Faraday Soc.*, **60**, 1146 (1964).
32. R. B. Cundall and W. Tippett, *Trans. Faraday Soc.*, **66**, 350 (1970).
33. H. Ishikawa and W. A. Noyes, *J. Chem. Phys.*, **37**, 583 (1962).
34. J. M. Bonnier and P. Jardon, *J. Chem. Phys.*, **67**, 577 (1970).
35. K. Sandros, *Acta Chem. Scand.*, **23**, 2815 (1969).
36. S. H. Jones and T. L. Brewer, *J. Amer. Chem. Soc.*, **94**, 6310 (1972).
37. C. A. Parker and C. G. Hatchard, *Trans. Faraday Soc.*, **59**, 284 (1963).
38. C. A. Parker and T. A. Joyce, *Chem. Comm.*, **1966**, 234 (1966).
39. T. Medinger and F. Wilkinson, *Trans. Faraday Soc.*, **61**, 620 (1965).
40. V. L. Ermolaev, E. B. Sveshnikova and E. A. Saenko, *Optics and Spectrosc.*, **22**, 86 (1967).
41. V. L. Ermolaev and E. B. Sveshnikova, *Optics and Spectrosc.*, **24** 153 (1968).
42. M. Nemoto, Y. Usui and M. Koizumi, *Bull. Chem. Soc. Japan*, **40**, 1035 (1967).
43. M. Nemoto, H. Kokubun and M. Koizumi, *Bull. Chem. Soc. Japan*, **42**, 123, (1969).
44. A. R. Horrocks and F. Wilkinson, *Proc. Roy. Soc.*, **A 306**, 257 (1968).
45. A. R. Horrocks, A. Kearvell, K. Tickle and F. Wilkinson, *Trans. Faraday Soc.*, **62**, 3393 (1966).
46. A. R. Horrocks, T. Medinger and F. Wilkinson, *Photochem. Photobiol.*, **6**, 21 (1967).
47. T. Medinger and F. Wilkinson, *Trans. Faraday Soc.*, **62**, 1785 (1966).
48. H. G. Lewis and E. D. Owen, *J. Chem. Soc.* (*B*), 422 (1967).

49. K. Gollnick, *Adv. Photochem.*, **6**, 1 (1969).
50. B. Stevens and B. E. Algar, *J. Phys. Chem.*, **72**, 3468 (1968).
51. B. Stevens and B. E. Algar, *Chem. Phys. Letters*, **1**, 58, 219 (1967).
52. G. Porter and M. R. Topp, *Proc. Roy. Soc.*, **A 315**, 163 (1970).
53. J. T. Richards, G. West and J. K. Thomas, *J. Phys. Chem.*, **74**, 4137 (1970).
54. K. Gollnick and G. O. Schenk, *Pure Appl. Chem.*, **9**, 507 (1964).
55. R. Potashnik, C. R. Goldschmidt and M. Ottolenghi, *Chem. Phys. Letters*, **9**, 424 (1971).
56. J. B. Callis, M. Goutermann and J. D. S. Danielson, *Rev. Sci. Instr.*, **40**, 1599 (1969).
57. P. G. Seybold, M. Goutermann and J. B. Callis, *Photochem. Photobiol.*, **9**, 229 (1969).
58. I. V. Aleksandrov and K. K. Pukhov, *Optics and Spectrosc.*, **17**, 513 (1964).
59. M. Gueron, J. Eisinger and R. G. Shulman, *Molec. Phys.*, **14**, 111 (1968). *J. Chem. Phys.*, **47**, 4077 (1967).
60. A. Beckett and G. Porter, *Trans. Faraday Soc.*, **59**, 2051 (1963).
61. K. Tickle and F. Wilkinson, *Trans. Faraday Soc.*, **61**, 1981 (1965).
62. J. Nafisi Movaghar and F. Wilkinson, *Trans. Faraday Soc.*, **66**, 2257, 2268 (1970).
63. W. Siebrand, *The Triplet State*, p. 31 (Ed. A. B. Zahlan), Cambridge University Press, 1967.
64. R. B. Cundall, L. C. Pereira and D. A. Robinson, *J. Chem. Soc. Faraday II*, **69**, 710 (1973).
65. A. Kearvell and F. Wilkinson, *Chem. Phys. Letters*, **11**, 472 (1971).
66. E. Vander Donckt and D. Lietaer, *J. Chem. Soc. Faraday I*, **68**, 112 (1972).
67. W. R. Dawson and J. L. Kropp, *J. Phys. Chem.*, **73**, 693 (1969).
68. W. Siebrand and D. F. Williams, *J. Chem. Phys.*, **49**, 1860 (1968).
69. J. B. Birks, *Photophysics of Aromatic Molecules*, Wiley–Interscience, London and New York, 1970.
70. P. F. Jones and A. R. Calloway, *Transitions Non-Radiatives dans les Molecules*, Paris, 1969; *J. Chim. Phys.*, **1970**, 110 (1970).
71. C. E. Easterly, L. G. Christophorou and J. G. Carter, *J. Chem. Soc. Faraday II*, **69**, 471 (1973).
72. R. G. Bennett and P. J. McCartin, *J. Chem. Phys.*, **44**, 1969 (1966).
73. Y. H. Meyer, R. Astier and J. M. LeClerq, *J. Chem. Phys.*, **56**, 801 (1972).
74. A. Kearvell and F. Wilkinson, *Transitions Non-Radiatives dans les Molecules*, Paris, 1969; *J. Chim. Phys.*, **1970**, 125 (1970).
75. E. C. Lim and H. R. Bhattacharjee, *Chem. Phys. Letters*, **9**, 249 (1971).
76. P. F. Jones and S. Siegel, *Chem. Phys. Letters*, **2**, 486 (1968).
77. E. C. Lim, J. D. Laposa and J. M. H. Yu, *J. Molec. Spectrosc.*, **19**, 412 (1966).
78. R. P. Widman and J. R. Huber, *J. Phys. Chem.*, **76**, 1524 (1972).
79. B. R. Henry and W. Siebrand, *J. Chem. Phys.*, **54**, 1072 (1971).
80. T. F. Hunter and R. F. Wyatt, *Chem. Phys. Letters*, **6**, 221 (1970).
81. W. Siebrand, *J. Chem. Phys.*, **50**, 1040 (1969).
82. E. W. Schlag, S. Schneider and S. F. Fischer, *Ann. Rev. Phys. Chem.*, **1971**, 465 (1971).
83. P. F. Jones and S. Siegel, *J. Chem. Phys.*, **50**, 1134 (1969).
84. W. E. Graves, R. H. Hofeldt and S. P. McGlynn, *J. Chem. Phys.*, **56**, 1460 (1972).
85. C. A. Parker and C. G. Hatchard, *Trans. Faraday Soc.*, **57**, 1894 (1961).

86. J. L. Kropp and W. R. Dawson, *J. Phys. Chem.*, **71**, 4499 (1967).
87. V. L. Ermolaev and K. K. Svitashev, *Optics and Spectrosc.*, **7**, 399 (1959).
88. B. Stevens and M. F. Thomaz, *Chem. Phys. Letters*, **1**, 549 (1968).
89. I. Barradas, J. A. Ferreira and M. F. Thomaz, *J. Chem. Soc. Faraday II*, **69**, 384 (1973).
90. G. Karamos, T. Cole, P. Scribe, J. C. Dalton and N. J. Turro, *J. Amer. Chem. Soc.*, **93**, 1032 (1971).
91. B. Stevens, *Adv. Photochem.*, **8**, 161 (1971).
92. D. S. McClure, *J. Chem. Phys.*, **17**, 905 (1949).
93. H. H. Jaffé and M. Orchin, *Theory and Application of Ultraviolet Spectroscopy*, Wiley, New York, 1962.
94. M. A. El-Sayed, *J. Chem. Phys.*, **38**, 2834 (1963).
95. M. A. El-Sayed, *J. Chem. Phys.*, **36**, 573 (1962).
96. S. Dym and R. M. Hochstrasser, *J. Chem. Phys.*, **51**, 2458 (1969).
97. P. F. Jones and A. R. Calloway, *J. Amer. Chem. Soc.*, **92**, 4997 (1970).
98. B. J. Cohen and L. Goodman, *J. Chem. Phys.*, **46**, 713 (1967).
99. Y. H. Li and E. C. Lim, *J. Chem. Phys.*, **57**, 605 (1972).
100. S. G. Hadley, *J. Phys. Chem.*, **75**, 2083 (1971).
101. E. C. Lim and J. M. H. Yu, *J. Chem. Phys.* **45**, 4742 (1966).
102. Y. H. Li and E. C. Lim, *J. Chem. Phys.*, **56**, 1004 (1972).
103. V. L. Alvarez and S. G. Hadley, *J. Phys. Chem.*, **76**, 3937 (1972).
104. M. Goutermann, *J. Chem. Phys.*, **30**, 1139 (1959); *J. Molec. Spectrosc.*, **6**, 138 (1961).
105. A. T. Gradyushko and M. P. Tsvirko, *Optics and Spectrosc.*, **27**, 291 (1971).
106. P. G. Bowers and G. Porter, *Proc. Roy. Soc.*, **A 296**, 435 (1967).
107. E. B. Sveshnikova and M. I. Snegov, *Optics and Spectrosc.*, **27**, 265 (1970).
108. M. Martin and L. Lindqvist, *Transitions Non-Radiatives dans les Molecules*, Paris, 1969; *J. Chim. Phys.*, **1970**, 144 (1970).

4 *Excited molecular π-complexes in solution*

Henk Beens and Albert Weller

Introduction

During the last few decades the phenomenon of molecular complex formation, both in the ground and in the electronically excited states, has received considerable attention. Research in this area led to and, to a large extent, was stimulated by the Mulliken theory of intermolecular charge-transfer (CT) interaction.[1] It has been found that complexes are generally formed between electron-donor molecules, D, and electron-acceptor molecules, A, with the charge-transfer state, A^-D^+, between them being of crucial importance. Complexes of this type which are stable in the ground state are denoted as electron–donor–acceptor (EDA) complexes.[2]

Molecular complexes which are stable in the excited (singlet) state, but not in the ground state, have been studied since Förster and Kasper[3] observed the concentration-dependent fluorescence transformation of pyrene solutions and ascribed this phenomenon to formation of an excited pyrene dimer (excimer) which is unstable in the ground state. The importance of CT states in the description of the electronic structure of excimers was recognized by Ferguson.[4]

The first example of excited state complex formation between two different molecules (hetero-excimer) was provided by Leonhardt and Weller,[6] who observed the complex formation between perylene in its excited singlet state and dimethylaniline in the ground state. Since then numerous other examples of hetero-excimers have been reported and their CT character has been demonstrated in several ways.

The terms excimer and hetero-excimer are reserved here for homo-molecular and heteromolecular excited double molecules, respectively, which in general (but not necessarily) are unstable in the ground state. The general term exciplex, as used here, not only includes excimers and hetero-excimers but also comprises excited triple (and higher) complexes as well as excited electron–donor–acceptor (EDA) complexes which are not dissociated in their electronic ground states.†

† This differs from the usual definition[68] of exciplex (editor).

The discussion will be confined to π-complexes, i.e. to exciplexes whose components are substituted and unsubstituted aromatic compounds.

4.1 Experimental results

4.1.1 *Molecular π-complexes formed in the ground state*

The formation of molecular complexes in the ground state

$$A + D \rightarrow (AD) \tag{4.1}$$

can be observed in the electronic *absorption* spectrum, in which one or more new, generally broad and structureless, absorption bands are found, which often occur at longer wavelengths than those of the components. The longest wavelength band of the complex (AD) corresponds to an electronic transition, which, in the first approximation, can be described as a CT transition

$$(A^-D^+) \leftarrow (AD) \tag{4.2}$$

in which one electron is transferred from D to A. It is found that the frequency of this absorption band is correlated with the electron donating and accepting ability of the components; this corresponds to the fact that the energy of the lowest charge-transfer excited state, $E(A^-D^+)$, can be expressed by

$$E(A^-D^+) = IP_D - EA_A - C \tag{4.3}$$

in which IP_D and EA_A are the ionization potential of D and the electron affinity of A, respectively, while C is the Coulomb energy which is gained when the two ions A^- and D^+ are brought together. Measurements of the absorption spectrum as a function of the concentration of the components give the complex stoichiometry, the dissociation constant and the extinction coefficient.

For some of these complexes the structure in the crystalline state has been determined; with few exceptions the two components are found to lie in parallel planes at a distance of 3·2–3·5 Å.[8] It is under these conditions that the interaction between the π-orbitals of the components, which is at least in part responsible for the stability of the complex, can become important. It is therefore expected that parallel orientation with a similar intermolecular distance will also apply to π-complexes in solution.

With few exceptions EDA complexes do not show emission in liquid solution, though many do luminesce in frozen solutions and in the crystalline state,[2] the broad and structureless fluorescence spectrum showing a mirror-image relationship to the absorption spectrum, which indicates that this fluorescence is also of the CT type.[2] The phosphorescence spectra are often almost identical with those of the free aromatic hydrocarbon components, though recently also the phenomenon of CT phosphorescence, manifesting itself in broad, structureless, delayed emissions, has been found.[10,11]

4.1.2 *Singlet state exciplexes*

4.1.2.1 *Formation of exciplexes*

Exciplex formation in the excited singlet state manifests itself in the *fluorescence* spectrum. An example is given in Figure 4.1. The pyrene (D) fluorescence is quenched by *p*-dicyanobenzene (A) (the latter is not excited by the 313 nm Hg-line) and a new broad and structureless emission band appears at longer wavelengths. This new emission is ascribed to an exciplex, formed in the excited singlet state according to

$$^{1}D^{*} + {}^{1}A \rightarrow {}^{1}(AD)^{*} \tag{4.4}$$

As there is no corresponding change in the absorption spectrum, the complex, evidently, is not formed in the ground state.

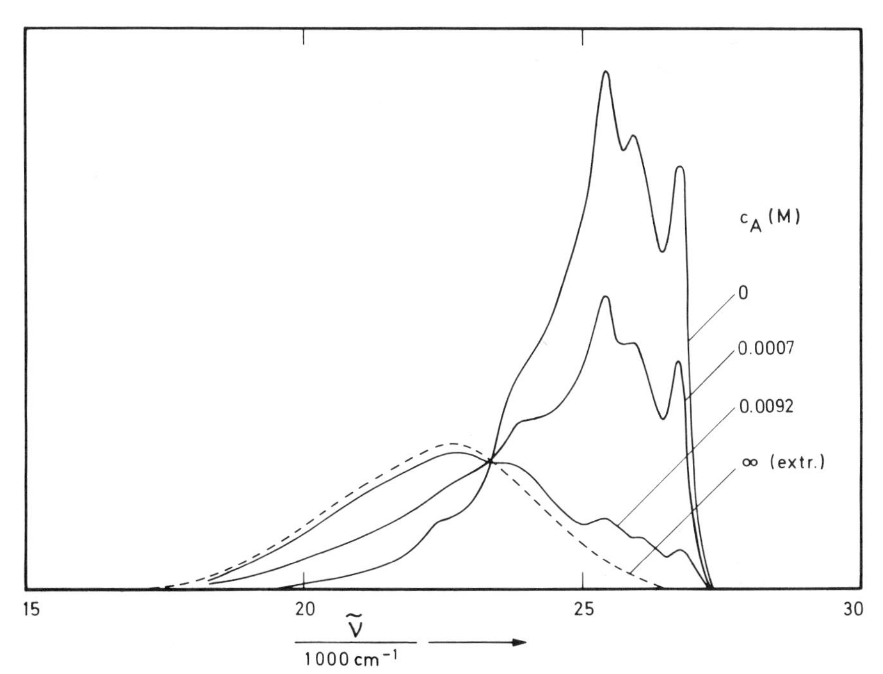

Figure 4.1 Fluorescence spectra in toluene at room temperature of $5{\cdot}1 \times 10^{-5}$ M pyrene alone and with increasing amounts of 1,4-dicyanobenzene added (excitation with Hg–313 nm)

The dependence of the intensities of the pyrene and exciplex emissions, I and I', on the concentration, c, of the non-excited species can be described by Stern–Volmer type equations

$$I = I(0)\frac{1}{1 + \kappa c} \quad \text{and} \quad I' = I'(\infty)\frac{\kappa c}{1 + \kappa c} \tag{4.5}$$

with the constant κ being given by

$$\kappa = \frac{k_a \tau_0}{1 + k_d \tau'_0} \tag{4.6}$$

where k_a is the rate constant for the exciplex formation reaction, while k_d refers to the back-reaction; τ_0 and τ'_0 are the lifetimes of the primary and exciplex emissions, respectively.

By extrapolation $c \to \infty$ the emission spectrum ($I'(\infty)$) of the exciplex is obtained (dashed curve in Figure 4.1).

From this type of experiment quantitative information with respect to exciplex quantum yield (comparison of $I(0)$ and $I'(\infty)$) and the kinetics of the complex formation reaction can be obtained.[5,14,22,30,33] From the quantum yield and lifetime of the exciplex the magnitude of the corresponding transition dipole moment can be calculated. For two different hetero-excimers, recently, the relative direction of this transition moment has also been determined; this has been done by utilizing the liquid crystal orientation technique.[9]

The phenomenon of hetero-excimer formation has been observed with numerous aromatic hydrocarbons, including alternant as well as non-alternant ones[13,14] as acceptors, and several aromatic and aliphatic amines[14,19b] or aromatic methoxy compounds[20] as electron donors. The formation of hetero-excimers with aromatic hydrocarbons as donors has also been observed, *viz.*, with acceptors like *p*-dicyanobenzene, terephthalic dimethylester and other hydrocarbons substituted with one or more acceptor groups.[12,15,21,22]

Finally it may be mentioned that evidence has been presented for triple complex formation in the excited state, e.g. between *p*-dicyanobenzene (A) and naphthalene dimer (D).[22]

4.1.2.2 *Charge-transfer hetero-excimers*

As already anticipated by Leonhardt and Weller,[6] exciplexes such as those formed between pyrene and dimethylaniline or between pyrene and *p*-dicyanobenzene are preferably represented by $^1(A^-D^+)$, expressing the fact that the excited state is a singlet CT state; the emission of the complex, therefore, corresponds to the CT transition

$$^1(A^-D^+) \rightarrow {}^1(AD) \tag{4.7}$$

the reverse of equation (4.2). The singlet character of the emissive state is indicated by lifetime measurements, which yield values of 10^{-7}–10^{-8} s.[14,19]

Indirect evidence in favour of the CT character of hetero-excimers comes from flash photolysis experiments in highly polar solvents, where transient absorption spectra due to the radical ions A^- and D^+ have been observed.[6,30,33,63]

The energy of the pure CT state, (A^-D^+), in the gas phase relative to the ground state, A + D, is given by equation (4.3). Therefore, in the first approximation, the energy of the CT emission band, $h\nu_e^{max}$, should be linearly correlated with IP_D and EA_A which, in turn, are related to the polarographic oxidation potential of the donor, E_D^{ox}, and to the polarographic reduction potential of the acceptor, E_A^{red}, respectively. Such linear correlations between $h\nu_e^{max}$ and E_A^{red} for a series of complexes with a common donor and between $h\nu_e^{max}$ and E_D^{ox} for another series of complexes with a common acceptor, indeed, have been observed.[12,13,21] It has even been found that in cases, as in Table 4.1 where relatively strong acceptors are combined with relatively strong donors (so that virtually pure charge-transfer complexes are produced in the excited state), the exciplex emission frequencies in *n*-hexane are directly related to the differences of the polarographic oxidation and reduction potentials measured in polar solvents like acetonitrile or dimethylformamide:[14,15]

$$h\nu_e^{max}(\text{hexane}) = E_D^{ox} - E_A^{red} - \Delta \tag{4.8}$$

with

$$\Delta = 0{\cdot}10 \pm 0{\cdot}08 \text{ eV}$$

The applicability of this relation which holds within 0·08 eV (the combined error of frequency and potential measurements) is evidently due to the fact that the solvation energy in a polar solvent of the separate ions, D^+ and A^-, is of the same order and depends on the size of the ions in the same way as the Coulomb term C.

One further characteristic of typical CT emission bands is their wavelength sensitivity to the polarity of the solvent. The red shift of $h\nu_e^{max}$ which is observed when the solvent polarity is increased can be related to the dipole moment, μ_c, of the exciplex by:

$$h\nu_e^{max} = h\nu_e^{max}(0) - \frac{2\mu_c^2}{\rho^3}\left(\frac{\varepsilon - 1}{2\varepsilon + 1} - \frac{1}{2}\frac{n^2 - 1}{2n^2 + 1}\right) \tag{4.9}$$

Table 4.1 Data on charge–transfer exciplexes $^1(A^-D^+)$ [15]

No.	Acceptor	E_A^{red} vs. SCE (V)	Donor	E_D^{ox} vs. SCE (V)	$h\nu_e^{max}$ in hexane (eV)	Δ (eV)	μ_c^2/ρ^3 (eV)
1	Benzonitrile	−2·35	Diethylaniline	0·76	3·08	0·03	1·17
2	1-Cyanonaphthalene	−2·00	Dimethyl-2-naphthylamine	0·67	2·63	0·04	1·08
3	Naphthalene	−2·58	Diethylaniline	0·76	3·21	0·13	0·90
4	*p*-Dicyanobenzene	−1·80	Naphthalene	1·60	3·30	0·10	0·89
5	Biphenyl	−2·65	Diethylaniline	0·76	3·24	0·17	0·88
6	*p*-Dicyanobenzene	−1·80	Pyrene	1·20	2·92	0·08	0·83
7	Pyrene	−2·10	Diethylaniline	0·76	2·76	0·10	0·70
8	1:2-Benzanthracene	−2·03	Diethylaniline	0·76	2·68	0·11	0·61
9	Anthracene	−1·96	Diethylaniline	0·76	2·59	0·13	0·58

where ε is the dielectric constant, n the refractive index, and ρ the equivalent sphere radius of the complex volume (cf. §4.3.4).

Plots of $h\nu_e^{max}$, measured in solvents of different polarity, versus $f(\varepsilon, n)$ (cf. equation (4.9)) exhibit the usual scatter due to specific solute–solvent interactions. These effects, however, largely cancel out when $h\nu_e^{max}$ values of two different molecular complexes are plotted against each other. Such plots, exhibiting surprisingly little deviation from a straight line, yield relative values of μ_c^2/ρ^3 that reliably can be compared. For the most thoroughly investigated system, anthracene-diethylaniline, which has been chosen for comparison throughout, the $h\nu_e^{max}$ versus $f(\varepsilon, n)$ plot yields by least-square evaluation $\mu_c^2/\rho^3 = 0{\cdot}58 \pm 0{\cdot}04$ eV. Thus the μ_c^2/ρ^3 values given in Tables 4.1 and 4.2 are at best reliable within the same limits of $\pm 0{\cdot}04$ eV

The variation within a factor of two of μ_c^2/ρ^3 in Table 4.1 can easily be accounted for by a 25 % increase in ρ in going from system 1 to system 9. With reasonable estimates for ρ, dipole moments μ_c of 12 ± 2 Debye are obtained.

4.1.2.3 *Mixed excimers and excimers*

The exciplexes listed in Table 4.2 have Δ values $> 0{\cdot}20$ eV and dipole moments which are smaller than those of the CT heteroexcimers of comparable dimensions: $\mu_c(i)$ in the last column of Table 4.2 refers to the dipole moment of the ith hetero-excimer in Table 4.1.

These deviations from 'normal' CT exciplex behaviour are most pronounced and evidently typical for excimers (systems 18–20 in Table 4.2) and it is for this reason that exciplexes of intermediate character such as systems 10–17 of Table 4.2 are termed mixed excimers. Their behaviour with respect to Δ and μ_c can be understood in terms of interactions between the CT state $^1(A^-D^+)$ and non-polar (locally) excited complex states such as $^1(A^*D)$ and $^1(AD^*)$ leading to stabilization of the CT state and to lowering of the dipole moment. Clearly, these interactions are even more important with typical excimers where $\Delta > 0{\cdot}6$ eV and $\mu_c = 0$.

4.1.2.4 *EDA complexes*

When, as in the case of EDA complexes, the interaction between the CT state $^1(A^-D^+)$ and the ground state $^1(AD)$ leads to stabilization of the ground state and destabilization of the CT state, negative values of Δ are to be expected. Δ values between $-0{\cdot}07$ and $-0{\cdot}21$ eV have, indeed, been found for typical EDA complexes.[15]

4.1.3 *Triplet state exciplexes*

Entirely analogous to fluorescence from the singlet CT state, phosphorescence from the triplet CT state may be expected if this state,

Table 4.2 Data on mixed excimers and excimers[15]

No.	Acceptor	E_A^{red} *vs.* SCE (V)	Donor	E_D^{ox} *vs.* SCE (V)	$h\nu_e^{max}$ in hexane (eV)	Δ (eV)	μ_c^2/ρ^3 (eV)	μ_c
10	Anthracene	−1·96	1,2,4-Trimethoxybenzene	1·12	2·73	0·35	0·48	0·91 $\mu_c(9)$
11	1:2 Benzanthracene	−2·03	1,2,4-Trimethoxybenzene	1·12	2·77	0·38	0·50	0·90 $\mu_c(8)$
12	1-Cyanonaphthalene	−2·00	Durene	1·59	3·36	0·23	0·51	0·75 $\mu_c(3)$
13	1-Cyanonaphthalene	−2·00	1,8-Dimethylnaphthalene	1·34	3·04	0·30	0·40	0·61 $\mu_c(2)$
14	Biphenyl	−2·65	Hexamethylbenzene	1·46	3·59	0·52	0·30	0·58 $\mu_c(5)$
15	1-Cyanonaphthalene	−2·00	Naphthalene	1·60	3·08	0·52	0·27	0·50 $\mu_c(2)$
16	1-Cyanonaphthalene	−2·00	*p*-Xylene	1·86	3·38	0·48	0·22	0·50 $\mu_c(4)$
17	1-Cyanonaphthalene	−2·00	*m*-Xylene	1·90	3·43	0·47	0·11	0·35 $\mu_c(4)$
18	1-Cyanonaphthalene	2·00	1-Cyanonaphthalene	1·60	2·98	0·62	0·0	0
19	Pyrene	2·10	Pyrene	1·20	2·60	0·70	0·0	0
20	Naphthalene	2·58	Naphthalene	1·60	3·12	1·06	0·0	0

$^3(A^-D^+)$, is energetically below the locally excited triplet states, $^3(A^*D)$ and $^3(AD^*)$. This condition implies a relatively large energy gap between CT singlet and locally excited singlet states, so that their mixing will be small or even negligible. As the two unpaired electrons in a typical excited CT complex occupy different orbitals which are more or less localized in the complex components, the singlet–triplet splitting of the CT state will be small and, in fact, will be zero for the zero-order CT state (cf. §4.2.2.8).

Generally, phosphorescence is observable only in rigid media where diffusion is extremely slow. Hence hetero-excimer phosphorescence can be studied only with complexes which are present in the ground state already, and it is under these conditions that phosphorescence from such complexes, indeed, has been observed.[10,11] ESR measurements, which have been carried out with two of these complexes to determine the zero-field-splitting parameters $|D|$ and $|E|$, establish their triplet state character.[23,24] The values of $|D|$ obtained for these complexes are smaller by a factor of four than those of the triplet states of the components and, thus, confirm the concept of triplet CT states with considerable spin separation.

Equation (4.8) and its corollaries with respect to variations in Δ are applicable also to exciplex phosphorescence and can be used to arrive at an interpretation of triplet exciplexes, similar to that obtained for singlet exciplexes.[11] Thus, positive deviations ($\Delta > 0{\cdot}18$ eV) which have been found in the phosphorescence of complexes with $E_D^{ox} - E_A^{red} > 2{\cdot}75$ eV are ascribed to the stabilizing interaction between the triplet CT state and (energetically higher) locally excited triplet states. There is, of course, no energetically effective interaction between the triplet CT state and the (singlet) ground state. However, the interaction between $^1(A^-D^+)$ and $^1(AD)$ because of its stabilizing effect on the ground state can cause negative deviations ($\Delta < 0{\cdot}02$ eV) as observed in the CT phosphorescence of complexes with $E_D^{ox} - E_A^{red} < 2{\cdot}50$ eV. Apparently it is only in the case of complexes with $E_D^{ox} - E_A^{red}$ between 2·50 and 2·75 eV that the positive and negative deviations tend to cancel each other so that the 'normal' value of $\Delta = 0{\cdot}10 \pm 0{\cdot}08$ eV is observed.

Triplet excimer formation is assumed to occur in cooled alcoholic solutions of phenanthrene and naphthalene, where besides the monomer phosphorescence a structureless, red-shifted, long-lived emission has been observed.[25]

Hetero-excimer formation in the triplet state has been detected in fluid solution with the system A = *p*-dicyanobenzene and D = diethylaniline.[11] The observation of this complex formation is made possible by the fact that the singlet and triplet CT levels (which are the lowest excited states in this case) are close enough together, the singlet CT state being only about 0·1 eV higher than the corresponding triplet state, so that after formation of the complex in the triplet state according to

$$^3D^* + A \rightarrow {}^3(D^+A^-) \tag{4.10}$$

an E-type delayed emission due to the endothermic reaction

$$^3(D^+A^-) \rightarrow {}^1(D^+A^-) \tag{4.11}$$

can occur.

The results of spectroscopic measurements which have been carried out with this system in methylcyclohexane at room temperature are shown in Figure 4.2. In addition to the ultraviolet fluorescence of the

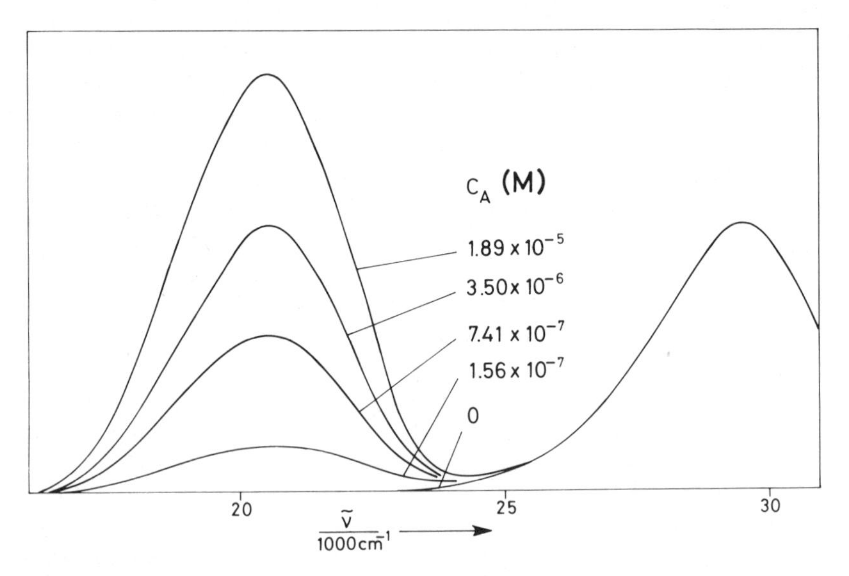

Figure 4.2 Emission spectra of $1{\cdot}3 \times 10^{-4}$ M *N,N*-diethylaniline (donor) in methylcyclohexane (de-aerated) at room temperature with increasing amounts of 1,4-dicyanobenzene (acceptor) added (excitation 313 nm)

donor (maximum at 29,600 cm^{-1}), which is primarily excited and does not change in intensity, a new blue-green emission occurs around 20,400 cm^{-1} whose intensity, I, increases with increasing acceptor concentration, c_A, according to the Stern–Volmer-type relation

$$I(c_A) = \frac{I(\infty)}{1 + c_h/c_A} \tag{4.12}$$

with a half-value concentration $c_h = 1{\cdot}3 \times 10^{-6}$ M. These findings, in particular the constant intensity of the diethylaniline fluorescence and the low half-value concentration, can only be accounted for by the assumption that diethylaniline in the triplet state, $^3D^*$, which is generated in constant yield from the primarily excited singlet state, reacts with 1,4-dicyanobenzene, A, to form the triplet CT complex according to equation (4.10). This is followed by the fast thermal reaction (4.11) and subsequent emission from the excited singlet CT state.

Other observations, which corroborate the proposed mechanism, are:

(i) Lifetime measurements give values of about 10^{-5} s for the new emission at room temperature, proving the delayed character of this fluorescence.[26]

(ii) Added substances, such as biphenyl, with the triplet state below that of $^3D^*$, but above that of $^3(D^+A^-)$, have a tremendous quenching effect on the emission.[11]

(iii) At temperatures below 0°C the emission intensity at infinite acceptor concentration (4.12) decreases with decreasing temperature.[21,26] This confirms the idea that the (emitting) singlet CT state is the one with the higher energy (cf. §4.2.2.8).

4.1.4 *Solvent effects on exciplex emission properties*

The maximum exciplex emission intensity, $I'(\infty)$, (4.5) decreases with increasing solvent polarity. This intensity drop is strong for hetero-excimers,[19a,30,32] less for mixed excimers,[14,21] and virtually zero for excimers. The lifetime, τ'_0, of the complex emission determined for two hetero-excimers (anthracene-diethylaniline[32] and pyrene-dimethylaniline[19a]) in different solvents, also decreases with increasing polarity of the solvent; however, less strongly than $I'(\infty)$.

This peculiar difference in solvent dependence of $I'(\infty)$ and τ'_0 (which does not occur with excimers) has led to the supposition (Mataga *et al.*[19]) that the radiative rate constant, k_f, of the hetero-excimer decreases with increasing solvent polarity. In order to rationalize this it was assumed that the complex electronic and geometrical structure depended on the solvent, a more polar solvent inducing more polar character in the hetero-excimer. This latter assumption is of some importance, though several conclusions derived from this model can be shown to be incorrect (cf. §4.3). Furthermore, it does not give a plausible explanation for the decrease in lifetime.

An alternative explanation,[32] assuming k_f to be virtually constant was based on the kinetic scheme

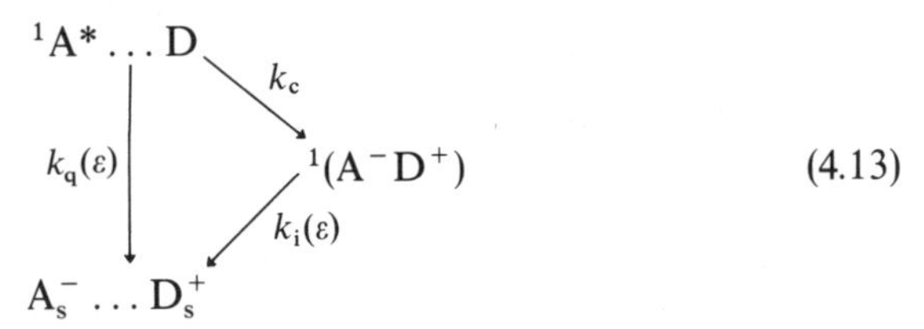

(4.13)

and suggests that the additional process that necessarily competes with hetero-excimer formation (k_c) involves the direct production of a (solvated) radical ion pair by electron transfer occurring in the encounter complex prior to hetero-excimer formation and increasing in rate with increase in solvent polarity (ε). This process with the rate constant $k_q(\varepsilon)$ reduces the quantum yield of hetero-excimer fluorescence by the solvent-dependent factor $k_c/(k_c + k_q(\varepsilon))$ without affecting the hetero-excimer lifetime. Ionic dissociation with the rate constant $k_i(\varepsilon)$ increasing with increase in solvent polarity is, accordingly, responsible for the decrease in lifetime.

Energy considerations based on polarographic oxidation and reduction potentials show that radical ion formation from $^1A^* \ldots D$ (or $A \ldots {}^1D^*$) is sufficiently exothermic to become the predominant quenching process in highly polar solvents such as acetonitrile, where, indeed, radical ions, A^- and D^+ have been observed in flash photolysis experiments.[6,30,33,63] Encounter distances evaluated from molecular quenching rate constants in this solvent are between 6 and 7 Å showing that the electron transfer occurs over twice the interplanar separation in the exciplex and does not require any stringent geometry of the transition state.[33] These findings supply strong evidence in favour of scheme (4.13).

4.1.5 *Thermodynamic data on exciplexes*

By measuring the temperature and pressure dependence of I and I' (4.5) the following thermodynamic parameters of exciplex formation equilibria have been obtained:

(a) The dissociation enthalpies, ΔH_d, have been determined for some excimers in different solvents[5,7] and for a number of hetero-excimers in alkane solutions.[14] When the energy of the $0 - 0$ transition of the primarily excited molecular species is denoted as $E_{0,0}$, the repulsion energy in the Franck–Condon ground state, δE_{rep}, can be expressed as

$$\delta E_{rep} = E_{0,0} - \Delta H_d - h\nu_e^{max} \tag{4.14}$$

The values of δE_{rep} ranging between 0·65 and 0·13 eV are in accordance with the idea that the ground state of these exciplexes is repulsive.

(b) The dissociation entropies, ΔS_d, varying around 19 $\pm$ 3 e.u. point to a relatively rigid structure of these exciplexes.

(c) The volume contraction on exciplex formation, $-\Delta V$, has been evaluated from the pressure dependence of the equilibrium constants in the case of some exciplexes, for which the equilibrium in the excited state is established. The naphthalene excimer[7] has a value of 16 cm^3/mole and for a series of hetero-excimers in hexane values between 12 cm^3/mole (1:2-benzanthracene-1,2,4-trimethoxybenzene) and 29 cm^3/mole (pyrene-diethylaniline) have been obtained.[67] On the basis of the naphthalene data and assuming a sandwich structure, an interplanar distance of 3·0 Å has been calculated for the naphthalene excimer.[7] A similar value may be assumed to apply to the other exciplexes as well.

4.2 The electronic structure of molecular π-complexes

4.2.1 *General considerations*

In §4.1 the experimental results have been interpreted tentatively in terms of mixing between charge-transfer (CT) and locally excited states (including the ground state). Now the nature of the electronic states of molecular π-complexes will be investigated by making a more theoretical approach. The purpose of this will be to give an understanding of the factors which govern the character of the relevant electronic states of the complex and especially of the contribution of the structure (A^-D^+), rather than to try to make as precise as possible calculations.

For the description of molecular π-complexes, especially for excimers, several approaches with MO-theory have been made. The following treatment will be more like the valence-bond (resonance-structure) method[1] because the binding energies in the various states are small. An attempt will be made to estimate the magnitude of the various interactions and their importance with respect to experimentally accessible quantities such as energy, dipole moment and radiative ability of the excited state. Thereby use will be made of the theory developed for excimers[37,40,41] and of SCF–MO theory.[46,47] In a simple approach the number of states which are considered is limited rather drastically. It has become common practice to treat the ground state of EDA complexes (wavefunction Ψ_g) as a resonance-hybrid of the no-bond ground state (AD) and one dative structure, the CT state (A^-D^+), only:[2]

$$\Psi_g = b_0\Psi(AD) + c_0\Psi(A^-D^+) \tag{4.15}$$

In this simple picture the excited state is characterized by

$$\Psi_a = s_0\Psi(A^-D^+) + r_0\Psi(AD) \tag{4.16}$$

with $b_0 \approx s_0 \approx 1$ and $c_0 \approx -r_0 \approx 0$. Though this description can account for several empirical findings such as the stabilization of the ground state, the energy of the absorption band and the increased polar character of the ground state,[1,2] it is clear that, generally, more states have to be included in the wavefunctions. So it has been shown recently[39] that for weakly bonded systems, such as the pyrene-tetracyanoethylene complex, the lowest CT state (A^-D^+) contributes only a fraction to the stabilization of the ground state, the higher excited CT states being equally important. Furthermore, for such weak complexes the contributions of classical electrostatic interactions to ground state stabilization and electric dipole moment can be of comparable magnitude.[64,65] In addition, the final structure of the complex will be determined by the equilibrium between the attractive and repulsive forces between the components. However, although the repulsion energy (predominantly caused by exchange of electrons of equal spin in overlapping orbitals) is a very important factor for the ultimate structure of the complex, there are experimental indications[14] that its contribution at the equilibrium distance in different excited π-complexes does not vary very much.

In principle, for the lowest excited states of molecular complexes interactions with many other states also have to be considered. So, in order to explain the unexpectedly high absorption intensities of the so-called contact-charge-transfer complexes, Murrell considered also the interaction with locally excited states.[45] The necessity of taking these into account was emphasized by the theoretical description of the excimers ($A = D$), where neither the pure charge-resonance picture

$$\Psi^{\pm} = 2^{-\frac{1}{2}}\{\Psi(A^-D^+) \pm \Psi(A^+D^-)\} \tag{4.17}$$

nor the exciton picture

$$\Psi_e^{\pm} = 2^{-\frac{1}{2}}\{\Psi(AD^*) \pm \Psi(A^*D)\} \tag{4.18}$$

alone is satisfactory. It is only after including the interaction between the states corresponding to (4.17) and (4.18) that a full account of the properties of excimers can be given.[37,40,41]

4.2.2 *Resonance interactions*

4.2.2.1 *Introduction*

The molecular complexes which are considered here are composed of electron donors and acceptors, which, in turn, are aromatic hydrocarbons,

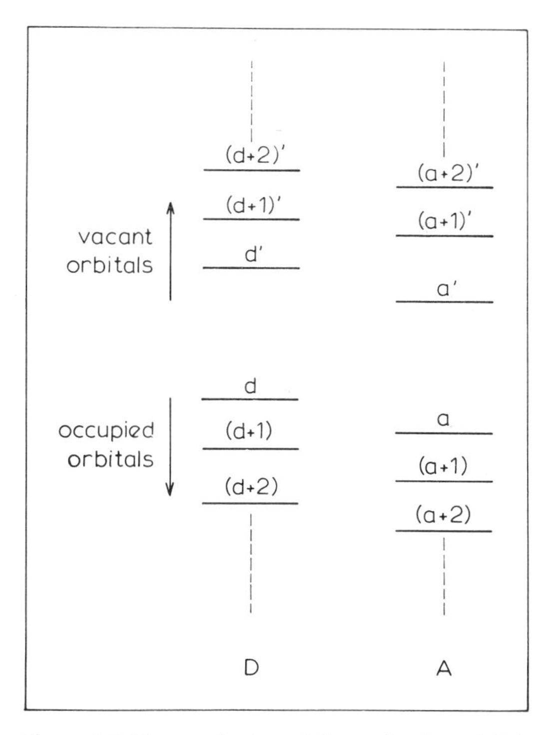

Figure 4.3 The numbering of the molecular orbitals

possibly substituted with donor groups like $-NR_2$ and $-OR$, or acceptor substituents like $-CN$ and $-COOR$, respectively, with R corresponding to alkyl.

If the molecular orbitals of the free components, of which only the π-orbitals are considered, are numbered according to Figure 4.3, the wavefunction of the CT state (A^-D^+) can be denoted as $[d, a']$, expressing the fact that one electron has been transferred from d, the highest occupied orbital of D, to a', the lowest vacant orbital of A. It is assumed that A and D are chosen in such a way that (A^-D^+) is the energetically lowest CT state of the complex; in view of equations (4.34) and (4.56) this corresponds to

$$(IP + EA)_D \leqslant (IP + EA)_A \tag{4.19}$$

It is further assumed that this state is relatively low-lying, energetically comparable to the locally excited states of the components. As the influence of the solvent will be discussed later (§4.3), the complex will be considered to be in vacuum.

4.2.2.2 *Zero-order wavefunctions*

In zero-order (in the absence of intermolecular overlap) the wavefunction of the CT state (A^-D^+) can be formulated as

$$^{1,3}\psi(A^-D^+) = {}^{1,3}[d, a'] = 2^{-\frac{1}{2}}\{|\ldots a\bar{a}d\bar{a}'| \pm |\ldots a\bar{a}a'\bar{d}|\} \qquad (4.20)$$

in which the upper and lower sign refer to the singlet and triplet state ($S_z = 0$), respectively.

Analogously,

$$^{1,3}\psi(A^+D^-) = {}^{1,3}[a, d'] = 2^{-\frac{1}{2}}\{|\ldots d\bar{d}a\bar{d}'| \pm |\ldots d\bar{d}d'\bar{a}|\} \qquad (4.21)$$

The ground state (AD) is represented as

$$^{1}\psi(AD) = |\ldots a\bar{a}d\bar{d}| \qquad (4.22)$$

The wavefunctions of the lowest excited singlet and triplet states of substituted and unsubstituted alternant aromatic hydrocarbons often can be approximated (e.g. for D) by

$$^{1,3}\psi(D_p^*) = {}^{1,3}[d, d'] \qquad (4.23)$$

$$^{1}\psi(D_\alpha^*) = {}^{1}\lambda \,.\, {}^{1}[d+1, d'] - (1 - {}^{1}\lambda^2)^{\frac{1}{2}} \,.\, {}^{1}[d, (d+1)'] \qquad (4.24a)$$

$$^{3}\psi(D_\beta^*) = {}^{3}\lambda \,.\, {}^{3}[d+1, d'] - (1 - {}^{3}\lambda^2)^{\frac{1}{2}} \,.\, {}^{3}[d, (d+1)'] \qquad (4.24b)$$

where p, α and β refer to the classification by Clar;[44] p and α correspond to ^1L_a and ^1L_b, respectively, in the nomenclature of Platt.[43] In equations (4.24) λ depends on the nature of the substituents and on the position at which the substituent is introduced; in the case of no substituent $\lambda = 2^{-1/2}$.

Thus the complex locally singly excited states are given by, e.g.,

$$^{1,3}\psi(AD_p^*) = 2^{-\frac{1}{2}}\{|\ldots a\bar{a}d\bar{d}'| \pm |\ldots a\bar{a}d'\bar{d}|\} \qquad (4.25)$$

$$^{1,3}\psi(A_p^*D) = 2^{-\frac{1}{2}}\{|\ldots d\bar{d}a\bar{a}'| \pm |\ldots d\bar{d}a'\bar{a}|\} \qquad (4.26)$$

$$^{1}\psi(A_\alpha^*D) = \lambda \,.\, 2^{-\frac{1}{2}}\{|\ldots d\bar{d}(a+1)\bar{a}'| + |\ldots d\bar{d}a'\overline{(a+1)}|\} - 2^{-\frac{1}{2}}(1-\lambda^2)^{\frac{1}{2}}\{|\ldots d\bar{d}a\overline{(a+1)'}| + |\ldots d\bar{d}(a+1)'\bar{a}|\} \qquad (4.27)$$

The energetically lowest doubly excited states are $^{1,3}(^3A_p^{*3}D_p^*)$, the wavefunctions of which are given by

$$^{1}\psi(^3A_p^{*3}D_p^*) = (12)^{-\frac{1}{2}}\{|\ldots d'\bar{d}a'\bar{a}| - |\ldots d'\bar{d}a\bar{a}'| - 2|\ldots \bar{d}\bar{d}'aa'| - 2|\ldots dd'\bar{a}\bar{a}'| - |\ldots d\bar{d}'a'\bar{a}| + |\ldots d\bar{d}'a\bar{a}'|\} \qquad (4.28)$$

$$^{3}\psi(^3A_p^{*3}D_p^*) = 2^{-\frac{1}{2}}\{|\ldots \bar{d}\bar{d}'aa'| - |\ldots dd'\bar{a}\bar{a}'|\} \qquad (4.29)$$

The wavefunctions of excited CT states are, for example, given by

$$^{1,3}\psi(A^-D^{*+}) = {}^{1,3}[d+1, a'] = 2^{-\frac{1}{2}}\{|\ldots a\bar{a}d\bar{d}(d+1)\bar{a}'| \pm |\ldots a\bar{a}d\bar{d}a'\overline{(d+1)}|\} \quad (4.30)$$

4.2.2.3 *Zero-order energies*

The zero-order energies can be expressed in terms of the matrix elements of the F-operator[46,47,48] as

$$E_0^{1,3}(A^-D^+) = F(a', a') - F(d, d) + \langle da'|da'\rangle \pm \langle da'|da'\rangle - \langle dd|a'a'\rangle \quad (4.31)$$

if the zero-order energy of the ground state is taken as the energy-zero and

$$F(p, q) = \langle p|\mathscr{H}_c|q\rangle + \sum_n \{2\langle pq|(a+n)(a+n)\rangle - \langle (a+n)p|q(a+n)\rangle\} + \sum_m \{2\langle pq|(d+m)(d+m)\rangle - \langle (d+m)p|q(d+m)\rangle\} \quad (4.32)$$

in which $\mathscr{H}_c$ is the one-electron core Hamiltonian operator. Here $n = 0, 1, \ldots, (N-1)$ and $m = 0, 1, \ldots, (M-1)$, where N and M are the number of occupied orbitals of A and D, respectively. In zero-order there is no overlap between the orbitals of A and D, so, in order to be consistent, integrals which involve electron densities between A and D have to be omitted. Equation (4.31), therefore, has to be written as

$$E_0^{1,3}(A^-D^+) = F(a', a') - F(d, d) - \langle dd|a'a'\rangle \quad (4.33)$$

which is equivalent to (4.3):[48]

$$E_0^{1,3}(A^-D^+) = IP_D - EA_A - C \quad (4.34)$$

The other wavefunctions, of course, also give the familiar expressions, e.g.

$$E_0^{1,3}(A_p^*D) = F(a', a') - F(a, a) + \langle aa'|aa'\rangle \pm \langle aa'|aa'\rangle - \langle aa|a'a'\rangle \quad (4.35)$$

4.2.2.4 *Zero-order interactions*

In the absence of overlap there is no interaction between (A^-D^+) and locally excited states and the ground state.

However, there are non-zero interaction elements with other CT states, *viz.*, CT states such as $^{1,3}[d, (a+1)'] = {}^{1,3}\psi(A^{*-}D^+)$, with charge-transfer in the same direction as in (A^-D^+),[49] e.g.

$$\langle {}^{1,3}\psi(A^-D^+)|\mathscr{H}|{}^{1,3}\psi(A^{*-}D^+)\rangle = -\langle dd|a'(a+1)'\rangle \quad (4.36)$$

and also between the locally excited states, e.g.

$$\langle {}^{1,3}\psi(\mathrm{A}_p^*\mathrm{D})|\mathscr{H}|{}^{1,3}\psi(\mathrm{AD}_p^*)\rangle = \langle dd'|aa'\rangle \pm \langle dd'|aa'\rangle \tag{4.37}$$

These interactions are presumably small relative to the interactions which are introduced by the intermolecular overlap: equation (4.36) corresponds to an interaction between a charge density (d, d), localized on D, and a transition charge density $(a', (a+1)')$ localized on A, which, owing to the orthogonality of a' and $(a+1)'$, consists of an equal amount of positive and negative contributions, so that the interaction will be quite small. The interactions according to (4.37) are higher. In the dipole–dipole approximation these integrals are given by

$$\langle dd'|aa'\rangle = \frac{1}{2}\frac{\mathbf{M}_\mathrm{D}\cdot\mathbf{M}_\mathrm{A}}{R^3} \tag{4.38}$$

in which $\mathbf{M}_\mathrm{A}$ and $\mathbf{M}_\mathrm{D}$ are the transition dipole moments of the p-transitions, while R is the interplanar distance. It seems that this expression overestimates[37,40] the interaction energy at actual values of R (3·0–3·5 Å): for A = D = pyrene (4.38) gives about 0·3 eV at 3·0 Å (maximum value for $\mathbf{M}_\mathrm{A}$ and $\mathbf{M}_\mathrm{D}$ parallel), whereas 0·05–0·10 eV is probably a more realistic value.[37,40]

4.2.2.5 *First-order energies*

As a result of the overlap between the π-orbitals of D and A the energies of the states are changed. The expressions for the first-order energies are rather complicated, but in order to see in which direction the energies are changed, it is sufficient[40] to inspect the diagonal-elements of the overlap matrix. It is easily derived that, to second order in the overlap, for the ground state and the CT state

$$\langle {}^1\psi(\mathrm{AD})|{}^1\psi(\mathrm{AD})\rangle = 1 - 2\sum_n\sum_m S^2(d+m, a+n) \tag{4.39}$$

and

$$\begin{aligned}\langle {}^{1,3}\psi(\mathrm{A}^-\mathrm{D}^+)|{}^{1,3}\psi(\mathrm{A}^-\mathrm{D}^+)\rangle = \langle {}^1\psi(\mathrm{AD})|{}^1\psi(\mathrm{AD})\rangle &+ \sum_n S^2(d, a+n)\\ &- \sum_m S^2(a', d+m)\\ &+ S^2(d, a') \pm S^2(d, a')\end{aligned} \tag{4.40}$$

where

$$S(p, q) = \langle p|q\rangle \tag{4.41}$$

is the intermolecular overlap integral between the orbitals p and q.

From (4.39) it follows that the ground state is destabilized by the overlap of $M \cdot N$ pairs of occupied orbitals (exchange repulsion energy).

From the first sum on the right-hand side of (4.40) it follows that as a result of the overlap between d and the occupied orbitals of A there is an increase of the electron density between A and D, so that $^{1,3}(A^-D^+)$ is stabilized relative to $^1(AD)$; in the same way the second sum corresponds to a destabilization. The last two terms indicate that the singlet CT state is stabilized relative to the triplet CT state, so that in first order the triplet CT state is *above* the singlet CT state.

It should be noted here that this result is different from the conclusions drawn from an MO-description: when complex orbitals are used, then, in first order the triplet state is *lower* than the singlet state, the splitting being given by twice the exchange integral (cf. equation (4.31), in which d and a' now have to be looked at as orbitals of the whole complex). This contradiction already disappears in second order (cf. §4.2.2.8).

The intermolecular overlap, S, between the π-orbitals of the components which are 3·0–3·5 Å apart, is rather small, in any case small enough to neglect S^2 relative to unity. This implies that the zero-order wavefunctions can still be used in the presence of overlap.

4.2.2.6 *Interactions in the presence of overlap*

In order to consider the interactions between the CT states and other complex states it is convenient to inspect the off-diagonal terms of the overlap matrix: the Hamiltonian matrix elements are to a good approximation proportional to these overlap matrix elements. Those which involve p-states (L_a-states) are given in Table 4.3, correct to first order in the overlap. Elements with other states are entirely analogous.

From the table it can be concluded that:

(i) The interaction between the triplet CT state and other states is analogous to (and, as will be shown later, of the same order of magnitude as) the corresponding interactions in the singlet manifold, with the exception that the triplet CT state does not interact with the ground state.

(ii) The interaction between $^1(A^-D^+)$ and (AD), which is partly responsible for the stabilization of the ground state, is governed by another overlap than is the interaction with doubly excited states, so that taking into account the interaction between (A^-D^+) and (AD) does not necessarily imply that the interactions with doubly excited states also have to be taken into account.

The Hamiltonian matrix elements between the various states in the case of the orbitals of A and D being mutually orthogonal can be obtained

Table 4.3 Overlap elements between the lowest CT states (A^-D^+) and (A^+D^-) and some *p*-states (the other elements between states occurring in the table are zero)

	$^1(AD)$	$^1(AD_p^*)$	$^1(A_p^*D)$	$^1(^3A_p^*\,^3D_p^*)$	$^1(A_p^*D_p^*)$
$^1(A^-D^+)$	$\sqrt{2}\langle d\vert a'\rangle$	$\langle d'\vert a'\rangle$	$-\langle d\vert a\rangle$	$-\frac{1}{2}\sqrt{6}\langle d'\vert a\rangle$	$-\langle d'\vert a\rangle$
$^1(A^+D^-)$	$\sqrt{2}\langle d'\vert a\rangle$	$-\langle d\vert a\rangle$	$\langle d'\vert a'\rangle$	$-\frac{1}{2}\sqrt{6}\langle d\vert a'\rangle$	$-\langle d\vert a'\rangle$
		$^3(AD_p^*)$	$^3(A_p^*D)$	$^3(^3A_p^*\,^3D_p^*)$	
$^3(A^-D^+)$		$\langle d'\vert a'\rangle$	$-\langle d\vert a\rangle$	$-\langle d'\vert a\rangle$	
$^3(A^+D^-)$		$-\langle d\vert a\rangle$	$\langle d'\vert a'\rangle$	$\langle d\vert a'\rangle$	

in the usual way,[50] e.g.

$$\langle {}^{1,3}(\mathrm{A}^-\mathrm{D}^+)|\mathscr{H}|{}^{1,3}(\mathrm{A}_p^*\mathrm{D})\rangle = -F(a,d) + \langle aa'|da'\rangle \pm \langle aa'|da'\rangle - \langle ad|a'a'\rangle \quad (4.42)$$

$$\langle {}^{1,3}(\mathrm{A}^-\mathrm{D}^+)|\mathscr{H}|{}^{1,3}(\mathrm{A}\mathrm{D}_p^*)\rangle = F(a',d') + \langle dd'|da'\rangle \pm \langle dd'|da'\rangle - \langle dd|a'd'\rangle \quad (4.43)$$

$$\langle {}^{1}(\mathrm{A}^-\mathrm{D}^+)|\mathscr{H}|{}^{1}(\mathrm{AD})\rangle = \sqrt{2}F(a',d) \quad (4.44)$$

Because the orbitals of D and A are not orthogonal, these equations have to be corrected. The corrections are rather complicated expressions, however. As the overlap is rather small, terms which involve more than one intermolecular overlap may safely be neglected. By the reduction of the many-electron integrals according to the usual methods,[50] the corrections which have to be added to the right-hand sides of (4.42)–(4.44), when the energy of the ground state is taken as the energy-zero,† are found to be

$$-S(a,d)[E_0^{1,3}(\mathrm{A}_p^*\mathrm{D}) - F(d,d) + \langle dd|aa\rangle - \langle dd|a'a'\rangle] \quad (4.42\mathrm{a})$$

$$-S(a',d')F(d,d) \quad (4.43\mathrm{a})$$

$$-\sqrt{2}S(d,a')F(d,d) \quad (4.44\mathrm{a})$$

Corrections for other Hamiltonian matrix elements, which depend on the first power of the overlap, are entirely analogous.

In order to be able to make some numerical estimates it is necessary to find an approximation for terms like $\langle p|\mathscr{H}|q\rangle$ as they appear in (4.32) and therefore also in (4.42–4.44a). To this purpose use is made of the approximations by Konijnenberg[37] and McGlynn.[41]

The Hamiltonian operator can be expressed as

$$\mathscr{H} = \sum_{\nu} \mathscr{H}_\mathrm{c}(\nu) + \sum_{\nu<\mu} \frac{e_0^2}{r_{\mu,\nu}} \quad (4.45)$$

in which the one-electron core Hamiltonian operator, $\mathscr{H}_\mathrm{c}$, is given by

$$\mathscr{H}_\mathrm{c} = T + U_\mathrm{A} + U_\mathrm{D} \quad (4.46)$$

the kinetic operator, T, plus the potential energy operators U_A and U_D for the electron in the field of the cores of A and D deprived of their π-electrons. If U_A^* and U_D^* are the potential fields of the neutral cores and

† Equations (4.42a)–(4.44a) are, in fact, a little more complicated. Instead of $\langle\psi_i|\mathscr{H}|\psi_j\rangle$ expressions are given for $\langle\psi_i|\mathscr{H}|\psi_j\rangle - E_0\langle\psi_i|\psi_j\rangle$, where E_0 is the energy of the first-order ground state. Since this energy is taken as the energy-zero, terms which involve E_0 are omitted, however.

$\mathscr{H}_c^*$ is the analogue of $\mathscr{H}_c$ in (4.46), any intermolecular matrix element $\langle p|\mathscr{H}_c|q\rangle$ can be written as

$$\langle p|\mathscr{H}_c|q\rangle = \langle p|\mathscr{H}_c^*|q\rangle - \sum_n 2\langle pq|(a+n)(a+n)\rangle - \sum_m 2\langle pq|(d+m)(d+m)\rangle \tag{4.47}$$

The MO's of A and D are written as LCAO-MO's, e.g.

$$(a+n) = \sum_i a_i^n \phi_i^{\mathrm{A}} \quad \text{and} \quad (d+m) = \sum_j d_j^m \phi_j^{\mathrm{D}} \tag{4.48}$$

whereas the potential fields U_{A}^* and U_{D}^* are split up as

$$U_{\mathrm{A}}^* = \sum_i U_{\mathrm{A},i}^* \quad \text{and} \quad U_{\mathrm{D}}^* = \sum_j U_{\mathrm{D},j}^* \tag{4.49}$$

Applying the Mulliken approximation[51] to the Coulomb penetration integrals, matrix elements such as $\langle (a+n)|\mathscr{H}_c^*|(d+m)\rangle$ can be approximated by

$$\begin{aligned}\langle (a+n)|\mathscr{H}_c^*|(d+m)\rangle &= \sum_i \sum_j a_i^n d_j^m \langle \phi_i^{\mathrm{A}}|T + U_{\mathrm{A}}^* + U_{\mathrm{D}}^*|\phi_j^{\mathrm{D}}\rangle \\ &\approx \sum_i \sum_j a_i^n d_j^m \langle \phi_i^{\mathrm{A}}|\phi_j^{\mathrm{D}}\rangle [\langle \phi_j^{\mathrm{D}}|T + U_{\mathrm{D},j}^*|\phi_j^{\mathrm{D}}\rangle + \tfrac{1}{2}\langle \phi_j^{\mathrm{D}}|U_{\mathrm{D},j+1}^*|\phi_j^{\mathrm{D}}\rangle \\ &\quad + \tfrac{1}{2}\langle \phi_j^{\mathrm{D}}|U_{\mathrm{D},j-1}^*|\phi_j^{\mathrm{D}}\rangle + \tfrac{1}{2}\langle \phi_i^{\mathrm{A}}|U_{\mathrm{A},i}^*|\phi_i^{\mathrm{A}}\rangle + \tfrac{1}{2}\langle \phi_i^{\mathrm{A}}|U_{\mathrm{A},i+1}^*|\phi_i^{\mathrm{A}}\rangle \\ &\quad + \tfrac{1}{2}\langle \phi_i^{\mathrm{A}}|U_{\mathrm{A},i-1}^*|\phi_i^{\mathrm{A}}\rangle].\end{aligned} \tag{4.50}$$

In expression (4.50) terms of the type $\langle \phi_j^{\mathrm{D}}|T + U_{\mathrm{D},j}^*|\phi_j^{\mathrm{D}}\rangle$ appear, which is the negative of the electron affinity of carbon in its valence state ($-\mathrm{A_C}$), and terms of the type $(k|ii) = \langle \phi_i^{\mathrm{A}}|U_{\mathrm{A},k}^*|\phi_i^{\mathrm{A}}\rangle$, Coulomb penetration integrals. These latter quantities have high values only for $i = k$; they are much smaller for $i = k \pm 1$, and are negligible, and therefore omitted in (4.50), for other values of k. Integrals such as $\langle \phi_i^{\mathrm{A}}|U_{\mathrm{D},k}^*|\phi_i^{\mathrm{A}}\rangle$ are negligible, too.

Within this approximation it does not seem unreasonable to write

$$\langle (a+n)|\mathscr{H}_c^*|(d+m)\rangle \approx -\gamma \,.\, S(a+n, d+m) \tag{4.51a}$$

with

$$\gamma = \mathrm{A_C} - \tfrac{1}{2}(i|ii) - 2(ii|(i+1)(i+1)) \tag{4.51b}$$

This implies that one single hetero-atom occurring in A or D is treated as a carbon atom and, in addition, that every carbon atom in A or D is considered as having two neighbour carbon atoms (intramolecular).

It is useful to define quantities like $F^{\mathrm{D}}(p, q)$ which are expressions analogous to (4.32) except that they are defined for isolated molecules D and A, e.g.

$$F^{\mathrm{D}}(p, q) = \langle p|T + U_{\mathrm{D}}|q\rangle + \sum_{m} [2\langle pq|(d + m)(d + m)\rangle - \langle (d + m)p|(d + m)q\rangle] \quad (4.52)$$

Within the approximations used,

$$F^{\mathrm{D}}(d, d) \approx F(d, d) \quad (4.53)$$

If the other quantities are approximated as follows

$$\langle dd|dd\rangle \approx \langle dd|d'd'\rangle = J_{\mathrm{D}} \quad (4.54)$$

$$\langle aa|aa\rangle \approx \langle aa|a'a'\rangle = J_{\mathrm{A}} \quad (4.55)$$

$$\langle dd|a'a'\rangle \approx \langle d'd'|aa\rangle \approx \langle dd|aa\rangle = C \quad (4.56)$$

$$F^{\mathrm{D}}(d, d) \approx -IP_{\mathrm{D}} \quad \text{and} \quad F^{\mathrm{A}}(a, a) \approx -IP_{\mathrm{A}} \quad (4.57)$$ [46,50]

When the Mulliken approximation[51] for the two-electron integrals is applied, the following expressions for the Hamiltonian matrix elements are found:

$$\langle {}^{1,3}(\mathrm{A^-D^+})|\mathscr{H}|{}^{1,3}(\mathrm{A}_p^*\mathrm{D})\rangle \approx -S(d, a)[\Gamma_{\mathrm{D}} + E_0^{1,3}(\mathrm{A}_p^*\mathrm{D})] \quad (4.58)$$

$$\langle {}^{1,3}(\mathrm{A^-D^+})|\mathscr{H}|{}^{1,3}(\mathrm{AD}_p^*)\rangle \approx S(d', a') . \Gamma_{\mathrm{D}} \quad (4.59)$$

$$\langle {}^{1}(\mathrm{A^-D^+})|\mathscr{H}|{}^{1}(\mathrm{AD})\rangle \approx \sqrt{2}S(d, a') . \Gamma_{\mathrm{D}} \quad (4.60)$$

$$\langle {}^{1}(\mathrm{A^-D^+})|\mathscr{H}|{}^{1}({}^3\mathrm{A}_p^{*3}\mathrm{D}_p^*)\rangle \approx -\tfrac{1}{2}\sqrt{6}S(d', a)[\Gamma_{\mathrm{D}} + E_0^1(\mathrm{A}_p^*\mathrm{D})] \quad (4.61)$$

$$\langle {}^{3}(\mathrm{A^-D^+})|\mathscr{H}|{}^{3}({}^3\mathrm{A}_p^{*3}\mathrm{D}_p^*)\rangle \approx -S(d', a)[\Gamma_{\mathrm{D}} + E_0^3(\mathrm{A}_p\mathrm{D})] \quad (4.62)$$

The matrix elements with $(\mathrm{A^+D^-})$ are entirely analogous, e.g.

$$\langle {}^{1,3}(\mathrm{A^+D^-})|\mathscr{H}|{}^{1,3}(\mathrm{A}_p^*\mathrm{D})\rangle \approx S(d', a') . \Gamma_{\mathrm{A}} \quad (4.63)$$

Interaction elements with other states are, e.g.,

$$\langle {}^{1}(\mathrm{A^-D^+})|\mathscr{H}|{}^{1}(\mathrm{A}_\alpha^*\mathrm{D})\rangle \approx -{}^{1}\lambda S(d, a + 1)[\Gamma_{\mathrm{D}} + \tfrac{1}{2}(E_0^1(\mathrm{A}_\alpha^*\mathrm{D}) + E_0^1(\mathrm{A}_\beta^*\mathrm{D}))] \quad (4.64)$$

$$\langle {}^{3}(\mathrm{A^-D^+})|\mathscr{H}|{}^{3}(\mathrm{A}_\beta^*\mathrm{D})\rangle \approx -{}^{3}\lambda S(d, a + 1)[\Gamma_{\mathrm{D}} + \tfrac{1}{2}(E_0^3(\mathrm{A}_\alpha^*\mathrm{D}) + E_0^3(\mathrm{A}_\beta^*\mathrm{D}))] \quad (4.65)$$

In these equations Γ_{A} and Γ_{D} are given by

$$\Gamma_{\mathrm{D}} = -\gamma + IP_{\mathrm{D}} - \tfrac{1}{2}J_{\mathrm{D}} - \tfrac{1}{2}C \quad (4.66)$$

$$\Gamma_{\mathrm{A}} = -\gamma + IP_{\mathrm{A}} - \tfrac{1}{2}J_{\mathrm{A}} - \tfrac{1}{2}C \quad (4.67)$$

The Hamiltonian matrix elements in (4.58)–(4.65) thus appear to be proportional to the corresponding overlap matrix elements. This is the basis of discussing these interactions in terms of overlap integrals. Finally, some of the most important interaction elements which (in this approximation) do not depend on the intermolecular overlap are given:

$$\langle {}^{1,3}(\mathrm{A^-D^+})|\mathscr{H}|{}^{1,3}(\mathrm{A^+D^-})\rangle \approx 0 \tag{4.68}$$

$$\langle {}^{1,3}(\mathrm{A}_p^*\mathrm{D})|\mathscr{H}|{}^{1,3}(\mathrm{AD}_p^*)\rangle \approx \langle dd'|aa'\rangle \pm \langle dd'|aa'\rangle \tag{4.69}$$

(cf. (4.37)).

With the aid of analytical expressions given in the literature for the Coulomb penetration integrals in (4.51b) and the experimental value of A_C, γ is readily calculated.[41] With a value of 3·18–3·25 for the effective nuclear charge of carbon, γ is found to be 16 eV. The factor Γ then becomes about -12 eV.

It may now be concluded that, with the orbital overlap integrals equal, the interaction element of $^1(\mathrm{A^-D^+})$ with the ground state (AD) has the highest value. Next comes the interaction with the $p(=L_a)$-states and finally that with the $\alpha(=L_b)$-states. In addition it is observed that the interaction elements in the triplet manifold are at least equal to those in the singlet manifold (except for (4.69)).

In this place some further remarks about intermolecular overlap should be made. When special symmetry relations between the orbitals are absent it can be stated that:

(i) This overlap will strongly depend on the matching of donor and acceptor. Perfect matching, of course, can occur for the excimer with equivalent atoms opposite. In the case of A and D being rather different in size, the overlap integral is not expected to be high.

(ii) The more modal planes there are in the pertinent orbitals, the higher will be the probability that positive and negative contributions cancel and the smaller the overlap. So, generally, antibonding orbitals will give lower overlap than bonding orbitals.

The numerical value of the overlap also depends on the choice of the AO's. Murrell[40] used Slater-orbitals with an exponent of 1·4 instead of the 'normal' value of 1·59–1·625; the object of this was to obtain values similar to the $p\sigma$–$p\sigma$-overlap of SCF-carbon-orbitals. On the other hand, the choice of the interplanar distance also affects the magnitude of the overlap.

The values chosen here for Slater-exponent and interplanar distance, 1·625 and 3·0 Å, respectively, are of course, somewhat arbitrary, and should be justified by the results of the calculations.

4.2.2.7 *The resulting wavefunctions and the transition dipole moment*

In view of the results obtained, the wavefunctions for the complex ground state, ${}^1\Psi_g$, and for the excited state, ${}^{1,3}\Psi_a$, can be written as

$$ {}^1\Psi_g = \sum_i b_i \psi_i(\text{loc}) + \sum_j c_j \psi_j(\text{CT}) \tag{4.70} $$

$$ {}^{1,3}\Psi_a = \sum_v r_v \psi_v(\text{loc}) + \sum_w s_w \psi_w(\text{CT}) \tag{4.71} $$

in which the $\psi_j(\text{CT})$ correspond to the CT states of the complex, including those with $A^{\pm}$ or $D^{\pm}$ excited ($\psi_0(\text{CT}) = \psi(A^-D^+)$), while the $\psi_i(\text{loc})$ characterize the locally excited states, including the ground state (AD) ($\psi_0(\text{loc}) = \psi(\text{AD})$). The coefficients b_i, c_j, r_v and s_w depend on the magnitude of the Hamiltonian matrix elements as well as on the energy separations between the states; they can be obtained with the variation principle, or, when the energy differences are sufficiently high, with first-order perturbation theory.

The probability of a transition between $|\Psi_g\rangle$ and $|\Psi_a\rangle$ is determined by the value of the transition dipole moment, $\mathbf{M}$, which is given by

$$ \mathbf{M} = \langle \Psi_a | \boldsymbol{\mu}_{op} | \Psi_g \rangle = \sum_{i,v} b_i r_v \mathbf{M}_{i,v}(1,1) + \sum_{j,w} c_j s_w \mathbf{M}_{j,w}(c,c) + \sum_{i,w} b_i s_w \mathbf{M}_{i,w}(1,c) + \sum_{j,v} c_j r_v \mathbf{M}_{j,v}(c,1) \tag{4.72} $$

where $\boldsymbol{\mu}_{op}$ is the electric dipole moment operator. The different types of contributions to $\mathbf{M}$ are given by:

(i) $M_{i,v}(1,1)$. For $i = 0$, e.g.

$$ \mathbf{M}_{0,v}(1,1) = \langle {}^1\psi(\text{A*D}) | \boldsymbol{\mu}_{op} | {}^1\psi(\text{AD}) \rangle = \langle {}^1\psi(\text{A*}) | \boldsymbol{\mu}_{op} | {}^1\psi(\text{A}) \rangle \tag{4.73} $$

which is the dipole moment of the local transition ${}^1\text{A*} \leftarrow {}^1\text{A}$. These dipole moments can have values as high as 5 Debye (${}^1L_a$-transitions). For $i = v = 0$:

$$ \mathbf{M}_{0,0}(1,1) = \langle {}^1\psi(\text{AD}) | \boldsymbol{\mu}_{op} | {}^1\psi(\text{AD}) \rangle \tag{4.74} $$

which is the expectation value of the dipole moment in the ground state. This is often negligible compared to other dipole moments discussed in this section.

(ii) $M_{j,w}(c,c)$. This integral is negligible for $j \neq w$, but has a high value for $j = w$: $\mathbf{M}_{j,j}(c,c)$ is the expectation value, $\boldsymbol{\mu}_0$, of the dipole moment in the state $|\psi_j(\text{CT})\rangle$, which can be up to 15 Debye.

(iii) $M_{i,w}(1, c)$ and $M_{j,v}(c, 1)$. For $i = w = 0$ or $j = v = 0$:

$$\mathbf{M}_{0,0}(1, c) = \langle \psi(\mathrm{AD}) | \boldsymbol{\mu}_{\mathrm{op}} | \psi(\mathrm{A^-D^+}) \rangle \tag{4.75a}$$

which in the Mulliken-approximation becomes:[1a]

$$\mathbf{M}_{0,0}(1, c) = \tfrac{1}{2}\sqrt{2}S(d, a')\{\langle \psi(\mathrm{AD}) | \boldsymbol{\mu}_{\mathrm{op}} | \psi(\mathrm{AD}) \rangle + \langle \psi(\mathrm{A^-D^+}) | \boldsymbol{\mu}_{\mathrm{op}} | \psi(\mathrm{A^-D^+}) \rangle\} \tag{4.75b}$$

Since the first term in (4.75b) is small compared to the second one, the polarization due to this contribution is in the direction of the charge-transfer (perpendicular to the aromatic planes). Murrell, however, showed that (4.75b) does not give the correct limit for $S = 0$: when the orbitals d and a' do overlap, but $S = 0$ because of symmetry, the transition dipole moment, in fact, can still be finite;[45] in that case the polarization of the transition is perpendicular to the direction of the charge-transfer. However, its absolute value is quite small. Also, as will be shown below, the contibution due to (4.75) is generally unimportant.

The overlap between the orbitals d and a' gives rise to a finite transition moment $M_{0,0}(1, c)$. On the other hand it induces the interaction between $(\mathrm{A^-D^+})$ and (AD) and so contributes to the transition moment via $M_{j,j}(c, c)$. According to equation (4.72) $M_{0,0}(1, c)$ and $M_{j,j}(c, c)$ have still to be multiplied with the corresponding coefficients.

Now, often $b_0 \approx s_0 \approx 1$. From perturbation theory

$$|c_0| \approx |r_0| \approx (\langle \psi(\mathrm{AD}) | \mathscr{H} | \psi(\mathrm{A^-D^+}) \rangle - E \,.\, \langle \psi(\mathrm{AD}) | \psi(\mathrm{A^-D^+}) \rangle)/E(\mathrm{A^-D^+}) \tag{4.76}$$

from which, with $E(\mathrm{A^-D^+}) = 3$ eV and (4.60) it follows that

$$|c_0| \approx |r_0| \approx 7S(d, a') \tag{4.77}$$

The two contributions to **M** are

$$\text{type (ii):} \qquad c_0 s_0 M_{0,0}(c, c) \approx 105S(d, a') \text{ (Debye)} \tag{4.78}$$

$$\text{type (iii):} \qquad b_0 s_0 M_{0,0}(1, c) \approx 11S(d, a') \text{ (Debye)} \tag{4.79}$$

It is seen that the contribution of type (iii) is small compared to that of type (ii).

So, in general, the emissive ability of the exciplex will be due to only contributions by $M_{0,v}(1, 1)$, with $v \neq 0$, and by $M_{j,j}(c, c)$. As the former gives a polarization in the aromatic plane while the latter is perpendicular to it, polarization studies can lead to experimental verification of these considerations. Some corroborating experimental evidence has already been given (cf. §4.1.2).

4.2.2.8 *The triplet charge-transfer state*

By the method used in §4.2.2.6, retaining terms to order S^2, however, (when necessary also in the normalization factors of the wavefunctions), the first-order singlet–triplet splitting, Δ_{ST}, for the CT state (A^-D^+) is found to be

$$\begin{aligned}\Delta_{ST} = E_1^1(A^-D^+) - E_1^3(A^-D^+) \approx 2[\langle da'|da'\rangle \\ + 2S(d,a')\{F(d,a') - \langle da'|dd\rangle\} \\ - S^2(d,a')\{2F(d,d) + E_0(A^-D^+) - \langle dd|dd\rangle\}]\end{aligned} \tag{4.80}$$

when the energy of the ground state (AD) is taken as the energy-zero. Equation (4.80) can be written as

$$\Delta_{ST} \approx -(4\gamma - 4IP_D + 2E_0(A^-D^+) + \tfrac{3}{2}J_D - \tfrac{1}{2}J_A + 3C) \,.\, S^2(d,a') \tag{4.81}$$

or, after substituting $E_0(A^-D^+) = 3$ eV (cf. §4.2.2.1) and reasonable values for the other quantities (cf. §4.2.3)

$$\Delta_{ST} \approx -53S^2(d,a')\,(\text{eV}) \tag{4.82}$$

So, in first order, as was noticed already in §4.2.2.5, the overlap between d and a' destabilizes the triplet CT state relative to the singlet state.

On the other hand this overlap is also responsible for the interaction between $^1(A^-D^+)$ and (AD), which destabilizes the *singlet* CT state (the triplet state is not affected) by an amount Δ'_{ST}, which according to perturbation theory is given by

$$\Delta'_{ST} = \frac{(\langle\psi(AD)|\mathscr{H}|\psi(A^-D^+)\rangle - E\,.\,\langle\psi(AD)|\psi(A^-D^+)\rangle)^2}{E(A^-D^+) - E(AD)} \tag{4.83}$$

With $E_0(A^-D^+) = 3$ eV, (4.83) and (4.60) lead to

$$\Delta'_{ST} \approx 180\, S^2(d,a')(\text{eV}) \tag{4.84}$$

Comparison of (4.82) and (4.84) shows that the second-order interaction is the most important, so that finally the triplet CT state is again below the singlet state. The doubly excited states probably have a comparable (stabilizing) effect on both the triplet and the singlet CT state.

Of course, the interaction with singly locally excited states could again change the sequence. However, when the CT state is the lowest excited state of the complex, the final triplet CT state is below the singlet state. This is, for example, the case for the system D = diethylaniline and A = *p*-dicyanobenzene (cf. reference 11 and §4.1.3).

4.2.3 *Applications*

In this section the preceding results will be used to make some numerical estimates of the contributions of the various complex states to the emitting state wavefunction of a number of actual complexes.

First some remarks have to be made with respect to the complex geometry. Whereas for the excimers of aromatic hydrocarbons, such as naphthalene and pyrene, it is easy to make a plausible guess as to the orientation of the two components (*viz.*, it is likely that the carbon atoms of A are opposite to the equivalent ones of D, since this makes the Coulomb term in (4.34) as well as the important overlap integrals maximum), this is not so in general if A ≠ D. The situation is even more difficult in the latter case, because configurations which correspond to a maximum Coulomb energy are not necessarily the same as those which lead to a maximum stabilization by means of resonance interaction. Of course, a full calculation would supply the energetically most favourable configuration, but as yet such calculations are not sufficiently reliable. Moreover, the van der Waals attraction energy and the repulsion energy due to intermolecular overlap are very difficult to estimate quantitatively.

The following calculations, which make use of as many empirical data as possible, are therefore meant to give no more than the order of magnitude of the interactions.

In succession the following types of complexes will be considered:

(i) the excimer,

(ii) the charge-transfer hetero-excimer,

(iii) the mixed excimer,

(iv) the EDA complex.

4.2.3.1 *The excimer*

The overlap integrals between the important orbital-pairs for the pyrene excimer in its most symmetrical configuration and with an interplanar distance of 3·0 Å are given in Table 4.4. These integrals have been calculated according to

$$S(a+n, d+m) = \sum_i \sum_j a_i^n d_j^m \langle \phi_i^A | \phi_j^D \rangle \tag{4.85}$$

where ϕ_i^A and ϕ_j^D are the $(2p_z\text{-})$AO's of the two components; they are assumed to be Slater orbitals with exponent 1·625. The AO-overlap integrals are taken from the table in reference 52. The coefficients a_i^n and d_j^m are approximated with the Hückel method.[53]

Table 4.4 Intermolecular overlap integrals for some A, D systems

Donor	Acceptor	Configuration[a]	$S(d, a+1)$	$S(d, a)$	$S(d, a')$	$S(d', a)$	$S(d', a')$	$S((d+1)', a')$	C(eV)[b]
Pyrene	Pyrene	(iii)	0	0·059	0	0	0·025	0	3·29[41]
Aniline[d]	Anthracene	(i)[c] see text	0	0·014	0·013		0·010	0	3·39
Aniline[d]	Anthracene	(ii) Figure 4.5B	0	0	0·010	0	0·012	0	3·59
Aniline[d]	Pyrene	(i) Figure 4.5A	0	+[e]	0	+	0	+	3·40
Aniline[d]	Pyrene	(ii) Figure 4.5A	+	0	0	0	0	+	3·43
(a) Naphthalene	α-Cyano-naphthalene[f]	(iii)	0	0·046	0·010	0·002	0·023	0·001	
(b) Naphthalene	Naphthalene	(iii)	0	0·056	0	0	0·030		
(c) α-Amino-naphthalene[g]	Naphthalene	(iii)		0·038			0·020		

[a] (i) Long axes of A and D perpendicular; (ii) long axes of A and D parallel; (iii) equivalent atoms of A and D opposite.
[b] Coulomb energy between A^- and D^+ at 3·0 Å interplanar separation.
[c] This configuration does not correspond to an energy minimum.
[d] $h_N = 1{\cdot}2$, $k_{CN} = 1{\cdot}0$.
[e] + indicates non-zero overlap.
[f] $h_N = 0{\cdot}5$, $k_{CN} = 1{\cdot}0$.
[g] $h_N = 1{\cdot}0$, $k_{CN} = 1{\cdot}0$.

It follows from Tables 4.3 and 4.4 that interaction occurs between the CT states (A^-D^+), (A^+D^-) and the L_a-states (A_p^*D) and (AD_p^*).

The interaction elements are estimated as follows: With help of the data: $IP = 7{\cdot}53$ eV;[2] $EA = 0{\cdot}59$ eV;[55] $E(^1A_p^*) = 3{\cdot}72$ eV;[54] $E(^3A_p^*) = 2{\cdot}09$ eV[54] and $C = 3{\cdot}29$ eV,[41] the following quantities are calculated:

$$J = 4{\cdot}85 \text{ eV (from (4.35), (4.53), (4.54) and (4.57))};^{54}$$

$$E(A^-D^+) = E(A^+D^-) = 3{\cdot}65 \text{ eV (from (4.34))};$$

$$\Gamma = -12{\cdot}5 \text{ eV (from (4.66))}.$$

With (4.58) and (4.59) one obtains:

$$\langle ^{1,3}(A^-D^+)|\mathscr{H}|^{1,3}(AD_p^*)\rangle = \langle ^{1,3}(A^+D^-)|\mathscr{H}|^{1,3}(A_p^*D)\rangle \approx -0{\cdot}31 \text{ eV}$$

$$\langle ^{1,3}(A^-D^+)|\mathscr{H}|^{1,3}(A_p^*D)\rangle = \langle ^{1,3}(A^+D^-)|\mathscr{H}|^{1,3}(AD_p^*)\rangle$$

$$\approx 0{\cdot}53 \text{ eV (singlet)}$$

$$\approx 0{\cdot}62 \text{ eV (triplet)}$$

Assuming a value for $\langle aa'|dd'\rangle$ in (4.69) of 0·05 eV (cf. §4.2.2.4) the energies of the lowest excited states are found with the variation principle, by solving the secular equations

$$|\mathscr{H} - E\,.\,\mathscr{S}| = 0 \qquad (4.86)$$

The excited states are found at

$$E\,(\text{singlet}) \approx 2{\cdot}58 \text{ eV} \quad \text{and} \quad E\,(\text{triplet}) \approx 1{\cdot}55 \text{ eV},$$

while experimentally the emissions are found at

$$E\,(\text{singlet}) = 2{\cdot}60 \text{ eV}^{7} \quad \text{and} \quad E\,(\text{triplet}) = 1{\cdot}61 \text{ eV},^{25}$$

The good agreement, of course, is also caused by a fortunate choice of the interplanar distance and the exponent for the Slater orbital.

The coefficients which appear in the wavefunctions of the excited states are evaluated as usual.[50] The wavefunctions are found to be

$$^1\Psi_a = 0{\cdot}50\,\{^1\psi(A_p^*D) - {}^1\psi(AD_p^*)\} - 0{\cdot}50\,\{^1\psi(A^-D^+) - {}^1\psi(A^+D^-)\} \qquad (4.87)$$

$$^3\Psi_a = 0{\cdot}63\,\{^3\psi(A_p^*D) - {}^3\psi(AD_p^*)\} - 0{\cdot}33\,\{^3\psi(A^-D^+) - {}^3\psi(A^+D^-)\} \qquad (4.88)$$

For the singlet state the transition dipole moment with the ground state is (cf. (4.71))

$$\mathbf{M} = 0{\cdot}50\,(\mathbf{M}_A - \mathbf{M}_D) \qquad (4.89)$$

where $\mathbf{M}_A$ and $\mathbf{M}_D$ are the dipole moments of the ^1L_a-transitions of A and D. It is easily verified that, with the signs chosen for the overlap in Table 4.4, $\mathbf{M}_A$ and $\mathbf{M}_D$ point in the same direction. For the symmetrical dimer, therefore, $M = 0$. In order to explain the relatively short lifetime of pyrene excimer emission (70 ns), Konijnenberg, therefore, assumed that the emission originates from a slightly distorted species.[37]

4.2.3.2 *The charge-transfer hetero-excimer*

When the size of A and D is rather different, the intermolecular overlap cannot reach high values. This means that the contribution of the CT state (A^-D^+) to the exciplex wavefunction will be very high, and in that case the complex equilibrium configuration will be one in which the energy of the CT state is minimum, i.e. with the Coulomb energy, C, maximum.

For this reason C has been calculated for all possible configurations (interplanar distance kept fixed at 3·00 Å; aromatic planes parallel) of the complexes between A = anthracene, pyrene, biphenyl, naphthalene and D = aniline. Thereby the point-charge approximation has been used, the charges on the nuclei having been obtained from the Hückel

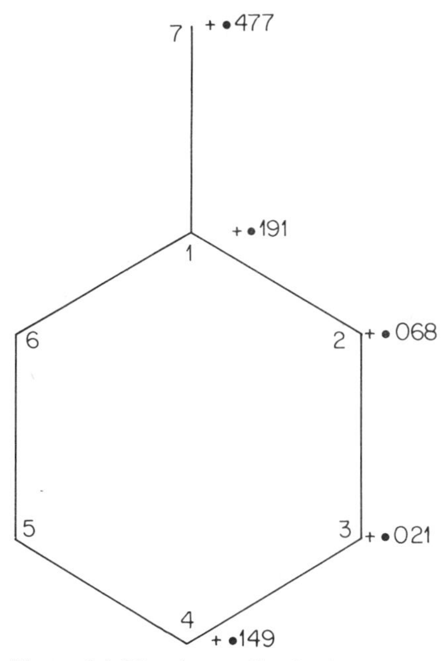

Figure 4.4 The charge distribution of the aniline radical cation

coefficients, taking $h_N = 1{\cdot}2$ and $k_{CN} = 1{\cdot}0$ for aniline;[42] the charge distribution obtained for the aniline positive ion is shown in Figure 4.4.

The configurations for which C has been found to be maximum are given in Figure 4.5. For pyrene-aniline two different energy minima for the zero-order CT state were found in this way, the energy difference between the two configurations being only 0·03 eV.

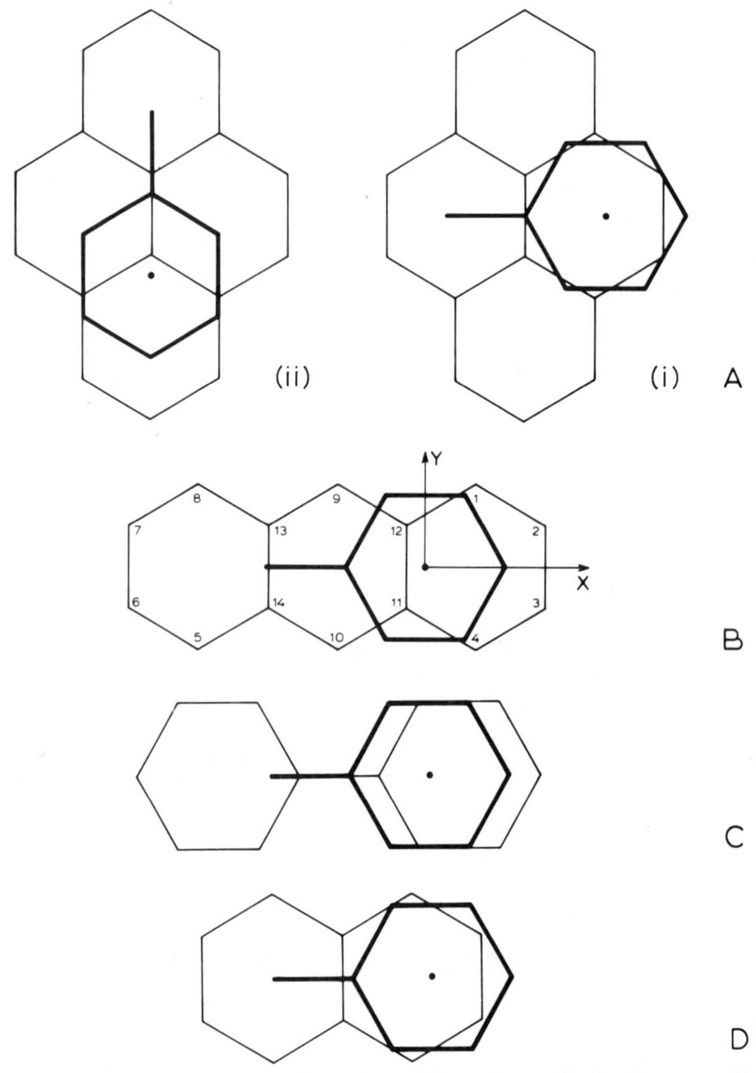

Figure 4.5 Hetero-excimer configurations with maximum Coulomb energy. (The dot indicates the centre of the aniline hexagon)

These configurations will correspond to the geometry of the exciplex only, when the interactions with other states are small. For the complexes of Figure 4.5 this is probably the case, since the matching is poor (small overlap) and the energy difference between the CT states and the other states rather high.

(*a*) *The anthracene* (*A*)-*diethylaniline* (*D*) *exciplex.* Some experimental data are given in Table 4.5. It is seen that the transition dipole moment is about 1 Debye. This is also the case for the system with A = pyrene. As the contribution according to equation (4.78) gives maximally 100 % of the emission probability, the following inequality holds:

$$105\, S(d, a') \lesssim 1 \text{ (Debye)} \tag{4.90}$$

This implies that the overlap integral between d and a' must be smaller than ~0·01, so that (with (4.77)) $|r_0|$ and $|c_0|$ are smaller than 0·07. The corresponding interaction energy is calculated with perturbation theory (4.83), which gives an energy smaller than 0·02 eV. Thus, even in the case where the transition probability is entirely due to the interaction between (A^-D^+) and (AD), the corresponding energy changes are so small that they are not of experimental interest.

Now the interactions as they follow from inspection of the MO's will be discussed. The pertinent Hückel orbitals are shown in Figure 4.6.

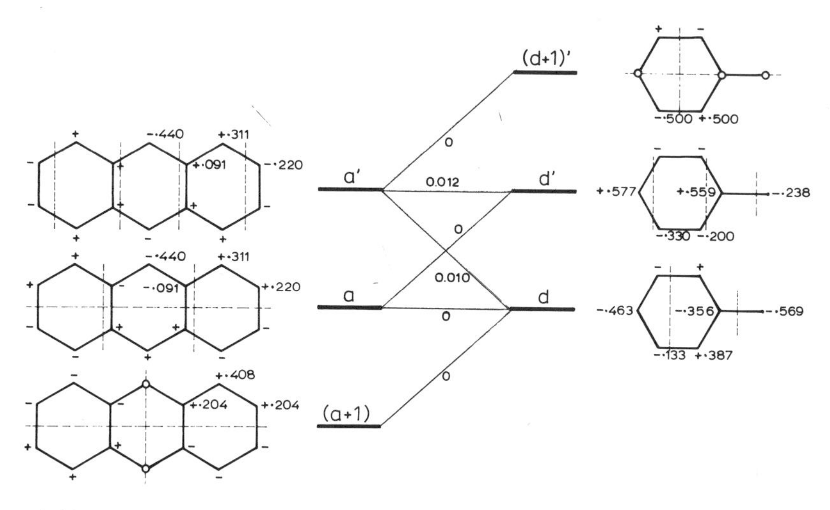

Figure 4.6 Hückel orbitals pertinent to the anthracene-aniline hetero-excimer. Intermolecular overlap integrals are indicated (middle of the figure) for the configuration of Figure 4.5B

Table 4.5 Values pertinent to the anthracene–diethylaniline exciplex

State	$^1A^*_p$	$^1D^*_p$	$^3A^*_p$	$^3D^*_p$	$^1A^*_\alpha$	$^1D^*_\alpha$	Exciplex
0,0-transition energy (eV)	3·28	4·45	1·82	3·33	(3·48)	3·84	2·67[a]
Transition dipole moment (Debye)	2·9	4·5				1·4	$1{\cdot}1 \pm 0{\cdot}1$[b]

[a] Emission maximum; extrapolated to vacuum with (4.142).
[b] Value in *n*-hexane, calculated from the lifetime of the emission,[32] assuming a quantum yield of $0{\cdot}8 \pm 0{\cdot}2$.

Their overlap integrals have been calculated for two different configurations, (i) one chosen naively because of an expected high overlap (carbon atoms 1 and 4 of aniline opposite to 9 and 10 of anthracene) and (ii) the one corresponding with a maximum Coulomb energy. It should be noted that the former configuration does not correspond to a (local) maximum of C (Table 4.4).

Comparison of Tables 4.3 and 4.4 shows that for (i) the interaction is between the states (A^-D^+), (A_p^*D), (AD_p^*) and (A^+D^-), while interaction with the ground state also occurs. A calculation analogous to the one in §4.2.3.1 shows that as a result of this interaction the CT state (A^-D^+), which, undoubtedly, gives by far the most important contribution to the exciplex fluorescent state, is stabilized by less than 0·05 eV. This is much smaller than the difference in Coulomb energy with configuration (ii). The conclusion, therefore, is that configuration (ii) most likely corresponds to that of the actual (singlet) exciplex.

Figure 4.7 shows the changes in Coulomb energy of configuration (ii) (cf. Figure 4.5B) for small changes in θ (θ is the angle between the X-axis and the line joining carbon atoms 1 and 4 of aniline) for small displacements in the X-direction and for small displacements in the Y-direction. Although these curves show that a distortion in the X- or Y-direction is accompanied by relatively small changes in C, the gain in overlap, and thus in resonance stabilization, is also small.

The Hamiltonian matrix elements for the complex in configuration (ii) are estimated as before, making use of the data in Tables 4.4 and 4.5 and taking $J_D = 6$ eV and $IP_D = 7$ eV:

$$\langle {}^{1,3}(A^-D^+)|\mathscr{H}|{}^{1,3}(AD_p^*)\rangle = -0{\cdot}15\ \text{eV}$$

$$\langle {}^{1}(A^-D^+)|\mathscr{H}|{}^{1}(AD)\rangle = -0{\cdot}17\ \text{eV}$$

The other interactions are negligible. With perturbation theory it follows that the exciplex wavefunction is given by

$$ {}^1\Psi_a = 0{\cdot}99\ {}^1\psi(A^-D^+) - 0{\cdot}10\ {}^1\psi(AD_p^*) - 0{\cdot}08\ {}^1\psi(AD) \qquad (4.91)$$

while the corresponding ground-state wavefunction is given by

$$ {}^1\Psi_g = 1{\cdot}00\ {}^1\psi(AD) + 0{\cdot}08\ {}^1\psi(A^-D^+) \qquad (4.92)$$

The transition dipole moment follows from (4.71) as

$$\mathbf{M} = -0{\cdot}10\ \mathbf{M}(1, 1) + 0{\cdot}08\ \mathbf{M}(c, c) \qquad (4.93)$$

so that, since both contributions are perpendicular, $|\mathbf{M}| = 1{\cdot}2$ Debye, which is in agreement with the experimental value (Table 4.5). The total

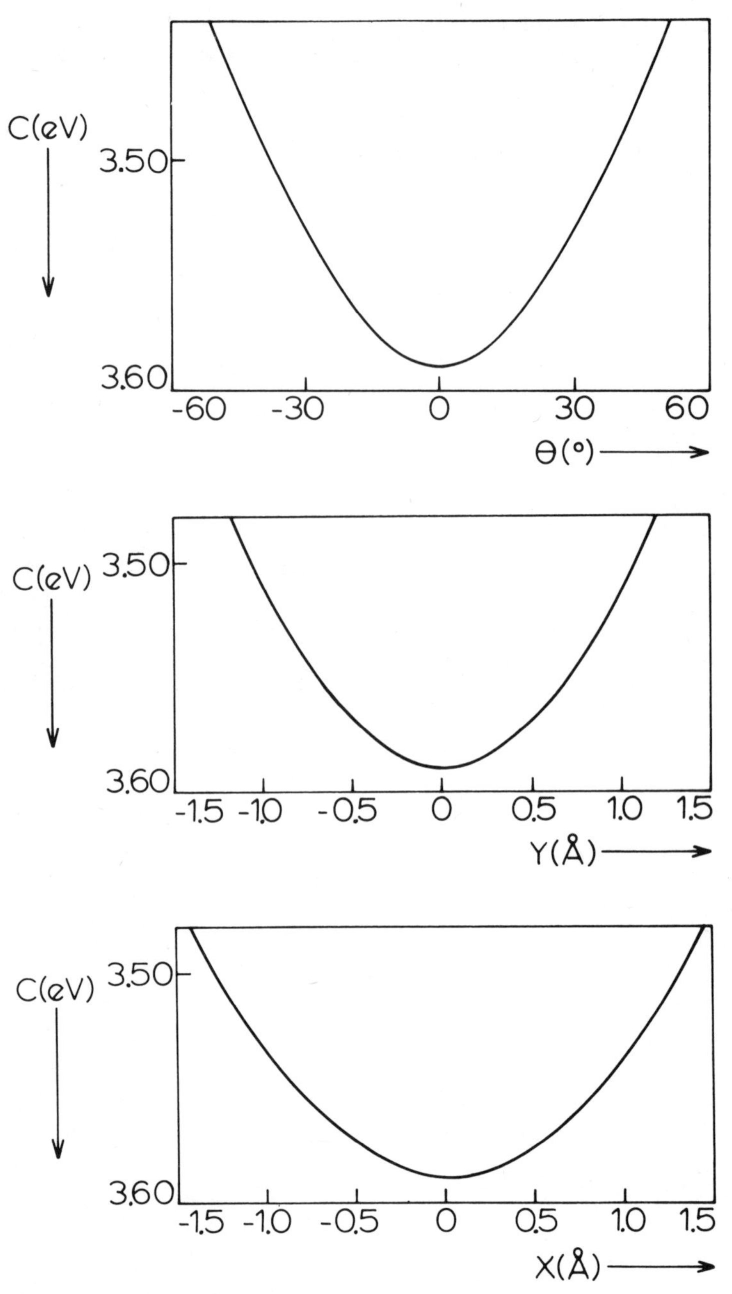

Figure 4.7 The Coulomb energy for the anthracene-aniline heteroexcimer as a function of X, Y and θ

resonance stabilization energy of the emitting state is calculated to be less than one hundredth of an electron-volt.

(*b*) *The pyrene* (*A*)-*aniline* (*D*) *system.* The overlap integrals for the two configurations of Figure 4.5 are indicated in Table 4.4; some experimental data are given in Table 4.6. In configuration (i) (A^-D^+) interacts with (A_p^*D) and (AD_α^*); in (ii) with (A_α^*D) and (AD_α^*). Because the overlap integrals are small again, and as the α ($= {}^1L_b$)-transitions have small dipole moments, while those of *p*-states are high, it may be assumed that, though perhaps different exciplexes (exciplex isomers) may exist in this case, the complex corresponding to configuration (i) is actually observed in emission. The exciplex transition dipole moment probably originates almost exclusively from the pyrene ${}^1L_a$-state; since, when the corresponding orbital overlap integrals are equal, the ${}^1L_b$-state ($\lambda < 1$) gives rise to a smaller interaction element than the ${}^1L_a$-state (cf. (4.64)).

(*c*) *A = naphthalene and biphenyl; D = aniline.* Proceeding as before it is found that for A = naphthalene, (A^-D^+) interacts predominantly with (AD_α^*) and (AD_p^*) (not with (AD)); for A = biphenyl with (A_p^*D) and (AD).

Although the estimations used here are rather crude, they probably give the right order of magnitude, as for example is indicated by the results for pyrene excimer. In addition, they illustrate the important difference between the excimer and the polar exciplex: the former being characterized by relatively high values of the overlap integrals (and, therefore, high values of the interaction elements) and small energy separations between the interacting states, the latter showing small overlaps and relatively large energy gaps.

Furthermore it is seen that, while from an energetic point of view the interaction between (A^-D^+) and (AD) is always negligible, this is not so with respect to the transition dipole moment of the exciplex. For anthracene-diethylaniline, for example, it is this interaction which appears to be predominantly responsible for the radiative ability.

4.2.3.3 *The mixed excimer*

A typical example is the complex between naphthalene (D) and α-cyanonaphthalene (A) (complex (a) Table 4.4).

It is to be expected that the cyano-group, being an acceptor substituent, will affect primarily the anti-bonding orbitals of the hydrocarbon. This implies that EA_A and $E({}^1A_p^*D)$ are lowered by about an equal amount, while $IP_A \approx IP_D$. While for the excimers of alternant hydrocarbons (cf. §4.2.3.1)

$$E(A^-D^+) = E(A^+D^-) \approx E^1(A_p^*D) = E^1(AD_p^*) \qquad (4.94)$$

Table 4.6 Values pertinent to the pyrene–dimethylaniline exciplex

State	$^1A_p^*$	$^1D_p^*$	$^3A_p^*$	$^3D_p^*$	$^1A_\alpha^*$	$^1D_\alpha^*$	Exciplex
0,0-transition energy (eV)	3·70	4·45	2·09	3·33	3·34	3·84	2·83[a]
Transition dipole moment (Debye)	5·0	4.5			0·7	1·4	0·97[b,19a]

[a] Emission maximum; extrapolated to vacuum with (4.142).
[b] Value in *n*-hexane.

(so that the CT states and the locally excited structures become about equally mixed), the relations for this exciplex will be

$$(E_1 =)\, E^1(A^-D^+) \approx E^1(A_p^*D) \quad \text{and} \quad (E_2 =)\, E^1(A^+D^-) \approx E^1(AD_p^*) \tag{4.95}$$

The intermolecular overlap integrals for this exciplex and the naphthalene excimer (complex (b)) have been calculated as before (Table 4.4), assuming that the configurations of both complexes are similar: equivalent carbon atoms of D and A opposite each other.

It appears that for complex (a) $S(d, a)$ and $S(d', a')$ are both smaller than for the excimer (complex (b)), while $S(d, a')$ and $S(d', a)$ are both greater. Interaction with ${}^1L_b$-states (via $S((d + 1)', a')$ or $S(d, a + 1))$ is negligible in both cases.

In order to calculate the contributions to the complex excited singlet state wavefunction, the following linear combinations are constructed:

$$\Phi_1^{\pm} = 2^{-\frac{1}{2}}\{\psi(A^-D^+) \pm \psi(A_p^*D)\} \tag{4.96}$$

$$\Phi_2^{\pm} = 2^{-\frac{1}{2}}\{\psi(A^+D^-) \pm \psi(AD_p^*)\} \tag{4.97}$$

The Hamiltonian matrix elements between these functions are

$$\langle\Phi_1^{\pm}|\mathscr{H}|\Phi_1^{\pm}\rangle = E_1 \pm \alpha_1 \quad \text{and} \quad \langle\Phi_1^{\pm}|\mathscr{H}|\Phi_1^{\mp}\rangle = 0 \tag{4.98}$$

$$\langle\Phi_1^{\pm}|\mathscr{H}|\Phi_2^{\pm}\rangle = \tfrac{1}{2}\beta \pm \alpha_2 \tag{4.99}$$

$$\langle\Phi_1^{\pm}|\mathscr{H}|\Phi_2^{\mp}\rangle = -\tfrac{1}{2}\beta \tag{4.100}$$

$$\langle\Phi_2^{\pm}|\mathscr{H}|\Phi_2^{\pm}\rangle = E_2 \pm \alpha_1 \quad \text{and} \quad \langle\Phi_2^{\pm}|\mathscr{H}|\Phi_2^{\mp}\rangle = 0 \tag{4.101}$$

in which β is the interaction element corresponding to (4.69), α_1 is that of (4.58), while α_2 corresponds to equation (4.63); Γ_A and Γ_D are taken to be equal ($-12{\cdot}7$ eV).

Since β is small (§4.2.2.4), the mixing of the states with different sign is neglected (equation (4.100)). α_1 and α_2 are calculated as before; they are given in Table 4.8, together with the values used for E_1 and E_2.

With the variation principle the following wavefunctions and energies are found:

Complex (a)

$$\Psi_a = 0{\cdot}59\, \psi(A^-D^+) - 0{\cdot}38\, \psi(A^+D^-) - 0{\cdot}59\, \psi(A_p^*D) + 0{\cdot}38\, \psi(AD_p^*)$$

$$E = 3{\cdot}10 \text{ eV} \tag{4.102}$$

Table 4.7 Energies (in eV) pertinent to the exciplexes of §4.2.3.3

Donor	Acceptor	$E(^1A_p^*)$	$E(^1D_p^*)$	$h\nu_e^{max}$ [a]
(a) Naphthalene	α-Cyanonaphthalene	3·94	4·33	3·09
(b) Naphthalene	Naphthalene	4·33	4·33	3·11
(c) α-Dimethylaminonaphthalene	Naphthalene	4.33	3·7[58]	—[b]

[a] Emission maximum in n-hexane.
[b] Exciplex not stable.

Complex (b)

$$\Psi_a = 0{\cdot}50\,\psi(A^-D^+) - 0{\cdot}50\,\psi(A^+D^-) - 0{\cdot}50\,\psi(A_p^*D) + 0{\cdot}50\,\psi(AD_p^*)$$

$$E = 3{\cdot}15\text{ eV} \tag{4.103}$$

The calculated energies are in good agreement with the experimental ones (Table 4.7). The values for the excited-state dipole moments follow from

$$\boldsymbol{\mu} = \langle \Psi_a | \boldsymbol{\mu}_{op} | \Psi_a \rangle \tag{4.104}$$

as:

$$\text{complex (a)}: \ \mu = 0{\cdot}21\,\mu_0 \quad \text{and} \quad \text{complex (b)}: \ \mu = 0$$

where μ_0 is the dipole moment in the state (A^-D^+). The experimental values are 0·5 μ_0 and 0, respectively.[15] The discrepancy in the former case is to be expected, not only because of the approximations in (4.95), but also in view of the fact that the present calculation is for the gas phase. As will be shown in §4.3, the solvent is able to increase the polar character of the exciplex.

An interesting feature of the wavefunction (4.102) is that as distinct from (4.103), the coefficients denoting the contributions of the locally excited states are no longer equal. As a result the transition dipole moment will have a finite contribution from these states. With $|\mathbf{M}_A|$ and $|\mathbf{M}_D|$ about 3·5 Debye, the contribution is found to be 0·7 Debye.

The other part comes from the mixing with the ground state, which has been omitted up to now. Taking the ground state into account one can write the excited-state wavefunction as

$$\Psi'_a = \Psi_a + r_0\psi(AD) \tag{4.105}$$

From perturbation theory $r_0 = -0{\cdot}07$ is obtained. An equal amount of (A^-D^+) is mixed into the ground state: $c_0 = 0{\cdot}07$. The contribution of

(A^+D^-) to the ground state is negligibly small ($S(d', a)$ is small; Table 4.4). Analogous to (4.78) the contribution to M is found to be $c_0 s_0 M_{0,0}(c, c) \approx 1$ Debye. Since both contributions are perpendicular: $|\mathbf{M}| = 1{\cdot}2$ Debye. Even in this case the main contribution to the radiative ability of the exciplex comes from the interaction with the ground state.

The complex between A = naphthalene and D = α(dimethylamino)-naphthalene (complex (c)), which is not stable in the excited state[15a] will be considered now. Since one of the components carries a donor substituent, it may be expected that in this case $EA_A \approx EA_D$ and

$$(E_1' =)\, E^1(A^-D^+) \approx E^1(AD_p^*) \quad \text{and} \quad (E_2' =)\, E^1(A^+D^-) \approx E^1(A_p^*D) \tag{4.106}$$

Therefore the following linear combinations are used:

$$\Phi_1'^{\pm} = 2^{-\frac{1}{2}}\{\psi(A^-D^+) \pm \psi(AD_p^*)\} \tag{4.107}$$

$$\Phi_2'^{\pm} = 2^{-\frac{1}{2}}\{\psi(A^+D^-) \pm \psi(A_p^*D)\} \tag{4.108}$$

Some Hamiltonian matrix elements are:

$$\langle \Phi_1'^{\pm} | \mathscr{H} | \Phi_1'^{\pm} \rangle = E_1' \pm \alpha_2 \tag{4.109}$$

$$\langle \Phi_2'^{\pm} | \mathscr{H} | \Phi_2'^{\pm} \rangle = E_2' \pm \alpha_2 \tag{4.110}$$

$$\langle \Phi_1'^{\pm} | \mathscr{H} | \Phi_2'^{\pm} \rangle = \tfrac{1}{2}\beta \pm \alpha_1 \tag{4.111}$$

Proceeding as before, the energy of the lowest excited singlet state is calculated as 3·18 eV.

Table 4.8 Energy values (in eV) used in the calculations of §4.2.3.3

Complex	E_1	E_1'	E_2	E_2'	α_1	α_2	β	Γ
(a)	3·94		4·33		0·39	−0·29	0·10	−12·7
(b)	4·33	4·33	4·33	4·33	0·47	−0·38	0·10	−12·7
(c)		3·7		4·33	0·32	−0·25	0·10	−12·7

The resonance stabilization energy relative to (A^-D^+), $E(A^-D^+) - E$, is for complex (a) 0·84 eV; for (b) 1·18 eV; for (c) 0·52 eV. It is clear that for these complexes the geometry will not be determined by the principle of maximum Coulomb energy, although the configuration which has been assumed (equivalent carbon atoms opposite to each other), presumably giving maximum overlap, does not necessarily deviate much from one with maximum Coulomb energy.

The repulsion energy, δE_{rep} (cf. §4.1.5), for the naphthalene excimer is about 0·50 eV.[62] Assuming a similar value for (c), the enthalpy of complex formation in the excited state is calculated to be −0·02 eV (in this case $^1(AD_p^*)$ is the lowest excited singlet state). This small value is in accordance with the experimental fact that complex (c) is not stable in the excited state.

The main difference between (a) and (c) is, that with (a) the interaction via $S(d, a)$ is the most important one (first-order interaction; cf. (4.98), (4.101) and (4.58)), whereas with (c) the first-order interaction is governed by the overlap between a' and d'. Since, generally, $S(d, a) > S(d', a')$, complexes of type (a) are more likely to be stable than those of type (c).

4.2.3.4 *The EDA complex*

As an example the complex between naphthalene (D) and trinitrobenzene (A) will be discussed. In this case both the donor as well as the acceptor have no permanent dipole moments.

It is assumed that the ground state and the excited state are represented by the wavefunctions (4.15) and (4.16), respectively. From the experimental value of the dipole moment of the complex in the ground state ($\mu_g = 0{\cdot}88$ Debye[2]) it follows with $\mu_0 = 15$ Debye, and

$$\mu_g = c_0^2 \mu_0 \tag{4.112}$$

that $c_0 = -r_0 = 0{\cdot}24$ and $b_0 = s_0 = 0{\cdot}97$. With (4.60) and (4.76), substituting $\Gamma = -12$ eV and $E(A^-D^+) = 2{\cdot}82$ eV (the experimental transition energy[2]), the value of $S(d, a')$ is found to be

$$S(d, a') = 0{\cdot}035.$$

Similarly to (4.78) the CT transition dipole moment is given by

$$\mathbf{M} = c_0 s_0 \mathbf{M}_{0,0}(c, c) = c_0 s_0 \boldsymbol{\mu}_0$$

so that $|\mathbf{M}| = 3{\cdot}5$ Debye (experiment: 3·3 Debye[2]).

The stabilization of the ground state resulting from perturbation theory (equations (4.76) and (4.83)) is $\Delta E = 0{\cdot}16$ eV, whereas the experimentally obtained enthalpy of complex formation is −0·19 eV.[2] The good agreement in the latter case is rather fortuitous, however, because a first-order repulsion energy has not been accounted for.

On the whole, the experimental values seem rather well to be connected by theory in this case also; the value found for the overlap integral also fits well into the picture developed in this chapter.

4.2.4 *Concluding remarks*

Summarizing it can be stated that, with respect to the emissive ability, the exciplex may be considered as the limiting case of the EDA complex:

even in the case of mixed excimers the main contribution to the transition dipole moment comes from the interaction between the CT state and the ground state. With respect to other exciplex properties like energy and dipole moment, however, the interactions with locally excited states, generally, are of more importance.

In addition, the classification of the exciplexes as made on the basis of experimental data (§4.1) emerges clearly from the results of the calculations.

As for the triplet CT state, it is found that the first-order interaction caused by the overlap between the donating (d) and the accepting orbital (a'), which destabilizes the triplet CT state relative to the singlet CT state, is generally exceeded by the second-order interaction, which destabilizes the singlet state. This means that, interactions with locally excited states being absent, the triplet CT state is below the singlet CT state.

Though interactions with locally excited states could change this sequence, there is as yet no single experimental indication that this has ever been realized. Since very small energy differences are involved, it would be difficult to make any statement on the basis of theoretical results in the latter case.

4.3 The role of the solvent

4.3.1 *General considerations*

The formulas, with the aid of which the excited state dipole moments are usually evaluated from the solvent-dependent emission frequencies[14-16] (cf. (4.9)), all assume a solvent-independent dipole moment. However, as is easily recognized, the exciplex wavefunction and, therefore, the electric dipole moment will be a function of the solvent polarity: suppose that the excited state may be considered as a resonance hybrid of the (polar) CT state (A^-D^+) mixed with a small amount of a (non-polar) locally excited state (AD)*, with either A or D excited. The mixing of (AD)* into (A^-D^+) results in an increased electron density between A and D, so that the binding energy increases. At the same time, however, the electric dipole moment of the complex decreases. When the complex is in solution, solvation of the dipole can occur, and as the solvation energy increases with increasing dipole moment (cf. (4.118)), the solvation will oppose this delocalization tendency. As a result the dipole moment will increase with increasing polarity of the solvent.

In addition, there could be an effect possibly allied to this: the solvent-induced change of the complex geometry should also result in a change of the dipole moment and other exciplex properties.

In order to get an idea of the importance of these phenomena an attempt will be made to treat these effects in a more quantitative way, making use of a continuum model for the solvent. As will be shown, it follows from these more quantitative considerations that the above-mentioned experimental dipole moments, though not being constants, retain their significance, with the numerical values to be regarded as average values.

The competition between the delocalization tendency and the solvation has also been recognized by Mataga.[19]

4.3.2 *The solvation of a permanent point-dipole*

When a dipole m is brought from the gas phase into solution, solvation occurs, i.e., if only dipole–dipole interactions are considered, the solvent dipoles orient themselves relative to m, so that as a result the free energy of the total system is decreased.

The orientation of the solvent dipoles creates an electric field, the reaction field,[59] **F**, at the location of the dipole m. In a simple model (point-dipole in the centre of a sphere of radius ρ, surrounded by a homogeneous dielectric with dielectric constant ε) the reaction field at m, which has a dipole moment $\boldsymbol{\mu}$, is given by the relation[59]

$$\mathbf{F} = \boldsymbol{\mu} \, . \, f_\varepsilon \tag{4.113}$$

in which f_ε is

$$f_\varepsilon = \frac{2}{\rho^3} \frac{\varepsilon - 1}{2\varepsilon + 1} \tag{4.114}$$

The energy of a dipole in an electric field **F** is

$$E_m = -\boldsymbol{\mu} \, . \, \mathbf{F} \tag{4.115}$$

The amount of work required to bring the dielectric into its polarized state is[59]

$$E_{\text{pol}} = \tfrac{1}{2}\boldsymbol{\mu} \, . \, \mathbf{F} \tag{4.116}$$

The free solvation energy, E_s, therefore, is given by

$$E_s = E_m + E_{\text{pol}} = -\tfrac{1}{2}\boldsymbol{\mu} \, . \, \mathbf{F} \tag{4.117}$$

which also can be written as (cf. (4.114))

$$E_s = -\frac{\mu^2}{\rho^3} \frac{\varepsilon - 1}{2\varepsilon + 1} \tag{4.118}$$

4.3.3 *The influence of the solvent on the electronic and geometrical structure of a dipolar species*

When a molecule or complex, which in the gas phase is described by the Hamiltonian $\mathscr{H}^0$ with eigenfunctions ψ_i^0, is brought into a solvent, the reaction field, caused by the solvation, acts as a perturbation, which can mix the ψ_i^0s. The perturbation, $\mathscr{H}'$, by virtue of (4.115) is given by

$$\mathscr{H}' = -\boldsymbol{\mu}_{op} \cdot \mathbf{F} \tag{4.119}$$

where $\boldsymbol{\mu}_{op}$ is the electric dipole moment operator, and $\mathbf{F}$ the reaction field. The wavefunctions of the perturbed states are given by

$$\Psi = \sum_i c_i \psi_i^0 \tag{4.120}$$

It is clear that in calculations, the aim of which is to get the final wavefunctions in the solvated state, the value which has to be substituted for $\mathbf{F}$ is the *final* reaction field. As this field depends on $\boldsymbol{\mu}$ (4.113) and as $\boldsymbol{\mu}$ depends on the final wavefunction through

$$\boldsymbol{\mu} = \langle \Psi | \boldsymbol{\mu}_{op} | \Psi \rangle \tag{4.121}$$

it is seen that the problem of calculating Ψ is a self-consistent field problem.

When the molecule or complex may not be considered as rigid, the equilibrium values λ_j^0 of its geometrical parameters λ_j will depend on the polarity of the solvent:

$$\lambda_j^0 = \lambda_j^0(\varepsilon) \tag{4.122}$$

The equilibrium values of λ_j are determined by an additional condition:

$$\left(\frac{\partial E_a(\lambda_j)}{\partial \lambda_j}\right)_{\varepsilon, \lambda_i^0} = 0 \tag{4.123}$$

for all j, where λ_i^0 is the equilibrium value of λ_i, while E_a is the (free) energy of molecule plus solvent. The latter quantity can be expressed as (cf. (4.117))

$$E_a = E + E_{pol} \tag{4.124}$$

with E being

$$E = \langle \Psi(\lambda_j) | \mathscr{H}^0(\lambda_j) + \mathscr{H}'(\lambda_j) | \Psi(\lambda_j) \rangle \tag{4.125}$$

and E_{pol}:

$$E_{pol} = \tfrac{1}{2}\boldsymbol{\mu}(\lambda_j) \cdot \mathbf{F}(\lambda_j) \tag{4.126}$$

In the following these principles are used in order to investigate the influence of the solvent on the absorption and emission characteristics of excited π-complexes. Two borderline cases will be considered:

(i) an exciplex in which there is only one (locally) excited state (AD)* which can interact with the CT state (A^-D^+); and

(ii) an EDA complex which shows only interaction between (A^-D^+) and (AD).

4.3.4 *The influence of the solvent on exciplex properties*

4.3.4.1 *The energy*

The excimer case has been treated elsewhere.[18] On the basis of a simple model it could be explained why even the emission originating from excimers (and other species consisting of two identical and weakly interacting moieties) can show a considerable solvent dependence. Later, more extensive calculations for one particular case, *viz.*, 9,9′-dianthryl, appeared to lead to essentially similar conclusions.[66]

Here another simple case will be investigated: the exciplex in which there is only one locally excited state (AD)* which can interact with the CT state (A^-D^+).

Let the wavefunction of that exciplex in the solvent be

$$\Psi = c_1\psi(A^-D^+) + c_2\psi(AD)^* \tag{4.127}$$

Neglect of intermolecular overlap means that the matrix element

$$\langle\psi(A^-D^+)|\boldsymbol{\mu}_{op}|\psi(AD)^*\rangle = \mathbf{M}_{0,0}(1, c) \tag{4.128}$$

(cf. (4.75)) vanishes. If the dipole moment in the state (AD)* is zero, the final dipole moment, $\boldsymbol{\mu}$, is

$$\boldsymbol{\mu} = c_1^2\langle\psi(A^-D^+)|\boldsymbol{\mu}_{op}|\psi(A^-D^+)\rangle = c_1^2\boldsymbol{\mu}_0 \tag{4.129}$$

with $\boldsymbol{\mu}_0$ being the dipole moment of (A^-D^+) and the normalization condition requiring that

$$c_1^2 + c_2^2 = 1 \tag{4.130}$$

The Hamiltonian matrix elements of the complex in the gas phase are denoted by

$$\langle\psi(A^-D^+)|\mathscr{H}^0|\psi(A^-D^+)\rangle = E_c \tag{4.131}$$

$$\langle\psi(AD)^*|\mathscr{H}^0|\psi(AD)^*\rangle = E_e \tag{4.132}$$

$$\langle\psi(A^-D^+)|\mathscr{H}^0|\psi(AD)^*\rangle = \alpha \tag{4.133}$$

The influence of the solvent is accounted for by the matrix element (cf. (4.113), (4.119) and (4.129))

$$\langle\psi(A^-D^+)|\mathscr{H}'|\psi(A^-D^+)\rangle = -c_1^2\mu_0^2 f_\varepsilon \tag{4.134}$$

In the approximation used here, the role of the solvent consists in changing the energy difference between the zero-order states (A^-D^+) and (AD)*.

Application of the variation principle and subsequent substitution of (4.131)–(4.134) leads to the secular equations

$$c_1(E_c - c_1^2\mu_0^2 f_\varepsilon - E) + c_2\alpha = 0 \tag{4.135a}$$

$$c_1\alpha + c_2(E_e - E) = 0 \tag{4.135b}$$

which can be combined to give

$$\left(E_c - E - \frac{\alpha^2}{E_e - E}\right)\left(1 + \frac{\alpha^2}{(E_e - E)^2}\right) = \mu_0^2 f_\varepsilon \tag{4.136}$$

where E is the energy of the resulting complex states.

The polarization energy, E_{pol}, stored in the solvent is from (4.113), (4.116) and (4.129):

$$E_{pol} = \tfrac{1}{2}c_1^4\mu_0^2 f_\varepsilon \tag{4.137}$$

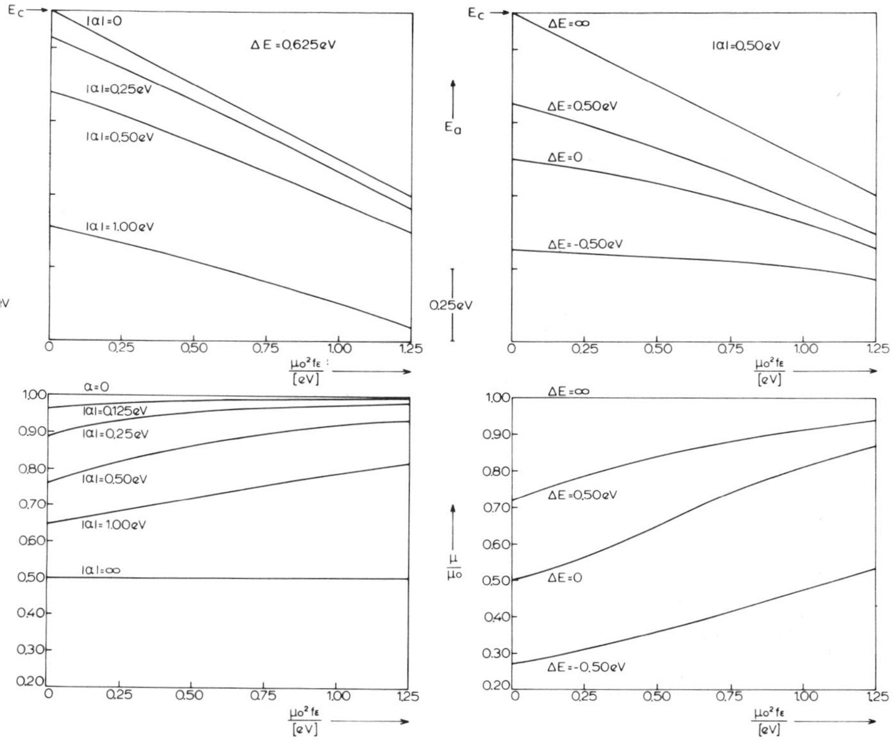

Figure 4.8 The (free) energy, E_a, of a rigid exciplex (upper parts) and the relative dipole moment, μ/μ_0, (lower parts) as a function of $\mu_0^2 f_\varepsilon$ with $\Delta E = E_e - E_c$ and α as parameters

The total (free) energy of complex plus solvent is obtained from (4.126). When E has been calculated, c_1^2 follows from (4.130) and (4.135), which give

$$c_1^2 = \left\{1 + \frac{\alpha^2}{(E_e - E)^2}\right\}^{-1} \tag{4.138}$$

As an illustration, the free energy of a rigid complex, E_a, calculated with the aid of (4.136)–(4.138) and (4.124) is given in Figure 4.8 as a function of $\mu_0^2 f_\varepsilon$ (with $\Delta E = E_e - E_c$ and α as parameters), together with the function μ/μ_0 (from (4.129) and (4.138)).

Since the (experimentally obtained) values of μ_0^2/ρ^3 are about 1 eV at the most[14–16], the value of $\mu_0^2 f_\varepsilon$ for the exciplex in actual solvents runs from hexane ($f = 0{\cdot}19$) to acetonitrile ($f = 0{\cdot}48$) over a range of less than 0·6 eV. It appears that the dipole moment μ approaches the value μ_0 for the pure CT state the closer, the higher $\mu_0^2 f_\varepsilon$ and $E_e - E_c$ and the smaller α^2.

4.3.4.2 *The emission frequency*

After the introduction of the quantities f and f'

$$f = \tfrac{1}{2}\rho^3 f_\varepsilon = \frac{\varepsilon - 1}{2\varepsilon + 1} \qquad f' = \tfrac{1}{2}\rho^3 f_n = \frac{n^2 - 1}{2n^2 + 1} \tag{4.139}$$

the energy of the Franck–Condon (FC) ground state, E_g^{FC}, to which the exciplex transition occurs, can be written[18,21]

$$E_g^{FC} = E_g^0(1 - f'/f)E_{pol} \tag{4.140}$$

where E_{pol} corresponds to solvation of the excited-state dipole moment and E_g^0 is the gas-phase energy of the ground state. The frequency of the exciplex emission is given by

$$h\nu_e = E_a - E_g^{FC} \tag{4.141}$$

As can be easily shown,[21] equations (4.113), (4.124), (4.126) and (4.139)–(4.141), applied to an emitting species with a solvent-independent excited-state dipole moment μ_a, lead to the following expression for the emission frequency

$$h\nu_e = h\nu_e(0) - \frac{2\mu_a^2}{\rho^3}(f - \tfrac{1}{2}f') \tag{4.142}$$

if it may be assumed that the ground-state dipole moment is negligible compared to μ_a.

If the more general case of the exciplex defined by (4.127) is considered, it is found by application of (4.124), (4.129) and (4.137)–(4.141) that[21]

$$\delta(hv_e) = -\frac{2\mu^2}{\rho^3}\delta(f - \tfrac{1}{2}f') - (f - f')\frac{d(\mu^2/\rho^3)}{df}\delta f \qquad (4.143)$$

for small variations of the independent variables f and f', since because of (4.123)

$$\delta E_a = \left(\frac{\partial E_a}{\partial f}\right)_{\lambda_f^{\rho}} \delta f \qquad (4.144)$$

According to (4.143) a plot of hv_e against $(f - \frac{1}{2}f')$ gives a straight line (compare (4.142)) only if μ^2/ρ^3 does not depend on f. This is the case if two conditions are satisfied: (i) c_1^2 does not depend on f, which is the case only if $c_1 = 1$; (ii) μ_0^2/ρ^3 does not depend on f, which holds only if the geometry of the complex is not affected by the solvent.

When these conditions are not fulfilled, the second term in (4.143) corresponds to a scattering in the hv_e *vs.* $(f - \frac{1}{2}f')$ plots (since f and $(f - \frac{1}{2}f')$ for the usual solvents are badly correlated) and the solvent dependence of μ^2/ρ^3 gives rise to a curvature. Because of this scattering, still further enlarged by specific solvent–solute interactions, the curvature corresponding to a variation of μ^2/ρ^3 can never be detected experimentally. The values of μ^2/ρ^3 obtained from the slopes of the straight lines drawn through the experimental points, therefore, have to be looked at as average values.

4.3.4.3 *The emission probability for a rigid complex*

The expression for the transition dipole moment between the excited state and the ground state (AD) is given by

$$\mathbf{M} = \langle \Psi | \boldsymbol{\mu}_{op} | \Psi(AD) \rangle = c_2 \mathbf{M}(1, 1) \qquad (4.145)$$

(cf. §4.2.2.7 and equation (4.127)) with $M(1, 1)$ being the transition dipole moment of the pertinent local transition ($D^* \rightarrow D$ or $A^* \rightarrow A$).

The radiative rate constant, k_f, for spontaneous emission in a solution with refractive index n is[48,57]

$$k_f = \frac{64\pi^4}{3hc^3} \cdot v_e^3 \cdot n^3 \cdot |\mathbf{M}|^2 \qquad (4.146)$$

The equations (4.145) and (4.146) indicate that k_f for the solvated exciplex, characterized by the wavefunction (4.127), decreases when f increases: v_e and c_2^2 become smaller.

As was shown in §4.2, however, the emission probability in case of an actual exciplex mostly originates from the interaction between (A^-D^+) and (AD). For such a case M^2 will increase with increasing solvent polarity (cf. §4.3.5).

4.3.4.4 *The geometrical parameters*

For a complex consisting of two rigid aromatic frames, one of which is held fixed, the most important types of distortion are: (i) variation of the interplanar distance, $\lambda = r$; (ii) rotation around the Z-axis (which is perpendicular to the aromatic planes), $\lambda = \theta$; (iii) displacements in the X- or Y-direction, $\lambda = x, y$. It is assumed throughout that A and D are parallel oriented aromatic planes without solvent molecules between them and that this holds true when the λ_j are changed. At a certain distance, r, solvent molecules, of course, do penetrate between A and D. Then, however, a situation is reached which is more like a solvated ion-pair and in that case the change in the complex corresponds to a chemical conversion rather than to a small change in the exciplex properties.

The conditions which govern the equilibrium values of the geometrical parameters follow from (4.123), (4.124) and (4.135) as

$$\frac{\partial E_c}{\partial \lambda_j} - \frac{1}{E_e - E} \cdot \frac{\partial \alpha^2}{\partial \lambda_j} = c_1^2 f \frac{\partial(\mu_0^2/\rho^3)}{\partial \lambda_j} \qquad (4.147)$$

for all j. In (4.147) the first term indicates the energy change of the pure CT state, the second one the change in resonance stabilization and the right-hand term approximately the change in solvation energy with changing λ_j.

From (4.147) it is clear that when the exciplex is brought from the gas phase ($f = 0$) into a solvent, the change in λ_j^0 will be such that μ^2/ρ^3 increases. This means that the interplanar distance, for example, increases with increasing solvent polarity. Since α^2 decreases with increasing interplanar distance and because for the lowest excited state $E_e - E > 0$, the resonance stabilization, evidently, counteracts the change of the interplanar distance. Although this is only a minor effect, this counteraction is the smaller the more polar the solvent ($E_e - E$ increases with f).

Because the geometrical changes caused by the solvent are maximum in this case, the magnitude of the distortions has been estimated for a complex in the pure CT state (A^-D^+).

(*i*) *Variation of the interplanar distance.* The free energy estimated for the aniline-anthracene exciplex[21] is given in Figure 4.9 as a function of the interplanar distance (dielectric constant of the solvent ε as parameter). It is seen that the equilibrium distance increases by as little as 2%, when

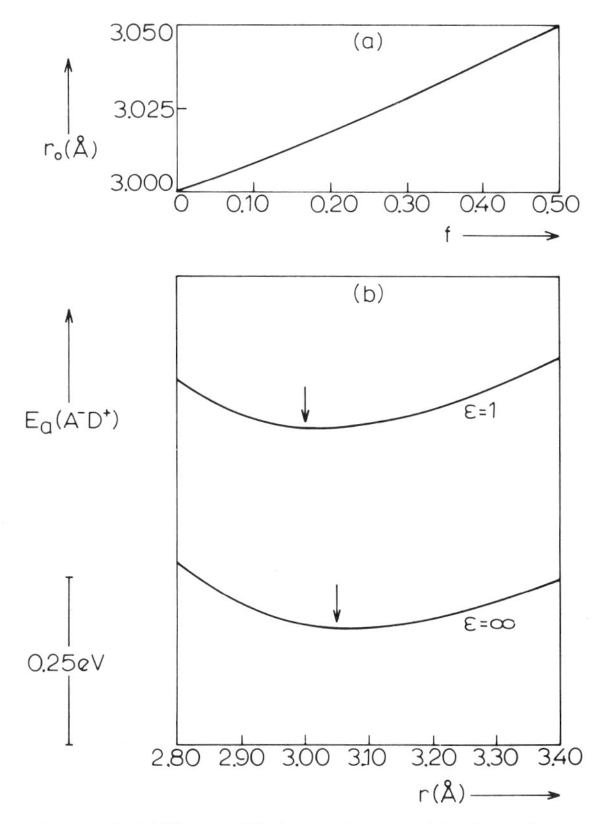

Figure 4.9 (a) The equilibrium value, r_0, of the interplanar distance as a function of the solvent polarity f. (b) The (free) energy of an exciplex in the CT state (A^-D^+) as a function of the interplanar distance, r, for dielectric constant 1 and ∞. Arrows indicate energy minima

ε increases from one to infinity. Furthermore, the (free) energy change caused by this distortion amounts to less than 0·01 eV.

(*ii*) *Rotation around the Z-axis*, $\lambda = \theta$. Since this does not change μ_0^2/ρ^3, the solvent will have no significant effect on θ^0 and the equilibrium value can be determined from

$$\frac{\partial E_c}{\partial \theta} - \frac{1}{E_e - E} \frac{\partial \alpha^2}{\partial \theta} = 0 \tag{4.148}$$

(cf. (4.147)), so that the complex in the lowest excited state tends to adjust to that value of θ^0 which makes the resonance integral $|\alpha|$ maximum.

Here, too, the geometrical change is counteracted by the Coulomb term (cf. Figure 4.7).

(*iii*) *Displacements perpendicular to the Z-axis.* The free energy of the aniline-anthracene complex in the CT state has been calculated as a function of the displacement, x, of one of the components along the X-axis ($x = 0$ for the configuration of Figure 4.5B), by making use of a very simple model.[21]

The results are shown in Figure 4.10. Energy minima appear at $x = 0$ (undistorted complex) and at high x-values ($x > 4$ Å), the latter corresponding to the solvated ion pair.

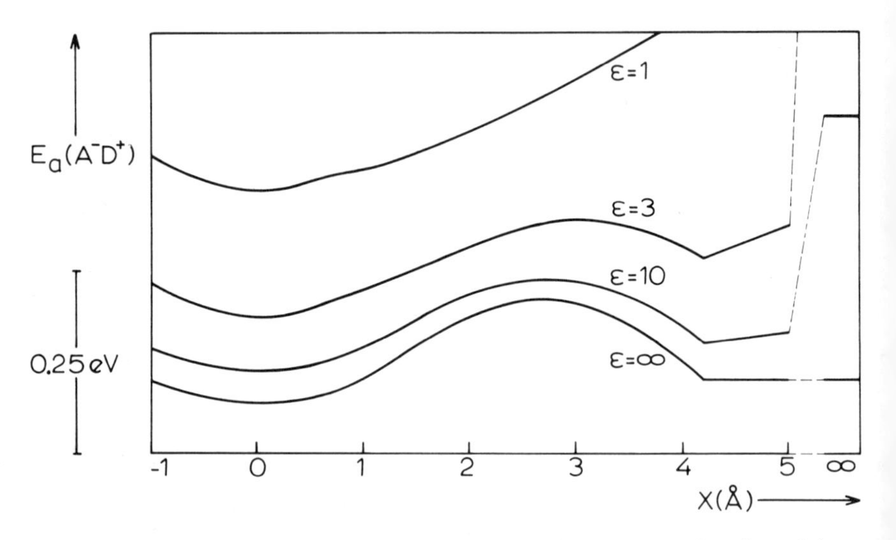

Figure 4.10 The (free) energy of an exciplex in the CT state as a function of the displacement x with the dielectric constant, ε, as parameter

So, although the energy curves are still more shallow than the ones in Figure 4.9, the solvent does not change the complex structure. At sufficiently high polarity, however, conversion into the solvated ion-pair is possible, owing to a relatively low (free) energy barrier. It should be noted that one entropy term has been neglected up to now, *viz.*, the entropy connected with the complex itself. Since for the conversion of the solvated complex into the solvated ion-pair this entropy, certainly, will increase, the free energy corresponding to the solvated ion-pair will be lower than indicated in the figure.

4.3.5 *The excited state of an EDA complex*

In this section attention will be paid to the excited state of an EDA complex, in which there is only interaction between (A^-D^+) and (AD). Such a complex can be described by the wavefunctions (4.15) and (4.16), which may be written as

$$\Psi_g = \psi(AD) + c_0\psi(A^-D^+) \quad (4.149)$$

and

$$\Psi_a = \psi(A^-D^+) - r_0\psi(AD) \quad (4.150)$$

if the gas phase energies $E_c = E(A^-D^+)$ and $E_n = E(AD)$ are not too close. It is noted that Ψ_g and Ψ_a are the wavefunctions for the complex being in the *equilibrium* ground state and the *equilibrium* excited state, respectively.

Since in a solvent the energy difference between these two states is given by (cf. (4.115))

$$E_c - E_n - \boldsymbol{\mu}_0 \cdot \mathbf{F},$$

perturbation theory gives

$$c_0 = \frac{|\beta|}{E_c - E_n - \boldsymbol{\mu}_0 \cdot \mathbf{F}_g} \quad \text{and} \quad r_0 = \frac{|\beta|}{E_c - E_n - \boldsymbol{\mu}_0 \cdot \mathbf{F}_a} \quad (4.151)$$

where β is the Hamiltonian matrix element between $\psi(AD)$ and $\psi(A^-D^+)$. These equations show that, since $\mathbf{F}$ increases with f_ε (cf. (4.113)), both c_0 and r_0 increase with increasing solvent polarity. In view of (4.149) and (4.150) this means that the contribution of the CT state to the excited state wavefunction decreases with increasing solvent polarity, whereas it increases for the ground state.

The probability of emission from $|\Psi_a\rangle$ to the FC ground state $|\Psi_g^{FC}\rangle$ is governed by

$$\mathbf{M} \approx c_0^{FC} \cdot \boldsymbol{\mu}_0 \quad (4.152)$$

(cf. (4.78)). The coefficient c_0^{FC} is given by (4.151), with $\mathbf{F}_g$ being the reaction field in the FC ground state:

$$\mathbf{F}_g^{FC} \approx \boldsymbol{\mu}_0(f_\varepsilon - f_n) \quad (4.153)$$

The emission probability for a rigid EDA complex thus increases with increasing $(f_\varepsilon - f_n)$.

Since the reaction field in the equilibrated ground state is small (c_0 is small), the solvent dependence of the absorption probability for a rigid EDA complex will be negligible.

The expressions of §4.3.4.4 can be applied directly to the excited EDA complex when E_e, α and c_1^2 in these formulas are replaced by E_n, β and $(1 + r_0^2)^{-1}$, respectively. Equation (4.147), then, gives:

$$\frac{\partial E_c}{\partial \lambda_j} - \frac{1}{E_n - E} \cdot \frac{\partial \beta^2}{\partial \lambda_j} = \frac{1}{1 + r_0^2} \cdot f \cdot \frac{\partial(\mu_0^2/\rho^3)}{\partial \lambda_j} \tag{4.154}$$

In this case $E_n - E < 0$, so that, in contrast to the exciplex in §4.3.4.4, the interaction in the excited state cooperates with the solvent to change the complex geometry: the excited state is stabilized when β^2 decreases. As a consequence the geometrical changes will be more important than for the pure CT state. The angle θ, for example, will adjust to that value, which makes β^2 minimum. This implies a decrease of **M** (cf. (4.151) and (4.152)).

Because for the ground state of the EDA complex $E_n - E > 0$, the situation here is similar to that of the exciplex of §4.3.4.1: the interaction between (AD) and (A^-D^+) opposes solvent-induced structural changes and θ will tend to a value which makes β^2 maximum (and **M** maximum).

It is this behaviour of θ, being different for the ground state and the equilibrated excited state of an EDA complex which, very probably, can account for some observations by Mataga[19e]: the lifetime (60 ns) of the excited state of the EDA complex between toluene (D) and tetracyanobenzene (A) as evaluated from the absorption spectra at room temperature, is in accordance with the value found for the emission at 77°K. At this low temperature the environmental rigidity evidently prevents an adjustment to the most favourable θ-value (which would correspond to a decrease of M). This hindrance, however, is no longer present at room temperature in liquid solution, where, accordingly, a value almost three times as high ($\sim$150 ns) is measured.

It is to be expected that this decrease of the emissive ability of the excited state is particularly pronounced with a complex in which both D and A are benzenoid systems: the Coulomb term C will not depend as much on θ and, therefore, will not counteract the rotation as much as would be the case when A and D are derived from different aromatic hydrocarbons (cf. Figure 4.7).

4.3.6 *Concluding remarks*

As shown, the role of the solvent considered only insofar as dipole–dipole interactions are concerned, consists in solvation of the exciplex dipole moment, whereby the resulting reaction field decreases the weight of the locally excited states in the exciplex wavefunction and increases the contribution of the ground state, the former effect being enlarged, the latter diminished by a change of the complex equilibrium geometry. However, as these effects are opposing each other with respect to the emission probability, the radiative rate constant of hetero-excimers is expected to show very little solvent dependence.

With respect to the complex structure the results show that the idea of a continuous change of the intermolecular equilibrium distance with changing solvent polarity has to be rejected insofar as it is meant to explain the experimentally observed relation between the solvent-dependent exciplex quantum yield and lifetime. Calculations in §4.3.4.4 indicate a maximum variation of the interplanar distance of 0·05 Å, which at 3·0 Å could account for a change in intermolecular overlap of less than 5 %,[52] and this corresponds to a change in M^2 of only about 10% (M is proportional to the overlap).

4.4 References

1. (a) R. S. Mulliken, *J. Amer. Chem. Soc.*, **72**, 600 (1950); *J. Amer. Chem. Soc.*, **74**, 811 (1952); *J. Phys. Chem.*, **56**, 801 (1952); (b) R. S. Mulliken, *J. Chim. Phys.*, **61**, 20 (1963); (c) R. S. Mulliken and W. B. Person, *Ann. Rev. Phys. Chem.*, **13**, 107 (1962).
2. G. Briegleb, *Elektronen–Donator–Acceptor–Komplexe*, Springer-Verlag, Heidelberg, 1961.
3. Th. Förster and K. Kasper, *Z. Phys. Chem.* (*NF*), **1**, 275 (1954).
4. J. Ferguson, *J. Chem. Phys.*, **28**, 765 (1958).
5. B. Stevens, *Adv. Photochem.*, **8**, 161 (1971); and references therein.
6. (a) H. Leonhardt and A. Weller, *Z. Phys. Chem.* (*NF*), **29**, 277 (1961); *Ber. Bunsenges. Phys. Chem.*, **67**, 791 (1963); (b) H. Leonhardt, Thesis, Stuttgart (1962).
7. Th. Förster, *Angew. Chem.*, **81**, 364 (1969); and references therein.
8. C. K. Prout and J. D. Wright, *Angew. Chem.*, **80**, 688 (1968).
9. H. Beens, H. Möhwald, D. Rehm, E. Sackmann and A. Weller, *Chem. Phys. Letters*, **8**, 341 (1971).
10. S. Iwata, J. Tanaka and S. Nagakura, *J. Chem. Phys.*, **47**, 2203 (1967).
11. H. Beens and A. Weller, *Molecular Luminescence*, p. 203 (Ed. E. C. Lim), W. A. Benjamin, Inc., New York, 1969; and unpublished results.
12. E. A. Chandross and J. Ferguson, *J. Chem. Phys.*, **47**, 2557 (1967).
13. H. Knibbe, D. Rehm and A. Weller, *Z. Phys. Chem.* (*NF*), **56**, 96 (1967).
14. (a) H. Knibbe, D. Rehm and A. Weller, *Ber. Bunsenges. Phys. Chem.*, **73**, 839 (1969); (b) J. Guttenplan, D. Rehm and A. Weller, to be published.
15. (a) H. Beens and A. Weller, *Acta Phys. Polon.*, **34**, 593 (1968); (b) D. Rehm and A. Weller, *Z. Phys. Chem.* (*NF*), **69**, 183 (1970).
16. H. Beens, H. Knibbe and A. Weller, *J. Chem. Phys.*, **47**, 1183 (1967).
17. (a) A. K. Chandra and E. C. Lim, *J. Chem. Phys.*, **48**, 2589 (1968); (b) *J. Chem. Phys.*, **49**, 5066 (1968).
18. H. Beens and A. Weller, *Chem. Phys. Letters*, **3**, 666 (1969).
19. (a) N. Mataga, T. Okada and N. Yamamoto, *Chem. Phys. Letters*, **1**, 119 (1967); (b) N. Mataga and K. Ezumi, *Bull. Chem. Soc. Japan*, **40**, 1355 (1967); (c) N.

Mataga, T. Okada and N. Yamamoto, *Bull. Chem. Soc. Japan*, **39**, 2562 (1966); (d) T. Okada, H. Matsui, H. Oohari, H. Matsumoto and N. Mataga, *J. Chem. Phys.*, **49**, 4717 (1968); (c) N. Mataga and Y. Murata, *J. Amer. Chem. Soc.*, **91**, 3144 (1969).
20. D. Rehm, *Z. Naturforsch.*, **25a**, 1442 (1970).
21. H. Beens, Thesis, Free University, Amsterdam (1970).
22. H. Beens and A. Weller, *Chem. Phys. Letters*, **2**, 140 (1968).
23. H. Hayashi, S. Iwata and S. Nagakura, *J. Chem. Phys.*, **50**, 993 (1969).
24. H. Beens, J. de Jong and A. Weller, *Colloque Ampère*, *XV*, p. 289, North-Holland, Amsterdam, 1969.
25. J. Langelaar, R. P. H. Rettschnick, A. M. F. Lambooy and G. J. Hoytink, *Chem. Phys. Letters*, **1**, 609 (1967).
26. H. Staerk, Thesis, Technical University, Munich (1969).
27. E. A. Chandross, *J. Chem. Phys.*, **43**, 4175 (1965).
28. C. A. Parker and C. G. Hatchard, *Nature*, **190**, 165 (1961).
29. A. Weller and K. Zachariasse, *J. Chem. Phys.*, **46**, 4984 (1968); *Molecular Luminescence*, p. 895 (Ed. E. C. Lim), W. A. Benjamin, Inc., New York, 1969; *Chem. Phys. Letters*, **10**, 197; 424; 590 (1971).
30. A. Weller, *Nobel Symposium*, **5**, p. 413 (Ed. S. Claesson), Stockholm, 1967.
31. E. A. Chandross, J. W. Longworth and R. E. Visco, *J. Amer. Chem. Soc.*, **87**, 3259 (1965).
32. H. Knibbe, K. Röllig, F. P. Schäfer and A. Weller, *J. Chem. Phys.*, **47**, 1184 (1967).
33. H. Leonhardt and A. Weller, *Luminescence of Organic and Inorganic Materials*, p. 74 (Eds. Kallman and Spruch), Wiley, New York, 1962; H. Knibbe, D. Rehm and A. Weller, *Ber. Bunsenges. Phys. Chem.*, **72**, 257 (1968); D. Rehm and A. Weller, *Ber. Bunsenges. Phys. Chem.*, **73**, 834 (1969); *Israel J. Chem.*, **8**, 259 (1970).
34. A. Streitwieser, *Molecular Orbital Theory for Organic Chemists*, Wiley, New York and London, 1961.
35. M. J. S. Dewar and A. R. Lepley, *J. Amer. Chem. Soc.*, **83**, 4560 (1960).
36. R. L. Flurry, *J. Phys. Chem.*, **69**, 1927 (1965).
37. E. Konijnenberg, Thesis, University of Amsterdam (1963).
38. S. Azumi and T. Azumi, *Bull. Chem. Soc. Japan*, **40**, 279 (1967).
39. H. Kuroda, I. Ikemoto and H. Akamatu, *Bull. Chem. Soc. Japan*, **39**, 1842 (1965).
40. J. N. Murrell and J. Tanaka, *Molec. Phys.*, **7**, 363 (1964).
41. S. P. McGlynn, A. T. Armstrong and T. Azumi, *Modern Quantum Chemistry*, Part 3, p. 203 (Ed. O. Sinanoglu), Academic Press, New York, 1965.
42. (a) H. Baba, *Bull. Chem. Soc. Japan*, **34**, 76 (1960); (b) H. Baba and S. Suzuki, *Bull. Chem. Soc. Japan*, **34**, 82 (1960).
43. J. R. Platt, *J. Chem. Phys.*, **17**, 484 (1949).
44. E. Clar, *Aromatische Kohlenwasserstoffe*, Springer-Verlag, Berlin, 1952.
45. J. N. Murrell, *Quart. Revs.*, **15**, 191 (1961).
46. C. C. J. Roothaan, *Rev. Mod. Phys.*, **23**, 69 (1951).
47. J. A. Pople, *Proc. Phys. Soc.*, **A68**, 81 (1955).
48. J. N. Murrell, *The Theory of the Electronic Spectra of Organic Molecules*, Methuen, London, 1963.
49. H. C. Longuet-Higgins and J. N. Murrell, *Proc. Phys. Soc.*, **A68**, 601 (1955).
50. R. G. Parr, *The Quantum Theory of Molecular Electronic Structure*, W. A. Benjamin, Inc., New York, 1963.

51. R. S. Mulliken, *J. Chim. Phys.*, **46**, 497 (1949).
52. R. S. Mulliken, C. A. Rieke, D. Orloff and H. Orloff, *J. Chem. Phys.*, **17**, 1248 (1949).
53. E. Heilbronner and P. A. Straub, *Hückel Molecular Orbitals*, Springer-Verlag, Berlin, 1966.
54. J. Michl and R. S. Becker, *J. Chem. Phys.*, **46**, 3889 (1967).
55. R. S. Becker and E. Chen, *J. Chem. Phys.*, **45**, 2403 (1966).
56. J. Czekalla and K. O. Meyer, *Z. Phys. Chem. (NF)*, **27**, 185 (1961).
57. J. B. Birks and I. H. Munro, *Progress in Reaction Kinetics*, **4**, 239 (1967).
58. E. A. Steck and G. W. Ewing, *J. Amer. Chem. Soc.*, **70**, 3397 (1948).
59. C. J. F. Böttcher, *Theory of Electric Polarization*, Amsterdam, 1952.
60. *Handbook of Chemistry and Physics*, 49th ed. (Ed. R. C. West).
61. E. Lippert, *Z. Elektrochem.*, **61**, 962 (1957).
62. B. Stevens and M. I. Ban, *Trans. Faraday Soc.*, **60**, 1515 (1964).
63. K. H. Grellmann, A. R. Watkins and A. Weller, *J. Luminescence*, **1**, **2**, 678 (1970); *J. Phys. Chem.*, **76**, 469 (1972).
64. J. L. Lippert, M. W. Hanna and P. J. Trotter, *J. Amer. Chem. Soc.*, **91**, 4035 (1969).
65. E. G. Cook and J. C. Schug, *J. Chem. Phys.*, **53**, 723 (1970).
66. F. Schneider and E. Lippert, *Ber. Bunsenges. Phys. Chem.*, **74**, 624 (1970).
67. P. Pollmann, D. Rehm and A. Weller, to be published.
68. *Molecular Luminescence* (Ed. E. C. Lim), W. A. Benjamin, Inc., New York, 907 (1969); *Molec. Photochem.* **1**, 157 (1969).

5 *Theory of molecular decay processes*

René Voltz

5.1 Introduction

Investigation of the temporal and spectral properties of molecular systems by photophysical methods can, in principle, adopt one of the two general and extreme viewpoints, also widely used in comparable studies of atomic and nuclear resonances.[1–5] The first concerns resonance fluorescence, which is a single scattering process where a distinction of excited state preparation and decay as independent stages cannot be made. Incident photon energy must be sufficiently well defined for the corresponding coherence time to be larger than the characteristic molecular decay times. The response of the material system to the radiation field then essentially involves forced oscillations; observation is made under steady-state conditions, measuring resonance scattering cross-sections as functions of incident photon energy. General scattering theory obviously provides here the natural descriptive framework.[2–6] If, on the other hand, the spectral width of the light beam is larger than the natural decay width of the energy levels, one has the complementary situation where it is excitation time, rather than energy, which is well defined. Under these conditions, the photon–molecule interaction phenomenon can be regarded as two independent processes of pulse excitation and subsequent excited level decay. The problem is thus reduced to relaxation studies, where one follows the decay of initially prepared unstable levels. In practice, most of the available information on the molecular properties come from such experiments, with broad band excitation, rather than from photon-scattering techniques—which generally require very stringent narrow band excitation conditions in the ultraviolet energy range. Molecular photophysical and chemical studies thus naturally tend to focus on time dependence, kinetics and relaxation of excited species.[7,8] The theoretical tools appropriate to treat this class of problems are those of general damping theory, extensively developed to describe atomic level or elementary particle decay.[1–4]

In the last years, relaxation of molecular excitation energy has become a field of very active research, the progress of which has been repeatedly reviewed.[7,9–12] Progressively, the simple pictures of molecular rate processes—essentially considering independent relaxation channels, each characterized by a rate constant in a golden rule form—were recognized to be insufficient. It became abundantly clear that a full quantum mechanical treatment is required to understand and properly describe all the essential features of molecular state decay. It is the purpose of this article, to present such a comprehensive description, as applied to the relatively general case of an excited molecule, isolated in an 'inert' condensed host material.[13]

Our review has two special aims. The first is to present the theoretical treatment in a form from which the basic physical features and concepts of the molecular relaxation processes can be both simply and naturally abstracted; within the formal framework thus designed, we will then take up a unified discussion of the various aspects of the physical problem, regarding, in particular, the nature, interrelations and practical significance of the different molecular electronic relaxation modes, together with their influence on fluorescence emission. All along this study, we necessarily recover the essential results established in former work by Robinson and Frosch,[13] Mies and Kraus,[14] Jortner *et al.*,[15–18] Rhodes[19] and others,[20–26] who often adopted very different formalisms and viewpoints. But, in addition to providing coherence in presentation, the present review also presents some more general and new results.

In order to present the basic concepts as clearly as possible, the treatment is restricted to the simplest possible physical models having the required features of real systems. The electronically excited molecule is taken as embedded in an inert solid host material. In practice, we assume a highly diluted solution, where the solvent has electronic states notably higher than those of the solute, so that mixing of guest and host configurations does not take place. Intermolecular resonance interactions, either directly or via virtual states of the host material, can thus be neglected. Under these conditions, the host environment only contributes to the decay of the solute excited levels by vibrational relaxation. The general problem to be solved is that of any decay theory: at some given finite time, taken as $t = 0$, the solute molecule is prepared, by a short electromagnetic perturbation, in a well-defined excited state, and we ask for its subsequent evolution. Such suddenly excited states can generally be identified as dipole allowed vibronic states, defined in the Born–Oppenheimer adiabatic approximation.[19,26] For simplicity, it is assumed that only a single vibronic level is initially prepared. The results can thus be related to the 'single vibronic level' fluorescence studies.[27,28] We accordingly discuss in some detail experimentally significant quantities like fluorescence intensity decay and quantum yield. But the treatment will concentrate on the theoretical description, rather than to review the experimental results which only serve an illustrative purpose.

We now summarize the contents of the report. The treatment necessarily begins with a discussion of the Hamiltonian which governs the dynamical development of the molecular system in the radiation field (§5.2). In most of the existing treatments, the attention was focused on molecules in the gas phase, so that coupling with the surroundings played a minor role; it thus appears as appropriate to re-examine the case of molecules in the condensed phase, and consider in some detail the question of vibrational relaxation. As usual, the total energy operator H is separated in a 'Model Hamiltonian', H^0, and the corresponding residual interaction, $H' = H - H^0$. By successive applications of adiabatic approximations, the operator H^0 is displayed as a sum of terms, which respectively define the zero-order electronic, vibrational, lattice phonon and photon subsystems. The different parts of H', coupling these subsystems, describe the various radiative and non-radiative, electronic and vibrational relaxation transitions.

In §5.3, the formalism adopted to describe the time evolution of the excited molecular system is presented. It must be based on one of the various equivalent theories of damping phenomena.[1–4] For the present purpose, the methods considering the evolution and resolvent operators appear as most appropriate, since they naturally emphasize the time-dependent aspects of the problem.[29–31] We require, on the other hand, that the formal results interpret as clearly and literally as possible the fundamental features of the physical picture. This is achieved by using the 'partitioning techniques' to introduce explicitly in the formalism the natural distinctions that exist between the energy states of the system.[29,32] One begins with a partition of the photon and phonon continua from the discrete molecular levels, which allows one to formalize the basic notions of irreversible and reversible transitions. In a second step, the idea of natural hierarchy of molecular levels is introduced, by partitioning the subsets of 'radiant' and 'non-radiant' states, i.e., the molecular configurations that are strongly or weakly coupled with the radiation field; this leads to the exact formalization of the other fundamental concepts of radiative and radiationless transitions. At this point, it is in order to note that the present approach has many formal analogies with the treatments developed by Feshbach, Bloch and others in nuclear reaction theory: Feshbach's 'Unified Theory of Nuclear Reactions' rests on the partition between nuclear bound states (closed channels) and one-nucleon continua (open channels).[33] The notion of 'nuclear state hierarchy' determines the description of intermediate structure in the energy dependence of cross-sections:[33–36] a clear distinction is made between the important nuclear states that are strongly coupled with the incident continuum, and the other ones; the former configurations are given the picturesque name of 'doorway' states and play the same role as the 'radiant' states in the molecular systems. §5.3 is concluded by deriving the formal expressions for the decay law and for light-emission intensity and quantum yield.

In §5.4—which forms the central part of the review—we are then in a position to concentrate more explicitly on the actual physical problem. The main subject to be examined here is, of course, that of 'radiationless transitions', since the specific features of radiant molecular level decay, and those of fluorescence, are known to be determined essentially by mixing of the radiant configuration with the non-radiant ones. In most cases, the non-radiant configurations are practically unknown, so that one must resort to very general and crude models for the molecular level schemes and interstate coupling. We first consider the extreme case, expectantly valid in many situations involving large molecules, where the non-radiant level manifold may be regarded as very dense, and too complex to allow an exact treatment. Simple statistical assumptions can be used, with the result that the non-radiant manifold is reduced to an effective continuum. This means that coupling of the radiant and non-radiant states leads to irreversible intramolecular energy dissipation, which adds to damping by interaction with the radiation and phonon continua: in this limiting case, decay of the initially excited vibronic state is strictly exponential (§5.4.1). Next, we turn to the more general situation, where the non-radiant manifold still contains configurations for which the foregoing statistical treatment is sufficient, but involves in addition a set of discrete special levels that must be treated exactly, considering details of level distribution and interstate coupling. The former statistical non-radiant continuum, together with the photon and phonon fields, is still responsible for irreversible radiant energy dissipation; but interaction of the radiant state with the special non-radiant set gives rise to coherent energy exchange: depending on the relative strength of intramolecular coupling and of natural damping of the discrete molecular levels, several types of decay, with eventual periodic and non-exponential features, may here be expected. This is discussed in detail in §5.4.2, with results that are in general accord and, on some aspects, complementary, with the other most recent studies.[18,24] Here again, close analogies with physical situations in nuclear resonance reaction theory must be noted: the statistical limit in §5.4.1 corresponds, in essence, to the uniform model of compound nuclei considered in the original treatment by Breit and Wigner;[44] more recent developments concerned with the optical model and intermediate resonances[34–36] have introduced physical pictures that are similar to those presented in §5.4.2 of this work.

§5.5 is devoted to a discussion of the practical significance of the theoretical results. As the treatment rests on an idealization of the excitation process, we have to examine in some detail the hidden approximations thus involved. Similar consideration is given to the usually neglected role of the detector. The discussion is closed by some concluding remarks on the most interesting physical features exhibited by this treatment, and on their experimental observability.

5.2 Hamiltonian and choice of representation

The basis of the theory is the separation of the total Hamiltonian of the system in a 'model part' H_0 and a 'residual interaction' $H' = H - H^0$. The model Hamiltonian H^0 must be chosen so that its set of eigenstates describes the position of the energy levels as determined by spectroscopy; in addition, the corresponding residual interaction H' must be a perturbation accounting for the observed linewidths, or for the transitions between eigenstates of H^0 responsible for the rate processes amenable to experimental observation. For isolated molecules, it was shown that the definition of the model molecular Hamiltonian which satisfies best these constraining conditions is that resulting from the adiabatic Born–Oppenheimer approximation: the model part H_{BO} interprets the energy levels while the non-adiabaticity operator H_{NA} is responsible for radiationless electronic relaxation.[9] The choice of the model Hamiltonian for the present system, i.e., for the polyatomic molecule plus the host lattice in the electromagnetic radiation field, must follow similar general lines.

The total energy operator reads:

$$H = H_m + H_r^0 + H'_{int} \tag{5.1}$$

denoting by H_r^0 and H'_{int} the terms for the radiation field and its interaction with the material system. H_m characterizes the material subsystem.

5.2.1 *Hamiltonian for the material system*

The system of the organic molecule with the host lattice, is characterized by the Hamiltonian:

$$H_m = T(\mathbf{q}) + T(\mathbf{Q}) + T(\mathbf{R}) + U(\mathbf{q}, \mathbf{Q}, \mathbf{R}) \tag{5.2}$$

where $\mathbf{q}$, $\mathbf{Q}$ and $\mathbf{R}$ respectively represent the sets of molecular electronic, molecular vibrational and lattice vibrational coordinates. T and U are the kinetic and potential energy operators.

The internal vibrations of an organic molecule have generally much higher frequencies (equivalent to *c.* 1000 cm^{-1}) than the relative motions of the molecules in the condensed medium: in molecular solids, the wavenumbers for lattice vibrations are typically of the order of 100 cm^{-1} or less.[37] This shows that the two classes of motions are separable, which allows to consider the whole matter Hamiltonian in a 'double adiabatic approximation':[38] after separating the lattice vibrations (slow subsystem) from the motions of the intramolecular degrees of freedom (fast subsystem), the adiabatic approximation is used again for the latter, to separate the intramolecular vibrational and electronic modes, according to the familiar Born–Oppenheimer procedure. The molecular and lattice vibrations are treated in the harmonic approximation.

We thus define a model part H^0_{m} of the matter Hamiltonian, with eigenfunctions and eigenvalues of the form:

$$|ev, p\rangle = |\phi_{ev}(\mathbf{q}, \mathbf{Q}, \mathbf{R})\Lambda_{evp}(\mathbf{R})\rangle \tag{5.3}$$

$$\phi_{ev} = \eta_e(\mathbf{q}, \mathbf{Q}, \mathbf{R})\chi_{ev}(\mathbf{Q}, \mathbf{R}) \tag{5.3'}$$

$$E^0_{evp} = \hbar(\omega_e + \omega_v + \omega_p) \tag{5.4}$$

where e, v and p are the quantum numbers respectively specifying the electronic (wavefunction η) vibrational (wavefunction χ) and phonon (wavefunction Λ) states; in other words:

$$H^0_{\mathrm{m}} = \sum |evp\rangle E^0_{evp}\langle evp| \tag{5.5}$$

The residual interaction term H'_{m} is related to the non-adiabaticity operators neglected in the two steps of the double adiabatic approximation. It is determined by a matrix with the following significant non-diagonal elements:

$$\langle e', v', p|H'_{\mathrm{m}}|e, v, p\rangle = \langle e', v'|H'_{\mathrm{m}}|e, v\rangle$$

$$= -\sum_i \frac{\hbar^2}{m_i}\left\langle \chi_{e'v'}\left|\left\langle \eta_{e'}\left|\frac{\partial}{\partial Q_i}\right|\eta_e\right\rangle\frac{\partial}{\partial Q_i}\right|\chi_{ev}\right\rangle \tag{5.6}$$

$$\langle e, w, p'|H'_{\mathrm{m}}|e, v, p\rangle = \langle w, p'|H'_{\mathrm{m}}|v, p\rangle$$

$$= -\sum_j \frac{\hbar^2}{M_j}\left\langle \Lambda_{ewp'}\left|\left\langle \chi_{ew}\left|\frac{\partial}{\partial R_j}\right|\chi_{ev}\right\rangle\frac{\partial}{\partial R_j}\right|\Lambda_{evp}\right\rangle \tag{5.7}$$

m_i and M_j are the masses of the vibrating nuclei and molecules.

The first of these expressions represents the electronic matrix element of the $T(\mathbf{Q})$ operator known to be responsible for the radiationless transitions between the electronic states $|e\rangle$ and $|e'\rangle$ (electronic relaxation). The summation is over the N intramolecular vibrational normal modes. The properties of the various terms are usually discussed by considering the electronic functions in the form of a Herzberg–Teller expansion, and expressing the vibrational functions χ_{ev} as the product of the N factors $\gamma_{ek}(v_k)$ (kth normal mode, with quantum number v_k) characteristic of the individual normal modes of molecular vibrations.[9–11] Each term in (5.6) contains an 'electronic' and a 'vibrational Franck–Condon' factor: the electronic part is of the form:

$$J_i(e', e) = \frac{i\hbar}{\sqrt{m_i}} \frac{\langle \eta_{e'} | A_i(\mathbf{q}) | \eta_e \rangle}{\hbar(\omega_e - \omega_{e'})} \tag{5.8}$$

where $A_i(\mathbf{q})$ is given, in terms of the electronic potential energy term, as $(\partial U/\partial Q_i)_0$, taken at the equilibrium configurations of the nuclei and the molecules in the lattice. $J_i(e', e)$ is non-negligible for the so-called promoting modes, which in general correspond to C–C skeletal vibrations of the organic molecule.[9–11] The vibrational factor is expressed as:

$$F_i(e', e) = \frac{i\hbar}{\sqrt{m_i}} \left\langle \gamma_{e'i}(v_i') \left| \frac{\partial}{\partial Q_i} \right| \gamma_{ei}(v_i) \right\rangle \prod_{k \neq i} \langle \gamma_{e'k}(v_k') | \gamma_{ek}(v_k) \rangle \tag{5.9}$$

and contains overlap integrals $\langle \gamma_{e'k} | \gamma_{ek} \rangle$ for vibrational modes different from the considered promoting mode i; large overlap integrals characterize modes accepting most of the electronic energy in the electronic relaxation process; such 'acceptor' modes essentially correspond to C–H vibrations.[9–11]

The other expression (5.7) to be considered as a matrix element of the residual interaction has a structure similar to that of (5.6); it represents molecular vibrational matrix elements of the lattice kinetic energy operator $T(\mathbf{R})$, assumed to be responsible for vibrational relaxation in the condensed medium. Just as electronic relaxation, vibrational relaxation is a multiphonon process; but here the transitions lead from a vibronic molecular state $|ev\rangle$ to another vibrational level $|ew\rangle$ ($\hbar\omega_w < \hbar\omega_v$) of the same electronic configuration, with a concomitant excitation of several lattice phonons. The summation in (5.7) is carried out over the lattice modes. By arguments similar to those used in the discussion of electronic relaxation, it is seen that each term of the sum is the product of two factors, $K_j(w, v)$ and $H_j(w, v)$, which can here be denoted as 'molecular vibrational' and 'lattice vibrational' respectively. The molecular vibrational part has a form similar to that of $J_i(e', e)$ (5.8), i.e.:

$$K_j(w, v) = \frac{i\hbar}{\sqrt{M_j}} \frac{\langle \chi_{ew} | B_j(\mathbf{Q}) | \chi_{ev} \rangle}{\hbar(\omega_v - \omega_w)} \tag{5.10}$$

The operator $B_j(\mathbf{Q})$ is given by $[\partial U(\mathbf{Q}, \mathbf{R})/\partial R_j]_0$, taken at the equilibrium lattice positions; $U(\mathbf{Q}, \mathbf{R})$ is the term describing the nucleus–molecule interactions in the expression of the potential energy in (5.2). On the other hand, expressing the lattice functions $\Lambda_{evp}(\mathbf{R})$ and $\Lambda_{ewp'}(\mathbf{R})$ as the products of the individual normal lattice mode factors $\lambda_{ev,j}(p_j)$ (jth normal mode with quantum number p_j, or $\lambda_{ew,j}(p'_j)$, the lattice vibrational part $H_j(w, v)$ is found to be similar in form to the Franck–Condon factor $F_i(e', e)$ defined by equation (5.9):

$$H_j(w, v) = \frac{i\hbar}{\sqrt{M_j}} \left\langle \lambda_{ew,j}(p'_j) \left| \frac{\partial}{\partial R_j} \right| \lambda_{ev,j}(p_j) \right\rangle \prod_{h \neq j} \langle \lambda_{ew,h}(p'_h) | \lambda_{ev,h}(p_h) \rangle \quad (5.11)$$

The formal analogy between the various expressions describing electronic and vibrational relaxations (equations (5.6) and (5.7), (5.8) and (5.10), (5.9) and (5.11)) indicates that the characteristics of vibrational relaxation should be similar in many respects to those of electronic relaxation: distinction between promoting lattice phonon modes, such as

$$\langle \chi_{ew} | \partial/\partial R_i | \chi_{ev} \rangle \neq 0,$$

and acceptor modes characterized by large phonon overlap integrals; importance of high-frequency lattice vibrations as acceptor modes; energy-gap law, ...

In addition to the terms (5.6) and (5.7) just discussed, a detailed analysis of the non-adiabatic effects should also consider the electronic matrix elements of the operator $T(\mathbf{R})$. But in most of the systems of interest in the present context, we can safely assume that $\partial/\partial R_i|\eta_e\rangle \ll \partial/\partial Q_i|\eta_e\rangle$ and neglect the promotion of electronic relaxation by the lattice phonons.

Besides the non-adiabaticity operators neglected in the two steps of the double adiabatic approximation, the residual interaction term related to the breakdown of the harmonic approximation for the molecular vibrations may be important in some instances. The corresponding non-diagonal matrix elements are of the form:

$$\langle e, v', p | H'_{\mathrm{m}} | e, v, p \rangle = \frac{1}{6} \sum_{i,j,k} B_{i,j,k} \langle \chi_{ev'} | Q_i Q_j Q_k | \chi_{ev} \rangle \quad (5.12)$$

They represent anharmonic coupling between different normal modes of intramolecular vibrations, and are responsible for redistribution of energy among isoenergetic vibrational states.[10,38] The effects of anharmonicity are expected to be important for large excess vibrational energies in a given electronic configuration, which is the case for the vibronic levels populated by internal conversion between electronic states separated by a large energy gap.[9–11] For the vibrational levels populated by light absorption, anharmonic intrastate mixing should be less important.

5.2.2 *Total Hamiltonian and residual interaction*

Having defined the terms H^0_m and H'_m for the material subsystem, we can write, for the total system:

$$H^0 = H^0_\mathrm{m} + H^0_\mathrm{r} \tag{5.13}$$

$$H' = H'_\mathrm{m} + H'_\mathrm{int} \tag{5.14}$$

The eigenvectors of the model Hamiltonian H^0 are thus of the form $|e, v, p, k\rangle = |e, v, p\rangle|k\rangle$, where $|e, v, p\rangle$ was previously defined and $|k\rangle$ describes the radiation state: $|k\rangle = |\mathbf{k}, \hat{e}_\lambda\rangle$, wavevector $\mathbf{k}$, polarization $\hat{e}_\lambda$. The corresponding eigenvalues include four terms:

$$E^0_{e,v,p,k} = \hbar(\omega_e + \omega_v + \omega_p + \omega_k) \tag{5.15}$$

Among the full set of eigenvectors of H^0, we will distinguish several subsets, which play distinct and characteristic roles in the evolution of the molecular states under the radiative and non-radiative transitions:

(1)
$$|V\rangle \equiv |e, v\rangle|p\rangle|vac\rangle \tag{5.16}$$

the molecule is excited in a vibronic state $|e, v\rangle$ connected to the ground state by an allowed radiative transition: in the following such a state will be designated as '*radiant*'; the lattice is unexcited and no photon is present.

(2)
$$|V'\rangle \equiv |e', v'\rangle|p\rangle|vac\rangle \tag{5.17}$$

the molecule is here in an excited state $|e', v'\rangle$ from, or to which radiative transitions are forbidden ('*non-radiant*' state):[39,40]

(3)
$$|K\rangle \equiv |0, v_0\rangle|p\rangle|\mathbf{k}\hat{e}_\lambda\rangle \tag{5.18}$$

the molecule is in its ground state, with eventual excitation of vibrations ($|0, v_0\rangle$); the lattice remains unexcited ($|p\rangle$), but there is one photon in the radiation field.

(4)
$$|P\rangle \equiv |e, w\rangle|p'\rangle|vac\rangle \tag{5.19}$$

the molecule is in the vibronic state $|e, w\rangle$, with the lattice excited $|p'\rangle$, but without photons.

In order to introduce explicitly this natural distinction between the states in the general formulation, it is convenient to define the projection operators for the different types of eigenvectors:[41]

$$M = \sum|V\rangle\langle V|, \quad M' = \sum|V'\rangle\langle V'| \tag{5.20}$$

$$R = \sum|K\rangle\langle K|, \quad L = \sum|P\rangle\langle P| \tag{5.21}$$

The summations include all the vectors of the appropriate subset. The projectors M and M' select the discrete manifolds of states in which the electronic energy is located on the solute molecule, respectively in a radiant and non-radiant configuration. The operators R and L select those subsets of states where the radiation field and the lattice are respectively excited; since the corresponding spectra are continuous, the summation symbols in the expressions (5.21) involve integrations: for R, $\sum$ represents a summation over the quantum numbers v_0, together with a summation over the polarization directions (λ) and an integration over the photon propagation directions (solid angle Ω) and over the continuous spectrum of the energies $E_k = \hbar ck = \hbar\omega_k$,[1]

$$\sum_K = \sum_{v_0}\sum_{\lambda}\int d\Omega \int \rho(E_k)\,dE_k,\ \rho(E_k) = E_k^2/(2\pi\hbar c)^3 \tag{5.22}$$

In the definition of projector L, $\sum$ has a similar meaning and involves an integration over the continuous phonon distribution:

$$\sum_P = \sum_w \int \rho(E_{p'})\,dE_{p'} \tag{5.23}$$

where $\rho(E_{p'})$ represents the density of phonon states around the energy $E_{p'} = \hbar\omega_{p'}$; supposing that the phonon distribution in the host material containing the excited solute can be treated in the Debye approximation, one has for example:

$$\begin{aligned} \rho(E_{p'}) &= 3E_{p'}^2/2\pi^2(\hbar\bar{u})^3 \quad \text{for } E_{p'} \leqslant \hbar\omega_{\max} \\ &= 0 \qquad\qquad\qquad\quad \text{for } E_{p'} > \hbar\omega_{\max} \end{aligned} \tag{5.24}$$

where $\bar{u}$ denotes a mean velocity of sound, as averaged over the longitudinal and the two transversal waves; $\omega_{\max}$ is Debye's cut-off frequency.

The operators defined in (5.20) and (5.21) satisfy the characteristic properties of projection operators.[41] Assuming that the subsets of vectors defined by the relations (5.16) to (5.19) represent all the states of the system, we can write:

$$M + M' + R + L = 1 \tag{5.25}$$

Following the orthogonal and idempotent properties, one has also:

$$MM' = M'M = 0, \qquad MR = RM = 0, \quad \text{etc.}\ldots \tag{5.26}$$

$$M^2 = M, \qquad M'^2 = M', \quad \text{etc.}\ldots \tag{5.27}$$

This allows to display the model Hamiltonian of the full system under the form of the following spectral representation:

$$H^0 = H^0_{MM} + H^0_{M'M'} + H^0_{RR} + H^0_{LL} \tag{5.28}$$

with the definitions:

$$H^0_{MM} = MH^0M, \qquad H^0_{M'M'} = M'H^0M', \quad \text{etc.}\ldots \tag{5.29}$$

Each term in expression (5.28) corresponds to the typical subsystems defined above; the complete energy spectrum as resulting from the various terms, is illustrated in Figure 5.1.

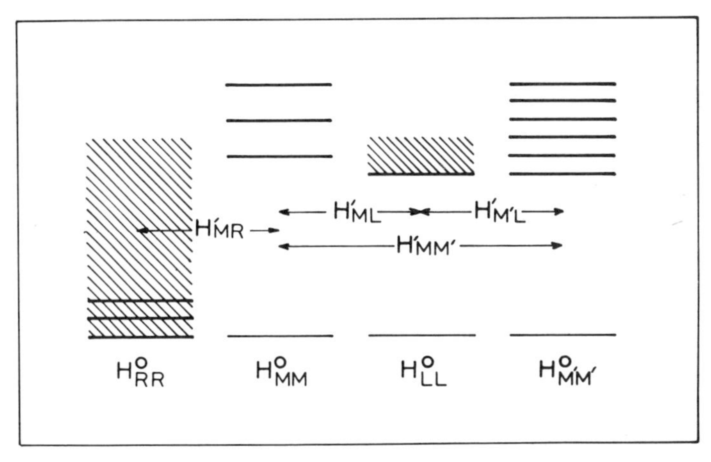

Figure 5.1 Decomposition of the total zero-order energy operator H^0 into the natural reduced Hamiltonians: H^0_{MM} and $H^0_{M'M'}$ respectively define the discrete spectra of radiant and non-radiant vibronic levels; H^0_{RR} and H^0_{LL} define the continua where the radiation and lattice phonon fields are respectively excited. Also represented are the residual interaction terms responsible for the transitions: H'_{MR} accounts for radiative processes; $H'_{MM'}$ for intramolecular non-radiative transitions; H'_{ML} and $H'_{M'L}$ for vibrational relaxation

The residual interaction H' couples the zero-order subsets of states, and accounts for the transitions between the different blocks of the spectrum in Figure 5.1. Using again the projection operators and writing that $H' = (M + M' + R + L)\,H'(M + M' + R + L)$, we can separate the various terms of interest. Among these, the 'diagonal' expressions ($MH'M$, $M'H'M'$, ...) describe weak energy level shifts and may generally be neglected; keeping only the non-diagonal terms, we get:

$$H' = H'_{MR} + H'_{MM'} + H'_{ML} + H'_{M'L} \tag{5.30}$$

if we set:

$$H'_{MR} = MH'R + RH'M \tag{5.31}$$

and similar definitions for the other terms.

Operator H'_{MR} describes the *radiative transitions* between the molecular ground and excited radiant states. The terms $MH'R$ and $RH'M$ respectively account for photon absorption and emission; examining for example $MH'R = \sum\sum |V\rangle\langle V|H'|K\rangle\langle K|$, it is seen to be determined by the matrix elements:

$$\begin{aligned}\langle V|H'|K\rangle &= \langle vac; ev|H'_{\text{int}}|0v^0; \mathbf{k}\hat{e}_\lambda\rangle \\ &= \text{i}(2\pi\hbar ck)^{\frac{1}{2}}\hat{e}_\lambda\langle\eta_e|\mathbf{d}|\eta_0\rangle\langle\chi_{ev}|\chi_{0v^0}\rangle \qquad (5.32)\end{aligned}$$

where $\langle\eta_e|\mathbf{d}|\eta_0\rangle$ is the non-negligible transition dipole moment, and $\langle\chi_{ev}|\chi_{0v^0}\rangle$ a Franck–Condon overlap integral. Similar considerations apply to $RH'M$. For the non-radiant excited molecular states, per definition, the quantities like $\langle V'|H'|K\rangle = \langle vac; e'v'|H'_{\text{int}}|0v^0; \mathbf{k}\hat{e}_\lambda\rangle$ are very small, either because the transition moment vanishes (forbidden transition), or because the overlap integral is small; so that $H'_{M'R} = 0$.

The second term, $H'_{MM'} = MH'M' + M'H'M$, in the expression (5.30) of the residual interaction operator, accounts for the radiationless transitions between the radiant and non-radiant Born–Oppenheimer states of the solute molecule, i.e. for the *intramolecular non-radiative electronic relaxation* processes in competition with radiation. It is characterized by matrix elements

$$\langle V'|H'|V\rangle = \langle e'v'|H'_{\text{m}}|ev\rangle \qquad (5.33)$$

of the non-adiabaticity operator, like those examined in detail before (equations (5.6), (5.8), (5.9)).

The remaining operators H'_{ML} and $H'_{M'L}$ in (5.30) respectively represent the *vibrational relaxation processes* from the molecular radiant and non-radiant vibronic levels. They are determined by the matrix elements

$$\begin{aligned}\langle P|H'|V\rangle &= \langle p'; ew|H'_{\text{m}}|ev; p\rangle \\ \langle P'|H'|V'\rangle &= \langle p'; e'w'|H'_{\text{m}}|e'v'; p\rangle \qquad (5.34)\end{aligned}$$

typical of the multiphonon transitions through which quanta of intramolecular vibrations are transformed in lattice phonon excitation energy (equations (5.7), (5.10), (5.11)). At this place, it is interesting to note that, among the vibrational modes of organic molecules, the C–H vibrations are more strongly coupled to lattice phonons than the skeletal C–C vibrations;[37] this suggests that the non-radiant levels involving the C–H accepting modes, as populated by the radiationless electronic transitions, are characterized by larger vibrational relaxation matrix elements than those for the radiant levels which essentially involve C–C modes.

The different terms of the residual interaction, inducing the transitions between the typical subsystems of states in the spectrum of the system are represented in Figure 5.1. On general grounds (cf. the theorem of Fock and Krylov),[42] it is expected that the intramolecular coupling term, $H'_{MM'}$, connecting discrete states, may lead to recurrent processes; but the other terms in (5.30) that couple the molecular states to the photon and lattice–phonon continua lead to irreversible damping. This will now be examined more precisely by analysing the decay of an excited molecular vibronic state.

5.3 Decay of an excited vibronic state

The solute molecule, isolated in the inert host material, is supposed to be excited in a radiant vibronic state $|ev\rangle$ by a very short electromagnetic perturbation, due for example to a broad-band light pulse or to the interaction with a fast electron. At time $t = 0$, the full system is thus prepared in a non-stationary state of its Hamiltonian H (5.1), which is represented by an eigenvector $|V\rangle$ of the model energy operator H^0 (equations (5.13) and (5.16)). Under this initial condition, we have to solve the equations of motion.

This is most rigorously and conveniently achieved using a general formalism based on the resolvent operator,[2–4] which will be recalled in §5.3.1. In the present problem, it will further be illuminating to introduce the techniques of partitioning that are widely used in other fields; this allows the formal separation of the typical classes of states and transitions utilizing the projection operators defined in §5.2.2. By means of the partitioning methods, we first take into account the natural distinction between the bound and the continuous subsets of states; this allows one to define precisely the true dissipative phenomena responsible for irreversible damping of the bound vibronic levels through coupling to the photon and phonon continua (§5.3.2). The next step is to introduce explicitly in the formalism the natural hierarchy in the set of molecular states, and to distinguish between the radiant states and the non-radiant ones; this allows one to define and to discuss in detail the intramolecular energy transfer phenomena (§5.3.3). In a final paragraph of this formal section, we consider the theoretical expressions for the 'decay law' and experimentally significant quantities like the intensity and quantum yield of fluorescence emission.[46]

5.3.1 *Evolution and resolvent operators*

The evolution of the system, from $t = 0$ to positive times, is conveniently described in terms of the evolution operator

$$U_+(t) = \theta(t)U(t) = \begin{cases} U(t) & (t > 0) \\ 0 & (t < 0) \end{cases} \tag{5.35}$$

$U(t)$ is the unitary operator satisfying the usual Schrödinger equation; as H does not contain the variable t explicitly, we can take:

$$U(t) = \exp(-\mathrm{i}Ht)$$

Operator $U_+(t)$ is seen to satisfy:

$$\left(\mathrm{i}\frac{\partial}{\partial t} - H\right)U_+(t) = \mathrm{i}\delta(t),\ U_+(0) = 1$$

and may therefore be considered as the *time-dependent Green operator.*[4,45] In order to get an integral representation of $U_+(t)$, one introduces the Fourier transform:

$$U_+(t) = \frac{\mathrm{i}}{2\pi}\int_{-\infty}^{+\infty} \mathrm{d}E\, \mathrm{e}^{-\mathrm{i}Et} G_+(E) \tag{5.36}$$

$$G_+(E) = \lim_{\varepsilon \to 0^+} (E - H + \mathrm{i}\varepsilon)^{-1} \tag{5.37}$$

The operator $G_+(E)$ is the *propagator in energy space.* It may be related to the *resolvent* $G(z)$ of the Hamiltonian H, defined as

$$G(z) = (z - H)^{-1} \tag{5.38}$$

by the equation:

$$G_+(E) = G(E + \mathrm{i}\varepsilon) \equiv G(E^+) \tag{5.39}$$

In terms of the general operators thus defined, the state vector of the system at time t,

$$|t\rangle = U_+(t)|0\rangle$$

may be expressed as:

$$|t\rangle = \frac{\mathrm{i}}{2\pi}\int_{-\infty}^{+\infty} \mathrm{d}E\, \mathrm{e}^{-\mathrm{i}Et} G(E^+)|0\rangle \tag{5.40}$$

Also, the probability amplitude $A_n(t) = \langle n|t\rangle$ to find the system in state $|n\rangle$ at time t is given by the integral representation:

$$A_n(t) = \frac{\mathrm{i}}{2\pi}\int_{-\infty}^{+\infty} \mathrm{d}E\, \mathrm{e}^{-\mathrm{i}Et} \langle n|G(E^+)|0\rangle \tag{5.41}$$

The details of the dynamical behaviour of the system are thence represented by the analytical properties of the resolvent operator, or its matrix elements. Formally the evolution in time of the initially prepared unstable states appears as determined by the distributions of poles and residues of $G(E^+)$ in the complex energy plane, when the expressions (5.40) and (5.41) are evaluated by the residue method.

5.3.2 *Partition of bound and continuous states: irreversible bound state damping*

As demonstrated by Fock and Krylov,[42] true irreversible damping of a discrete bound state is due to coupling with dissipative continua. A description of the decay of an unstable state must therefore begin by introducing the natural distinction between the discrete and continuous manifolds of states. Explicitly this may be built in the formalism by a partitioning method.[32]

In the present problem, we thus define the projection operators

$$B = M + M', \qquad C = R + L \tag{5.42}$$

which respectively separate the discrete and continuous parts of the model Hamiltonian spectrum (cf. Figure 5.1). Operator B projects on the subspaces of states $\{|V\rangle\}$ and $\{|V'\rangle\}$ corresponding to the excited bound vibronic levels of the solute molecule; C selects the continuous subspaces spanned by the vectors $|K\rangle$ and $|P\rangle$, representing the states where photons and lattice phonons are present. In addition to their characteristic Hermitian and idempotent properties:

$$B = B^{+} = B^{2}, \qquad C = C^{+} = C^{2} \tag{5.43}$$

these projectors satisfy the relations

$$BC = CB = 0, \qquad B + C = 1 \tag{5.44}$$

The operators B and C allow to partition the state vector of the system into 'bound' and 'continuous' components:

$$|t\rangle = B|t\rangle + C|t\rangle \tag{5.45}$$

Since, in the decay problems, the initial state is in the bound manifold, one has:

$$|0\rangle = B|0\rangle \tag{5.46}$$

The equations of motion can now be solved, considering separately the two parts of $|t\rangle$:

$$B|t\rangle = \frac{\mathrm{i}}{2\pi}\int_{-\infty}^{+\infty} \mathrm{d}E\, \mathrm{e}^{-\mathrm{i}Et} BG(E^{+})B|0\rangle \tag{5.47}$$

$$C|t\rangle = \frac{\mathrm{i}}{2\pi}\int_{-\infty}^{+\infty} \mathrm{d}E\, \mathrm{e}^{-\mathrm{i}Et} CG(E^{+})B|0\rangle \tag{5.48}$$

The state of the system is thus seen to be governed by the reduced resolvents $BG(E)B$ and $CG(E)B$.

They are obtained by operating with B and C on equation (5.49), which is equivalent to the definition of the resolvent $G(E)$:

$$(E - H)G(E) = (E - H^0 - H')G(E) = 1 \tag{5.49}$$

With standard operator algebra, bearing in mind the projector properties (5.43) and (5.44), the following coupled equations are obtained:

$$B(E - H)BBG(E)B - BH'CCG(E)B = B \tag{5.50a}$$

$$-CH'BBG(E)B + C(E - H)CCG(E)B = 0 \tag{5.50b}$$

These expressions actually represent a set of algebraic equations for the individual matrix elements of the resolvent, and must be used as such in practice. But in the formal theory, one solves (5.50b) to obtain:

$$CG(E)B = \frac{1}{C(E - H)C}CH'BBG(E)B \tag{5.51}$$

which substituted in (5.50a) yields:

$$\left[E - BHB - BH'C\frac{1}{C(E - H)C}CH'B\right]BG(E)B = B \tag{5.52}$$

This equation is usually written as:

$$[E - \mathscr{H}_B(E)]BG(E)B = B \tag{5.53}$$

with the following definitions:

$$\mathscr{H}_B(E) = B[H^0 + H' + W_B(E)]B \tag{5.54}$$

$$W_B(E) = BH'C\frac{1}{C(E - H)C}CH'B \tag{5.55}$$

The reduced resolvent $BG(E)B$ can thus be considered under the form:

$$BG(E)B = B[E - \mathscr{H}_B(E)]^{-1}B \tag{5.56}$$

which, inserted in (5.51), also determines the operator $CG(E)B$ controlling the transitions to the continua.

These relations are the basic results of the projection operator formalism. They suggest that $\mathscr{H}_B(E)$ and $BG(E)B$ be interpreted as effective Hamiltonian and resolvent operators acting only in the subspace of the bound discrete states. The remaining parts of the spectrum, i.e. the photon and phonon continua, are treated as a field generating an additional, non-local, energy dependent '*optical potential*' given by $W_B(E)$. A simple interpretation of expression (5.55) of $W_B(E)$ could consider, first a transition from subspace B to the continua C (hence the factor $CH'B$), followed by propagation in the C subspace (described by the propagator $[C(E - H)C]^{-1}$) and ended by the transition back to the bound manifold of states (submatrix $BH'C$). The major role of operator $W_B(E)$ will be to account for the details of damping of the bound excited molecular levels.

The integral representations of the state vectors (equations (5.47) and (5.48)) depend on $G(E^+)$, with $E^+ = E + i\varepsilon$. From the $i\varepsilon$ term, it follows that $W_B(E^+)$ is complex and has an imaginary part which describes the influence of damping of the bound excited molecular levels by interaction with photons and phonons. As mentioned before, the perturbation H' does not couple the continuous subspaces, i.e., $CH'C = 0$. The optical potential operator $W_B(E^+)$ can then be considered in the form of a sum of two separate terms corresponding to the photon (projector R) and phonon (projector L) continua:

$$W_B(E^+) = BH'R\frac{1}{e_R^+}RH'B + BH'L\frac{1}{e_L^+}LH'B \tag{5.57}$$

where:

$$e_R^+ = R(E^+ - H^0)R, \qquad e_L^+ = L(E^+ - H^0)L \tag{5.58}$$

For the matrix elements that represent $W_B(E^+)$, one has, for the initially excited radiant state, $|0\rangle \equiv |V\rangle$:

$$\langle V|W_B(E^+)|V\rangle = \sum_K \frac{\langle V|H'|K\rangle\langle K|H'|V\rangle}{E - E_K + i\varepsilon} + \sum_P \frac{\langle V|H'|P\rangle\langle P|H'|V\rangle}{E - E_P + i\varepsilon} \tag{5.59}$$

Per definition of the non-radiant states, the corresponding matrix elements do not contain the first term, and we get:

$$\langle V'_k|W_B(E^+)|V'_h\rangle = \sum_P \frac{\langle V'_k|H'|P\rangle\langle P|H'|V'_h\rangle}{E - E_P + i\varepsilon} \tag{5.60}$$

$$\langle V'_k|W_B(E^+)|V_i\rangle = \sum_P \frac{\langle V'_k|H'|P\rangle\langle P|H'|V_i\rangle}{E - E_P + i\varepsilon} \tag{5.61}$$

together with a similar expression for $\langle V_j|W_B(E^+)|V'_h\rangle$. It may however be argued that only the diagonal elements of the phonon term $[BH'Le_L^{-1}LH'B]$ should be retained. Vibrational relaxation from different vibrational levels of the molecule embedded in the host material are indeed completely uncorrelated processes, so that we can introduce a 'statistical assumption' stating that the various matrix elements $\langle V_i|H'|P\rangle$

and $\langle V'_k|H'|P\rangle$ are independent random quantities with equal probabilities for being positive or negative;[34,35] as a consequence, the non-diagonal elements for the expressions (5.60) and (5.61) are expected to be small. Remembering the significance of

$$\sum_K \text{ and } \sum_P$$

(cf. equations 5.22), (5.23)) and since:

$$(x + i\varepsilon)^{-1} = \mathscr{P}/x - i\pi\,\delta(x)$$

($\mathscr{P}$: principal part), we can write the matrix elements of $W_B(E^+)$ as sums of a real and an imaginary part. The real terms are given by the integrals with the principal part; they represent weak energy-level shifts which are of minor importance in the present discussion. The imaginary terms characterize the level width and decay phenomena due to coupling with the radiation field and the lattice vibrations. If we set:

$$\gamma_R = 2\pi \sum_K \langle V|H'|K\rangle \delta(E - E_K)\langle K|H'|V\rangle \tag{5.62}$$

$$\gamma_L = 2\pi \sum_P \langle V|H'|P\rangle \delta(E - E_P)\langle P|H'|V\rangle \tag{5.63}$$

$$\gamma'_L = 2\pi \sum_P \langle V'_k|H'|P\rangle \delta(E - E_P)\langle P|H'|V'_h\rangle \delta_{kh} \tag{5.64}$$

we have for the matrix elements of the optical potential $W_B(E^+)$, the expressions:

$$\langle V|W_B|V\rangle = (\Delta_R + \Delta_L) - i(\gamma_R + \gamma_L)/2 \tag{5.65}$$

$$\langle V'_k|W_B|V'_h\rangle = (\Delta'_L - i\gamma'_L/2)\delta_{kh} \tag{5.66}$$

$$\langle V'_k|W_B|V\rangle = 0 \tag{5.67}$$

where Δ_R, Δ_L and Δ'_L represent the generally negligible energy corrections. As usual, the energy shifts Δ and width γ will be taken as constant, i.e., their dependence on energy E may be neglected.[1]

5.3.3 *Partition of radiant and non-radiant states: intramolecular radiationless transitions*

As the next step in the general formulation, we now explicitly introduce the natural heierarchy of the molecular bound states, and separate the radiant states (projector M) from the non-radiant ones (projector M'). This allows one to define and to discuss precisely the radiationless intramolecular energy transfer processes that are initiated or terminated by the radiative transitions in the molecules.

Writing the vector $B|t\rangle$ in the form:

$$B|t\rangle = M|t\rangle + M'|t\rangle \tag{5.68}$$

the two components are given by:

$$M|t\rangle = \frac{\mathrm{i}}{2\pi}\int_{-\infty}^{+\infty} \mathrm{d}E\, \mathrm{e}^{-\mathrm{i}Et} MG(E^+)M|0\rangle \tag{5.69}$$

$$M'|t\rangle = \frac{\mathrm{i}}{2\pi}\int_{-\infty}^{+\infty} \mathrm{d}E\, \mathrm{e}^{-\mathrm{i}Et} M'G(E^+)M|0\rangle \tag{5.70}$$

if we remember that the initial state pertains to the radiant manifold: $|0\rangle \equiv |V\rangle$. Whereupon, the reduced resolvents of interest are $M'G(E)M$ and $MG(E)M$.

They are determined using methods similar to those of the previous paragraph: operating with M and M' on equation (5.53), one obtains the system of equations:

$$[E - M\mathscr{H}_B(E)M]MG(E)M - [M\mathscr{H}_B(E)M']M'G(E)M = M \tag{5.71a}$$

$$-[M'\mathscr{H}_B(E)M]MG(E)M + [E - M'\mathscr{H}_B(E)M']M'G(E)M = 0 \tag{5.71b}$$

From (5.71b), we get the relation between $M'GM$ and MGM:

$$M'G(E)M = \frac{1}{M'[E - \mathscr{H}_B(E)]M'}[M'\mathscr{H}_B(E)M]MG(E)M \tag{5.72}$$

which, inserted into (5.71a), leads to the equation:

$$[E - \mathscr{H}_M(E)]MG(E)M = M \tag{5.73}$$

Here $\mathscr{H}_M(E)$ is defined by:

$$\mathscr{H}_M(E) = M[\mathscr{H}_B(E) + W_M(E)]M \tag{5.74}$$

with:

$$W_M(E) = M\mathscr{H}_B(E)M'\frac{1}{M'[E - \mathscr{H}_B(E)]M'}M'\mathscr{H}_B(E)M \tag{5.75}$$

From this, one obtains the expression of $MG(E)M$:

$$MG(E)M = M[E - \mathscr{H}_M(E)]^{-1}M \tag{5.76}$$

together with that of $M'G(E)M$, using (5.73). All these operators are determined by the reduced elements of the effective Hamiltonian $\mathscr{H}_B(E)$

in the bound subspace (equation (5.54)), which are of the form:

$$M\mathscr{H}_B(E)M = M(H^0 + H')M + MW_B(E)M \tag{5.77}$$

$$M\mathscr{H}_B(E)M' = MH'M' \tag{5.78}$$

$$M'\mathscr{H}_B(E)M = M'H'M \tag{5.79}$$

$$M'\mathscr{H}_B(E)M' = M'(H^0 + H')M' + M'W_B(E)M' \tag{5.80}$$

considering the properties (5.65)–(5.67) of the $W_B(E)$ operator.

Again, we can interpret $\mathscr{H}_M(E)$ and $MG(E)M$ as effective operators, only acting in the subspace of radiant states, on which our attention is centred in the present section. The influence of the non-radiant manifold is described by the new optical potential (5.75):

$$W_M(E) = MH'M'\frac{1}{E - M'(H_0 + H')M' - M'W_B(E)M'}M'H'M \tag{5.75'}$$

This energy-dependent operator plays a central role in our discussion since it accounts for all the features of radiationless energy exchange between the molecular radiant and non-radiant states.

5.3.4 *Decay law and light emission*

From the general formal results obtained by the partitioning techniques, it is easy to derive the time-dependent probability amplitudes of finding the molecular system in the various subspaces of states, and so to express all the possible experimentally significant quantities. In the present discussion, it will however be sufficient to consider the decay law for the initially excited molecular radiant state, together with the measurable characteristics of light emission, which provides the best experimental means to study the intramolecular energy transfer processes.

Under the assumption that only one radiant state—the 'isolated' initially excited level $|0\rangle = |V\rangle$—needs to be taken into account, according to equation (5.69), one gets for the probability amplitude $\langle V|t\rangle = A_v(t)$ of the molecule to remain in the initial state at time t, the integral representation:

$$A_v(t) = \frac{\mathrm{i}}{2\pi}\int_{-\infty}^{+\infty} \mathrm{d}E\, \mathrm{e}^{-\mathrm{i}Et}\langle V|MG(E^+)M|V\rangle \tag{5.81}$$

The function $\langle V|MG(E^+)M|V\rangle$ is determined from the expressions in the previous section. According to the equations (5.76) and (5.74), we can write:

$$\langle V|MG(E^+)M|V\rangle = [E - \langle V|\mathscr{H}_M(E^+)|V\rangle]^{-1} \tag{5.82}$$

with:

$$\langle V|\mathscr{H}_M(E^+)|V\rangle = E_v - \mathrm{i}\tfrac{1}{2}\Gamma_v + \Phi_v(E) \tag{5.83}$$

In relation (5.83), we have set:

$$E_v = E_v^0 + \langle V|H'|V\rangle + \Delta_R + \Delta_L \tag{5.84}$$

this is the energy of the radiant state as corrected by the first-order perturbation term and the previously mentioned level shifts due to the interaction with the photon and phonon continua (cf. equation (5.65)). The quantity

$$\Gamma_v = \gamma_R + \gamma_L \tag{5.85}$$

characterizes damping and width of the $|V\rangle$-level, induced by coupling to the dissipative photon and phonon continua (equation (5.65)). The function $\Phi_v(E)$ is defined by:

$$\Phi_v(E) = \langle V|W_M(E)|V\rangle$$
$$= \sum_{V'} \frac{\langle V|H'|V'\rangle\langle V'|H'|V\rangle}{E - E_{v'} + \mathrm{i}\tfrac{1}{2}\Gamma_{v'}} \tag{5.86}$$

where

$$E_{v'} = E_{v'}^0 + \langle V'|H'|V'\rangle + \Delta_L' \tag{5.87}$$

and

$$\Gamma_{v'} = \gamma_L' \tag{5.88}$$

are respectively the corrected energies and the damping constants of the non-radiant levels $|V'\rangle$ (cf. equation (5.66)).

The integral representation of the probability amplitude $A_v(t)$ may thus be explicitly written in the form:

$$A_v(t) = \frac{\mathrm{i}}{2\pi}\int_{-\infty}^{+\infty} \frac{\mathrm{d}E\,\mathrm{e}^{-\mathrm{i}Et}}{E^+ - E_v + \mathrm{i}\tfrac{1}{2}\Gamma_v - \Phi_v(E^+)} \tag{5.89}$$

which can be evaluated by the residue method, closing the integration contour by a semicircle in the lower energy halfplane. The function $\langle V|MG(E^+)M|V\rangle$ may be considered in the usual meromorphic form:

$$[E^+ - E_v + \mathrm{i}\tfrac{1}{2}\Gamma_v - \Phi_v(E^+)]^{-1} = \sum_p R_p/(E^+ - E_p) \tag{5.90}$$

where the poles E_p within the integration contour satisfy the equation:

$$E_p - E_v + \mathrm{i}\tfrac{1}{2}\Gamma_v - \Phi_v(E_p) = 0 \tag{5.91}$$

R_p denotes the corresponding residues, given by:

$$R_p = \left(1 - \frac{\mathrm{d}\Phi_v}{\mathrm{d}E}\right)^{-1}\Bigg|_{E=E_p} \tag{5.92}$$

From the form of these equations, it is clear that E_p and R_p are complex. In terms of these quantities, the probability amplitude $A_v(t)$ becomes:

$$A_v(t) = \sum_p R_p \, \mathrm{e}^{-\mathrm{i}E_p t} = \sum_p R_p \, \mathrm{e}^{-E''_p t} \, \mathrm{e}^{-\mathrm{i}E'_p t} \tag{5.93}$$

where we have explicitly introduced the real and imaginary parts of the poles, setting $E_p = E'_p - \mathrm{i}E''_p$ ($E''_p > 0$).

From expression (5.93), the *decay law*:

$$P_v(t) = |A_v(t)|^2 \tag{5.94}$$

which gives the probability for finding the molecule in the decaying excited radiant state, is easily determined.

But the decay law *per se* has only limited practical interest. Experimentally, the decay of excited molecular states is generally studied by following light emission. This is obviously governed by the amplitude $\langle K|t\rangle = A_K(t)$ of finding the system in a state $|K\rangle$ (equation (5.18)):

$$A_K(t) = \frac{\mathrm{i}}{2\pi}\int_{-\infty}^{+\infty} \mathrm{d}E \, \mathrm{e}^{-\mathrm{i}Et} \langle K|RG(E^+)M|V\rangle \tag{5.95}$$

Operating with M and R respectively at right and left of $CG(E)B$ in equation (5.51), one gets for the reduced resolvent of interest:

$$RG(E)M = \frac{1}{R(E-H)R}(RH'M)MG(E)M \tag{5.96}$$

and for the matrix element in (5.95)

$$\langle K|RG(E^+)M|V\rangle = \frac{\langle K|H'|V\rangle}{E - E_K + \mathrm{i}\varepsilon}\langle V|MG(E^+)M|V\rangle \tag{5.97}$$

Hence:

$$A_K(t) = \frac{\mathrm{i}}{2\pi}\int_{-\infty}^{+\infty} \mathrm{d}E \, \mathrm{e}^{-\mathrm{i}Et} \frac{\langle K|H'|V\rangle}{E - E_K + \mathrm{i}\varepsilon}\langle V|MG(E^+)M|V\rangle \tag{5.98}$$

Also evaluating this amplitude in terms of poles and residues of the integrand, one obtains:

$$A_K(t) = \langle K|H'|V\rangle \, \langle V|MG(E_K)M|V\rangle \, \mathrm{e}^{-\mathrm{i}E_K t} + \sum_p \frac{R_p}{E_p - E_K} \mathrm{e}^{-E''_p t} \, \mathrm{e}^{-\mathrm{i}E'_p t} \tag{5.99}$$

In the optical relaxation studies, the experimentally significant quantities are *the intensity* $I(t)$ (photons per unit time) and *the quantum yield* Q *of fluorescence*; they are given by:

$$I(t) = \frac{\mathrm{d}}{\mathrm{d}t} \sum_K |A_K(t)|^2$$

$$Q = \sum_K |A_K(\infty)|^2$$

and can be evaluated by calculating

$$\sum_K |A_K(t)|^2$$

from (5.98):

$$\sum_K |A_K(t)|^2 = \iint_{-\infty}^{+\infty} \frac{\mathrm{d}E\,\mathrm{d}E'}{(2\pi)^2} \mathrm{e}^{-\mathrm{i}(E-E')t} \langle V|MG(E^+)M|V\rangle\langle V|MG(E'^-)M|V\rangle$$

$$\times \left\{ \sum_K \frac{|\langle K|H'|V\rangle|^2}{(E - E_K + \mathrm{i}\varepsilon)(E' - E_K - \mathrm{i}\varepsilon)} \right\}$$

The expression in the bracket may be split up by partial fractions:

$$\sum_K \frac{|\langle K|H'|V\rangle|^2}{(E - E_K + \mathrm{i}\varepsilon)(E' - E_K - \mathrm{i}\varepsilon)}$$

$$= \sum_K \left[\frac{|\langle K|H'|V|^2}{E' - E_K - \mathrm{i}\varepsilon} - \frac{|\langle K|H'|V\rangle|^2}{E - E_K + \mathrm{i}\varepsilon} \right] \left[\frac{1}{E - E' + 2\mathrm{i}\varepsilon} \right]$$

Using the definition (5.62) of γ_R and Δ_R (cf. equation (5.65)), one then has:

$$\sum_K \frac{|\langle K|H'|V\rangle|^2}{(E - E_K + \mathrm{i}\varepsilon)(E' - E_K - \mathrm{i}\varepsilon)} = \mathrm{i}\gamma_R \frac{1}{(E - E' + 2\mathrm{i}\varepsilon)}$$

Hence:

$$\sum_K |A_K(t)|^2 = \mathrm{i}\gamma_R \iint_{-\infty}^{+\infty} \frac{\mathrm{d}E\,\mathrm{d}E'}{(2\pi)^2} \frac{\mathrm{e}^{-\mathrm{i}(E-E')t}}{(E - E' + 2\mathrm{i}\varepsilon)}$$

$$\times \langle V|MG(E^+)M|V\rangle\langle V|MG(E'^-)M|V\rangle$$

Bearing in mind the distribution relation $(E - E')(E - E' + 2i\varepsilon)^{-1} = 1$, we first verify that:

$$\begin{aligned} I(t) &= \frac{d}{dt} \sum_K |A_K(t)|^2 \\ &= \gamma_R \left| \int_{-\infty}^{+\infty} \frac{dE}{2\pi} e^{-iEt} \langle V|MG(E^+)M|V\rangle \right|^2 \\ &= \gamma_R P_v(t) \end{aligned} \tag{5.100}$$

where $P_v(t)$ is given by the decay law (5.94). On the other hand, if we note that:

$$\lim_{t=\infty} (E - E' + 2i\varepsilon)^{-1} e^{-i(E-E')t} = -2\pi i \delta(E - E')$$

we obtain:

$$\begin{aligned} Q &= \sum_K |A_K(\infty)|^2 \\ &= \frac{\gamma_R}{2\pi} \int_{-\infty}^{+\infty} dE |\langle V|MG(E^+)M|V\rangle|^2 \end{aligned} \tag{5.101'}$$

Following equation (5.90), this can be expressed in terms of the poles E_p and the residues R_p as:

$$Q = \frac{\gamma_R}{2\pi} \sum_p R_p \int_{-\infty}^{+\infty} dE \frac{\langle V|MG(E)M|V\rangle^*}{E - E_p}$$

$$= -i\gamma_R \sum_p R_p \langle V|MG(E)M|V\rangle^*_{E=E_p} \tag{5.101''}$$

$$= -i\gamma_R \sum_{p,p'} \frac{R_p R^*_{p'}}{(E_p - E^*_{p'})} \tag{5.101'''}$$

In some instances, the exact *energy distribution function of the emitted photons* is measured. This energy profile is described by $J(E_K) = |A_K(\infty)|^2$. Due to the finite values of E''_p in the expression (5.99) of $A_K(t)$, the first term only remains as $t \to +\infty$; hence:

$$J(E_K) = |\langle K|H'|V\rangle|^2 |\langle V|MG(E_K)M|V\rangle|^2 \tag{5.102'}$$

or, introducing the expansion (5.90):

$$J(E_K) = |\langle K|H'|V\rangle|^2 \sum_{p,p'} \frac{R_p R^*_{p'}}{(E_K - E_p)(E_K - E^*_{p'})} \tag{5.102''}$$

5.4 Intramolecular energy transfer modes

Coupling of the initially excited radiant state of the solute molecule with the discrete manifold of the non-radiant molecular vibronic states should be responsible for recurrent processes, with the molecule returning to its initial state after a finite time. On the other hand, interaction of the bound radiant and non-radiant molecular states with the photon and lattice phonon continua leads to energy dissipation. As a consequence of the cooperation of these reversible and irreversible transitions, and depending on their relative importance, various modes of radiationless intramolecular decay can thus be encountered. This problem has been considered in the recent years, using various methods.[13–26] It is interesting to note that it bears a close analogy with a number of line-broadening problems treated in nuclear physics, from where many concepts and theoretical techniques may be borrowed.[33–36,48]

In the present section, the different modes of intramolecular energy exchange will be discussed with the help of the general results obtained in §5.3. The decay law (5.94), together with the expressions of intensity (5.100), quantum yield (5.101) and spectral distribution (5.102) of light emission, all depend on the poles and residues of $\langle V|MG(E^+)M|V\rangle$ through equation (5.90); our primary problem will therefore be to solve the equations (5.91) and (5.92) for the different physically significant situations. We begin with the practically most common case of true irreversible exponential decay (§5.4.1), where the initially excited molecular radiant level can be considered as interacting with a non-radiant continuum of molecular states.

In §5.4.2, we then examine the less frequently encountered—but in principle more interesting—examples where at least one part of the non-radiant molecular manifold consists of quasi-discrete levels. A variety of possible non-exponential and oscillatory decay modes may thus appear, which are classified by a detailed analysis of the influence of intramolecular mixing and of damping.

5.4.1 *Exponential decay and irreversible energy transfer*

For polyatomic molecules in the condensed phase, exponential non-radiative decay is the rule rather than the exception. This can be justified using simple and reasonable statistical assumptions, that are similar to those extensively used in the theory of nuclear resonance reactions.[5,34–36]

In most polyatomic systems, the manifold of non-radiant levels $\{|V'\rangle\}$ interacting with the excited radiant state is very complicated. The non-radiant states are known to be densely spaced; in addition they are generally characterized by relatively high excess vibrational energies, so that they have strong natural decay (vibrational relaxation, anharmonic interstate mixing) and merge into each other. Under these conditions $\Phi_V(E)$, defined by equation (5.86), becomes a smooth function of E and an averaging of the matrix elements over V' takes place. Assuming a random distribution of the complicated $|V'\rangle$ levels, we then get:

$$\begin{aligned}\Phi_V(E) &= \sum_{V'} \frac{|\langle V|H'|V'\rangle|^2}{E - E_{V'} + \mathrm{i}\tfrac{1}{2}\Gamma_{V'}} \\ &\approx \frac{|\langle V|H'|V'\rangle|^2_{\mathrm{Av}}}{D} \int \frac{\mathrm{d}E_{V'}}{E - E_{V'} + \mathrm{i}\tfrac{1}{2}\Gamma_{V'}} \\ &= -\mathrm{i}\frac{\pi}{D}|\langle V|H'|V'\rangle^2_{\mathrm{Av}}\end{aligned} \tag{5.103}$$

where $|\langle V|H'|V'\rangle|^2_{\mathrm{Av}}$ and D represent average values of $|\langle V|H'|V'\rangle|^2$ and of the non-radiant level spacing in the energy region around E. The result is therefore that an additional constant imaginary potential appears, which describes dissolution of the excited $|V\rangle$ state in the complex manifold of the molecular non-radiant configurations.

Equation (5.91) has a single solution $E_p = E'_p - \mathrm{i}E''_p$, with $E'_p = E_V$ and $E''_p = \frac{1}{2}(\Gamma_V + \Gamma'_V)$, where:

$$\Gamma'_V = \frac{2\pi}{D}|\langle V|H'|V'\rangle|^2_{\mathrm{Av}} \tag{5.104}$$

Also $R_p = 1$, so that (equation (5.93)):

$$A_V(t) = \mathrm{e}^{-\mathrm{i}E_V t}\,\mathrm{e}^{-\frac{1}{2}(\Gamma_V + \Gamma'_V)t} \tag{5.105}$$

which leads to the pure exponential decay law (equation (5.94)):

$$P_V(t) = \mathrm{e}^{-(\Gamma_V + \Gamma'_V)t} \tag{5.106}$$

Following (5.100), the fluorescence intensity is accordingly given by:

$$I(t) = \gamma_R\,\mathrm{e}^{-(\Gamma_V + \Gamma'_V)t} \tag{5.107}$$

The corresponding quantum yield Q is easily obtained from one of the forms of equation (5.101):

$$Q = \gamma_R/(\Gamma_V + \Gamma'_V) \tag{5.108}$$

Finally, the energy distribution of the emitted photons, as given by (5.102), has a Lorentzian form with a total line width $\frac{1}{2}(\Gamma_V + \Gamma'_V)$ centred around $E_K = E_V$.

It is seen that, in the present case, the general approach recovers—and justifies—the results obtained by the usual, intuitive arguments, taking the golden rule of perturbation theory to describe the rate processes. According to equation (5.104), in particular, a constant rate of transition, from the initial radiant state to the final continuum of the dense and broadened manifold of non-radiant levels, can be defined. The statistical procedures used to derive equation (5.103) imply that density and width of the non-radiant levels are high enough for the condition:

$$\tfrac{1}{2}\Gamma_{V'} \gg D \tag{5.109}$$

to be satisfied; also it is assumed that no individual level(s) in the $|V'\rangle$ manifold has an anomalously strong interaction with the initial radiant state. These conditions define the '*statistical*' limiting case, which generally applies to radiationless transitions in polyatomic molecules between electronic configurations separated by a large energy gap.[9,11] Less frequent, but more interesting in the present context, are the cases where these conditions are not satisfied, and where deviations from the simple results of this section are expected.

5.4.2 *Non-exponential decay and recurrent energy exchange*

5.4.2.1 *The molecular model*

General physical features. In many cases, concerning for instance small molecules, or polyatomic molecules with a discrete set of strongly coupled zero-order states, the conditions defining the 'statistical limit' are only partly, or not at all, satisfied.[9–11] The characteristic behaviour of such systems must then be investigated by distinguishing, within the non-radiant manifold, the subset of states—say $|\overline{V}'\rangle$—for which the simple statistical assumptions do not apply, from all the other states that can be treated as in §5.4.1 (cf. Figure 5.2). Expression (5.86) of the function $\Phi_V(E)$ will consequently contain two terms:

$$\Phi_V(E) = -\mathrm{i}\tfrac{1}{2}\Gamma'_V + \overline{\Phi}_V(E) \tag{5.110}$$

where Γ'_V has the same meaning as before (cf. equation (5.104)), while the second term must be explicitly considered as the sum:

$$\overline{\Phi}_V(E) = \sum_{\overline{V}'} \frac{|\langle \overline{V}'|H'|V\rangle|^2}{E - E_{\overline{V}'} + \mathrm{i}\tfrac{1}{2}\Gamma_{\overline{V}'}} \tag{5.111}$$

over the $|\overline{V}'\rangle$ states. In general, these states may be supposed to form a correlated subset of levels characterized by a coupling strength $\langle \overline{V}'|H'|V\rangle$ larger than that for the $|V'\rangle$ levels entering in the definition of Γ'_V in equation (5.110) (selection rules); on the other hand, the level spacing $\overline{D}$ is large enough, and the level widths $\Gamma_{\overline{V}'}$ are small enough, for a condition

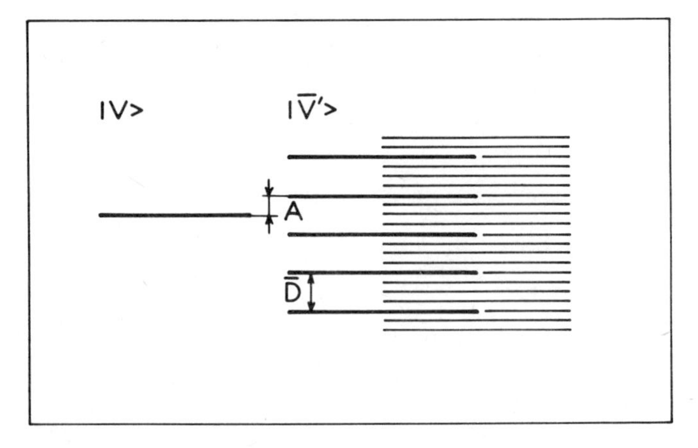

Figure 5.2 Molecular vibronic energy-level scheme. The radiant level $|V\rangle$ is separated from the non-radiant manifold. Among the non-radiant configurations, a special subset $\{|\bar{V}'\rangle\}$ is distinguished by particular selection rules of interstate coupling with the radiant state. Quantum mechanical mixing of the radiant state with the special non-radiant $|\bar{V}'\rangle$ states is responsible for the 'intermediate' or 'small molecule' type decay features. Interaction of the radiant state with the non-special, non-radiant configurations is treated under general statistical assumptions

inverse of (5.109) to be satisfied, i.e.:

$$\tfrac{1}{2}\Gamma_{\bar{V}'} \ll \bar{D} \tag{5.112}$$

This means that $\bar{\Phi}_V(E)$ cannot be considered as slowly varying, and will show a succession of peaks, as a function of E. Solution of equation (5.91) then defines a discrete distribution for the poles E_p over which the summations in expansion (5.90) and in the expressions of the probability amplitudes (equation (5.93), for example) are carried out. The recurrent processes arise from interference of the different terms in these sums.

Properties of poles and residues. The poles E_p are given by an equation of the form (5.91), that is written as

$$E - E_V + \mathrm{i}\tfrac{1}{2}\Gamma_V^{\updownarrow} - \bar{\Phi}_V(E) = 0 \tag{5.91$'$}$$

if one introduces the constant:

$$\Gamma_V^{\updownarrow} = \Gamma_V + \Gamma_V' \tag{5.113}$$

which accounts for damping of the initially excited radiant state $|V\rangle$ by coupling with the photon and lattice phonon continua (Γ_V), together with the eventual intramolecular non-radiant continuous manifolds

(Γ'_V).[47] If N represents the number of non-radiant $|\overline{V}'\rangle$ states to be considered in expression (5.111) of $\overline{\Phi}_V(E)$, equation (5.91′) is easily seen to be equivalent to a polynomial equation of degree $N + 1$, with the first terms given by:

$$E^{N+1} - E^N[(E_V - \mathrm{i}\tfrac{1}{2}\Gamma_V^{\updownarrow}) + \sum_{\overline{V}'}(E_{\overline{V}'} - \mathrm{i}\tfrac{1}{2}\Gamma_{\overline{V}'})] + \ldots = 0 \quad (5.114)$$

The $N + 1$ roots of this equation define the poles $E_p = E'_p - \mathrm{i}E''_p$. According to the well-known relation between the sum of the $N + 1$ roots and the coefficient of E^N in equation (5.114), we can write:

$$\sum_p E_p \equiv \sum_p (E'_p - \mathrm{i}E''_p) = (E_V - \mathrm{i}\tfrac{1}{2}\Gamma_V^{\updownarrow}) + \sum_{\overline{V}'}(E_{\overline{V}'} - \mathrm{i}\tfrac{1}{2}\Gamma_{\overline{V}'})$$

which leads to the following *sum rules*, of some interest in the later discussion:

$$\sum_p E'_p = E_V + \sum_{\overline{V}'} E_{\overline{V}'} \quad (5.115)$$

$$\sum_p E''_p = \tfrac{1}{2}\Gamma_V^{\updownarrow} + \sum_{\overline{V}'} \tfrac{1}{2}\Gamma_{\overline{V}'} \quad (5.116)$$

The residues R_p in expansion (5.90) are given, following (5.92) and (5.110), by:

$$R_p = \left[1 - \frac{\mathrm{d}\overline{\Phi}_V(E)}{\mathrm{d}E}\right]^{-1}\Bigg|_{E=E_p} = \left[1 + \sum_{\overline{V}'} \frac{|\langle \overline{V}'|H'|V\rangle|^2}{(E_p - E_{\overline{V}'} + \mathrm{i}\tfrac{1}{2}\Gamma_{\overline{V}'})^2}\right]^{-1} \quad (5.117)$$

According to the initial condition $A_V(0) = 1$, one has, from equation (5.93), the sum rule:

$$\sum_p R_p = 1 \quad (5.118)$$

which supplements (5.115) and (5.116).

The detailed distributions of poles E_p and residues R_p obviously depend on the exact form of the characteristic function $\overline{\Phi}_V(E)$. But solving equations (5.91′) and (5.117) in terms of the general form (5.111) is difficult, so that one usually introduces simplifying models.

Simplifying models. In general, as suggested by the previously mentioned characteristics of the $|\overline{V}'\rangle$ states, it appears as reasonable to assume that the coupling matrix elements in (5.111) are approximately equal, and to set:

$$|\langle \overline{V}'|H'|V\rangle|^2 = u^2 \quad (5.119)$$

In addition, the levels $|\overline{V}'\rangle$ are supposed to have the same width $\Gamma_{\overline{V}'}$ and to be equally spaced between two energy limits given by $E_0 \pm N\overline{D}$:

$$E_{\overline{V}'} = E_0 + n\overline{D}, \qquad n = 0, \pm 1, \ldots, \pm N \tag{5.120}$$

The number of the $|\overline{V}'\rangle$ states is thus $2N + 1$ and may be large. Under these conditions the function (5.111) is seen to depend on the complex variable $E + \mathrm{i}\frac{1}{2}\Gamma_{\overline{V}'}$:

$$\overline{\Phi}_V(E + \mathrm{i}\tfrac{1}{2}\Gamma_{\overline{V}'}) = u^2 \sum_{n=-N}^{+N} (E - E_0 - n\overline{D} + \mathrm{i}\tfrac{1}{2}\Gamma_{\overline{V}'})^{-1} \tag{5.121}$$

In many cases, the effects of the edges of the distribution (5.120) may be disregarded. It is then convenient to assume an infinite distribution, with $N = \infty$. For this idealized model, often referred to as 'picket-fence model',[36] one gets for $\overline{\Phi}_V$ the analytical form:

$$\overline{\Phi}_V(E + \mathrm{i}\tfrac{1}{2}\Gamma_{\overline{V}'}) = \pi\frac{u^2}{\overline{D}}\,\mathrm{cotg}\left[\frac{\pi}{\overline{D}}(E - E_0 + \mathrm{i}\tfrac{1}{2}\Gamma_{\overline{V}'})\right] \tag{5.122}$$

which is easier to handle than the physically more significant expression (5.121).

The transcendental equation (5.91′) for the poles $E_p = E'_p - \mathrm{i}E''_p$ then becomes:

$$E_p - E_V + \mathrm{i}\tfrac{1}{2}\Gamma_V^{\uparrow} - \tfrac{1}{2}\Gamma_V^{\downarrow}\,\mathrm{cotg}\left[\frac{\pi}{\overline{D}}(E_p - E_0 + \mathrm{i}\tfrac{1}{2}\Gamma_{\overline{V}'})\right] = 0 \tag{5.123}$$

where we have defined the quantity:[47]

$$\Gamma_V^{\downarrow} = 2\pi u^2/\overline{D} \tag{5.124}$$

This parameter represents an intramolecular energy transfer width, which characterizes the mixing of the radiant state $|V\rangle$ with the non-radiant $|\overline{V}'\rangle$ manifold (cf. *infra*).

Introducing for later convenience the dimensionless quantities:

$$\begin{aligned}
&\varepsilon'_p = \frac{\pi}{\overline{D}}(E'_p - E_0), \quad \varepsilon_V = \frac{\pi}{\overline{D}}(E_V - E_0), \quad \varepsilon''_p = \frac{\pi}{\overline{D}}(E''_p - \tfrac{1}{2}\Gamma_{\overline{V}'})\\
&\rho^{\uparrow} = \frac{\pi}{\overline{D}}\tfrac{1}{2}(\Gamma_V^{\uparrow} - \Gamma_{\overline{V}'}), \quad \rho^{\downarrow} = \frac{\pi}{\overline{D}}\tfrac{1}{2}\Gamma_V^{\downarrow}
\end{aligned} \tag{5.125}$$

the equation (5.123) becomes:

$$\varepsilon'_p - \varepsilon_V - \mathrm{i}(\varepsilon''_p - \rho^{\uparrow}) - \rho^{\downarrow}\,\mathrm{cotg}\,(\varepsilon'_p - \mathrm{i}\varepsilon''_p) = 0 \tag{5.126′}$$

or equivalently:

$$\begin{cases} \varepsilon'_p - \varepsilon_V = \rho^{\downarrow}(\mathrm{tg}\,\varepsilon'_p)(1 - \mathrm{th}^2\,\varepsilon''_p)/(\mathrm{tg}^2\,\varepsilon'_p + \mathrm{th}^2\,\varepsilon''_p) \\ \varepsilon''_p - \rho^{\uparrow} = -\rho^{\downarrow}(\mathrm{th}\,\varepsilon''_p)(1 + \mathrm{tg}^2\,\varepsilon'_p)/(\mathrm{tg}^2\,\varepsilon'_p + \mathrm{th}^2\,\varepsilon''_p) \end{cases} \quad (5.126'')$$

We note that in some cases, the imaginary parts of the poles may be small enough, so that $\varepsilon''_p \ll 1$. The general equations (5.126″) can then be considered in the simplified form obtained by setting $\mathrm{th}\varepsilon''_p \approx \varepsilon''_p$ and neglecting the second-order terms:

$$\begin{aligned} \varepsilon'_p - \varepsilon_V &= \rho^{\downarrow}\,\mathrm{cotg}\,\varepsilon'_p \\ \varepsilon''_p - \rho^{\uparrow} &= -\rho^{\downarrow}\varepsilon''_p(1 + \mathrm{cotg}^2\,\varepsilon'_p) \end{aligned} \quad (5.127')$$

or:

$$\varepsilon''_p = \frac{\rho^{\uparrow}\rho^{\downarrow}}{\rho^{\downarrow} + (\rho^{\downarrow})^2 + (\varepsilon'_p - \varepsilon_V)^2} \quad (5.127'')$$

The residue R_p for the pole E_p is, according to equation (5.117), given by:

$$\begin{aligned} R_p &= \left\{1 + \left(\frac{\pi u}{\bar{D}}\right)^2 \left[1 + \mathrm{cotg}^2 \frac{\pi}{\bar{D}}(E_p - E_0 + \mathrm{i}\tfrac{1}{2}\Gamma_{\bar{V}'})\right]\right\}^{-1} \\ &= \{1 + \rho^{\downarrow}[1 + \mathrm{cotg}^2\,(\varepsilon'_p - \mathrm{i}\varepsilon''_p)]\}^{-1} \end{aligned} \quad (5.128')$$

Following equations (5.123), this may also be considered under the forms:

$$\begin{aligned} R_p &= \frac{u^2}{u^2 + (\tfrac{1}{2}\Gamma_V^{\downarrow})^2 + [E_p - E_V + \mathrm{i}\tfrac{1}{2}\Gamma_V^{\uparrow}]^2} \\ &= \frac{\rho^{\downarrow}}{\rho^{\downarrow} + (\rho^{\downarrow})^2 + [(\varepsilon'_p - \varepsilon_V) - \mathrm{i}(\varepsilon''_p - \rho^{\uparrow})]^2} \end{aligned} \quad (5.128'')$$

These simplified models will be useful to analyse the details of the intramolecular energy transfer processes. But the general solution of the problem remains difficult, so that one resorts to approximation methods, which depend on the extent of mixing of the radiant with the non-radiant molecular states. The distinction of weak and strong mixing will further be necessary to define the different decay types.

Molecular interstate mixing. Per definition, the *weak mixing* case is assumed to apply when a close one-to-one correspondence exists between the $2N + 2$ levels of the zero-order spectrum (1 radiant state $|V\rangle$ + $2N + 1$ non-radiant states $|\bar{V}'\rangle$), and the $2N + 2$ roots of the generalized eigenvalue equation (5.91′), which is solved by standard perturbation techniques. For each of the $2N + 1$ non-radiant levels (5.120), one looks for a pole $E_n = E'_n - \mathrm{i}E''_n$, with $E'_n \approx E_0 + n\bar{D}$, $E'' \approx \frac{1}{2}\Gamma_{\bar{V}'}$. The remaining pole, denoted by $E_{2N+2} = E'_{2N+2} - \mathrm{i}E''_{2N+2}$, is related to the initially excited radiant state, i.e., $E'_{2N+2} \approx E_V$ and $E''_{2N+2} \approx \frac{1}{2}\Gamma^{\updownarrow}_V$; this means, in particular, that the $|V\rangle$ state retains its special, radiant character.

According to equation (5.91′), with $\bar{\Phi}_V(E)$ of the general form (5.121), the pole E_n corresponding to the zero-order level $E_0 + n\bar{D}$ does satisfy the equation:

$$E_n - E_V + \mathrm{i}\tfrac{1}{2}\Gamma^{\updownarrow}_V - \frac{u^2}{E_n - E_0 - n\bar{D} + \mathrm{i}\frac{1}{2}\Gamma_{\bar{V}'}}$$

$$- u^2 \sum_{n' \neq n} \frac{1}{E_n - E_0 - n'\bar{D} + \mathrm{i}\frac{1}{2}\Gamma_{\bar{V}'}} = 0$$

Weak mixing clearly means that:

$$|E_n - E_0 - n\bar{D} + \mathrm{i}\tfrac{1}{2}\Gamma_{\bar{V}'}| \ll \bar{D}$$

On the other hand, for the terms involved in the summation, one has:

$$|E_n - E_0 - n'\bar{D} + \mathrm{i}\tfrac{1}{2}\Gamma_{\bar{V}'}| \gtrsim \bar{D}$$

so that they may be neglected. Hence:

$$E_n - E_V + \mathrm{i}\tfrac{1}{2}\Gamma^{\updownarrow}_V - u^2[E_n - E_0 - n\bar{D} + \mathrm{i}\tfrac{1}{2}\Gamma_{\bar{V}'}]^{-1} = 0$$

or equivalently:

$$E_n - E_0 - n\bar{D} + \mathrm{i}\tfrac{1}{2}\Gamma_{\bar{V}'} = \frac{u^2}{E_n - E_V + \mathrm{i}\frac{1}{2}\Gamma^{\updownarrow}_V}$$

Inserting the zero-order approximation value for E_n in the right-hand side of this equation, one gets:

$$(E'_n - E_0 - n\bar{D}) + \mathrm{i}[\tfrac{1}{2}\Gamma_{\bar{V}'} - E''_n] = \frac{u^2}{(E_0 + n\bar{D} - E_V) + \mathrm{i}\frac{1}{2}(\Gamma^{\updownarrow}_V - \Gamma_{\bar{V}'})} \tag{5.129}$$

The quantitative expression defining weak mixing is obtained by writing that the correction term in the right-hand side of equation (5.129) has a modulus less than the mean spacing $\bar{D}$:

$$u^2\left[(E_0 + n\bar{D} - E_V)^2 \times \left(\frac{\Gamma^{\updownarrow}_V - \Gamma_{\bar{V}'}}{2}\right)^2\right]^{-\frac{1}{2}} \ll \bar{D} \tag{5.130′}$$

or alternatively, introducing the mixing width $\Gamma^{\downarrow}_V$ (5.124):

$$\left(\frac{\Gamma^{\downarrow}_V}{2\pi}\right)\left[(E_V - E_0 - n\bar{D})^2 + \left(\frac{\Gamma^{\updownarrow}_V - \Gamma_{\bar{V}'}}{2}\right)^2\right]^{-\frac{1}{2}} \ll 1 \tag{5.130″}$$

Expressed in terms of the dimensionless quantities (5.125), this reads:

$$(\rho^{\downarrow}/2\pi([(\varepsilon_V - \pi n)^2 + (\rho^{\uparrow})^2]^{-\frac{1}{2}} \ll 1 \qquad (5.130''')$$

The weak mixing condition is seen to depend on n: it always applies for the $|\bar{V}'\rangle$ levels sufficiently far from the radiant level.

In the following, the '*weak mixing cases*' are defined by the condition that (5.130) is satisfied throughout the whole distribution (5.120) of the $|\bar{V}'\rangle$ states. This implies that it must be fulfilled for the $|\bar{V}'\rangle$ level nearest to the radiant level. Denoting by $A(\leqslant \bar{D}/2)$ the energy difference between this $|\bar{V}'\rangle$ and the $|V\rangle$ state (cf. Figure 5.2), the defining criterion for the weak mixing cases thus becomes:

$$\left(\frac{\Gamma_V^{\downarrow}}{2\pi}\right)\left[A^2 + \left(\frac{\Gamma_V^{\uparrow} - \Gamma_{\bar{V}'}}{2}\right)^2\right]^{-\frac{1}{2}} \ll 1 \qquad (5.131')$$

or equivalently

$$(\rho^{\downarrow})^2 \ll \alpha^2 + (\rho^{\uparrow})^2, \qquad \alpha = \pi A/\bar{D} \qquad (5.131'')$$

On the other hand, the '*strong mixing cases*' are defined to correspond to systems for which conditions inverse of (5.130) hold for at least one level of the non-radiant $|\bar{V}'\rangle$ manifold. The characteristic condition is accordingly:

$$\frac{\Gamma_V^{\downarrow}}{2\pi}\left[A^2 + \left(\frac{\Gamma_V^{\uparrow} - \Gamma_{\bar{V}'}}{2}\right)^2\right]^{-\frac{1}{2}} \gg 1 \qquad (5.132')$$

or:

$$(\rho^{\downarrow})^2 \gg \alpha^2 + (\rho^{\uparrow})^2 > (\rho^{\uparrow})^2 \qquad (5.132'')$$

By contrast to the weak mixing case, the whole set (5.120) of $|\bar{V}'\rangle$ levels cannot necessarily be treated on the same footing, since the mixing strength decreases for increasing energy gaps with the radiant level and may get weak. The number of non-radiant $|\bar{V}'\rangle$ levels which satisfy

$$(\rho^{\downarrow}/2\pi([(\pi n)^2 \times (\rho^{\uparrow})^2]^{-\frac{1}{2}} > 1$$

and may thus be considered to experience strong mixing, is seen to be of the order $2\mathfrak{N}$, with:

$$\mathfrak{N} = \rho^{\downarrow}/2\pi^2 = \frac{1}{2}\left(\frac{u}{\bar{D}}\right)^2 \qquad (5.133)$$

If $\mathfrak{N} < N$ (cf. equation (5.120)) strong mixing does not apply throughout the whole distribution of the non-radiant $|V'\rangle$ manifold; if $\mathfrak{N} > N$, all the levels are thoroughly mixed; but here particular effects associated with the edges of (5.120) may be of importance, as will be seen below.

We next turn to the derivation of the distribution of poles and residues, which determine the expression (5.93) of the probability amplitude $A_V(t)$ and consequently the decay law $P_V(t) = |A_V(t)|^2$ together with the light intensity $I(t) = \gamma_R |A_V(t)|^2$ (cf. equations (5.94) and (5.100)).

5.4.2.2 *Distributions of poles and residues*

The two limiting cases are treated separately, beginning with that of weak mixing.

Weak mixing. We use standard perturbation techniques to solve the generalized eigenvalue equation (5.91′). The poles remain in the neighbourhood of the complex zero-order energies valid in the absence of intramolecular coupling. They are therefore located within the energy interval $(E_0 - N\bar{D}, E_0 + N\bar{D})$ spanned by the $|\bar{V}'\rangle$ manifold, as well as in the immediate exterior vicinity of the edges where the physical $\bar{\Phi}_V(E)$ function (5.121) is well approximated by the analytical form (5.122); we may thus safely proceed with the picket-fence model.

The $2N + 1$ poles related to the $|\bar{V}'\rangle$ states are given by:

$$E_n = E'_n - \mathrm{i}E''_n, \qquad n = 0, \pm 1, \ldots, \pm N$$
$$E'_n = E_0 + n\bar{D} + \Delta E'_n, \qquad E''_n = \tfrac{1}{2}\Gamma_{\bar{V}'} + \Delta E''_n, \tag{5.134′}$$

or equivalently, in terms of the reduced quantities, by:

$$\varepsilon'_n = n\pi + \Delta\varepsilon'_n, \qquad \varepsilon''_n = \Delta\varepsilon''_n \tag{5.134″}$$

where $\Delta\varepsilon'_n = (\pi\Delta E'_n/\bar{D}) \ll 1$ and $\Delta\varepsilon''_n = (\pi\Delta E''_n/\bar{D}) \ll 1$ are the corrections to be determined. Inserting (5.134″) in the equation (5.126″) and keeping only the terms up to first-order for the corrections, one gets the required results:

$$\Delta\varepsilon'_n = \rho^{\downarrow}\frac{(\pi n - \varepsilon_V)}{(\pi n - \varepsilon_V)^2 + (\rho^{\uparrow})^2} \tag{5.135}$$

$$\Delta\varepsilon''_n = \rho^{\downarrow}\frac{\rho^{\uparrow}}{(\pi n - \varepsilon_V)^2 + (\rho^{\uparrow})^2} \tag{5.136}$$

The corresponding residues are given by expressions (5.128″) as:

$$R_n = \frac{\rho^{\downarrow}}{\rho^{\downarrow} + (\rho^{\downarrow})^2 + [(n\pi - \varepsilon_V) + \mathrm{i}\rho^{\uparrow}]^2} \tag{5.137}$$

For the special pole $E_{2N+2} = E'_{2N+2} - \mathrm{i}E''_{2N+2}$, corresponding to the radiant state, we proceed in a similar way and write:

$$E'_{2N+2} = E_V + \Delta E'_{2N+2}, \qquad E''_{2N+2} = \tfrac{1}{2}\Gamma_V^{\uparrow} + \Delta E''_{2N+2} \tag{5.138′}$$

$$\varepsilon'_{2N+2} = \varepsilon_V + \Delta\varepsilon'_{2N+2}, \qquad \varepsilon''_{2N+2} = \rho^{\uparrow} + \Delta\varepsilon''_{2N+2} \tag{5.138″}$$

to obtain from (5.126″):

$$\Delta\varepsilon'_{2N+2} = \rho^{\downarrow}\,\mathrm{tg}\,\varepsilon_V \frac{1 - \mathrm{th}^2\,\rho^{\uparrow}}{\mathrm{tg}^2\,\varepsilon_V + \mathrm{th}^2\,\rho} \tag{5.139}$$

$$\Delta\varepsilon''_{2N+2} = -\rho^{\downarrow}\,\mathrm{th}\,\rho^{\uparrow} \frac{1 + \mathrm{tg}^2\,\varepsilon_V}{\mathrm{tg}^2\,\varepsilon_V + \mathrm{th}^2\,\rho^{\uparrow}} \tag{5.140}$$

Following equation (5.128″), the residue is of the form:

$$R_{2N+2} = \frac{\rho^{\downarrow}}{\rho^{\downarrow} + (\rho^{\downarrow})^2 + (\Delta\varepsilon'_{2N+2} - \mathrm{i}\Delta\varepsilon''_{2N+2})^2} \tag{5.141}$$

For later reference, we may now specialize these general results to particular, physically significant situations controlled by the weak mixing conditions (5.131). More precisely, two limiting cases are distinguished, depending on the extent of natural decay (i.e. of $\rho^{\uparrow}$) of the initially excited radiant state.

1. Consider first the case where weak mixing can be uniquely ascribed *to weak interstate coupling*. This applies, if one has:

$$\rho^{\uparrow} \ll \alpha < 1 \quad \text{or} \quad \tfrac{1}{2}(\Gamma_V^{\updownarrow} - \Gamma_{\bar{V}'}) \ll A \tag{5.142'}$$

   either because the widths $\Gamma_V^{\updownarrow}$ and $\Gamma_{\bar{V}'}$ of the molecular levels are all very small, or/and because these widths are accidentally identical ($\Gamma_V^{\updownarrow} \approx \Gamma_{\bar{V}'}$). The weak mixing condition (5.131′) then reduces to

$$\rho^{\downarrow} \ll \alpha < 1 \quad \text{or} \quad u^2/A \ll \bar{D} \tag{5.142''}$$

   This is the trivial condition of validity of non-degenerate bound-state perturbation theory: in the present case, it is equivalent to the criterion of applicability of the BO approximation in the study of stationary molecular states. Bearing in mind that $(\rho^{\uparrow}, \rho^{\downarrow}) \ll \alpha \lesssim 1$, the expressions for the poles and residues are seen to simplify to:

$$\begin{cases} \Delta\varepsilon'_n = \rho^{\downarrow}(\pi n - \varepsilon_V)^{-1}, & \varepsilon'_n = n\pi + \rho^{\downarrow}(\pi n - \varepsilon_V)^{-1} \\ \Delta\varepsilon'_{2N+2} = \rho^{\downarrow}\,\mathrm{cotg}\,\alpha, & \varepsilon'_{2N+2} = \varepsilon_V + \rho^{\downarrow}\,\mathrm{cotg}\,\alpha \\ \Delta\varepsilon''_n = \varepsilon''_n = 0 & \\ \Delta\varepsilon''_{2N+2} = 0, \quad \varepsilon''_{2N+2} = \rho^{\uparrow} & \\ R_n = \rho^{\downarrow}(n\pi - \varepsilon_V)^{-2} & \\ R_{2N+2} = \rho^{\downarrow}[\rho^{\downarrow} + (\rho^{\downarrow})^2 + (\rho^{\downarrow})^2\,\mathrm{cotg}^2\,\alpha]^{-1} & \end{cases} \tag{5.143'}$$

or equivalently:

$$E'_n = E_0 + n\bar{D} + \frac{u^2}{E_0 - E_V + n\bar{D}}, \quad E'_{2N+2} = E_V + \tfrac{1}{2}\Gamma_V^{\downarrow} \operatorname{cotg}\left(\frac{\pi A}{\bar{D}}\right);$$

$$E''_n = \tfrac{1}{2}\Gamma_{\bar{V}'}, \quad E''_{2N+2} = \tfrac{1}{2}\Gamma_V^{\updownarrow} \qquad (5.143'')$$

$$R_n = \frac{u^2}{(E_0 - E_V + n\bar{D})^2}, \quad R_{2N+2} = \frac{1}{1 + (\pi u/\bar{D})^2[1 + \operatorname{cotg}^2(\pi A/\bar{D})]}$$

Note that $\rho^{\downarrow} \ll 1$ implies that the denominator of R_{2N+2} only weakly exceeds unity, so that R_{2N+2} is only slightly smaller than one; on the other hand, one has from (5.143'), $R_n \leqslant \rho^{\downarrow}/\alpha^2 \ll (\rho^{\downarrow}/\alpha)^2 \ll 1$. This shows that

$$R_n \ll R_{2N+2} \lesssim 1 \qquad (5.144)$$

meaning that, in the expansion (5.93), which determines the dynamical behaviour of the excited system, the principal term corresponds to the pole E_{2N+2}.

2. For the second case of particular interest to be considered, weak mixing must be essentially ascribed to *rapid decay of the initially excited radiant state.* It is characterized by the condition:

$$\rho^{\uparrow} \gg 1 > \alpha \quad \text{or} \quad \tfrac{1}{2}(\Gamma_V^{\updownarrow} - \Gamma_{\bar{V}'}) \gg \bar{D} > A \qquad (5.145')$$

Here, situations where a criterion inverse from (5.142) would indicate strong mixing (e.g. by breakdown of the BO approximation) in the absence of natural damping, can however be converted in a weak mixing one, if:

$$\rho^{\downarrow} \ll \rho^{\uparrow} \quad \text{or} \quad \Gamma_V^{\downarrow} \ll (\Gamma_V^{\updownarrow} - \Gamma_{\bar{V}'}) \qquad (5.145'')$$

Physically this means that, if the radiant state quickly decays in the continua—as implied by a large $\Gamma_V^{\updownarrow}$—there is no time for mixing with the discrete $|\bar{V}'\rangle$ states to act efficiently. In this case, some of the foregoing general expressions for the poles and residues are simplified if we note that $\rho^{\uparrow} \gg 1$ implies that th $\rho^{\uparrow} = 1$: while the corrections $\Delta\varepsilon'_n$ and $\Delta\varepsilon''_n$ must still be taken in their full form (5.135) and (5.136), the others reduce to

$$\begin{cases} \Delta\varepsilon'_{2N+2} = 0, & \Delta\varepsilon''_{2N+2} = -\rho^{\downarrow} \\ R_n = \rho^{\downarrow}(n\pi - \varepsilon_V + i\rho^{\uparrow})^{-2}, & R_{2N+2} = 1 \end{cases} \qquad (5.146')$$

We thus have:

$$\begin{cases} \varepsilon_n' = n\pi + \rho^{\downarrow}\dfrac{(n\pi - \varepsilon_V)}{(n\pi - \varepsilon_V)^2 + (\rho^{\uparrow})^2}, & \varepsilon_{2N+2}' = \varepsilon_V \\ \varepsilon_n'' = \rho^{\downarrow}\dfrac{\rho^{\uparrow}}{(n\pi - \varepsilon_V)^2 + (\rho)^2}, & \varepsilon_{2N+2}'' = \rho^{\uparrow} - \rho^{\downarrow} \end{cases} \tag{5.146''}$$

or:

$$\begin{cases} E_n' = E_0 + n\bar{D} + u^2\dfrac{(E_0 - E_V + n\bar{D})}{(E_0 - E_V + n\bar{D})^2 + (\frac{1}{2}[\Gamma_V^{\updownarrow} - \Gamma_{\bar{V}'}])^2} \\ E_n'' = \frac{1}{2}\Gamma_{\bar{V}'} + u^2\dfrac{\frac{1}{2}(\Gamma_V^{\updownarrow} - \Gamma_{\bar{V}'})}{(E_0 - E_V + n\bar{D})^2 + (\frac{1}{2}[\Gamma_V^{\updownarrow} - \Gamma_{\bar{V}'}])^2} \\ E_{2N+2}' = E_V, \qquad E_{2N+2}'' = \frac{1}{2}(\Gamma_V^{\updownarrow} - \Gamma_V^{\downarrow}) \end{cases} \tag{5.146'''}$$

$$R_n = \frac{u^2}{[(E_0 - E_V + n\bar{D}) + i\frac{1}{2}(\Gamma_V^{\updownarrow} - \Gamma_{\bar{V}'})]^2};\ R_{2N+2} = 1$$

A schematic representation of the pole distribution in the complex ($\varepsilon = \varepsilon' - i\varepsilon''$) plane is given in Figure 5.3. The open circles correspond

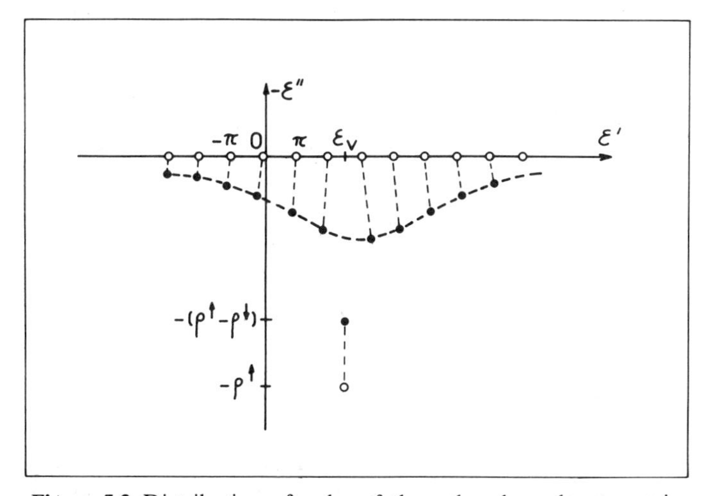

Figure 5.3 Distribution of poles of the reduced resolvent matrix element $\langle V|MGM|V\rangle$ in the complex energy plane: solutions of the generalized eigenvalue equation (5.126), $\varepsilon_p - \varepsilon_V + i\rho^{\uparrow} - \rho^{\downarrow} \operatorname{cotg} \varepsilon_p = 0$. The open circles represent the zero-order values of the spectrum; due to the residual interaction, the poles move to the corrected values given by the full circles. Case I ($1 < \rho^{\downarrow} < \rho^{\uparrow}$): weak mixing by fast radiant state decay

to the zero-order model system, in the absence of intramolecular coupling. The corrected values $\varepsilon_p = \varepsilon'_p - i\varepsilon''_p$ accounting for the influence of intramolecular coupling are represented by the full circles. For the poles ε_n, the distribution is seen to be symmetric about $\varepsilon' = \varepsilon_V$, the reduced line widths being located on a Lorentzian curve of width at half maximum equal to $\rho^{\uparrow}$. Under the present weak mixing conditions, the pole ε_{2N+2} is well separated from the $2N + 1$ poles corresponding to the non-radiant set (5.120); it therefore retains its special radiant character. But if intramolecular coupling $\rho^{\downarrow}$ increases, the pole ε_{2N+2} moves towards the real axis and tends to approach the $2N + 1$ other poles and to merge with them: it then loses its special identity and the radiant character becomes spread over many of the $2N + 2$ compound states (cf. *infra*). We further note that the modulus of the residues R_n is given by the Lorentzian:

$$|R_n| = \rho^{\downarrow}[(n\pi - \varepsilon_V)^2 + (\rho^{\uparrow})^2]^{-1}$$

which, according to the weak mixing conditions (5.131) is very much smaller than the residue $R_{2N+2} = 1$. This shows that the term related to the radiant level in the basic meromorphic expansion (5.90) is the principal one: the dynamical behaviour of the excited system essentially involves the radiant level; the non-radiant levels to be eventually taken into account are distributed within an interval $\Delta\varepsilon' \approx \rho^{\uparrow}$ about the radiant state.

Strong mixing. The present discussion will essentially be concerned with systems where natural damping is larger for the radiant state than for the non-radiant $|\bar{V}'\rangle$ states: $\Gamma_V^{\updownarrow} > \Gamma_{\bar{V}'}$, i.e., $\rho^{\uparrow} > 0$. The cases where inverse conditions ($\rho^{\uparrow} < 0$) hold appear to be physically less interesting in molecular systems; they do not present essentially new features and are easily discussed in the light of the results in the following sections (cf. *infra*).

We begin to examine physical situations where the distribution (5.120) of the non-radiant $|\bar{V}'\rangle$ levels is sufficiently extended, so that $N > \mathfrak{N}$ (cf. equation (5.133)). Not all the $|\bar{V}'\rangle$ levels are then submitted to strong mixing; the boundaries of the set (5.120), in particular, are only weakly mixed, and no special edge effects are expected. Under such conditions, the results of the picket-fence model may be used (§5.4.2.1). Also, without significant loss of generality, we can assume that $E_V = E_0$ ($\varepsilon_V = 0$).

For the $2(N - \mathfrak{N})$ non-radiant levels that do not satisfy the strong mixing condition, $(\rho^{\downarrow}/2\pi)[\pi^2 n^2 + (\rho^{\uparrow})^2]^{-\frac{1}{2}} > 1$, all the results of the beginning of this section are valid (weak mixing).

In order to determine the remaining $2\mathfrak{N} + 2$ poles for the strongly mixed molecular states, we first note that, according to the second of equations (5.126″), written as:

$$\frac{\rho^{\uparrow} - \varepsilon''_p}{\rho^{\downarrow}} = \operatorname{th} \varepsilon''_p \frac{1 + \operatorname{tg}^2 \varepsilon'_p}{\operatorname{tg}^2 \varepsilon'_p + \operatorname{th}^2 \varepsilon''_p} \tag{5.147}$$

the quantities $\rho^{\uparrow} - \varepsilon''_p$ and ε''_p have the same sign.

Following the sum rule (5.116), taken under the equivalent form:

$$\sum_p \varepsilon''_p = \rho^{\uparrow} \tag{5.148}$$

the sign is plus since, as mentioned before, we take $\rho^{\uparrow} > 0$. As a consequence, we can write:

$$\rho^{\uparrow} \geqslant \varepsilon''_p \geqslant 0$$

and, remembering the strong mixing condition (5.132″):

$$\rho^{\downarrow} \gg \rho^{\uparrow} \geqslant \varepsilon''_p \tag{5.149'}$$

These relations (5.149′) indicate that the two sides of equation (5.147) are both very small, so that:

$$\operatorname{th} \varepsilon''_p \approx \varepsilon'_p \ll 1 \tag{5.149''}$$

The use of the simplified equations (5.127) to determine the poles $\varepsilon'_p - i\varepsilon''_p$ is thus justified.

The real parts ε'_p are obtained by solving the transcendental equation:

$$\varepsilon'_p = \rho^{\downarrow} \operatorname{cotg} \varepsilon'_p \tag{5.150}$$

by numerical or graphical methods. Figure 5.4 shows that the solutions are sandwiched between the zero-order values $\varepsilon' = n\pi$ corresponding to (5.120). A one-to-one correspondence still connects the solutions ε'_p to the zero-order spectrum; but near $\varepsilon' = 0$, the energy differences $\Delta\varepsilon'_n = \varepsilon'_n - n\pi$ cannot be considered as very much smaller than unity by contrast to the weak mixing cases. The imaginary parts are obtained from equation (5.127″) as:

$$\varepsilon''_p = \rho^{\uparrow} \frac{\rho^{\downarrow}}{(\varepsilon'_p)^2 + (\rho^{\downarrow})^2 + (\rho^{\downarrow})} \tag{5.151}$$

They are distributed around $\varepsilon' = 0$ with a width determined by $\rho^{\downarrow} \gg 1$. No essential difference exists between the pole ε_{2N+2} related to the radiant level and the poles related to the neighbouring non-radiant levels: while ε''_{2N+2} has the greatest possible value, $\varepsilon''_{2N+2} = \rho^{\uparrow}/(\rho^{\downarrow} + 1)$, the other widths remain of the same order in the central part of the distribution:

$$\varepsilon''_p \leqslant \varepsilon''_{2N+2} = \rho^{\uparrow}/(\rho^{\downarrow} + 1) \tag{5.152}$$

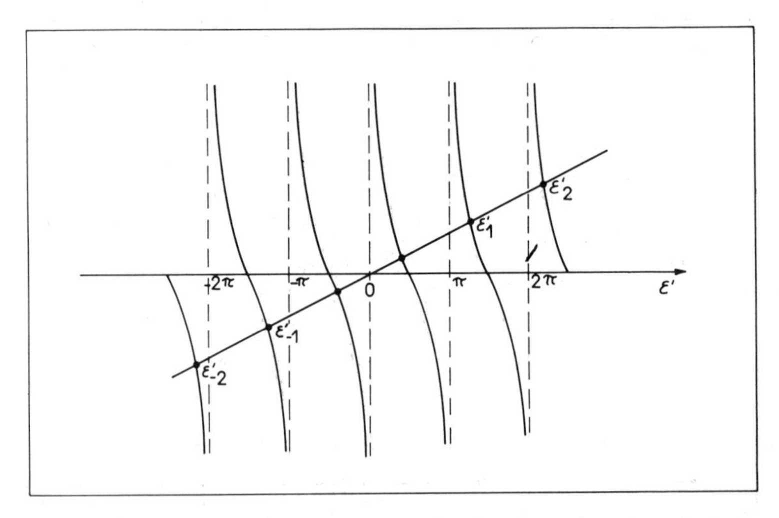

Figure 5.4 Graphic solution of equation (5.150), $\varepsilon'_p = \rho^{\downarrow} \operatorname{cotg} \varepsilon'_p$: the intersections give the real parts ε'_p of the poles of $\langle V|MGM|V\rangle$. Strong mixing conditions ($\rho^{\uparrow} < \rho^{\downarrow}$)

These characteristic features of the pole distribution are represented in Figure 5.5, where the relation between the zero-order spectrum (open circles) and the poles ε_p (full circles) clearly appears; a comparison with Figure 5.3 stresses the differences with the weak mixing situations.

The residues R_p are given by relation (5.128″):

$$R_p = \frac{\rho^{\downarrow}}{(\varepsilon'_p + i\rho^{\uparrow})^2 + (\rho^{\uparrow})^2 + \rho^{\downarrow}}$$

which, under the presently valid condition $\rho^{\downarrow} \gg \rho^{\uparrow}$, simplifies to:

$$R_p = \frac{\rho^{\downarrow}}{(\varepsilon'_p)^2 + (\rho^{\downarrow})^2 + \rho^{\downarrow}} \tag{5.153}$$

This expression is seen to be related to (5.151) as:

$$R_p = \varepsilon''_p/\rho^{\uparrow} \tag{5.154}$$

so that the distribution of residues is similar to that of the widths ε''_p. If $\rho^{\downarrow}$ is large enough, it is relatively smooth, and the central value R_{2N+2} does not significantly exceed the residues for the non-radiant levels in the immediate vicinity of the radiant state:

$$R_p \leqslant R_{2N+2} = (\rho^{\downarrow} + 1)^{-1} \tag{5.155}$$

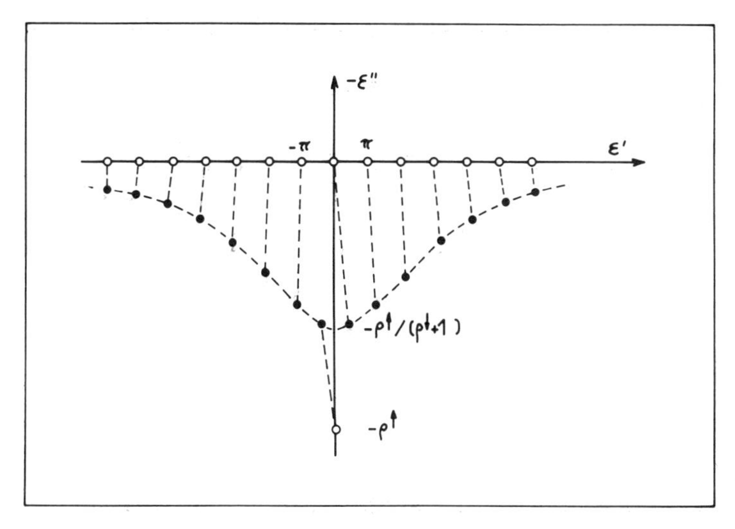

Figure 5.5 Distribution of poles of $\langle V|MGM|V\rangle$ for strong mixing conditions (cf. Figure 5.3, for the symbols used). Cases IV and V $(1, \rho^{\uparrow} < \rho^{\downarrow})$

Together with (5.152), this expresses the fact that the original radiant character is now smoothly redistributed over all the strongly mixed zero-order states.

As stated at the beginning of this section devoted to strong mixing, the results obtained suppose that the boundaries $E_0 \pm N\bar{D}$ of the non-radiant $|\bar{V}'\rangle$ manifold (equation (5.120)) are but weakly mixed. We now consider the cases where one or both edges of the distribution (5.120) are submitted to strong mixing.

Strong mixing: edge effects. The special poles, related to boundary levels submitted to strong mixing, cannot be determined using the simple picket-fence model, as mentioned in §5.4.2.1. Instead of expression (5.122) for the function $\bar{\Phi}_V(E + i\frac{1}{2}\Gamma_{\bar{V}'})$, one has to take the more realistic form (5.121), and write the basic equation for the poles as:

$$(\varepsilon_p' - \varepsilon_V) - i(\varepsilon_p'' - \rho^{\uparrow}) - \rho^{\downarrow} \sum_{n=-N}^{+N} \frac{1}{(\varepsilon_p' - n\pi) - i\varepsilon_p''} = 0 \qquad (5.156)$$

in place of equation (5.126). The transcendental equation replacing (5.150), for the real parts, likewise reads:

$$\varepsilon_p' - \varepsilon_V = \rho^{\downarrow} \sum_{n=-N}^{+N} (\varepsilon_p' - n\pi)^{-1} \qquad (5.157)$$

and may also be solved by graphical methods. For the other remaining poles—those not related to a strongly mixed edge level—no essential difference is expected with the previous picket-fence results, which can therefore be used.

1. We first examine the cases where a *single edge* of the $|\bar{V}'\rangle$ manifold has some influence. Usually this may be expected to be the lowest one, which is the most sharply defined as the lowest vibrational level of the strongly coupled non-radiant electronic configuration. One may thus consider a molecular system, where the radiant level E_V is far from one of the boundaries, say the upper $E_0 + N\bar{D}$, of the $|\bar{V}'\rangle$ distribution, so that:

$$E_0 + N\bar{D} - E_V \gg \tfrac{1}{2}\Gamma_V^{\downarrow} \quad \text{or} \quad N\pi - \varepsilon_V \gg \rho^{\downarrow} \tag{5.158'}$$

On the other hand, the radiant level is assumed to be sufficiently near to the other boundary (the lowest, $E_0 - N\bar{D}$), with an energy gap $\delta E_V = E_V - (E_0 - N\bar{D})$ such that:

$$\delta E_V = E_V - E_0 + N\bar{D} \lesssim \tfrac{1}{2}\Gamma_V^{\downarrow} \quad \text{or} \quad \delta\varepsilon_V = \varepsilon_V + N\pi \leqslant \rho^{\downarrow} \tag{5.158''}$$

(cf. Figure 5.6). Under these conditions, only the lower boundary level $E_0 - N\bar{D}$ is strongly mixed.

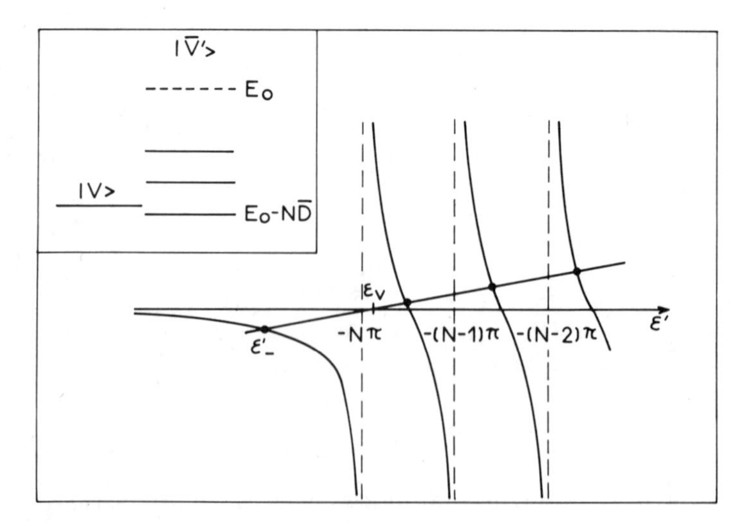

Figure 5.6 Graphic solution of equation (5.157),

$$\varepsilon_p' - \varepsilon_V = \rho^{\downarrow} \sum_{n=-N}^{+N} (\varepsilon_p' - n\pi)^{-1}$$

under strong mixing conditions, with the indicated level scheme. Note the special solution separating below the lower edge of the energy band

The special pole $E_- = E'_- - iE''_-$ related to this level accounts for the particular edge effects we are looking for. From the graphical representation of equation (5.157), given in Figure 5.6, it is seen that the real part separates below the edge $E_0 - N\bar{D}$ so much more as the coupling strength (or $\rho^\downarrow$) is larger. To determine ε'_-, we keep only the dominant term in the right-hand side of (5.157), and write:

$$(\varepsilon'_- + N\pi) - \delta\varepsilon_V = \rho^\downarrow/(\varepsilon'_- + N\pi) \tag{5.159}$$

which yields:

$$\varepsilon'_- = -N\pi - \tfrac{1}{2}\delta\varepsilon_V\left\{\left[1 + \frac{4\rho^\downarrow}{(\delta\varepsilon_V)^2}\right]^{\frac{1}{2}} - 1\right\} \tag{5.160'}$$

or

$$E'_- = E_0 - N\bar{D} - \tfrac{1}{2}\delta E_V\left\{\left[1 + \left(\frac{2u}{\delta E_V}\right)^2\right]^{\frac{1}{2}} - 1\right\} \tag{5.160''}$$

Note that, when the radiant level tends to coincide with the distribution edge, i.e., when δE_V becomes very small compared to $2u$, one gets the simple result:

$$E'_- = E_0 - N\bar{D} - u \tag{5.161}$$

The imaginary part ε''_- can be evaluated from the imaginary part of equation (5.156), in which the sum is approximated by its principal term ($n = -N$):

$$\varepsilon''_- - \rho^\uparrow + (\varepsilon''_-)\rho^\downarrow[(\varepsilon'_- + N\pi)^2 + (\varepsilon''_-)^2]^{-1} = 0 \tag{5.162}$$

This equation can be transformed by noting that, following equations (5.159) and (5.160), one has:

$$(\varepsilon'_- + N\pi)^2 = \rho^\downarrow - \tfrac{1}{2}(\delta\varepsilon_V)^2\{[1 + 4\rho^\downarrow/(\delta\varepsilon_V)^2]^{\frac{1}{2}} - 1\}$$

Also $(\varepsilon''_-)^2$ may be neglected, since the sum rule (5.148) implies that $\varepsilon''_- < \rho^\uparrow$, so that: $\varepsilon''_- \ll \rho^\downarrow$, bearing in mind the strong mixing criterion $\rho^\uparrow \ll \rho^\downarrow$. Equation (5.162) thus yields:

$$\varepsilon''_- = \rho^\uparrow\left\{1 + \frac{\rho^\downarrow}{\rho^\downarrow - \tfrac{1}{2}(\delta\varepsilon_V)^2\{[1 + 4\rho^\downarrow/(\delta\varepsilon_V)^2]^{\frac{1}{2}} - 1\}}\right\}^{-1} \tag{5.163}$$

Note that, in the limit $\delta\varepsilon_V \to 0$, the reduced linewidth tends to:

$$\varepsilon''_- = \rho^\uparrow/2 \tag{5.164}$$

Figure 5.7 illustrates, for the particular case of vanishing energy gap, the positions of the poles ε_p. Strong mixing of the lowest boundary $\varepsilon' = -N\pi$ of the $|\bar{V}'\rangle$ manifold pushes the lowest level to still lower energies,

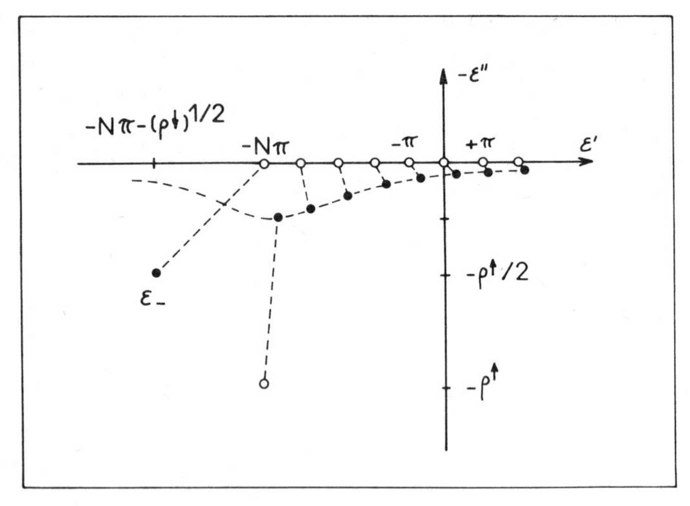

Figure 5.7 Distribution of poles of $\langle V|MGM|V\rangle$ for strong mixing and the level scheme of Figure 5.6: the pole ε_- is responsible for the edge effects considered under Case IV′. The symbols are the same as in Figure 5.3

$-\pi N - (\rho^{\downarrow})^{\frac{1}{2}}$, and endows it with about one half of the existent decay width $\rho^{\uparrow}$. When molecular interstate coupling is large (i.e., if $\rho^{\downarrow} \gg 1$), the special pole ε_- is seen to be well separated from the $2N + 1$ remaining ones, which are described by equations (5.150)–(5.152).

To evaluate the residue R_- of $\langle V|MG(E)M|V\rangle$ at the special pole, one best takes the general expression (5.117), keeping, as before, the sole term with $n = -N$ in the sums over n. R_- is then found to be given by an expression like (5.154), i.e.:

$$R_- = (\varepsilon''_-/\rho^{\uparrow}) \tag{5.165}$$

with ε''_- of the form (5.163). If the radiant level is near the edge ($\delta E_V = 0$), one has again a simple result:

$$R_- = \tfrac{1}{2} \tag{5.166}$$

The residues for the remaining, non-special poles are given by the equations (5.153)–(5.155), and are seen to be much smaller than R_-, if $\rho^{\downarrow}$ is large.

All these results mean that a large fraction—possibly approaching one half of the original radiant character—concentrates on the single pole related to the strongly coupled edge level; the remaining part is smoothly distributed over the $2N + 1$ other levels. This indicates that the special pole will be responsible for significant new features in the characteristics of light emission. The other possible cases, where single edge effects are due to the upper boundary level $E_0 + N\bar{D}$, instead of the lower one, can be treated in a similar way, with similar results: the real part of the special pole, $E_+ = E'_+ - iE''_+$, here separates above the upper edge by an absolute amount equal to that in equation (5.160); E''_+ and R_+ are obtained under the same forms as E''_- and R_- in equations (5.163)–(5.166).

2. Effects that can be ascribed to the *two edges* of the $|\bar{V}'\rangle$ distribution (5.120) have to be considered under extreme strong mixing conditions, when the totality of the $2N + 2$ zero-order radiant and non-radiant states ($|V\rangle + \{|\bar{V}'\rangle\}$) are strongly coupled together. This implies that the number N of $|\bar{V}'\rangle$ levels must be sufficiently small and interstate coupling u sufficiently strong, so that (cf. equation (5.133)):

$$\mathfrak{N} > N$$

Here we have to determine the special poles ε_+ and ε_-, respectively related to the two edges $E_0 + N\bar{D}$ and $E_0 - N\bar{D}$ of distribution (5.120), by taking the equations (5.156) and (5.157); for simplicity, we may assume that $E_V = E_0$ ($\varepsilon_V = 0$). The $2N$ other poles, different from the special ones, are given by the foregoing picket-fence results (cf. equations (5.150)–(5.155)).

Due to the symmetry properties of the zero-order energy spectrum (cf. Figure 5.8), the real part of the poles $\varepsilon_p = \varepsilon'_p - i\varepsilon''_p$ are distributed

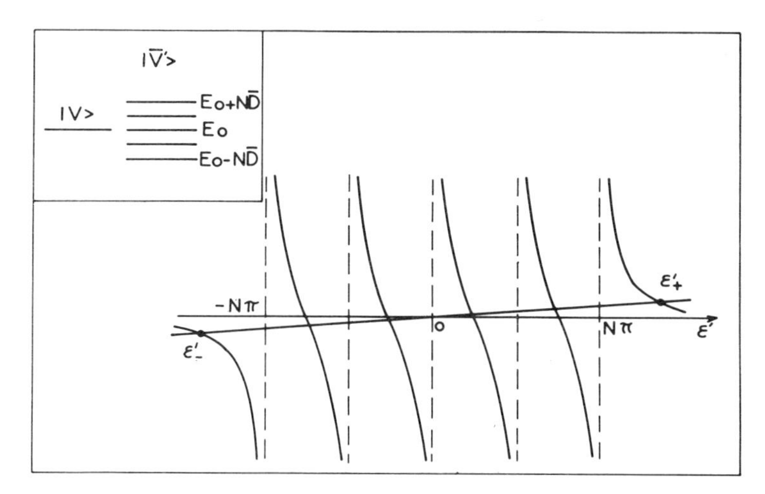

Figure 5.8 Graphic solution of equation (5.157) under strong mixing conditions, for the indicated level scheme. Note the two special roots ε'_- and ε'_+ which separate below and above the energy level band

symmetrically about $\varepsilon' = 0$; on the other hand ε''_p and R_p only depend on $(\varepsilon'_p)^2$. For the special poles and residues, we thus have:

$$\varepsilon'_+ = -\varepsilon'_-, \qquad \varepsilon''_+ + \varepsilon''_-, \qquad R_+ = R_- \tag{5.167}$$

Figure 5.8 shows that the real part ε'_+ and ε'_- do indeed separate symmetrically out of the zero-order spectrum, in an extent determined by $\rho^\downarrow$, i.e., the interstate coupling strength. To determine one of these quantities, ε'_- for example, we keep only the dominant term in the sum of equation (5.157), and write:

$$\varepsilon'_- = \rho^\downarrow/(\varepsilon'_- + N\pi)$$

with the result:

$$\varepsilon'_- = -\frac{N\pi}{2}\left\{1 + \left[1 + \frac{4\rho^\downarrow}{(N\pi)^2}\right]^{\frac{1}{2}}\right\}$$

Hence:

$$\varepsilon'_\pm = \pm\frac{N\pi}{2}\left\{1 + \left[1 + \left(\frac{2u}{N\bar{D}}\right)^2\right]^{\frac{1}{2}}\right\} \tag{5.168}$$

If interstate coupling is so large that $(2u/N\bar{D}) \gg 1$, one gets the simple result that:

$$\varepsilon'_\pm = \pm(\pi u/\bar{D}) \quad \text{or} \quad E'_\pm = E_0 \pm u \tag{5.169}$$

The imaginary part $\varepsilon''_+ = \varepsilon''_-$ can be simply evaluated with the help of the sum rule (5.148), which reads here:

$$\varepsilon''_+ + \varepsilon''_- + \sum_p{}' \varepsilon''_p = \rho^\uparrow$$

where

$$\sum_p{}'$$

is understood to represent a summation over the $2N$ non-special poles for which ε''_p is given by equation (5.151). One thus gets:

$$2\varepsilon''_+ = \rho^\uparrow\left[1 - \sum_p{}' \frac{\rho^\downarrow}{(\varepsilon'_p)^2 + (\rho^\downarrow)^2 + \rho^\downarrow}\right] \tag{5.170}$$

where the sum is necessarily smaller than $2N(\rho^\downarrow + 1)$, which itself is very small in the extreme strong coupling conditions. The reduced line widths are hence seen to be:

$$\varepsilon''_+ = \varepsilon''_- \approx (\rho^\uparrow/2) \tag{5.171}$$

To determine the residues $R_+ = R_-$, one proceeds in a similar way, considering now the sum rule (5.118), to obtain:

$$2R_+ = 1 - \sum_p{}' \frac{\rho^\downarrow}{(\varepsilon'_p)^2 + (\rho^\downarrow)^2 + \rho^\downarrow} \tag{5.172}$$

and

$$R_+ = R_- \approx \tfrac{1}{2} \tag{5.173}$$

These results show that extreme strong interstate coupling, including the two edges of the non-radiant $|\bar{V}'\rangle$ manifold, mixes the zero-order states in such a way as to push two of the states significantly out of the primitive spectrum and to endow them with almost all the radiant character, while the others are practically denuded. Interstate coupling is here responsible for drastic changes in the optical properties of the excited system.

Having determined the distributions of poles and residues of the reduced resolvent matrix element $\langle V|MG(E^+)M|V\rangle$, we are now in position to analyze in detail the decay law and the characteristic features of light emission as defined in §5.3.4.

5.4.2.3 *Decay and light emission modes*

General expressions. Decay law $P_V(t)$ and *light intensity* $I(t) = \gamma_R P_V(t)$ are determined by the probability amplitude $A_V(t)$, given by expansion (5.93) in terms of the poles E_p and the residues R_p. Introducing the dimensionless quantities (5.125), together with the reduced time variable

$$\tau = \bar{D}t/\pi \tag{5.174}$$

we may equivalently take:

$$A_V(t) = e^{-\frac{1}{2}\Gamma_{\bar{V}'}t} e^{-iE_0 t} \left[\sum_p R_p\, e^{-\varepsilon''_p \tau} e^{-i\varepsilon'_p \tau} \right] \tag{5.175}$$

The general expressions of $P_V(t) = |A_V(t)|^2$ and $I(t) = \gamma_R |A_V(t)|^2$ are separated into two naturally distinct parts:

$$P_V(t) = P'_V(t) + P''_V(t); \quad I(t) = I'(t) + I''(t) \tag{5.176}$$

respectively representing normal monotonic decay and interference effects. Using (5.93) and (5.175), we write:

$$I'(t) = \gamma_R P'_V(t)$$

$$= \gamma_R \left[\sum_p |R_p|^2 e^{-2E''_p t} \right] \tag{5.177'}$$

$$= \gamma_R\, e^{-\Gamma_{\bar{V}'}t} \left[\sum_p |R_p|^2 e^{-2\varepsilon''_p \tau} \right] \tag{5.177''}$$

obtaining a sum of different exponentially decaying components, each with its own decay constant $(2E''_p)^{-1}$. The interference part reads:

$$
\begin{aligned}
I''(t) &= \gamma_R P''_V(t) \\
&= \gamma_R \left\{ \sum_{p \neq p'} R_p R^*_{p'} \, \mathrm{e}^{-(E''_p + E''_{p'})t} \, \mathrm{e}^{-\mathrm{i}(E'_p - E'_{p'})t} \right\} \\
&= \gamma_R \left\{ \sum_{p < p'} [R_p R^*_{p'} \, \mathrm{e}^{-(E''_p + E''_{p'})t} \, \mathrm{e}^{-\mathrm{i}(E'_p - E'_{p'})t} + C.c] \right\}
\end{aligned}
\tag{5.178'}
$$

or equivalently:

$$
I''(t) = \gamma_R \, \mathrm{e}^{-\Gamma_{\bar{V}'} t} \left\{ \sum_{p < p'} [R_p R^*_{p'} \, \mathrm{e}^{-(\varepsilon''_p + \varepsilon''_{p'})\tau} \, \mathrm{e}^{-\mathrm{i}(\varepsilon'_p - \varepsilon'_{p'})\tau} + C.c] \right\} \tag{5.178''}
$$

Each of these terms adds an oscillatory component with a frequency $(E'_p - E'_{p'})$, which superimposes on the smooth envelope represented by $I'(t)$. In actuality, the available experimental information on light intensity decay is obviously limited by finite time resolution of the photon detectors; the recurrent term, in particular, will be averaged to zero in many practical circumstances.

The *energy distribution of fluorescence emission* is described by the distribution function $J(E_K)$ which, following equation (5.102''), can be considered under the general form:

$$
J(E_K) = |\langle K|H'|V\rangle|^2 \sum_{p,p'} \frac{R_p R^*_{p'}}{(E_K - E'_p + \mathrm{i}E''_p)(E_K - E'_{p'} - \mathrm{i}E''_{p'})} \tag{5.179}
$$

Here again, it is in order to distinguish the normal and the interference components; one thus writes:

$$
J(E_K) = J'(E_K) + J''(E_K) \tag{5.180}
$$

with:

$$
J'(E_K) = |\langle K|H'|V\rangle|^2 \sum_p \frac{|R_p|^2}{(E_K - E'_p)^2 + (E''_p)^2} \tag{5.181}
$$

and:

$$
J''(E_K) = |\langle K|H'|V\rangle|^2 \sum_{p \neq p'} \frac{R_p R^*_{p'}}{(E_K - E'_p + \mathrm{i}E''_p)(E_K - E'_{p'} - \mathrm{i}E''_{p'})} \tag{5.182}
$$

The normal component represents a spectrum of pure Lorentzian lines, each centred at the energy E'_p with a width E''_p. The lines form a well resolved pattern if $E''_p < \bar{D}$; in such cases, the various terms in the interference component $J''(E_K)$ are verified to be small. On the other hand, if $E''_p > \bar{D}$

(overlapping resonances), a non-negligible contribution from the interference terms (5.182) must be added to the normal component $J'(E_K)$, which results in deviation from the simple spectrum of overlapping Lorentzian distributions.

The other experimentally significant quantity is *light emission quantum yield*. Its expression may be considered under the form (5.101″):

$$Q = -\mathrm{i}\gamma_R \sum_p R_p \langle V|MG(E)M|V\rangle^*_{E_p} \tag{5.101″}$$

where:

$$\langle V|MG(E)M|V\rangle^*_{E_p} = [E_p - E_V - \mathrm{i}\tfrac{1}{2}\Gamma_V^{\updownarrow} - \overline{\Phi}_V(E_p - \mathrm{i}\tfrac{1}{2}\Gamma_{\overline{V}'})]^{-1}$$

or, if we remember that the energy E_p satisfies equation (5.91′):

$$\langle V|MG(E)M|V\rangle^*_{E_p} = [-\mathrm{i}\Gamma_V^{\updownarrow} + \overline{\Phi}_V(E_p + \mathrm{i}\tfrac{1}{2}\Gamma_{\overline{V}'}) - \overline{\Phi}_V(E_p - \mathrm{i}\tfrac{1}{2}\Gamma_{\overline{V}'})]^{-1}$$

Now, considering condition (5.112) and the definition (5.117) of R_p, we can successively write:

$$\overline{\Phi}_V(E_p + \mathrm{i}\tfrac{1}{2}\Gamma_{\overline{V}'}) - \overline{\Phi}_V(E_p - \mathrm{i}\tfrac{1}{2}\Gamma_{\overline{V}'}) \approx \mathrm{i}\Gamma_{\overline{V}'}\left(\frac{\mathrm{d}\overline{\Phi}_V}{\mathrm{d}E}\right)_{E_p} = \mathrm{i}\Gamma_{\overline{V}'}(1 - R_p^{-1})$$

Hence:

$$\langle V|MG(E)M|V\rangle^*_{E_p} = [-\mathrm{i}\Gamma_V^{\updownarrow} - \mathrm{i}\Gamma_{\overline{V}'}(R_p^{-1} - 1)]^{-1}$$

and:

$$Q = \gamma_R \sum_p \frac{R_p}{\Gamma_V^{\updownarrow} + \Gamma_{\overline{V}'}(R_p^{-1} - 1)} \tag{5.183}$$

We may note at once that, for negligible broadening of the non-radiant $|\overline{V}'\rangle$ levels, with $\Gamma_{\overline{V}'} \approx 0$, one has:

$$Q = (\gamma_R/\Gamma_V^{\updownarrow}) \sum_p R_p = \gamma_R/\Gamma_V^{\updownarrow} \tag{5.184}$$

where use was made of the sum rule (5.118). The physical interpretation of this general result is obvious, since natural decay of the radiant state, as described by (5.113), provides the unique dissipation channels in this extreme case.

The general expressions (5.176)–(5.183) allow one to translate the previously discussed features of the distributions of poles and residues (cf. §5.4.2.2), into physically significant information. We are now in position for a comprehensive analysis of the decay and light emission characteristics in the various particular cases of interest. We begin with the weak mixing situations, first considering the case of rapidly decaying radiant states.

Case I: Weak mixing by fast decay of the radiant level. The defining conditions are given by the relations (5.145) and were discussed before. It should be noted that this case is expected to apply quite frequently for the molecular systems in the condensed phase, as soon as radiant levels with high vibrational or electronic excess energies are considered.

The distributions of poles and residues are represented by the equations (5.146), and illustrated in Figure 5.3. We already noted the special role of the pole E_{2N+2} related to the initially excited radiant level; in all the expressions displayed as expansions over the poles, the term corresponding to E_{2N+2} is found to be dominant.

Consider firstly decay or *the variations of light intensity* $I(t)$, which was separated into its naturally distinct components, $I'(t)$ and $I''(t)$, describing respectively normal decay and interference effects (equations (5.176)–(5.178)). For the normal monotonically decaying intensity component, we write:

$$I'(t) = I'_1(t) + I'_2(t) \tag{5.185'}$$

with:

$$I'_1(t) = \gamma_R |R_{2N+2}|^2 \exp(-2E''_{2N+2}t) \tag{5.185''}$$

$$I'_2(t) = \gamma_R \sum_{n=-N}^{+N} |R_n|^2 \exp(-2E''_n t) \tag{5.185'''}$$

In these relations, one has $|R_n|^2 \ll |R_{2N+2}|^2 = 1$ and $E''_n \ll E''_{2N+2} = \frac{1}{2}(\Gamma_V^{\uparrow} - \Gamma_V^{\downarrow})$. In the practically most frequent examples, where the decay is studied for the initial times, $t \lesssim (\Gamma_V^{\uparrow})^{-1}$, the term $I'_1(t)$ is thus the principal one. Due to the presence of the factors $|R_n|^2 \ll 1$, the other term $I'_2(t)$ is generally very weak at the beginning; but, decaying much more slowly than $I'_1(t)$, it gains some relative importance at longer times. In actual situations, we may however expect that $I'(t) \approx I'_1(t)$, i.e.:

$$I'(t) \approx \gamma_R \exp[-(\Gamma_V^{\uparrow} - \Gamma_V^{\downarrow})t] \tag{5.186}$$

For the interference component, we can proceed in a similar way, and argue that only the terms involving $R_{2N+2} = 1$ need to be taken into account in equation (5.178):

$$I''(t) \approx \gamma_R \sum_{n=-N}^{+N} [R_n R^*_{2N+2} e^{-i(E'_n - E'_{2N+2})t} e^{-(E''_n + E''_{2N+2})t} + C.c]$$

$$= 2\gamma_R \sum_{n=-N}^{+N} \operatorname{Re}[R_n R^*_{2N+2} e^{-i(E'_n - E'_{2N+2})t}] e^{-(E''_n + E''_{2N+2})t} \tag{5.187}$$

Taking the values (5.146″), we obtain more precisely:

$$I''(t) = \gamma_R \, e^{-\Gamma_V^\uparrow t} \left\{ e^{-(\rho^\uparrow - \rho^\downarrow)\tau} \sum_{n=-N}^{+N} \frac{\rho^\downarrow}{[(n\pi - \varepsilon_V) + i\rho^\uparrow]^2} e^{-i(n\pi - \varepsilon_V)\tau} + C.c \right\}$$

or:

$$I''(t) = \gamma_R \, e^{-\Gamma_V^\uparrow t} \, e^{\frac{1}{2}\Gamma_V^\downarrow t} \rho^\downarrow \left\{ \sum_n \frac{e^{-i[(n\pi - \varepsilon_V) + i\rho^\uparrow]\tau}}{[(n\pi - \varepsilon_V) + i\rho^\uparrow]^2} + C.c \right\}$$

Since, according to the defining conditions of the present case, one has $\rho^\uparrow \gg 1$, the factors R_n vary slowly with n; the sum over n can then be replaced by an integral:

$$\sum_n \rightarrow \int dx/\pi$$

Assuming further that the edges of the non-radiant manifold (5.120) are sufficiently far from the radiant level, the integration can be carried out over the whole energy axis. Hence:

$$I''(t) = \gamma_R \, e^{-\Gamma_V^\uparrow t} \, e^{\frac{1}{2}\Gamma_V^\downarrow t} \frac{\rho^\downarrow}{\pi} \int_{-\infty}^{+\infty} \frac{e^{-i(x + i\rho^\uparrow)\tau}}{(x + i\rho^\uparrow)^2} dx + C.c$$

By the residue method, the integral is evaluated as $-2\pi\tau$, so that one gets:

$$I''(t) = -2\gamma_R(\Gamma_V^\downarrow t) \, e^{-\Gamma_V^\uparrow t} \, e^{\frac{1}{2}\Gamma_V^\downarrow t} \qquad (5.188)$$

In the present case, the interference intensity component is also characterized by monotonic variation because important damping prevents true recurrent processes. Intramolecular coupling is however seen to be responsible for deviation from pure exponential decay. As the damping factor $\exp(-\Gamma_V^\uparrow t)$ restricts light emission to times less than, or of the order of $(\Gamma_V^\uparrow)^{-1}$, for which $\Gamma_V^\downarrow t \ll 1$, we need only to consider quantities of the lowest order of $(\Gamma_V^\downarrow t)$; the intensity $I(t) = I'(t) + I''(t)$ then becomes:

$$I(t) = \gamma_R[1 - (\Gamma_V^\downarrow t) - \tfrac{1}{2}(\Gamma_V^\downarrow t)^2 + \ldots] \, e^{-\Gamma_V^\uparrow t} \qquad (5.189)$$

We note that the initial slope of the decay curves is the same as that of the exponential $\exp[-(\Gamma_V^\uparrow + \Gamma_V^\downarrow)t]$, but subsequently the decay rate increases (cf. Figure 5.9).

Turning to the *emission energy spectrum* $J(E_K)$, as described by equations (5.180)–(5.182), we likewise have to distinguish the normal and interference parts, $J'(E_K)$ and $J''(E_K)$. Again, we note that the major terms in the expansions representing these quantities are those involving the special pole E_{2N+2}. For the normal component, we thus write:

$$J'(E_K) = J'_1(E_K) + J'_2(E_K) \qquad (5.190')$$

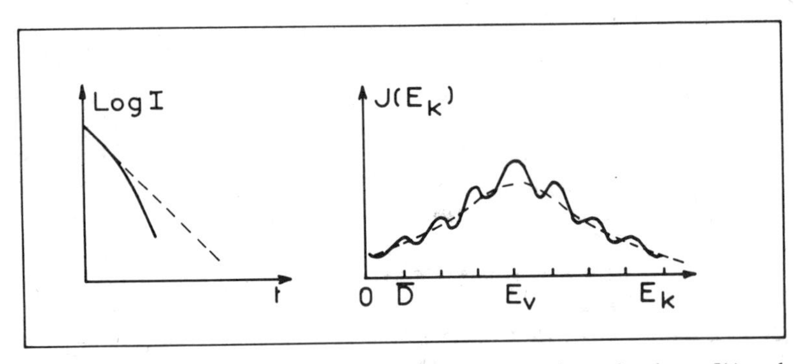

Figure 5.9 Schematic representation of fluorescence intensity decay $I(t)$ and spectral distribution $J(E_K)$. Case I ($1 < \rho^{\downarrow} < \rho^{\uparrow}$): weak mixing through fast decay of the radiant state

with:

$$J_1'(E_K) = |\langle K|H'|V\rangle|^2 \frac{|R_{2N+2}|^2}{(E_K - E_{2N+2}')^2 + (E_{2N+2}'')^2} \qquad (5.190'')$$

$$J_2'(E_K) = |\langle K|H'|V\rangle|^2 \sum_{n=-N}^{+N} \frac{|R_n|^2}{(E_K - E_n')^2 + (E_n'')^2} \qquad (5.190''')$$

In the present case, one has:

$$J_1'(E_K) = |\langle K|H'|V\rangle|^2 \frac{1}{(E_K - E_V)^2 + [\frac{1}{2}(\Gamma_V^{\uparrow} - \Gamma_V^{\downarrow})]^2} \qquad (5.191)$$

which describes an important and broad Lorentzian line extending over a large number of level spacing $\bar{D}$. Since $|R_n|^2 \ll 1$ and $E_N' \ll \bar{D}$, the remaining term $J_2'(E_K)$ (equation (5.190''')) represents a set of very weak, narrow lines, separated by the level spacing $\bar{D}$. The special resonance E_{2N+2} is seen to overlap, and therefore to interfere, with all the others; as the latter, non-special poles do not overlap, we neglect their mutual interference effects. The component $J''(E_K)$ may therefore be considered as:

$$J''(E_K) = |\langle K|H'|V\rangle|^2 \sum_{n=-N}^{+N}$$

$$\times \left[\frac{R_n}{(E_K - E_{2N+2}' - iE_{2N+2}'')(E_K - E_n' + iE_n'')} + C.c\right] \qquad (5.192)$$

Since $|R_n| \ll 1$, it is seen to be weaker than (5.191), but stronger than $J_2'(E_K)$; the expression (5.192) thus describes the essentials of the fine structure which modulates the broad Lorentzian (5.191) in the emission spectrum. It rapidly varies around the energies $E_K = E_0 + n\bar{D}$, with line shapes that are similar to those ascribed to interference between potential and resonance scattering in nuclear elastic scattering theory,[5] or the Fano-type profiles in atomic spectroscopy.[6]

As a final point, we consider *fluorescence quantum yield.* Its expression (5.183) is also displayed as the sum of two terms, with the first corresponding to $p = 2N + 2$:

$$Q = Q_1 + Q_2 \tag{5.193'}$$

$$Q_1 = \gamma_R \frac{R_{2N+2}}{\Gamma_V^{\updownarrow} + \Gamma_{\bar{V}'}(R_{2N+2}^{-1} - 1)} \tag{5.193''}$$

$$Q_2 = \gamma_R \sum_{n=-N}^{+N} \frac{R_n}{\Gamma_V^{\updownarrow} + \Gamma_{\bar{V}'}(R_n^{-1} - 1)} \tag{5.193'''}$$

Q_1 represents the principal term which, since $R_{2N+2} \approx 1$, is approximately equal to $\gamma_R/\Gamma_V^{\updownarrow}$. The expression of Q_2 shows that:

$$|Q_2| < (\gamma_R/\Gamma_V^{\updownarrow})\left(\sum_{n=-N}^{+N} R_n\right) \ll \gamma_R/\Gamma_V^{\updownarrow}$$

if one remembers that

$$\sum_n R_n = 1 - R_{2N+2} \ll 1$$

We thus conclude that, even if $\Gamma_{\bar{V}'} \neq 0$, one has:

$$Q \approx Q_1 \approx \gamma_R/\Gamma_V^{\updownarrow} \tag{5.194}$$

The physical meaning is that dissipation of radiant energy through the non-radiant damping channels is prevented by too weak mixing of the $|V\rangle$ state with the non-radiant manifold.

The analytical results obtained on emission intensity decay $I(t)$ (equation (5.189)) and energy spectrum $J(E_K)$ are schematically represented in Figure 5.9; they appear to be in accord with numerical computation in reference 49.

Case II: Weak mixing per weak interstate coupling. In practice, the defining conditions (5.142) of this case correspond to the extreme small molecule limit, with widely spaced, non-overlapping levels. The dynamical development of such excited molecular systems is that of well separated quasi-stationary states, and in many respects some similarity with the better known behaviour of excited atoms is expected; quantum beats of molecular states, for example, could in principle lead to optical manifestations like those, extensively studied, due to interference of non-degenerate atomic levels.[50,51]

As in Case I, the formal expansions over the distributions of poles and residues, which represent the experimentally significant quantities, are dominated by the special term corresponding to the pole E_{2N+2}. Fluorescence *intensity decay* is likewise represented by the set of equations (5.185), in which the expressions (5.143) are to be substituted. The normally decaying component thus reads:

$$I'(t) = \gamma_R \left\{1 + \left(\frac{\pi u}{\bar{D}}\right)^2 \left[1 + \mathrm{cotg}^2 \left(\frac{\pi A}{\bar{D}}\right)\right]\right\}^{-2} \exp(-\Gamma_V^{\updownarrow} t) \quad (5.195')$$

or, if we approximate R_{2N+2} by unity:

$$I'(t) \approx \gamma_R \exp(-\Gamma_V^{\updownarrow} t) \quad (5.195'')$$

For the interference component, we substitute (5.143) in expression (5.187), to obtain:

$$I''(t) = 2\gamma_R \left\{ \sum_{n=-N}^{+N} \frac{u^2}{(E_0 - E_V + n\bar{D})^2} \cos(E_0 - E_V + n\bar{D})t \right\} \mathrm{e}^{-\frac{1}{2}(\Gamma_V^{\updownarrow} + \Gamma_{\bar{V}'})t} \quad (5.196')$$

Since $R_n = u^2/(E_0 - E_V + n\bar{D})^2$ is very much smaller than unity, the interference component contributes much less to intensity than the normal one. It decays with approximately twice the normal decay time appearing in equation (5.195). We further notice that the leading term of expansion (5.196') corresponds to the $|\bar{V}'\rangle$ level nearest to the radiant level $(E_0 - E_V + n\bar{D} = A)$, and in many instances may be sufficient to approximate the expression of $I''(t)$, i.e.:

$$I''(t) \approx 2\gamma_R \left(\frac{u}{A}\right)^2 \cos(At) \exp[-\tfrac{1}{2}(\Gamma_V^{\updownarrow} + \Gamma_{\bar{V}'})t] \quad (5.196'')$$

Note that under the defining conditions of Case II, one has: $u \ll A$; on the other hand, the damping constant $\frac{1}{2}(\Gamma_V^{\updownarrow} + \Gamma_{\bar{V}'})$ is smaller than the frequency A, which is one of the necessary conditions for the quantum beats to be actually observable.

Considering the *fluorescence energy spectrum*, as represented by equations (5.180)–(5.182), we remember that all the individual widths E_p'' are smaller than the level spacing $\bar{D}$. As a consequence, the interference component (5.182) is in general negligible and only the normal component (5.181) needs to be taken into account. The energy profile of emitted lights is thus represented by approximately isolated Lorentzian lines, each related to one of the $2N + 2$ compound states of the system (cf. equation (5.190)). The dominant line, as given by equation (5.190''), is centred at $E_K = E_V + \frac{1}{2}\Gamma_V^{\updownarrow} \mathrm{cotg}(\pi A/\bar{D})$ with the width $\frac{1}{2}\Gamma_V^{\updownarrow}$ determined by the natural irreversible decay processes of the radiant state. The set of other lines (equation (5.190‴)) provides less intense lateral fine structure in the emission spectrum.

Regarding *fluorescence quantum yield*, arguments essentially similar to those applied to Case I lead to the same results, with the same physical meaning (cf. equations (5.193)–(5.194)).

Figure 5.10 schematically illustrates the variations of light intensity $I(t)$ and the emission spectrum $J(E_K)$, expected in Case II.

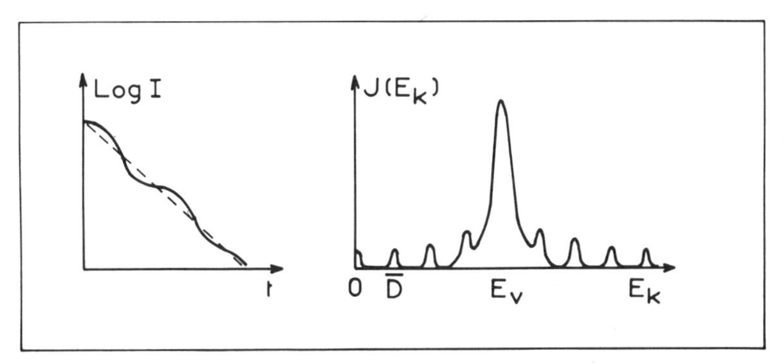

Figure 5.10 Schematic representation of fluorescence intensity decay $I(t)$ and spectral distribution $J(E_K)$. Case II ($\rho^{\downarrow} < \rho^{\uparrow}, 1$): weak mixing through weak coupling

While being physically very different, Cases I and II share the property that the initially excited level retains its special radiant character, and remains largely responsible for the main features of light emission; molecular interstate coupling only introduces perturbation effects. When the weak mixing conditions (5.131) break down, i.e., when the $|V\rangle$ state is strongly coupled to one or several $|\bar{V}'\rangle$ levels, the radiant character is redistributed over the whole set of strongly coupled states, which together control the characteristics of light emission. We next turn to the simplest possible case where strong mixing takes place.

Case III: Two-level resonance. Only one single non-radiant $|\bar{V}'\rangle$ level is assumed to strongly interact with the initially excited $|V\rangle$ level; the weak mixing condition (5.131) is supposed to hold for all the other states of the non-radiant $|\bar{V}'\rangle$ manifold. For the strongly mixed $|\bar{V}'\rangle$ level—which is necessarily the nearest to the radiant level—one has:

$$(\Gamma_V^{\downarrow}/2\pi)\{A^2 + [(\Gamma_V^{\uparrow} - \Gamma_{\bar{V}'})/2]^2\}^{-\frac{1}{2}} \gtrsim 1$$

which implies that: $\Gamma_V^{\downarrow} \geqslant (\Gamma_V^{\uparrow} - \Gamma_{\bar{V}'})$ and $(u^2/\bar{D}) \gtrsim A$, or: $(\alpha, \rho^{\uparrow}) \lesssim \rho^{\downarrow}$. On the other hand, validity of the weak mixing criterion for the other $|\overline{V}'\rangle$ states mean that $u \ll \overline{D}$, or $\rho^{\downarrow} \ll 1$. The present case is consequently characterized by:

$$(\alpha, \rho^{\uparrow}) \lesssim \rho^{\downarrow} \ll 1 \tag{5.197}$$

i.e., by $A \ll \overline{D}$ and very small values of the damping width $(\Gamma_V^{\updownarrow} - \Gamma_{\overline{V}'})$. We thus recover here the somewhat trivial case of resonance between two isoenergetic bound states, which is the immediate generalization of the foregoing Case II, when the energy difference A tends to vanish. Practical examples where the conditions (5.197) are satisfied might be found among small molecules with large level spacings and strong vibronic interaction between pairs of accidentally degenerate levels.[52]

Without loss of generality, we may set $A = 0$ $(\alpha = 0)$ and $E_V = E_0$ $(\varepsilon_V = 0)$. The $|\overline{V}'\rangle$ level in resonance with the radiant level then corresponds to $n = 0$ in the distribution (5.120). For the $2N$ weakly coupled levels, the values of the poles and residues are those given by equations (5.143), i.e.:

$$E_n' = E_0 + n\overline{D} + u^2/n\overline{D}, \qquad E_n'' = \tfrac{1}{2}\Gamma_{\overline{V}'}$$

$$R_n = (u/\overline{D})^2 n^{-2}; \qquad n = \pm 1, \pm 2, \ldots, \pm N. \tag{5.198}$$

For the two remaining poles, which characterize the strongly interacting level pair, the strong mixing relations (5.150)–(5.152) must be used: in equation (5.150), the cotangent function can be approximated by $(\varepsilon_p')^{-1}$, so that we obtain for the two poles, denoted by E_+ and E_-, the equivalent results:

$$\varepsilon_{\pm}' = \pm(\rho^{\downarrow})^{\frac{1}{2}}, \qquad E_{\pm}' = E_V \pm u \tag{5.199}$$

According to equation (5.151), the imaginary parts are:

$$\varepsilon_{\pm}'' = \rho^{\uparrow}(2 + \rho^{\downarrow})^{-1} \approx \tfrac{1}{2}\rho^{\uparrow} \tag{5.200'}$$

$$E_{\pm}'' = \tfrac{1}{4}(\Gamma_V^{\updownarrow} + \Gamma_{\overline{V}'}) \tag{5.200''}$$

Inserting the values (5.198) and (5.199), in the general expressions (5.128) of the residues, one gets:

$$R_{\pm} = \frac{\rho^{\downarrow}}{\rho^{\downarrow} + (\rho^{\downarrow})^2 + [\pm(\rho^{\downarrow})^{\frac{1}{2}} + \mathrm{i}(\frac{1}{2}\rho^{\uparrow})]^2}$$

$$R_{\pm} = \frac{1}{1 + \{1 \pm \mathrm{i}[(\Gamma_V^{\updownarrow} - \Gamma_{\overline{V}'})/4u]\}^2} \tag{5.201}$$

where $\rho^{\downarrow} \ll 1$ has been neglected. These quantities are seen to depend on the relative values of $\frac{1}{4}(\Gamma_V^{\updownarrow} - \Gamma_{\overline{V}'})$ and u. When $u \gg \frac{1}{4}(\Gamma_V^{\updownarrow} - \Gamma_{\overline{V}'})$, the most simple result is obtained: $R_+ = R_- = \frac{1}{2}$. We note that the other limiting case, where damping largely exceeds intermolecular energy exchange, satisfies the defining conditions of weak mixing Case I, and is therefore described by the previously derived results.

The values obtained for the two special poles E_+ and E_- show that they practically exhaust the sum rules $\sum R_p = 1$ and $\sum \varepsilon_p'' = \rho^\uparrow$. This means that the radiant character is essentially distributed between them: in the expansions defining light intensity decay, energy spectrum and quantum yield, the corresponding two terms represent the chief contributions.

Beginning with the two components of *light intensity decay*, $I(t) = I'(t) + I''(t)$, we write the normally decaying part as:

$$I'(t) = I'_1(t) + I'_2(t) \tag{5.202}$$

with, following (5.177′):

$$I'_1(t) = \gamma_R[|R_+|^2 \, e^{-2E''_+ t} + |R_-|^2 \, e^{-2E''_- t}]$$

$$I'_1(t) = 2\gamma_R|R_+|^2 \exp\left[-\tfrac{1}{2}(\Gamma_V^\updownarrow + \Gamma_{\overline{V}'})t\right] \tag{5.203′}$$

Following expression (5.201), one has:

$$|R_+|^2 = \{4 + [(\Gamma_V^\updownarrow - \Gamma_{\overline{V}'})/4u]^4\}^{-1} \tag{5.204}$$

which under the condition $(\Gamma_V^\updownarrow - \Gamma_{\overline{V}'}) \ll 4u$, is of the order of $\frac{1}{4}$; expression (5.203′) then simplifies to:

$$I'_1(t) \approx \tfrac{1}{2}\gamma_R \exp\left[-\tfrac{1}{2}(\Gamma_V^\updownarrow + \Gamma_{\overline{V}'})t\right] \tag{5.203″}$$

The second term in equation (5.202) is very much smaller. It reads (cf. equation (5.177″)):

$$I'_2(t) = \gamma_R \, e^{-\Gamma_{\overline{V}'} t} \sum_n |R_n|^2, \qquad n = \pm 1, \ldots, \pm N \tag{5.205}$$

with R_n given by (5.198). If we successively write:

$$\sum_n |R_n|^2 = 2 \sum_{n=1}^{-N} |R_n|^2 = 2\left(\frac{u}{D}\right)^4 \sum_{n=1}^{N} n^{-4}$$

$$< 2\left(\frac{u}{\overline{D}}\right)^4 \sum_{n=1}^{\infty} (n^{-4}) = \frac{2}{90}\left(\frac{\pi u}{\overline{D}}\right)^4 = \frac{2}{90}(\rho^\downarrow)^2$$

and remember that $\rho^\downarrow \ll 1$ (equation (5.197)), we indeed verify that $I'_2(t)$ is negligible. For the interference component, we likewise distinguish the principal term $I''_1(t)$ corresponding to interference of the two special resonances, and write:

$$I''(t) = I''_1(t) + I''_2(t) \tag{5.206}$$

$$I''_1(t) = \gamma_R[R_+(R_-)^* \, e^{-i(E_+ - E_-)t} + C.c]$$

$$I''_1(t) = 2\gamma_R \exp[-\tfrac{1}{2}(\Gamma_V^\updownarrow + \Gamma_{\overline{V}'})t] \, \mathrm{Re}\,[(R_+)^2 \exp(-2iut)] \tag{5.207}$$

In the limit of weak damping, with $(\Gamma_V^\updownarrow - \Gamma_{\overline{V}'}) \ll 4u$, one gets the simpler form:

$$I''_1(t) = \tfrac{1}{2}\gamma_R \cos(2ut) \exp[-\tfrac{1}{2}(\Gamma_V^\updownarrow + \Gamma_{\overline{V}'})t] \tag{5.207''}$$

Note that the intensity (5.207) is comparable to (5.203). The other term $I''_2(t)$ in equation (5.206) essentially accounts for the effects of interference of the two special resonances with the $2N$ remaining resonances; the term describing mutual interference of the latter ones is easily shown to be, like $I_2(t)$, of the order of $(\rho^\downarrow)^2 \ll 1$, and can be neglected. We then take:

$$I''_2(t) \approx \gamma_R \left(\frac{u}{\overline{D}}\right)^2 e^{-\frac{1}{4}(\Gamma_V^\updownarrow + \Gamma_{\overline{V}'})t} \, e^{-\frac{1}{2}\Gamma_{\overline{V}'}t}$$
$$\times \left\{(R_+ \, e^{-iut} + R_- \, e^{iut})\left(\sum_{n=1}^{N} \frac{1}{n^2} \cos n\overline{D}t\right) + C.c\right\} \tag{5.208'}$$

and obtain for the weak damping case, where: $R_+ = R_- = \frac{1}{2}$, the expression:

$$I''_2(t) \approx 2\gamma_R \left(\frac{u}{\overline{D}}\right)^2 e^{-\frac{1}{4}(\Gamma_V^\updownarrow + \Gamma_{\overline{V}'})t} \, e^{-\frac{1}{2}\Gamma_{\overline{V}'}t}(\cos ut)\left(\sum_{n=1}^{N} \frac{\cos n\overline{D}t}{n^2}\right) \tag{5.208''}$$

Due to the factor $(u/\overline{D})^2 \ll 1$, $I''_2(t)$ is a small perturbation as compared to $I''_1(t)$ and $I'_1(t)$; we may further note that the function defined by the sum over n is responsible for rapid intensity oscillations, with a period $2\pi(\overline{D})^{-1}$ short compared to the period of energy exchange $(2\pi u^{-1})$, between the strongly coupled zero-order levels. Summing up, intensity decay $I(t)$ is essentially represented by (5.203) and (5.207); taking the simplest forms (5.203″) and (5.207″), one has:

$$I(t) \approx I'_1(t) + I''_1(t)$$
$$= \tfrac{1}{2}\gamma_R(1 + \cos 2ut) \exp[-\tfrac{1}{2}(\Gamma_V^\updownarrow + \Gamma_{\overline{V}'})t]$$
$$= \gamma_R \cos^2 ut \exp[-\tfrac{1}{2}(\Gamma_V^\updownarrow + \Gamma_{\overline{V}'})t] \tag{5.209}$$

This may be interpreted as indicating that excitation energy oscillates back and forth between the two strongly coupled degenerate states, with a period $(2\pi/u)$; damping is described by the exponential factor, with a

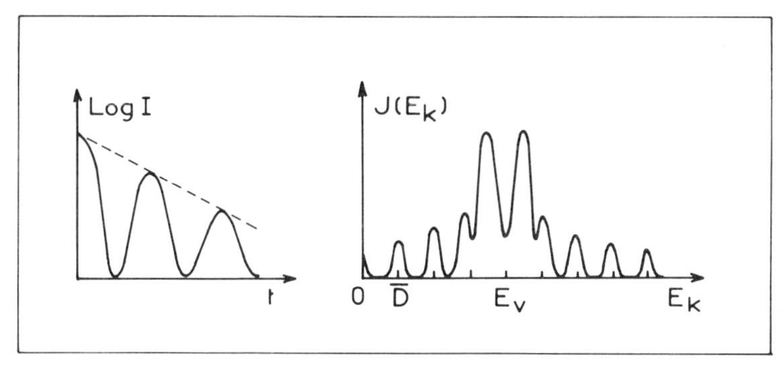

Figure 5.11 Schematic representation of fluorescence intensity decay $I(t)$ and spectral distribution $J(E_K)$. Case III ($\rho^{\uparrow} < \rho^{\downarrow} < 1$): two-level resonance

decay time twice the 'natural decay time' $(\Gamma_V^{\updownarrow} + \Gamma_{\bar{V}'})^{-1}$, which is often referred to as the 'Douglas effect'[53] (cf. Figure 5.11).

Turning our attention to the *energy spectrum* of emitted light (equations (5.179)–(5.182)), the spectral pattern is found to be similar to that of Case II, with the exception of two central intense lines instead of one. The resonances represented by E_n are all non-overlapping and of minor intensity. In the repeatedly considered particular case, where $\frac{1}{4}(\Gamma_V^{\updownarrow} - \Gamma_{\bar{V}'}) \ll u$, the two central lines corresponding to E_+ and E_- are two equally intense, well separated Lorentzians centred at $E_0 \pm u$ and having the same width $\frac{1}{4}(\Gamma_V^{\updownarrow} + \Gamma_{\bar{V}'})$. When $\frac{1}{4}(\Gamma_V^{\updownarrow} + \Gamma_{\bar{V}'}) \approx 2u$, the two resonances tend to overlap; interference of the two amplitudes must then be taken into account, and a more complicated shape is to be expected for the central lines (cf. Figure 5.11).

Light emission quantum yield Q may be estimated by inserting, in expression (5.183), the values (5.198) and (5.201) of the residues R_n and $R_{\pm}$. Only the terms involving the two principal poles need to be taken into account, since the others lead to a negligible contribution. If one takes the most simple and significant case, where $\frac{1}{4}(\Gamma_V^{\updownarrow} - \Gamma_{\bar{V}'}) \ll u$, $R_+ = R_- = \frac{1}{2}$, it is found that:

$$Q = \gamma_R/(\Gamma_V^{\updownarrow} + \Gamma_{\bar{V}'}) \tag{5.210}$$

We recover the intuitively obvious result that light emission efficiency is here determined by the radiationless processes involving both the two strongly coupled states, which share the radiant character.

Characteristic features of light emission for the two-level resonance case are illustrated in Figure 5.11. As the following point in the discussion, we proceed with the more general case where several of the zero-order molecular states are strongly mixed.

Case IV: Few-level resonance. More than a single, but still a relatively limited number, of the levels in the non-radiant distribution (5.120) do strongly mix with the initially excited radiant state. The number of strongly coupled $|\bar{V}'\rangle$ levels was stated above to be equal to $2\mathfrak{N}$, with $\mathfrak{N}$ defined by equation (5.133). The characteristic condition for the present case then reads:

$$2\mathfrak{N} = \rho^{\downarrow}/\pi^2 = (u/\bar{D})^2 > 1 \tag{5.211}$$

and must be added to the general strong mixing criterion (5.132), here written as:

$$\rho^{\downarrow} \gg \rho^{\uparrow} > 0 \tag{5.212}$$

if we assume, for simplicity, that $\alpha = 0$ (or $A = 0$). We also begin to neglect possible edge effects, which implies that the radiant level is sufficiently separated from the borders $E_0 \pm N\bar{D}$ of the non-radiant distribution (5.120). Note that the two conditions (5.211) and (5.212) mean that $\rho^{\uparrow}$ must not be too large.

Case IV thus defined was termed 'Sparse Intermediate Case' and studied by Jortner *et al.*[15–18] Typical examples where it applies are the first singlet states of small, triatomic molecules (SO_2, CO_2, CS_2, ...). Radiant states of polyatomic molecules, with small energy separation from the next lower energy level, may also eventually fall in this category.

The distributions of poles and residues for the special poles, related to the strongly mixed molecular levels, are given by the previously derived and discussed equations (5.149)–(5.155). For the remaining poles, the general results obtained under the weak mixing conditions must be used. Inspection of expression (5.137) for R_n however shows that it is similar to (5.153), so that we may take, for the whole spectrum, the common form for the residues:

$$R_n = \rho^{\downarrow}[(\varepsilon_n')^2 + (\rho^{\downarrow})^2 + \rho^{\downarrow}]^{-1}$$

The distribution of the values of ε_n', as illustrated for example in Figure 5.4, is symmetric about $\varepsilon' = 0$; for the energies around $\varepsilon' = 0$, one has approximately $|\varepsilon_n'| = (n + \frac{1}{2})\pi$; but for the present purpose, it will be convenient to take $|\varepsilon'| = |n|\pi$ for the whole energy range, the errors being negligible if $\rho^{\downarrow} > 1$. We may then write R_n as:

$$R_n = (\rho^{\downarrow} + 1)^{-1}[1 + (n/n_0)^2]^{-1} \tag{5.213}$$

with:

$$n_0 = [(\rho^{\downarrow})^2 + \rho^{\downarrow}]^{\frac{1}{2}}/\pi \tag{5.214}$$

Regarding the reduced widths ε''_n, we must turn to the expressions (5.151) or (5.154) for the strongly mixed levels. But again, we verify that the same expressions also approximately apply for the levels with $|\varepsilon'_n|$ large enough for interstate mixing to be weak. Hence, we may take, for all the poles $\varepsilon''_n = \rho^{\uparrow} R_n$ (cf. equation (5.154)) with R_n defined by (5.213):

$$\varepsilon''_n = \frac{1}{[1 + (n/n_0)^2]} \frac{\rho^{\uparrow}}{(\rho^{\downarrow} + 1)} \tag{5.215}$$

We now consider the theoretical expressions representing *fluorescence intensity decay*. Beginning with the normal component $I'(t)$, we insert the values (5.213) and (5.215) in equation (5.177″) to get:

$$I'(t) = \frac{2\gamma_R \, e^{-\Gamma_{\bar{V}'}t}}{(\rho^{\downarrow} + 1)^2} \left\{ \sum_{n \geq 0} \frac{\exp[-2\rho^{\uparrow}\tau/\{[1 + (n/n_0)^2](\rho^{\downarrow} + 1)\}]}{[1 + (n/n_0)^2]^2} \right\} \tag{5.216}$$

Since we work under condition (5.212), we have $\Gamma_V^{\updownarrow} \gg \Gamma_{\bar{V}'}$. Thus equation (5.216) represents non-exponential decay, which is described by a discrete, decreasing distribution of lifetimes ranging from the value obtained by setting $n = 0$ in (5.216) to that corresponding to $n = \infty$. At the beginning of decay, dynamics is controlled by the poles most distant from the real axis (cf. Figure 5.5), i.e., by $\{E_n\}$ with $n < n_0$; here the 'decay time constants' are of the order of:

$$(2E'')^{-1} \gtrsim [(\Gamma_V^{\updownarrow} + \rho^{\downarrow}\Gamma_{\bar{V}'})/(1 + \rho^{\downarrow})]^{-1} \tag{5.217}$$

where the right-hand side involves a width which can be considered as the weighted mean of the individual widths for the strongly coupled levels. As time development proceeds the poles nearest to the real axis ($n > n_0$) gain more and more importance and the decay law tends to a pure exponential with a decay time $(\Gamma_{\bar{V}'})^{-1}$. We note that, in the particular limiting case where $\Gamma_{\bar{V}'} \approx 0$, we recover the results of Jortner *et al.*;[15–16] in such cases, (5.217) becomes:

$$(2E'')^{-1} \geqslant [\Gamma_V^{\updownarrow}/(1 + \rho^{\downarrow})]^{-1}$$

and the often mentioned characteristic 'lengthening' of the observed fluorescence lifetime with respect to the natural lifetime $(\Gamma_V^{\updownarrow})^{-1}$ is then most emphasized. One may further note that in the particular cases where $\rho^{\uparrow} = 0$, decay is purely exponential with the characteristic lifetime $(\Gamma_V^{\updownarrow})^{-1} = (\Gamma_{\bar{V}'})^{-1}$. Considering next the interference component $I''(t)$ of fluorescence intensity, we find, from equation (5.178″), an expansion into harmonics with frequencies given in terms of the distances $(n - n')\bar{D}$ between the levels in the zero-order spectrum:

$$I''(t) = 2\gamma_R \, e^{-\Gamma_{\bar{V}'}t} \left\{ \sum_{n < n'} R_n R_{n'} \, e^{-(\varepsilon''_n + \varepsilon''_{n'})\tau} \cos[(n - n')\pi\tau] \right\} \tag{5.218}$$

The most important contributions come from the terms where n' is not too different from n. The pre-exponential and damping factors are comparable to those for the normal component $I'(t)$. But, in practice, limited experimental time resolution is expected to make actual observation of these interference components difficult, even for the lowest frequencies, that are of the order of $\bar{D}$ (cf. Figure 5.12).

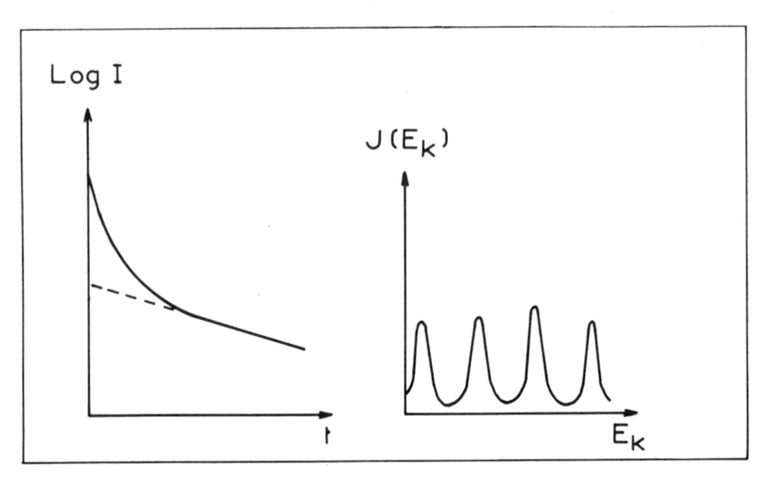

Figure 5.12 Schematic representation of fluorescence intensity decay $I(t)$ and spectral distribution $J(E_K)$. Case IV $(1, \rho^{\uparrow} < \rho^{\downarrow})$: few-level resonance

The *energy spectrum* of emitted photons is found to be very simple, if one remembers that the imaginary parts of the poles ε_n are all such as $\varepsilon''_n \ll 1$ (cf. equations (5.149″) and (5.109)). Since the widths E''_n are smaller than spacing $\bar{D}$, the distribution function $J(E_K)$ is essentially given by the normal term $J'(E_K)$. One thus gets a set of $2N + 2$ narrow Lorentzian lines, regularly centred at the energies $E_K = E'_n$ with the widths E''_n, given by the previously discussed expressions:

$$J(E_K) = |\langle K|H'|V\rangle|^2 \sum_n \frac{|R_n|^2}{(E_K - E'_n)^2 + (E''_n)^2}$$

(cf. Figure 5.12).

In order to evaluate *fluorescence quantum yield*, we use the general expression (5.183), written as:

$$Q = \frac{\gamma_R}{\Gamma_{\bar{V}'}} \sum_n \frac{R_n}{R_n^{-1} + (\Gamma_V^{\updownarrow} - \Gamma_{\bar{V}'})/\Gamma_{\bar{V}'}}$$

with R_n defined by (5.213)–(5.214): $R_n = \rho^{\downarrow}/[(\varepsilon'_n)^2 + (\pi n_0)^2]$. One thus obtains:

$$Q = \frac{\gamma_R}{\Gamma_{\bar{V}'}} \sum_n \frac{(\rho^{\downarrow})^2}{[(\varepsilon_n')^2 + (\pi n_0)^2][(\varepsilon_n')^2 + (\pi n_0)^2 + \rho^{\downarrow}(\Gamma_V^{\updownarrow} - \Gamma_{\bar{V}'})/\Gamma_{\bar{V}'}]} \tag{5.219'}$$

As $\pi n_0 > \pi$, which is the spacing between the discrete values of ε_n', we may approximate the sum by an integral:

$$Q = \frac{\gamma_R}{\Gamma_{\bar{V}'}} \int_{-\infty}^{+\infty} \frac{\mathrm{d}x}{\pi} \frac{(\rho^{\downarrow})^2}{[x^2 + (\pi n_0)^2]\{x^2 + (\pi n_0)^2[1 + (\Gamma_V^{\updownarrow} - \Gamma_{\bar{V}'})/(\rho^{\downarrow} + 1)\Gamma_{\bar{V}'}]\}}$$

which is evaluated and finally yields:

$$Q = \frac{\rho^{\downarrow}}{\pi n_0} \frac{\gamma_R}{\Gamma_V^{\updownarrow} + \rho^{\downarrow}\Gamma_{\bar{V}'}} \left\{1 + \left[\frac{(\rho^{\downarrow} + 1)\Gamma_{\bar{V}'}}{\Gamma_V^{\updownarrow} + \rho^{\downarrow}\Gamma_{\bar{V}'}}\right]^{\frac{1}{2}}\right\}^{-1} \tag{5.219''}$$

In many cases, the width $\Gamma_{\bar{V}'}$ may be expected to be not too different from $\Gamma_V^{\updownarrow}$, with the consequence that, in (5.219), $(\rho^{\downarrow} + 1)\Gamma_{\bar{V}'} \lesssim (\Gamma_V^{\updownarrow} + \rho^{\downarrow}\Gamma_{\bar{V}'})$ and:

$$Q \approx \gamma_R[2(\Gamma_V^{\updownarrow} + \rho^{\downarrow}\Gamma_{\bar{V}'})]^{-1} \tag{5.220}$$

where we have further assumed that $\rho^{\downarrow}$ is large enough to have $\rho^{\downarrow} \approx \pi n_0$. Under these conditions, light emission is largely quenched through the non-radiant levels, which are the most numerous in the strongly coupled manifold. On the other hand, if $\Gamma_{\bar{V}'}$ is so small that, in (5.219), the second term is very much smaller than unity, the effect of natural decay of the initially excited radiant level becomes more important; the approximate form here obtained reads:

$$Q \approx \frac{\gamma_R}{\Gamma_V^{\updownarrow} + \rho^{\downarrow}\Gamma_{\bar{V}'}} \left\{1 - \left[\frac{(\rho^{\downarrow} + 1)\Gamma_{\bar{V}'}}{\Gamma_V^{\updownarrow} + \rho^{\downarrow}\Gamma_{\bar{V}'}}\right]^{\frac{1}{2}}\right\} \tag{5.221}$$

and, for $\Gamma_{\bar{V}'} \to 0$, tends to the trivial result (5.184).

These general results must now be supplemented by a study of the special features arising when the previously mentioned edge effects cannot be neglected.

Case IV′: Few-level resonance, edge effects.

1. We first examine the cases where a *single edge* can possibly be made responsible for the particular effects to be studied. Practical examples of potential interest are molecules where the electronic configurations of the interacting radiant and non-radiant manifolds are only separated by a small energy gap; new features due to the presence of the lower edge of the $|\bar{V}'\rangle$ distribution (5.120) may then indeed be notable. Thus, we direct special attention to such situations where, besides the conditions (5.211) and (5.212) defining Case IV, the conditions (5.158)

are satisfied. For the special features to be as apparent as possible, we further suppose that the energy gap, $\delta E_V = E_V - (E_0 - N\bar{D})$, between the $|V\rangle$ level and the lowest $|\bar{V}'\rangle$ level is much smaller than the intramolecular coupling energy:

$$\delta E_V \ll 2u \tag{5.222}$$

The distribution of poles, for the case thus defined, was derived before, and is represented in Figures 5.6 and 5.7. For the special pole related to the lowest edge, the expressions to be taken are (5.161), (5.164) and (5.165), i.e.:

$$\varepsilon'_- = -N\pi - (\rho^{\downarrow})^{\frac{1}{2}}, \qquad \varepsilon''_- = \tfrac{1}{2}\rho^{\uparrow}, \qquad R_- = \tfrac{1}{2} \tag{5.223'}$$

The values for the $2N + 1$ remaining poles are the normal ones, considered in the previous Case IV. But here we cannot set $\varepsilon_V = 0$ but have to take: $\varepsilon_V = -N\pi$. It then appears convenient to redefine the integer over which the sums in the pole expansions are carried out, as:

$$n' = n + N$$

n' is seen to range between 0 and $2N$. The expressions of poles and residues then read:

$$\varepsilon'_{n'} = (n' - N)\pi, \quad \varepsilon''_{n'} = \rho^{\uparrow} R_{n'}, \quad R_{n'} = \frac{\rho^{\downarrow}}{(n'\pi)^2 + (\rho^{\downarrow})^2 + \rho^{\downarrow}} \tag{5.223''}$$

To examine the special features of *light intensity decay*, one separates the condition from the special pole E_- in expansion (5.175) of the probability amplitude $A_V(t)$:

$$A_V = (A_V)_1 + (A_V)_2 \tag{5.224'}$$

$$\begin{aligned} (A_V)_1 &= \mathrm{e}^{-\frac{1}{2}\Gamma_{\bar{V}'}t}\,\mathrm{e}^{-\mathrm{i}E_0 t}[R_-\,\mathrm{e}^{-\varepsilon''_-\tau}\,\mathrm{e}^{-\mathrm{i}\varepsilon'_-\tau}] \\ &= \tfrac{1}{2}\,\mathrm{e}^{-\mathrm{i}(E_0 - N\bar{D})t}\,\mathrm{e}^{\mathrm{i}ut}\,\mathrm{e}^{-\frac{1}{4}(\Gamma_V^{\uparrow} + \Gamma_{\bar{V}'})t} \end{aligned} \tag{5.224''}$$

$$\begin{aligned} (A_V)_2 &= \mathrm{e}^{-\frac{1}{2}\Gamma_{\bar{V}'}t}\,\mathrm{e}^{-\mathrm{i}E_0 t}\left[\sum_{n'=0}^{2N} R_{n'}\,\mathrm{e}^{-\varepsilon''_{n'}\tau}\,\mathrm{e}^{-\mathrm{i}\varepsilon'_{n'}\tau}\right] \\ &= \mathrm{e}^{-\frac{1}{2}\Gamma_{\bar{V}'}t}\,\mathrm{e}^{-\mathrm{i}(E_0 - N\bar{D})t}\left[\sum_{n'=0}^{2N} \frac{\rho^{\downarrow}}{(n'\pi)^2 + (\rho^{\downarrow})^2 + \rho^{\downarrow}}\,\mathrm{e}^{-R_{n'}\rho^{\uparrow}\tau}\,\mathrm{e}^{-\mathrm{i}n'\pi\tau}\right]. \end{aligned} \tag{5.224'''}$$

For the normal intensity component, one then likewise distinguishes two terms I_1 and I_2:

$$I'(t) = I'_1(t) + I'_2(t) \tag{5.225'}$$

$$I'_1(t) = \tfrac{1}{4}\gamma_R \exp\left[-\tfrac{1}{2}(\Gamma_V^{\uparrow} + \Gamma_{\bar{V}'})t\right] \tag{5.225''}$$

$$I_2'(t) = \frac{\gamma_R \, e^{-\Gamma_{\bar{V}'}t}}{(\rho^{\downarrow} + 1)^2}\left\{\sum_{n' \geq 0} \frac{\exp\left[-2\rho^{\uparrow}\tau/\{[1 + (n'/n_0)^2](\rho^{\downarrow} + 1)\}\right]}{[1 + (n'/n_0)^2]^2}\right\} \quad (5.225''')$$

where n_0 is defined by equation (5.214). The second term is seen to be half of intensity $I'(t)$ observed in the absence of any edge effect (cf. equation (5.216)). This relative decrease of the characteristic slowly varying, non-exponential component is accompanied by the appearance of the exponential term $I_1'(t)$ (5.225″), responsible for faster intensity decrease, with a decay time twice the natural constant $(\Gamma_V^{\uparrow} + \Gamma_{\bar{V}'})^{-1}$; the relative importance of this new term is expectantly largest at the beginning of decay. These results hold under the extreme condition (5.222). When the energy gap δE_V increases so that condition (5.222) is progressively relaxed, the special intensity term $I_1'(t)$ loses importance, which is transferred to the normal component $I_2(t)$; if one approaches a condition inverse of (5.222), only the slowly decaying behaviour described under Case IV remains. Regarding the interference intensity component, defined by equation (5.178), one notes that the terms involving the special pole E_- contribute most significantly in the characteristic cases where $R_- = \frac{1}{2} \gg R_{n'}$. One then considers:

$$\begin{aligned} I''(t) &\approx \gamma_R[(A_V)_1^*(A_V)_2 + C\,.\,c] \\ &= \gamma_R \, e^{-\Gamma_{\bar{V}'}t} \sum_{n' \geq 0} R_{n'} \, e^{-(\varepsilon''_- + \varepsilon''_{n'})\tau} \cos(\varepsilon'_{n'} - \varepsilon'_-)\tau \\ &= \gamma_R \left[\sum_{n' \geq 0} \frac{\rho^{\downarrow}}{(\pi n')^2 + (\rho^{\downarrow})^2 + \rho^{\downarrow}} \cos(u + n'\bar{D})t\right] e^{-\frac{1}{2}(\Gamma_V^{\uparrow} + \Gamma_{\bar{V}'})t} \end{aligned} \quad (5.226)$$

where we neglected $\varepsilon''_{n'} \ll 1$ as compared to $\varepsilon''_- = \frac{1}{2}\rho^{\uparrow}$. The most important contributions are given by the smallest values of n'. But, for the same reasons as before (Case IV), experimental study of the interference intensity component should generally be impossible: the frequencies $(\omega \geqslant u \gg \bar{D})$ in equation (5.226) are so large, that one cannot follow instantaneous variations of $I''(t)$, which thus average to zero (cf. Figure 5.13).

New features also appear in the *energy spectrum* of fluorescence. They are best considered by writing the normal contributions $J'(E_K)$ as a sum of the special term determined by R_- and E_- and a term determined by the non-special poles:

$$J'(E_K) = J_1'(E_K) + J_2'(E_K) \quad (5.227')$$

$$J_1'(E_K) = |\langle K|H'|V\rangle|^2 \frac{(\frac{1}{4})}{(E_K + N\bar{D} + u)^2 + [\frac{1}{4}(\Gamma_V^{\uparrow} + \Gamma_{\bar{V}'})]^2} \quad (5.227'')$$

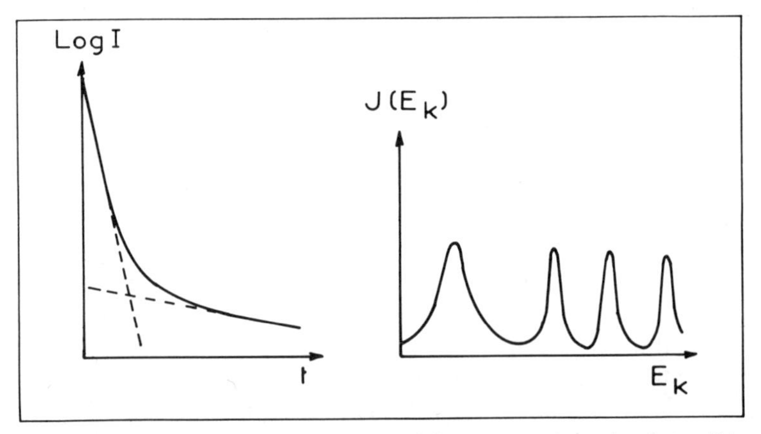

Figure 5.13 Schematic representation of fluorescence intensity decay $I(t)$ and spectral distribution $J(E_K)$. Case IV′ (1, $\rho^{\uparrow} < \rho^{\downarrow}$): few-level resonance with edge effects

$$J_2'(E_K) = |\langle K|H'|V\rangle|^2 \sum_{n'=0}^{2N} \frac{|R_{n'}|^2}{(E_K - E_{n'}')^2 + (E_{n'}'')^2} \tag{5.227‴}$$

the latter contribution (5.227‴) is entirely comparable to the spectrum for Case IV; it represents a set of $2N + 1$ narrow Lorentzian lines centred near the $|\bar{V}'\rangle$ levels. The former Lorentzian term (5.227″) is new; it lies below the lowest line of distribution (5.227‴) at an energy distance u, and has a width equal to half the normal damping width $\frac{1}{2}(\Gamma_V^{\updownarrow} + \Gamma_{\bar{V}'})$. If coupling u is strong enough so that $[(\Gamma_V^{\updownarrow} + \Gamma_{\bar{V}'})/4u] < 1$, interference of the special resonance with the non-special ones may be neglected ($J'' = 0$); this is expectantly the most frequent situation under the present strong mixing conditions (cf. Figure 5.13).

For the *quantum yield* of light emission, one proceeds similarly, and writes expression (5.183) as:

$$Q = Q_1 + Q_2 \tag{5.228′}$$

thus separating the term Q_1 corresponding to the special pole E_-. It is easily verified to be given by:

$$Q_1 = \frac{1}{2} \frac{\gamma_R}{\Gamma_V^{\updownarrow} + \Gamma_{\bar{V}'}} \tag{5.228″}$$

The other term describes the condition coming from the $2N + 1$ other poles $E_{n'}$ ($n' = 0, \ldots, 2N$); it is of the form:

$$Q_2 = \frac{\gamma_R}{\Gamma_{\bar{V}'}} \sum_{n'=0}^{2N} \frac{R_{n'}}{R_{n'}^{-1} + (\Gamma_V^{\updownarrow} - \Gamma_{\bar{V}'})/\Gamma_{\bar{V}'}} \tag{5.228‴}$$

which is similar to expression (5.219) for Case IV, except that the sum here only involves the positive values of the integer n'. Using the same arguments as in Case IV, Q_2 is evaluated as one half of the values (5.219)–(5.221) for the quantum yield in the absence of edge effects. Under condition (5.222), the new resonance related to the lowest edge of the zero-order level spectrum thus appears to be endowed by a large fraction of the total radiant quantum yield, while the others are notably denuded. If (5.222) is progressively relaxed, these special features tend to disappear: the radiant character is redistributed more evenly among all the compound levels so that the characteristic behaviour of Case IV is recovered.

2. To be complete, one has also to examine briefly the cases where *the two edges* $E_0 \pm N\bar{D}$ of the zero-order spectrum give rise to special effects. As mentioned above, this corresponds to rather extreme situations where all the zero-order levels, $|V\rangle + \{|\bar{V}'\rangle\}$, are strongly coupled.

   The optical properties of the system, still assuming initial excitation of the unique $|V\rangle$ level, are controlled by the two special poles $E_\pm$. Taking the characteristic values given by equations (5.169), (5.171) and (5.173):

$$E'_\pm = E_0 \pm u, \quad E''_\pm = \tfrac{1}{4}(\Gamma_V^\updownarrow + \Gamma_{\bar{V}'}); \quad R_\pm = \tfrac{1}{2} \tag{5.229}$$

one gets:

$$\begin{aligned} I(t) &= \tfrac{1}{2}\gamma_R(1 + \cos 2ut)\exp[-\tfrac{1}{2}(\Gamma_V^\updownarrow + \Gamma_{\bar{V}'})t] \\ &= \gamma_R \cos^2(ut)\exp[-\tfrac{1}{2}(\Gamma_V^\updownarrow + \Gamma_{\bar{V}'})t] \end{aligned} \tag{5.230}$$

$$\begin{aligned} J(E_K) = \frac{|\langle K|H'|V\rangle|^2}{4}\bigg\{&\frac{1}{(E_K - E_0 - u)^2 + [\frac{1}{4}(\Gamma_V^\updownarrow + \Gamma_{\bar{V}'})]^2} \\ &+ \frac{1}{(E_K - E_0 + u)^2 + [\frac{1}{4}(\Gamma_V^\updownarrow + \Gamma_{\bar{V}'})]^2}\bigg\} \end{aligned} \tag{5.231}$$

$$Q = \gamma_R/(\Gamma_V^\updownarrow + \Gamma_{\bar{V}'}) \tag{5.232}$$

These expressions all reflect typical two-level resonance behaviour also found in Case III. But here the whole set of vibronic $|\bar{V}'\rangle$ levels act together as a unique entity. The two resonant states need only to be characterized by the electronic quantum numbers. The quantum numbers for vibrations that distinguish the various $|\bar{V}'\rangle$ levels do not play any role because electronic energy exchange between the radiant and non-radiant configurations is faster than vibrational motion.[54]

After the molecular systems where only a relatively small number of $|\bar{V}'\rangle$ states are strongly coupled to the initially excited radiant state, one must turn to the extreme case where strong mixing involves a very large number of the non-radiant $|\bar{V}'\rangle$ levels.

Case V: Giant resonance. The number of non-radiant $|\bar{V}'\rangle$ states, strongly mixed with the radiant $|V\rangle$ state, is assumed to be very large. In addition to the general strong mixing conditions (5.132) or (5.212):

$$\rho^{\downarrow} \gg \rho^{\uparrow} \tag{5.233'}$$

we thus have:

$$2\mathfrak{N} = \rho^{\downarrow}/\pi^2 = (u/\bar{D})^2 \gg 1 \tag{5.233''}$$

On the other hand, it is first supposed that the non-radiant manifold is sufficiently wide, so that no special effects associated with the boundaries $E_0 \pm N\bar{D}$ of distribution (5.120) are notable: $N > \mathfrak{N}$. It is interesting to note that the molecular model thus defined is identical to the general 'giant resonance' model, first proposed for the interpretation of the optical model resonances in nuclear reaction studies.[56] For the molecular systems, the defining conditions of this case are expected to be best satisfied in the extreme limit of large molecules, with densely spaced (but non-overlapping) vibronic levels. The term 'statistical limit', proposed by Jortner *et al.* to denote the present case[15,16] does not seem to be very appropriate; it better applies to the previously discussed case of overlapping levels ($\frac{1}{2}\Gamma_{\bar{V}'} \gg \bar{D}$, cf. §5.4.1) where statistical assumptions are explicitly introduced, which is not the case here.

Formally, no essential difference exists with the foregoing few-level resonance case. In practice the Cases IV and V are distinguished because different time scales are physically important. The poles and residues of $\langle V|MG(E)M|V\rangle$ to be taken are those previously derived for strong mixing in the absence of edge effects: they are described by equations (5.149)–(5.155); Figure 5.4 represents the set of the real ε_p' values, while Figure 5.5 gives the distributions of ε_p'' and $R_p = \varepsilon_p''/\rho^{\uparrow}$.

To examine the *intensity decay law* in the giant resonance case, it is convenient to start with the probability amplitude $A_V(t)$ under one of its general forms, (5.93) or (5.175), in terms of the poles and the residues. We first focus our attention to *relatively small times t after excitation*, such that:

$$t \lesssim (\Gamma_V^{\downarrow})^{-1} \quad \text{or} \quad \tau \lesssim (\rho^{\downarrow})^{-1} \tag{5.234}$$

Since, in the present case, one has: $\varepsilon_p'' \ll 1 \ll \rho^{\downarrow}$, the damping factors $\exp(-\varepsilon_p''\tau)$ in expression (5.175) are seen to be negligible; further noting that $(\rho^{\downarrow})^2 \gg \rho^{\downarrow}$, one can write equation (5.175) as:

$$A_V(t) = e^{-iE_0 t} \sum_p \frac{\rho^{\downarrow}}{(\varepsilon'_p)^2 + (\rho^{\downarrow})^2} e^{-i\varepsilon'_p \tau} \tag{5.235}$$

where the successive values of ε'_p are approximately separated by π (cf. Figure 5.4). Now, combining condition (5.233′) together with (5.234) implies that $\exp(-i\varepsilon'_p\tau)$ varies slowly on the τ-interval of interest, so that we can replace the sum in (5.235) by an integral:

$$A_V(t) = e^{-iE_0 t} \frac{1}{\pi} \int_{-\infty}^{+\infty} dx \frac{\rho^{\downarrow}}{x^2 + (\rho^{\downarrow})^2} e^{-ix\tau}$$

Hence, one gets:

$$A_V(t) = e^{-iE_0 t} e^{-\rho^{\downarrow}\tau} = e^{-iE_0 t} e^{-\frac{1}{2}\Gamma_V^{\downarrow} t} \tag{5.236}$$

For times restricted by condition (5.234), the giant resonance model thus yields a pure exponential decay law:

$$I(t) = \gamma_R \exp(-\Gamma_V^{\downarrow} t) \tag{5.237}$$

This is just the result obtained for the systems satisfying the conditions of the statistical limit as defined in §5.4.1, where the non-radiant manifold is a continuum. There is however an essential difference in the nature of time development for the two cases: while in the statistical limit, decay in the non-radiant continuum is truly irreversible, the presence of the discrete $|\bar{V}'\rangle$ levels in giant resonance is responsible for recurrent energy exchange, which can in principle be observed at times longer than $(\Gamma_V^{\downarrow})^{-1}$. For *times t long enough*, so that:

$$t \geqslant \bar{D}^{-1} \gg (\Gamma_V^{\downarrow})^{-1} \quad \text{or} \quad \tau > 1 \gg (\rho^{\downarrow})^{-1} \tag{5.238}$$

the probability amplitude $A_V(t)$ must be explicitly considered as the trigonometric series (5.235): it is seen to be a function periodic in time with a period $2\pi/\bar{D}$, which can be considered as a characteristic recurrence time for energy exchange between the $|V\rangle$ and $|\bar{V}'\rangle$ levels.[25,82] Coupling between these states cannot be made responsible for true irreversible damping; but, in practice, the recurrence time $2\pi/\bar{D}$ may be so large—for large molecules—that only the initial stage of decay described by equation (5.237) is actually observed, with apparent irreversibility.

The energy spectrum of the emitted photons presents essentially the same characteristics as in Case IV. The widths E''_p are all smaller than the spacing $\bar{D}$, so that one again gets a pattern of narrow, non-overlapping Lorentzian lines centred at E'_p. But it should be noted here, that actual observation of this fine-structure spectrum requires generally higher energy resolution conditions than in the few-level resonance case.

Fluorescence quantum yield Q is in principle represented by the same expressions (5.219)–(5.221) as in the previous few-level resonance situations. But this requires photon emission to be followed during an infinite time period. In practice, observation of the system is generally restricted to time intervals larger than $(\Gamma_V^{\downarrow})^{-1}$, but smaller than the first recurrence time $2\pi/\bar{D}$, so that the foregoing expressions cannot be taken; one then uses:

$$Q = \int_0^{t_0} I(t)\,\mathrm{d}t, \qquad (\Gamma_V^{\downarrow})^{-1} \ll t_0 < 2\pi/\bar{D}$$

with $I(t)$ given by equation (5.237). The result:

$$Q \approx (\gamma_R/\Gamma_V^{\downarrow}) \tag{5.239}$$

is also what would be obtained under the extreme statistical conditions, where the non-radiant manifold is continuous (§5.4.1).

As for the few-level resonance case, interesting edge effects may be in principle expected under giant resonance conditions. This will now be briefly examined.

Case V′: Giant resonance, edge effects. In principle, no essential difference exists between Cases IV′ and V′. Under the extreme strong coupling conditions, where the radiant character concentrates on the two compound states $E_{\pm}$, the results (5.229)–(5.232) also apply for giant resonance systems. Some specific giant resonance features however appear when single edge effects are examined. As before, we concentrate attention on the case where the radiant level is near the lowest edge of the non-radiant $|\bar{V}'\rangle$ distribution ($E_V \approx E_0 - N\bar{D}$). The analysis proceeds as for Case IV′, leading to the same general expressions (cf. for example (5.222)–(5.224)). But the discussion of the experimentally significant quantities must account for the fact that, under giant resonance conditions, the practical time scales are generally smaller than the natural recurrence period $2\pi/\bar{D}$.

For fluorescence intensity decay, one considers equations (5.224). Amplitude $(A_V)_1$ related to the special pole retains the form (5.224″). But the other term $(A_V)_2$ is re-examined with the same arguments as for equation (5.235) in Case V: for the physically significant time periods $\tau \leqslant (\rho^{\downarrow})^{-1} \ll 1$, the sum in expression (5.224‴) is approximately given by the integral:

$$\begin{aligned} F(\tau) &= \frac{1}{\pi}\int_0^{\infty} \mathrm{d}x \frac{\rho^{\downarrow}}{x^2 + (\rho^{\downarrow})^2}\,\mathrm{e}^{-\mathrm{i}x\tau} \\ &= \int_{-\infty}^{\infty} \mathrm{d}x\, \theta(x)\frac{(\rho^{\downarrow}/\pi)}{x^2 + (\rho^{\downarrow})^2}\,\mathrm{e}^{-\mathrm{i}x\tau} \end{aligned} \tag{5.240}$$

Formally, the specific features in decay, due to the edge effects, are ascribed to the distribution $\theta(x)$ (Heaviside's step function) in (5.240). As noted by Khalfin[57] the practical consequence is a deviation from the pure exponential law (5.237), with slower decay. This may be analysed in more detail by noting that $F(\tau)$ is the Fourier transform of the product of two functions, $\theta(x)$ and $(\rho^{\downarrow}/\pi)[x^2 + (\rho^{\downarrow})^2]^{-1}$, whose transforms are respectively $\pi\delta(\tau) + (\mathrm{i}\tau)^{-1}$ and $\exp(-\rho^{\downarrow}|\tau|)$. Equation (5.240) can then be expressed as the convolution integral[58]

$$F(\tau) = \frac{1}{2\pi}\int_{-\infty}^{+\infty}\left[\pi\delta(y) + \frac{1}{\mathrm{i}y}\right]\mathrm{e}^{-\rho^{\downarrow}|\tau - y|}\,\mathrm{d}y$$

$$= \tfrac{1}{2}\,\mathrm{e}^{-\rho^{\downarrow}\tau} + \frac{1}{2\pi\mathrm{i}}\int_{-\infty}^{+\infty}\frac{\mathrm{e}^{-\rho^{\downarrow}|\tau - y|}}{y}\,\mathrm{d}y$$

Introducing the exponential integral functions[59]

$$E_1(x) = \int_x^{\infty}(\mathrm{e}^{-t}/t)\,\mathrm{d}t, \qquad Ei(x) = ⨍_{-\infty}^{x}(\mathrm{e}^{t}/t)\,\mathrm{d}t, \qquad x > 0$$

one hence gets the form:

$$F(\tau) = \tfrac{1}{2}\,\mathrm{e}^{-\rho^{\downarrow}\tau} + \frac{1}{2\pi i}[\mathrm{e}^{-\rho^{\downarrow}\tau}Ei(\rho^{\downarrow}\tau) + \mathrm{e}^{\rho^{\downarrow}\tau}E_1(\rho^{\downarrow}\tau)],$$

where the first term represents normal giant resonance decay behaviour; the second term is responsible for slower variations: according to the typical asymptotic expansions of $Ei(x)$ and $E_1(x)$[59] it decays as $(\mathrm{i}\pi\rho^{\downarrow}\tau)^{-1}$ if $(\rho^{\downarrow}\tau) \gg 1$. Regarding the physically most significant normal intensity component $I'(t)$, one has a first term given by (5.225″); the other contribution, $I_2'(t) = \gamma_R|(A_V)_2|^2$, is given by $\gamma_R\,\mathrm{e}^{-\Gamma_{\bar{V}'}t}|F(\tau)|^2$, so that:

$$I'(t) = I_1'(t) + I_2'(t)$$

$$I_1'(t) = \tfrac{1}{4}\gamma_R \exp[-\tfrac{1}{2}(\Gamma_V^{\updownarrow} + \Gamma_{\bar{V}'})t] \tag{5.241}$$

$$I_2'(t) = \tfrac{1}{4}\gamma_R \exp(-\Gamma_V^{\updownarrow}t) + \frac{\gamma_R}{(4\pi)^2}[\mathrm{e}^{-\rho^{\downarrow}\tau}Ei(\rho^{\downarrow}\tau) + \mathrm{e}^{\rho^{\downarrow}\tau}E_1(\rho^{\downarrow}\tau)]^2\,\mathrm{e}^{-\Gamma_{\bar{V}'}t}$$

At the beginning, the normal intensity decay is essentially controlled by the exponential contribution to $I_2'(t)$, which is similar to decay in the absence of the edge effects. Later, the slower components, $I_1'(t)$ and the second term of $I_2'(t)$, become important; at very long times, decay is mainly controlled by the asymptotic form of the non-exponential part of $I_2'(t)$:

$$I'(t) \sim \frac{\gamma_R}{\pi^2}(\Gamma^{\downarrow}t)^{-2}\,\mathrm{e}^{-\Gamma_{\bar{V}'}t}$$

The interference component $I''(t)$ of light intensity involves oscillating factors with a frequency equal to u: for the same reasons as before, $I''(t)$ is expected to be difficult to follow experimentally, and will not be further discussed.

Fluorescence energy spectrum has the same general structure as in Case IV′. It is likewise represented by equations (5.227), with a special line $J'_1(E_K)$ separating below a set, $J'_2(E_K)$, of narrow Lorentzians centred near the energy levels of the $|\bar{V}'\rangle$ manifold (5.120). The special line is still well separated from the others, and should be easy to observe in many practical cases. On the other hand, the set of the other lines is expected to be in general so dense, that experimental resolution of the fine structure in $J'_2(E_K)$ is very difficult.

Light emission quantum yield is similarly considered as the sum, $Q = Q_1 + Q_2$, of two terms respectively related to the special pole E_- and the remaining ones (cf. equations (5.228)). As before, the contribution from the special line is given by (5.228″): $Q_1 = \frac{1}{2}[\gamma_R/(\Gamma_V^{\updownarrow} + \Gamma_{\bar{V}'})]$; note that it could be directly obtained by integrating expression (5.241) of $I'_1(t)$. In the light of arguments identical to those used to derive equation (5.239), the expression (5.228‴) of the second term Q_2 is seen to have no practical significance, under the giant resonance conditions, where the observation time t_0 is such that $(\Gamma_V^{\downarrow})^{-1} < t_0 < 2\pi/\bar{D}$. Instead, one can take:

$$Q_2 = \int_0^{t_0} I_2(t)\,\mathrm{d}t \geqslant \tfrac{1}{2}(\gamma_R/\Gamma_V^{\downarrow})$$

remembering the variations of intensity $I'_2(t)$ given by (5.241). Hence:

$$Q \gtrsim \frac{1}{2}\frac{\gamma_R}{\Gamma_V^{\downarrow}} + \frac{\gamma_R}{\Gamma_V^{\updownarrow} + \Gamma_{\bar{V}'}} \tag{5.242}$$

which is larger than the quantum yield (5.239) in Case V: the appearance of edge effects is accompanied by an increase of light emission efficiency.

5.5 Discussion

The foregoing treatment forms a general framework for the description of molecular electronic relaxation and its study by fluorescence after impulsive excitation. In order to draw practically significant conclusions, a better reference to the actual experimental situation seems desirable. The whole discussion assumes initial impulsive excitation, followed by decay of a single molecular vibronic level: the former step is considered in a delta-function approximation, so that the latter step plays the major role; clearly, a more detailed analysis of the excitation process is necessary to assert the validity of so crude a picture (§5.5.1). On the other hand, the influence of the process of detection should also be included in the description, since actual observation of the derived laws for fluorescence intensity decay and spectral distribution obviously depends on the complementary temporal and spectral resolution properties of the photon counters (§5.5.2). The essential features emerging from a quantum-theoretical treatment are often conveniently discussed in terms of a language founded on classical analogues, one of which is briefly considered in §5.5.3. The general discussion is then closed by a summary and some concluding remarks on the main physical aspects of the present problem, as they are stressed by this time-dependent treatment.

5.5.1 *Initial excitation*

For the present purpose, it will be sufficient to take an elementary model for the incident light beam, supposed to be well collimated in one direction and polarized. The energy distribution is centred around E_I with a characteristic bandwidth I, so that the coherence time is accordingly equal to I^{-1}. We may then represent the incident radiation beam by wave packets constructed with the vectors $|K\rangle$ defined by equation (5.18),[60–62]

$$|\psi_I\rangle = \int \mathrm{d}E_K F_I(E_K - E_I, I)|K\rangle \qquad (5.243')$$

For analytical convenience, we choose a normalized spectral distribution amplitude of the Lorentzian form:

$$F_I = \frac{(I/2\pi)^{\frac{1}{2}}}{(E_K - E_I) + \mathrm{i}(I/2)} \qquad (5.243'')$$

To discuss the molecular excitation process, one starts with the system initially in the state

$$|0\rangle = |\psi_I\rangle = R|\psi_I\rangle$$

and looks, for later times, $t > 0$, at the population of the excited radiant molecular levels, described by the projection $M|t\rangle$ of the state vector $|t\rangle$. The reduced resolvent of interest is here $MG(E^+)R$ which, in a similar way as in §5.3, is easily shown to be given by:

$$MG(E^+)R = MG(E^+)MMH'RR(E - H + \mathrm{i}\varepsilon)^{-1}R$$

The pertinent matrix elements are hence of the form:

$$\begin{aligned}\langle V|MG(E^+)R|K\rangle &= \langle V|MG(E^+)M|V\rangle\langle V|H'|K\rangle\frac{1}{E - E_K + \mathrm{i}\varepsilon}\\ &= \langle V|H'|K\rangle\sum_p\frac{R_p}{(E - E_p + \mathrm{i}\varepsilon)(E - E_K + \mathrm{i}\varepsilon)}\end{aligned} \qquad (5.244)$$

where the meromorphic expansion (5.90) of $\langle V|MG(E^{+})M|V\rangle$ is taken into account. We may now write the probability amplitude of excitation in the vibronic state $|V\rangle$, $A_V(t) = \langle V|t\rangle$ as:

$$A_V(t) = \frac{\mathrm{i}}{2\pi}\int_{-\infty}^{+\infty} \mathrm{d}E\, \mathrm{e}^{-\mathrm{i}Et}\langle V|MG(E^{+})R|\psi_I\rangle$$

Inserting the expressions (5.243) and (5.244), and integrating, we get the expression:

$$A_V(t) = \sum_p \frac{\langle V|H'|K\rangle (I/2\pi)^{\frac{1}{2}}}{(E'_p - E_I) + \mathrm{i}\frac{1}{2}I} R_p[\mathrm{e}^{-\mathrm{i}E'_pt}\,\mathrm{e}^{-E''_pt} - \mathrm{e}^{-\mathrm{i}E_It}\,\mathrm{e}^{-\frac{1}{2}It}] \quad (5.245)$$

which allows one to discuss the influence of excitation band width on the overall process of photon–molecule interaction.[60]

It clearly appears that the optimal conditions for molecular relaxation to be observable correspond to 'broad-band excitation', i.e.,

$$\tfrac{1}{2}I \gg E''_p \quad (5.246)$$

for all the poles $E_p = E'_p - \mathrm{i}E''_p$ involved. After initial transients, during the short correlation time I^{-1} of the incident light, only the first term in equation (5.245) remains, describing free molecular relaxation:

$$A_V(t) \sim \sum_p \frac{\langle V|H'|K\rangle (I/2\pi)^{\frac{1}{2}}}{(E'_p - E_I) + \mathrm{i}\frac{1}{2}I} R_p\,\mathrm{e}^{-\mathrm{i}E'_pt}\,\mathrm{e}^{-E''_pt} \quad (5.247)$$

This must be compared with our previous result:

$$A_V(t) = \sum_p R_p\,\mathrm{e}^{-\mathrm{i}E'_pt}\,\mathrm{e}^{-E''_pt} \quad (5.93)$$

derived by standard decay theory, under the 'initial condition':

$$A_V(0) = \sum_p R_p = 1 \quad (5.118)$$

It is seen that, in actuality, each of the residues R_p in the sums (5.118) and (5.93) must be multiplied by a weighting factor $\langle V|H'|K\rangle F_I(E'_p - E_I, I)$ dependent on the spectral distribution function of the exciting light beam: the initial probability amplitude in decay theory should be taken as the superposition:

$$A_V(0) = \sum_p \langle V|H'|K\rangle F_I(E'_p - E_I, I)R_p \quad (5.248)$$

resulting from a 'filtering' of the molecular compound resonances $\{E_p, R_p\}$ by the incident electromagnetic field (cf. reference 63, where similar conclusions are apparently reached from a different viewpoint). The real time-dependent behaviour, as given by equation (5.247), is essentially the same as that described by expression (5.93), provided the spectral distribution function F_I is nearly constant over the energy range spanned by the distributions of residues discussed and illustrated in §5.4.2. This implies that, for the general results of §5.4 to be valid, the maxima of the residue distributions (E_V, in general) and of the incident spectrum (E_I) should approximately coincide; in addition, the spectral bandwidth I should at least be of the same order as the width characterizing the pole distributions, $R_p(E'_p)$.

A last limiting condition on the spectral distribution of the exciting light source comes from the requirement that only one vibronic radiant level $|V\rangle$ be mainly excited; this means, of course, that the spectral distribution width I must be smaller than the spacing of the radiant levels for excitation energies around E_I.

5.5.2 *Detection*

Quite generally, the detection of the fluorescence wave packets emitted by the impulsive excited molecules is accomplished through a photo-absorption process which brings the atoms of the detector from their ground state to an excited state; the latter may be part of a continuum, which ensures a broad-band response. It will here be sufficient to take a simple two-level model for the detector.[62] We further assume that the experimental set-up is designed so that detection only begins after fluorescence emission is completed. The formal problem to be solved is hence the following: at a given time $t_0 \gg (E''_p)^{-1}$, one starts with an initial wavepacket, $|t_0\rangle = |\psi_{\mathrm{fl}}\rangle = R|\psi_{\mathrm{fl}}\rangle$, the form of which is seen to be:

$$
\begin{aligned}
|\psi_{\mathrm{fl}}\rangle &= \sum_K A_K(t_0)|K\rangle \\
&= \langle K|H'|V\rangle \sum_K \langle V|MG(E_K)M|\mathbf{V}\rangle\, \mathrm{e}^{-\mathrm{i}E_K t_0}|K\rangle
\end{aligned}
\tag{5.249}
$$

if one remembers equation (5.99); what is, under these conditions, the probability of photodetection at later times, $t = t_0 + t'$?

The amplitude of detection probability is written as:

$$
\langle D|t\rangle = \frac{\mathrm{i}}{2\pi}\int_{-\infty}^{+\infty} \mathrm{d}E\, \mathrm{e}^{-\mathrm{i}Et'}\langle D|DG(E^+)R|\psi_{\mathrm{fl}}(t_0)\rangle \tag{5.250}
$$

if $|D\rangle$, and $D = |D\rangle\langle D|$, respectively denote the state vector, and the corresponding projector of the system with the counter excited. $|D\rangle$ is obviously a radiant state; the reduced resolvent operator $DG(E^+)R$, and the matrix elements $\langle D|DG(E^+)R|K\rangle$ are thus of the same form as $MG(E^+)R$

and $\langle V|MGR|K\rangle$ in §5.5.1, *viz.*,

$$\langle D|DG(E^+)R|K\rangle = \langle D|DG(E^+)D|D\rangle\langle D|H'|K\rangle(E - E_K + i\varepsilon)^{-1} \quad (5.251)$$

Inserting the expressions (5.249) and (5.251) in equation (5.250), one gets:

$$\langle D|t\rangle = \sum_K \langle D|H'|K\rangle\langle K|H'|V\rangle\langle V|MG(E_K)M|V\rangle\, e^{-iE_K t}$$
$$\times \left\{\frac{i}{2\pi}\int_{-\infty}^{+\infty} dE \frac{e^{-i(E-E_K)t'}}{E - E_K + i\varepsilon}\langle D|DG(E^+)D|D\rangle\right\} \quad (5.252)$$

The response characteristics of the detector are described by the function $\langle D|DG(E^+)D|D\rangle$. As the most natural choice in the present context, we may take a single Lorentzian resonance amplitude centred around the central energy E_Δ with the bandwidth $\frac{1}{2}\Delta$:

$$\langle D|DG(E)D|D\rangle = \frac{(\Delta/2\pi)^{\frac{1}{2}}}{(E - E_\Delta) + i\frac{1}{2}\Delta} \quad (2.253)$$

The width Δ determines the spectral and temporal resolution properties, that are required in actual experiments.

If one concentrates on *fluorescence intensity measurements* to study excited state relaxation, a sharp definition of time requires that the energy spectrum of light be determined with correspondingly low resolution; the 'resolution time' Δ^{-1} must be smaller than the decay times $(E_p'')^{-1}$, which corresponds to the 'broad-band detection' conditions:

$$\Delta \gg E_p'' \quad (5.254)$$

Evaluation of (5.252) in terms of (5.253), then shows that, except for transients during the short resolution time, the time behaviour recorded by the counter is given by:

$$\langle D|t\rangle = \sum_K \langle D|H'|K\rangle\langle K|H'|V\rangle\langle V|MG(E_K)M|V\rangle\ \langle D|DG(E_K)D|D\rangle\, e^{-iE_K t}$$

or, if the summation is replaced by an integration (cf. equation (5.22)):

$$\langle D|t\rangle = \langle D|H'|K\rangle\rho(E_K)\langle K|H'|V\rangle \int_{-\infty}^{+\infty} dE_K\, e^{-iE_K t}\langle V|MG(E_K)M|V\rangle$$
$$\times\ \langle D|DG(E_K)D|D\rangle \quad (5.255)$$

where the slowly varying quantities are taken out of the integral as average quantities. Contour integration yields, if we still disregard the terms decaying as $\exp(-\frac{1}{2}\Delta t)$:

$$\langle D|t\rangle = (-2\pi\mathrm{i})\langle D|H'|K\rangle\rho(E_K)\langle K|H'|V\rangle$$
$$\times\left\{\sum_p \frac{(\Delta/2\pi)^{\frac{1}{2}}}{(E'_p - E_\Delta) + \mathrm{i}\frac{1}{2}\Delta} R_p \,\mathrm{e}^{-\mathrm{i}E'_p t}\,\mathrm{e}^{-E''_p t}\right\} \tag{5.256}$$

Detection probability amplitude is hence seen to exhibit a time dependence which compares with that of the basic quantity,

$$A_V(t) = \sum_p R_p \,\mathrm{e}^{-\mathrm{i}E'_p t}\,\mathrm{e}^{-E''_p t}$$

with the difference that each term is weighted according to the characteristic detector system function (5.253): in a similar way, and in addition to excitation, the detection process thus filters those terms in the expansions over the poles E_p, for which the real values E'_p lie within the bandwidth, $E_\Delta \pm \frac{1}{2}\Delta$, of the photocounter. If the width Δ is sufficiently large to overlap the whole energy interval spanned by the distributions of residues $R_p(E'_p)$ in §5.4.2, the two amplitudes are practically proportional:

$$\langle D|t\rangle = -2\pi\langle D|H'|K\rangle\rho(E_K)\langle K|H'|V\rangle(\pi\Delta)^{-\frac{1}{2}}A_V(t) \tag{5.257}$$

this corresponds to the best possible conditions for the relaxation measurements.

If one is interested in a precise *study of the fluorescence emission spectrum*, i.e., in the fine structure of the energy distribution function $J(E_K)$, the experimental situation and requirements are complementary to the preceding ones: sharp definition of energy by the photon counter implies here 'narrow band conditions', the inverse of (5.254). Analysis of spectral distributions generally involves the probability amplitudes taken in the limit of large times (cf. §5.3.4 and reference 1): if $t' \to \infty$, the integral in equation (5.252) becomes $\langle D|DG(E_K)D|D\rangle$, so that we recover expression (5.255) for the detection probability amplitude. But here, evaluation in the long time limit, $t \gg (E''_p)^{-1}$, only keeps the pole of the detector function (5.253), that is now the nearest to the real axis:

$$\langle D|t\rangle \sim -(2\pi\mathrm{i})\langle D|H'|K\rangle\rho(E_K)\langle K|H'|V\rangle(\Delta/2\pi)^{\frac{1}{2}}$$
$$\times\langle V|MG(E_\Delta - \mathrm{i}\tfrac{1}{2}\Delta)M|V\rangle\,\mathrm{e}^{-\mathrm{i}E_\Delta t}\,\mathrm{e}^{-\frac{1}{2}\Delta t}$$

The detection probability $|\langle D|t\rangle|^2$ is thus indeed found to be proportional to $J(E_\Delta) = |\langle K|H'|V\rangle|^2|\langle V|MG(E_\Delta)M|V\rangle|^2$, i.e., the value of the fluorescence distribution function (5.102) at the energy defined by the detector.

5.5.3 *A classical analogue*

As considered in this treatment, the molecular system is the analogue of a set of damped oscillators, which includes a single central one $|V\rangle$, driven by the external forces, and coupled to a number of other oscillators ($\{|V'\rangle\}$). The radiation-mediated processes of excitation and observation are exclusively focused on the special oscillator $|V\rangle$, which is therefore the radiant 'doorway' through which the motion of the system of coupled oscillators must necessarily be excited and analysed. The presently considered relaxation studies involve impulsive excitation of the special oscillator, which in turn produces vibrations in the others, so that the full range of the free normal modes of the system is excited; these are then continuously observed by the electromagnetic signal emitted by the special radiant oscillator. The standard treatment of the classical problem proceeds along the same lines as the present quantum-mechanical approach (cf. for example reference 65). The central problem is likewise the search of the complete set of damped normal modes, each characterized by a complex frequency $\omega_p = \omega'_p - i\omega''_p$, where ω'_p is the normal frequency of vibration, and ω''_p a damping factor due to the frictional forces. In a similar way as for the poles E_p of the reduced resolvent matrix element $\langle V|MG(E^+)M|V\rangle$ (cf. equation (5.91)), the complex frequencies ω_p are solutions of a generalized eigenvalue problem; also the amplitude, say $X_V(t)$, of the central oscillator is found under a form identical to that of probability amplitude $A_V(t)$:

$$X_V(t) = \sum_p C_p \, e^{-i\omega'_p t} \, e^{-\omega''_p t}$$

C_p: complex pre-exponential factors. This formal equivalence of time-dependence of molecular excited states with classical motion of coupled oscillators justifies the use of common concepts like the various typical damped oscillation modes, beats, Poincaré recurrences, etc....

5.5.4 *Concluding remarks*

The theoretical treatment used in this study is, of course, equivalent to other general perturbation theoretical approaches using other variants of the resolvent or Green's function formalisms,[17,24,25,30,66,67] Fano's configuration interaction method,[15,26,68,69] extended Wigner–Weisskopf schemes,[68] Heitler's damping theory,[23] etc.... Conceptually, the present approach has the advantage to concentrate on notions like effective non-Hermitian Hamiltonians and 'optical' potentials, which allow a proper definition of molecular resonances and decaying states (cf. for example references 70, 71); also, the extension to collision theory—required, for instance, to describe resonance fluorescence—is straightforward.[72] Practically, in the specialized field of molecular photophysics (where relaxation techniques are dominant), this formalism has the merit of emphasizing time dependence at each step of the development. In addition, it systematically distinguishes the natural classes of energy states, the natural elementary dynamical stages, that play the major role on the overall molecular relaxation phenomenon. This quantum treatment hence appears as particularly suitable for a comparison with, and a discussion of, the elementary descriptions using the standard methods of chemical kinetics.[7,8]

The theoretical analysis is based on a very elementary model, which essentially distinguishes between three classes of states of the molecular system. First, the radiant states are separated from the others; hence, the special role of the radiant levels as doorways, through which the intramolecular dynamical modes are excited and observed per coupling with the external radiation field, is explicitly recognized; this leads, in particular, to the formalization of the basic concepts of radiative and non-radiative transitions that follow molecular broad-band excitation. A second partition is made within the non-radiant manifold, in order to account for the possible existence of a limited subset of non-radiant configurations playing a special role in intramolecular dynamics. On very general grounds, one accordingly distinguishes:

(1) a special class of well separated and correlated vibronic levels, which must be treated exactly in the quantum-mechanical description;

(2) the remaining non-radiant states, forming in general a complicated and dense set, for which a statistical treatment is sufficient.

A necessary condition implied by the definition of the special levels $\{|\bar{V}'\rangle\}$, is that they do not overlap (cf. equation (5.112)). Non-radiant subsets that do not satisfy this condition form an effective continuum, and should be classified in the second, non-special class of molecular states. By interacting with the radiant state, the special and non-special non-radiant configurations are respectively responsible for the coherent and incoherent intramolecular components of motion, which determine the general types of periodic and aperiodic decay modes.

In practice, the relative importance of the two non-radiant level classes and, consequently, of the characteristic decay modes, depends upon the molecular system and the excitation energy. Two extreme limiting cases appear as trivial. For relatively small molecules, the non-special manifold is generally unimportant because of the reduced density of vibronic levels: here, the special features of radiative decay (quantum beats, lengthening of fluorescence decay times, ...) due to eventual molecular state interferences should be best exhibited. For large molecules, on the other hand, the mean level density is generally so large that the non-radiant manifold is usually best considered as dominated by the previously defined attributes of non-special levels: this 'statistical limit' is characterized by familiar exponential decay; for organic molecular systems, it may be regarded as the rule rather than the exception. Between the two extreme 'small and large molecule' cases a number of situations may be envisaged, which must be treated without simplifying the foregoing molecular model; these 'intermediate' cases appear, by far, as most interesting to gain insight into non-trivial details of intramolecular electronic relaxation.

The primary condition to be satisfied in order to observe the special fluorescence features related to such intermediate situations is given by equation (5.112), or:

$$(\Gamma_{\bar{V}'}/2\bar{D}) < 1 \qquad (5.112')$$

remembering that $\bar{D}$ and $\Gamma_{\bar{V}'}$ denote width and spacing of the special non-radiant sequence of levels $\{|\bar{V}'\rangle\}$. The favourable large values for $\bar{D}$ may be found, either in small or medium-sized molecules, or in large molecules when the radiant $|V\rangle$ and non-radiant $|\bar{V}'\rangle$ manifolds are separated by a small energy gap.[18,24] Illustrative, recently studied examples involve the second excited singlet state (S_2) of organic molecules, where the energy gap with the S_1-state is small (3000–4000 cm^{-1}): 3:4-benzopyrene,[73] naphthalene,[74] quinoxaline;[75] other examples concern the interaction of the lowest singlet state $|V\rangle$ with a nearby triplet $|\bar{V}'\rangle$ manifold: benzophenone,[76,77] biacetyl,[78] pyrazine.[79] Most of these experiments were performed with vapours, under 'collision free' conditions: as required by condition (5.112'), the width $\Gamma_{\bar{V}'}$ is accordingly the smallest possible, being determined by the sole intramolecular mechanisms (nonadiabaticity, non-harmonicity, . . .). In dense materials, an additional contribution to $\Gamma_{\bar{V}'}$ comes from vibrational relaxation per interaction with the lattice phonons, as determined by the operator $H'_{M'L}$ in expression (5.30) (cf. also equation (5.34)), or by alternative processes considered in reference 80. Here, condition (5.112') is thus more difficult to satisfy than in the free-molecule limit; no reason is however apparent to believe that it necessarily breaks down in any condensed system. Favourable circumstances may, for example, be expected with medium-sized molecules dissolved in an inert host material with a spectrum of small energy phonons; vibrational relaxation, as a multiphonon transition (cf. §5.2.2), may then be notably slowed down, according to the general energy gap law. Solid rare-gas host matrices appear as particularly relevant for this purpose, since their Debye frequency ω_{max} (cf. equation (5.24)) is around 60–80 cm^{-1},[81] while, in organic solids, one has $\omega_{max} \approx 200$ cm^{-1}.[38]

For a proper classification of the characteristic types of decay modes in the 'intermediate cases', it is indispensable to distinguish between strength of coupling and strength of mixing of the $|V\rangle$ with the special $|\overline{V}'\rangle$ non-radiant states. Intramolecular coupling strength is measured by the dimensionless quantity $\rho^{\downarrow}$, defined by equation (5.125):

$$\rho^{\downarrow} = (\pi\Gamma_V^{\downarrow}/2\overline{D}) = (\pi u/\overline{D})^2 \qquad (5.125')$$

where u is the coupling energy $\langle\overline{V}'|H'|V\rangle$ and $\Gamma_V^{\downarrow}$ the characteristic width introduced by equation (5.124). But actual mixing of the $|V\rangle$ and $|\overline{V}'\rangle$ levels is a balance between interstate coupling which promotes it, and irreversible level damping which prevents it. The influence of damping must hence also be taken into account: it depends on the dimensionless quantity $\rho^{\uparrow}$, also defined by equations (5.125):

$$\rho^{\uparrow} = \pi(\Gamma_V^{\updownarrow} - \Gamma_{\overline{V}'})/2\overline{D} \qquad (5.125'')$$

where it should be remembered that the radiant level width $\Gamma_V^{\updownarrow}$ includes, in addition to the natural radiation and lattice phonon terms, the contribution from interaction with the non-special class of molecular non-radiant configurations (cf. equation (5.113)). The extent of interstate mixing depends on the relative values of the coupling and damping strengths, $\rho^{\downarrow}$ and $\rho^{\uparrow}$: the two natural limits to be considered as those corresponding to 'weak mixing', with $\rho^{\downarrow}\langle[1 + (\rho^{\uparrow})^2]^{\frac{1}{2}}$, and 'strong mixing', with $\rho^{\downarrow} > \rho^{\uparrow}$. If this is combined with coupling strength ($\rho^{\downarrow} > 1$: 'strong' coupling; $\rho^{\downarrow} < 1$: 'weak' coupling) one obtains the general classification, in Table 5.1, of the typical cases examined in §5.4.

Table 5.1 Classification scheme based on the extent of molecular interstate coupling and mixing

Mixing \ Coupling	Weak: $\rho^{\downarrow} < 1$	Strong: $1 < \rho^{\downarrow}$
Weak: $\rho^{\downarrow} < [1 + (\rho^{\uparrow})^2]^{1/2}$	Case II: $\rho^{\downarrow} < (\rho^{\uparrow}, 1)$	Case I: $1 < \rho^{\downarrow} < \rho^{\uparrow}$
Strong: $\rho^{\uparrow} < \rho^{\downarrow}$	Case III: $\rho^{\uparrow} < \rho^{\downarrow} < 1$	Cases IV, V: $(1, \rho^{\uparrow}) < \rho^{\downarrow}$

Coupling strength $\rho^{\downarrow}$ exclusively depends upon the molecular attributes u and $\overline{D}$. Within a given category of molecules, the interstate non-adiabaticity interaction u may be expected to vary slowly; this means that the coupling strength is mainly determined by the spacing $\bar{D}$ of the special non-radiant vibronic levels or also, crudely speaking, by the molecular size. The weak coupling Cases II and III hence certainly apply better to relatively small molecules than to polyatomic molecular systems. Their distinctive features, like quantum beat behaviour, therefore appear of limited practical interest in organic materials. In these systems 'intermediate type' decay features are expectedly determined by the strong coupling cases I and (IV, V).

In Case I, the physical situation is dominated by fast irreversible decay of the impulsively excited radiant state, which reduces the extent of intramolecular mixing with the special non-radiant manifold. This should be quite common for radiant vibronic levels with high electronic and vibrational quantum numbers: for highly electronically excited radiant states, the number and density of the dissipative non-radiant configurations increase; also other continuous decay channels (molecular dissociation, isomerization, autoionization,...) may gain importance; all these electronic relaxation processes contribute to increase the radiant level width $\Gamma_V^{\updownarrow}$. Further contributions from vibrational relaxation in dense media, anharmonic mixing of molecular vibration normal modes,... appear for vibrationally excited molecules. Despite its large potential importance, Case I has apparently not yet been examined in detail by previous workers (see however reference 24). The gross features of fluorescence emission, and of decay of the radiant state, are controlled by the broad resonance of width $\Gamma_V^{\updownarrow}$ centred at the initial level. Particular effects may however be observable, that are due to the special narrow non-radiant resonances interfering with the principal broad resonance in which they are imbedded (cf. Figure 5.9). According to §§5.5.1 and 5.5.2, the conditions of observability of the theoretical results derived for this case are that excitation (I) and detection (Δ) bandwidths be larger than the width $\Gamma_V^{\updownarrow}$ of the principal resonance ($\Gamma_V^{\updownarrow} < I, \Delta$). Remembering further the condition of single vibronic level excitation stated at the end of §5.5.1 ($I < D_{\text{rad}}$, D_{rad}: spacing of the radiant vibronic levels), we may summarize all the necessary conditions implied by the present analysis of Case I:

$$\Gamma_{\bar{V}'} < (2\bar{D}/\pi) < \Gamma_V^{\downarrow} < \Gamma_V^{\updownarrow} - \Gamma_{\bar{V}'} < (I, \Delta) < D_{\text{rad}}$$

The other 'intermediate type' category, where both coupling and mixing are strong, involves the two Cases IV and V (together with IV′ and V′) which, in previous work, have often been more or less artificially distinguished. From the physical viewpoint, no essential difference exists between these situations that are all governed by thorough mixing of the radiant state with a number of the special non-radiant states; the radiant character is then shared by a collection of the molecular levels: the dynamical behaviour and light emission of the impulsively excited molecules are determined by the cooperative behaviour of these strongly mixed configurations. The most notable practical consequence in the relaxation studies is then the characteristic giant resonance decay mode, discussed under Case V (cf. equations (5.236)–(5.237)). The general significance of this initial exponential component, with the time constant $(\Gamma_V^{\downarrow})^{-1}$, in the strongly mixed intermediate situations has been particularly stressed in the recent work of Tric and Tramer.[24,79,83] The distinction between the Cases IV (few-level resonance) and V (giant resonance) appears to be only based on different practical time scales that are imposed by the necessary, but adjunct, excitation and/or detection parts of the experimental system (§§5.5.1 and 5.5.2).

Case IV is best related to circumstances where at least one of the widths I (excitation) and Δ (detection) is smaller than the bandwidth of the strongly mixed molecular vibronic states ($I, \Delta < \Gamma_V^{\downarrow}$). Following equations (5.247) and (5.256), the excitation and/or detection devices then act as filters with reduced band-pass properties, the response of which only accounts for the influence of a limited number of molecular resonances. The full spectrum-dependent giant resonance behaviour can hence not be observed: the fast initial decay component, in particular, is not recorded; fluorescence intensity decay is best described by the slowly decaying law (5.216), where the sum includes only the levels 'filtered' by the excitation/detection parts of the physical system. The characteristic conditions for Case IV may be summarized as:

$$\Gamma_{\overline{V}'} < (2\bar{D}/\pi, \Gamma_V^{\uparrow} - \Gamma_{\overline{V}'}) < (I, \Delta) < (\Gamma_V^{\downarrow}, D_{\mathrm{rad}})$$

Case V, and the corresponding giant resonance results, should apply quite generally to the strong mixing intermediate situations, whenever no spectral limitations are introduced by the excitation and detection systems, i.e., when: $(I, \Delta) > \Gamma_V^{\downarrow}$. Initial intensity decay is then dominated by the characteristic giant resonance component, decreasing exponentially with the lifetime $(\Gamma_V^{\downarrow})^{-1}$ as described by equation (5.237). This crude expression accounts for the gross features; but it is easily generalized, just by dropping some unnecessary simplifying assumptions made on the quantities R_p and ε_p'' in expansion (5.175) of $A_V(t)$; one hence shows that, for the times (5.234):

$$A_V(t) \approx \mathrm{e}^{-\mathrm{i}E_0 t}\,\mathrm{e}^{-\frac{1}{2}\Gamma_{\overline{V}'}t}\,\mathrm{e}^{-[\rho^{\uparrow}/(\rho^{\downarrow}+1)]\tau}\int_{-\infty}^{+\infty}\frac{\mathrm{d}x}{\pi}\,\frac{\rho^{\downarrow}\,\mathrm{e}^{-\mathrm{i}x\tau}}{(x+\mathrm{i}\rho^{\uparrow})^2+(\pi n_0)^2}$$

and:

$$I(t) \approx \gamma_R \exp\left[-\frac{\Gamma_V^{\uparrow} - \Gamma_{\overline{V}'}}{(\rho^{\downarrow}+1)}t\right]\exp\left[-(\Gamma_V^{\downarrow} + \Gamma_V^{\uparrow})t\right] \qquad (5.237')$$

In addition to this rapidly decaying component, a slower decaying part generally appears at longer times. This must be accounted for by the discrete sum in the basic expansion (5.175) of the amplitude $A_V(t)$, since the approximation by an integral is no longer valid. Problem and arguments are then similar to those of Case IV, which means that the slowly decaying component is represented by expression (5.216) (cf. also reference 82). Recent experimental results[75–79] seem to agree qualitatively with such a general decay law involving two components. It should be noted however that the edge effects considered under Cases IV′ and V′ also tend to introduce extra fast and slow components. They may well play a non-negligible role in particular cases, which should be experimentally studied by considering, for example, the influence of excitation wavelengths on the characteristics of fluorescence emission. The conditions supposed to be verified in Case V are:

$$\Gamma_{\bar{V}'} < (2\bar{D}/\pi, \Gamma_V^{\uparrow} - \Gamma_{\bar{V}'}) < \Gamma_V^{\downarrow} < (I, \Delta) < D_{\text{rad}}$$

We close by noting that many of the problems mentioned in this review still appear as practically unexplored. The current experimental and theoretical activity in the field will, however, very rapidly lead to substantial progress in our understanding of the molecular decay processes.

5.6 References

1. W. Heitler, *The Quantum Theory of Radiation*, Chapters 16 and 20, 3rd edition, Clarendon Press, Oxford, 1966.
2. M. L. Goldberger and K. M. Watson, *Collision Theory*, Chapter 8, John Wiley, New York, 1967.
3. A. Messiah, *Mécanique Quantique*, Chapters 19 and 21, Dunod, Paris, 1965.
4. R. D. Levine, *Quantum Mechanics of Molecular Rate Processes*, Clarendon Press, Oxford, 1969.
5. I. E. McCarthy, *Introduction to Nuclear Theory*, John Wiley, New York, 1968.
6. B. W. Shore, *Rev. Mod. Phys.*, **39**, 439 (1967).
7. J. B. Birks, *Photophysics of Aromatic Molecules*, John Wiley, New York, 1970.
8. J. G. Calvert and J. N. Pitts, *Photochemistry*, John Wiley, New York, 1966.
9. B. R. Henry and W. Siebrand, *Organic Molecular Photophysics*, Vol. 1, Chapter 4 (Ed. J. B. Birks), John Wiley, New York, 1972.
10. E. W. Schlag, S. Schneider and S. F. Fischer, *Ann. Rev. Phys. Chem.*, **22**, 465 (1971).
11. J. Jortner, S. A. Rice and R. M. Hochstrasser, *Advances in Photochemistry*, **7**, 149 (1969).
12. B. R. Henry and M. Kasha, *Ann. Rev. Phys. Chem.*, **19**, 161 (1968).
13. G. W. Robinson and R. P. Frosch, *J. Chem. Phys.*, **37**, 1962 (1962); **38**, 1187 (1963).
14. F. H. Mies and M. Kraus, *J. Chem. Phys.*, **45**, 4455 (1966).
15. M. Bixon and J. Jortner, *J. Chem. Phys.*,**48**, 715 (1968); **50**, 3284 (1969); **50**, 4061 (1969).

16. J. Jortner and R. S. Berry, *J. Chem. Phys.*, **48**, 2757 (1968).
17. K. F. Freed and J. Jortner, *J. Chem. Phys.*, **50**, 2916 (1969).
18. A. Nitzan, J. Jortner and P. M. Rentzepis, *Proc. Roy. Soc.*, **A 328**, 367 (1972); *Molec. Phys.*, **22**, 585 (1971).
19. W. Rhodes, *J. Chem. Phys.*, **50**, 2885 (1969); *J. Chim. Phys.* (special issue), *Transitions non-radiatives dans les molécules*, 40 (1970).
20. M. Berrondo and K. Weiss, *J. Chem. Phys.*, **52**, 807 (1970).
21. I. Riess, *J. Chem. Phys.*, **52**, 871 (1970).
22. K. F. Freed, *J. Chem. Phys.*, **52**, 1345 (1970).
23. R. Voltz, *Molec. Phys.*, **19**, 881 (1970).
24. C. Tric, *Chem. Phys. Letters*, **21**, **83** (1973); *J. Chem. Phys.*, **55**, 4303 (1971).
25. R. A. van Santen, *Physica*, **62**, 84 (1972).
26. J. A. Beswick, R. Lefèbvre and A. M. Plumejeau, *J. Chem. Phys.*, **56**, 4011 (1972).
27. C. S. Parmenter, *Adv. Chem. Phys.*, **22**, 365 (1972).
28. M. Stockburger, *Organic Molecular Photophysics*, Vol. 1, Chapter 2 (Ed. J. B. Birks), John Wiley, New York, 1972.
29. A. Messiah, *Mécanique Quantique*, Chapter 21, Section 13, Dunod, Paris, 1965.
30. R. A. Harris, *J. Chem. Phys.*, **39**, 978 (1963).
31. L. Mower, *Phys. Rev.*, **142**, 799 (1966).
32. R. D. Levine, *Quantum Mechanics of Molecular Rate Processes*, Part 3, Clarendon Press, Oxford, 1969.
33. H. Feshbach and F. S. Levine, *Reaction Dynamics*, Part II, Gordon and Breach, New York, 1973. A complete and accessible account of Feshbach's work.
34. H. Feshbach, A. K. Kerman and R. H. Lemmer, *Ann. Phys.* (*N.Y.*), **41**, 230 (1967).
35. C. Bloch, *Many Body Description of Nuclear Structure and Reactions* (Ed. C. Bloch), Academic Press, New York, 1966.
36. C. Mahaux and H. A. Weidenmüller, *Shell-Model Approach to Nuclear Reactions*, Chapter 9, North-Holland, Amsterdam, 1969.
37. O. Schnepp and N. Jacobi, *Adv. Chem. Phys.*, **22**, 205 (1972).
38. K. K. Rebane, *Impurity Spectra of Solids*, Plenum Press, New York, 1970.
39. J. A. Beswick and G. Lochak, *J. Physique*, **32**, 11 (1971). Discussion of the concepts of 'radiant' and 'non-radiant' states; cf. also reference 40.
40. G. Lochak, *J. Physique*, **31**, 871 (1970).
41. A. Messiah, *Mécanique Quantique*, Chapter 7, Dunod, Paris, 1965.
42. V. A. Fock and S. N. Krylov, *J. Exp. Theoret. Phys.* (*USSR*), **17**, 93 (1947). This work is reviewed in reference 43.
43. A. S. Davydov, *Quantum Mechanics*, Chapter 9, Section 80, Pergamon Press, Oxford, 1965.
44. G. Breit and E. Wigner, *Phys. Rev.*, **49**, 519 (1936).
45. P. Roman, *Advanced Quantum Theory*, Section 4–5, Addison-Wesley, Reading, Mass., 1965.
46. In the following sections, we set $\hbar = 1$.
47. The notations $\Gamma_V^{\uparrow}$ and $\Gamma_V^{\downarrow}$ are borrowed from the studies of nuclear doorway states (references 33–36), where they denote widths having essentially the same meaning as in the present problem.
48. E. M. Lane, *Isospin in Nuclear Physics* (Ed. D. H. Wilkinson), Chapter 11, North-Holland, Amsterdam, 1969.
49. C. A. Langhoff and G. W. Robinson, *Molec. Phys.*, **26**, 249 (1973).
50. E. B. Aleksandrov, *Sov. Phys. Uspekhi*, **15**, 436 (1973).

51. U. Fano and J. H. Macek, *Rev. Mod. Phys.*, **45**, 553 (1973).
52. H. E. Radford and H. P. Broida, *J. Chem. Phys.*, **38**, 644 (1963); *Phys. Rev.*, **128**, 231 (1963). An example of two level resonance: $^2\Sigma - ^2\pi$ mixing in CN.
53. A. E. Douglas, *J. Chem. Phys.*, **45**, 1007 (1966).
54. The physical situation is similar to that of the 'strong-coupling' limit for the intermolecular energy transfer processes in dimers.[55]
55. W. T. Simpson and D. L. Peterson, *J. Chem. Phys.*, **26**, 588 (1957).
56. A. M. Lane, R. G. Thomas and E. P. Wigner, *Phys. Rev.*, **98**, 693 (1955).
57. L. A. Khalfin, *Sov. Phys. JETP*, **6**, 1053 (1958).
58. A. Papoulis, *The Fourier Integral and its Applications*, Chapter 2, McGraw-Hill, New York, 1962.
59. *Handbook of Mathematical Functions*, Chapter 5, National Bureau of Standards, Washington, 1966.
60. C. Cohen-Tannoudji, *Compléments de Mécanique Quantique*, Université de Paris, 1966 (unpublished).
61. I. E. McCarthy, *Introduction to Nuclear Theory*, Chapter 9, John Wiley, New York, 1968.
62. R. J. Glauber, *Optique et Electronique Quantiques*, p. 63 (Eds. C. De Witt, A. Blandin and C. Cohen-Tannoudji), Gordon and Breach, New York, 1965.
63. W. Rhodes, *Chem. Phys. Letters*, **11**, 179 (1971). See also reference 64.
64. A. Nitzan and J. Jortner, *Chem. Phys. Letters*, **14**, 177 (1972).
65. H. Goldstein, *Classical Mechanics*, Chapter 10, Addison-Wesley, Reading, Mass., 1959.
66. D. P. Chock, J. Jortner and S. A. Rice, *J. Chem. Phys.*, **49**, 610 (1968).
67. K. Nishikawa and S. Aono, *Progr. Theor. Phys.*, **50**, 345 (1973).
68. M. Bixon, J. Jortner and Y. Dothan, *Molec. Phys.*, **17**, 109 (1969).
69. R. Lefèbvre and J. A. Beswick, *Molec. Phys.*, **23**, 1223 (1972).
70. L. Rosenfeld, *Spectroscopic and Group Theoretical Methods in Physics*, p. 203 (Eds. F. Bloch *et al.*), North-Holland, Amsterdam, 1968.
71. M. M. Sternheim and J. F. Walker, *Phys. Rev. C*, **6**, 114 (1972).
72. A. Villaeys, to be published.
73. P. Wannier, P. M. Rentzepis and J. Jortner, *Chem. Phys. Letters*, **10**, 102 (1971).
74. P. Wannier, P. M. Rentzepis and J. Jortner, *Chem. Phys. Letters*, **10**, 193 (1971).
75. J. R. McDonald and L. E. Brus, *Chem. Phys. Letters*, **23**, 87 (1973).
76. G. E. Busch, P. M. Rentzepis and J. Jortner, *J. Chem. Phys.*, **56**, 361 (1972).
77. R. M. Hochstrasser and J. E. Wessel, *Chem. Phys. Letters*, **19**, 156 (1973).
78. A. Nitzan, J. Jortner, J. Kommandeur and E. Drent, *Chem. Phys. Letters*, **9**, 273 (1971).
79. F. Lahmani, A. Frad and A. Tramer, *Chem. Phys. Letters*, **14**, 337 (1972).
80. A. Nitzan and J. Jortner, *J. Chem. Phys.*, **58**, 2412 (1973).
81. G. A. Cook, *Argon, Helium and the Rare Gases*, Interscience, New York, 1961.
82. R. Lefèbvre and J. Savolainen, *J. Chem. Phys.*, **60**, 2509 (1974).
83. J. M. Delory and C. Tric, *Chem. Phys.*, **3**, 54 (1974).

6 *Radiationless transitions—a postscript*

Bryan R. Henry and Willem Siebrand

6.1 Discussion

This postscript concerns developments in the period early-1971–early-1973. We shall be mostly concerned with those developments that either clarify or alter the picture sketched in Chapter 4 of Volume 1. We shall use the same ordering as before and omit (mathematical) details for reasons of space and readability. We also refer the reader to the above article for many of the original references.

In §4.2 (Vol. 1) we introduced the notion of a recurrence of the initial state and used it to describe radiationless processes in terms of Poincaré cycles. It should be noted, however, that Poincaré's recurrence theorem does not predict a conservative system (i.e. a system without energy dissipation) to be involved in a periodic motion in the ordinary sense.[1] Thus the theorem does not state that the system will return exactly to the initial state, but only that it will approach this state closely in a finite time. In general the recurrence period will depend on how closely we wish to approach it. Thus the more exacting are our requirements, the longer we will have to wait for a recurrence to occur, except in special cases. Such a special case is the Bixon–Jortner model[2] with its constant level spacing. Since the recurrence pattern is the Fourier transform of the level spacing, an irregular level spacing will produce an irregular recurrence pattern. On the other hand, any regularity observed in the recurrence pattern will indicate a regular spacing of a group of (directly coupled) levels in the vibrational manifold. Since the directly coupled levels are generally associated with a limited number of specific normal or local modes, the observation of such regularities would lead to information about the part played by these modes in the transition.[3] The matrix element V_{mn} would then be analysable in terms of components associated with specific levels or groups of levels in the manifold of final states, $\{\Psi_m\}$.

Not much progress has been made thus far with the experimental observation, let alone the analysis of such quasi-periodic phenomena. One of the main difficulties is the requirement of coherent excitation of all the levels involved. This is not difficult when only a single molecule is to be excited, but to get a strong enough signal, one is forced to deal with a large number of molecules in a large container, so that the coherent excitation of all the relevant levels in all these molecules can be a major problem. Another major problem is the complicated form of recurrence patterns involving more than two levels. They are hard to recognize and hard to distinguish from 'noise' or experimental artifacts. Their Fourier transformation into an energy-level pattern is not easy to carry out unambiguously unless the observation time is much longer than the recurrence time.

Once such an energy-level pattern is obtained, the assignment of these levels becomes the next step in the physical interpretation of the observed periodicity. An assignment in terms of approximate rather than exact molecular quantum numbers will often be most amenable to physical interpretation, e.g., an interpretation in terms of specific normal modes of vibration. Without such an assignment the information to be gained from a quantum-beat observation is of questionable validity.

This assignment problem is a generalization of that encountered for radiationless transitions leading to exponential decay which was discussed and solved in principle in §4.5 (Vol. 1). In view of its central importance for the interpretation of radiationless decay data, we return to it here and give a simple and general prescription for matrix elements governing exponentially decaying radiationless transitions.[4] For convenience we assume the experiment to involve excitation of a molecule in the ground state by a photon $h\nu$ where ν varies uniformly from $\nu - \frac{1}{2}\Delta\nu$ to $\nu + \frac{1}{2}\Delta\nu$. The radiative and radiationless transitions following the excitation are monitored via an emission signal. These transitions are formulated in terms of an arbitrarily chosen complete set of states $\{\Psi_a\}$, defined by

$$(H_M - V)\Psi_a = E_a^0\Psi_a \tag{6.1}$$

where H_M is the molecular Hamiltonian and V is arbitrary for the time being. Now we divide the set $\{\Psi_a\}$ into two subsets $\{\Psi_j\}$ and $\{\Psi_k\}$ such that

$$\begin{gathered} \nu - \tfrac{1}{2}\Delta\nu \leqslant E_k^0/h \leqslant \nu + \tfrac{1}{2}\Delta\nu \\ \nu - \tfrac{1}{2}\Delta\nu > E_j^0/h \quad \text{or} \quad \nu + \tfrac{1}{2}\Delta\nu < E_j^0/h \end{gathered} \tag{6.2}$$

We now try to separate the set $\{\Psi_k\}$, directly populated by photon absorption, into a set of initial states $\{\Psi_n\}$ and a set of final states $\{\Psi_m\}$. Only the states $\{\Psi_n\}$ will have transition dipole moments to the ground state. To facilitate this separation we choose V and thus $\{\Psi_n\}$ so that good optical selection rules are obtained. More specifically, we favour the representation in which the number of states in the set $\{\Psi_n\}$ is minimized. In the special case where reduction to one state Ψ_n is possible, radiative and radiationless processes are independent, i.e., no interference occurs, and the radiationless matrix element

$$V_{mn} = \langle\{\Psi_m\}|V|\Psi_n\rangle + \sum_i{}'(E_i^0 - E_n^0)\langle\{\Psi_m\}|V|\Psi_i\rangle\langle\Psi_i|V|\Psi_n\rangle + \dots \quad (6.3)$$

is independent of the representation, provided the following conditions are met: (i) The intermediate states $\{\Psi_i\}$ are taken exclusively from the set $\{\Psi_j\}$; (ii) the expansion is continued until convergence is achieved; (iii) the linewidths $|V_{mn}|$ are small compared to $h\Delta\nu$.

This special case reduces to the usually observed case of exponential decay when $\{\Psi_m\}$ is a dense manifold. When $\{\Psi_m\}$ is a coarsely structured manifold, recurrence beats may be observable if an observation time much longer than the recurrence time is possible.[3] Quantum beats may also arise when $\{\Psi_n\}$ cannot be reduced to one state or when $|V_{mn}| \geqslant h\Delta\nu$. These two cases are related and may often be transformed into each other by a change of basis. It must be kept in mind, however, that the linewidth of the level n is governed not only by $|V_{mn}|$ but also by the radiative linewidth. If $h\Delta\nu$ is smaller than the latter, one can no longer speak of a prepared state Ψ_n, since emission from Ψ_n will take place while there is still light being absorbed, so that emission and absorption can no longer be treated as separate processes. Hence the amount of information obtainable by using highly monochromatic light is limited by the long duration of the absorption process. These limits are, in fact, just a manifestation of the uncertainty principle. Nitzan and Jortner[5] have made a detailed study of the interference effects that may arise in the limit of highly monochromatic exciting light beams.

Returning now to exponential decay, we note that equation (6.3) and the associated conditions (i)–(iii) allow a direct test of the matrix elements used to calculate radiationless transition probabilities. It is readily seen that the matrix elements proposed on the basis of the adiabatic Born–Oppenheimer approximation and analyzed in §4.6 (Vol. 1) are in agreement with this basic rule. Note also, that some alternative schemes[6,7] apparently leading to different results, are incorrect, either because they clearly violate the basic conditions (i)–(iii) for the validity of equation (6.3), or because an expansion has been truncated prematurely.[4]

Unfortunately, the ability to formulate a matrix element unambiguously does not necessarily entail the ability to calculate it. The vibronic integrals remain the main stumbling block. The most complete attempt thus far to include the nuclear-coordinate dependence of the electronic wavefunctions into the calculation has recently been made by Nitzan and Jortner.[8] However, even this calculation involves too many drastic assumptions to be entirely convincing especially for vibronic coupling between states separated by a large energy gap.[9]

A recent attempt to improve the quantitative evaluation of radiationless decay rates has been made by Christie and Craig.[10] These authors attempt to account for the differing magnitudes of coupling to the various states in the vibrational quasicontinuum with a method based on Laplace transformation. However, an application of their approach to the first excited singlet to triplet intersystem crossing in benzene[11] requires many serious approximations so that a test of the relative accuracy of this method must await further work.

The problem of calculation of matrix elements has hitherto prevented the satisfactory evaluation of decay rates for internal conversion. The experimental data about $S_1 \rightsquigarrow S_0$ conversion in aromatic hydrocarbons remain meagre, but there is some evidence[12–14] that their Franck–Condon factors follow the same empirical relation as the $T_1 \rightsquigarrow S_0$ rate constants (see §4.7 (Vol. 1)). In addition the spin-prohibition factor of 10^8 seems to be the same for the radiative as for the non-radiative rate constants.[12,13] Recent observations of several instances of fluorescence from S_2, contrary to Kasha's rule, has led to a few $S_2 \rightsquigarrow S_1$ conversion rate constants.[15–24] A full theoretical analysis of these results must await further experimental data.

A significant contribution to the evaluation of intersystem crossing matrix elements has been made by Metz *et al.*,[25] who included two-centre vibronic integrals in their calculation. On this basis they reversed the order of two of the terms in §4.6 (Vol. 1); $|H_{mn}^{(2)}| > |H_{mn}^{(3)}|$, without changing the basic mechanism; i.e., $\pi\sigma$ spin–orbit coupling induced by out-of-plane vibrations. The quantitative assessment of this claim must await the publication of the details of their calculation. The available new experimental data[26–32] do not clearly distinguish between the two mechanisms. It should be noted that cross-terms of the form $H_{mn}^{(2)}H_{mn}^{(3)}$ will contribute to intersystem crossing if both matrix elements are governed by $\pi\sigma$ spin–orbit coupling and are of comparable magnitude. Thus in that case $H_{mn}^{(2)}$ and $H_{mn}^{(3)}$ cannot be strictly identified with separate mechanisms. This may, for instance, complicate the interpretation of the isotope effect. Inclusion of two-centre vibronic integrals also complicates matters in that it requires introduction of the normal modes of vibration into the calculation. In the one-centre scheme, on the other hand, one can get away with local modes of vibration. In general not much is known of the normal-mode properties in excited electronic states. It is also worth noting that ground state normal-modes belonging to the same symmetry species may mix in the excited state. This phenomenon, sometimes referred to as the Duschinsky effect,[33] is well-known in spectroscopy. It makes it difficult to interpret isotope effects due to inducing modes if more than one inducing mode of a given species is active.[34]

Recent observations[26–30] of position-dependent deuterium effects in molecules other than naphthalene have generally confirmed the overall picture outlined in §4.8 (Vol. 1). They show convincingly that intersystem crossing in aromatic hydrocarbons is vibrationally induced, the vibrations being out-of-plane modes, as follows from the resulting spin alignment. Ideally one should combine both measurements, i.e., measure the position-dependent isotope effect for each of the three triplet components separately. Although efforts are being made in this direction,[30] it is not easy to get the necessary accuracy. The one additional feature suggested by these measurements is that the τ_z (i.e., $m_z = 0$) component which can be (de) populated by $\pi\pi$ spin–orbit coupling, either directly or via vibrational induction, is instead likely to be dominated by $\pi\sigma$ spin–orbit coupling, just like the τ_x and τ_y ($m_z = \pm 1$) components; although for the τ_z component this requires a third-order matrix element consisting of one spin–orbit integral and two vibronic coupling integrals.[25]

Work on the temperature dependence of intersystem crossing continues[35–37] and the existence of a temperature-independent and a temperature-dependent rate of singlet to triplet crossing is now well established in several molecules. However, the simple Arrhenius behaviour is not always exhibited and in some cases the density of states must be included to describe the observed behaviour.[36]

The study of the effect of vibronic excitation on radiationless transitions has made rapid progress.[38–47] The investigation of the single vibronic level (often abbreviated as SVL) fluorescence of benzene has been extended to deuterobenzene and substituted benzenes.[38] There has also been new theoretical work;[39–42] the most detailed study of the benzene data is that by Heller, Freed and Gelbart[39] who interpreted the observed radiationless rate constants in terms of the equilibrium distances and frequencies of the corresponding modes in the triplet state. Since these parameters are not known *a priori*, their analysis can only show that the observations are in agreement with a reasonable and mutually consistent set of vibrational parameters. However, one should not lose track of the basic aim of such an analysis, which is to shed light on the mechanism of the transition. The main questions to be answered are: Which modes induce the $S_1 \rightsquigarrow T_1$ crossing? How is the energy distributed among the accepting modes? What is the nature of the process that quenches the fluorescence of levels 3000 cm^{-1} or more above S_1? These questions remain essentially unanswered. As directly shown by the simple analysis of §4.9 (Vol. 1), the available data do not allow such answers since none of the SVLs brought to fluorescence contains quanta of an inducing or a dominant accepting mode without being influenced by the quenching process active above $E(S_1) + 3000\ cm^{-1}$.

The latter process, sometimes referred to as channel 3, remains unidentified. Earlier attempts to identify it with predissociation have met with the objection that apparently no hydrogen is set free.[48] There are also theoretical arguments against this possibility.[49] An argument against a photochemical reaction is the absence of products. It has also been suggested[48] that there may be an as yet unidentified electronic state, perhaps a $\sigma\pi^*$ state, at this energy, but no strong independent evidence for this has been found thus far. Hence, the channel 3 problem remains unsolved.

The effect of such embedded electronic states in the S_1 vibrational manifold of several other molecules, e.g. naphthalene, has been noted. Under low resolution a plot of the radiationless decay rate constant against excess energy may show an inflection at the position of such a state.[43] It is difficult to achieve true SVL fluorescence in molecules larger and less symmetric than benzene, because of the high density of vibrational levels. However, some progress is being made, notably in the case of naphthalene, where it has been found that $\nu_8(b_{1g})$ is not an inducing mode for intersystem crossing and that the previously reported anomalous deuterium effect does not occur under SVL fluorescence conditions.[47] As expected, the situation intermediate between SVL excitation and excitation of a dense manifold leads to signals that are more difficult to interpret than those corresponding to the two limiting cases.[50] Interesting work under SVL conditions has also been done on pyrazine[46] which seems to be small enough to approach the dense intermediate category for intersystem crossing from S_1. Thus in this molecule, as the pressure is decreased to small values, the phosphorescence disappears.

Although there has been some recent activity[51–53] in relating the theories of radiationless transitions and photochemical reactions the situation is far from resolved. Most of the recent investigations of photochemical reactivity pursue a different approach and follow a Woodward–Hoffmann type pattern and all but ignore the nuclear motions. Typically, adiabatic potentials are calculated to predict reaction paths and velocities, so that the computational problems, which may be formidable, are those of Hartree–Fock theory and electron correlation. Thus it is assumed that the dynamics of the nuclei plays no independent part in determining the course of the reaction. It is not known at present under which precise conditions this adiabatic approximation holds. For some simple inorganic reactions a diabatic description, i.e. one based on wavefunctions cruder than adiabatic Born–Oppenheimer states, has been found to be more appropriate.[54–56] No general attempt has as yet been made to find the 'best' first-order description of a photochemical reaction, analogous to the 'best' first-order description of a radiationless transition, discussed in §4.5 (Vol. 1). One of the fundamental differences in the two types of processes is that radiationless transitions are monitored via the decay of an initially prepared state and photochemical reactions via the generation of products.[57]

Medium effects of radiationless transitions have obtained a good deal of recent attention. As usual when a complicated phenomenon is given detailed scrutiny, a number of generalizations were overturned. Several examples have been found of highly specific effects caused by so-called inert media. For instance, deuteration of such a medium has been shown repeatedly to slow down radiationless decay for specific solute–solvent pairs.[58,59] In these systems the internal vibrations of the solvent molecules apparently, to some degree, act as accepting modes. This would imply that the electronic excitation of the solute molecule has a relatively strong effect on the vibrations of one or more neighbouring solvent molecules. Also we mention that specific matrix site effects on triplet state lifetimes have once again recently been noted.[60,61] These also imply some participation of the solvent in molecular relaxation.

To the extent that these observations concern intermolecular energy transfer, they are outside the scope of the article, but it may not always be easy to disentangle their intramolecular and intermolecular aspects. In this connection we mentioned the recently observed intermolecular intersystem crossing, the precise mechanism of which is still subject to discussion.[62–64]

For reasons of space and coherence we mention rather than discuss the recent activity in the area of vibrational relaxation.[65–69] These developments are expected to contribute a great deal to our understanding of radiationless phenomena in the near future.

6.2 References

1. See, e.g. M. Kac, *Probability and Related Topics in Physical Sciences*, Chapter 3, Interscience, London, 1959.
2. M. Bixon and J. Jortner, *J. Chem. Phys.*, **48**, 715 (1968).
3. W. Siebrand, *Chem. Phys. Letters*, **14**, 23 (1972).
4. G. Orlandi and W. Siebrand, *Chem. Phys. Letters*, **14**, 19 (1972); **21**, 217 (1973).
5. A. Nitzan and J. Jortner, *J. Chem. Phys.*, **57**, 2870 (1972).
6. B. Sharf, *Chem. Phys. Letters*, **14**, 315, 475 (1972).
7. T. Azumi, *Chem. Phys. Letters*, **17**, 211 (1972).
8. A. Nitzan and J. Jortner, *J. Chem. Phys.*, **56**, 3360 (1972).
9. A. Marconi and G. Orlandi, *Chem. Phys. Letters*, **21**, 89 (1973).
10. J. R. Christie and D. P. Craig, *Molec. Phys.*, **23**, 345 (1972).

11. J. R. Christie and D. P. Craig, *Molec. Phys.*, **23**, 353 (1972).
12. J. B. Birks, *Photophysics of Aromatic Molecules*, John Wiley and Sons, Inc., New York, 1970.
13. W. Siebrand and D. F. Williams, *J. Chem. Phys.*, **49**, 1860 (1968).
14. A. Kearvell and F. Wilkinson, *Chem. Phys. Letters*, **11**, 472 (1971).
15. P. Wannier, P. M. Rentzepis and J. Jortner, *Chem. Phys. Letters*, **10**, 102, 193 (1971).
16. P. A. Geldof, R. P. Rettschnick and G. J. Hoytink, *Chem. Phys. Letters*, **4**, 59 (1969).
17. C. E. Easterly, L. G. Christophorou, R. P. Blaunstein and J. G. Carter, *Chem. Phys. Letters*, **6**, 579 (1970).
18. W. R. Dawson and J. L. Kropp, *J. Phys. Chem.*, **73**, 1752 (1969).
19. H. Baba, A. Nakajima, M. Aoi and K. Chihara, *J. Chem. Phys.*, **55**, 2433 (1971).
20. A. Nakajima, *Bull. Chem. Soc. Japan*, **45**, 1687 (1972).
21. T. Deinum, C. J. Werkhoven, J. Langelaar, R. P. H. Rettschnick and J. D. W. van Voorst, *Chem. Phys. Letters*, **12**, 189 (1971); **19**, 29 (1973).
22. P. C. Johnson and H. W. Offen, *J. Chem. Phys.*, **57**, 336 (1972).
23. C. E. Easterly, L. G. Christophorou and J. G. Carter, *J. Chem. Soc. Faraday Trans.*, **II 4**, 471 (1973).
24. J. B. Birks, *Chem. Phys. Letters*, **17**, 370 (1972).
25. F. Metz, S. Friedrich and G. Hohlneicher, *Chem. Phys. Letters*, **16**, 353 (1972).
26. G. Heinrich, G. Holzer, H. Blume and D. Schulte-Frohlinde, *Z. Naturforsch.*, **25b**, 496 (1970).
27. G. Heinrich, H. Güstenn, F. Mark, G. Olbrich and D. Schulte-Frohlinde, *Ber. Bunsenges. Physik Chem.*, **77**, 103 (1973).
28. J. D. Simpson, H. W. Offen and J. G. Burr, *Chem. Phys. Letters*, **2**, 383 (1968).
29. J. L. Charlton and B. R. Henry, *J. Amer. Chem. Soc.*, **95**, 2782 (1973).
30. D. A. Antheunis, J. Schmidt and J. H. van der Waals, *Molec. Phys.*, **27**, 1521 (1974).
31. R. Li and E. C. Lim, *J. Chem. Phys.*, **57**, 605 (1972).
32. C. S. Parmenter and M. D. Schuh, *Chem. Phys. Letters*, **13**, 120 (1972).
33. F. Duschinsky, *Acta Physicochem. USSR*, **7**, 551 (1937).
34. G. Orlandi and W. Siebrand, *J. Chem. Phys.*, to be published.
35. S. H. Lin, *J. Chem. Phys.*, **56**, 2648 (1972).
36. E. R. Pantke and H. Labhart, *Chem. Phys. Letters*, **16**, 255 (1972).
37. M. F. Thomas, M. I. Barradas and B. Stevens, *Chem. Phys. Letters*, **17**, 160 (1972).
38. A. S. Abramson, K. G. Spears and S. A. Rice, *J. Chem. Phys.*, **56**, 2291 (1972).
39. D. F. Heller, K. F. Freed and W. M. Gelbart, *J. Chem. Phys.*, **56**, 2309 (1972).
40. A. Nitzan and J. Jortner, *Chem. Phys. Letters*, **13**, 466 (1972).
41. A. Nitzan and J. Jortner, *J. Chem. Phys.*, **56**, 2079 (1972).
42. A. Nitzan and J. Jortner, *J. Chem. Phys.*, **56**, 5200 (1972).
43. E. C. Lim and C. S. Huang, *J. Chem. Phys.*, **58**, 1247 (1973).
44. G. S. Beddard, G. R. Fleming, O. L. J. Gijzeman and G. Porter, *Chem. Phys. Letters*, **18**, 481 (1973).
45. E. W. Schlag, S. Schneider and D. W. Chandler, *Chem. Phys. Letters*, **11**, 474 (1971).
46. F. Lahmani, A. Frad and A. Tramer, *Chem. Phys. Letters*, **14**, 337 (1972).
47. A. E. W. Knight, B. K. Selinger and I. G. Ross, *Aust. J. Chem.*, **26**, 1159 (1973).

48. J. H. Callomon, J. E. Parkin and R. Lopez-Delgado, *Chem. Phys. Letters*, **13**, 125 (1972).
49. V. Lawetz, W. Siebrand and G. Orlandi, *Chem. Phys. Letters*, **16**, 448 (1972).
50. A. Nitzan, J. Jortner and P. M. Rentzepis, *Proc. Roy. Soc.*, **A 327**, 367 (1972).
51. K. G. Kay and S. A. Rice, *J. Chem. Phys.*, **57**, 3041 (1972).
52. R. Lefebvre and J. A. Beswick, *Molec. Phys.*, **23**, 1223 (1972).
53. R. A. van Santen, *Chem. Phys. Letters*, **15**, 621 (1972).
54. W. Lichten, *Phys. Rev.*, **131**, 229 (1963); **139**, A27 (1965); **164**, 131 (1967).
55. F. T. Smith, *Phys. Rev.*, **179**, 111 (1969).
56. G. L. Hofacker and R. D. Levine, *Chem. Phys. Letters*, **15**, 165 (1972).
57. R. W. Yip and W. Siebrand, *Chem. Phys. Letters*, **13**, 209 (1972).
58. N. Hirota and C. A. Hutchinson, *J. Chem. Phys.*, **46**, 1561 (1967).
59. J. P. Simons and A. L. Smith, *Chem. Phys. Letters*, **16**, 536 (1972).
60. A. Bernas and T. B. Truong, *Photochem. Photobiol.*, **15**, 311 (1972).
61. B. Meyer and J. L. Metzger, *Spectrochim. Acta*, **28A**, 1563 (1972).
62. G. Vaubel, *Chem. Phys. Letters*, **9**, 51 (1971).
63. G. Vaubel and H. Baessler, *Chem. Phys. Letters*, **11**, 613 (1971).
64. H. Zimmermann, D. Stehlik and K. H. Hausser, *Chem. Phys. Letters*, **11**, 496 (1971).
65. E. R. Menzel, K. E. Rieckhoff and E. M. Voight, *Chem. Phys. Letters*, **13**, 604 (1972); *Phys. Rev. Letters*, **28**, 261 (1972).
66. P. A. M. van der Bogaardt, R. P. H. Rettschnick and J. D. W. van Voorst, *Chem. Phys. Letters*, **18**, 351 (1973).
67. S. F. Fischer, *Chem. Phys. Letters*, **17**, 25 (1972).
68. A. Nitzan and J. Jortner, *J. Chem. Phys.*, **58**, 2412 (1973).
69. D. F. Heller and K. F. Freed, to be published.

7 *Theory of charge transport processes in organic molecular solids*

Stephen D. Druger

7.1 Introduction

Organic crystals are systems of notable complexity in which a wide variety of phenomena might reasonably be discussed under the general heading of 'conductivity'. First, there are experimental and theoretical problems of indirect relevance to the actual transport of charge. These include such phenomena as carrier generation, carrier recombination, the effects of irradiation, and the interaction of carriers with excitons and with defects. Secondly, the more restrictive heading of 'charge transport phenomena' might be taken to include a number of theoretical and experimental problems that do not arise when only the transport of isolated carriers in the bulk of the ideal crystal is considered. Examples of such matters are the effects of point defects and surfaces on the observed transport, as well as the overlapping problem of space-charge-limited carrier transport. Many of these problems have been reviewed elsewhere by Pope and Kallman,[1] Gutmann and Lyons,[2] LeBlanc[3] and Sharp and Smith.[4] The present review concerns itself almost entirely with the fundamental transport mechanisms for an isolated carrier in the bulk of an ideal organic molecular crystal at finite temperatures, and emphasizes the effects of the electron–phonon interaction on the transport of charge. We discuss mainly theoretical aspects of this subject. Furthermore, we confine our discussion of specific systems almost exclusively to the aromatic hydrocarbons.

The aromatic hydrocarbons have often been cited as the prototype of organic molecular crystals. Their structure is relatively simple compared with many organic systems, yet in contrast with some still simpler systems such as the rare-gas solids, they retain the main features characteristic of an array of interacting organic molecules. Anthracene, in particular, is both relatively stable and relatively easy to purify, and among the aromatic hydrocarbons has probably been subjected to the most extensive and detailed experimental study. A brief summary of some of its relevant properties is therefore an appropriate starting point.

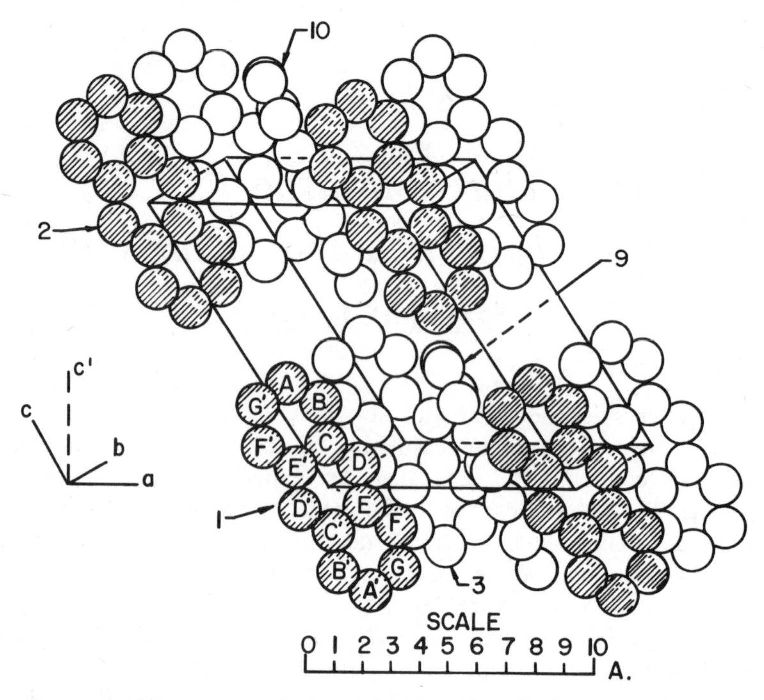

Figure 7.1 Molecular packing in crystalline anthracene. (J. M. Robertson, *Rev. Mod. Phys.*, **30**, 155 (1958); reproduced by permission of the American Institute of Physics)

Crystalline anthracene is monoclinic with two molecules per unit cell, and with the molecular arrangement shown in Figure 7.1. The distance of closest approach between neighbouring molecules is 3·60 Å between atoms F1 and C9. The molecular packing arrangement suggests that electronic interactions between neighbouring molecules in the *ab* plane might tend to be larger than for molecules in different *ab* planes. This tendency is manifested to varying degrees in a number of calculated results, as exemplified by the intermolecular overlaps in Table 7.1. It is also seen in some experimental data, such as the anisotropy of triplet excition diffusion, observed by Ern.[5] As is typical for van der Waals crystals, the intermolecular interactions are generally weak, and the molecular properties tend to persist in the crystalline environment. In particular, the neutral excited states of the solid can be classified in terms of those of the isolated molecule (Avakian and Merrifield[6]).

It has been suggested that the lowest lying ionized state of the anthracene crystal might involve the transfer of an electron from the highest bonding orbital of an anthracene molecule to the lowest antibonding orbitals of its

Table 7.1 Some typical intermolecular electronic overlaps in anthracene. The overlap integrals are taken between the wavefunctions for the excess carrier localized on the molecules numbered according to Figure 7.1. Each of the values given below is to be multiplied by 10^{-4}. (From J. N. Murrell, *Molec. Phys.*, **4**, 205 (1961))

Overlap	*ab* Planes	Hole	Electron
1 and 3	Same	15·5	−12·7
1 and 9	Same	−13·1	−19·3
1 and 2	Different	−0·05	−0·22
1 and 10	Different	3·3	0·7

neighbours (Lyons,[7] Pope, Burgos and Giachino[8]), requiring about 3·45 eV of excitation energy. The ionized states involved in the transport of charge, on the other hand, are those in which the electron and hole are isolated from each other, and the lowest such conduction state of anthracene requires about 3·9 eV of energy to be excited directly from the ground state (Castro and Hornig,[9] Chaiken and Kearns,[10] Geacintov and Pope[11]). Although the conduction states can be populated by intrinsic photo-generation at about this energy, extrinsic and two-photon carrier-generation processes requiring lower energies are also well known (Pope and Kallmann[1]). Theoretical treatments have often considered the conduction states to be band-like, and describable by wavefunctions constructed from the molecular orbitals in the tight-binding approximation. The calculated results then suggest that the bands should be narrow, with bandwidths of the order of $k_B T$ (with k_B the Boltzmann constant) and should show substantial anisotropy. Some experimental data have also been reported suggesting the existence of additional wide bands separated from the conventional narrow bands by about 0·5 eV (Dresner[12]).

In considering carrier transport, it is convenient to define the mobility tensor for the excess carriers operationally as

$$\boldsymbol{\mu} = \frac{\boldsymbol{\sigma}}{ne} \tag{7.1}$$

where n is the volume density of carriers, $-e$ is the electron charge, and $\boldsymbol{\sigma}$ is the conductivity tensor defined so that the current density $\mathbf{J}$ is given by

$$\mathbf{J} = \boldsymbol{\sigma} . \mathbf{E} \tag{7.2}$$

in an applied field. Consistently with the weak intermolecular coupling between excess electron states and with the correspondingly narrow bandwidths, the mobility components in anthracene tend to be small, of the order of 1 cm^2/V s. The narrow bands and low mobilities only marginally satisfy the conditions under which a physical description of transport in terms of approximately stationary Bloch states might be appropriate (e.g. Glaeser and Berry[13]). The measured mobility components indicate observable anisotropy, but not to the extent that the overlap integrals of Table 7.1 would suggest. The observed characteristics of the mobility and its behaviour as a function of temperature, pressure and other experimental parameters are discussed in greater detail in the review that follows.

The theory of charge transport considered here is a study of the consequences of the electronic and vibrational properties of the molecular solid. The adiabatic approximation allows a separation of the electronic and vibrational dynamics of the neutral ground state crystal. When the excess carrier is present, the variation of its energy with nuclear displacements corresponds to an electron–phonon coupling that can, in principle, affect the vibrational dynamics of the environment as well as the dynamical behaviour of the carrier. The basic formulation of the problem is then analogous to that for exciton transport (e.g. Grover and Silbey[14]). The state of the excess carrier initially localized on a single molecule is coupled by non-zero resonance transfer integrals to states on neighbouring sites; the transport then arises from the interplay between the resonance transfer interactions and the electron–phonon interaction. The detailed behaviour of the carrier depends in part on the nature of the electron–phonon coupling and on the specific vibrational modes to which coupling occurs.

7.2 The Born–Oppenheimer separation

The adiabatic,† or Born–Oppenheimer, separation[15] is of central importance in what follows. Since it has been discussed many times before (Chester,[16] Born and Huang,[17] Sham and Ziman[18]) we only briefly describe some of the results.

A molecular crystal composed of a set of mutually interacting electrons nuclei may in principle be described by a wavefunction $\Psi(r, R)$ that satisfies the Schrödinger equation

$$[T_e + T_N + U(r, R)]\Psi(r, R) = E\Psi(r, R) \tag{7.3}$$

where r represents schematically the set of electron coordinates $\{\mathbf{r}_i\}$ and R the set of nuclear coordinates $\{\mathbf{R}_I\}$. The first term in (7.3) is the sum of

† It is frequently useful to distinguish between the 'adiabatic' approximation of separating the nuclear and electronic motion according to the physical picture given above, and the Born–Oppenheimer approximation of representing the system by a single vibronic wavefunction in (7.5).

electron kinetic energies

$$T_e = \sum_i \frac{p_i^2}{2m_i}$$

while the second is the sum of nuclear kinetic energies

$$T_N = \sum_I \frac{P_I^2}{2M_I}$$

with $\mathbf{p}_i$ and m_i respectively the momentum and mass of electron i, and $\mathbf{P}_I$ and M_I the corresponding quantities for nucleus I. The term $U(r, R)$ is the total potential energy, and in the absence of externally applied fields is the sum of electron–electron, electron–nucleus and nucleus–nucleus interactions. In principle, spin-dependent terms may also be included in U.

The great disparity between electronic and nuclear masses usually allows a separation of (7.3) into an electronic and a nuclear eigenvalue problem. In the adiabatic approximation, the electronic wavefunction is calculated using the instantaneous nuclear positions, while the nuclei are treated as moving in a potential given by the electronic energy eigenvalue as a function of nuclear configuration. The electronic wavefunction then deforms continuously with the nuclear motion, without transitions occurring between different electron states.

More precisely, consider $\psi_m(r, R)$ as the mth solution of the electronic eigenvalue equation

$$[T_e + U(r, R)]\psi_m(r, R) = \mathscr{E}_m(R)\psi_m(r, R) \tag{7.4}$$

with eigenvalue $\mathscr{E}_m(R)$ obtained for each set of values of the nuclear coordinates. If solutions to (7.3) of the form

$$\Psi(r, R) = \sum_n \psi_n(r, R)\chi_n(R) \tag{7.5}$$

are considered, there results the equation

$$[\mathscr{E}_m(R) + T_N]\chi_m = \sum_j C_{mj}\chi_j \tag{7.6}$$

with

$$C_{mj} = A_{mj} + B_{mj} \tag{7.7}$$

The contributions to the right-hand side of (7.7) are operators that involve matrix elements of the nuclear momentum between various electronic states and are given by

$$A_{mj} = \sum_J -\frac{\hbar^2}{M_J}\int [\psi_m(r, R)\nabla_J\psi_j(r, R)] \,.\, \nabla_J \, \mathrm{d}^{3N}r \tag{7.8}$$

and

$$B_{mj} = \sum_J -\frac{\hbar^2}{2M_j} \int [\psi_m(r, R)\nabla_J^2 \psi_j(r, R)] \, \mathrm{d}^{3N}r \tag{7.9}$$

The diagonal terms C_{mm} are not of important physical consequence in the present discussion. The Born–Oppenheimer approximation consists of neglecting also the coupling terms C_{mj} between different vibronic wavefunctions in (7.6). The solutions to (7.6) are then simple products

$$\Psi_m = \psi_m(r, R)\chi_m(R)$$

with the nuclear wavefunction $\chi_m(R)$ satisfying the Schrödinger equation for motion in the 'potential' $\mathscr{E}_m(R)$

$$[\mathscr{E}_m(R) + T_N]\chi_m = E\chi_m \tag{7.10}$$

This leads to the physical interpretation already described.

The C_{mj} operators generally have non-zero matrix elements between different electronic states, however. If the matrix element $\langle\chi_m|C_{mj}|\chi_j\rangle$ is small, the first-order correction to the wavefunction is roughly

$$\delta\Psi \simeq \sum_j \frac{\langle\chi_m|C_{mj}|\chi_j\rangle}{\mathscr{E}_m - \mathscr{E}_j} \Psi_j \tag{7.11}$$

Detailed treatment (Chester[16]) shows that $\langle\chi_m|C_{mj}|\chi_j\rangle$ has magnitude much less than $\hbar/\tau \simeq \hbar\omega$, where τ is a characteristic time for the nuclei to have enough energy to change Ψ_m significantly, and ω is a vibrational frequency for the nuclear motion. Thus, the mixing will not be serious if $\hbar\omega$ is small compared to the electronic energy difference between vibronic states connected by C_{mj}.

The breakdown of the Born–Oppenheimer separation occurs to varying degrees when the energies of different vibronic states are degenerate or quasidegenerate, and is associated with a number of important phenomena in solids and in molecules. The Jahn–Teller effect[19] occurs when the vibronic states of interest are degenerate, and the mixing between them leads to a lowering of the energy for nuclear displacements from what would otherwise have been the equilibrium configuration of the system. The non-adiabatic correction terms of (7.7) also play an important role by coupling an electronically excited vibronic state of a molecule to vibronic states of lower electronic energy with which it is, on the whole, quasi-degenerate. The coupling can then lead to vibrationally induced non-radiative decay of the electronic part of the excitation (e.g. Nitzan and Jortner[20]). In the case of charge transport, it is often possible, as a zero-order approximation, to describe the crystal with an excess carrier by a single Born–Oppenheimer product. In the next order of approximation, the non-adiabatic coupling terms of (7.7) lead to transitions between different vibronic states; they determine the effect of the electron–phonon coupling on the charge transport when the coupling is weak enough for a description in these terms to be appropriate (Sham and Ziman[18]). On the other hand, the small-polaron band picture that sometimes applies for strong electron–phonon coupling is based on a zero-order description of the system as a superposition of Born–Oppenheimer products, each product describing the system with the carrier localized on a given site and the crystal environment relaxed around it (Holstein[21]).

7.3 Vibrational modes of molecular crystals

Theoretical treatments of excess carrier transport have generally considered coupling of the excess carrier to a relatively small number of vibrational degrees of freedom of a specific physical character; the nature of the modes considered determines to some degree the consequences of the coupling. We outline in this section some features of the vibrational dynamics of molecular crystals that are relevant to the theory of charge transport.

7.3.1 *Molecular vibrations*

The vibrational properties of a molecular solid may be considered from the viewpoint that the electronic energy eigenvalue $\mathscr{E}_m(R)$ in the adiabatic separation acts as the potential governing the nuclear motion. The usual procedure may be followed of expanding the potential for the nuclear Schrödinger equation to second order in displacements from equilibrium. The transformation that decouples the resultant system of equations then defines the normal modes of vibration, whose physical properties may then be considered. A more detailed discussion of the elementary aspects of the theory than we present may be found in the standard references on molecular vibrations (e.g. Wilson *et al.*[22]) and on lattice dynamics (e.g. Maradudin *et al.*[23]).

For an isolated molecule, the adiabatic potential $\mathscr{E}_0(R)$ is a function of the $3n_N$ nuclear cartesian coordinates. When the adiabatic potential is expanded in displacements $X_{i\xi}$ of the nuclei from their equilibrium positions, and terms beyond those bilinear in the $X_{i\xi}$ are ignored in the harmonic approximation, the Hamiltonian governing the nuclear motion in equation (7.10) becomes

$$H_N = \sum_{i\xi} \frac{P_{i\xi}^2}{2m_i} + \sum_{ij\xi\eta} \left(\frac{\partial^2 \mathscr{E}_0}{\partial X_{i\xi} \partial X_{j\eta}} \right)_0 X_{i\xi} X_{j\eta} \qquad (\xi, \eta = 1-3) \quad (7.12)$$

which may be conveniently written as

$$H_N = \sum_{i\xi} p_{i\xi}^2 + \sum_{ij\xi\eta} D_{i\xi,j\eta} q_{i\xi} q_{j\eta} \tag{7.13}$$

in terms of the 'mass weighted' coordinates

$$\begin{aligned} q_{i\xi} &= (m_i)^{\frac{1}{2}} X_{i\xi} \\ p_{i\xi} &= (m_i)^{-\frac{1}{2}} P_{i\xi} \end{aligned} \tag{7.14}$$

with m_i the mass of nucleus i and with

$$D_{i\xi,j\eta} = (m_i m_j)^{-\frac{1}{2}} \left(\frac{\partial^2 \mathscr{E}_0}{\partial X_{i\xi} \partial X_{j\eta}} \right)_0$$

evaluated in the equilibrium configuration. The equations of motion for the nuclear positions follow by regarding the coordinates and momenta as classical quantities and applying Newton's second law or Hamilton's equations to (7.13); an alternative procedure is to treat the coordinates and momenta as quantum mechanical operators in the Heisenberg picture, and to apply the commutation relations

$$[q_{i\xi}, p_{j\eta}] = i\hbar \delta_{ij} \delta_{\xi\eta}$$

$$\frac{i\hbar\, \mathrm{d}p_{j\eta}}{\mathrm{d}t} = [p_{j\eta}, H_N]$$

and

$$\frac{i\hbar\, \mathrm{d}q_{j\eta}}{\mathrm{d}t} = [q_{j\eta}, H_N]$$

Either procedure gives the same $3n_N$ differential equations describing a system of coupled harmonic oscillators

$$\frac{\mathrm{d}^2 q_{i\xi}}{\mathrm{d}t^2} + \sum_{j,\eta} D_{i\xi,j\eta} q_{j\eta} = 0 \tag{7.15}$$

with $3n_N$ corresponding equations for the momenta

$$\frac{\mathrm{d}^2 p_{i\xi}}{\mathrm{d}t^2} + \sum_{j,\eta} D_{i\xi,j\eta} p_{j\eta} = 0$$

If the vibrational problem is formulated in a set of coordinates other than the cartesian $X_{i\xi}$, and if the new displacement coordinates are linear combinations of the old, with the particular linear combinations given by some non-singular transformation

$$Q_\lambda = \sum_{j\eta} T_{\lambda,j\eta} q_{j\eta} \tag{7.16}$$

equations of the same general form as (7.15) should result. In particular, the transformation could be chosen so that, when D is constructed in terms of the new coordinates, only the diagonal matrix elements $D_{\lambda\lambda}$ are non-zero. The equations (7.15) are then decoupled, and each has a solution in which the normal coordinate Q_λ oscillates independently with angular frequency $\omega_\lambda = (D_{\lambda\lambda})^{\frac{1}{2}}$. The solution for Q_λ in turn corresponds to vibration of the cartesian coordinates at a single frequency ω and with relative amplitudes given by the inverse of (7.16). The requirement that the cartesian coordinates oscillate wih a single frequency ω implies that each $q_{i\xi}$ satisfies

$$\frac{d^2 q_{i\xi}}{dt^2} = -\omega^2 q_{i\xi} \tag{7.17}$$

which in turn reduces (7.17) to a system of algebraic equations whose non-trivial solution requires that the characteristic equation

$$\det\left[D_{i\xi,j\eta} - \omega^2 \delta_{ij}\delta_{\xi\eta}\right] = 0 \tag{7.18}$$

be satisfied. Solution of (7.18) for ω gives the normal frequencies, and use of the solutions in the system of algebraic equations resulting from (7.17) and (7.15) gives the transformation to the normal coordinates $\{Q_\lambda\}$.

We list some of the relevant properties of the solutions to (7.18):

(1) Since D is Hermitian, ω_λ^2 is real. Since the change in potential energy $\omega_\lambda^2 Q_\lambda^2/2$ for displacement Q_λ from equilibrium is physically required to be non-negative in order that the motion be stable, ω_λ^2 is also non-negative; hence ω_λ is real.

(2) Three independent solutions for Q_λ describe uniform translations of the entire molecule, with the corresponding value of ω_λ equal to zero. This follows since the potential $\mathscr{E}_0(R)$ is unaffected by a uniform infinitesimal translation, and there is no restoring force for such a displacement.

(3) By the same reasoning as for (2), three independent solutions for Q_λ describe small uniform rotations of the molecular axes,† and correspond to $\omega_\lambda = 0$.

Formulation of the molecular vibrational problem in terms of internal coordinates, such as various bond stretchings and bendings,‡ eliminates

† The true separability of the dynamical motion into a rotational part and an internal vibrational part that is independent of rotation requires, however, that the moments of inertia be unaffected by the rotational motion.

‡ In this case the transformation (7.16) is often singular, since the set of internal coordinates is likely to include redundancies.

the six superfluous coordinates, and is often convenient in such applications as the experimental determination of the molecular vibrational force constants. It is often also convenient to use symmetry-adapted internal coordinates, constructed as particular linear combinations of internal displacements chosen so that the symmetry coordinates transform into one another according to the various irreducible representations of the molecular point group. In particular, the normal coordinates are each a sum only over symmetry-adapted internal coordinates that transform according to a single irreducible representation; the use of symmetry coordinates partially factorizes (7.18) into products of subdeterminants each associated with a particular representation. Frequently, it happens that some set of internal coordinates, such as C–H bond stretchings for example (Table 7.2), are associated with frequencies well separated from

Table 7.2 Typical vibrational frequencies for the benzene molecule in its ground state (S_0) and first excited singlet state (S_1). The modes are labelled according to the irreducible representations of the molecular point group D_{6h} [a]

Mode	Symmetry	$\hbar\omega(\text{cm}^{-1})$	
		$S_0(^1A_{1g})$	$S_1(^2B_{2u})$
C—C stretch	a_{1g}	995	923
C—H stretch	a_{1g}	3073	3130
C—C bend	e_{2g}	608	522
C—H bend	e_{2g}	1178	1130
Out-of-plane	b_{2g}	707	365
	a_{2u}	674	513
	e_{2u}	399	243

[a] From R. W. Munn and W. Siebrand, *J. Chem. Phys.*, **52**, 47 (1970), and based on A. O. Best, F. M. Garforth, C. K. Ingold, H. G. Poole and C. L. Wilson, *J. Chem. Soc.*, **1948**, 406 (Part I) and 516 (Part XII) (1948).

those of other internal coordinate displacements. If the other frequencies are crudely regarded as either zero or infinite by comparison, then appropriate linear combinations of only the displacements in question would be normal coordinates. The true normal coordinates will therefore separate into those that exhibit mainly the physical character of the displacement in question and the remainder that involve only a small contribution from such displacements.

7.3.2 *Molecular crystal vibrations*

The vibrations of a neutral molecular crystal in its electronic ground state can be treated similarly to those of an isolated molecule. The adiabatic potential for the crystal, $\mathscr{E}_0(R)$, can be expanded to second order in displacements from equilibrium. For convenience, we assume the expansion to be in the cartesian displacement components $X_{Ii\xi}$ for each atom i in each unit cell I, although an (equivalent) expansion in terms of a combination of internal and cartesian coordinates might be more practical for actual calculation. An equation of the same form as (7.13) is then obtained in terms of the mass-weighted coordinates $q_{Ii\xi}$ defined analogously to equation (7.14). The requirement that the coordinates oscillate with a single frequency ω leads to the characteristic equation

$$\det\left[D_{Ii\xi,Jj\eta} - \omega^2 \delta_{IJ}\delta_{ij}\delta_{\xi\eta}\right] = 0 \tag{7.19}$$

for the normal frequencies. By virtue of the lattice periodicity, each matrix element $D_{Ii\xi,Jj\eta}$ that appears in (7.19) can depend on the location $\mathbf{R}_I$ of unit cell I and on the location $\mathbf{R}_J$ of unit cell J only through the displacement $\mathbf{R}_I - \mathbf{R}_J$ between them. As a consequence, it is straightforward to show that the transformation

$$U_{\mathbf{k}i\xi} = \sum_I \exp\left[i\mathbf{k} \,.\, (\mathbf{R}_J - \mathbf{R}_I)\right] q_{Ii\xi} \tag{7.20}$$

factors the characteristic determinant (7.19) into as many subdeterminants as there are unit cells in the region over which the usual periodic boundary conditions on $\mathbf{k}$ are applied. The solution of the factorized equations

$$\det\left[D_{i\xi,j\eta}(\mathbf{k}) - \omega^2 \delta_{ij}\delta_{\xi\eta}\right] = 0 \tag{7.21}$$

with

$$D_{i\xi,j\eta}(\mathbf{k}) = \sum_I D_{Ii\xi,Jj\eta} \exp\left[i\mathbf{k} \,.\, (\mathbf{R}_J - \mathbf{R}_I)\right]$$

gives the normal mode frequencies for each wavevector $\mathbf{k}$.

Some features of the solutions to (7.21) may be briefly summarized:

(1) For each value of $\mathbf{k}$ there are $3s$ solutions for $\omega(\mathbf{k})$, where s is the number of atoms per unit cell. The total number of solutions is therefore $3sN$ with N the number of unit cells over which periodic boundary conditions are defined. The functions $\omega_\lambda(\mathbf{k})$ for $\lambda = 1$ to $3s$ can be regarded as branches of a multivalued function; each $\omega_\lambda(\mathbf{k})$ is the 'dispersion relation' for the corresponding branch λ.
(2) Since D is Hermitian and since the motion is assumed stable, the solutions for $\omega_\lambda(\mathbf{k})$ are real.
(3) Three solutions for the angular frequency in (7.19) are zero and correspond to three independent uniform translations of the entire crystal.

The solutions that correspond to a uniform translation of the system occur for zero wavevector, since the cartesian displacements involved in such a translation occur with the same phase. There therefore exist three acoustic branches that approach zero frequency at zero wavenumber. The acoustic modes for $\mathbf{k} = 0$ correspond to displacement of each unit cell as a whole, and this correspondence is maintained for small but non-zero wavevectors as well. The long-wavelength acoustic modes correspond roughly to elastic waves in a continuous medium. All other vibrational modes are classified as 'optical.'

The weak intermolecular forces characteristic of van der Waals solids allow a second classification of the vibrational modes of the crystal. For each molecule composed of t atoms, there are $3t - 6$ internal coordinates that specify the molecular deformation, and six external coordinates that specify the position and orientation of the molecule. In calculating the vibrational modes of the crystal, it is often convenient to treat the motion of the internal molecular coordinates by ignoring the effects of changes in external coordinates, and to treat the motion of the external coordinates as though the restoring forces for the internal molecular displacements were infinite and the molecules perfectly rigid. This can be done by assuming that the restoring forces for internal molecular deformations are much greater than the restoring forces for changes in intermolecular separations. In this approximation the internal and external coordinates of the free molecule retain their character in the crystal. The lowest branches of the vibrational spectrum will be 'lattice' branches whose modes correspond to librations and translations of rigid molecules, and whose normal coordinates are essentially linear combinations of the external molecular coordinates. The lattice branches can, of course, be subdivided into acoustic and optical branches. The remaining branches of the spectrum are intramolecular in character. Their normal modes describe internal molecular vibrations, and their normal coordinates are essentially linear combinations of the internal molecular coordinates. The frequencies of the intramolecular branches should generally lie close to some normal frequency of the free molecule, and should show a weak $\mathbf{k}$-dependence arising from the coupling between neighbouring molecules.

A third classification of each lattice mode as either a libration or a translation might be useful, but would hold rigorously only under special conditions.

Specifically quantum mechanical aspects of the crystal vibrations can be treated by noting that a normal coordinate transformation separates the vibrational Hamiltonian (7.12) into a sum of simple harmonic oscillator Hamiltonians, each corresponding to a normal mode of the system. It is then possible to study the system formally by defining $a_{\mathbf{k}\lambda}$ and $a^{+}_{\mathbf{k}\lambda}$ respectively as the operators for annihilation and creation of a vibrational quantum in the mode $(\mathbf{k}\lambda)$. When the usual commutation relations between the creation and annihilation operators are applied, the vibrational

Hamiltonian can be shown to take the form (e.g. Maradudin *et al.*[23])

$$H_N = \sum_{\mathbf{k}\lambda} \hbar\omega_\lambda(\mathbf{k})[a^+_{\mathbf{k}\lambda}a_{\mathbf{k}\lambda} + \tfrac{1}{2}] \tag{7.22}$$

The details of the formalism are not essential for our purposes, however. We only need note that the separation of the Hamiltonian into a sum over normal modes implies that an appropriate set of nuclear eigenfunctions for (7.10) are the products of simple harmonic oscillator wavefunctions, each such wavefunction involving a single normal coordinate of the crystal. The energy of each normal mode is quantized. It is found that the thermal probability for a mode of frequency ω to have quantum number n and therefore energy $(n + \frac{1}{2})\hbar\omega$ at temperature T is

$$\exp - \frac{n\hbar\omega}{k_B T} \bigg/ \left[\sum_m \exp - \frac{n\hbar\omega}{k_B T}\right]$$

and that the thermal population of the mode is

$$\langle n \rangle = \left[\exp\frac{\hbar\omega}{k_B T} - 1\right]^{-1}$$

It is useful to define the quantity $g(\omega)\,d\omega$ as the fraction of the total number of modes that lie between ω and $\omega + d\omega$. Thermal averages over the vibrational modes can then be carried out using the expressions for $\langle n \rangle$ once $g(\omega)$ is calculated. The treatment of Debye,[24] for example, considers the crystal as an elastic continuum, with the result

$$g(\omega) \propto \omega^2$$

The assumption of a Debye $g(\omega)$ for acoustic modes is ubiquitous in the theory of solids. Its deficiencies were pointed out in detail by Blackman.[25–6]

7.3.3 *The separation approximation*

The separability of the crystal vibrations into intramolecular and lattice modes is tacitly assumed in a number of treatments of the electron–phonon interaction. The degree to which this assumption is justified can be discussed in terms of existing calculations of the vibrational properties of molecular crystals.

Detailed calculation of vibrational frequencies $\omega_\lambda(\mathbf{k})$ for non-zero values of $\mathbf{k}$ have been carried out for such inorganic molecular crystals as solid nitrogen (Schnepp and Ron[27]) and the rare-gas solids (see, e.g. Horton[28]). For organic molecular crystals, vibrational frequencies have been calculated as a function of wavevector for the lattice modes of the highly symmetric crystal hexamethylenetetramine (Cochran and Pawley[29]), for the lattice modes of naphthalene and anthracene (Pawley[30]), and for the vibrational modes of benzene (Bonadeo and Taddei[31]). A greater number of calculations have considered the vibrational frequencies only for zero wavevector (e.g. Taddei *et al.*,[32] Harada and Shimanouchi[33]) where considerable simplification of the characteristic equation occurs when the problem is formulated in terms of internal molecular coordinates.

In the early calculations of Cochran and Pawley[29] and Pawley[30] the lattice frequencies are calculated by regarding the molecules as rigid. It is this assumption that Pawley calls the 'separation approximation.' The force constants needed for the calculation can then be evaluated, in principle at least, from measured elastic data (Cochran and Pawley[29]). A more general procedure is, however, to assume that the intermolecular potential is a sum of interactions between each pair of atoms, and that each atom–atom interaction is given by an empirical potential (Kitaigorodskii,[34] Pawley[30]) in terms of the interatomic distance r

$$V(r) = -\frac{A}{r^6} + B\exp(-\alpha r)$$

with a single set of parameters for all C–C interactions, a second set for all C–H interactions, and a third for all H–H interactions. An appropriately defined dynamical matrix $D(\mathbf{k})$ (cf. equation (7.21)) can then be evaluated numerically for each wavevector of interest, and the characteristic equation solved for the frequencies $\omega_\lambda(\mathbf{k})$. The results for anthracene are shown in Figure 7.2.

It can be shown (Cochran and Pawley,[29] Sándor[35]) that, for zero wavevector in a centrosymmetric crystal, each mode is either a translation or a libration. A mode corresponding to non-zero wavevector, however, can have both a librational and a vibrational contribution, with the two components oscillating $\frac{1}{2}\pi$ out of phase. It is seen in Figure 7.2 that the mixing between librational and translational displacements can be substantial, and that a branch whose modes are purely of one kind when $\mathbf{k} = 0$ may assume characteristics predominantly of the other before the zone boundary is reached. The acoustic modes near $\mathbf{k} = 0$ have frequencies fairly well separated from those of librational modes, and so their character remains predominantly translational in this region. Of particular importance is the fact that the lowest lying intramolecular frequency of about 200 cm^{-1} for anthracene lies dangerously close to the highest lattice frequency, suggesting that the separation approximation might not be justified.

The validity of the separation approximation has since been examined in greater detail for the aromatic hydrocarbons. Non-separability should

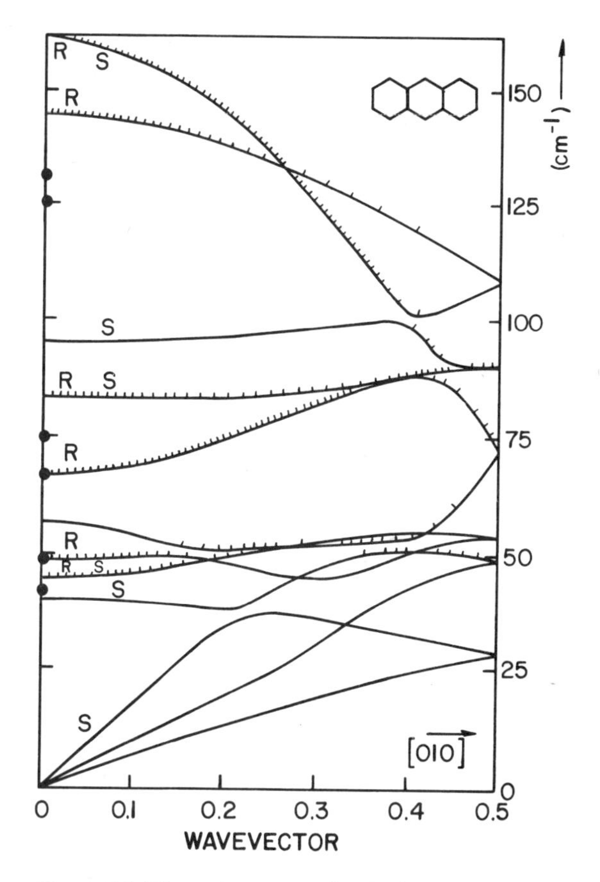

Figure 7.2 Dispersion curves for the lattice vibrations of anthracene based on a rigid-molecule approximation. The cross-hatchings indicate rotational character; *R* denotes a rotational mode and *S* a symmetric mode. (Pawley;[30] reproduced by permission of Akademie-Verlag Berlin)

be expected to have the greatest effect on those internal modes involving the hydrogen atom displacements, considering the peripheral location of the hydrogen atoms in the molecule, and the high frequencies associated with their light mass. Pawley[36] has shown that when neutron diffraction data for anthracene, naphthalene and pyrene are analysed using a model that allows some non-rigid displacement of the hydrogen atoms, improved agreement with theory is obtained. Pawley and Cyvin[37] have also calculated the vibrational dispersion along **k**-directions of high symmetry in naphthalene, taking into account the full set of fifty-four cartesian displacement coordinates per molecule. Relaxing the assumption of separability is equivalent to increasing the intermolecular restoring forces from zero when calculating the intramolecular branches of $\omega(\mathbf{k})$, and to decreasing the intramolecular restoring forces to a finite value when calculating the lattice frequencies. It may be argued on physical grounds that the frequencies are thereby increased for the intramolecular modes and decreased for the lattice modes; we note that the result is also suggested by Rayleigh's theorems (see §7.3.4). When the separability assumption is relaxed, the calculated lattice frequencies are lowered by 5 to 10 cm^{-1}, with the acoustic branches less strongly affected than the optical branches. The intramolecular frequencies are shifted by about 1 to 5%, the larger shifts occurring for the lower frequencies. The small but definite effects that the mixing† of lattice and molecular modes has on the normal frequencies is in agreement with the recent calculations of Taddei *et al.*[32] and Bonadeo and Taddei,[31] who obtain a less than 3% shift in the lattice frequencies of benzene when intermode mixing is allowed.

7.3.4 *Rayleigh's theorems*

It is frequently useful to know how the normal-mode frequencies of a system of coupled oscillators are affected when a single force constant or a single mass is changed. Munn and Siebrand,[38] for example, suggest that the presence of an excess carrier might modify the force constants for the internal vibrations of the molecule on which it is localized. In the theory of impurities in solids, the changes in vibrational spectrum produced changes of the interatomic force constants and atomic masses are frequently considered (Klein[39]).

The effects of altering a mass or a force constant are described qualitatively by Rayleigh's theorems (Maradudin *et al.*[23]). It can be shown that increasing a single mass in a system of coupled oscillators can only decrease those normal frequencies that are changed. In each case, the decrease is no greater than the distance to the next lower original frequency. Correspondingly, a decrease in the mass of one oscillator can only increase the normal frequencies that are changed, but each by no more than the distance to the next higher unperturbed frequency. The increase of a single force constant has the same qualitative effect as the decrease of a single mass, and can only increase the normal frequencies that are affected, while the decrease of a single force constant can only decrease the normal frequencies that are affected. The usual proof (Maradudin *et al.*[23]) involves

† 'Mixing' in the theory of vibrational spectra generally refers to the specific coupling of different modes by the anharmonic contributions to the vibrational Hamiltonian. The sense in which the term is used here is of course rather different.

assumptions inconsistent with the invariance of the system under uniform infinitesimal translations. Nevertheless, the qualitative results of Rayleigh's theorems hold also for more generally realistic models of impurities in solids (Klein[39]).

7.4 Excess carrier bands in molecular solids

7.4.1 *General considerations*

A one-electron band picture is well justified on experimental and theoretical grounds for a variety of inorganic solids. In particular it provides a useful model for treating charge transport phenomena in normal metals and high-mobility semiconductors, although its applicability as a physical description of low-mobility transport is less certain. Considerable effort has been devoted in developing a detailed understanding of excess carrier transport in organic solids on the basis of a band picture.

The one-electron bands of conventional inorganic crystals arise from the commonly assumed theoretical model in which every electron in the solid moves under the influence of a single effective potential $V(\mathbf{r})$. Furthermore, $V(\mathbf{r})$ is usually assumed, as a first approximation, to be the effective potential that obtains when the crystal is in its perfect lattice configuration. The lattice periodicity then implies that the one-electron Schrödinger equation

$$\left[-\frac{\hbar^2}{2m} + V(\mathbf{r})\right]\psi_{\nu\mathbf{k}}(\mathbf{r}) = \varepsilon_\nu(\mathbf{k})\psi_{\nu\mathbf{k}}(\mathbf{r}) \tag{7.23}$$

is satisfied by a set of Bloch orbitals $\psi_{\nu\mathbf{k}}(\mathbf{r})$, where each $\psi_{\nu\mathbf{k}}$ is characterized by a wavevector $\mathbf{k}$ and satisfies

$$\psi_{\nu\mathbf{k}}(\mathbf{r} + \boldsymbol{\xi}) = \exp(i\mathbf{k} \,.\, \boldsymbol{\xi})\psi_{\nu\mathbf{k}}(\mathbf{r})$$

Here $\boldsymbol{\xi}$ is a vector connecting the origins of different unit cells, and m is the free electron mass. The energy eigenvalues $\varepsilon_\nu(\mathbf{k})$ in (7.23) are the one-electron energy bands of the solid, and obtain whenever the simple model on which they are based can be assumed. It is a separate question whether the abstractly constructed band-like orbitals and energies are useful as a physical description for solids characterized by very small electronic overlaps between their constituent units (see, e.g., Knox and Teegarden,[40] Ziman[41] p. 109).

It is important to distinguish between the bare bands of the neutral crystal calculated as we have described and the energies that become involved when an excess charge is actually added to a conduction band or removed from a valence band. The excess carrier might in principle modify the potential $V(\mathbf{r})$, and its unbalanced charge is likely to introduce substantial polarization effects whose treatment lies beyond the simple one-electron approximation. The view that the carrier is simply placed in an unmodified bare-band state is frequently useful, and can often be justified if the polarization energies can be ignored (see, e.g., Knox and Teegarden[40] pp. 20–23). In the aromatic hydrocarbons, the one-electron band calculations have generally been carried out specifically for an excess carrier, although in practice some of the special features introduced by the presence of an actual charge are often ignored. It is also possible to generalize the energy band calculation to include some of the effects of polarization, as has been done by Glaeser and Berry.[13]

The wavefunction $\Psi_\nu(\mathbf{r}, t)$ of an excess carrier localized within some region of the crystal may be considered as a wavepacket formed from excess carrier Bloch functions; each Bloch function of wavevector $\mathbf{k}$ is then associated with a velocity

$$\mathbf{v}(\mathbf{k}) = \hbar^{-1}\nabla_{\mathbf{k}}\varepsilon_\nu(\mathbf{k})$$

(Jones and Zener[42]). Wannier's theorem then describes the behaviour of the wavepacket under the influence of an applied perturbing potential $W_\mathrm{p}(\mathbf{r})$

$$[\varepsilon_\nu(-\mathrm{i}\nabla) + W_\mathrm{p}(\mathbf{r})]\Psi_\nu = \mathrm{i}\hbar\frac{\partial\Psi_\nu}{\partial t} \tag{7.24}$$

provided the potential varies slowly over many lattice sites and the crystal is rigid (Wannier,[43] Slater,[44] Pekar[45]). Here $\varepsilon_\nu(-\mathrm{i}\nabla)$ is obtained by expanding $\varepsilon_\nu(\mathbf{k})$ in a Taylor series and replacing $\mathbf{k}$ by $-\mathrm{i}\nabla$. When the band has a minimum at $\mathbf{k} = 0$ and the wavepacket consists mainly of Bloch functions corresponding to small wavenumbers, equation (7.24) reduces to the usual result of the effective-mass theorem for an anisotropic crystal

$$\left\{\left[\frac{\hbar^2}{2}\sum_{\xi\eta}\left(\frac{1}{m^*}\right)_{\xi\eta}\left(\frac{\partial^2}{\partial r_\xi \partial r_\eta}\right)_0\right] + W_\mathrm{p}(\mathbf{r})\right\}\Psi_\nu(\mathbf{r}) = \mathrm{i}\hbar\frac{\partial\Psi_\nu}{\partial t} \tag{7.25}$$

where

$$\left(\frac{1}{m^*}\right)_{\xi\eta} = \hbar^{-2}\left[\frac{\partial^2\varepsilon_\nu(\mathbf{k})}{\partial k_\xi \partial k_\eta}\right]_{\mathbf{k}=0}$$

is the reciprocal effective mass tensor. The usual form of the effective mass equation (7.25) does not apply when the excess carriers are thermally distributed in a narrow band of width comparable to $k_\mathrm{B}T$, since each carrier then has a substantial probability of occupying states with wavenumbers not close to zero, and the expansion used to obtain (7.25) from (7.24) is no longer valid.

7.4.2 *Theory of excess carrier bands in the aromatic hydrocarbons*

The assumption of weak coupling between excess carrier states of neighbouring molecules simplifies considerably the calculation of approximate one-electron energies and orbitals for the aromatic hydrocarbons. The tight binding approximation can be applied in its simplest form, and the Bloch states of the crystal constructed from the ionic states of the molecules themselves.

The orbitals $\phi_\nu(\mathbf{r})$ of the isolated molecule obey the Schrödinger equation

$$\left[-\frac{\hbar^2\nabla^2}{2m} + V_n(\mathbf{r})\right]\phi_\nu(\mathbf{r}) = \varepsilon_\nu\phi_\nu(\mathbf{r})$$

for an appropriate molecular potential V_n. The interaction of the excess carrier with the crystal can then be taken as the sum of interactions with each molecule separately

$$V(\mathbf{r}) = \sum_{Ii} V_n(\mathbf{r} - \mathbf{R}_{Ii}) \tag{7.26}$$

where $\mathbf{R}_{Ii}$ is the position of molecule i in unit cell I. Since V has non-vanishing matrix elements between molecular states of neighbouring molecules, an excess carrier localized on a given site is not in a stationary state of the one-electron Hamiltonian $[V(\mathbf{r}) + (\hbar^2/2m)\nabla^2]$. In particular, the potential $V(\mathbf{r})$ could, in principle, mix different states of the same molecule through matrix elements of the form

$$\langle\phi_\nu(\mathbf{r} - \mathbf{R}_{Ii})|V_n(\mathbf{r} - \mathbf{R}_{Jj})|\phi_\nu(\mathbf{r} - \mathbf{R}_{Ii})\rangle$$

as well as mixing the molecular states of neighbouring molecules through the transfer integrals of the form

$$\langle\phi_\nu(\mathbf{r} - \mathbf{R}_{Ii})|V_n(\mathbf{r} - \mathbf{R}_{Jj})|\phi_\nu(\mathbf{r} - \mathbf{R}_{Jj})\rangle$$

Two equivalent procedures have been described for calculating the Bloch states of the crystal. One method is to first form a separate Bloch sum for each molecule of the unit cell (Friedman[46]).

$$\phi_{\mathbf{k}i\nu}(\mathbf{r}) = N^{-\frac{1}{2}} \sum_I \exp(i\mathbf{k}\,.\,\mathbf{R}_I)\phi_\nu(\mathbf{r} - \mathbf{R}_{Ii}) \tag{7.27}$$

Each Bloch sum in (7.27) is then coupled to other Bloch sums of the same $\mathbf{k}$ by non-zero matrix elements of V, and the one-electron eigenfunctions of the crystal are linear combinations of the Bloch sums (7.27). Alternatively, linear combinations of the orbitals within each unit cell can first be formed, and the Bloch sums then taken over the various unit cells (Katz *et al.*,[47] Glarum[48]).

$$\psi_{\mathbf{k}} = N^{-\frac{1}{2}} \sum_{I} \left\{ \sum_{iv} \lambda_{iv} \phi_v(\mathbf{r} - \mathbf{R}_{Ii}) \right\} \exp(i\mathbf{k} \cdot \mathbf{R}_I)$$

In either case, minimization of $\langle \psi_{\mathbf{k}} | H_{\text{oe}} | \psi_{\mathbf{k}} \rangle$ subject to the restriction

$$\langle \psi_{\mathbf{k}} | \psi_{\mathbf{k}} \rangle = 1$$

would lead to the equations

$$\sum_{Iv} \mathrm{e}^{-i\mathbf{k}.\mathbf{R}_I} [\langle \phi_v(\mathbf{r} - \mathbf{R}_{Ii}) | H_{\text{oe}} | \phi_{v'}(\mathbf{r} - \mathbf{R}_{oj}) \rangle - \varepsilon_v(\mathbf{k}) \langle \phi_v(\mathbf{r} - \mathbf{R}_{Ii}) | \phi_{v'}(\mathbf{r} - \mathbf{R}_{0j}) \rangle] \lambda_{iv} = 0 \tag{7.28}$$

where

$$H_{\text{oe}} = \frac{p^2}{2m} + V(\mathbf{r})$$

is the one-electron Hamiltonian. In practice, overlaps between different molecules and mixing between different states of the same molecule are neglected. Equation (7.28) then reduces to the secular equation for diagonalizing the Hamiltonian matrix.

Anthracene, for example, has two molecules per unit cell, so that solution of the secular equation

$$\begin{vmatrix} \mathrm{H}_{11}(\mathbf{k}) - \varepsilon(\mathbf{k}) & H_{12}(\mathbf{k}) \\ H_{21}(\mathbf{k}) & H_{22}(\mathbf{k}) - \varepsilon(\mathbf{k}) \end{vmatrix} = 0 \tag{7.29}$$

with $H_{ij} = \langle \phi_{\mathbf{k}iv} | H_{\text{oe}} | \phi_{\mathbf{k}jv} \rangle$ yields two energy bands. At zero wavevector and along directions of high symmetry in the Brillouin zone, $H_{11}(\mathbf{k})$ and $H_{22}(\mathbf{k})$ are equal. The two Bloch functions that diagonalize the Hamiltonian matrix are then a symmetric and an antisymmetric sum of the two $\phi_{\mathbf{k}iv}$ that each correspond to a different molecule of the unit cell. This holds only for special **k**-vectors. Solution of (7.29) using matrix elements calculated from the potential in (7.26) leads to an expression for the two $\varepsilon(\mathbf{k})$ for a general value of **k**

$$\begin{aligned} \varepsilon_{\pm} = {} & \text{constant} + 2\varepsilon_b \cos(\mathbf{k} \cdot \mathbf{b}) \\ & \pm 2\varepsilon_a \{\cos[\mathbf{k} \cdot (\mathbf{a} + \mathbf{b})/2] + \cos[\mathbf{k} \cdot (\mathbf{a} - \mathbf{b})/2]\} \\ & \pm 2\varepsilon_\gamma \{\cos \mathbf{k} \cdot [(\mathbf{a} + \mathbf{b})/2 + \mathbf{c}] + \cos \mathbf{k} \cdot [(\mathbf{a} - \mathbf{b})/2 + \mathbf{c}]\} \end{aligned} \tag{7.30}$$

in terms of appropriately defined transfer integrals ε_a, ε_b and ε_γ between molecules, with **a**, **b** and **c** the usual lattice vectors for a monoclinic lattice (Friedman[46]).

The procedure outlined so far does not include a number of special features arising from the presence of an actual excess charge:

(1) The excess charge localized on a single molecule polarizes the electron distributions of its neighbours. When delocalized in a Bloch state it carries its polarization with it. The composite carrier and polarization field, which Toyozawa[49] calls an 'electronic polaron', it still one-electron-like (Kohn[50]) but has modified effective mass and modified band structure. Such polarization effects are ignored in the simple one-electron approximation. They can, however, be included heuristically in the present formulation by replacing the Bloch sums of (7.26) by the analogous sums over the many-electron wavefunctions Φ_{Iiv}, where each Φ_{Iiv} describes the carrier localized on site (Ii) with the environment polarized around it (Glaeser and Berry[13]).

(2) The orbitals of a monopositive or mononegative molecular ion differ somewhat from those of the neutral molecule, and should be used in constructing $\psi_{v\mathbf{k}}(\mathbf{r})$.

(3) The presence of an excess carrier may modify the equilibrium nuclear positions, as well as other local vibrational properties of the crystal (e.g. Silbey *et al.*[51]). In particular, the vibrational wavefunction may be changed for the molecule on which the carrier is localized, possibly introducing vibrational overlap factors into the effective values of the transfer integrals.

7.4.3 *Calculated excess carrier bands*

The excess hole and electron bands of anthracene were calculated by LeBlanc.[52] Thaxton, Jarnagin and Silver[53] later reported similar calculations for naphthalene, anthracene, tetracene and pentacene. In each calculation excess electron and excess hole Bloch states are constructed using, respectively, the lowest unfilled and highest filled π orbitals of the neutral molecule, following the treatment of Balk *et al.*[54] for the anthracene molecular ion. The molecular orbitals are taken as linear combinations of Slater-type $2p_z$ atomic orbitals, and the potential V_n is taken as the sum of the Goeppert-Mayer and Sklar[55] neutral carbon-atom Hartree terms. The various matrix elements of V_n are then complicated sums of many-centre integrals; all overlap integrals, and integrals involving either more than two centres or separations greater than 5 Å, are ignored.

The calculations have been extended in a number of ways. First, Katz, Rice, Choi and Jortner[56] corrected three shortcomings of the earlier calculations:

(1) The transfer integrals depend sensitively on the tail of the molecular wavefunction; SCF wavefunctions rather than Slater orbitals should be used.

(2) The simplifying assumption of considering one molecule per unit cell is unjustified.

(3) Those three-centre integrals that involve two atoms on the same molecule are not small, and there are many of them. Some of these are therefore to be included.

Silbey *et al.*[51] include intermolecular exchange in their calculated transfer integrals, and consider some effects arising from the intramolecular vibrational degrees of freedom. The wavefunction for the crystal with a localized carrier is regarded as a product of vibrational and electronic wavefunctions for each molecule, with the vibrational wavefunction of one molecule modified by the presence of the charge localized on it. Each of the earlier electronic bands now becomes a series of vibronic bands. Since, however, the electronic bandwidth is much smaller than the vibrational quantum of energy for an intramolecular mode, and since the energy spacing to the first vibrationally excited band is much greater than $k_B T$, only the lowest vibrational band need be considered. The calculated transfer integrals then include the overlap between the ground state vibrational wavefunctions of the ion and the neutral molecule. The same overlap occurs for both molecules involved in the transfer integral, and the transfer matrix element is the purely electronic integral multiplied by the vibrational overlap squared

$$|\langle \chi_{\text{ion}} | \chi_{\text{neutral}} \rangle|^2$$

Some of the effects of electronic polarization have been considered by Glaeser and Berry.[13] The wavefunctions used in calculating the transfer integrals are formally taken as an antisymmetrized product of the one-electron orbitals for each molecule; one molecule has an extra orbital occupied by the excess charge, and the orbitals of the other molecules are polarized around it. Detailed calculations suggest that the inclusion of polarization might substantially reduce the intermolecular transfer integrals, and could lead to a considerable narrowing of the band. The effect is found to be sensitive to the detailed assumptions made in its calculation, however.

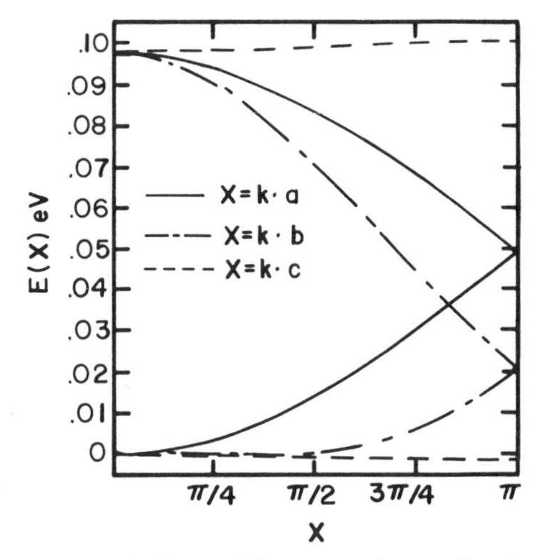

Figure 7.3 Shape of the excess electron band of anthracene in the $\mathbf{a}^{-1}$, $\mathbf{b}^{-1}$ and $\mathbf{c}^{-1}$ directions. (Katz *et al.*;[47] reproduced by permission of the American Institute of Physics)

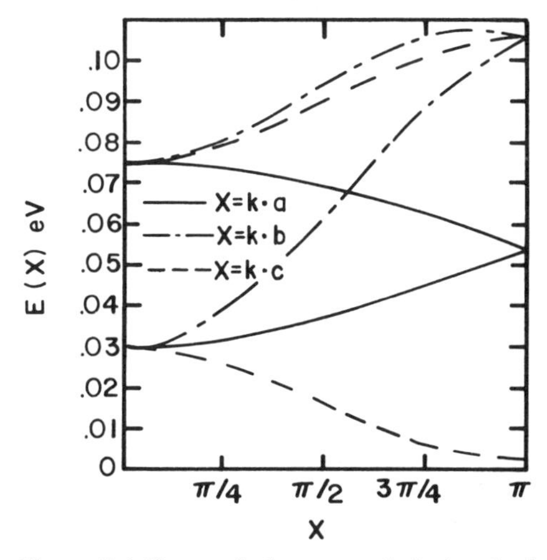

Figure 7.4 Shape of the excess hole band of anthracene in the $\mathbf{a}^{-1}$, $\mathbf{b}^{-1}$ and $\mathbf{c}^{-1}$ directions. (Katz *et al.*;[47] reproduced by permission of the American Institute of Physics)

Some typical results for anthracene are shown in Figures 7.3 and 7.4. The bands calculated by the various authors all tend to be rather narrow, ranging in width from the order of $k_B T$ to several times $k_B T$. For the excess electron in anthracene, the actual minimum of $\varepsilon(\mathbf{k})$ occurs for some non-zero wavevector parallel to the b-axis. The calculated bands also tend to be rather anisotropic. For anthracene, the symmetry of the crystal is such that the Bloch orbitals for wavevectors parallel to the b-axis are a symmetric and antisymmetric sum over the two molecules of the unit cell; along this direction, the symmetric and antisymmetric bands transform smoothly one into the other across the zone boundary in an extended zone scheme. For wavevectors along the c'-direction, on the other hand, there is a substantial splitting at the zone boundary. LeBlanc[52] notes, in particular, that his transfer integrals in the c'-direction for electrons in anthracene arise from the near cancellation of two large terms, and might therefore depend sensitively on such details as the intermolecular separation. This does not hold for holes in the same direction.

7.5 Band theory of charge transport

7.5.1 *Electron–phonon interaction for Bloch electrons*

Under the assumption of a rigorously periodic crystal structure, a band-picture treatment would suggest that an excess carrier in an applied field could eventually acquire a velocity of magnitude

$$|\mathbf{v}(\mathbf{k})| \equiv \hbar^{-1}|\nabla_{\mathbf{k}}\varepsilon(\mathbf{k})|$$

limited only by the bandshape.† In practice, thermal vibrations produce deviations from periodicity; the deviations cause scattering between Bloch states, restricting the lifetime of each Bloch state to a finite value, and thereby limiting transport.

The scattering of Bloch electrons has been discussed in terms of the adiabatic separation by Sham and Ziman[18] who regarded the one-electron Bloch orbital for the excess carrier as the electronic part of the Born–Oppenheimer vibronic wavefunction (7.9). Because of the quasi-continuous distribution of Bloch state energies, there are generally thermally accessible vibrational modes with energy comparable to the energy differences between pairs of Bloch states. The non-adiabatic corrections to the Born–Oppenheimer separation might then induce transitions between Born–Oppenheimer vibronic states constructed with

† A possible effect of this limitation on charge transport in rare gas solids is discussed by Spear.[57]

different Bloch orbitals. On the other hand, a description of charge transport in these terms is useful only if the scattering is sufficiently weak that the adiabatic separation is approximately maintained and the Bloch orbitals are approximate electronic stationary states to begin with; with this same qualitative assumption, the scattering might be treated in first-order perturbation theory and the scattering rate calculated from the matrix elements of the non-adiabatic terms of §7.2. The matrix elements of A_{mj} are of particular interest, since they connect Born–Oppenheimer states with different vibrational wavefunctions; they can therefore lead to processes in which both vibrational and electronic energies change, but in such a way that the total energy is conserved. With the crystal vibronic wavefunction taken as

$$\psi_{\mathbf{k}}(r, R)\chi(\{n_{\mathbf{q}}\}, R) \tag{7.31}$$

where $n_{\mathbf{q}}$ is the vibrational occupation number for mode $\mathbf{q}$, and with H_{ep} taken to denote (very schematically)

$$-\sum_I \frac{\hbar^2}{M_I}[\nabla_I \psi_{\mathbf{k}}(r, R)] \cdot [\nabla_I \chi(\{n_{\mathbf{q}}\}, R)] = H_{\text{ep}} \psi_{\mathbf{k}}(r, R)\chi(\{n_{\mathbf{q}}\}, R)$$

the scattering rate between $|\mathbf{k}\{n_{\mathbf{q}}\}\rangle$ is, in the first order,

$$\frac{2\pi}{\hbar}|\langle \mathbf{k}, \{n_{\mathbf{q}}\}|H_{\text{ep}}|\mathbf{k}', \{n_{\mathbf{q}}\}\rangle|^2 \, \delta[E(\mathbf{k}, \{n_{\mathbf{q}}\}) - E(\mathbf{k}', \{n'_{\mathbf{q}}\})] \tag{7.32}$$

The matrix elements (7.32) are defined somewhat differently in the Bloch representation. A change in vibrational coordinate Q from its equilibrium value changes the electronic interaction potential U of (2.4)

$$\Delta U \simeq \left(\frac{\partial U}{\partial Q}\right)_0 \Delta Q$$

It can then be shown (Ziman,[58] Sham and Ziman[18]) that the matrix elements of $\Delta U(r, R)$ give the same scattering rate as the matrix elements of A_{mj}, at least to first order in perturbation theory. The system is then governed by the Hamiltonian

$$H = H_L + H_{\text{e}} + H_{\text{ep}} \tag{7.33}$$

where H_L is the vibrational Hamiltonian of the crystal without the carrier (cf. (7.12)), H_{oe} is the one-electron Hamiltonian for the excess electron with the crystal in its perfect lattice configuration and H_{ep} is now the change in U with crystal deformation. The Bloch states are eigenstates when H_{ep} is ignored, and H_{ep} causes scattering in the next order of approximation. In calculating the scattering rate, the matrix elements are generally evaluated between the Bloch states of the perfect lattice. An expansion of H_{ep} to terms linear in displacements is almost always assumed.

For an excess carrier having small wavenumber and interacting with long-wavelength acoustic phonons, it is particularly useful to assume that the scattering results from local changes in the electronic energy eigenvalue as a function of the local deformation. The effective scattering potential is then (to first order in the local deformation Δ)

$$\Delta\varepsilon(\mathbf{k} \approx 0) \approx \varepsilon_1 \Delta$$

where Δ is the dilatation defined as the divergence of the elastic displacement $\mathbf{u}(\mathbf{r})$, and ε_1 is a phenomenological 'deformation potential' constant. Arguments justifying this approximation are given by Bardeen and Shockley.[59] The method is useful in wide-band semiconductors, but excess carriers in the aromatic hydrocarbons are not restricted to have small wavenumbers at reasonable temperatures (see Friedman[60]).

In the second-order scattering, difficulties arise from considering the transitions to be between perfect lattice Bloch states. Intuitive considerations suggest that the tight-binding Bloch functions might best be constructed from localized orbitals that follow the displacements of the molecules, whereas only variations in the potentials are considered in the matrix elements of (7.32). More generally, Herring[61] has argued that the use of the Bloch representation gives very large contributions to the scattering from the electron–lattice interaction terms bilinear in the displacements, and that these contributions are largely cancelled by the second-order interband scattering. The spurious contributions are attributed to the inability of the Bloch functions to follow the displacements. The merits of an approach based on a representation that deforms locally with the lattice have been emphasized by Blount[62] and a systematic theory on this basis has been presented by Whitfield.[63] From this viewpoint, the electrons follow the lattice strain in the first approximation, and the scattering is an additional effect. The method is discussed by Sham and Ziman[18] (p. 241 ff). A treatment of the electron–phonon interaction that includes molecular orbital displacement in first order for the aromatic hydrocarbons has been presented by Friedman.[60]

7.5.2 *Transport coefficients*

The calculation of macroscopic transport coefficients from the microscopic details of scattering requires that the statistical aspects of charge transport be considered. Several methods have been applied in considering the charge transport coefficients of solids. These include (1) approximate solution of the Boltzmann equation (e.g. Ziman[64] p. 257 ff., Wilson[65] p. 193 ff.); (2) application of density matrix methods (Kohn and Luttinger,[66]

Friedman[46]) and (3) the application of linear response theory (Kubo and Tomita,[67] Kubo[68]). Methods (1) and (3) have been applied explicitly to charge transport coefficients in the aromatic hydrocarbons, and we discuss them in this section. Also, we briefly describe some of the results that can be obtained for the aromatic hydrocarbons without detailed knowledge of the scattering mechanism. The present discussion is complemented by §7.5.3, where the detailed treatment of the electron–phonon coupling is considered.

7.5.2.1 *The Boltzmann equation*

The calculation of transport coefficients from a band model for the organic molecular solids has generally been formulated in terms of the Boltzmann equation. Applied to band-like transport, the method assumes a carrier distribution function $f(\mathbf{k}, \mathbf{r})$ for the density of carriers in the state $\mathbf{k}$ at the point $\mathbf{r}$ that obeys the equation

$$\mathbf{v}(\mathbf{k}) \, . \, \mathbf{\nabla}_r f + \frac{e}{\hbar}\left(\mathbf{E} + \frac{1}{c}\,\mathbf{v} \times \mathbf{H}\right) . \, \mathbf{\nabla}_{\mathbf{k}} f = \left[\frac{\partial f}{\partial t}\right]_{\text{coll}} = -\frac{f - f_0}{\tau(\mathbf{k})} \tag{7.34}$$

Here $[\partial f(\mathbf{k}, \mathbf{r})/\partial t]_{\text{coll}}$ is the rate at which $f(\mathbf{k}, \mathbf{r})$ is changed by scattering out of Bloch states with wavevector $\mathbf{k}$, $f_0(\mathbf{k}, \mathbf{r})$ is the equilibrium distribution function, $\tau(\mathbf{k})$ is the collision relaxation time, $\mathbf{v}(\mathbf{k})$ is the velocity $\hbar^{-1}\mathbf{\nabla}_{\mathbf{k}}\varepsilon(\mathbf{k})$, and $\mathbf{E}$ and $\mathbf{H}$ are respectively electric and magnetic fields acting on the system. The Boltzmann equation (7.34) arises by requiring the net rate of change of f to be zero in the steady state, taking into account the changes arising from scattering, diffusion and the effects of the applied fields; it is based on intuitive considerations. Conditions sufficient for it to follow from basic principles are known (Kohn and Luttinger[66]):

(1) The interactions causing the scattering are either weak or highly localized.

(2) The density matrix of the system (by virtue of the assumption of random phases) is diagonal at every instant of time.

The current density $\mathbf{J}$ is given in terms of $f(\mathbf{k}, \mathbf{r})$ by

$$\mathbf{J}(\mathbf{r}) = \int e\mathbf{v}(\mathbf{k}) f(\mathbf{k}, \mathbf{r}) \frac{\mathrm{d}^3 k}{4\pi^3} \tag{7.35}$$

An approximate formal solution of (7.34) for $f(\mathbf{k}, \mathbf{r})$ in terms of the applied fields then leads to a formal expression for such observables as $\mathbf{J}$. From these expressions, charge transport coefficients that relate the currents to the applied fields can be obtained. The transport coefficients for various charge transport phenomena are discussed by Blatt,[69] and their calculation in the narrow band limit by Friedman.[70] The coefficients of the greatest relevance to reported experimental studies of transport in the aromatic hydrocarbons are those that involve the Hall effect and the conductivity at zero magnetic field.

The solution of (7.34) for zero magnetic fields gives an expression for drift mobility. A standard approximate method (e.g. Ziman[64]) is to express $f(\mathbf{k}, \mathbf{r})$ as $f_0(\mathbf{k}, \mathbf{r}) + g(\mathbf{k}, \mathbf{r})$ and to keep in (7.34) only those terms of no higher than first order in $g(\mathbf{k}, \mathbf{r})$ and in the field $\mathbf{E}$. An expression for $g(\mathbf{k}, \mathbf{r})$ is then obtained in terms of $\tau(\mathbf{k})$, and can be used in (7.35) to obtain a formal expression for $\mathbf{J}$. By comparing the result with

$$\mathbf{J} \equiv ne\boldsymbol{\mu} \cdot \mathbf{E}$$

it is seen that the mobility tensor is given by

$$\boldsymbol{\mu} = \frac{e}{k_B T} \langle \tau(\mathbf{k})\mathbf{v}(\mathbf{k})\mathbf{v}(\mathbf{k}) \rangle_0 \tag{7.36a}$$

$$= \frac{e}{k_B T} \langle \mathbf{v}(\mathbf{k})\mathbf{v}(\mathbf{k})/[\lambda(\mathbf{k})|\mathbf{v}(\mathbf{k})|] \rangle_0 \tag{7.36b}$$

Here n is the density of carriers and the brackets denote the thermal average over the band, given schematically as

$$\langle \ldots \rangle_0 = \frac{2}{(2\pi)^3} \int [\ldots] \exp\left[-\frac{\varepsilon(\mathbf{k})}{k_B T} \right] d^3k \tag{7.37}$$

also $\tau(\mathbf{k})$ is a mean free path defined as

$$\lambda(\mathbf{k}) = |\mathbf{v}(\mathbf{k})| \tau(\mathbf{k})$$

Although actual evaluation of the drift mobility components using (7.36) requires sufficient knowledge of the scattering to determine $\tau(\mathbf{k})$, a crude analysis is often made without this information by assuming either an average isotropic mean free path λ_0, or an average isotropic relaxation time τ_0. In particular, it is then possible to predict the relative values of the mobility tensor components from the band structure alone by evaluating $\langle \mathbf{v}(\mathbf{k})\mathbf{v}(\mathbf{k}) \rangle$ for the constant relaxation time approximation, or $\langle \mathbf{v}(\mathbf{k})\mathbf{v}(\mathbf{k})/|\mathbf{v}(\mathbf{k})| \rangle$ for the constant mean free path approximation. Such an analysis generally predicts a rather anisotropic mobility tensor for anthracene with the relative mobility components depending on the specific assumptions made in calculating the transfer integrals. Generally, $\mu_{c'c'}$ is predicted to be substantially smaller than μ_{aa} or μ_{bb} for electrons.

The anisotropies are generally reported to be about the same for the average mean free path and average relaxation time approximations. These results are discussed in greater detail in §7.5.4.

The Hall effect involves crossed electric and magnetic fields which are usually taken to be $(E_x, E_y, 0)$ and $(0, 0, H_z)$ respectively. The response of the carriers to the fields is the net current density

$$\begin{aligned} J_x &= \sigma_{xx}^{(H)} E_x + \sigma_{xy}^{(H)} E_y \\ J_y &= \sigma_{yx}^{(H)} E_x + \sigma_{yy}^{(H)} E_y \end{aligned} \tag{7.38}$$

where the coefficients $\sigma_{\xi\eta}^{(H)}$ are magnetic-field dependent. We assume that the transverse components $\sigma_{xy}^{(H)}$ and $\sigma_{yx}^{(H)}$ vanish at zero magnetic field. To prevent a current from flowing in the y-direction, the field $\mathscr{E}_y$ must be applied, and the Hall coefficient R is then defined in terms of $\mathscr{E}_y$ as

$$R = \frac{\mathscr{E}_y}{HJ_x}$$

so that, from (7.38)

$$R = -\frac{\sigma_{yx}^{(H)}}{H\sigma_{xx}^{(0)}\sigma_{yy}^{(0)}} \tag{7.39}$$

where the superscript '(0)' denotes zero magnetic field. The Hall mobility is then defined in terms of R as

$$\begin{aligned} \mu_{\mathrm{H}} &= (necR)\mu_{yy} \\ &= -\frac{c\sigma_{yx}^{(H)}}{H\sigma_{xx}} = \frac{c\sigma_{xy}^{(H)}}{H\sigma_{xx}} \end{aligned}$$

and has the same dimensions as μ. More generally, the applied voltage in the x-direction together with the applied magnetic field in the z-direction might lead to an electric field component E_z as well as E_y. Solution of the system of equations

$$\begin{aligned} J_x &= \sigma_{xx}E_y + \sigma_{xy}E_y + \sigma_{xz}E_z \\ 0 &= \sigma_{yx}E_x + \sigma_{yy}E_y + \sigma_{yz}E_z \\ 0 &= \sigma_{zx}E_x + \sigma_{zy}E_y + \sigma_{zz}E_z \end{aligned}$$

leads to a definition of the Hall coefficient

$$\begin{aligned} R &= \frac{E_y}{HJ_x} \\ &= \frac{\Sigma_{yz}}{H \det \sigma} \end{aligned}$$

where σ is the determinant of the σ_{ij} matrix and Σ_{yz} is the determinant

$$\begin{vmatrix} \sigma_{xx}\sigma_{xy} \\ \sigma_{yx}\sigma_{yy} \end{vmatrix}$$

This procedure gives the same result as (7.38) whenever $\sigma_{yx}\sigma_{zy} - \sigma_{zx}\sigma_{yy}$ is zero, which does not necessarily occur for an anisotropic crystal. The usual analysis based on (7.38) sometimes leads to difficulties in this case (Coldwell-Horsfall and ter Haar,[71] Beer[72]).

An expression for the Hall mobility can be obtained by finding an approximate formal solution of the Boltzmann equation for the particular field arrangement of interest. An appropriate procedure is first to express (7.34) in terms of

$$f(\mathbf{k},\mathbf{r}) - f^0(\mathbf{k},\mathbf{r}) = \phi_{\mathbf{k}}\partial f_{\mathbf{k}}/\partial\varepsilon(\mathbf{k})$$

next to neglect terms beyond those linear in $\phi_{\mathbf{k}}$, and then to follow the Jones and Zener method (Wilson[65] p. 194 ff.) of developing $\phi_{\mathbf{k}}$ as an expansion in powers of H. Determination of the part of J_y linear in H gives σ_{yx}/H, and therefore determines R and μ_{H}. The result obtained by Friedman[70] using essentially (7.38) can be written in the form

$$\left(\frac{\mu_{\mathrm{H}}}{\mu_{xx}}\right) = k_{\mathrm{B}}T\frac{\langle\tau[v_y^2M_{xx}^{-1} - v_yv_xM_{yx}^{-1}]\rangle}{\langle\tau v_x^2\rangle\langle\tau v_y^2\rangle} \tag{7.40}$$

where x and y respectively correspond to the a and b crystallographic axes, τ and $\mathbf{v}$ are the usual functions of $\mathbf{k}$, and the $M_{\xi\eta}^{-1}$ are the components of the reciprocal effective mass tensor as a function of $\mathbf{k}$ (cf. (7.25)).

In order to treat the narrow band limit for anthracene, Friedman[70] expands the expressions for the Hall coefficient and drift mobility in powers of the intermolecular transfer integral divided by $k_{\mathrm{B}}T$, and assumes the band structure (7.30). The resultant expressions for the diagonal drift mobility components are of the form

$$\mu_{\xi\xi} = \frac{e\tau_0}{\hbar^2}\left(\frac{c_\xi}{k_{\mathrm{B}}T}\right)\left[1 - \frac{d_\xi}{k_{\mathrm{B}}T} + \dots\right]$$

where c_ξ and d_ξ are coefficients that depend on the various transfer integrals in (7.30) differently for each crystallographic direction ξ. The $(k_{\mathrm{B}}T)^{-1}$ dependence at large T is expected, since all band states averaged over in (7.36) are about equally populated when $k_{\mathrm{B}}T$ is greater than the bandwidth.

LeBlanc[73] and Friedman[70] have both considered the expected behaviour of (μ_{H}/μ). LeBlanc specifically calculates, as a function of bandwidth divided by $k_{\mathrm{B}}T$, what is essentially the average of (7.40) over the **a** and **b** crystallographic directions. For large bandwidth, the conventional result

$(\mu_H/\mu) \approx +1$ is obtained. For narrower bandwidths, however, the sign of the ratio can change and both authors suggest a negative ratio μ_H/μ for carriers in anthracene. The positive sign of μ_H/μ in conventional semiconductors arises from the respective curvatures of the conduction and valance bands near their band edges, where the carriers in the wide band for the most part lie; under standard sign conventions this leads to a positive effective mass for both holes and electrons. When the bands are narrow and all states about equally populated, negative effective-mass states near the zone boundary contribute to (7.40) as well. If the bands are also anisotropic, contributions associated with positive and negative effective masses can contribute unequally. The overall sign of μ_H/μ then depends on the detailed bandshape, of which Hall effect measurements should, in principle, be a sensitive probe.

7.5.2.2 *Linear response theory*

The calculation of transport coefficients from the Boltzmann equation generally involves the assumption of weak electron–phonon coupling and the evaluation of only its lowest-order effect on transport. A formal approach applicable at low fields is to regard the excess carriers interacting with the crystal together as the initial system, and to consider the linear part of the system's response to the applied fields (Kubo[68]). The response is related through the generalized fluctuation–dissipation theorem to appropriately defined correlation functions for the system, and the formal evaluation of the correlation functions, when it is possible, leads to expressions for the transport coefficients. In practice, the correlation functions can be evaluated only with substantial approximation, so that the linear response formulation does not necessarily provide a 'better' practical means of calculation than the solution of a single particle transport equation such as (7.34).

The response of the system to the applied field may be regarded as a linear superposition of responses to a series of weak electric field pulses between $t = -\infty$ and $t = 0$. The electric field pulse E_ξ at time t' and the current J_η it produces at a later time t are related through the response function $\phi_{\eta\xi}(t - t')$,

$$J_\eta = \int_{-\infty}^{t} \phi_{\eta\xi}(t - t')E_\xi(t')\,\mathrm{d}t'$$

The conductivity is therefore given in terms of $\phi_{\eta\xi}(t')$ by

$$\sigma_{\eta\xi}(\omega) = \int_{-\infty}^{0} \phi_{\eta\xi}(t')\exp(\mathrm{i}\omega t')\,\mathrm{d}t' \tag{7.41}$$

for an oscillatory applied field $\mathbf{E}_0 \exp(\mathrm{i}\omega t')$.

An expression for the conductivity can be obtained straightforwardly (see, e.g., ter Haar[74]). The Hamiltonian for the system may be divided into a zero field part $\hat{H}_0$ and a perturbation $\hat{H}'$ produced by the applied field

$$\hat{H} = \hat{H}_0 + \hat{H}'(t)$$

where for convenience $\hat{H}'(t)$ may be taken to oscillate with frequency ω. Also, H' is assumed to be turned on infinitely slowly from $t = -\infty$

$$H'(t) = \mathrm{e}^{-\varepsilon|t|}\,\mathrm{e}^{\mathrm{i}\omega t}\hat{V} \qquad \varepsilon \to 0^+$$

For an applied field **E**, the potential operator has the form

$$\hat{V} = -e\sum_j \mathbf{E}\,.\,\mathbf{r}_j \tag{7.42}$$

where $\mathbf{r}_i$ is the coordinate of electron i.

The density operator ρ for the system has the property that for an observable F (and with $\{|i\rangle\}$ a set of basis states)

$$\begin{aligned}\langle F\rangle &= \mathrm{Tr}\,\{\rho F\}\\ &\equiv \sum_i \langle i|\rho F|i\rangle\end{aligned} \tag{7.43}$$

is the statistical mechanical average of the quantum mechanical expectation value of F. The effect of the field is then reflected in the change of the density operator ρ from its equilibrium ρ_0,

$$\rho = \rho_0 + \Delta\rho$$

where

$$\rho_0 = \frac{\exp(-\beta\hat{H})}{Z} \tag{7.44}$$

with $\beta = (k_B T)^{-1}$, and where

$$Z = \mathrm{Tr}\,\{\exp[-\beta H_0]\}$$

is the quantum statistical mechanical partition function. Also the evolution of ρ (in the Schrödinger picture) is governed by

$$\mathrm{i}\hbar\frac{\partial\rho}{\partial t} = [\hat{H}, \rho] \tag{7.55}$$

The response of the system to the applied field is expected to be a current density **j**, defined as the time derivative of

$$\frac{\left(e\sum_i \mathbf{r}_i\right)}{V}$$

where V is the volume. The current operator is therefore

$$\mathbf{j} = (i\hbar V)^{-1}\left[e\sum_i \mathbf{r}_i, \hat{H}\right] \tag{7.56}$$

An approximate solution for $\Delta\rho$ can be found, and $\langle j \rangle$ evaluated, by expressing (7.55) in terms of (7.51) and (7.53), and keeping only terms linear in $\Delta\rho$ and $\hat{H}'(t)$. The procedure leads directly to an expression for the conductivity at $\omega = 0$

$$\sigma_{\mu\nu} = Z^{-1}\int_0^\infty \mathrm{d}\tau \int_0^\beta \langle j_\nu^H(-i\hbar\lambda) j_\mu^H(\tau)\rangle \,\mathrm{d}\lambda \tag{7.57}$$

in terms of a current–current correlation function. Here $j_\nu^{\hat{H}}$ is the νth component of $\mathbf{j}$ in the Heisenberg picture.

Gosar and Choi[75] have treated the effects of acoustic phonon coupling on charge transport in anthracene using the linear response formulation. Their results are discussed in the next section.

7.5.3 *Mechanism of electron coupling to lattice modes*

The constant relaxation time and constant-mean-free path approximations yield only mobility ratios, and can predict only the anisotropy arising from the band structure, but not the anisotropy contributed by the scattering itself. Actual calculation of the transport coefficients requires more detailed consideration of the scattering mechanism.

The viewpoint generally followed in band theory calculations has been to regard the intramolecular frequencies as high enough so that the carrier moves in the lowest intramolecular vibrational band with the effective transfer integrals perhaps modified by the appropriate vibrational overlap factor. The scattering between different Bloch states is then produced by the coupling to lattice modes. Friedman[60] and Gosar and Choi[75] have examined the nature and consequences of the acoustic mode coupling in some detail. While the treatment of Gosar and Choi is not a band model approach in the sense used here, we discuss it in this context because of its relevance to the nature of the coupling between excess carriers and the lattice vibrational modes.

7.5.3.1 *Electron–phonon coupling for Bloch electrons in molecular crystals*

The narrow bandwidths and small transfer integrals characteristic of aromatic hydrocarbons and of many other molecular crystals† lead to

† The rare-gas solids are an almost pathological exception to this rule. Their lowest unfilled bands are several eV wide (Fowler,[76] Reilly,[77] Knox and Bassani,[78] Mattheiss[79]), the overlap integrals between excess-electron orbitals on neighbouring sites can sometimes be large ($\langle\phi_{4s}|\phi_{4s}\rangle = 0{\cdot}55$ for argon; Knox[80] p. 36) and observed excess electron drift mobilities are of the order of 10^3 cm^2/V s (Miller, Howe and Spear,[81] Pruett and Broida[82]).

difficulties in treating the electron–phonon interaction by conventional means. One difficulty considered by Friedman[60] is that the scattering rate in the conventional picture would normally be calculated in first order by evaluating the matrix elements

$$\langle \psi_{\mathbf{k}} | H_{\mathrm{ep}} | \psi_{\mathbf{k}'} \rangle = \sum_{I} \langle \psi_{\mathbf{k}} | \sum_{I} \nabla V \, . \, \delta \mathbf{R}_I | \psi_{\mathbf{k}'} \rangle$$

between Bloch states of the undeformed crystal. Intuitive considerations suggest, especially for tight binding, that the Bloch functions should be allowed to follow the displacements of the molecules. A second difficulty is that, with all band states about equally occupied, the effective-mass theorem cannot be applied in its simplest form (7.25).

These problems have been discussed by Friedman[60] who considers the electron–phonon interaction for acoustic modes in the aromatic hydrocarbons from a microscopic viewpoint. The system with an excess carrier is described by a superposition of vibronic wavefunctions, each corresponding to an electron localized on a given site

$$\Psi(\mathbf{r}, \ldots \mathbf{R}_g \ldots) = \sum_{g'} C_{g'}(\ldots \mathbf{R}_g \ldots) \phi(\mathbf{r} - \mathbf{R}_{g'}) \tag{7.58}$$

where $\phi(\mathbf{r} - \mathbf{R}_g')$ is a localized molecular wavefunction that deforms instantaneously with displacement of $\mathbf{R}_{g'}$ and satisfies

$$\left[-\frac{\hbar^2}{2m} \nabla_r^2 + U(\mathbf{r} - \mathbf{R}_g) \right] \phi(\mathbf{r} - \mathbf{R}_g) = \varepsilon \phi(\mathbf{r} - \mathbf{R}_g) \tag{7.59}$$

The wavefunction Ψ then satisfies (cf. (7.3))

$$i\hbar \frac{\partial \Psi}{\partial t} = [H_{\mathrm{e}}(\ldots \mathbf{R}_g \ldots) + H_{\mathrm{L}}(\ldots \mathbf{R}_g \ldots)] \Psi \tag{7.60}$$

where

$$H_{\mathrm{e}}(\ldots \mathbf{R}_g \ldots) = -\frac{\hbar^2 \nabla_r^2}{2m} + \sum_{g} U(\mathbf{r} - \mathbf{R}_g) \tag{7.61}$$

is the one-electron Hamiltonian for the excess carrier and H_{L} is the vibrational Hamiltonian of the crystal. By requiring the expansion of ψ in (7.58) to satisfy (7.60) and by keeping terms up to first order in the displacements $\mathbf{u}_g = \mathbf{R}_g - \mathbf{R}_g^0$ and in the two-centre overlap integrals, an expression is obtained for the C_l coefficients as a function of time. The result, which we write in the form

$$\left[\frac{i\hbar \partial}{\partial t} - (\varepsilon + W_0 + \delta W + H_{\mathrm{L}}) \right] C_l$$

$$= \sum [J_h^0 + (\nabla_h J_h) \, . \, (\mathbf{u}_{l+h} - \mathbf{u}_l)] C_{l+h} + F_1 + F_2 + F_3 \tag{7.62}$$

with

$$J_{lg} = \int d^3r\, \phi(\mathbf{r} - \mathbf{R}_l) U(\mathbf{r}, \mathbf{R}_l) \phi(\mathbf{r} - \mathbf{R}_g)$$

$$W \equiv \int d^3r\, |\phi(\mathbf{r} - \mathbf{R}_l)|^2 \sum_{g'' \neq l} U(\mathbf{r} - \mathbf{R}_{g''})$$

and

$$\delta W = (\nabla W)_0 \cdot \mathbf{u}_l$$

is rather complicated, but we note that the various terms may be identified straightforwardly.

The left-hand side of (7.62) describes the adiabatic motion of the lattice in a potential given by the total electronic energy in the instantaneous lattice configuration, without transitions between localized electron states. The terms on the right couple states localized on different sites. These terms involve the usual transfer integral $J_{l,l+h} \equiv J_h$ expanded to first order in intermolecular displacements; the variation of J with displacement is the conventional source of electron–phonon scattering for acoustic modes. The additional terms F_1, F_2 and F_3 arise from the constraint that the origin of the wavefunction $\phi(\mathbf{r} - \mathbf{R}_g)$ follow the molecular displacements. The expressions for F_1, F_2 and F_3 involve the matrix elements of kinetic energy operators similar to A_{mj} and B_{mj} of (7.7), where the spatial derivatives are now taken only with respect to the explicit dependence of $\phi(\mathbf{r} - \mathbf{R}_g)$ on $\mathbf{R}_g$, and not on the implicit dependence of ϕ on $\mathbf{R}_g$ through the coordinate dependence of U in (7.59).† In the localized representation, rather different roles are played by the variation of Coulomb integrals W and the variation of transfer integrals J. We also note that the treatment to this point is not restricted by a choice between localized functions and tight-binding Bloch functions as a physical description of the system.

An essential step in deriving the final form of the matrix elements is to take the solutions to (7.62) in the absence of displacements

$$\Psi_{\mathbf{k}} = N^{-\frac{1}{2}} \sum_l \phi(\mathbf{r} - \mathbf{R}_l^0) \exp(i\mathbf{k} \cdot \mathbf{R}_l^0) \chi_{\{N\}} \tag{7.63}$$

as zero-order states between which the remaining (displacement dependent) terms cause transitions. (In (7.63), $\{N\}$ is the set of vibrational quantum numbers for the many-mode vibrational wavefunction χ.) Since the scattering is, in practice, calculated only to first order, the use of the perfect-crystal Bloch states (7.63) assumes a physical description of the system in terms of weak scattering between approximately stationary band states.

† The latter dependence is already included in (7.62) through variations in W and J.

As a physical consequence, the role of thermal fluctuations of the transfer integrals is to increase scattering between Bloch states, which would then be expected to decrease the mobility by decreasing $\tau(\mathbf{k})$ in (7.34). The electron–phonon matrix elements are found to be the sum of three terms

$$\langle \mathbf{k}\{N\}|V_1|\mathbf{k}'\{N'\}\rangle = -N^{-1}\sum_l \exp[-i(\mathbf{k}-\mathbf{k}')\,.\,\mathbf{R}_l^0]\exp(i\mathbf{k}\,.\,\mathbf{R}_h^0) \times \langle\{N\}|(\mathbf{u}_{l+h}-\mathbf{u}_l)|\{N\}\rangle\,.\,(\nabla_h J_h)_0 \qquad (7.64)$$

$$\langle \mathbf{k}\{N\}|V_2|\mathbf{k}'\{N\}\rangle = -N^{-1}\sum_l \exp[-i(\mathbf{k}-\mathbf{k}')\,.\,\mathbf{R}_l^0]\sum_h \times \langle\{N\}|(\mathbf{u}_{l+h}-\mathbf{u}_l)|\{N'\}\rangle\,.\,(\nabla_h\langle\phi|U(\mathbf{r}-\mathbf{R}_h|\phi\rangle)$$

$$\langle \mathbf{k}\{N\}|V_3|\mathbf{k}'\{N\}\rangle = -(\hbar^2/M)N^{-1}\sum_l \exp[-i(\mathbf{k}-\mathbf{k}')\,.\,\mathbf{R}_l^0] \times \sum_I \exp(i\mathbf{k}\,.\,\mathbf{R}_I^0)\langle\{N\}|\nabla_{l+h}|\{N'\}\rangle \times \langle\phi(\mathbf{r})|\nabla_r|\phi(\mathbf{r}-\mathbf{R}_h^0)\rangle \qquad (7.66)$$

The first term arises from fluctuations in transfer integrals, the second from variations in Coulomb integrals, and the third from the local orbitals moving with molecular displacement. Detailed consideration then shows that the scattering from (7.65) and (7.66) is negligible compared with the scattering from transfer integral variations of (7.64). When expressed in terms of the phonon coordinates of the crystal, (7.64) then takes the form

$$-2i\sum_{\mathbf{q},j,\pm}(\hbar/2MN\omega_{\mathbf{q}})^{\frac{1}{2}}\delta(\mathbf{k},\mathbf{k}'\pm\mathbf{q}+\mathbf{K})\delta(N_{\mathbf{q},j},N_{\mathbf{q}j}\pm 1) \times (N_{\mathbf{q}j'}+\tfrac{1}{2}\mp\tfrac{1}{2})\sum_h \hat{\varepsilon}_{\mathbf{q}j}\cdot(\nabla_h J_h)_0[\sin(\mathbf{k}\,.\,\mathbf{R}_h^0)-\sin(\mathbf{k}'\,.\,\mathbf{R}_h^0)] \qquad (7.67)$$

with $\omega_{\mathbf{q}j}$ and $\hat{\varepsilon}_{\mathbf{q}j}$ respectively the frequency and polarization of mode $(\mathbf{q}, j)$, and $\mathbf{K}$ a reciprocal lattice vector. Expression (7.67) involves the usual selection rules that the phonon number change by one, and that the total electron and phonon wavevector be conserved modulo a reciprocal lattice vector. The selection rule on the wavevectors arises directly in any of the expressions for linear electron–phonon-coupling matrix elements when the displacement is given by a phonon coordinate $U_{\mathbf{q}i\xi}$ of (7.21).

The application of these formal results to a one-dimensional crystal model has also been discussed by Friedman.[60] In this case elastic scattering of the excess carrier is assumed (i.e., the phonon energy is neglected). This leads to an expression for the Boltzmann collision term that is already explicitly of the form

$$\left[\frac{\partial f(k)}{\partial t}\right]_{\text{coll}} = -\frac{g(k)}{\tau(k)}$$

thereby establishing the existence of a relaxation time in this limit. Both normal processes (in which $k' = k \pm q$) and Umklapp processes ($k' = k + q + K$) are considered, since the initial state is not restricted to $k = 0$. The mobility is found to follow closely a $(1/k_B T)^2$ dependence.

With inelastic scattering included, the treatment does not lead explicitly to $(\partial f/\partial t)_{\text{coll}}$ in terms of a relaxation time. A variational treatment is therefore employed to bypass the calculation of a collision time. Again, a $1/T^2$ temperature dependence is found.

Finally, the theory is applied to a three-dimensional model for the anthracene crystal. The specific assumptions made include:

(1) Neglect of the electronic energy dependence in the Boltzmann factor when evaluating thermal averages over the narrow carrier band.

(2) Elastic scattering for the excess carrier.

(3) Neglect of Umklapp processes.

(4) Use of a Debye spectrum for the (acoustic) phonons and replacement of the phonon distribution by its high-temperature limit.

The band structure parameters of LeBlanc[52] are assumed and so effects of polarization are neglected. Application of a variational method to the Boltzmann equation then leads to expressions for the mobility-tensor components. The calculation gives predicted values for the anisotropy, once again predicts a $1/T^2$ temperature dependence and, in addition, gives calculated magnitudes of the mobility components. Comparison of these results with experiment is discussed in §7.5.4.

7.5.3.2 *Electron–phonon coupling mechanism for acoustic modes—phonon-assisted tunnelling and polarization coupling*

A detailed mechanism of electron coupling to acoustic modes has been studied by Gosar and Choi.[75] In addition to the short-range electron–phonon coupling arising from the variation of transfer integrals, the excess charge is coupled to the lattice displacements through the induced polarization of its environment. The polarization energy for a localized charge is substantial, about $-1{\cdot}7$ eV for anthracene and $-1{\cdot}3$ eV for naphthalene, and the variation of this energy with thermal fluctuations of intermolecular distances is estimated to be about $0{\cdot}03$ eV for nearest neighbour polarization alone. Gosar and Choi's treatment in effect compares the relative significance of the two coupling mechanisms; also, since the treatment is based on Kubo's linear response formalism and does not assume weak electron–phonon coupling, the calculation re-examines the usual result, based on the Boltzmann equation, that the fluctuation of transfer integrals generally decreases excess carrier mobilities.

It is convenient to describe the excess carrier in terms of localized (Wannier) basis states, although it is important to note that these are related to Bloch states by a unitary transformation; in itself, this is only a mathematical choice of a convenient set of basis functions, and not a physical assumption. It is also convenient to use the notation of second quantization, and to therefore express the Hamiltonian operator directly in terms of its matrix elements between localized basis states†

$$H = \sum_{i,j} w(i,j)a^+(i)a(j) + \sum_{\lambda} \omega(\lambda)b^+(\lambda)b(\lambda) + H_{\text{int}} \tag{7.68}$$

Here $w(i,j)$ is the transfer integral between i and site j, $a^+(i)$ is the creation operator for an electron on site i, $b^+(\lambda)$ is the creation operator for a vibrational quantum in mode λ with angular frequency $\omega(\lambda)$ and H_{int} gives the change in H to first order in nuclear displacements

$$\begin{aligned} H_{\text{int}} = &\sum_{i,\lambda} [u(i,\lambda)b(\lambda) + u(i,\lambda)^* b^+(\lambda)]a^+(i)a(i) \\ &+ \sum_{i,j,\lambda} [v(i,j,\lambda)b(\lambda) + v(i,j,\lambda)b^+(\lambda)]a^+(i)a(j) \end{aligned} \tag{7.69}$$

For each mode λ, the bracketed term involving the coupling constants $u(i,\lambda)$ gives the change in polarization energy to first order in the coordinate displacement

$$\mathbf{u}(\mathbf{r}) = \hat{\varepsilon}[\alpha(\lambda)b(\lambda) + \alpha^*(\lambda)b^+(\lambda)]$$

(with an appropriate $\hat{\varepsilon}$ and α for each λ). Similarly, the terms involving $v(i,j,\lambda)$ give, to the first order in displacement of the normal coordinate for mode λ, the change in the contribution of the transfer integral $w(i,j)$ to the Hamiltonian.

The dipole moment operator per unit volume V is

$$\mathbf{p} = \frac{e}{V}\sum_i \mathbf{r}_i a^+(i)a(i)$$

The operator for the current density is then the time rate of change of p,

† For convenience and consistency of notation with Gosar and Choi, we use the convention $\hbar = 1$.

and is given in the Heisenberg picture by

$$\mathbf{j} = \dot{\mathbf{p}} = (\mathrm{i}\hbar)^{-1}[\mathbf{p}, H] \tag{7.70}$$

or

$$\mathbf{j} = -\frac{\mathrm{i}e}{V}\sum_{i,j}(\mathbf{r}_j - \mathbf{r}_i)a^+(j)a(i)\left\{w(i,j) + \sum_{\lambda}[v(i,j,\lambda)b(\lambda) + v^*(i,j,\lambda)b^+(\lambda)]\right\} \tag{7.71}$$

from (7.68), (7.69) and (7.70). For any two sites i and j, there is not only a contribution to $\mathbf{j}$ in (7.71) that involves the transfer integral $w(i,j)$, but also a contribution that arises from the change in $w(i,j)$ with molecular displacement. The extra terms can be shown to increase the current. The increase arises physically from the transitory increases in transfer integrals which are, on the average, more effective in enhancing the transfer rate than the associated decreases are in inhibiting transfer.

The conductivity is given by the Kubo formula (7.57). By assuming that

$$\langle j_\nu^H(t_1)j_\xi^H(t_2)\rangle = \langle j_\nu^H(t_1 - t_2)j_\xi^H(0)\rangle$$

and by using the explicit time dependence for j_ν^H in the Heisenberg picture, (7.57) may be written in the form

$$\boldsymbol{\sigma} = \beta V \int_0^\infty \langle \mathbf{j}(t)\mathbf{j}(0)\rangle\, \mathrm{e}^{-\varepsilon t}\, \mathrm{d}t_{\varepsilon\to 0^+} \tag{7.72}$$

Evaluation of the correlation functions that result when (7.71) is substituted into (7.72) is a formidable task, and in practice requires extensive approximation. For example, terms involving $\langle a^+(j,t)a(i,t)a^+(k,0)a(l,0)\rangle$ arise from the $w(i,j)$ contributions to (7.71). The two-particle correlation functions of this form are

$$\begin{aligned} g_{ij}(t) &= \langle a^+(j,t)a(i,t)a^+(i,0)a(j,0)\rangle \\ &\equiv \langle \exp(\mathrm{i}Ht)a^+(j,0)\exp(-\mathrm{i}Ht) \\ &\quad \times \exp(\mathrm{i}Ht)a(i,0)\exp(-\mathrm{i}Ht)a^+(i,0)a(j,0)\rangle \end{aligned} \tag{7.73}$$

These are evaluated by assuming $g(t)$ to be gaussian in time

$$g_{ij}(t) = \frac{nV}{N}\exp\left(-\frac{\alpha_{ij}^2 t^2}{4\pi}\right)$$

Then Gosar and Choi expand (7.73) in powers of H, ignore the variation of *electron* energy in the Boltzmann factor (i.e. assume narrow electron bands) and use a Debye vibrational spectrum in order to evaluate the thermal average in (7.73). In general, the various contributions that arise in (7.72) from (7.71) can be classified into three groups, according to whether they involve two, three or four molecules. Only the two-molecule terms are found to contribute substantially to the current. It is also found that the fluctuation of polarization energy dominates over the transfer integral fluctuations in determining transport under the assumed approximations. The phonon-assisted transport is, however, substantial and constitutes about 10–30 % of the total current in anthracene. That the phonon assisted transport comprises so large a fraction of the total current results from the small values of transfer integrals and their relatively rapid increase with decreasing intermolecular distance. The effect is negligible in conventional wideband semiconductors (Gosar and Vilfan[83]). The calculated mobility components are in fairly good agreement with experiment, except for the c'-direction transport of electrons in anthracene (see §7.5.4).

7.5.4 *Applicability of the band picture*

The applicability of a band picture of charge transport to the aromatic hydrocarbons can be considered first of all with regard to how well its conclusions agree with experiment. A second question is whether the assumption of quasi-stationary Bloch states weakly coupled to one another through the electron–phonon interaction is itself physically reasonable for these solids, and in particular whether the experimentally fitted mean free paths and lifetimes allow a physical description in these terms.

7.5.4.1 *Comparison with experiment*

The experimental data that a satisfactory theory of charge transport should explain include:

(1) (a) observed mobility anisotropies, (b) experimental values of the actual mobility tensor components,

(2) the temperature dependence of the observed drift mobility ratios,

(3) pressure dependence of the mobility components,

(4) effects of deuteration on the transport coefficients and

(5) the ratios of Hall mobility components to the corresponding drift mobility components.

Some of the basic features of charge transport in aromatic crystals are well established experimentally (LeBlanc,[3,84,85] Kepler,[86,87] Many *et al.*,[88] Helfrich and Mark,[89,90] Nakada and Ishihara,[91] Raman and McGlynn,[92] Fourny and Delacôte,[93] Mey and Hermann[94]). Typically, the diagonal drift mobility components are of the order of 1 cm^2/V s and show a T^{-n} temperature dependence, with n usually between 1 and 2. An exception is electron transport in the c' direction of anthracene; $\mu_{c'c'}$ in this case is somewhat smaller than the other components, and increases gently with temperature. Deviations from the $1/T^n$ dependence have also been reported for electron transport in the **a** and **b** crystallographic directions of naphthalene, where the corresponding mobility tensor components are observed to be essentially independent of temperature (Mey and Hermann[94]). Some typical experimental results (Kepler[86]) for the temperature dependence of the electron mobility are shown in Figure 7.5.

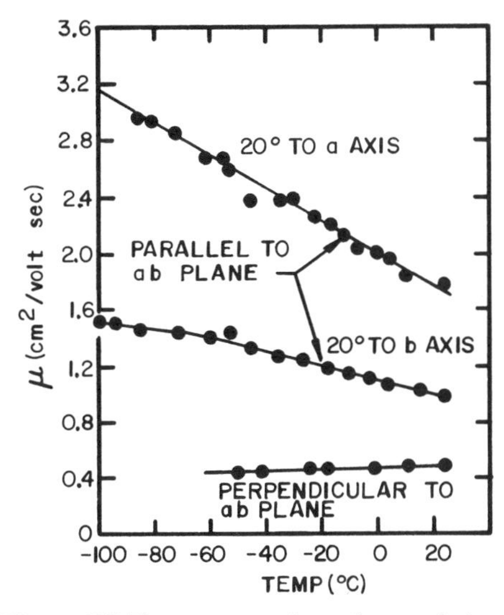

Figure 7.5 Temperature dependence of the electron mobility in anthracene. (Kepler;[87] reproduced by permission of the Illinois Institute of Technology.) Additional temperature dependence measurements are reported in the references cited in text

The pressure dependence shown in Figure 7.6 is seen to be substantial; however, the change in mobility is found to be greater when the volume is altered by changing the temperature rather than by changing the pressure (Kepler[87]). The pressure dependence of $\mu_{c'c'}$ is especially small. The effect of deuteration on charge transport has been observed only fairly recently, and a rather substantial 10% effect is found for the c'-direction mobility of electrons in anthracene (Munn *et al.*,[95] Morel and Hermann,[96]

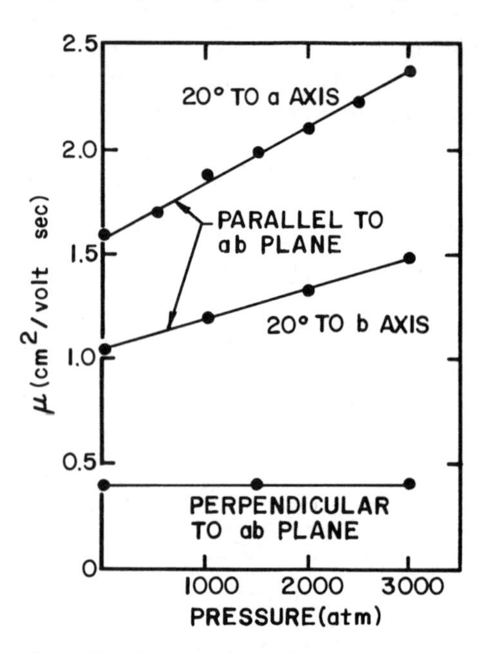

Figure 7.6 Pressure dependence of the electron mobility in anthracene. (Kepler;[87] reproduced by permission of the Illinois Institute of Technology)

Mey *et al.* [97]). The change is reportedly a decrease in $\mu_{c'c'}$ upon deuteration (Mey *et al.*[97]) contrary to earlier interpretations, with the other mobility components in anthracene showing no effect.

Hall mobility data, on the other hand, are inconsistent among the various authors (Spielberg *et al.*,[98] Korn *et al.*,[99] Schadt and Williams,[100] Smith[101]) particularly with respect to the sign of the Hall coefficient. The measured Hall mobilities generally indicate $\mu_H > \mu$, with different signs for (μ_H/μ) often observed for different crystallographic directions in the same crystal. The observation of Dresner[12] are of particular interest. He concludes first of all that the Hall mobility for holes in anthracene is not anomalous in sign. Partly on the basis of Hall mobility data, he further suggests that there is, in addition to the usual narrow hole band, a lower lying wide band separated from it by about $2k_BT$. Wide bands of this sort have also been suggested for electrons in anthracene (Pope and Burgos,[102] Dresner[103]).

These general experimental features may be compared with the calculated results of band theory. The degree of consistency between observed and calculated anisotropies depends to some extent on the particular band-structure calculation considered. Glaeser and Berry,[13] in particular, base their calculations on a critical assessment of the approximations involved in earlier calculations; their results, some of which are included in Table 7.3, take some account of exchange and polarization, but do not

Table 7.3 Ratios of mobility tensor components in anthracene

	μ_{aa}/μ_{bb}	$\mu_{c'c'}/\mu_{bb}$	$\mu_{c'c'}/\mu_{aa}$	Source
		(Electron)		
Constant relaxation time	1·5	0·0	0·0	Katz *et al.*[47]
	2·1	0·05	0·02	Silbey *et al.*[51]
	2·25	0·008	—	Glaeser and Berry[13 a]
Constant mean free path	1·5	0·01	0·01	Katz *et al.*[47]
	1·9	0·07	0·04	Silbey *et al.*[51]
	2·16	0·03	—	Glaeser and Berry[13 a]
Hopping model	1·0	0·11	0·11	Glaeser and Berry[13 a]
Experimental	1·7	0·4	0·24	Kepler[86]
		(Hole)		
Constant relaxation time	0·41	0·11	0·26	Katz *et al.*[47]
	0·07	0·19	2·5	Silbey *et al.*[51]
Constant mean free path	0·45	0·14	0·31	Katz *et al.*[47]
	0·08	0·25	2·9	Silbey *et al.*[51]
Experimental	0·5	0·4	0·8	Kepler[86]

[a] Calculated with polarization oscillator strength = 2·0; see Glaeser and Berry.[13]

include the intramolecular vibrational narrowing that is sometimes considered in such calculations. The order of magnitude disagreement with experiment for mobility ratios involving electron transport in the c' direction of anthracene occurs in a number of calculations. For other directions, agreement with experiment is reasonable considering the approximations necessarily involved and the fact that many of the calculated results are based on the average mean-free-path and average relaxation-time approximation, and therefore ignore the anisotropy of the scattering itself. Caution is required in interpreting the overall agreement between calculated and measured anisotropy ratios as evidence for band-like transport, however, since the same transfer integrals that determine the band structure play a role in determining the rate at which an actually localized carrier transfers to a neighbouring site, and the same anisotropy could easily arise when transport is by a series of uncorrelated near-neighbour hops (Glaeser and Berry[13]). Gosar and Choi[75] (Table 7.4) explicitly consider the electron–phonon coupling in their calculation and obtain absolute as well as relative magnitudes of the mobility components. Their results are in good agreement with experiment, except for the order of magnitude discrepancy for electron transport in the c' direction of anthracene. Since their treatment assumes neither weak scattering nor Bloch states, the general agreement with experiment is not evidence for a band picture. Their results do indicate that the electron–phonon interaction mechanism of polarization coupling to acoustic modes is, with one particular exception, capable of explaining the actual magnitude of the mobility components.

Table 7.4 Some calculated values of the mobility components of carriers in anthracene (units of cm^2/V s)

	μ_{aa}	μ_{bb}	$\mu_{c'c'}$	μ_{aa}/μ_{bb}
	(Electron)			
Calculated (constant τ)[60]	—	—	—	1·13
Calculated (Friedman)[60]	—	0·85	—	1·33
Calculated (Gosar and Choi[75])	1·64	1·35	0·003	1·21
Experimental (Kepler[86])	1·70	—	0·4	—
	(Hole)			
Calculated (constant τ)[60]	—	—	—	0·61
Calculated (Friedman[60])	—	0·72	—	0·61
Calculated (Gosar and Choi[75])	0·95	2·34	0·65	0·41
Experimental (Kepler[86])	1·0	2·0	0·8	0·5

The usually observed T^{-n} temperature dependence of the mobility is rather suggestive of band-like transport. The temperature dependence of the electron mobility for the c' direction of anthracene cannot be explained straightforwardly on a simple band model, however, since the increase of scattering between Bloch states with increasing temperature would be expected to decrease $\mu_{c'c'}$. It is reported that the relevant transfer integral for electron transport in this case does increase with changes in unit cell dimension at increasing temperature, but a straightforward explanation on this basis appears to be inadequate (Delacôte,[104] Chojnacki[105]). The possibility that the anomalous temperature dependence might be explained in terms of phonon-assisted tunneling also has been mentioned (Delacôte[104]), while another explanation (discussed later in this article) has been proposed in terms of electron–phonon coupling to intramolecular modes (Munn and Siebrand[38]).

The comparison between observed and calculated Hall mobilities is of potential value in assessing the relative merits of different band structure calculations if a band model description is physically appropriate (Korn *et al.*,[99] Spielberg *et al.*[98]). It is important to note, however, that both anomalous and non-anomalous signs of the Hall coefficient are also possible when transport is by random jumps from site to site (Munn and Siebrand[106,107]).

Finally, we note that the deuterium effect has not apparently been treated on the basis of a band model. Also, to the present author's knowledge, no theoretical explanation has yet been developed for the origin of Dresner's wide bands.

7.5.4.2 *Consistency of a band picture for the aromatic hydrocarbons*

In assessing the applicability of the band model as a physical description, it is important to distinguish between the use of either Bloch or localized states as merely a convenient choice of basis functions, and the use in some essential way of approximations that depend on one or the other physical picture. For the present discussion, the band picture is the physical description of excess carrier transport in terms of nearly stationary electronic Bloch states (with the system approximately described by a single Born–Oppenheimer product) and with only weak scattering induced by the electron–phonon interaction. Friedman's[60] treatment of the electron coupling to acoustic modes involves this assumption, for example, while Gosar and Choi do not limit themselves to a localized physical description merely in their choice of basis functions.

The difficulties that arise when a band picture is applied to low-mobility transport have been discussed a number of times (Fröhlich and Sewell,[108] Joffé,[109] Glarum[48]). For the band model to be a consistent physical description, the electron–phonon coupling must be weak enough compared to the intermolecular electronic coupling that the energy broadening caused by the finite lifetime of the Bloch states is less than the electron bandwidth, or $W < \hbar\tau_0$ where W is the bandwidth and τ_0 the relaxation time. By setting

$$\mu \approx \frac{e\tau_0}{k_B T}\langle \mathbf{v}(\mathbf{k})\mathbf{v}(\mathbf{k})\rangle$$

and

$$\mathbf{v}(\mathbf{k}) = \hbar^{-1}\frac{\partial\varepsilon(\mathbf{k})}{\partial\mathbf{k}} \approx \frac{Wa}{\hbar}$$

the condition for band transport to be a reasonable description becomes

$$\mu \gtrsim \frac{ea^2}{\hbar} \frac{W}{k_B T}$$

where a is an appropriate lattice constant. With $W \approx k_B T$ and $a \approx 3$ Å, the borderline region for the mobility is very roughly 1 cm^2/V s—a value typical of the aromatic hydrocarbons. Frequently, the applicability of the band model is discussed in terms of the requirement that the mean free path λ_0 be greater than a lattice distance. By noting that

$$\mu \approx \frac{e}{k_B T} \lambda_0 \langle \mathbf{v}(\mathbf{k})\mathbf{v}(\mathbf{k})/|\mathbf{v}(\mathbf{k})| \rangle$$

and by replacing v by $v_{max} = Wa/\hbar$, it is seen that

$$\mu \approx \frac{ea^2}{\hbar} \frac{\lambda}{a} \frac{W}{k_B T}$$

For $\lambda/a \approx 1$, roughly the same lower limit for the band picture is obtained as before. It should be noted (Klinger[110]) that the failure of the scattering to satisfy these conditions argues only against the specific kind of Bloch state being considered, and does not preclude the possibility of a quasi-particle description in terms of the other band-like states (§7.6).

From calculated bandwidths and measured mobility components, τ_0 and λ_0 can be estimated for each direction of transport, and can be compared with $W/\hbar$ and with the lattice separation, respectively. Except for the early calculations of LeBlanc,[52] this procedure generally yields mean free paths and relaxation times uncomfortably close to their lower limits. Glaeser and Berry,[13] for example, find λ_0 significantly smaller than the intermolecular spacing, and t only about twice as large as $W/\hbar$. It would then appear that the conditions for the simple band model to apply are barely satisfied by the aromatic hydrocarbons.

7.5.5 *Transport through localized states—an alternative to the band picture*

An intuitively reasonable picture in the limit of small electronic transfer integrals is to regard the carriers as localized on particular molecules in zero order, and to consider the transitions to neighbouring molecules produced by the resonance interactions in the next order of approximation. In the extreme limit where the transfer is infrequent compared to the rate at which the lattice relaxes around the localized carrier, the motion is a series of uncorrelated hops. Since the transfer then occurs preferentially for nuclear configurations in which the carrier would have the same energy on either site, the transport is likely to exhibit an activation energy temperature dependence, reflecting the minimum energy needed to deform the configuration to achieve energy coincidence (Holstein[21]). Such a temperature dependence is in contrast with that observed for the aromatic hydrocarbons. Nonetheless, some simple considerations suggest that a model intermediate between hopping and band-like transport might be a viable alternative to the band picture.

First of all, the temperature dependence in the intermediate cases could differ from that of either the hopping or the band limits as, for example, can occur when the vibrational relaxation time permits coherent electron transport over several sites (Emin[111]). The detailed behaviour in the intermediate region depends on the nature and magnitude of the electron–vibrational coupling. In the alkali halides, for example, the effective mass of an excess electron moving with its associated deformation changes rapidly but continuously as its coupling to optical lattice modes is increased (Feynman[112]), while Toyozawa[113,114] argues that the short-range coupling to acoustic modes might produce a nearly discontinuous change to a self-trapped state at some critical value of the coupling constant. The rare-gas solids (Druger and Knox,[115] Song[116]) and the alkali halides (Castner and Kanzig[117]) are unusual in that holes are self-trapped and exhibit behaviour in the opposite extreme to that of electrons under the same conditions. It is particularly interesting to note that the positive carrier mobility in solid xenon decreases with increasing temperature (Spear[57]), although both theoretical (Druger and Knox[115]) and experimental (Miller *et al.*[81]) evidence suggests that the carriers are, indeed, quasilocalized.†

Furthermore, Glaeser and Berry[13] have considered a simple diffusion model in which one-step uncorrelated jumps occur at rates determined by the intermolecular transfer integrals, while the effects of the electron–phonon interactions are not treated. The simplified model not only predicts anisotropies to within a factor of 2, but predicts the magnitude of the mobility to well within a factor of 10, and thus compares favourably with the band model.

Other simplified models based on localized states have also been considered in the past (Eley and Spivey,[118] Nelson,[119] Keller and Rast,[120] Keller,[121] Tredgold,[122] Kearns[123]). Such models are often of considerable value for complex organic systems that are not amenable to a microscopically detailed treatment.

† It has been suggested that the temperature dependence in this case can be explained by assuming that the transport is by hopping with a low activation energy (Spear[57]).

7.6 Small-polaron theory of charge transport

Some experimental aspects of charge transport in the aromatic hydrocarbons have not, at present, been adequately explained on the basis of a band picture; at the same time, the relative magnitudes of the electron–phonon and intermolecular–electronic couplings appear to only marginally satisfy the conditions for the applicability of a simple band model. An alternative approach is to include the nuclear displacements in zeroth order by considering the slowly moving carrier to be accompanied by the distortion that the electron–lattice coupling induces around it. The carrier and its induced distortion may be regarded as a quasiparticle, which has generally been called a 'polaron'.

The size of the polaron is measured by the extent of the region over which the distortion is induced. In the alkali halides, the electron–phonon coupling arises mainly from the long-range Coulomb interaction between the carrier and the optical lattice modes; the distortion covers a region of radius much larger than a lattice constant, and the quasiparticle is a large polaron. A well-known approximation in this case is to regard the crystal as a continuum (Fröhlich[124]). On the other hand, the coupling may be strong and of short range, and the distortion may occur predominantly within the order of a lattice distance around the carrier. The quasiparticle in this case is a small polaron.

A number of theoretical treatments of the small polaron have been presented (Holstein,[21,125] Yamashita and Kurosawa,[126,127] Sewell[128]) and lead to similar physical results. Models similar to that of Holstein[21,125] have been applied to both organic (Munn and Siebrand[38]) and inorganic molecular crystals (Gibbons and Spear[129]), and we particularly emphasize this treatment in what follows. The theory of polarons has been reviewed by Appel[130] and has been discussed with specific emphasis on transition metal oxides by Austin and Mott.[131]

7.6.1 *Electron–phonon coupling for localized states*

A commonly adopted procedure for treating a localized excess electron interacting with the lattice is to describe the system by a Hamiltonian of the form

$$H = H_e + H_L + H_{int} \tag{7.74}$$

In (7.74),

$$H_e = \left[\frac{p^2}{2m} + V(\mathbf{r})\right]$$

is the one-electron Hamiltonian for the excess carrier in the undeformed crystal and H_L is the vibrational Hamiltonian for the crystal without the carrier. The lattice potential contained in H_L is given in the harmonic approximation by an expression of the form

$$V_L = \sum_{\lambda} \tfrac{1}{2} m_{\lambda} \omega_{\lambda}^2 Q_{\lambda}^2$$

in terms of some set of (real) normal coordinates $\{Q_{\lambda}\}$. The remaining term H_{int} is the interaction between the excess electron and its environment as a function of the electron coordinate and the nuclear displacements, and may be expanded in normal coordinates

$$H_{\text{int}} = \sum_{\lambda} \alpha_{\lambda}(\mathbf{r}) Q_{\lambda} + \sum_{\lambda\eta} \beta_{\lambda\eta}(\mathbf{r}) Q_{\lambda} Q_{\eta} \tag{7.75}$$

Because of the H_{int} contribution to (7.74), the presence of the electron modifies the local vibrational properties of the crystal, while the change in vibrational properties acts also through H_{int} in (7.74) to alter the dynamical behaviour of the carrier.

If only the linear term in (7.75) is kept and the adiabatic behaviour of the system is considered when the electron is in a state (s), then the vibrational potential energy for the mode λ is given by the total electron energy

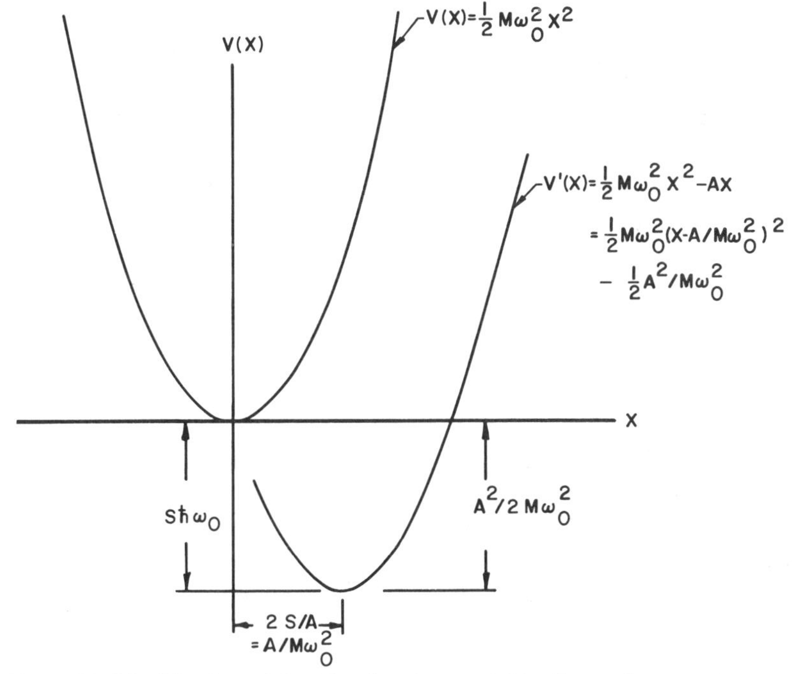

Figure 7.7 Modification of the vibrational potential by linear electron–phonon coupling. The presence of the excess carrier leads to the additional term $-Ax$ in the vibrational potential $\tfrac{1}{2}M\omega_0^2 X^2$

eigenvalue which is, to first order in $A_\lambda Q_\lambda$,

$$\mathscr{E}_s(Q_\lambda) = \tfrac{1}{2}m_\lambda\omega_\lambda^2 Q_\lambda^2 + A_\lambda^{(\xi)}Q_\lambda + E_0^{(\xi)} \tag{7.76}$$

(with appropriate constants $A_\lambda^{(\xi)}$ and $E_0^{(\xi)}$). The added term $A_\lambda^{(\xi)}Q_\lambda$ in the potential (7.76) serves as a linear electron–phonon interaction and leads to a change in equilibrium position without a change in frequency (Figure 7.7). The relaxation of the coordinate Q_λ to its modified equilibrium point then lowers the electron energy by

$$\frac{A_\lambda^2}{2m_\lambda\omega_\lambda^2}$$

The quadratic coupling in (7.75) is more complicated. If the terms coupling different modes are ignored, and only the quadratic coupling for a single mode is considered, the potential for the vibrational motion becomes

$$\mathscr{E}_s(Q_\lambda) = \tfrac{1}{2}m\omega_\lambda^2 Q_\lambda^2 + B_{\lambda\lambda}^{(s)}Q_\lambda^2 + E_0^{(s)} \tag{7.77}$$

(with appropriate constants $B_{\lambda\lambda}^{(s)}$ and $E_0^{(s)}$). The coupling then modifies the vibrational frequency without a change in equilibrium positions. Quadratic coupling might be of interest, for example, when the presence of an excess carrier modifies the vibrational frequency of some molecular mode (as has been suggested by Munn and Siebrand[38] for anthracene). It is then of interest to know how the crystal normal modes are affected. Rayleigh's theorems suggest that the only changes in normal frequencies are in the same direction as the change in molecular frequency, and are no greater than the distance to the next unperturbed frequency, provided the appropriate equilibrium positions are held fixed.† The changes in frequency within any vibrational branch are then infinitesimal (of the order N^{-1} for periodic boundary conditions over N unit cells), except possibly for the highest or lowest frequency of each vibrational branch. For an increase in the molecular spring constant, the highest frequency of each branch can split off from the quasicontinuum of frequencies. The same can happen for the lowest frequencies with a decrease in coupling constant. The corresponding mode is localized and involves cartesian displacements that die off rapidly with distance from the molecule whose vibrational frequency has been changed; since the mode is not degenerate with the phonon modes of the crystal, its energy does not propagate. If the modified intramolecular frequency, on the other hand, is degenerate with the vibrational bands of crystal, the corresponding vibrational state is not a localized mode; however, it is then a pseudolocalized mode if

† See Maradudin *et al.*[23] p. 357.

its coupling to the rest of the system is weak enough so that the relaxation of its vibrational energy occurs only slowly.

Localized and pseudolocalized modes are frequently considered in the theory of impurities in solids (Klein[39]), and in this general case can be associated with translational motion of the impurity itself, or with the motion of its immediate neighbours, or with its own internal motion. The vibrational properties of a molecular solid with a substitutional impurity have been studied for an array in which each molecule has one rotational and one translational degree of freedom (Dettmann and Ludwig[132]) and for the lattice and intramolecular modes of a linear chain of diatomic molecules (Munn[133]). For the linear chain of diatomic molecules, application of appropriate Green's function methods confirms the existence of a localized mode either in the bandgap or above the optical vibrational band when a single intramolecular spring constant differs from those of the remaining molecules. We note that in the more general case of a three-dimensional crystal, a change of spring constant leads to a localized mode only under specific, and sometimes stringent, conditions (Maradudin *et al.*[23] p. 373).

7.6.2 *Small-polaron states*

Small polarons have frequently been treated in a generalized tight binding approximation (Tiablikov,[134] Nettel,[135] Yamashita and Kurosawa,[126,127] Holstein[21,125]). Zero-order wavefunctions are constructed as the product of an electronic wavefunction for the carrier localized on a given molecule and a vibrational wavefunction for the crystal deformed around it. The small but non-vanishing transfer integrals between wavefunctions localized on different sites then ensures that the basis states are not stationary. The stationary state wavefunctions are then formed as linear combinations of the localized wavefunctions.

At zero temperature, the zero-order basis states for different sites are degenerate when the transfer integrals are ignored. The stationary states when the transfer integrals are taken into account are then Bloch sums over the localized states, and the transport is band-like. However, each transfer matrix element now includes a vibrational overlap as well as a purely electronic factor. As the temperature is increased, the probability of non-zero phonon occupation numbers increases, and the rate of scattering between polaron band states increases. At the same time, if linear electron–phonon coupling is assumed, the thermal average of each vibrational overlap decreases, the effective transfer integrals decrease, and the band becomes narrower. At some temperature, the band picture of small polaron transport becomes inapplicable. At still higher temperatures, the transport is by hopping between localized states. Holstein,[21] in particular, has considered the transport of small polarons in the low- and high-temperature limits. We outline the main features of his treatment and summarize his conclusions.

Holstein's[21,125] theory is based on a model of the crystal as a one-dimensional chain of N equally spaced molecules. Each molecule n has a single vibrational coordinate x_n and a molecular vibrational frequency ω_0, with the vibrations of neighbouring molecules weakly coupled. With the excess carrier present, the Hamiltonian of the system is

$$H = H_e + H_L + H_{int}$$

where

$$H_L = \sum_{n=1}^{N} \left(-\frac{\hbar^2}{2m} \frac{\partial^2}{\partial x_n^2} + \tfrac{1}{2} m\omega_0^2 x_n^2 + \tfrac{1}{2} m\omega_1^2 x_n x_{n+1} \right) \tag{7.78}$$

is the vibrational Hamiltonian of the crystal in the absence of the carrier, and where

$$H_e + H_{int} = -\frac{\hbar^2}{2m}\nabla^2 + \sum_{n=1}^{N} U(r - R_n, x_n) \tag{7.79}$$

The wavefunction of the system is assumed to be a superposition of molecular wavefunctions

$$\Psi = \sum_n a_n(x_1, \ldots, x_N)\phi_n(r, x_n) \tag{7.80}$$

where each $\phi_n(r, x_n)$ is a molecular orbital that follows the instantaneous configuration of the molecule on which it is localized

$$\left[-\frac{\hbar^2}{2m}\nabla^2 + U(r - R_n, x_n) \right] \phi(r - R_n, x_n) = \mathscr{E}(x_n)\phi(r - R_n, x_n) \tag{7.81}$$

The electron energy eigenvalue $\mathscr{E}(x_n)$ then serves as the electron–phonon coupling between the lattice and carrier localized on the nth molecule.

A number of simplifying assumptions can now be made:

1. As in almost all polaron theories, $\mathscr{E}_n(x)$ is expanded only to the term linear in displacements

$$\mathscr{E}(x_n) = \text{const.} + Ax_n$$

2. In the spirit of the tight-binding approximation in its simplest form, all transfer integrals

$$J_l(x_n, x_m) = \int \phi^*(r - R_n, x_n)U(r - R_l, x_l)\phi(r - R_m, x_m)\, d^3r$$

are assumed to vanish except for those that are either one-centre integrals or two-centre integrals between nearest-neighbour sites.

3. The matrix elements of the form

$$-\frac{\hbar^2}{2m}\iint \phi_m^* a_n^* \frac{\partial \phi_m}{\partial x_m}\frac{\partial a_m}{\partial x_m}\,\mathrm{d}r\,\mathrm{d}R$$

are ignored.

4. Those transfer integrals $J(x_n, x_{n\pm 1})$ that just do not vanish are assumed independent of the vibrational coordinates.

With these assumptions, the time dependence of the $\{a_n\}$ is seen from (7.78) to be given by

$$i\hbar\frac{\partial a_n}{\partial t} = H_{\mathrm{L}} a_n - J(a_{n+1} + a_{n-1}) - Ax_n \tag{7.82}$$

The behaviour of the small polaron then depends on the following parameters:

(1) the molecular vibrational frequency ω_0,

(2) the intermolecular vibrational coupling constant ω_1,

(3) the electronic transfer integral $J = -J(x_n, x_{n\pm 1})$ defined above, and

(4) the electron–phonon coupling constant A.

Holstein considers first the conditions for the solutions of (7.82) to exhibit small-polaron-like behaviour. The nature of the solutions can be examined by assuming an adiabatic approximation in which the nuclear kinetic energy terms of (7.82) are ignored. Equation (7.82) then becomes

$$\left(\tfrac{1}{2}\sum_m M\omega_0^2 x_m^2 - Ax_n\right)a_n - J(a_{n+1} + a_{n-1}) = \mathscr{E}(x_1, \ldots, x_N)a_n \tag{7.83}$$

for the energy eigenvalue $\mathscr{E}$. We note that if the normal coordinates were fixed in their ideal crystal positions, the usual Bloch solution $a_n = a_0 \exp(ikn)$ would give†

$$\mathscr{E}(k) = 2J\cos k \tag{7.84}$$

for purely electronic band, as expected. On the other hand, polaron-like solutions can be found by letting the x_m relax into a configuration that minimizes the total energy subject to the constraint

$$\sum_n |a_n|^2 = 1$$

† Following Holstein, we assume for convenience that the intermolecular spacing is $a = 1$.

A relation is then obtained between the amplitude of the electron on the molecule m and the displacement of x_m that is thereby induced (see Figure 7.7)

$$x_m^{(0)} = |a_m^{(0)}|^2(A/M\omega_0^2) \tag{7.85}$$

The corresponding energies can be found by applying a continuum expansion to the $\{a_n\}$ coefficients.

$$a_{n+1} = a_n + \frac{\partial a_n}{\partial n} + \frac{1}{2}\frac{\partial^2 a_n}{\partial n^2}$$

so that (7.83) reduces to a differential equation

$$\left(\frac{A^2}{M\omega_0^2}|a_n|^2 - \lambda\right)a_n + J\frac{\partial^2 a_n}{\partial n^2} = 0 \tag{7.86}$$

with

$$-\lambda = \mathscr{E}(x_1^{(0)}, \ldots, x_N^{(0)}) - \tfrac{1}{2}\sum_m M\omega_0^2 x_m^{(0)\,2} + 2J$$

Two physically different solutions to (7.86) are possible. If the electron is localized within a finite region, so that

$$\lim_{n\to\infty} |a_n|^2 = 0,$$

then the solutions are

$$a_n = \left(\frac{A^2}{8M\omega_0^2 J}\right)^{\frac{1}{2}} \operatorname{sech}\left[\frac{A^2}{4M\omega_0^2 J}(n - n_0)\right] \tag{7.87}$$

$$\mathscr{E}(x_1^{(0)}, \ldots, x_N^{(0)}) = -2J - \frac{1}{48J}\frac{A^2}{M\omega_0^2} \tag{7.88}$$

where n_0 is the site around which the polaron is centred. On the other hand, band-like solutions for (7.86) can be found by letting

$$a_n = N^{-\frac{1}{2}} \exp(ikn)$$

so that from (7.86),

$$\mathscr{E}(x_1^{(0)}, \ldots, x_N^{(0)}) \approx -2J + Jk^2 \tag{7.89}$$

The terms involving relaxation of the nuclear coordinates do not appear in (7.89) since their contribution is of order $1/N$ or less, and the calculated energy agrees with (7.84) when the cos k factor is expanded in powers of k.

Comparison of (7.88) and (7.89) shows that the bound state is lower in energy than the lowest band-like state by

$$E_b = \frac{1}{48J}\left(\frac{A^2}{M\omega_0^2}\right)^2$$

and E_b is the polaron binding energy in the approximation of ignoring the nuclear kinetic energy operators. This binding energy differs, however, from the energy to dissociate the polaron optically. For thermal dissociation, the energy of deforming the lattice against the restoring forces of the crystal is recovered, and the thermal dissociation energy is E_b. According to the Franck–Condon principle on the other hand, the lattice deformation energy still resides with the crystal immediately after the optical transition.

It can be seen from (7.87) that the polaron has linear dimensions of the order

$$L \approx a\left(\frac{4J}{A^2/M\omega_0^2}\right) \tag{7.90}$$

for a lattice separation a. The condition for a large polaron to obtain is then that L be much greater than a lattice distance, or from (7.90)

$$2J \lesssim \frac{A^2}{2M\omega_0^2} \equiv S\hbar\omega_0$$

For a small polaron, L must be the order of a or less, and

$$2J \gtrsim \frac{A^2}{2M\omega_0^2} \equiv S\hbar\omega_0 \tag{7.91}$$

Here $S\hbar\omega_0$ is the energy gained by allowing the lattice to relax around the carrier localized on a given site, and $2J$ is the half-width of the electron band (and is also the energy spent in localizing the electron in the perfect rigid crystal).

7.6.3 *Small-polaron transport*

Equation (7.82) is the starting point for considering the transport of small polarons in Holstein's model. It is convenient to reformulate (7.82) in terms of phonon coordinates of the crystal

$$Q_q = \left(\frac{2}{N}\right)^{\frac{1}{2}} \sum_{n=1}^{N} x_n \sin(qn + \pi/4)$$

When this is done (7.82) reduces to a set of N equations, each of the form

$$i\hbar\frac{\partial a_n}{\partial t} = \sum_q \left[\left(-\frac{\hbar^2}{2m}\frac{\partial^2}{\partial Q_q^2} + \tfrac{1}{2}m\omega_q^2 Q_q^2\right) - \left(\frac{2}{N}\right)^{\frac{1}{2}} A Q_q \sin(qn + \pi/4)\right] a_n - J(a_{n+1} + a_{n-1}) \quad (7.92)$$

where

$$\omega_q^2 = \omega_0^2 + \omega_1^2 \cos q$$

is the angular frequency for the phonon mode having wavenumber q. It is then seen that each of the N equations (7.92) involves a vibrational Hamiltonian for a system of N independent oscillators, with the equilibrium position of the oscillator having wavenumber q displaced by

$$Q_q^{(n)} = \left(\frac{A}{M\omega_q}\right)^2 \left(\frac{2}{N}\right)^{\frac{1}{2}} \sin(kn + \pi/4)$$

for the nth equation of (7.92). As we have already described, the zero-order solutions to (7.92) are obtained by ignoring J. The resultant solutions then correspond each to a state in which the carrier is localized on a given site; in the next order of approximation the localized solutions are coupled to each other by the transfer integral J. The zero-order solutions for the $\{a_n\}$ are found from (7.92) to be products of displaced oscillator wavefunctions

$$a_n(p, \{N_q\}) = \delta_{np} \prod_q \chi_{\{N_q\}}\left[\left(\frac{M\omega_q}{\hbar}\right)^{\frac{1}{2}} (Q_p - Q_p^{(p)})\right] \quad (7.93)$$

with total energy

$$E(\{N_q\}) = \sum_q \hbar\omega_q(N_q + \tfrac{1}{2}) - E_b$$

where

$$E_b = \frac{1}{48J}\left(\frac{A^2}{M\omega_0^2}\right)^2.$$

In equation (7.93), χ_{N_q} is the normalized simple harmonic oscillator wavefunction with vibrational quantum number N_q, δ_{np} is 1 for $n = p$ and zero otherwise, and n labels each of the N coefficients $\{a_n(p, \{N\})\}$ that together are the solution of (7.93) describing a carrier localized on site p when $J = 0$.

For $J \neq 0$, the solutions to (7.92) may be constructed as a superposition of the zero-order solutions

$$a_n(\{N_q\}) = \sum_{p', \{N_{q'}\}} C(p', \{N_{q'}\}) a_n(p', \{N_{q'}\}) \quad (7.94)$$

It can then be seen from (7.92) and (7.93) that the coefficients $C(p; \{N_q\})$ satisfy

$$i\hbar \frac{\partial C(p, \{N_q\})}{\partial t} = \sum_{p', \{N_q\}} \langle p, \{N_q\} | V | p', \{N_{q'}\} \rangle C(p', \{N'_q\}) \times \exp\left\{ \frac{it}{\hbar} [E(\{N_q\}) - E(\{N'_q\})] \right\} \quad (7.95)$$

where in (7.95) the matrix element coupling the states localized on sites p and p' is given by

$$\langle p\{N_q\} | V | p'\{N'_q\} \rangle = -J\delta_{p,p' \pm 1} \langle \chi(p, \{N_q\}) | \chi(p', \{N'_q\}) \rangle \quad (7.96)$$

The effective transfer matrix element (7.96) is zero except for nearest neighbours, and when it is non-zero is given by the product of the electronic transfer integral and the overlap between the initial and final vibrational wavefunctions of the system. The many-mode vibrational overlap that appears in (7.96)

$$\langle \chi(p, \{N_q\}) | \chi(p', \{N'_q\}) \rangle$$

$$= \prod_q \int_{-\infty}^{\infty} \chi_{N_q} \left[\left(\frac{M\omega_q}{\hbar} \right)^{\frac{1}{2}} (Q_q - Q_q^{(p)}) \right] \chi_{N'_q} \left[\left(\frac{M\omega_q}{\hbar} \right)^{\frac{1}{2}} (Q_q - Q_q^{(p')}) \right] \frac{M\omega_q}{\hbar} \, dQ_q \quad (7.97)$$

is itself the product of single oscillator vibrational overlaps, each term involving the initial and final vibrational occupation number of a single mode of the crystal. The transitions induced by the coupling between localized basis functions in (7.96) are then of two kinds, and may be classified as 'diagonal' if each mode has the same vibrational quantum number in the final as it has in the initial state, and as 'non-diagonal' if some of the phonon quantum numbers N_q change. The two kinds of transitions play different physical roles in different temperature ranges.

7.6.3.1 *Low temperatures*

At zero temperature all initial N_q are zero and energy conservation requires that all transitions be diagonal. The diagonal transitions can then be expected to dominate for some finite range of temperature above zero.

In considering (7.94) under these circumstances, Holstein notes that the displacement $Q_q^{(p)}$ for each mode is infinitesimal. The overlaps can then be evaluated by expanding each factor on the right-hand side of (7.94) to second order in $Q_q^{(p)} - Q_q^{(p')}$. If, in addition, it is assumed that $N_q = N_{-q}$, then the diagonal coupling terms are found to be

$$\langle p, \{N_q\}|V|p', \{N_q\}\rangle = -J\delta_{p,p'\pm 1} \exp[-S(\{N_q\})] \tag{7.98}$$

where

$$S(\{N_q\}) = \sum_q (1 + 2N_q)\frac{\gamma_q}{N} \tag{7.99}$$

Here the coupling parameter γ_q is given by

$$\gamma_q = \frac{A^2}{2M\omega_q^2 \hbar\omega_q}(1 - \cos q) \tag{7.100}$$

and is seen, apart from the factor $(1 - \cos q)$, to be the ratio of lattice relaxation energy to the vibrational quantum $\hbar\omega_q$, given by the quantity S in Figure 7.7. The particular diagonal vibrational overlaps that occur in (7.94) occur also in the theory of F-centres in alkali halides, and the behaviour of their thermal averages with temperature is well known in this context (Huang and Rhys[136]).

If stationary solutions to (7.95) are sought by considering the coefficients $C(p)$, where

$$C(p, \{N_q\}) = \exp[-iE(\{N_q\})t/\hbar]C(p),$$

then for the special case where all non-diagonal matrix elements are ignored, (7.95) takes the time-independent form

$$EC(p) = -J[C(p + 1) + C(p - 1)] \exp(-S) \tag{7.101}$$

(with the same $\{N_q\}$ implicit in each term). The choice of $C(p)$ as

$$C(p) = \exp(ipk)$$

satisfies (7.101), and therefore describes a stationary state of the system. This particular choice of $C(p)$ corresponds in (7.94) to forming generalized tight-binding Bloch states in order to eliminate the non-stationary character of the localized states when $J \neq 0$. It is worth emphasizing that the time dependence drops out of the exponential in (7.95) and the Bloch-like state is stationary precisely because of the exact energy conservation for the diagonal transitions considered. The resulting polaron band energy is then

$$\mathscr{E}(k, \{N_q\}) = E(\{N_q\}) - 2J \cos k \exp[-S(\{N_q\})] \tag{7.102}$$

with halfwidth

$$\Delta\mathscr{E} = 2J \exp[-S(\{N_q\})] \tag{7.103}$$

where $S(\{N_q\})$ is given by (7.99). The bandwidth is therefore dependent on phonon number. As the temperature is increased, the average number of phonons for each mode of increases, and the bandwidths decrease rapidly (Huang and Rhys[136]).

At non-zero temperatures non-diagonal transitions can also occur, and their effect must be considered. Since the non-diagonal matrix elements of (7.96) connect a given polaron band state $(k, \{N_q\})$ of the crystal to polaron states having a different set of phonon and electron wavenumbers, their effect is to induce scattering and thereby to limit transport. From any initial polaron state, there are a wide range of possible combinations of changes in the wavenumber and vibrational quantum numbers for which energy is conserved. The transitions can then develop uniformly in time, and the transition rate is given in first-order perturbation theory by

$$W(k, \{N_q\}|k', \{N_q'\}) = \frac{2\pi}{\hbar} |\langle k\{N_q\}|V|k'\{N_q\}\rangle|^2 \, \delta[\mathscr{E}(k, \{N_q\}) - \mathscr{E}(k', \{N_q'\})] \tag{7.104}$$

It is then possible to calculate the relaxation time $\tau(k)$ for any given polaron band state as the reciprocal of the total transition rate to all other states.

For the band picture of small-polaron transport to be applicable, the energy broadening produced by scattering must be less than the polaron bandwidth. If the relaxation time is $\tau(k, \{N_q\})$ a band description of the small polaron will be reasonable so long as

$$\begin{aligned} \frac{\hbar}{\tau(k, \{N_q\})} &\ll \Delta E(k, \{N_q\}) \\ &= 2J \exp[-S(\{N_q\})] \end{aligned} \tag{7.105}$$

As the temperature increases from zero, the number of initial N_q that are non-zero increases. The rate at which non-diagonal transitions can occur then also increases, and so $\hbar/\tau$ in (7.105) increases. At the same time, the thermal average of the bandwidth in (7.105) becomes narrower. Near some temperature T_t, the condition for the polaron band picture to be physically reasonable is only marginally satisfied. The transition region, in view of the exponential temperature dependences involved, should be rather narrow.

7.6.3.2 *High temperatures*

Above the temperature T_t, non-diagonal transitions dominate, and an appropriate description of the transport is in terms of transitions between localized polaron states, rather than between polaron band states. It is then of interest to consider the thermally averaged transition probability

$$W(p \to p', T) = \frac{\sum_{\{N_q\}} W(p \to p'; \{N_q\}) \exp\left[-\sum_q \beta\hbar\alpha_q(N_q + \tfrac{1}{2})\right]}{\sum_{\{N_q\}} \exp\left[-\sum_q \beta\hbar\omega_q(N_q + \tfrac{1}{2})\right]} \tag{7.106}$$

where $W(p \to p'; \{N_q\})$ is the total probability for a transfer from the site p with the crystal in the vibrational state $\{N_q\}$, to site p' with any allowed final vibrational state. The transition rate is then given by

$$W(p, \{N_q\} \to p', \{N'_q\}) = \frac{2}{\hbar^2}|\langle p, \{N_q\}|V|p', \{N'_q\}\rangle|^2 \cdot \frac{\partial}{\partial t}\Omega\left[\sum_q \hbar\omega_q(N_q - N'_q)\right] \tag{7.107}$$

where

$$\Omega(x) = \frac{1 - \cos(xt/\hbar)}{x^2/\hbar^2}$$

is the characteristic resonance transfer function of time-dependent perturbation theory, whose time derivative may be expressed in the integral form

$$\frac{\partial}{\partial t}\Omega(x) = \tfrac{1}{2}\int_{-t}^{t} \exp\left(\frac{ixt'}{\hbar}\right) dt' \tag{7.108}$$

Holstein evaluates the transition matrix element in (7.107) by the earlier procedure of expanding the vibrational overlaps in (7.107) to second order in the displacements of the equilibrium positions. Transitions in which any one phonon number changes by two or more are ignored. Combining the resultant expression for the transfer matrix elements with (7.107) and (7.108) leads to the transition rate in terms of an integral over the time t. The transition rate is found to increase linearly with t in the limit of large t, and the anomalous behaviour is attributed to the implicit inclusion of diagonal contributions. Because of the rigorously exact energy conservation involved in the diagonal transitions, they do not develop uniformly in time, and their contributions to the transition rate are therefore subtracted off. The remaining contributions arise from non-diagonal transitions, and give a transition rate constant in time,

$$W(p \to p \pm 1; T) = \frac{J^2}{\hbar^2}\left[\frac{2\pi^2}{\int_0^\pi 2\gamma_q\omega_q^2 \operatorname{csch}(\tfrac{1}{2}\beta\hbar\omega_q)\, dq}\right]^{\frac{1}{2}} \times \exp\left[\pi^{-1}\int_0^\pi 2\gamma_q \tanh(\beta\hbar\omega_q/4)\, dq\right] \tag{7.109}$$

For the classical limit of $\hbar\omega_0 \ll k_B T$, the transition rate is found to be

$$W(p \to p \pm 1; T) = \frac{J^2}{\hbar}\left(\frac{\pi}{4k_B T E_a}\right)^{\frac{1}{2}} \exp\left(-\frac{E_a}{k_B T}\right) \tag{7.110}$$

with activation energy

$$E_a = \pi^{-1} \int_0^{\pi} \frac{A^2}{4M\omega_q^2}(1 - \cos q)\, dq \tag{7.111}$$

The activation energy is seen to be smaller than the thermal dissociation energy

$$E_b = \pi^{-1} \int_0^{\pi} \frac{A^2}{2M\omega_q^2}\, dq \tag{7.112}$$

From (7.109) the lifetime of a localized state can then readily be evaluated

$$\tau(p)^{-1} = W(p \to p + 1; T) + W(p \to p - 1; T)$$

7.6.4 *Drift mobility and diffusion coefficient*

A number of procedures are available for calculating transport coefficients for small polaron transport. These include solution of an appropriate Boltzmann equation, application of density matrix methods and the application of the Kubo linear response formalism. Friedman,[46] in particular, has shown using a density matrix formulation that the total mobility is additive in the band and hopping contributions

$$\mu = \mu_{\text{band}} + \mu_{\text{hop}}$$

Less formal methods can, however, be used to evaluate the transport coefficients in the limits Holstein considers.

The mobility and diffusion constant can be obtained straightforwardly for $T < T_t$ and $T > T_t$ when J is small. For $T > T_t$, the transport is by random hopping from site to site. The diffusion coefficient is then given by

$$D = a^2 W_T(p \to p \pm 1)$$

and is related to the mobility through the Einstein relation

$$\mu = \frac{eD}{k_B T}$$

From equation (7.109) for W_T, the mobility is then

$$\mu_{\text{hop}} = \frac{ea^2}{k_B T} \frac{J^2}{\hbar^2 \omega_0} \left[\frac{\pi}{\gamma \operatorname{csch}(\beta\hbar\omega_0/2)}\right]^{\frac{1}{2}} \exp\left(-2\gamma \tanh\frac{\beta\hbar\omega_0}{4}\right) \tag{7.113}$$

for the narrow band limit of $\omega_q = \omega_0$, and with $\gamma = A^2/2M\omega_0^3\hbar$.

For $T < T_t$, the usual band theory treatment applies, and the mobility can be obtained from

$$D = \langle v^2 \tau \rangle_{Av} \tag{7.114}$$

where

$$\begin{aligned} v &= \frac{a}{\hbar} \frac{\partial \mathscr{E}(k, T)}{\partial k} \\ &= \frac{2J}{\hbar} \sin k \exp[-S(T)] \end{aligned} \tag{7.115}$$

from equation (7.102). In evaluating (7.114), the usual assumption is then made that all band states are about equally populated, by virtue of the small bandwidth. The mobility is then found to be

$$\mu_{\text{band}} = \frac{e}{k_B T} \frac{2J^2}{\hbar} a^2 \tau(k, T) \, e^{-2S(T)} \tag{7.116}$$

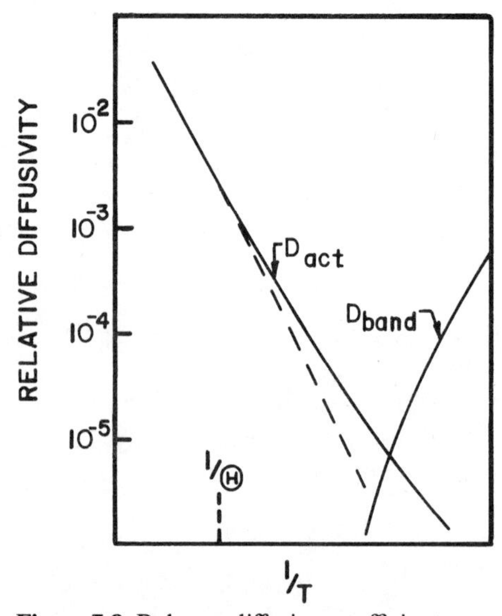

Figure 7.8 Polaron diffusion coefficient as a function of temperature. The curves D_{act} and D_{band} are respectively the temperature dependences for transport by random jumps and by polaron band motion. The solid line is the classical approximation obtained by assuming narrow bandwidth and high temperature. (Holstein;[21] reproduced by permission of Academic Press, Inc.)

in terms of a lifetime $\tau(k, T)$ that can be obtained by calculating the scattering rate between polaron band states. Formal evaluation of the transition rate then verifies that the lifetimes of the localized and band states are equal

$$\tau_{k,T}^{-1} = \tau_{p,T}^{-1}$$

and with (7.116) and (7.109) leads to an expression for the mobility

$$\mu_{\text{band}} = \frac{ea^2\omega_0}{k_B T}[\gamma\pi^{-1}\operatorname{csch}(\beta\hbar\omega_0/2)]\exp[-2\gamma\operatorname{csch}(\beta\hbar\omega_0/2)]$$

The temperature dependence of D in the two limits is shown in Figure 7.8.

7.6.5 *Adiabatic and correlated hopping of small polarons*

Equation (7.113) for the mobility in the polaron-hopping limit is based not only on the assumption that the small-polaron condition

$$2J < E_b \tag{7.117}$$

is satisfied, but also on the assumption that the electronic transfer integral is sufficiently small that it may be treated in first-order perturbation theory. These two conditions are different, as more detailed examination reveals, and it therefore becomes necessary to consider the small polaron behaviour in the case where the small-polaron condition (7.117) is satisfied while the conditions for applicability of the first-order perturbation treatment are not.

Holstein discusses the problem in terms of an adiabatic potential energy $\mathscr{E}(\{x_n\})$ for the system with an excess carrier. In calculating $\mathscr{E}(\{x_n\})$ from (7.82), the kinetic energy terms are again ignored, but the intermolecular vibrational coupling terms $\{m\omega_1^2 x_r x_{r+1}\}$ are now retained. It is then possible to examine the behaviour of the carrier in the region of site p by expanding (7.82) to second order in the perturbation J, which leads to

$$\mathscr{E}^{(p)}(\{x_n\}) = V_L(\{x_n\}) - Ax_p - \frac{J^2}{A}[(x_p - x_{p+1})^{-1} + (x_p - x_{p-1})^{-1}] + \dots \tag{7.118}$$

When the lattice is relaxed around the carrier, $|x_p - x_{p+1}| \approx A/M\omega_0^2$ is large. Since $J \ll A^2/M\omega_0^2$ for the small polaron, the adiabatic potential is then essentially just $V_L - Ax_p$ for the carrier localized on site p, and the effect of the J-dependent terms in (7.118) is small. When the configurations of molecule p and molecule $p + 1$ happen to be similar, the expansion (7.118) breaks down and J mixes the degenerate solutions for the carrier on neighbouring sites. In Figure 7.9 we show a highly schematic

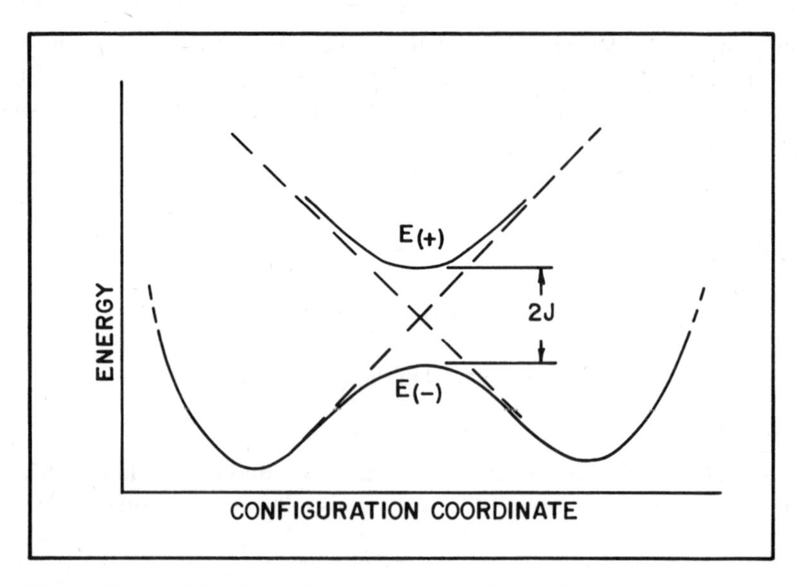

Figure 7.9 A highly schematic representation of the adiabatic energy of the two-site system

representation of the behaviour of $\mathscr{E}$ as a function of some hypothetical configuration coordinate chosen to reflect the relative values of the coordinates of the neighbouring molecules. When the carrier energy on each molecule differs substantially, the carrier is strongly localized on one of them. When the energies of the two configurations happen to be nearly equal, substantial mixing is induced by J, and the potential curve for localization on one site changes continuously into that for localization on the other. If J is small enough, then as the crossing point is passed the carrier may not be able to respond to the abrupt change, and may therefore 'jump' from $E_{(-)}$ to $E_{(+)}$. Eventually the crossing point is passed again in the opposite direction. The carrier may transfer to $E_{(-)}$ and return to its initial state, or else might remain on $E_{(+)}$ until approaching the crossing point again from the opposite direction. Detailed consideration of these possibilities leads to a rederivation of the classical (high-temperature) limit for the hopping probability (7.110).

On the other hand, if J is large enough, the splitting between $E_{(+)}$ and $E_{(-)}$ increases to the point that jumps between $E_{(+)}$ and $E_{(-)}$ become unlikely. The carrier then moves from site to site by adiabatically following the nuclear motion. These considerations lead to the conditions

$$J > E_a^{\frac{1}{4}} \left(\frac{2k_B T}{\pi} \right)^{\frac{1}{4}} \left(\frac{\hbar\omega_0}{\pi} \right)^{\frac{1}{2}}$$

for the adiabatic transport to obtain, and

$$J < E_a^{\frac{1}{4}} \left(\frac{2k_B T}{\pi}\right)^{\frac{1}{4}} \left(\frac{\hbar\omega_0}{\pi}\right)^{\frac{1}{2}}$$

for the non-adiabatic perturbation treatment to apply. The corresponding temperature dependences at high temperature are then found to be

$$W(p \to p+1) = \begin{cases} \dfrac{\omega_0}{2\pi} \exp\left(-\dfrac{E_a}{k_B T}\right) & \text{(adiabatic)} \\ \left(\dfrac{\pi}{4k_B T E_a}\right)^{\frac{1}{2}} \dfrac{J^2}{h} \exp\left(-\dfrac{E_a}{k_B T}\right) & \text{(non-adiabatic)} \end{cases}$$

Finally, it is important to note that the treatment of the small polaron hopping assumes that the lattice relaxes rapidly enough between jumps for successive hops to be uncorrelated. This in fact happens provided the intermolecular vibrational coupling is large enough so that (Emin[111,137])

$$\pi \frac{2\omega_1}{\omega_0} \frac{E_a k_B T}{\hbar\omega_0} \gg 1$$

If the condition is not satisfied, the rate at which a second hop occurs in the forward and reverse directions can be affected. Emin finds, in particular, that the temperature dependence may then be much milder than the usual activation energy result, and that the hopping rate can even decrease with temperature at high temperatures.

7.6.6 *Hall effect for small polarons*

A number of physically different cases may be distinguished in discussing the Hall effect for polarons:

(1) *'Wide' band (low-temperature) transport for large or small polarons.* When the polaron band is wide compared to $k_B T$, the usual treatment for wide-band semiconductors applies. The transport is described by a Boltzmann equation, and the corresponding Hall and drift mobility components are nearly equal.

(2) *Narrow band transport.* When the band is narrow compared to $k_B T$, the band states are all populated with about equal probability. If the band is unsymmetrical, various contributions associated with positive and negative effective masses contribute with different weights. Depending on which contributions dominate, the Hall effect may have either sign (Friedman[70]).

(3) *Non-adiabatic hopping of small polarons.* The effect of an applied magnetic field on the transfer matrix element between sites I and J is to multiply it by a phase factor $\exp[i\lambda \mathbf{R}_{IJ} . \mathbf{B}]$ with λ a constant. For a one-site jump, this has no effect on the jump rate. Interference can occur, however, between the amplitude for a direct jump from I to J, and the amplitude for an indirect jump from I to J through an intermediate site. The field-dependent interference can lead to a Hall effect (Friedman and Holstein[138]).

(4) *Adiabatic hopping of small polarons.* Emin and Holstein[139] have shown that when the carrier follows the lattice motion adiabatically, the motion through the applied magnetic field gives rise to additional terms in the Hamiltonian. These extra terms describe an effect on the lattice motion through the effect of the field on the carrier. The modification of the nuclear motion then leads in turn to a non-zero Hall effect.

Limits (3) and (4) are of particular interest. The theory of the Hall effect for small-polaron non-adiabatic hopping has been considered by a number of authors, whose results have not been in agreement. Friedman and Holstein[138] consider interference between one- and two-step jumps in a triangular lattice. Interference for a set of three neighbouring sites is allowed to contribute to the Hall effect whenever the three sites all have the same energy, and the results obtained explicitly reflect this requirement. In the high-temperature limit it is found that

$$\mu_{\mathrm{H}} = \frac{ea^2}{2\hbar} J \left[\frac{\pi}{4k_{\mathrm{B}}T\varepsilon_2}\right]^{\frac{1}{2}} \exp\left[-\frac{\varepsilon_3 - \varepsilon_2}{k_{\mathrm{B}}T}\right] \tag{7.119}$$

where already the drift mobility is

$$\mu = \left(\frac{3}{2}\right) \frac{ea^2}{k_{\mathrm{B}}T} \left(\frac{J^2}{\hbar}\right) \left(\frac{\pi}{4k_{\mathrm{B}}T\varepsilon_2}\right)^{\frac{1}{2}} \exp\left(-\frac{\varepsilon_2}{k_{\mathrm{B}}T}\right)$$

so that

$$\frac{\mu_{\mathrm{H}}}{\mu} = \frac{2}{3\sqrt{3}} \frac{k_{\mathrm{B}}T}{J} \exp\left(\frac{2\varepsilon_2 - \varepsilon_3}{k_{\mathrm{B}}T}\right) \tag{7.120}$$

Here ε_2 and ε_3 are the minimum energy of lattice deformation needed for two-site and three-site energy coincidence, respectively. Generally, $2\varepsilon_2 > \varepsilon_3$, and μ_{H} can be substantially larger than μ_{D}. Also, it can be shown that $\varepsilon_3 - \varepsilon_2 = \varepsilon_2/3$, so that the activation energy for μ_{H} in (7.119) is one third of the activation energy for μ_{D}. In the particular model considered, $\mu_{\mathrm{H}}/\mu_{\mathrm{D}}$ is positive.

Friedman and Holstein obtained these results following a semi-intuitive approach. Other authors (Klinger,[140] Kurosawa,[141] Firsov,[142] Schnakenberg[143]) applied the linear response formalism to the same model and arrived at conclusions inconsistent with these. Holstein and Friedman[144] have subsequently presented a linear response treatment of their model, in which they rederive their earlier results, while Firsov[145] has more recently presented an alternate derivation of these same results.

The positive sign of μ_H in (7.119) results from considering a triangular lattice where all the nearest-neighbour transfer integrals J_{ij} are equal to $-J$, with J positive for electrons and negative for holes. For an anisotropic lattice the problem is more complex. For three-site jumps, the Hall mobility is given by (Munn and Siebrand[106,107])

$$\mu_H = \frac{3q|q|}{4\hbar k_B T} \mu W_T(i \to j \to k)$$

where a is the lattice spacing, q is the charge of the carrier, i, j and k label the sites of interest, and W_T is the thermal average of

$$W(i \to j \to k) = (4\pi^2 V_{kj} V_{ji} V_{ik}/\hbar)\delta(E_k - E_i)\delta(E_j - E_i)$$

Here E_i is the electron energy on site i, and the transfer matrix element V_{ij} between sites i and j is the product of an electronic transfer integral J_{ij} and a Franck–Condon factor. For an anisotropic lattice each J can have either sign, and it is not clear without detailed calculation which sign dominates when the effects of all the transitions are considered. Munn and Siebrand emphasize that either sign for μ_H/μ is possible both for linear electron–phonon coupling and for quadratic coupling.

The Hall effect in the small polaron adiabatic limit has been discussed by Emin and Holstein,[139] who show that μ_H is given by

$$\mu_H = \tfrac{1}{2} e a^2 \frac{\omega_0 J}{2\pi k_B T} f(T) \exp\left[-\frac{(\varepsilon_3/3 - J)}{k_B T}\right]$$

with $f(T) = 1$ for $J \ll k_B T$ and $f(T) < 1$ for $J \simeq k_B T$. This differs markedly from the result for non-adiabatic hopping; for some values of the parameters, the exponential coefficient can be quite small and the adiabatic μ_H can decrease with temperature. The absence of an experimental activation energy dependence for μ_H would therefore not in itself be proof that the transport is bandlike.

7.7 Small polarons in aromatic hydrocarbon solids

7.7.1 *Linear electron–phonon coupling*

We have already discussed the behaviour expected for charge transport phenomena when the small polaron limit applies with the usual linear electron–phonon coupling. We therefore confine the present discussion of linear coupling to the question of whether the values of the various coupling parameters for the aromatic hydrocarbons suggest strong intermolecular electronic coupling (as exemplified by the one-electron band picture) or weak coupling (as exemplified by the small-polaron picture).

Siebrand[146] considers the problem first for a two-molecule system. For strong coupling, the system is described by a single Born–Oppenheimer product

$$\psi_{\pm}^{(S)} = 2^{-\frac{1}{2}}[\psi(1) \pm \psi(2)]\chi(1)\chi(2)$$

where $\psi(i)$ denotes the wavefunction for the carrier localized on the molecule i, and χ is the vibrational wavefunction for the molecule without the excess charge. For weak coupling, on the other hand, the presence of the electron modifies the vibrational properties of the molecule on which it is localized and the overall wavefunction is

$$\psi_{\pm}^{(W)} = 2^{-\frac{1}{2}}[\psi(1)\chi'(1)\chi(2) \pm \psi(2)\chi(1)\chi'(2)]$$

with modified vibrational wavefunction χ'. In the case of intermediate coupling, each wavefunction is then approximated by a linear superposition of $\psi^{(S)}$ and $\psi^{(W)}$

$$\psi_{\pm} = \frac{\psi_{\pm}^{(S)} + \lambda\psi_{\pm}^{(W)}}{1 + \lambda^2 + 2\lambda\,\mathrm{Re}\,(\langle\psi_{\pm}^{(S)}|\psi_{\pm}^{(W)}\rangle)}$$

The coefficient λ can then be chosen to minimize the expectation value of the molecular Hamiltonian, and the variational energy of the system calculated as a function of the coupling parameters. For a fixed well depth E_b and for small values of $J/\hbar\omega_0$ the calculated dependence of the energy on $J/\hbar\omega_0$ is somewhat similar to that calculated for $\psi^{(W)}$ alone, and for large $J/\hbar\omega_0$ is somewhat similar to that for $\psi^{(S)}$ alone, with a smooth transformation from one characteristic behaviour to the other within the intermediate region. An analysis of the calculated energy dependence on coupling parameters leads Siebrand to suggest that thermal dissociation of small polarons in the intermediate-coupling regime might occur at high temperatures, thereby producing a more positive temperature dependence for the mobility than would otherwise follow. The application of these results to a linear chain of molecules has also been discussed (Siebrand[147]), but without the detailed variational calculations being carried out. The possible coexistence of small and large polaron states has also been considered by others (Toyozawa,[113] Sumi[148]), but the question has apparently not been studied in detail with regard to the aromatic hydrocarbons.

The question of which limit appropriately describes charge transport has been discussed by Siebrand[149] for a number of solids. For anthracene and naphthalene an estimate based on the value of the characteristic force constant k for C–C stretching frequencies and on the relation between bond lengths $\{x_i\}$ and bond orders (Coulson[150]) gives an estimated relaxation energy

$$E_r = -\tfrac{1}{2}k \sum_i (\Delta x_i)^2$$

$$\gtrsim 0{\cdot}1 \text{ eV}$$

for the molecule with the excess electron. This is not negligible compared with the bandwidth, in contrast to the estimate of Friedman.[70] Furthermore, the estimated E_r reduces the effective bandwidth by the factor

$$\exp - \frac{|E_r|}{\hbar\omega_0} = \exp(-S)$$

with $S \gtrsim 1$, and a reduction by a factor of 2 to 10 is thereby suggested. Siebrand suggests that mean free paths estimated from measured mobilities and calculated bandstructures might then be increased to values substantially larger than a lattice distance.

7.7.2 *Quadratic coupling to intramolecular modes*

Detailed theoretical treatments of the small-polaron have always assumed an electron–phonon interaction linear in the vibrational coordinates. Munn and Siebrand,[38,151] using a simplified model, have discussed some possible effects arising from the quadratic part of the coupling considered alone. Their theory includes a possible effect of deuteration on the excess carrier mobility; such an effect has been observed (Munn *et al.*,[95] Morel and Hermann,[96] Mey *et al.*[97]) and so it is of particular interest to examine the theory they present.

Munn and Siebrand present at least two semiqualitative arguments for the dominance of the quadratic contribution to the electron–phonon interaction. First of all, the quadratic coupling corresponds to a change in frequency. Such changes in fact occur for the vibrational modes of the benzene molecule when an electron is excited from a bonding to an antibonding π-orbital, and the out-of-plane-bending frequencies, in particular, may change by as much as 30% (Table 7.1). Intuitive considerations suggest that the frequency change when an excess electron is placed in an antibonding π-orbital might be only somewhat smaller than this and that the effect of adding an excess electron to a larger molecule such as anthracene could also be substantial. On the other hand, if the symmetry of the benzene molecular ion is the same as that of the neutral-ground-state molecule, then the linear coupling vanishes identically for the intra-molecular modes that exhibit the strongest quadratic interaction.

Secondly, Munn and Siebrand discuss the nature of the restrictions imposed by energy conservation for linear and for quadratic coupling. When a transfer occurs from site 1 to site 2, the initial and final vibrational energies are

$$n_{\mathrm{i}}^{(1)}\hbar\omega' + n_{\mathrm{i}}^{(2)}\hbar\omega \qquad \text{(initial)}$$

$$n_{\mathrm{f}}^{(1)}\hbar\omega + n_{\mathrm{f}}^{(2)}\hbar\omega' \qquad \text{(final)}$$

where $n_{\mathrm{i}}^{(j)}$ and $n_{\mathrm{f}}^{(j)}$ are respectively the initial and final vibrational quantum numbers for site j, and ω' and ω are respectively the molecular frequencies with and without the excess carrier. For linear coupling alone, $\omega' = \omega$, and energy conservation requires only that the change in phonon number for one molecule be the negative of the change for the other. For quadratic coupling considered alone, strict energy conservation would impose the more restrictive condition that the phonon numbers actually interchange, except for some special values of the frequencies. The more stringent requirements of energy conservation suggest that the quadratic interaction might be more effective in limiting transport. In particular, Munn and Siebrand suggest the possibility of a 'slow-phonon' limit (Rashba[152]) in which the intermolecular vibrational coupling between molecules is small, and the rate at which vibrational energy can transfer between molecules effectively limits transport. These considerations would apply also to the formally similar problem of triplet–exciton diffusion (Munn and Siebrand[151]).

The molecular crystal model assumed by Munn and Siebrand is similar to that of Holstein, with the essential difference that the linear electron–phonon interaction is replaced by one quadratic in the molecular vibrational coordinates. The interactions included are then:

(1) the vibrational coupling between neighbouring molecules

$$\sum_r \tfrac{1}{4}m\omega_1^2(x_{r+1} + x_{r-1})x_r$$

(2) the electronic intermolecular coupling, with transfer integrals equal to a constant $-J$ between nearest neighbours and zero between non-neighbouring molecules,

(3) the electron–phonon interaction, given by

$$-\tfrac{1}{2}m\omega_1^2 x_n^2$$

for the coupling to the nth molecule.

The wavefunction for the system is then

$$\Psi = \sum_n \psi(r - nd)a_n(x_1, \ldots, x_N) \tag{7.121}$$

(with d the intermolecular separation) and the equations of motion for the amplitudes $\{a_n\}$ become

$$i\hbar\frac{\partial a_n}{\partial t} = \left[\sum_r \left(-\frac{\hbar^2}{2m}\right)\frac{\partial^2}{\partial x_r^2} + \tfrac{1}{2}m\omega_0^2 x_r^2 + \tfrac{1}{4}m\omega_1^2 x_r(x_{r+1} + x_{r-1}) - \tfrac{1}{2}m\omega_2^2 x_n^2\right]a_n - J(a_{n+1} + a_{n-1}) \tag{7.122}$$

Munn and Siebrand then consider the behaviour of the solutions in a number of limiting cases for the relative values of the coupling parameters:

(1) (a) Slow-electron hopping,
 (b) Slow-electron coherent transport,

(2) (a) Slow-phonon hopping,
 (b) Slow-phonon coherent transport,

(3) Quasifree electron transport.

In the slow-electron and slow-phonon limits, the quadratic electron–phonon coupling is assumed to be more substantial than the intermolecular electronic and intermolecular vibrational couplings. The slow-phonon and slow-electron limits are then distinguished from each other according to whether the intermolecular vibrational or electronic coupling is weaker. In the quasifree electron limit, the intermolecular electronic coupling is the most substantial of the three interactions.

We describe each of these limits in turn.

7.7.2.1 *The slow-electron limit*

In this case the electron–phonon coupling is assumed to be the most substantial of the three interactions, followed in decreasing order by the intermolecular vibrational coupling and the intermolecular electronic coupling. Munn and Siebrand discuss two cases within this limit, the first in which the transport is by a series of hops between sites and the second in which the transport is band-like. We describe their treatment of the hopping transport first.

The zero-order solutions to (7.122) are obtained by setting $\hbar\omega_1 = J = 0$, so that the solutions for the $\{a_n\}$ are each the product of simple harmonic oscillator wavefunctions

$$|a_n^{(0)}(\{v_r\})| = \chi(v_n) \prod_{r \neq n} \chi_0(v_r)$$

Here v_r is the vibrational quantum number for site r, χ_0 is the vibrational wavefunction for the unperturbed molecule and χ is the vibrational wavefunction for the molecule on which the excess electron is localized. The overall wavefunction in (7.121) is then the vibronic wavefunction for the system with a carrier localized on a single site.

In the next order of approximation, the intermolecular electronic coupling causes transfer of the carrier to neighbouring molecules. In Holstein's treatment, the equation governing the time dependence of the $\{a_n\}$ is expressed in terms of phonon coordinates of the crystal. The transfer matrix elements are then the product of a purely electronic transfer integral and a many mode vibrational overlap between the initial states of the crystal for a nearest-neighbour carrier hop (equation (7.96)). The transition rate is developed in terms of the resonance transfer function of (7.107), and the contribution of the diagonal transitions is subtracted off. Because of the variety of transitions that can occur involving different changes in phonon number for the various crystal modes, the resultant transfer rate between neighbouring sites is then constant in time. Munn and Siebrand follow a more heuristic approach, on the other hand. They argue that in the slow-electron limit the weak electronic intermolecular coupling causes the transitions to neighbouring sites, while the stronger intermolecular vibrational coupling provides a final density of states, so that the transition develops uniformly in time. The transfer rate is then given by

$$W(i \rightarrow j) = \frac{2\pi}{\hbar}|H_{ij}|^2 \rho(E_j) \tag{7.123}$$

For the transitions induced by the transfer integral J, the matrix elements are found from equation (7.123) to be

$$|H_{n,n+1}|^2 = J\langle \chi(v_n)|\chi_0(v_n')\rangle\langle \chi_0(v_{n+1})|\chi(v_{n+1}')\rangle \times \prod_r \delta(v_r, v_r') \tag{7.124}$$

When the vibrational wavefunctions are altered through a change in the vibrational frequency, their orthonormality with respect to the unmodified wavefunctions is not strongly affected for reasonable values of the parameters, so in (7.124)

$$S(v_n, v_n') \equiv \langle \chi(v_n)|\chi_0(v_n')\rangle \simeq \delta(v_n, v_n') \tag{7.125}$$

At the same time, energy conservation requires that the vibrational occupation numbers of the two molecules interchange. The combined effect of these two restrictions is to allow transfers only between molecules with the same initial vibrational quantum numbers. Since the density of final states in a one-dimensional crystal exhibits van Hove singularities that would not occur as singularities in a three-dimensional system, an average density of states is used in (7.123). Furthermore, the average density of states is not calculated *a priori*, but is constructed heuristically as the average number of phonons excited divided by the phonon bandwidth. The resultant diffusion coefficient is

$$D_{\text{he}} = 2d^3 W_{\text{T}}(n \to n+1) = \frac{4\pi d^2 J^2 \omega_0}{\omega_1^2} \frac{\xi}{1+\xi} \tag{7.126}$$

where

$$\xi = \exp(-\beta\hbar\omega_0)$$

with

$$\beta = (k_{\text{B}}T)^{-1}$$

Both the diffusion coefficient (7.126), and the corresponding mobility (given by $eD/k_{\text{B}}T$) are found to approach zero at zero temperature. Also, the electron mobility can have a maximum at some temperature. We note that these features of the temperature dependence appear to derive entirely from the assumed density of states. In the approximation (7.124), the matrix element H_{ij} in (7.123) would be largest when all initial and final vibrational numbers are the same, and this occurs at $T = 0$ where the density of states, and therefore the hopping rate, vanish. Furthermore, the thermal average of $|H_{ij}|^2$ is proportional to

$$\langle S^4 \rangle_{\text{T}} = \tanh\left(\tfrac{1}{2}\beta\hbar\omega_0\right)$$

(Munn and Siebrand[38] equation (32)) which is a monotonically increasing function of β and therefore a monotonically decreasing function of temperature.

In addition to the slow electron hopping, Munn and Siebrand consider a slow-electron coherent-transport limit in which the velocity of the carrier is

$$w_\lambda = \frac{d}{\hbar}\frac{\partial E}{\partial q_\lambda}$$

where E is the band energy of the electron for wavenumber q_λ. Using this velocity, a mean free time can be defined in terms how far the electron travels before encountering a site with different vibrational number. The resultant 'coherent electron' diffusion coefficient

$$D_{\text{ce}} = \langle \tau_\lambda w_\lambda^2 \rangle = \frac{4Jd^2}{\pi\hbar} \frac{(1-\xi)^2(1+\xi)}{\xi} \tag{7.127}$$

decreases rapidly with increasing temperature, and it is argued that hopping transport should dominate at high temperature and coherent transport at low temperature, with only a small intermediate temperature range. Munn and Siebrand point out, however, that when the coupling is only to a single mode, the transport is coherent up to about $T \simeq \hbar\omega_0/k_B$, which is typically comparable to the melting point of the crystal. They therefore suggest that couplings to additional modes might restrict the transport to incoherent hopping and, in particular, that scattering from acoustic lattice modes might contribute to this effect.

7.7.2.2 *The slow-phonon limit*

In this case the electron–phonon coupling is again assumed dominant, and is followed in decreasing order by the intermolecular electronic and the intermolecular vibrational couplings. The treatment is based again on the golden rule formula (7.123), but it is assumed that the intermolecular electronic coupling produces the density of states, while the relatively weak intermolecular vibrational coupling determines the rate of transfer.

Localized basis states are constructed again by ignoring the J and ω_1 terms of (7.122) in zeroth order. In the next order of approximation, the intermolecular electronic coupling terms might be included and stationary solutions of (7.122) found by first letting $a_n = A_n \exp(iEt/\hbar)$ in (7.122), and then expressing the resultant equation in terms of Bloch sums over the A_n. The resultant band-like solutions would have energy

$$\begin{aligned} E &= -J[e^{i\lambda}S(v_n, v_n)S(v_{n+1}, v_{n+1}) + e^{-i\lambda}S(v_n, v_n)S(v_{n-1}, v_{n-1})] \\ &= -2J\cos\lambda \end{aligned}$$

for wavenumber λ, when the approximation (7.125) is invoked. The thermally averaged bandwidth would then be essentially temperature independent. Munn and Siebrand therefore take the density of final states as proportional to the reciprocal of the electronic bandwidth. When matrix elements of $m\omega_1^2 x_i(x_{i+1} + x_{i-1})$ are taken between the zero-order wavefunctions localized on different sites, the occurrence of such factors as $\langle\chi_0(v_n)|x_n|\chi(v_n')\rangle$ effectively allows only transitions in which the vibrational quantum number of each molecule changes by 1. Also, energy conservation requires that the vibrational quantum numbers of the two molecules interchange. The overall restriction then allows only transitions between those molecules whose vibrational quantum numbers differ by 1 initially. The resultant diffusion coefficient

$$D_{hp} = \frac{\pi\hbar^2 d^2}{16\hbar J\omega_0^2} \frac{\xi(1 + \xi^2)}{(1 - \xi)(1 + \xi)^3}$$

is zero for $T = 0$, when all vibrational quantum numbers are the same, and remains finite as the temperature increases.

For the slow-phonon limit, Munn and Siebrand also consider a coherent transport case. Their treatment is analogous to that for coherent transport in the slow-electron limit, but with the phonon velocity used instead of the electron velocity in calculating the mean free time. The resultant diffusion coefficient is

$$D_{\mathrm{cp}} = \frac{\hbar\omega_1^2 d^2}{\pi\hbar\omega_0}\frac{(1-\xi)^2(1+\xi)}{\xi}$$

and the temperature-dependent part of D diverges rapidly as the temperature approaches zero.

7.7.2.3 *Quasifree electron limit*

Munn and Siebrand note that the electron bandwidth might be comparable to the vibrational energy change produced by quadratic coupling to an excess electron. They therefore consider also the extreme case in which the intermolecular electronic coupling dominates, followed in decreasing order by the electron–phonon and intermolecular vibrational couplings. The zero-order solutions obtained by setting $\omega_1 = \omega_2 = 0$ in (7.122) are Bloch functions. The electron–phonon coupling is assumed to mix states of different wavenumber into new states, each new state labeled by a wavenumber λ. The transition matrix elements of the intermolecular vibrational coupling between these states can then be calculated. It is argued that the density of final states should increase with the root-mean-square deviation of the vibrational quantum number from its average value

$$\Delta = [\langle v^2\rangle_{\mathrm{T}} - \langle v\rangle_{\mathrm{T}}^2]^{\frac{1}{2}}$$

and $\rho(E_{\mathrm{f}})$ is taken heuristically to be proportional to Δ. The thermally averaged total rate of transition out of the state of wavenumber λ can then be calculated. Since the reciprocal of the total transition rate is the relaxation time, this procedure leads to a calculation of the mobility. The mobility is then found to decrease by several orders of magnitude over a temperature range that is a fraction of $\hbar\omega_0/k_{\mathrm{B}}$, and the results do not apply directly to anthracene.

7.7.3 *Application to the aromatic hydrocarbons*

We complete our outline of Munn and Siebrand's theory by summarizing some aspects of their discussion of real systems in terms of the one-dimensional model.

Munn and Siebrand suggest that the transport in different directions of the crystal might correspond to different limiting cases considered for the one-dimensional model. By assuming reasonable values of $\hbar\omega_0$, $\hbar\omega_1$ and $\hbar\omega_2$, it is possible to plot the calculated mobility as a function of J for each limit considered for the model crystal. The results may be compared with experimental data. The open circles in Figure 7.10 are obtained from experimental data combined with Glaeser and Berry's[13] estimate of J. The observed mobilities tend to lie in regions intermediate between those for which the various limits apply. Based on the results shown in Figure 7.10, and on comparison with observed temperature dependences and other data, Munn and Siebrand suggest that the c' electron transport is intermediate between that described by the slow-electron and slow-phonon limits, while the other diagonal components lie either between the slow-phonon hopping and coherent transport limits, or else correspond to slow-phonon coherent transport.

Next Munn and Siebrand's results may be compared with experiment. The theory is notably successful in two respects. If it is assumed that excess electron transport in the c' direction of anthracene falls somewhere between the slow-phonon and slow-electron hopping limits, the calculated temperature dependence agrees qualitatively with experiment. Secondly, Munn and Siebrand's treatment indicates a possible deuterium effect, whose existence has recently been confirmed.

Deuteration increases the vibrational mass of a molecule and therefore decreases the molecular frequencies of those modes that are affected. Taking the ratio of vibrational frequencies for the normal and deuterated molecule to be

$$r = \frac{\omega_0^{\mathrm{H}}}{\omega_0^{\mathrm{D}}} > 1$$

and assuming that the corresponding ratio of vibrational coupling constants is

$$\frac{\omega_1^{\mathrm{H}}}{\omega_1^{\mathrm{D}}} = r$$

it is found from the expressions for the diffusion coefficient in each of the limits considered that

$$D_{\mathrm{ce}}^{\mathrm{H}} = D_{\mathrm{ce}}^{\mathrm{D}} \qquad \text{(coherent slow electron)} \qquad (7.128)$$

$$D_{\mathrm{he}}^{\mathrm{H}} = r^{-1} D_{\mathrm{he}}^{\mathrm{D}} \qquad \text{(slow-electron hopping)} \qquad (7.129)$$

$$D_{\mathrm{hp}}^{\mathrm{H}} = r^{2} D_{\mathrm{hp}}^{\mathrm{D}} \qquad \text{(slow-phonon hopping)} \qquad (7.130)$$

$$D_{\mathrm{cp}}^{\mathrm{H}} = r D_{\mathrm{cp}}^{\mathrm{D}} \qquad \text{(coherent slow phonon)} \qquad (7.131)$$

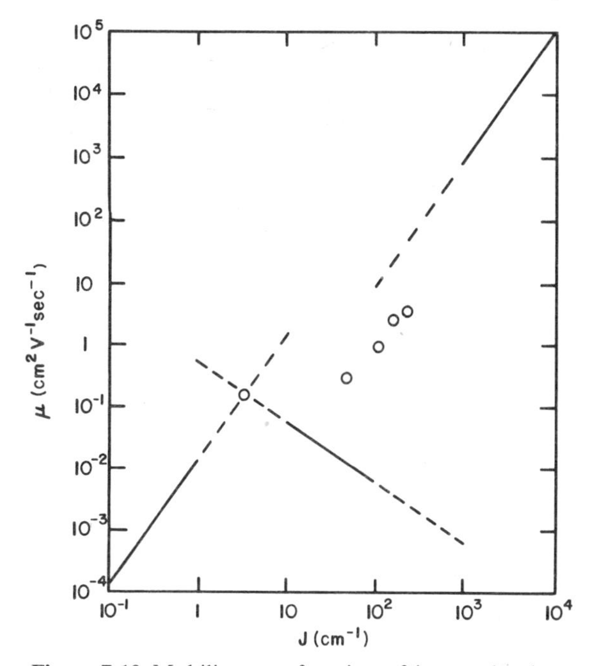

Figure 7.10 Mobility as a function of intermolecular electronic coupling J. The solid lines represent, in order of increasing J, the slow-electron hopping, slow-phonon hopping and slow-phonon coherent limits considered by Munn and Siebrand. The broken lines are extrapolations from the regions of applicability of the various limits. The open circles are deduced from experimental data. (Munn and Siebrand;[138] reproduced by permission of the American Institute of Physics)

Here the right-hand and left-hand sides of each equation are evaluated at different temperatures such that

$$\frac{\hbar\omega^{\mathrm{H}}}{k_{\mathrm{B}}T_{\mathrm{left}}} = \frac{\hbar\omega^{\mathrm{D}}}{k_{\mathrm{B}}T_{\mathrm{right}}}$$

The expression for D_{he} in (7.129) indicates an increase in mobility upon deuteration. We note that the r^{-1} part of the deuterium effect in the slow-electron limit arises from the assumed form of the density of final states, which introduces a factor $\zeta = 4J/(\hbar\omega_1^2/\omega_0)$ into D_{he}, where ζ is the ratio of free-electron to free-phonon bandwidths. In this limit, deuteration decreases the vibrational frequencies, thereby increasing the number of phonons excited at a given temperature; this in turn leads to an increased mobility, once it is accepted that the final density of states is proportional to the number of phonons excited, and that (by virtue of (7.125)) the matrix elements H_{ij} are not strongly modified by deuteration. For $r = 1{\cdot}15$ and $T = 300°K$, Munn and Siebrand estimate a 36% increase in $\mu_{c'c'}$ for anthracene in the slow-electron hopping limit (7.129), and a decrease by 10% for the slow-phonon hopping limit.

Experimental data were originally interpreted to indicate a 10% increase in $\mu_{c'c'}$ for electrons in deuterated as compared with normal anthracene. Munn and Siebrand suggest that such an effect can be explained if the transport in this case corresponds to either the slow-electron limit, or to behaviour intermediate between the slow-electron and slow-phonon limits. More recently, Mey *et al.*[97] have measured the mobilities of electrons and holes in both normal and perdeuterated anthracene. They conclude that both their own and earlier measurements indicate a 10% decrease in $\mu_{c'c'}$ for electrons upon deuteration, while the other mobility components appear to show no effect. They suggest that these results might be explained by assuming the slow-phonon limit for the c' transport of electrons.

In concluding this brief summary of some of the main points of Munn and Siebrand's theory, we make two comments. First, on the basis of theoretical considerations (Gosar and Choi,[75] Friedman[60]), other electron–phonon coupling mechanisms would appear to be substantial enough to explain the magnitude of most of the mobility components. It has not been clearly demonstrated how these couplings can be combined with the intramolecular quadratic interaction in such a way that the features resulting from the quadratic coupling persist, particularly in the case of the 'slow-electron' limit (see, however, Munn[153]). Secondly, as the authors point out, their theory assumes heuristic expressions for such quantities as the various densities of final states. We have emphasized that the predicted temperature dependence and deuterium effect, at least in the slow-electron limit, arise directly from the assumed behaviour of the density of final states. A treatment of the quadratic interaction that examines these features in greater detail would be of considerable value.

Acknowledgements

I am grateful to D. Konowalow and M. Eibert for a number of suggestions and to R. S. Knox for his comments on the manuscript.

7.8 References

1. M. Pope and H. Kallmann, *Disc. Faraday Soc.*, **51**, 7 (1971).
2. F. Gutmann and L. E. Lyons, *Organic Semiconductors*, Wiley, New York, 1967.

3. O. H. LeBlanc, Jr., *Physics and Chemistry of the Organic State*, Vol. 3, Ch. 3 (Eds. D. Fox, M. M. Labes and A. Weissberger), Wiley–Interscience, New York, 1967.
4. J. H. Sharp and M. Smith, *Organic Semiconductors. Physical Chemistry, an Advanced Treatise. Solid State Chemistry*, Vol. 10, Ch. 8, 1969.
5. V. Ern, *Phys. Rev. Letters*, **22**, 243 (1969).
6. P. Avakian and R. E. Merrifield, *Molec. Cryst.*, **5**, 37 (1968).
7. L. E. Lyons, *J. Chem. Soc.*, 5001 (1957).
8. M. Pope, J. Burgos and J. Giachino, *J. Chem. Phys.*, **43**, 3367 (1965).
9. G. Castro and J. F. Hornig, *J. Chem. Phys.*, **42**, 1459 (1965).
10. R. F. Chaiken and D. R. Kearns, *J. Chem. Phys.*, **45**, 3966 (1966).
11. N. E. Geacintov and M. Pope, *J. Chem. Phys.*, **45**, 3884 (1966).
12. J. Dresner, *Molec. Cryst. Liq. Cryst.*, **11**, 305 (1970).
13. R. M. Glaeser and R. S. Berry, *J. Chem. Phys.*, **44**, 3797 (1966).
14. M. K. Grover and R. Silbey, *J. Chem. Phys.*, **52**, 2099 (1970).
15. M. Born and R. Oppenheimer, *Ann. Physik*, **84**, 457 (1927).
16. C. V. Chester, *Adv. Phys.*, **10**, 357 (1961).
17. M. Born and K. Huang, *Dynamical Theory of Crystal Lattices*, Oxford University Press, 1954.
18. L. J. Sham and J. M. Ziman, *Solid State Phys.*, **15**, 221 (1963).
19. H. A. Jahn and E. Teller, *Proc. Roy. Soc.*, **A 161**, 220 (1937).
20. A. Nitzan and J. Jortner, *J. Chem. Phys.*, **56**, 3360 (1972).
21. T. Holstein, *Ann. Phys.*, **8**, 343 (1959).
22. E. B. Wilson, J. C. Decius and P. C. Cross, *Molecular Vibrations*, McGraw-Hill, New York, 1955.
23. A. A. Maradudin, E. W. Montroll and G. H. Weiss, *Theory of Lattice Dynamics in the Harmonic Approximation*, 2nd edn., *Solid State Physics*, Suppl. 3, Academic Press, New York, 1971.
24. P. Debye, *Ann. Phys.*, **39**, 789 (1912).
25. M. Blackman, *Proc. Roy. Soc.*, **A 148**, 365 (1935).
26. M. Blackman, *Proc. Camb. Phil. Soc.*, **33**, 94 (1937).
27. O. Schnepp and A. Ron, *Disc. Faraday Soc.*, **48**, 26 (1969).
28. G. K. Horton, *Amer. J. Phys.*, **36**, 93 (1968).
29. W. Cochran and G. S. Pawley, *Proc. Roy. Soc.*, **A 280**, 1 (1964).
30. G. S. Pawley, *Phys. Stat. Sol.*, **20**, 347 (1967).
31. H. Bonadeo and G. Taddei, *J. Chem. Phys.*, **58**, 979 (1973).
32. G. Taddei, H. Bonadeo, M. P. Marzocchi and S. Califano, *J. Chem. Phys.*, **58**, 966 (1973).
33. I. Harada and T. Shimanouchi, *J. Chem. Phys.*, **44**, 2016 (1966).
34. A. I. Kitaigorodskii, *J. Chem. Phys.*, **63**, 6 (1966).
35. E. Sàndor, *Acta Cryst.*, **15**, 463 (1962).
36. G. S. Pawley, *Acta Cryst.*, **A 27**, 80 (1971).
37. G. S. Pawley and S. J. Cyvin, *J. Chem. Phys.*, **52**, 4073 (1970).
38. R. W. Munn and W. Siebrand, *J. Chem. Phys.*, **52**, 6391 (1970).
39. M. V. Klein, *Physics of Color Centers*, p. 429 (Ed. W. B. Fowler), Academic Press, New York, 429, 1968.
40. R. S. Knox and K. J. Teegarden, *Physics of Color Centers*, Chapter 1 (Ed. W. B. Fowler), Academic Press, New York, 1968.
41. J. M. Ziman, *Principles of the Theory of Solids*, Cambridge University Press, 1964.

42. H. Jones and C. Zener, *Proc. Roy. Soc.*, **A 144**, 101 (1934).
43. G. H. Wannier, *Phys. Rev.*, **52**, 191 (1937).
44. J. C. Slater, *Phys. Rev.*, **76**, 1592 (1949).
45. S. Pekar, *J. Phys. USSR*, **10**, 431 (1946).
46. L. Friedman, *Phys. Rev.*, **135**, A 233 (1964).
47. J. L. Katz, S. A. Rice, S. I. Choi and J. Jortner, *J. Chem. Phys.*, **39**, 1683 (1963).
48. S. H. Glarum, *J. Phys. Chem. Solids*, **24**, 1577 (1963).
49. Y. Toyozawa, *Progr. Theor. Phys.*, **12**, 241 (1954).
50. W. Kohn, *Proc. Intern. Conf. Semiconductor Physics, Prague 1960*, p. 17, Czechoslovakian Academy of Sciences, Prague, 1961.
51. R. Silbey, J. Jortner, S. A. Rice and M. T. Vala, *J. Chem. Phys.*, **42**, 733 (1965); erratum *J. Chem. Phys.*, **43**, 2925 (1965).
52. O. H. LeBlanc, Jr., *J. Chem. Phys.*, **35**, 1275 (1961).
53. G. D. Thaxton, R. C. Jarnagin and M. Silver, *J. Phys. Chem.*, **66**, 2461 (1962).
54. P. Balk, S. deBruijn and G. J. Hoijtink, *Rec. Trav. Chim.*, **76**, 860 (1957).
55. M. Goeppert-Meyer and A. L. Sklar, *J. Chem. Phys.*, **6**, 648 (1938).
56. J. L. Katz, S. A. Rice, S. I. Choi and J. Jortner, *J. Chem. Phys.*, **39**, 1683 (1963).
57. W. E. Spear, *Solid State Comm.*, **9**, 65 (1971).
58. J. M. Ziman, *Proc. Camb. Phil. Soc.*, **51**, 707 (1955).
59. J. Bardeen and W. Shockley, *Phys. Rev.*, **80**, 72 (1950).
60. L. Friedman, *Phys. Rev.*, **140**, A 1649 (1965).
61. C. Herring, *Proc. Intern. Conf. Semiconductor Physics, Prague 1960*, p. 60, Czechoslovakian Academy of Sciences, Prague, 1961.
62. E. Blount, *Phys. Rev.*, **114**, 418 (1959).
63. G. Whitfield, *Phys. Rev.*, **121**, 720 (1961).
64. J. M. Ziman, *Electrons and Phonons*, Oxford University Press, 1960.
65. A. H. Wilson, *The Theory of Metals*, 2nd edn., Cambridge University Press, 1953.
66. W. Kohn and J. M. Luttinger, *Phys. Rev.*, **108**, 590 (1957).
67. R. Kubo and K. Tomita, *J. Phys. Soc. Japan*, **9**, 888 (1954).
68. R. Kubo, *J. Phys. Soc. Japan*, **12**, 570 (1957).
69. F. J. Blatt, *Solid State Physics*, Vol. 4, p. 199 (Eds. F. Seitz and D. Turnbull), Academic Press, New York, 1957.
70. L. Friedman, *Phys. Rev.*, **133**, A 1668 (1964).
71. R. A. Coldwell-Horsfall and D. ter Haar, *Physica*, **24**, 848 (1958).
72. A. C. Beer, *Galvanomagnetic Effects in Semiconductors*, *Solid State Physics*, Suppl. 5 (Eds. F. Seitz and D. Turnbull), Academic Press, New York, 1963.
73. O. H. LeBlanc, Jr., *J. Chem. Phys.*, **39**, 2395 (1963).
74. D. ter Haar, *Many-Body Problems*, p. 33 (Ed. S. F. Edwards), W. A. Benjamin, Inc., New York, 1969, reprinted from *Rep. Prog. Phys.*
75. P. Gosar and S. Choi, *Phys. Rev.*, **150**, 529 (1966).
76. W. B. Fowler, *Phys. Rev.*, **132**, 1591 (1963).
77. M. H. Reilly, *J. Phys. Chem. Solids*, **28**, 2067 (1967).
78. R. S. Knox and F. Bassani, *Phys. Rev.*, **124**, 652 (1961).
79. L. F. Mattheiss, *Phys. Rev.*, **133**, A 1399 (1964).
80. R. S. Knox, *Theory of Excitons*, *Solid State Physics*, Suppl. 5 (Eds. F. Seitz and D. Turnbull), Academic Press, New York, 1963.
81. L. S. Miller, O. Howe and W. E. Spear, *Phys. Rev.*, **166**, 871 (1968).
82. H. D. Pruett and H. P. Broida, *Phys. Rev.*, **164**, 1138 (1967).
83. P. Gosar and I. Vilfan, *Molec. Phys.*, **18**, 49 (1970).

84. O. H. LeBlanc, Jr., *J. Chem. Phys.*, **33**, 626 (1960).
85. O. H. LeBlanc, Jr., *J. Chem. Phys.*, **37**, 916 (1962).
86. R. G. Kepler, *Phys. Rev.*, **119**, 1226 (1960).
87. R. G. Kepler, *Organic Semiconductors* (Eds. J. J. Brophy and J. W. Buttrey), Macmillan, New York, 1962.
88. A. Many, S. Z. Weisz and M. Simhony, *Phys. Rev.*, **126**, 1989 (1962).
89. W. Helfrich and P. Mark, *Z. Phys.*, **166**, 370 (1962).
90. W. Helfrich, *Physics and Chemistry of the Organic Solid State*, Ch. 1 (Eds. D. Fox, M. M. Labes and A. Weissberger), Interscience, New York, 1967.
91. I. Nakada and Y. Ishihara, *J. Phys. Soc. Japan*, **19**, 695 (1964).
92. R. Raman and S. P. McGlynn, *J. Chem. Phys.*, **40**, 515 (1964).
93. J. Fourny and G. Delacôte, *J. Chem. Phys.*, **50**, 1028 (1969).
94. W. Mey and A. M. Hermann, *Phys. Rev.*, **B 7**, 1652 (1973).
95. R. W. Munn, J. R. Nicholson, W. Siebrand and D. F. Williams, *J. Chem. Phys.*, **52**, 6442 (1970).
96. D. L. Morel and A. M. Hermann, *Phys. Letters*, **36A**, 101 (1971).
97. W. Mey, T. J. Sonnonstine, D. L. Morel and A. M. Hermann, *J. Chem. Phys.*, **58**, 2542 (1973).
98. D. H. Spielberg, A. I. Korn and A. C. Damask, *Phys. Rev.*, **B3**, 2012 (1971).
99. A. I. Korn, R. A. Arndt and A. C. Damask, *Phys. Rev.*, **186**, 938 (1969).
100. M. Schadt and D. F. Williams, *Phys. Stat. Sol.*, **39**, 223 (1970).
101. G. C. Smith, *Phys. Rev.*, **185**, 1133 (1969).
102. M. Pope and J. Burgos, *Molec. Cryst.*, **1**, 395 (1966).
103. J. Dresner, *Phys. Rev. Letters*, **21**, 356 (1968).
104. G. Delacôte, *Molec. Cryst.*, **5**, 309 (1969).
105. H. Chojnacki, *Molec. Cryst.*, **3**, 275 (1968).
106. R. W. Munn and W. Siebrand, *J. Chem. Phys.*, **53**, 3343 (1970).
107. R. W. Munn and W. Siebrand, *Phys. Rev.*, **B2**, 3435 (1970).
108. H. Fröhlich and G. L. Sewell, *Proc. Phys. Soc.*, **74**, 643 (1959).
109. A. F. Joffé, *Canad. J. Phys.*, **34**, 1393 (1956).
110. M. J. Klinger, *Proc. Intern. Conf. Physics of Semiconductors* (1962).
111. D. Emin, *Phys. Rev. Letters*, **25**, 1751 (1970).
112. R. P. Feynmann, *Phys. Rev.*, **97**, 660 (1955).
113. Y. Toyozawa, *Progr. Theor. Phys.*, **26**, 29 (1961).
114. Y. Toyozawa, *J. Appl. Phys. Suppl.*, **33**, 340 (1962).
115. S. D. Druger and R. S. Knox, *J. Chem. Phys.*, **50**, 3143 (1969).
116. K. S. Song, *Canad. J. Phys.*, **49**, 26 (1971).
117. T. G. Castner and W. Kanzig, *J. Phys. Chem. Solids*, **3**, 178 (1957).
118. D. D. Eley and D. L. Spivey, *Trans. Faraday Soc.*, **56**, 1432 (1960).
119. R. C. Nelson, *J. Chem. Phys.*, **30**, 406 (1959).
120. R. A. Keller and H. E. Rast, Jr., *J. Chem. Phys.*, **36**, 2640 (1962).
121. R. A. Keller, *J. Chem. Phys.*, **38**, 1076 (1963).
122. R. H. Tredgold, *Proc. Phys. Soc.*, **80**, 807 (1962).
123. D. R. Kearns, *J. Chem. Phys.*, **35**, 2269 (1961).
124. H. Fröhlich, *Adv. Phys.*, **3**, 325 (1954).
125. T. Holstein, *Ann. Phys.*, **8**, 325 (1959).
126. J. Yamashita and T. Kurosawa, *J. Phys. Chem. Solids*, **5**, 34 (1958).
127. J. Yamashita and T. Kurosawa, *J. Phys. Soc. Japan*, **15**, 802 (1960).
128. G. L. Sewell, *Phil. Mag.*, **3**, 1361 (1958).
129. D. J. Gibbons and W. E. Spear, *J. Phys. Chem. Solids*, **27**, 1917 (1966).

130. J. Appel, *Solid State Physics*, **21**, 193 (Eds. F. Seitz and D. Turnbull), Academic Press, New York, 1968.
131. I. G. Austin and N. F. Mott, *Adv. Phys.*, **18**, 41 (1969).
132. K. Dettmann and W. Ludwig, *Phys. Stat. Sol.*, **10**, 689 (1965).
133. R. W. Munn, *J. Chem. Phys.*, **52**, 64 (1970).
134. S. V. Tiablikov, *Zh. Eksp. Teor. Fiz.*, **23**, 381 (1952).
135. S. J. Nettel, *Phys. Rev.*, **121**, 425 (1961).
136. K. Huang and A. Rhys, *Proc. Roy. Soc.*, **A 204**, 406 (1950).
137. D. Emin, *Phys. Rev.*, **B3**, 1321 (1971).
138. L. Friedman and T. Holstein, *Ann. Phys.*, **21**, 494 (1963).
139. D. Emin and T. Holstein, *Ann. Phys.*, **53**, 439 (1969).
140. M. I. Klinger, *Phys. Stat. Sol.*, **11**, 499 (1965).
141. T. Kurosawa, *Progr. Theor. Phys.*, **29**, 159 (1963).
142. Yu. A. Firsov, *Fiz. Tverd. Tela*, **5**, 2149 (1963); trans. *Sovt. Phys. Solid State*, **5**, 1566 (1964).
143. J. Schnakenberg, *Z. Phys.*, **208**, 165 (1968).
144. T. Holstein and L. Friedman, *Phys. Rev.*, **165**, 1019 (1968).
145. Yu. A. Firsov, *Fiz. Tverd. Tela*, **10**, 3027 (1968); trans. *Sovt. Phys. Solid State*, **10**, 2387 (1969).
146. W. Siebrand, *J. Chem. Phys.*, **40**, 2223 (1964).
147. W. Siebrand, *J. Chem. Phys.*, **40**, 2231 (1964).
148. H. Sumi, *J. Phys. Soc. Japan*, **33**, 327 (1972).
149. W. Siebrand, *J. Chem. Phys.*, **41**, 3574 (1964).
150. C. A. Coulson, *Proc. Roy. Soc.*, **A 169**, 413 (1939).
151. R. W. Munn and W. Siebrand, *J. Chem. Phys.*, **52**, 47 (1970).
152. E. I. Rashba, *Zh. Eksp. Teor. Fiz.*, **50**, 1964 (1966); trans. *Sovt. Phys. JETP*, **23**, 708 (1966).
153. R. W. Munn, *Disc. Faraday Soc.*, **51**, 73 (1971).

8 *Exciton interactions in organic solids—a postscript*

Nicholas E. Geacintov and Charles E. Swenberg

Since the first part of this article was written in 1972, several new discoveries and advances in the field of excitonic interactions in organic solids have been made. This chapter updates the review in Chapter 10, Vol. 1, which will be referred to as Part I. Avakian has also reviewed the effects of magnetic fields on luminescence involving triplet excitons in organic crystals.[1] The present review emphasizes important physical features and qualitative theoretical principles, rather than calculational details.

8.1 Effect of magnetic fields on certain photosensitized reactions

Gupta and Hammond[2] reported effects of magnetic fields (8–10kG) on the isomerization of piperylenes and stilbenes photosensitized by different ketones. The magnetic field affects the initial conversion rate from one isomeric form to the other and the composition in the photostationary state. The isomerization proceeds via the triplet state of the acceptor olefin (^3A), which is generated by a collision of the olefin in its ground state with the ketone donor in the triplet state (^3D). Decay of the triplet state of the acceptor molecule then occurs to give a characteristic mixture of *cis* and *trans* isomers. It was suggested that the decay of the intermediate triplet exciplex $^3(D\ldots A)^*$ via the singlet channel competes more effectively with decay of the exciplex into $^1D^0 + {}^3A^*$ in the presence of a magnetic field.[3] Atkins[3] has provided a tentative explanation of this effect by assuming that the intermediate exciplex has some radical character, i.e.

$$^3\psi[^3(D\ldots A)^*] = {}^3\psi[^3D\ldots {}^1A] + \lambda\,{}^3\psi[^2D^+\ldots {}^2A^-] \tag{8.1}$$

The second component of the exciplex wavefunction constitutes a charge-transfer contribution and consists of a caged radical pair. This radical pair is initially born as a triplet, but after some time (typically of the order of 10^{-8} s, the inverse of the Larmor frequency) this radical pair may acquire singlet character and thus the exciplex can decay into ground-state molecules rather than to $^3A^*$. Atkins suggests that the perturbation which brings about this triplet → singlet transition within the exciplex is the anisotropy in the g-factors of the two radicals. This gives rise to a non-vanishing matrix element between singlet and triplet states resulting in the observed effects of magnetic fields on the photosensitized isomerization of stilbenes and piperylenes. This mechanism is analogous to the one proposed by Frankevich[4] for the field-induced mixing of singlet- and triplet-type Wannier excitons in organic solids. Such effects are also discussed in connection with the Chemically Induced Dynamic Nuclear Polarization (CIDNP) effect.[5] In the latter phenomenon hyperfine modulation (HFM) is also operative, in which the Larmor precession frequency of the unpaired electrons is affected by electron–nuclei hyperfine interactions as well. HFM also induces mixing between singlet and triplet type radical pair states which leads to the observation of dynamic nuclear spin polarization in the recombination products.

To summarize, new types of magnetic-field effects can be observed whenever radical pairs are present as reactive intermediates. These effects are due to either or both

(a) a non-zero value of $\Delta g = g_1 - g_2$, where g_1 and g_2 are the g-factors of the two radicals,

(b) hyperfine modulation; this is due to an interaction between nuclear magnetic moments and the electron spin, and results in small differences in the effective magnetic field seen by the two radicals. This leads to different Larmor precession frequencies and thus to an oscillating singlet–triplet character of the radical pair.

Both of these effects have been postulated to explain recent observations of the influence of weak magnetic fields on the luminescence of organic crystals.[6,7] We now turn to a discussion of these effects.

8.2 Magnetic field effects on sensitized delayed fluorescence in organic crystals

In 1969, Nickel *et al.*[8] showed that delayed fluorescence in anthracene can be sensitized by photoexciting dye molecules (rhodamine B) absorbed on the anthracene crystal surface. More detailed accounts of studies of this phenomenon have been published subsequently.[9] The dye is photoexcited by light which is *not* absorbed by the anthracene, and the characteristic shorter wavelength delayed fluorescence of anthracene, which is due to triplet exciton fusion, is easily observed.[8,9] Photoexcitation of the dye (D) gives rise to the singlet excited state of the dye $^1D^*$. Delayed fluorescence of anthracene is generated by the following sequence of steps:

$$^1D^* + A \rightleftarrows {}^1[D^- \ldots A^+]^* \longrightarrow \begin{cases} D^- + A^+ & (8.2a) \\ ^1D^0 + {}^3A^* & (8.2b) \end{cases}$$

$$^3A^* + {}^3A^* \rightarrow \text{blue delayed anthracene fluorescence} \qquad (8.2c)$$

Energy transfer from the excited dye molecule to anthracene (A) gives rise to an intermediate charge–transfer state in which the dye molecule is the reduced species. This complex can decay to give a free hole (A^+) in the anthracene crystal. A sensitized hole photocurrent can be observed in the presence of an electric field of the proper polarity. The energy transfer step (8.2b) involves intersystem crossing and manifests itself indirectly as delayed fluorescence in anthracene which results from the fusion of two triplet excitons.

Groff *et al.*[6] observed that small magnetic fields (less than 100 gauss) decreased the rhodamine B sensitized delayed fluorescence in anthracene by up to 75%. This effect is completely different from the normal delayed fluorescence which is generated by exciting the anthracene crystal directly. The normal delayed fluorescence is modulated to a lesser extent and at higher magnetic fields. These phenomena have been interpreted in terms of magnetic-field dependent surface recombination of electrons and holes in the intermediate charge-transfer complex $[D^- \ldots A^+]^*$ in equation (8.2). According to the mechanism proposed by Groff *et al.*,[6] the charge-transfer (CT) state is born as a singlet. Because of the coulombic interaction between the electron on the rhodamine dye molecule and the hole on the anthracene molecule, the lifetime of this CT state is sufficiently long that significant spin motion can occur before either dissociation or recombination takes place. The spin motion is governed by the non-symmetric terms in the spin Hamiltonian:

$$\mathscr{H} = g_D \mu_B \mathbf{H}_D \cdot \mathbf{S}_D + g_A \mu_A \mathbf{H}_A \cdot \mathbf{S}_A + \sum_{l,m} a_l \mathbf{S}_m \cdot \mathbf{I}_l \qquad (8.3)$$

where the subscripts D and A refer to the dye and anthracene molecules respectively, μ_B is the Bohr magneton, $\mathbf{I}_l$ is the nuclear spin operator for all l protons interacting with the relevant electrons m, and the magnitude of the average hyperfine contact interaction is equal to a_l. Groff *et al.* assume that $g_D \approx g_A$ and explain the observed effects in terms of the relative magnitudes of the Zeeman and hyperfine terms in (8.3). In the absence of an applied magnetic field the hyperfine contact term mixes singlet and triplet spin states. The triplet type CT states can decay by geminate electron–hole recombination to yield an anthracene triplet exciton. The application of an external magnetic field reduces the relative importance of the hyperfine term and thereby the rate of singlet–triplet transitions. The CT state born originally as a singlet thus tends to remain as a singlet in the high field limit (singlet–triplet mixing can still occur when the Zeeman term dominates, but only between $M_S = 0$ states and only if $g_D \neq g_A$). The production of triplet type CT states is reduced and the generation rate of triplet excitons and thus the delayed fluorescence are also reduced.

Frankevich and Rumyantsev studied the fluorescence when tetracene, deposited as a thin film on the surface of an anthracene crystal, is photo-excited.[7a] The fluorescence spectrum is broad and structureless with a maximum at 18,200 cm^{-1} and differs from the spectrum of tetracene molecules dissolved in an anthracene matrix, and they attribute it to the emission from a tetracene-anthracene (TA)* exciplex with charge-transfer character (T^+A^-). This fluorescence decreases by 2–3% upon application of magnetic fields of about 20 gauss. Thus, the general aspects (but not the magnitude) of this effect are reminiscent of the one observed by Groff *et al.*[6] Moreover, the Russian authors explain their effect in terms of singlet–triplet mixing within the fluorescent exciplex $(T^+A^-)^*$ brought about by a non-zero Δg factor ($\Delta g = g_T - g_A \neq 0$). Mixing occurs only between the $M_S = 0$ spin components and the fluorescence intensity is decreased as the magnetic-field induced triplet character increases. Assuming $\Delta g \approx 10^{-1}$–10^{-2}, they calculate that the splitting ΔW between the singlet and triplet CT states is 10^{-7}–10^{-8} eV.

Similar effects with layers of rubrene, tetracene and perylene deposited on the surfaces of anthracene, naphthalene, stilbene and pyrene crystals were also observed and described by Frankevich.[7b] In some instances, however, the fluorescence increased upon application of a small magnetic field. Much more work is required before these interesting effects are fully understood.

Until now tetracene has been the only crystal in which fission of a thermally relaxed singlet exciton into two triplet excitons has been observed (see Part I). Recently, Frankevich has reported a similar effect at room temperature in solid rubrene.[7b]

Kaulatsch and Silinsh[10] have observed a magnetic effect on the photo-conductivity of thiotetracene crystals. The observed effects are temperature dependent and appear only at fields above 3000 gauss and there is a rapid rise and a saturation at ~5000–6000 gauss. Of all the possible magnetic-field effects discussed here and in Part I, only Case ii (Part I) is capable of predicting an initial lack of a magnetic-field effect and a subsequent rapid rise as the field strength is increased further. However, more experiments are needed (particularly at higher fields) before definite conclusions are possible.

8.3 Interaction of singlet excitons with metal surfaces

Baessler and his co-workers developed a method for quantitatively studying the interaction of excitons with metal layers.[11] The experimental technique involves a sandwich arrangement: metal | long chain fatty acid layer | anthracene crystal. The singlet excitons are efficiently quenched by the metal surface, but the efficiency decreases as the thickness of the fatty acid layer is increased. The thickness d of the fatty acid layers can be varied from about 20 to 100 Å. The anthracene crystals are excited with near UV light which is allowed to pass through the metal and fatty acid layer. The excitons created in the anthracene crystal diffuse towards the surface region and are quenched by interacting with the metal electrons via a long-range dipole–dipole coupling mechanism.[12] The quenching efficiency is monitored by measuring the fluorescence efficiency which decreases as the singlet excitons are quenched. The rate constant for energy transfer from an anthracene singlet exciton to an aluminium contact through a fatty acid monolayer is equal to $k_q \approx 10^{-9} d^{-3} \text{ s}^{-1}$.

Killesreiter and Baessler have also studied the dissociation of singlet excitons at an anthracene-metal contact. In solid anthracene this involves the tunnelling of an electron from the organic crystal to the metal leaving a hole in anthracene.[13,14] If a positive potential is applied to the illuminated contact (hole) photoconductivity can be observed. With *p*-chloranil crystals on the other hand, exciton decay at the metal contact results in electron injection.[14]

The properties of excitons in contact with metals have been considered theoretically in some detail by Agranovich and his co-workers.[15,16]

8.4 Exciton–carrier interaction

The detrapping of holes by singlet excitons and by photons in crystalline anthracene has been studied by Schott and Berrehar.[17] The rate constant of hole detrapping by singlet excitons was found to be $\gamma_S'' \sim 10^{-8} \text{ cm}^3 \text{ s}^{-1}$, somewhat larger than the value of $8 \times 10^{-10} \text{ cm}^3 \text{ s}^{-1}$ found by Pope *et al.*, which, however, they stated to be a lower limit.[18] The cross-section for detrapping of holes by photons ($4300 < \lambda < 5500$ Å) is of the order of 10^{-17} cm^2 and is roughly the same for photon detrapping of electrons.[19] Interaction of photons with trapped holes or electrons therefore appear to be similar processes.

The detrapping of carriers by excitons in anthracene was recently also studied by Frankevich *et al.*[20] The following rate constants for triplet exciton–hole interaction were measured: $\gamma''(h_t) = (0{\cdot}6 \pm 0{\cdot}2) \times 10^{-11}$ $\text{cm}^3 \text{ s}^{-1}$ and $\gamma''(h_f) = (2 \pm 1) \times 10^{-10} \text{ cm}^3 \text{ s}^{-1}$. (The notation of Part I is used, where h_f and h_t refer to free and trapped holes respectively.) These values are in good agreement with those of previous workers quoted in Part I. An upper limit for the singlet exciton-trapped hole interaction rate constant of $\gamma_S'' \lesssim 10^{-7}$ cm^3 s^{-1} was reported.[20] This also agrees with the values of Pope[18] and of Schott *et al.*,[17] but it is much smaller than the value measured by Wakayama and Williams.[21] Their large value of 5×10^{-5} cm^3 s^{-1}, however, has been attributed to a different process altogether[22] and is discussed more fully below.

The presence of charge carriers inside a crystal decreases the triplet lifetime. Pope and Weston have carried out precise measurements of the triplet exciton lifetime and thus determined the average total excess charge density and its voltage dependence in anthracene single crystals using 0·1 M Ce^{4+} in 7·5 M H_2SO_4 hole-injecting contacts.[23]

Using a sandwich type arrangement with gold electrodes on both sides, the $T_1 \leftarrow S_0$ absorption spectrum in single crystals of 9,10-dichloroanthracene was obtained.[24] As in anthracene, holes are probably injected by the gold electrodes, and are trapped and then detrapped by triplet excitons, giving rise to the observed photoenhanced currents. The same workers also observed photoconduction induced by a Q-switched ruby laser in 9,10-dichloroanthracene and 9,10-dibromoanthracene.[25] A third-power dependence on the light intensity was observed, as in anthracene (see Part I), and it can probably be explained in the same way, namely by the photoionization of singlet excitons with the latter being produced by a two-photon mechanism. Unlike anthracene, however, at low intensities a linear dependence of the photocurrent on the light intensity was observed. This can be explained by the triplet exciton–carrier detrapping mechanism.[24] It is observed in the 9,10-dichloroanthracene and 9,10-dibromoanthracene crystals because the triplet densities are much higher than in anthracene crystals using ruby laser excitation. This is due to the fact that in these two crystals the $0 - 0$ level of the $T_1 \leftarrow S_0$ transition is slightly below the ruby laser frequency, while in anthracene it is slightly above this frequency.[25]

8.5 Quartet and cage models

Wakayama and Williams have studied the quenching of the delayed fluorescence intensity, F, by injected charge carriers in anthracene crystals.[21] Their results indicate that for high excitation intensities, the observed fluorescence quenching is stronger than that predicted using standard triplet–triplet kinetics. Furthermore, this additional quenching is not due to direct singlet exciton carrier annihilation since the prompt fluorescence intensity is unaffected. At sufficiently high light intensities, injected charge carriers should have no effect on the quantum efficiency of delayed fluorescence, that is, F/F_0 should approach unity, where F_0 is the delayed fluorescence intensity in the absence of charge carriers. Thus one would *not* expect the delayed fluorescence intensity to vary linearly with absorbed light intensity and at the same time to depend on the injected charge carrier density as Wakayama and Williams[21] observed. As these authors noted, in order to explain these results, it is necessary to invoke a quenching process of an intermediate state. In addition, kinetic analysis implies that this intermediate state has a lifetime of at least 100 times the correlated lifetime of the triplet pair state.[26] Two possible mechanisms have been considered. Schwob and co-workers[22] have suggested the possibility that this third-order quenching process arises from triplet exciton quenching of a correlated triplet carrier pair (three $S = \frac{1}{2}$ unpaired spins on a given molecule). From spin selection rules the quartet spin component of the triplet–carrier pair has the necessary long lifetime implied by the kinetic analysis. From their analysis Schwob *et al.*[22] inferred a value of 10^4 s for the quartet lifetime and a bimolecular triplet–quartet annihilation rate of 10^{-9} cm^3 s^{-1}, both of which are reasonable. The difficulties with the model, however, are (1) the lack of independent evidence for quartet excitons and (2) the energy of the quartet state in solid anthracene is unknown.

The other mechanism suggested to explain the results of Wakayama and Williams involves the hypothesis of defect cages. It is well known that all crystals contain defects, e.g. line dislocations, screw dislocations, stacking faults, small and large angle grain boundaries, and thus defect concentrations of 10^{16} to 10^{18} cm^{-3} in bulk grown crystals are not unreasonable. Pope *et al.*,[27] hypothesized that for *some* of these defects, there is a region around the defect which is called a cage. This cage encompasses perhaps several hundred molecules, in which an exciton can become trapped in the sense that the exciton will tend to remain within the cage, being (partially) reflected from the cage boundary, but retaining its customary mobility. Two captured triplet excitons in a given defect cage will therefore be spatially *correlated* in the sense that they will (in general) either annihilate or collide with a quenching impurity within the cage. Furthermore, the spatial confinement enhances the fusion and triplet–carrier pair interaction rate constants. Pope *et al.*[27] have performed computer simulation analyses of exciton quenching within defect cages. Their results showed that the optically correlated triplet exciton pair within a cage is more efficiently quenched as the cage boundary reflection coefficient for triplet excitons is reduced from no reflection (perfect crystal) to total reflection (total caging of the triplet excitons). Computer fits to Wakayama and Williams' data were not performed. The cage model predicts that the delayed fluorescence intensity at high light intensity varies linearly with light intensity and at the same time is inversely proportional to the free carrier density, in agreement with experimental results. The saturation values of F/F_0 are predicted to vary with the applied voltage V in the manner $(F_0/F - 1)$ at (V_1): $(F_0/F - 1)$ at (V_2): $(F_0/F - 1)$ at $(V_3) = V_1 : V_2 : V_3$ in contrast to the quartet model where for all applied voltages F/F_0 approaches the free carrier limit. Pope[28] has suggested the possibility of distinguishing between the defect cage theory and the quartet theory by carrying out the carrier quenching experiment in low melting aromatic solids (e.g. phenanthrene or naphthalene) where presumably the defect cages would not exist. The validity of either the quartet or cage theory explanation of the quenching of the delayed fluorescence (under the conditions of high excitation intensity) awaits further experiments and more extensive theoretical developments. However there is little doubt that the study of defects (and the corresponding 'cages') may be an important avenue of future investigation.

8.6 Charge–transfer states

Jurgis and Silinsh[29] have calculated the energy of the nearest-neighbour charge-transfer (CT) state and the energy of the lowest conduction level by the self-consistent polarization field method. Better agreement between calculated and experimental values was obtained by these workers than by previous calculations of Hug and Berry.[30]

What may well prove to be the most direct experimental demonstration of a charge-transfer exciton in an organic molecular solid was provided in a preliminary report by Hanson.[31] A very sensitive electroreflectance technique was developed which is based on the Stark effect. The oscillator strength of the ground state–CT exciton transition is extremely low (10^{-4}–10^{-5}) and these transitions have not been observed before by standard spectroscopic techniques. The electroreflectance technique on the other hand is very sensitive and shows great promise for locating these elusive CT transitions. Hanson *et al.* have studied the Stark effect in 9,10-dichloroanthracene crystals and found three peaks with unusual properties between the first and second ${}^1\pi^* \leftarrow {}^1\pi$ transitions. The dipole moment was found to be about 23 Debye, which is very high, and was attributed to the CT state. It is hoped that this technique will prove useful in other crystals as well.

8.7 Exciton–exciton annihilation

Accurate measurements of triplet–triplet annihilation rate constants have been performed recently in solid pyrene and naphthalene. Pyrene is of special interest because of its dimer crystal structure[32] and because it displays both fluorescence[33] and phosphorescence[34] excimer emission. The dependence of the delayed fluorescence in pyrene crystals on magnetic field at room temperature has been analysed by Yarmus *et al.*[35] in terms of the exciton-pair model of Merrifield,[36] i.e.

$$\mathrm{T} + \mathrm{T} \underset{k_{-1}}{\overset{k_1}{\rightleftarrows}} (\mathrm{TT}) \overset{k_2}{\rightarrow} \mathrm{S}^* \qquad (8.4)$$

where S* denotes the fluorescent singlet state and the pair state (TT) represents an intermediate correlated state of the colliding triplets. The formation, scattering and annihilation rate constants are specified by k_1, k_{-1}, and k_2, respectively. Their analysis gave $k_{-1} = 7{\cdot}6 \times 10^9\ \mathrm{s}^{-1}$ ($\pm 10\%$), $k_2 = 3{\cdot}4 \times 10^9\ \mathrm{s}^{-1}$ ($\pm 10\%$), $D = 0{\cdot}066\ \mathrm{cm}^{-1}$ and $E = -0{\cdot}036\ \mathrm{cm}^{-1}$. The zero-field splitting parameters are in excellent agreement with those determined by the EPR spectrum of dilute pyrene-d_{10} solutions in single crystals of fluorene,[37] and k_{-1} and k_2 are approximately three times larger than the corresponding values in solid anthracene.[36] Furthermore, the results of Yarmus *et al.* can be explained in terms of monomeric triplet–triplet exciton annihilation only. This conclusion is further supported by Peter and Vaubel's measurements[38] of the bimolecular annihilation rate. They reported no significant change in the bimolecular decay curve between 160–300°K implying that triplet exciton migration requires little or no thermal activation. Thus it is essentially monomeric in nature in contrast to singlet exciton migration in solid pyrene.[39]

Peter and Vaubel[38] reported triplet exciton lifetimes in crystalline pyrene of 140 ± 10 ms when oxygen was carefully excluded. Triplet exciton lifetimes in solid pyrene previously reported are smaller, presumably due to quenching from photo-oxidation induced triplet trapping centres, e.g. pyrene quinone.[38] Values of $(7{\cdot}5 \pm 4{\cdot}0) \times 10^{-12}\ \mathrm{cm}^3\ \mathrm{s}^{-1}$ and $(2{\cdot}0 \pm 1) \times 10^{-12}\ \mathrm{cm}^3\ \mathrm{s}^{-1}$ for the room-temperature values of the total (γ_{tot}) and the radiative rate (γ_{rad}) constants, respectively, for triplet–triplet exciton annihilation in pyrene single crystals have been reported recently by Ern *et al.*[40] These values were inferred by measuring both the steady state delayed fluorescence intensity, F, of pyrene relative to anthracene (under the same experimental conditions) and the product $\alpha\gamma_{tot}$, by measuring the phase lag of the first harmonic of the delayed fluorescence[41] (α is the ground-state singlet–triplet absorption coefficient). From a least squares fit of $IF^{-\frac{1}{2}}$ versus $F^{\frac{1}{2}}$, where I is the steady-state excitation intensity, the authors inferred $f = \gamma_{rad}/\gamma_{tot} = 0{\cdot}25 \pm 0{\cdot}02$ and $R = 0{\cdot}10 \pm 0{\cdot}02$, where R is the fraction of triplet fusion events leading to singlets. This result implies equal electronic fusion matrix elements for the singlet and triplet channels, which is the case in solid anthracene (see Part I). Benz *et al.*[42] have studied the fluorescence and phosphorescence in highly purified naphthalene crystals (triplet exciton lifetimes up to 500 ms at room temperature) and reported $\gamma_{tot} = 3{\cdot}5 \times 10^{-12}$ cm^3 s^{-1} (assuming $f \approx 0{\cdot}4$) which is in good (but fortuitous) agreement with the theoretical prediction by Swenberg.[43]

Fourny *et al.*[44] have remeasured the singlet–singlet (γ_{ss}) and singlet–triplet (γ_{st}) quenching constants in anthracene by separating these two processes by a new kinetic method. The rate constants are $\gamma_{ss} = (1 \pm 0{\cdot}5) \times 10^{-8}$ and $\gamma_{st} = (5 \pm 3) \times 10^{-9}$ cm^3 s^{-1} which are in good agreement with the values obtained by other workers, quoted in Part I.

Prasad has discussed theoretically two mechanisms of interaction between excitons: (a) the direct coulombic interaction including exchange and (b) the indirect exciton–exciton phonon mediated interaction.[45] Case (a) has an R^{-6} dependence, where R is the distance between the excitons. This interaction is large (and therefore important) only when there is a large change in the polarizability of the molecule during excitation, in which case it can be attractive when the polarizability of the excited state of the molecule is larger than in the ground state. For optical phonons case (b) is extremely local (and for all practical purposes can be neglected) while for acoustic phonons it has a distance dependence of R^{-3} and is attractive in the case where the triplet exciton bandwidth is very narrow. The incorporation of these (attractive) exciton–exciton interactions into the theory of triplet–triplet annihilation remains to be carried out.

Fünfschilling and Zschokke-Gränacher[46] have reported preliminary measurements on triplet exciton annihilation in solid anthracene containing tetracene impurities (concentrations ranging from 5×10^{-9} to 10^{-5} M). From a kinetic analysis these authors determined a host–guest transfer rate of approximately $1{\cdot}8 \times 10^{-11}$ s^{-1}, and a minimum value for the heterofusion rate constant of $(f\gamma)_{AT} \gtrsim 0{\cdot}5\,(f\gamma)_{AA}$, where γ_{AA} is the host fusion rate constant.

8.8 Optically induced exciton fission

Klein, Voltz and Schott reported that the prompt fluorescence of anthracene crystals is magnetic-field sensitive.[47] The characteristic variation of the relative fluorescence quantum yield, $\Delta F/F$, in the presence of a magnetic field show that vibrationally excited molecular states can fission into two triplets. Since internal conversion within the singlet manifold is rapid, the effect is small, generally $\lesssim 1\,\%$. However its presence in crystalline anthracene shows the generality of the fission process. Recently Moller and Pope,[48] as well as Klein *et al.*[49] have observed the magnetic field enhancement of the prompt fluorescence of tetracene crystals at 77°K, where the thermally induced fission process is frozen out. This is in accordance with a suggestion of Vaubel and Baessler.[50]

The action spectrum of $\Delta F/F$ of Moller and Pope shows a steplike structure of approximately every 3000 cm^{-1} below 27,000 cm^{-1}. Swenberg *et al.*[51] have interpreted this structure as a progression in the totally symmetric C–H stretching mode which has a Franck–Condon vibrational coupling constant for the correlated pair singlet–triplet transition of the order of unity. A crucial test of this theory would be to observe the shift in the steplike structure in deuterated anthracene (or tetracene), since the C–H stretching mode frequency is expected to drop noticeably upon deuteration (for naphthalene this shift[52] is from 3055 cm^{-1} to 2272 cm^{-1}).

8.9 Exciton dynamics and diffusion

Preliminary measurements of the triplet exciton diffusion tensor in solid pyrene using the phase difference method of Ern[41,53] and laser excitation has been reported by Arnold *et al.*[54] The maximum diffusivity was found to occur along the *b*-axis with $D_{bb} = (1{\cdot}25 \pm 0{\cdot}3) \times 10^{-4}$ cm^2 s^{-1} and was nearly isotropic in the *ac*-plane with $D = (0{\cdot}3 \pm 0{\cdot}1) \times 10^{-4}$ cm^2 s^{-1}. These values are consistent with the measured bimolecular annihilation rate (assuming a diffusion-controlled reaction), although D_{aa} is larger than the value inferred by Arnold, Whitten and Damask from Davydov splitting measurements.[55] Static interactions, therefore, cannot account for the observed triplet diffusion coefficient which is probably dominated by non-local scattering.

Using the 4416 Å line of the He–Cd laser for direct singlet–triplet absorption and the phase shift method, Ern measured the triplet diffusion tensor at room temperature in naphthalene, *trans*-stilbene, and 1,4-dibromonaphthalene.[53] All crystallize into a monoclinic lattice structure and have 2, 4 and 8 molecules per unit cell respectively. At room temperature, Ern[53] reported $D_{aa} = (0{\cdot}9 \pm 0{\cdot}2) \times 10^{-5}$ cm^2 s^{-1}, $D_{bb} = (0{\cdot}7 \pm 0{\cdot}2) \times 10^{-4}$ cm^2 s^{-1} for solid stilbene, and $D_{aa} = (3{\cdot}3 \pm 0{\cdot}4) \times 10^{-5}$ cm^2 s^{-1}, $D_{bb} = (2{\cdot}7 \pm 0{\cdot}4) \times 10^{-5}$ cm^2 s^{-1} for crystalline naphthalene. Furthermore, Ern's results on exciton diffusion in crystalline 1,4-dibromonaphthalene show that triplet transport is primarily one-dimensional, i.e. $D_{cc} = (3{\cdot}5 \pm 0{\cdot}8) \times 10^{-4}$ cm^2 s^{-1} and D_{aa} and $D_{bb} \simeq 10^{-5}$ cm^2 s^{-1}. This is consistent with Hochstrasser and Whiteman's[56] results.

The effects of non-local scattering of excitons, in the theory of exciton diffusion and spin resonance line shapes has been considered by Haken and co-workers.[57–59] The reader is referred to the original literature for details. It may be noted that the only analysis of the temperature dependence of the triplet exciton diffusion coefficient is that of Ern *et al.*[60] These authors measured the diffusion coefficient from 118°K to 371°K along the *a*-axis for solid anthracene and inferred a temperature-insensitive non-local scattering rate which is approximately 0·1 cm^{-1}. This is contrary to the expectation that scattering due to phonons must decrease as the temperature is decreased.

8.10 Electron spin resonance

One of the most interesting developments in the field of excitons in organic solids are the recent magnetic resonance studies of tetracene triplet excitons by Yarmus *et al.*[61,62] which have revealed unexpected consequences of the exciton fission process. A careful study of the EPR spectrum of solid tetracene as a function of excitation intensity offers a test of existing theories of exciton spin relaxation.[26]

The fission process, in conserving spin angular momentum, selectively populates the triplet magnetic sublevels.[63] Relaxation mechanisms in crystalline tetracene were found sufficiently slow (T_1 is long, $\sim 10^{-5}$ s), so that during the average lifetime of the triplet state the system does not thermalize. This results in spin polarization throughout the *ab* plane.

Chopp[64] has solved the non-linear steady state equations describing the magnetic sublevel densities, n_i, which are given by

$$\frac{dn_i}{dt} = \alpha_i - \beta_i n_i + \sum_{j \neq i} (\omega_{ij} n_j - \omega_{ji} n_i) - \sum_j \gamma_{ij} n_i n_j = 0 \qquad (8.5)$$

Here α_i denotes the populating rate, ω_{ij} the relaxation rates and β_i and γ_{ij} the monomolecular and bimolecular decay rates respectively. Using known values of β_i and the populating rates α_i, Chopp inferred the relaxation rates $\omega_{ij}(\theta)$ for a given magnetic field orientation θ and bimolecular annihilation rates γ_{ij}. Values of $\omega_{ij}(\theta)$ were found to vary from 9×10^4 s^{-1} to a maximum of 5×10^6 s^{-1} and its angular dependence agrees favourably with Sternlicht and McConnell's mechanism of spin relaxation via exciton hopping between inequivalent sites.[65]

The EPR spectrum of triplet excitons in single crystals of 1,4-dibromonaphthalene (1,4-DBN) at 1·2°K has been reported by Schmidberger and Wolf.[66] Their results confirm the one-dimensionality of exciton migration in 1,4 DBN.

The effects of dimensionality on exciton dynamics are discussed in papers by Pearlstein[67] and by Harris[68] and their co-workers.

Acknowledgements

This work was in part supported by the Atomic Energy Commission. We thank L. Yarmus, J. Rosenthal and M. Chopp for providing us with results prior to publication.

8.11 References

1. P. Avakian, *Pure Appl. Chem.*, **32**, 1 (1974).
2. A. Gupta and G. S. Hammond, *J. Chem. Phys.*, **57**, 1789 (1972).
3. P. W. Atkins, *Chem. Phys. Letters*, **18**, 355 (1973).
4. B. A. Rusin and E. L. Frankevich, *Phys. Stat. Sol.*, **33**, 885 (1969).
5. See, e.g., H. R. Ward, *Accts. Chem. Res.*, **5**, 18 (1972); R. G. Lawler, *Accts. Chem. Res.*, **5**, 25 (1972).
6. R. P. Groff, R. E. Merrifield, A. Suna and P. Avakian, *Phys. Rev. Letters*, **29**, 429 (1972); *Abstracts VI Molec. Cryst. Symp.*, Elmau (1973).
7. (a) L. Frankevich and B. M. Rumyantsev, *Zh. Eksp. Teor. Fiz.*, **63**, 2015 (1972); (b) *Abstracts VI Molec. Cryst. Symp.*, Elmau (1973).
8. B. Nickel, H. Staerk and A. Weller, *Chem. Phys. Letters*, **4**, 27 (1969).
9. B. Nickel, *Molec. Cryst. Liq. Cryst.*, **18**, 227, 263 (1972).
10. I. S. Kaulatch and E. A. Silinsh, *Latv. PSR Zinat. Akad. Vest.*, **1972**, 69 (1972).
11. G. Vaubel, H. Baessler and D. Möbius, *Chem. Phys. Letters*, **10**, 334 (1971).
12. H. Killesreiter and H. Baessler, *Phys. Stat. Sol. (b)*, **51**, 657 (1972).
13. H. Killesreiter and H. Baessler, *Chem. Phys. Letters*, **11**, 411 (1971).
14. H. Killesreiter and H. Baessler, *Phys. Stat. Sol. (b)*, **53**, 193 (1972).
15. V. M. Agranovich, A. G. Mal'schukov and M. A. Mekhtiev, *Zh. Eksp. Teor. Fiz.*, **63**, 2274 (1972).
16. V. M. Agranovich, A. G. Mal'shukov and M. A. Mekhtiev, *Izv. Akad. Nauk. SSSR*, **37**, 324 (1973). (*Proceedings of the International Conference on Luminescence*, Leningrad, 1972.)
17. M. Schott and J. Berrehar, *Molec. Cryst. Liq. Cryst.*, **20**, 13 (1973).
18. M. Pope, J. Burgos and N. Wotherspoon, *Chem. Phys. Letters*, **12**, 140 (1971).
19. S. Z. Weisz, J. Levinson and A. Cobas, *Proc. 3rd Intern. Conf. on Photoconductivity, Stanford*, p. 297 (Ed. E. M. Pell), Pergamon, Oxford, 1969.
20. E. L. Frankevich, I. A. Sokolik and L. V. Lukin, *Phys. Stat. Sol. (b)*, **54**, 61 (1972).
21. N. Wakayama and D. F. Williams, *J. Chem. Phys.*, **57**, 1770 (1972).
22. H. P. Schwob, W. Siebrand, N. Wakayama and D. F. Williams, to be published.
23. M. Pope and W. Weston, *Molec. Cryst. Liq. Cryst.*, **25**, 205 (1974).
24. T. Izumi and U. Itoh, *J. Phys. Soc. Japan*, **34**, 1110 (1973).
25. T. Izumi and U. Itoh, *J. Phys. Soc. Japan*, **32**, 214 (1972).
26. A. Suna, *Phys. Rev.*, **B1**, 1716 (1970).
27. M. Pope, C. E. Swenberg and R. Selsby, unpublished.
28. M. Pope, private communication.
29. A. Jurgis and E. A. Silinsh, *Phys. Stat. Sol. (b)*, **53**, 735 (1972).
30. G. Hug and R. S. Berry, *J. Chem. Phys.*, **55**, 2516 (1971).
31. D. M. Hanson, *Abstracts VI Molec. Cryst. Symp.*, Elmau (1973).
32. J. M. Robertson and J. G. White, *J. Chem. Soc.*, 358 (1947).
33. J. B. Birks, A. A. Kazzaz and T. A. King, *Proc. Roy. Soc.*, **A 291**, 556 (1966).
34. O. L. J. Gijzeman, J. Langelaar and J. D. W. Van Voorst, *Chem. Phys. Letters*, **5**, 269 (1970).

35. L. Yarmus, C. Swenberg, J. Rosenthal and S. Arnold, *Phys. Letters*, **43A**, 103 (1973).
36. R. C. Johnson and R. E. Merrifield, *Phys. Rev.*, **B1**, 896 (1970).
37. S. W. Charles, P. H. H. Fischer and C. A. McDowell, *Mod. Phys.*, **9**, 517 (1965).
38. L. Peter and G. Vaubel, *Chem. Phys. Letters*, **18**, 531 (1973).
39. W. Klöpffer, H. Bauser, F. Dolezulek and G. Naundorf, *Molec. Cryst. Liq. Cryst.*, **16**, 229 (1972).
40. V. Ern, H. Bouchriha, M. Bisceglia, S. Arnold and M. Schott, *Phys. Rev.*, **B8**, 6038 (1973).
41. V. Ern, *Phys. Rev. Letters*, **22**, 343 (1969).
42. K. W. Benz, H. Port and H. C. Wolf, *Z. Naturforsch.*, **26a**, 5, 787 (1971).
43. C. E. Swenberg, *J. Chem. Phys.*, **51**, 1753 (1969).
44. J. Fourny, M. Schott and G. Delacôte, *Chem. Phys. Letters*, **20**, 559 (1973).
45. P. N. Prasad, *Chem. Phys. Letters*, **20**, 507 (1973).
46. J. Fünfschilling and I. Zschokke-Granacher, *VI Molec. Cryst. Symp.*, Elmau (1973).
47. G. Klein, R. Voltz and M. Schott, *Chem. Phys. Letters*, **16**, 340 (1972).
48. W. M. Moller and M. Pope, *J. Chem. Phys.*, **59**, 2760 (1973).
49. G. Klein, R. Voltz and M. Schott, *Chem. Phys. Letters*, **19**, 391 (1973).
50. G. Vaubel and H. Bassler, *Molec. Cryst.*, **15**, 15 (1971).
51. C. E. Swenberg, M. Ratner and N. E. Geacintov, *J. Chem. Phys.*, **60**, 2152 (1974).
52. A. I. Stamford and S. F. Fischer, *Chem. Phys.*, **1**, 99 (1973).
53. V. Ern, *J. Chem. Phys.*, **56**, 6259 (1972).
54. S. Arnold, J. L. Fave and M. Schott, to be published.
55. S. Arnold, W. B. Whitten and A. C. Damask, *Phys. Rev.*, **53**, 3452 (1971).
56. R. M. Hochstrasser and J. D. Whiteman, *J. Chem. Phys.*, **56**, 5945 (1972).
57. H. Haken and G. Strobl, *The Triplet State*, p. 311 (Ed. A. B. Zahlan), Cambridge University Press, 1967.
58. H. Haken and P. Reineker, *Z. Phys.*, **249**, 253 (1972).
59. P. Reineker and H. Haken, *Z. Phys.*, **250**, 300 (1972).
60. V. Ern, A. Suna, Y. Tomkiewicz, P. Avakian and R. F. Groff, *Phys. Rev.*, **B 5**, 3222 (1972).
61. L. Yarmus, private communication.
62. L. Yarmus, J. Rosenthal and M. Chopp, *Chem. Phys. Letters*, **16**, 482 (1972).
63. C. E. Swenberg, R. Van Metter and M. Ratner, *Chem. Phys. Letters*, **16**, 482 (1972).
64. M. Chopp, private communication.
65. H. Sternlicht and H. M. McConnell, *J. Chem. Phys.*, **35**, 1793 (1961).
66. R. Schmidberger and H. C. Wolf, *Chem. Phys. Letters*, **16**, 402 (1972).
67. R. M. Pearlstein, *J. Chem. Phys.*, **56**, 2431 (1972).
68. M. D. Fayer and C. B. Harris, *Phys. Rev.*, **B9**, 748 (1974).

9 *Photophysics of aromatic molecules—a postscript*

John B. Birks

9.1 Introduction

The manuscript of *Photophysics of Aromatic Molecules* (*PAM*) was completed in July 1969. Since then there has been significant progress in most areas of the subject. The present chapter aims to review this progress by considering relevant work published up till the end of 1972, and in some cases during 1973.

The task of updating *PAM* is greatly simplified by the other contributions to the present two-volume work. These have provided comprehensive accounts of many of the major growth areas of the subject, and it is only necessary to fit these into the general pattern. There are, however, several other topics considered in *PAM* which have not been discussed in earlier chapters, and these require more detailed treatment. For ease of cross-reference the section titles and numbers of this postscript correspond to the chapter titles and numbers of *PAM*, and all page number references to *PAM* are shown in italics.

9.2 Photophysical processes

9.2.1 *Unimolecular processes* (pp. *30–2*)

The unimolecular photophysical processes in aromatic hydrocarbons which may be used for the spectroscopy of their π-electronic states have been discussed previously (Chapter 1, Vol. 1).

The first example of $S_2 - S_1$ fluorescence (p. *32*) in an aromatic hydrocarbon was reported by Rentzepis, Jortner and co-workers,[1–3] who observed such emission, of quantum yield $\sim 2 \times 10^{-5}$, from azulene in solution and in the vapour phase. The photophysics of azulene is discussed in §9.5.3.1.

$S_2 - S_0$ fluorescence has been observed in benzene, toluene, *p*-xylene and mesitylene in the vapour and/or solution phases by Hirayama *et al.*,[4,5] in naphthalene vapour by Wannier *et al.*,[6] in azulene and its derivatives (pp. *166–8*), and in 1:2-benzanthracene, pyrene, 3:4-benzopyrene, 1:12-benzoperylene and ovalene (Chapter 1, Vol. 1). $S_3 - S_0$ fluorescence has been observed in *p*-xylene, mesitylene, naphthalene and pyrene in the vapour and/or solution phases by Hirayama *et al.*[4,5]

The list of aromatic hydrocarbons exhibiting multiple fluorescences is now sufficiently extensive that Kasha's rule (p. *144*) for spin-allowed radiative transitions can only be considered as an approximation. The improved detection techniques and more intense light sources, which permit the observation of fluorescence quantum yields as low as 10^{-6}, have penetrated the psychological barrier of the Kasha rule approximation and thus provided data on radiative and radiationless transitions from higher excited states, which are discussed in §9.5.1.

9.2.2 *Biphotonic processes* (pp. *32–4*)

Biphotonic ionization processes in condensed media have been discussed by Lesclaux and Joussot-Dubien (Chapter 9, Vol. 1). Related processes in aromatic crystals are described by Swenberg and Geacintov (Chapter 10, Vol. 1; Chapter 8, Vol. 2). Two-photon absorption is considered in §9.3.3.

9.2.3 *Bimolecular processes* (pp. *34–6*)

Swenberg and Geacintov (Chapter 10, Vol. 1) have enumerated the different types of singlet and triplet exciton interactions that may occur in aromatic crystals. Analogous interactions involving singlet ($^1M^*$) and triplet ($^3M^*$) excited molecules are to be expected in other environments. Possible bimolecular and biexcitonic interactions are listed in Table 9.1, which is an extended version of ones published previously (pp. *39–40*). The interacting excitons, molecules or molecular ions are distinguished by the suffixes A and B. In a homopolar interaction A and B belong to the same molecular species. In a heteropolar interaction they belong to different molecular species, and the suffixes A and B then refer to the species with the higher and lower excitation energy, respectively.

9.3 Absorption

9.3.1 $S_0 - S_p$ *absorption* (pp. *44–58*)

The $S_0 - S_p$ absorption spectra of aromatic hydrocarbons in solution at room temperature, and the assignment of their π-electronic states, have been discussed previously (Chapter 1, Vol. 1). The influence of pressure on the $S_0 - S_1$ absorption has been described by Offen (Chapter 3, Vol. 1).

Table 9.1 Interactions of triplet and singlet excited molecules or excitons

Reactants A	Reactants B		Products	Homopolar process (A = B)	Heteropolar process (A ≠ B)
$^3M_A^*$	^1M_B	(i)	$^1M_A + {^3M_B^*}$	Triplet energy migration	Triplet energy transfer
		(ii)	$^1M_A + {^1M_B^*}$	Triplet–singlet energy migration	Triplet–singlet energy transfer
		(iii)	$^3D^*$ ($^3E^*$)	Triplet excimer formation	Triplet exciplex formation
	$^2M_B^{\pm}$ (^2M_B)	(i)	$^1M_A + {^2M_B^{\pm}}$ (^2M_B)	Triplet quenching by ion (doublet)	Triplet quenching by ion (doublet)
	$^3M_B^*$	(i)	$^1M_A + {^1M_B^*}$	Homofusion	Heterofusion
		(ii)	$^1M_A + {^3M_B^{**}}$	Triplet quenching by triplet	Triplet quenching by triplet
		(iii)	$^2M_A^{\pm} + {^2M_B^{\mp}}$	Ion formation	Ion formation
		(iv)	$^1D^*$ ($^1E^*$)	Singlet excimer formation	Singlet exciplex formation
		(v)	$^3D^{**}$ ($^3E^{**}$)	Excited triplet excimer formation	Excited triplet exciplex formation
		(vi)	$^5D^*$ ($^5E^*$)	Quintet excimer formation	Quintet exciplex formation
$^1M_A^*$	^1M_B	(i)	$^1M_A + {^1M_B^*}$	Singlet energy migration	Singlet energy transfer
		(ii)	$^1M_A + {^3M_B^*}$	Singlet–triplet energy migration	Singlet–triplet energy transfer
		(iii)	$^1D^*$ ($^1E^*$)	Singlet excimer formation	Singlet exciplex formation
		(iv)	$^3M_A^* + {^3M_B^*}$	Homofission	Heterofission
		(v)	$^2M_A^{\pm} + {^2M_B^{\mp}}$	Ion formation	Ion formation
	$^2M_B^{\pm}$ (^2M_B)	(i)	$^3M_A^* + {^2M_B^{\pm}}$ (^2M_B)	Singlet quenching by ion (doublet)	Singlet quenching by ion (doublet)
		(ii)	$^1M_A + {^2M_B^{\pm}}$	Singlet quenching by ion (doublet)	Singlet quenching by ion (doublet)
	$^3M_B^*$	(i)	$^1M_A + {^3M_B^{**}}$	Singlet quenching by triplet	Singlet quenching by triplet
		(ii)	$^2M_A^{\pm} + {^2M_B^{\mp}}$	Ion formation	Ion formation
	$^1M_B^*$	(i)	$^1M_A + {^1M_B^{**}}$	Singlet quenching by singlet	Singlet quenching by singlet
		(ii)	$^1D^{**}$ ($^1E^{**}$)	Excited singlet excimer formation	Excited singlet exciplex formation
		(iii)	$^2M_A^{\pm} + {^2M_B^{\mp}}$	Ion formation	Ion formation
		(iv)	$^3M_A^{**} + {^3M_B^{**}}$	Excited homofission	Excited heterofission

9.3.1.1 *Benzene*

Pantos and Hamilton[7,8] have observed the $S_0 - S_2$ absorption band system of crystal benzene at low temperatures, and from a vibronic analysis they assigned it to the $^1B_{1u} \leftarrow {}^1A_{1g}$ ($^1L_a \leftarrow {}^1A$) transition. This assignment agrees with that of Katz *et al.*[9] who observed the absorption spectra of benzene in solid rare-gas matrices, and with that of Brith *et al.*[10] who observed the reflection and absorption spectra of crystal benzene at 4°K. The absorption spectra of the methyl-substituted benzenes (toluene, toluene-d_8, *p*-xylene and mesitylene) in solid krypton observed by Katz *et al.*[11] are also consistent with the $S_2 = {}^1B_{1u}$ assignment.

Although the different observers[7–10] agree on the assignment of the electronic transition, they differ in the analysis of the vibronic structure in terms of two or more progressions of totally symmetric $\nu_2(a_{1g})$ vibrations based on two or more origins. Katz *et al.*[9] interpreted the rare-gas matrix spectrum in terms of a weak progression on a $\nu_{18}(e_{2g})$ false origin, and a strong progression on a $\nu_{16}(e_{2g})$ false origin, the electronic (0 – 0) origin not being observed. Brith *et al.*[10] suggested two alternative assignments of the crystal spectrum: (a) a weak progression on a $\nu_{18}(e_{2g})$ false origin and a strong progression on a $\nu_{16}(e_{2g})$ false origin, as in the rare-gas matrix spectrum;[9] or (b) a weak progression on the electronic (0 – 0) origin and a strong progression on a $\nu_{16}(e_{2g})$ false origin. Pantos and Hamilton[7] interpreted the crystal spectrum in terms of a weak progression on the crystal-field-induced electronic (0 – 0) origin and a strong progression on a $\nu_{18}(e_{2g})$ false origin. The latter assignment avoids certain anomalies in those suggested by Brith *et al.*,[10] and the reader is referred to the original paper[7] for a discussion of these.

Overlap of the false origins and their associated progressions causes an apparent increase in linewidth $\Delta\bar{\nu}$, which contributes to the high values of $\Delta\bar{\nu} \sim 300\ \text{cm}^{-1}$ observed for the vibronic bands.[9,10] The relatively sharp crystal-field-induced electronic (0 – 0) band, which is the only one not affected by progression congestion, has $\Delta\bar{\nu} = 60\ \text{cm}^{-1}$ at 4·2°K.[12] Observations of $\Delta\bar{\nu}$ for this band, which provide data on the radiationless transition rate from the $^1B_{1u}(S_2)$ state of benzene, are discussed in §9.5.2.

The polarized absorption spectra of single benzene crystals in the region of the $S_0 - S_2$ transition have been observed by Brillante *et al.*[593] and Bird and Callomon.[606] The polarized spectra at 77°K observed by the two groups are similar, but they differ in the crystallographic description of their specimens. Brillante *et al.*[593] describe their specimens as *bc* (100) crystal plates, and they concluded that the corresponding *b*- and *c*-polarized spectra are inconsistent with $S_2 = {}^1B_{1u}$, but not with $S_2 = {}^1E_{2g}$. The sharp 46,565 cm^{-1} band, previously assigned[7] to the $^1B_{1u} \leftarrow {}^1A_{1g}$

electronic origin, is attributed either to a false origin of the $^1E_{2g} \leftarrow {}^1A_{1g}$ system or to a new transition, possibly of $\pi^* \leftarrow \sigma$ type, polarized normal to the molecular plane. In contrast Bird and Callomon[606] found that all their crystals grew consistently with the *ac* (010) face developed, and they concluded that the corresponding *a*- and *c*-polarized spectra are consistent with $S_2 = {}^1B_{1u}$, with the sharp band as the $^1B_{1u} \leftarrow {}^1A_{1g}$ electronic origin, in agreement with previous studies.[7–12]

Pantos *et al.*[616] have confirmed the $^1B_{1u} \leftarrow {}^1A_{1g}$ assignment in a further study of the absorption and fluorescence excitation spectra of crystal C_6H_6, *sym*-$C_6H_3D_3$ and C_6D_6 (wavenumbers are listed in that order below). The absorption spectra are analysed in terms of $\bar{\nu}_2(a_{1g})$ progressions on the following origins:

(i) the electronic 0 – 0 origin, $\bar{\nu}_0$ (46,500, 46,620, 46,740 cm^{-1})

(ii) the false origin, $\bar{\nu}_0 + \bar{\nu}_{18}(e_{2g})$ (510, 500, 480 cm^{-1})

(iii) the false origin, $\bar{\nu}_0 + \bar{\nu}_7(b_{2g})$ (760, 700, 660 cm^{-1})

(iv) the false origin, $\bar{\nu}_0 + \bar{\nu}_{16}(e_{2g})$ (—, —, 1250 cm^{-1})

(iii) is resolved from $\bar{\nu}_0 + \bar{\nu}_2(a_{1g})$ (910, 880, 870 cm^{-1}) in the *sym*-$C_6H_3D_3$ and C_6D_6 spectra, but not in the C_6H_6 spectrum, and (iv) appears as a shoulder in the C_6D_6 spectrum.

There is some evidence for a weak absorption band system in benzene originating about 2500 cm^{-1} below the $S_0 - S_2$ absorption. It will be described as the $S_0 - S_1'$ transition to avoid renumeration of the $S_2(^1B_{1u})$ and $S_3(^1E_{1u})$ states. Morris and Angus[13] first reported it in the phosphorescence excitation spectrum of benzene in solid krypton and xenon matrices. Taleb *et al.*[14] observed it in the absorption and phosphorescence excitation spectra of benzene in solid krypton and xenon matrices, and in the absorption spectrum of benzene in fluid perfluorohexane solution from room temperature to −85°C. Both groups[13,14] assigned the $S_0 - S_1'$ transition to the parity-forbidden $^1E_{2g} \leftarrow {}^1A_{1g}$ transition, vibronically induced by a non-totally-symmetric e_{1u} vibration. The observed vibronic transitions and their proposed assignments are listed in Table 9.2.

Birks *et al.*[415] observed no transition intermediate between the $S_0 - S_1$ and $S_0 - S_2$ absorption band systems in benzene vapour. Lipsky[602] has recently observed the absorption spectra of benzene in low-temperature fluid perfluorohexane and perfluoropentane solutions, but he found no evidence for the $S_0 - S_1'$ bands reported by Taleb *et al.*[14] The latter bands may be due to an impurity (e.g. toluene) or to benzene aggregates. The '$S_0 - S_1'$ bands' in solid krypton and xenon matrices may be heavy-atom-

Table 9.2 Energies (cm^{-1}) and tentative assignments of $^1E_{2g} \leftarrow {}^1A_{1g}$ vibronic bands of benzene

	Absorption spectrum			Phosphorescence excitation spectrum					
Solvent	Perfluorohexane	Xenon		Krypton		Xenon			
Solute	C_6H_6	C_6H_6	C_6D_6	C_6H_6	C_6D_6	C_6H_6	C_6D_5	Tentative assignment	Ref.
Band 1	46300	45770	45830	46170	46230	45770	45830	$0, 0 + e_{1u}$	14
				46350	—	45750	—	$0, 0 + e_{1u}$	13
Band 2	46950	46320	46430	46660	46750	46320	46430	$0, 0 + e'_{1u}$	14
				46860	—	46300	—	$0, 0 + e'_{1u}$	13
Band 3	47620	—	—	47170	—	—	—	$0, 0 + e_{1u} + a_{1g}$	14
				47300	—	46800	—	$0, 0 + e_{1u} + a_{1g}$	13

induced transitions to the adjacent triplet state $T_3(^3B_{1u})$. The present evidence for an excited singlet state S_1' between S_1 and S_2 in benzene is insufficient.

The strong $S_0 - S_3$ absorption in benzene corresponds to the allowed $^1E_{1u} \leftarrow {}^1A_{1g}$ transition. This absorption has been observed in the vapour,[15] liquid[16–18] and crystal[10,606] phases. Vibronic structure is observed in the vapour and crystal phases, but not in the liquid phase. Williams *et al.*[18] used reflectance measurements to determine the real and imaginary components of the complex refractive index of liquid benzene between 120 and 320 nm. In addition to the strong $^1E_{1u} \leftarrow {}^1A_{1g}$ absorption peak at 6·5 eV (190 nm), their results showed a maximum in the calculated electron energy loss spectrum at 7·3 eV (173 nm), which they attributed to a collective volume-plasma oscillation in liquid benzene. This interpretation currently lacks confirmation from other sources.

The absorption spectrum of benzene vapour from 6 eV (207 nm) to 35 eV (35·4 nm) has been measured by Koch and Otto[15] using synchrotron radiation. Above the $^1E_{1u} \leftarrow {}^1A_{1g}$ transition at 6·9 eV there is a broad absorption continuum starting from a minimum at 8·2 eV, reaching a maximum at about 17·8 eV, and then decreasing with increasing photon energy. This continuum, which is similar to that in the *n*-alkanes, is attributed to σ-electron excitation. Superimposed on this continuum there is a complex sequence of bands, which have been analysed in terms of Rydberg series. The photo-electron spectra[19–22] and the electron energy loss spectra[23] provide complementary data for the spectral analysis.

9.3.1.2 *Naphthalene*

The absorption spectrum of naphthalene vapour from 5 eV (248 nm) to 30 eV (41·3 nm) has been measured by Koch *et al.*[24] using synchrotron radiation. Above the bands at 5·89 and 7·7 eV, which are assigned to π–π^* transitions, there is a steep increase to a broad continuum with a maximum at about 16 eV, attributable to σ-electronic excitation. Superimposed on this continuum there is a complex sequence of bands, which are analysed as Rydberg series. The analysis is aided by complementary data from photoelectron spectroscopy.[19,25]

The absorption spectra of naphthalene in rare-gas matrices have been observed from 30,000–80,000 cm^{-1} (333–125 nm) by Angus and Morris.[26] Table 9.3 lists the experimental and theoretical values of the transition energies and oscillator strengths of the six transitions assigned to π-electronic states. The symmetry notation of the original paper[26] has been modified to conform to the Pariser[27] axis convention (x = long molecular axis; y = short molecular axis) used previously (Chapter 1, Vol. 1). It is of interest that all the assigned transitions are parity-allowed, and it is

Table 9.3 Energies (cm^{-1}) and oscillator strengths (f) of π-electronic transitions in naphthalene (after Angus and Morris[26])

Experimental				Theoretical				Assignment	
		Condition	Ref.	Pariser[27]		Hummel and Ruedenberg[28]			
32020	($f \sim 0{\cdot}02$)	Vapour	29	32400	($f \sim 0$)	33700	($f \sim 0{\cdot}03$)	$^1B_{3u}$	(^1L_b)
35910	($f \sim 0{\cdot}1$)	Vapour	30, 31	36200	($f \sim 0{\cdot}26$)	36500	($f \sim 0{\cdot}21$)	$^1B_{2u}$	(^1L_a)
47370	($f \sim 1{\cdot}0$)	Vapour	26	47900	($f \sim 2{\cdot}1$)	48900	($f \sim 2{\cdot}4$)	$^1B_{3u}$	(^1B_b)
49400	($f \sim 0{\cdot}3$)	Vapour	26	51000	($f \sim 0{\cdot}7$)	49200	($f \sim 0{\cdot}7$)	$^1B_{2u}$	(^1B_a)
57800	($f \sim 0{\cdot}1$)	Kr matrix	26	64200	($f \sim 0{\cdot}04$)	62000	($f \sim 0{\cdot}1$)	$^1B_{3u}$	(^1D_b)
61200	($f \sim 1$)	Kr matrix	26	66000	($f \sim 0{\cdot}85$)	66500	($f \sim 1{\cdot}0$)	$^1B_{2u}$	(^1D_a)

not necessary to postulate any parity-forbidden transitions (see Chapter 1, Vol. 1).

The lowest energy Rydberg transitions in naphthalene vapour have been reported[32] at 45,090 cm^{-1} and 45,390 cm^{-1}, within the π–π^* absorption region, and corresponding transitions are observed[26] in the krypton matrix spectrum of naphthalene. The peculiar lineshapes of certain absorption transitions in naphthalene have been interpreted[33–35] in terms of interference between Rydberg states and isoenergetic π^* vibronic states, leading to antiresonances in the absorption spectrum. Jortner and Morris[33] analysed the naphthalene vapour spectrum from 50,000–70,000 cm^{-1}, recorded photographically by Angus *et al.*,[36] and identified four such antiresonances. Scheps *et al.*[37] measured the vapour absorption spectra of benzene from 160 to 188 nm and of naphthalene from 150 to 180 nm. In naphthalene they confirmed three of the anti-resonances identified by Jortner and Morris,[33] the fourth being considered an artefact, and they reported other interference effects. No peculiar lineshapes were observed in the benzene absorption spectrum.

Although the naphthalene vapour absorption spectrum of Koch *et al.*[24] agrees closely with that of Scheps *et al.*,[37] the former are sceptical of the latter's interpretation of the three unusual absorption profiles in the 7–8 eV region as antiresonances. The analysis in this spectral range is difficult because of sequence congestion, and there are several accidental coincidences of presumed Rydberg bands and vibrational bands belonging to other Rydberg states. Koch *et al.*[24] consider that the uncertainties in the underlying background absorption due to the π–π^* band at approximately 7·7 eV make an unequivocal assignment impossible at present, and they suggest that some of the absorption profiles may be merely accidental. Further studies with improved resolution may settle these questions.

9.3.1.3 *Anthracene*

Magnetic circular dichroism (MCD) spectroscopy has been used to assign electronic transitions in a variety of molecules.[38–43] The MCD spectra of the following aromatic hydrocarbons have been studied: benzene,[42] toluene,[40] naphthalene,[40,41,43] anthracene,[40,41,43,608] phenanthrene,[41,43] tetracene,[40] 1:2-benzanthracene,[43] triphenylene,[42] 1:2:3:4-dibenzanthracene,[43] picene[43] and coronene.[42] In almost all cases the MCD spectra correlate with the optical absorption spectra.

Larkindale and Simkin[43] observed a prominent band at about 400 nm in the MCD spectrum of anthracene in solution at room temperature. The band lies 1400 cm^{-1} below the $^1A - {}^1L_a$ $0 - 0$ transition, and they assigned it to the $^1A - {}^1L_b$ transition in anthracene. Observations of the linear dichroism spectrum of anthracene in a stretched polyethylene film led to a similar assignment.[607] There are several reasons for rejecting this assignment.

(i) The majority of molecular orbital calculations, including those of Pariser,[27] find the ^1L_b energy of anthracene to lie above the ^1L_a energy, although two such calculations[44,45] have placed ^1L_b 1300–1800 cm^{-1} below ^1L_a.

(ii) The spectrum and radiative lifetime of the anthracene fluorescence correspond to the $^1L_a - {}^1A$ transition. If ^1L_b lay below ^1L_a, the anthracene fluorescence would represent a remarkable exception to Kasha's rule.

(iii) The $^1A - {}^1L_b$ absorption of unsubstituted aromatic hydrocarbons has a molar extinction coefficient $\varepsilon \sim 300$ (pp. *70–5*) which would be readily observable if ^1L_b lay below ^1L_a in anthracene.

(iv) Bergman and Jortner[46] have observed an electronic state at ~30,000 cm^{-1} in the two-photon absorption of anthracene which they assign to the ^1L_b ($^1B_{3u}$) state in agreement with the calculations of Pariser.[27]

(v) The 400 nm MCD band[43] coincides with the most prominent hot band in the anthracene $^1A - {}^1L_a$ absorption spectrum, and may be associated with it.

(vi) Recent measurements by Mason and Peacock[608] of the linear dichroism of anthracene in a stretched polyethylene film, its MCD in a range of solvents and its mesophase-induced circular dichroism in a cholesteric solvent, place the ^1L_b state near 28,000 cm^{-1}, about 1500 cm^{-1} above ^1L_a, contrary to the earlier assignment.[43,607]

Koch *et al.* have observed the absorption spectra of anthracene vapour[609] and crystal[610] in the vacuum ultraviolet from 5 to 8·5 eV photon energy. The positions and assignments of the most prominent π–π^* transitions are listed in Table 9.3A.

Table 9.3A Energies (eV) and assignments of most prominent π–π^* transitions of anthracene (after Koch *et al.*[609])

Vapour	*n*-Heptane solution[611]		Crystal[610]
3·45	3·27	^1L_a	3·13
5·24	4·84	^1B_b	4·56
~5·8 sh	5·62	^1C_b	5·56
~6·9	6·66	^1B_a	6·30
~8·1			7·75

9.3.2 *Flash photolysis and* $S_1 - S_p$ *absorption* (pp. *58–62*)

These topics have been treated in previous chapters. Flash photolysis and its application to triplet–triplet absorption spectroscopy have been described by Labhart and Heinzelmann (Chapter 6, Vol. 1). Dye lasers, including the pulsed and tunable versions which have opened up new experimental possibilities in nanosecond flash photolysis and multiphotonic absorption, have been discussed by Snavely (Chapter 5, Vol. 1). The applications of lasers to the study of molecular relaxation processes have been reviewed by Lippert (Chapter 1, Vol. 2). The $S_1 - S_p$ absorption spectra of aromatic hydrocarbons have been summarized by Birks (Chapter 1, Vol. 1).

9.3.3 *Multiphotonic absorption* (pp. *62–9*)

9.3.3.1 *Experimental methods*

Early measurements of two-photon absorption were limited by the monochromaticity of the laser light sources (pp. *67–9*; Chapter 1, Vol. 1). Observations were made using the single frequencies, or their harmonics, obtainable from a Q-switched neodymium laser (9450 cm^{-1}) or ruby laser (14,410 cm^{-1}). Various methods have been used to overcome this limitation.

(i) *The two-beam method* introduced by Hopfield *et al.*[47] involves the simultaneous absorption of photons from a Q-switched laser and from a conventional xenon flash lamp after passage through a monochromator or filter. The variation of energy of the photons from the conventional light source enables the two-photon absorption spectrum to be explored over a wider energy range. This method, which has been used by several observers,[47–53] has a relatively low sensitivity which prevents its use for the study of organic molecules in dilute solution, and it has been mainly applied to the study of pure organic liquids and crystals.

(ii) *The Raman method* utilizes the Raman emission stimulated from a series of organic liquids by a Q-switched laser to broaden the frequency range of the latter.[54,55] Webman and Jortner[55] used this technique to observe the two-photon excitation spectrum of the fluorescence of anthracene in solution and in the crystal phase from 3·10–3·56 eV. Provided Vavilov's law (p. *142*) is obeyed, the two-photon fluorescence excitation spectrum is proportional to the two-photon absorption spectrum. The method is limited in its frequency range, and it has a poor spectral resolution, determined by the separation of the Raman scattering frequencies.

(iii) *The high-intensity tunable dye laser*, described by Snavely (Chapter 5, Vol. 1), provides the ideal source for two-photon absorption spectrometry, and it does not suffer from the limitations of the previous two methods. Bergman and Jortner[46] have used tunable dye lasers to observe the two-photon fluorescence excitation spectrum of anthracene from 28,800 to 42,000 cm^{-1}, and their results are discussed in §9.3.3.4.

9.3.3.2 *Benzene*

Richards and Thomas[56] used the frequency-doubled (28,810 cm^{-1}) photons from a 15 ns-pulsed Q-switched ruby laser for the two-photon excitation of liquid benzene at 57,620 cm^{-1} (173·5 nm). Following excitation they observed a transient absorption, characteristic of the ${}^1B_{1g} - {}^1E_{1u}$ transition in the benzene excimer (p. *62*), and a permanent absorption attributable to a benzene photoproduct, probably fulvene (p. *173*). Richards and Thomas[56] concluded that the two-photon excitation occurs into the ${}^1E_{2g}$ state of benzene, which internally converts into the ${}^1B_{2u}(S_1)$ state of benzene and thence forms the benzene excimer.

Honig *et al.*[57] proposed that two-photon spectrometry be used to determine the symmetry of the S_2 state of benzene at 6·2 eV. If $S_2 = {}^1E_{2g}$, as proposed by Dunn and Ingold[58] and Brillante *et al.*,[593] the two-photon transition is parity-allowed and has a predicted absorption cross-section of 10^3–$10^4 \times 10^{-51}$ cm^4 s $photon^{-1}$. If $S_2 = {}^1B_{1u}$, the two-photon transition is parity-forbidden and has a predicted absorption cross-section of 0·4–20 $\times 10^{-51}$ cm^4 s $photon^{-1}$. Monson and McClain,[52] using the double-beam method, could not detect any two-photon absorption exceeding about 10^{-51} cm^4 s $photon^{-1}$ in the region of the $S_0 - S_2$ transition in liquid benzene. This result supports the $S_2 = {}^1B_{1u}$ assignment,[7–10] but it conflicts with the $S_2 = {}^1E_{2g}$ assignment.[58,593]

9.3.3.3 *1-Chloronaphthalene*

Monson and McClain[53] have shown that the two-photon absorption cross-section $\langle\delta\rangle$ for photons of any polarization (linear, circular or elliptical), averaged over all orientations of the absorbing molecule, is given by

$$\langle\delta\rangle = \delta_F F + \delta_G G + \delta_H H \tag{9.1}$$

where δ_F, δ_G and δ_H are molecular parameters, and F, G and H are simple functions of the polarization vectors. δ_F, δ_G and δ_H may be calculated theoretically and they may be determined experimentally if both linearly and circularly polarized light are used. Two rules are formulated for allowed two-photon transitions:

(i) in transitions which preserve the symmetry of the electronic wavefunction, $\delta_G = \delta_H$; and

(ii) in transitions which create two perpendicular nodal planes in the electronic wavefunction, $\delta_F = 0$.

The theory of the polarization of the fluorescence excited by two-photon absorption has been developed by McClain.[59–61]

Monson and McClain[52,53] used the two-beam method for polarization studies of the two-photon absorption spectrum of liquid 1-chloronaphthalene. They considered 1-chloronaphthalene as a slightly perturbed naphthalene molecule and assumed D_{2h} symmetry in the spectral analysis. Three maxima were observed[52] in the two-photon absorption spectrum at 42,600, 37,700 and 34,400 cm^{-1}, respectively. From its polarization characteristics the strong 42,600 cm^{-1} peak was assigned to a $^1A_g \rightarrow {}^1A_g$ transition, in agreement with an earlier assignment by Eisenthal *et al.*[50] who observed the unpolarized spectrum. The weak 34,400 cm^{-1} peak was assigned to a vibronically-induced two-photon-forbidden transition to the $^1B_{2u}(^1L_a)$ state. The 37,700 cm^{-1} peak, which does not correspond to any feature of the one-photon absorption spectrum, was assigned to a $^1A_g \rightarrow {}^1B_{1g}$ transition from its polarization characteristics.

Subsequently Monson and McClain[53] undertook a more complete polarization study, in which they determined δ_F, δ_G and δ_H and the ratio $R = (\delta_G - \delta_H)/\delta_F$ as a function of the two-photon energy for liquid 1-chloronaphthalene. The results confirm the $^1A_g \rightarrow {}^1A_g$ assignment of the 42,600 cm^{-1} peak, and the $^1A_g \rightarrow {}^1B_{2u}$ assignment of the 34,400 cm^{-1} peak and some adjacent maxima in the R spectrum, but they provide little or no evidence for the 37,700 cm^{-1} peak previously reported.[52] Monson and McClain[53] do not indicate which of their conflicting data are the more reliable. They state that 'the reader must decide for himself whether it (the 37,700 cm^{-1} peak) is there,' although they admit that 'the matter can only be settled in the laboratory. Until then one must regard our reported absolute absorptions as provisional values.'

Monson and McClain[53] noted that the 42,600 cm^{-1} peak, observed in the two-photon absorption spectrum of liquid 1-chloronaphthalene and assigned to the $^1A_g^-$ state, is of similar energy to the 43,500 cm^{-1} peak, observed in the threshold electron impact spectrum of naphthalene vapour and assigned by Birks *et al.*[62] to the $^3B_{2u}^+$ state. The differences in molecular structure, molecular environment and mode of excitation, the low spectral resolution and the relatively high electronic state density in

Table 9.4 Energies (cm^{-1}), absorption cross-sections σ (cm^4 s $photon^{-1}$) and assignments of two-photon transitions in anthracene (Bergman and Jortner[46])

Benzene solution	Cyclohexane solution	σ ($\times 10^{52}$)	Assignment	Theoretical (Pariser[27])	Crystal	σ ($\times 10^{51}$)	Assignment
31000	31200	5	$^1B_{3u}^- + \bar{\nu}_P$	29600	31600	40	^1A_u, ^1B_u ($^1B_{3u}^-$)
33200	33200	45	$^1B_{1g}^+$	36800	33400	4000	X
34600	34600	65	$^1B_{1g}^+ + \bar{\nu}_2$		34600	1000	$X + \bar{\nu}_2$
	35800	70	$^1B_{1g}^+ + 2\bar{\nu}_2$				
37600	37400	300	$^1B_{1g}^-$	39600	36600	700	^1A_g, ^1B_g ($^1B_{1g}^-$)
40000	40000	800	$^1A_{1g}^-$	40000	38800	1300	^1A_g, ^1B_g ($^1A_g^-$)

this spectral region, suggest that the approximate agreement is coincidental. It hardly justifies the conclusion of Monson and McClain[53] that 'their (electron impact) state is almost surely the same as the $^1A_g^-$ which we see', or its corollary that 'new theoretical work is needed to reassess the rules by which electron impact spectra are presently interpreted.'

9.3.3.4 *Anthracene*

Bergman and Jortner[46] observed the two-photon fluorescence excitation spectrum (assumed proportional to the two-photon absorption spectrum) of crystal anthracene and anthracene solutions from 28,800 cm^{-1} (347 nm) to 42,000 cm^{-1} (238 nm) using tunable dye lasers. Table 9.4 lists the energies and absorption cross-sections of the maxima in the two-photon absorption spectra of anthracene in benzene, anthracene in cyclohexane, and crystal anthracene, and the proposed assignments of the final states of the transitions.[46]

Transitions to five excited electronic states of anthracene have been identified in its two-photon solution absorption spectrum. Webman and Jortner[55] observed a weak two-photon absorption ($\delta \sim 10^{-52}$ cm^4 s $photon^{-1}$) at $\sim$28,000 cm^{-1} which they attributed to a vibronically induced, parity-forbidden transition to $^1B_{2u}^+(^1L_a)$. The 31,000 cm^{-1} transition (Table 9.4) is assigned[46] to the parity-forbidden transition to $^1B_{3u}^-(^1L_b)$, promoted by a vibrational mode $\bar{\nu}_P \sim 1000$ cm^{-1}, so that the $^1B_{3u}^- \leftarrow {}^1A_{1g}$ electronic origin is at $\sim$30,000 cm^{-1}. The 33,200, 37,600 and 40,000 cm^{-1} transitions (Table 9.4) are assigned to allowed two-photon transitions to $^1B_{1g}^+$, $^1B_{1g}^-$ and $^1A_{1g}^-$, respectively. The first of these exhibits a vibrational progression of the totally symmetric mode $\bar{\nu}_2$ ($\sim$1400 cm^{-1}).

In the two-photon crystal spectrum (Table 9.4) the 31,600, 36,600 and 38,800 cm^{-1} transitions are assigned to the unresolved Davydov components of $^1B_{3u}^-$, $^1B_{1g}^-$ and $^1A_{1g}^-$, respectively. The 33,400 cm^{-1} crystal absorption, designated X, presents difficulties because of its absorption intensity. Bergman and Jortner[46] discuss five possible assignments of X of which they favour either (a) $^1B_{1g}^+$ with crystal-field-induced interaction with higher intense transitions to g-states or (b) interference between zero-order bound exciton states and conduction band states. Further calculations and experiments are required to elucidate the matter.

9.3.3.5 *Organic crystals*

The two-photon excitation of several organic crystals has been studied by Kobayashi and Nagakura[63] using monochromatic (620 nm) photons from a N_2-laser-pumped dye laser. They observed the time-resolved fluorescence spectra excited by one-photon and two-photon absorption, respectively. From the similarity of the fluorescence lifetimes and spectra in the two cases they concluded that the fluorescence of anthracene, fluoranthene, chrysene, tetracyanobenzene (TCNB)-durene complex and TCNB-hexamethylbenzene (HMB) complex crystals occurs from the same excited state with both modes of excitation. The crystal self-absorption is, however, higher with two-photon excitation, because of the increased photon penetration. This reduces the relative intensity of the higher energy bands in the two-photon-excited fluorescence spectrum, and there is a red shift of its mean energy. The effect is particularly marked for the TCNB-durene and TCNB-HMB crystals, where there is a red shift of 10–20 nm in the maxima of the structureless fluorescence spectra with two-photon excitation.

From similar studies of the time-resolved fluorescence spectra of a mixed anthracene-tetracene crystal, Kobayashi and Nagakura[63] showed the energy transfer process to be the same for two-photon and one-photon excitation.

9.4 Fluorescence

9.4.1 *Determination of fluorescence lifetimes* (pp. *94–7*)

9.4.1.1 *Phase and modulation fluorometry*

The increasing popularity of the photon-sampling (single-photon) technique, which is probably related to the commercial availability of the instruments[64,65] has seen a decline in interest in phase and modulation fluorometry. Nevertheless the phase and modulation technique has advantages for some purposes over the usual single-photon method using a nanosecond flash lamp. No nanosecond flash lamp is required, the observed fluorescence lifetime does not require correction for the light pulse shape, and the minimum lifetime that can be observed is not limited by the flash duration. Phase and modulation fluorometry is probably the most accurate technique for the determination of simple exponential fluorescence decays. Observations of the phase θ and modulation m provide two independent determinations of the lifetime τ which agree only if the decay is exponential (pp. *94–5*). The instrument can also be used for the measurement of fluorescence response functions, which are the sum or difference of two exponential decays, such as occur in excimer or exciplex formation or in energy transfer (pp. *305–6*). In this case four independent parameters θ_1, θ_2, m_1 and m_2, corresponding to the two fluorescent components of the system, are observed, and these provide independent means of deriving the two characteristic lifetimes τ_1 and τ_2.

Spencer[66] has given a detailed account and analysis of the phase and modulation technique, which he has applied to studies of the fluorescence lifetimes of biological molecules and their complexes and the effects of Brownian rotation and energy transfer on fluorescence depolarization and lifetimes. For $\tau \leqslant 10$ ns the absolute standard deviation in his measurements is ± 30 ps, and for $\tau > 10$ ns it is $\leqslant 1\%$.[66] This accuracy is superior to the single-photon technique using nanosecond flash lamps, and comparable with that obtained using instantaneous excitation (§9.4.1.2).

9.4.1.2 *Single-photon fluorometry*

The single-photon technique was introduced in 1961 by Bollinger and Thomas,[67] who applied nucleonics techniques and instrumentation to measure the fluorescence (scintillation) decay of a *trans*-stilbene crystal excited by ionizing radiation. Certain features of this application may be noted

(i) the instantaneous (δ-function) excitation of the fluorescence;

(ii) the low and variable intensity ($\sim 10^4$ photons) of the fluorescence pulses;

(iii) the time-randomness of the pulses; and

(iv) the observation of the non-exponential decay of the fluorescence intensity by a factor of 10^7 over a time range from 1 to 10^7 ns.

Few of these features are to be found in most applications of the single-photon technique to the determination of photofluorescence lifetimes, suggesting that the potential of the technique is not yet being fully utilized.

Detailed reviews of the single-photon method have been given by Ware,[68,69] Knight and Selinger[70] and Lewis *et al.*[603] The statistics of single photon counting have been discussed by Morton,[71] and Tull[72] has compared photon counting and integrated current measuring techniques for the spectrophotometry of sources of low intensity. Experimental and theoretical methods of correction for photon pile-up in lifetime determinations by single-photon counting have been described by Davis and King.[73,74]

Various modes of fluorescence excitation have been used in conjunction with the single photon method:

(a) a recurrent pulsed light source, used either with or without a filter or monochromator;

(b) a recurrent pulsed monoenergetic beam of electrons;

(c) a random flux of individual monoenergetic electrons; and

(d) a random flux of individual ionizing radiations (γ-rays, α-particles, protons, etc.).

The pulsed light sources fall into two main groups: thyratron-triggered low-pressure gas discharges; and free-running (relaxation oscillator) high-pressure gas discharges operating at high voltages. The relative merits have been discussed by Ware[69] and Knight and Selinger.[70] If the lamp intensity and frequency are sufficiently high, the sensitivity of the single-photon method permits the use of an excitation monochromator before the specimen and/or an analysing monochromator after the specimen. The former has enabled the fluorescence lifetimes of single vibronic levels of aromatic molecular vapours to be determined (see Stockburger, Chapter 2, Vol. 1). The latter is used for nanosecond time-resolved emission spectroscopy.[75,76]

A pulsed beam (100 ns duration, <2 ns cut-off) of monoenergetic electrons (energy variable up to 50 eV, and stabilized within 30 meV) has been used by Bridgett *et al.*[77,78] in conjunction with the single-photon technique for the excitation and lifetime determination of excited states of helium and other gases.

Imhof and Read[79,80] have developed an original method for the determination of the lifetime of an excited state of energy E of a gas molecule, which emits photons of energy $\mathbf{h}\nu$. The technique involves the use of monoenergetic incident electrons of energy $E_0(>E)$, and the observation of delayed coincidences between the inelastically scattered electrons of energy $(E_0 - E)$ and the subsequent decay photons of energy $\mathbf{h}\nu$. In this manner the observations are restricted to the excited state of energy E and its characteristic photons $\mathbf{h}\nu$, and they are not influenced by the simultaneous excitation of other excited states and emissions therefrom, including cascade effects. The method has so far only been applied to atoms and simple molecules,[81–84] but it should also be useful for studies of aromatic molecular vapours.

The single-photon technique was originally developed for the measurement of the scintillation decay of organic molecular systems[67] and it has been widely used for this purpose.[85–94] A major advantage of excitation by ionizing radiation is the negligible duration of the excitation pulse compared with that of an optical flash lamp. The subsequent ion recombination and internal conversion into the first excited singlet state S_1 are also rapid, so that excitation by ionizing radiation is equivalent to δ-function excitation of S_1.[95]

The scintillation and photofluorescence decay times of anthracene and other organic crystals are identical, provided allowance is made for any self-absorption effects.[96,97] In a liquid solution scintillator, the aromatic solvent molecules are excited instantaneously into S_1, and the fluorescent solute is then excited via solvent–solute energy transfer. The solute fluorescence response function is the difference of two exponential decays, characteristic of the solvent lifetime (as reduced by energy transfer) and the solute fluorescence lifetime, respectively.[95] The solute fluorescence lifetime corresponds to the scintillation decay time at high concentrations and to the scintillation rise time at low concentrations. For compounds with short (~ns) fluorescence lifetimes and adequate solubility in aromatic solvents, the single-photon scintillation technique is a useful alternative method of lifetime determination.

9.4.2 *Determination of fluorescence spectra and quantum yields* (pp. *97–100*)

Shinitzky[98] has pointed out a potential source of error in fluorescence quantum yield and lifetime measurements. When a fluorescent system is excited with unpolarized light and its emission is detected without a polarizer, the emission intensity has a typical anisotropic distribution which is directly related to its degree of polarization. This effect can introduce an error of up to 20% in all steady state and dynamic fluorescence measurements, but it is eliminated when the fluorescence is detected at an angle of 55° or 125° to the direction of excitation.

Three different sets of parameters are used to describe the fluorescence properties of a compound in solution. Failure to distinguish them has led to much confusion in the literature.

(i) The molecular fluorescence spectrum $F_M(\bar{\nu})$, quantum efficiency $q_{FM}(=k_{FM}/k_M)$ and lifetime $\tau_M(=1/k_M)$ at infinite dilution are the fundamental parameters (pp. *88–9*).

(ii) The actual fluorescence spectrum $F(\bar{\nu})$, quantum yield Φ_{FM} and lifetime τ of a solution of molar concentration $[^1M]$ commonly differ from $F_M(\bar{\nu})$, q_{FM} and τ_M due to self-quenching. In general

$$\Phi_{FM} \leqslant q_{FM} \tag{9.2}$$

$$\tau \leqslant \tau_M \tag{9.3}$$

and $F(\bar{\nu})$ may include a component of the excimer fluorescence spectrum $F_D(\bar{\nu})$.

(iii) The technical or observed fluorescence spectrum $F^t(\bar{\nu})$, quantum yield Φ^t_{FM} and lifetime τ^t of a solution of concentration $[^1M]$ often differ from $F(\bar{\nu})$, Φ_{FM} and τ due to self-absorption, resulting from the overlap of $F(\bar{\nu})$ and the solution absorption spectrum $\varepsilon(\bar{\nu})\,[^1M]$. The difference between the technical and actual parameters depends on the magnitude of the latter, the spectral overlap, $[^1M]$, the excitation wavelength λ_{ex} and the method of observation.

$F(\bar{\nu})$ is attenuated in the spectral overlap region, but it is otherwise unaffected (pp. *99–100*). Most of the primary fluorescence photons, excited by the incident radiation, escape from the solution, but the remainder are absorbed by other solute molecules which then emit secondary fluorescence. Most of the secondary fluorescence photons escape, but the remainder are absorbed yielding tertiary fluorescence, and so on.[102] The observed fluorescence thus includes primary and higher-order components. The latter are delayed relative to the primary emission, so that in general

$$\tau^{t} \geqslant \tau \tag{9.4}$$

Self-absorption causes (a) a decrease in the yield of primary fluorescence, and (b) an increase in the yield of secondary and higher-order fluorescence. The overall effect on Φ_{FM}^{t} depends on the relative magnitudes of (a) and (b). For solutions observed in transmission, and for solutions of low Φ_{FM} observed in reflection, (a) $\geqslant$ (b), so that[99,102]

$$\Phi_{FM}^{t} \leqslant \Phi_{FM} \tag{9.5}$$

For solutions of high Φ_{FM} observed in reflection, (b) $\geqslant$ (a), so that[99,103]

$$1{\cdot}00 \geqslant \Phi_{FM}^{t} \geqslant \Phi_{FM} \tag{9.6}$$

This enhancement of Φ_{FM}^{t} is due to the secondary fluorescence produced by the reabsorption of primary fluorescence photons emitted into the hemisphere away from the surface, which would not otherwise be detected (p. *93*).

Four cases of self-quenching (Q) and self-absorption (A) may be distinguished.

(I) $Q = 0, \quad A = 0.$

(II) $Q = 0, \quad A \neq 0.$

(III) $Q \neq 0, \quad A = 0.$

(IV) $Q \neq 0, \quad A \neq 0.$

Phenanthrene ($Q = 0, A \simeq 0$) in solution is one compound which approximates to case I. 9,10-Diphenylanthracene (DPA) in solution corresponds to case II and quinine bisulphate in N H_2SO_4 solution to Case III.[99] Perylene in solution corresponds to Case IV. Experimental data on solutions of the last three compounds are listed in Table 9.5. The data, which are from nine different sources, are remarkably self-consistent. The experimental values of

$$k_{FM} = q_{FM}/\tau_{M} \tag{9.7}$$

Table 9.5 Fluorescence parameters of compounds in solution at room temperature

Solute	Solvent	[M]	Φ^{t}_{FM} [a]	Φ_{FM}	q_{FM}	τ_M(ns)	Ref.
DPA	Benzene	10^{-2} M	1·00	0·84	0·84	—	99
		$1{\cdot}5 \times 10^{-3}$ M	1·00	0·84	0·84	—	99
		$2{\cdot}2 \times 10^{-4}$ M	0·92	0·81	0·81	—	99
		5×10^{-5} M	0·87	0·83	0·83	—	99
		$1{\cdot}5 \times 10^{-5}$ M	0·82	0·82	0·82	—	99
		$\sim 10^{-5}$ M	—	—	0·85	7·3	104
		$\sim 10^{-5}$ M	—	—	—	7·4	105
		$\sim 10^{-5}$ M	—	0·81	0·81	—	106
		2×10^{-2} M	—	$\leqslant$0·87[b]	—	—	107
	Petroleum ether	10^{-3} M	1·00	0·83	0·83	—	99
	Ethanol	10^{-3} M	0·98	0·81	0·81	—	99
		$\sim 10^{-5}$ M	—	0·76	0·76	—	106
	Alcohol	10^{-5}–10^{-6} M	—	0·84	0·84	6·8	108
	Liquid paraffin	$\sim 10^{-5}$ M	—	$\leqslant$0·88[b]	0·81	—	106
	PMMA	10^{-3} M	0·99	0·83	0·83	—	100
Perylene	Benzene	2×10^{-3} M	0·95	0·80	0·89	—	99
		$\sim 10^{-6}$ M	—	—	0·89	4·9	104
		$\sim 10^{-6}$ M	—	—	—	5·0	105
		$\sim 10^{-6}$ M	—	—	—	4·8	109
		2×10^{-6} M	—	—	—	5·0	110
Quinine bisulphate	N H_2SO_4	5×10^{-3} M	0·51	0·51	0·55	—	99
		10^{-5} M	—	—	0·54	20·1	104
		10^{-5} M	—	—	—	19·4	105
		10^{-5} M	—	—	—	20·5	111

[a] Values of Φ^{t}_{FM} for $\lambda_{ex} = 366$ nm, front-surface observation.
[b] Values of Φ_{FM} from $\Phi_{FM} \leqslant 1 - \Phi_{TM}$, where Φ_{TM} is triplet quantum yield.

for the three compounds agree closely with the theoretical values derived from the absorption and fluorescence spectra.[104,105,109] Table 9.5 thus provides a reliable reference set of fluorescence parameters.

Fluorescence quantum yields are usually determined by comparison with a standard of known fluorescence quantum yield (pp. *97–8*). Melhuish[99,100] proposed quinine bisulphate in N H_2SO_4 solution as a fluorescence standard because $A = 0$, so that $\Phi^t_{FM} = \Phi_{FM}$ irrespective of the optical geometry. The scale of fluorescence quantum yields, based on this and related secondary standards, is self-consistent. It has been adopted by many observers and was used in *PAM* wherever possible.

Berlman[101] has made an extensive study of $F^t(\bar{\nu})$, Φ^t_{FM} and τ^t of aromatic compounds in solution. The observations were made in reflection, using monochromatic excitation for the determination of $F^t(\bar{\nu})$ and Φ^t_{FM}, and heterochromatic excitation for the observation of τ^t. For the former measurements λ_{ex} was chosen to coincide with a strong absorption peak to minimize the optical penetration depth d_{ex}. This procedure reduces the effect of self-absorption on $F^t(\bar{\nu})$, so that $F^t(\bar{\nu})$ approximates closely to $F(\bar{\nu})$ and, in the absence of excimer fluorescence, to $F_M(\bar{\nu})$. On the other hand, the reduction of d_{ex} increases the contribution of secondary fluorescence to Φ^t_{FM}. The increase of Φ^t_{FM} with increase in $[^1M]$, i.e. reduction in d_{ex}, for DPA in benzene (Table 9.5) is due to this effect, which corresponds to (9.6). Berlman[101] chose a $0{\cdot}3\ gl^{-1}$ ($\sim 10^{-3}$ M) solution of DPA in cyclohexane, observed in reflection at $\lambda_{ex} = 265$ nm, as his fluorescence standard, for which he assumed $\Phi^t_{FM} = 1{\cdot}00$. Since d_{ex} at $\lambda_{ex} = 265$ nm is less than at $\lambda_{ex} = 366$ nm, used by Melhuish,[99,100] this value of $\Phi^t_{FM} = 1{\cdot}00$ is fully consistent with the data of Table 9.5. Berlman's recent attempt[605] to discredit some of these data in favour of his own is therefore unnecessary.

Berlman[101] did not correct Φ^t_{FM} and τ^t for self-absorption to obtain Φ_{FM} and τ, or for self-quenching to obtain q_{FM} and τ_M. He made the simpler assumptions, only valid for Case I compounds, that

$$\Phi^t_{FM} = \Phi_{FM} = q_{FM} \tag{9.8.1a}$$

$$\tau^t = \tau = \tau_M \tag{9.8.1b}$$

and thence substituted in (9.7) to obtain τ_{FM} ($= 1/k_{FM}$). For Case II compounds, like DPA,

$$\Phi^t_{FM} \geqslant \Phi_{FM} = q_{FM} \tag{9.8.2a}$$

$$\tau^t \geqslant \tau = \tau_M \tag{9.8.2b}$$

Berlman's values[101] of $\Phi^t_{FM} = 1{\cdot}00$ and $\tau^t = 9{\cdot}35$ ns for DPA exceed q_{FM} and τ_M (Table 9.5) by 20% and 27%, respectively. For Case III compounds, like quinine bisulphate,

$$\Phi^{t}_{FM} = \Phi_{FM} \leqslant q_{FM} \tag{9.8.3a}$$

$$\tau^{t} = \tau \leqslant \tau_{M} \tag{9.8.3b}$$

For Case IV compounds, like perylene,

$$\Phi^{t}_{FM} \geqslant \Phi_{FM} \leqslant q_{FM} \tag{9.8.4a}$$

$$\tau^{t} \geqslant \tau \leqslant \tau_{M} \tag{9.8.4b}$$

Berlman's values[101] of $\Phi^{t}_{FM} = 0{\cdot}94$ and $\tau^{t} = 6{\cdot}9$ ns for perylene exceed q_{FM} and τ_{M} (Table 9.5) by 6% and 41%, respectively. Berlman[101] has provided a most useful *Handbook of Fluorescence Spectra*, but his other parameters Φ^{t}_{FM} and τ^{t}, although probably correct, do not generally correspond to the molecular parameters q_{FM} and τ_{M} which are of fundamental interest.

In *PAM* (pp. *103–4*, Tables *4.3*, *4.5*) Berlman's Φ^{t}_{FM} values[101] were multiplied by the DPA correction factor $q_{FM}/\Phi^{t}_{FM} = 0{\cdot}83$ in an attempt to estimate q_{FM} for his other compounds. This procedure is invalid, and it should be disregarded. The self-quenching and self-absorption corrections are specific to each solution, and the latter also depends on the excitation and observation conditions. The corrections can only be properly applied by the observer. If such corrections are not made, q_{FM} and τ_{M} should only be observed for very dilute solutions.

Improved methods of fluorescence spectrophotometry have been developed. Cundall and Evans[112] have designed an automatic instrument which yields corrected fluorescence and fluorescence excitation spectra. Lipsett *et al.*[113] have described a spectrofluorimeter with a digital recording system. Longworth[114] has described techniques for measuring the fluorescence and phosphorescence parameters of biological and other molecules with low luminescence yields.

9.4.3 *Fluorescence lifetimes and quantum efficiencies* (pp. *103–6*, *120–32*)

9.4.3.1 *Alkyl benzenes*

The fluorescence properties of benzene in fluid media have been described by Cundall (Chapter 2, Vol. 2) and, in the vapour phase, by Stockburger (Chapter 2, Vol. 1).

Two recent studies of the fluorescence properties of the alkyl benzenes in solution supplement those of Berlman[101] and others. Froehlich and Morrison[115] measured the fluorescence quantum efficiency q_{FM} and lifetime τ_{M} of twenty-six alkyl benzenes in $\sim 10^{-3}$ M hexane solution at room temperature, and evaluated therefrom k_{FM} and k_{IM}, the S_1 radiative

Table 9.6 Fluorescence parameters of methyl-substituted benzenes in room-temperature solution in cyclohexane (CX) or hexane (H)

Positions of substituents	Solvent	q_{FM}	Ref.	τ_M(ns)	Ref.	k_{FM} (10^6 s^{-1})	k_{IM} (10^6 s^{-1})	Ref.	Point group
0 (Benzene)	H	0·036	a	33·6	a	1·1	29	a	D_{6h}
	H	0·055	b	26	c	2·1	37	d	
	H	0·058	e	—					
	CX	0·063	f	29	g	2·0	33	f	
	CX	0·058	G	29	g	2·0	32·5	d	
	CX	0·07	h	28	h	2·6	33·4	h	
	CX	—		30·5	m				
1 (Toluene)	H	0·14	a	35·2	a	4·0	24	a	C_{2v}
	H	0·12	b	39·2	i	3·0	34·5	d	
	CX	0·14	f, G	34	g	4·2	27	f	
	CX	—		35	m				
1,2 (*o*-Xylene)	H	0·17	a	38·2	a	4·5	22	a	C_{2v}
	H	0·15	b	—					
	CX	0·16	G	32·2	g	4·9	26	d	
1,3 (*m*-Xylene)	H	0·13	a, b	32·7	a	4·0	27	a	C_{2v}
	CX	0·14	G	30·8	g	4·6	28	d	
1,4 (*p*-Xylene)	H	0·22	a	33·3	a	6·6	23	a	D_{2h}
	H	0·20	b	32·5	j	6·3	25	d	
	CX	0·33	G	30	g	11·1	23·3	d	
	CX	0·34	f	30	g	11·1	23	f	
	CX	—		34	m				
1,2,3	H	0·08	a, k	—					C_{2v}
	CX	0·15	f	—		4·2	24	f	
1,2,4	H	0·28	a	—		10·3	26	a	C_s
	CX	0·34	G	27·2	g	12·6	24	d	

1,3,5 (Mesitylene)	H	0·09	a	38·3	a	2·3	24	a	D_{3h}
	H	0·09	b	27	c	3·3	35	d	
	CX	0·14	G	36·5	g	3·9	24	d	
	CX	—		39·5	m				
1,2,3,4	H	0·11	a	33·8	a	3·3	26	a	C_{2v}
	CX	0·12	f	—		4·8	35	f	
1,2,3,5	H	0·13	a	27·9	a	4·7	31	a	C_{2v}
	CX	0·16	f	—		5·3	28	f	
1,2,4,5 (Durene)	H	0·30	a	28·6	a	10·5	24	a	D_{2h}
	CX	0·25	f	—		10·4	31	f	
1,2,3,4,5 (Pentamethyl)	H	0·075	a	15·7	a	4·8	59	a	C_{2v}
	CX	0·08	f	—		4·6	53	f	
1,2,3,4,5,6 (Hexamethyl)	H	0·01	a	0·6	a	2	1600	a	D_{6h}
	CX	0·015	f	—		3·2	240	f	
	CX	0·02	b, l	6	c	3·3	160	d	

For references see Table 9.7.

Table 9.7 Fluorescence parameters of alkyl-substituted benzenes in room temperature solution in cyclohexane (CX) or hexane (H)

Alkyl substituents	Solvent	q_{FM}	Ref.	τ_M(ns)	Ref.	k_{FM} (10^6 s^{-1})	k_{IM} (10^6 s^{-1})	Ref.	Point group
Ethyl	H	0·11	a	35·1	a	3·1	25	a	C_{2v}
	CX	0·15	G	31	g	4·9	27	d	
	CX	—		33	m				
n-Propyl	H	0·11	a	36·0	a	3·1	25	a	C_{2v}
	CX	0·12	G	32·4	g	3·7	27	n	
iso-Propyl	H	0·076	a	24·5	a	3·1	38	a	C_{2v}
	CX	0·10	G	22	g	4·5	41	n	
n-Butyl	H	0·11	a	35·1	a	3·1	25	a	C_{2v}
iso-Butyl	H	0·12	a	33·3	a	3·6	26	a	C_{2v}
sec-Butyl	H	0·088	a	29·2	a	3·0	31	a	C_{2v}
	CX	0·12	G	25	g	4·8	35	n	
tert-Butyl	H	0·032	a	10·0	a	3·2	97	a	C_{2v}
n-Hexyl	H	0·12	a	35·3	a	3·4	25	a	C_{2v}
n-Decyl	H	0·12	a	—					C_{2v}
n-Nonadecyl	H	0·11	a	34·2	a	3·2	26	a	C_{2v}
Cyclohexyl	H	0·089	a	29·2	a	3·0	31	a	C_{2v}
	CX	0·125	G	26·4	g	4·7	33	n	
1-Methyl, 4-ethyl	CX	0·23	G	30·8	g	7·5	25	n	
1,3-Di-*tert*-butyl	H	0·061	a	14·5	a	4·2	65	a	C_{2v}

1.3.5-Triethyl	CX	0·10	G	24	g	4·2	38	n	D_{3h}
1,3,5-Tri-*tert*-butyl	H	0·01	a	15·1	a	0·66	66	a	D_{3h}
Hexaethyl	H	0·01	a	—					D_{6h}

a. Froehlich and Morrison[115]
b. Lumb and Weyl[117]
c. Ivanova, Kudryashov and Sveshnikov[118]
d. Birks (*PAM* pp. *178–81*)
e. Eastman[110]
f. Reiser and Leyshon[116]
g. Berlman[101]
G. q_{FM} data of Berlman[101] renormalized (pp. *122–4*)
h. Helman[120]
i. Greenleaf, Lumb and Birks[121]
j. Greenleaf and King[122]
k. Berenfel'd and Krongauz[123]
l. Bowen and Williams[124]
m. Klein, Heisel, Lami and Laustriat[125]
n. Present work

and radiationless decay rates, respectively. Reiser and Leyshon[116] observed q_{FM} for several methyl benzenes in 10^{-4} M cyclohexane solution at room temperature, and they evaluated k_{FM} and k_{IM}, using calculated or published values of k_{FM}. The results of these and other studies are summarized in Tables 9.6 and 9.7.

The data show some interesting correlations with molecular structure.

(i) k_{FM} depends on the symmetry point group of the substituted benzene,[116] as explained by Petruska[126] and Murrell[127] in their theoretical treatment of the extinction coefficients of the alkyl benzenes. In the benzene D_{6h} point group the symmetry-forbidden $^1B_{2u} - {}^1A_{1g}$ $(S_1 - S_0)$ transition is made partially allowed by perturbations of e_{2g} symmetry (p. 9). Asymmetric substitution in benzene is equivalent to such a perturbation, and the relative content of e_{2g} symmetry in a substituted benzene determines the magnitude of k_{FM}.[127] The observed values of k_{FM} agree reasonably with the group-theoretical predictions (Figure 9.1).

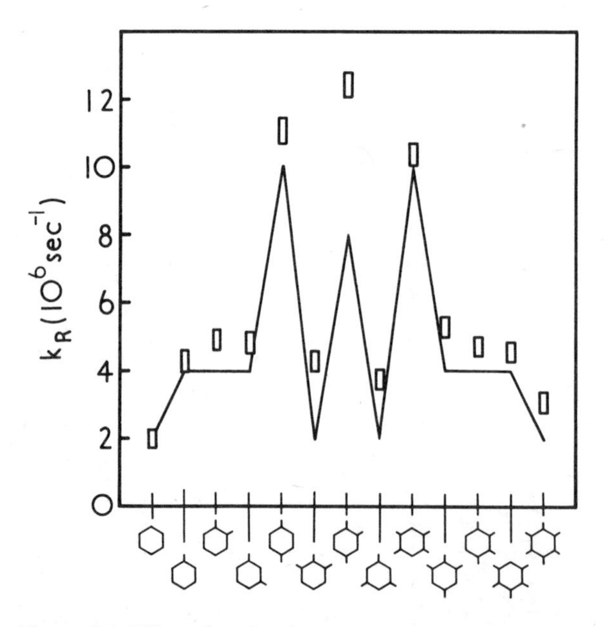

Figure 9.1 Effect of molecular symmetry on the $S_1 - S_0$ radiative decay rate k_R $(= k_{FM})$ in the methyl-substituted benzenes. The full line shows the group-theoretical predictions[127] and the rectangles the experimental values (after Reiser and Leyshon[116])

(ii) k_{IM} has similar values for benzene and its methyl derivatives, except pentamethyl and hexamethyl benzene. Since $k_{FM} \ll k_{IM} \simeq k_M$, q_{FM} is approximately proportional to k_{FM}, while τ_M is approximately constant, except for the last two compounds.

At room temperature k_{IM} is the sum of a temperature-independent component k_{IM}^0 and a temperature-dependent component k_{IM}^T. k_{IM}^0 is due to intersystem crossing from S_1 to the triplet manifold. In benzene and toluene k_{IM}^T corresponds to thermally activated internal conversion (channel 3) possibly to a physical isomer (Chapter 1, Vol. 1), but in other alkyl benzenes k_{IM}^T may be due to thermally activated intersystem crossing. Pending the separation and identification of the components of k_{IM}, it would be premature to speculate about its insensitivity to alkyl substitution in benzene at room temperature.

(iii) In the mono-alkyl benzenes k_{FM} and k_{IM} are insensitive to the extension of a linear alkyl chain substituent beyond two carbons, cf. ethyl, *n*-propyl, *n*-butyl, *n*-hexyl, *n*-nonadecyl (Table 9.7).[115] The $S_0 - S_1$ extinction coefficients show a similar insensitivity to linear alkyl chain length.[101]

(iv) Froehlich and Morrison[115] have observed an 'α-substitution effect' in the mono-alkyl benzenes. Alkyl substituents on the α-carbon attached to a benzene ring cause a reduction in q_{FM}, e.g. toluene (0·14), ethylbenzene (0·11), *iso*propylbenzene (0·07), *tert*-butylbenzene (0·03). A linear correlation between q_{FM} and the number of β-hydrogens is observed.[115] The prime effect is an increase in k_{IM} (Table 9.7). Similar behaviour was observed by Rabalais *et al.*[128] in studies of the radiationless decay of the triplet states of the alkyl benzenes. They concluded that 'CH stretching vibrations of the alkyl group appended to a benzene ring are efficient in modulating radiationless transitions. This quenching ability extends as far as carbon centres which are twice removed from the ring and not much further'. A similar explanation is applicable to the $S_1 - T_1$ intersystem crossing component (k_{IM}^0) of k_{IM}, but further experiments are required to determine whether k_{IM}^T is also influenced by the 'α-substitution effect'.

Phillips[129] has published a comprehensive review of the photophysics and photochemistry of methyl, fluoro and trifluoromethyl benzenes in the vapour phase. His article includes data on the observed and calculated singlet and triplet energy levels, fluorescence quantum yields and lifetimes, triplet quantum yields, fluorescence and triplet excitation spectra, photoisomerization quantum yields, vibrational relaxation parameters, and

Table 9.8 Photophysical parameters of methyl-substituted benzenes in the vapour phase (after Phillips[129])

Position of substituents	p (torr)	λ_{ex} (nm)	Φ_{FM}	Ref.	Φ_{TM}	Ref.	τ_M (ns)	Ref.	τ_{FM} (ns)	τ_{FM}(calc) (ns)	k_{FM} (10^6 s^{-1})	k_{IM} (10^6 s^{-1})	k_{TM} (10^6 s^{-1})	Point group
0	20	254	0·18	a	0·72	b	77	c	428	407	2·33	10·6	9·3	D_{6h}
1	3·5–18	264	0·30	d	0·70	d	56	e	180	141	5·36	12·5	12·5	C_{2v}
1,2	3–5·5	270	0·38	f	0·53	g	52	e	136	152	7·31	11·9	10·2	C_{2v}
1,3	3	271	0·35	f	0·40	g	49	e	140	167	7·14	13·3	8·2	C_{2v}
1,4	3	272	0·52	f	0·54	g	44	e	85	95	11·8	10·9	12·2	D_{2h}
1,2,3	1–2	270	(0·29)		—		70	e	—	238	(4·2)	(10·6)	—	C_{2v}
1,2,4	1·8	270	(0·55)		—		40	e	—	72	(13·8)	(11·2)	—	C_s
1,3,5	1·5–1·8	272	(0·24)		—		54	e	—	222	(4·49)	(14·1)	—	D_{3h}
1,2,3,4	0·04	280	(0·31)		—		64	e	—	206	(4·8)	(11·9)	—	C_{2v}
1,2,3,5	0·05	277	(0·36)		—		68	e	—	187	(5·3)	(9·4)	—	C_{2v}
1,2,4,5	0·03	280	(0·52)		—		37	e	—	71	(14·1)	(13·0)	—	D_{2h}
1,2,3,4,5	0·015	277	(0·25)		—		54	e	—	217	(4·6)	(13·8)	—	C_{2v}
1,2,3,4,5,6	—	—	—		—		—		—	312	(3·2)	—	—	D_{6h}

Notes

p = pressure; λ_{ex} = excitation wavelength; Φ_{FM} = fluorescence quantum yield; Φ_{TM} = triplet quantum yield; τ_M = fluorescence lifetime; τ_{FM} = radiative lifetime; τ_{FM}(calc) obtained from integrated absorption spectrum; k_{FM} = radiative decay rate; k_{IM} = radiationless decay rate; k_{TM} = intersystem crossing rate.
Bracketed values of Φ_{FM} from (τ_{FM}(calc)/τ_M).
Bracketed values of k_{FM} from τ_{FM}(calc)$^{-1}$.
Bracketed values of k_{IM} from ($\tau_M^{-1} - \tau_{FM}$(calc)$^{-1}$).

References

a. Noyes, Harter and Mulac[130]
b. Noyes and Harter[131] (corrected value)
c. Nishikawa and Ludwig[132]
d. Burton and Noyes[133]
e. Breuer and Lee[134]
f. Noyes and Harter[135]
g. Noyes and Harter[136]

impurity quenching of excited singlet and triplet states. Table 9.8 gives the S_1 photophysical parameters of the methyl benzenes in the vapour phase for comparison with Table 9.6. For the other data the reader is referred to the article by Phillips.[129]

9.4.3.2 *Methyl naphthalenes*

Wright and Reiser[137] have extended the studies of k_{FM} and k_{IM} for the methyl benzenes[116] to the monomethyl and dimethyl naphthalenes. In the latter compounds $\Phi_{FM} + \Phi_{TM} \simeq 1$ (p. *200*), so that $k_{IM} \simeq k_{TM}$, the intersystem crossing rate. The values of k_{FM} and k_{TM} for the methyl-substituted naphthalenes in room-temperature solution are listed in Table 9.9. Figure 9.2 compares the experimental values of k_{FM} with the

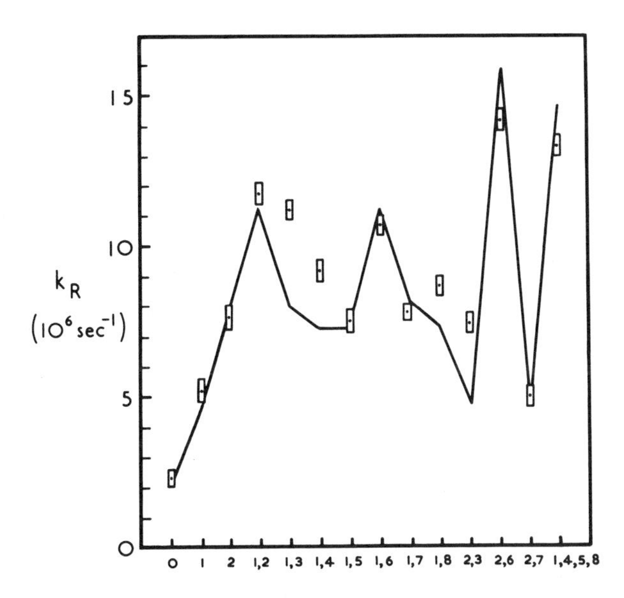

Figure 9.2 Effect of molecular symmetry on the $S_1 - S_0$ radiative decay rate k_R ($= k_{FM}$) in the methyl-substituted naphthalenes. The full line shows the group-theoretical predictions and the rectangles the experimental values (after Wright and Reiser[137])

predictions of a simple perturbation treatment[137] based on the method of Murrell and Longuet-Higgins.[138] The relatively high value of k_{TM} for 1,8-dimethylnaphthalene is attributed to steric effects between the adjacent methyl groups which distort the naphthalene planar configuration, and similar effects are considered to account for the high values of k_{IM} for pentamethyl- and hexamethyl-benzene.[137]

Table 9.9 Rate parameters of methyl-substituted naphthalenes in room-temperature solution (Wright and Reiser[137])

Positions of substituents	k_{FM} ($10^6\ s^{-1}$)	k_{TM} ($10^6\ s^{-1}$)
0	2·3	7·8
1	5·5	10·2
2	7·4	9·4
1,2	11·6	12·0
1,3	10·7	7·1
1,4	9·0	9·4
1,5	7·3	10·9
1,6	10·5	9·4
1,7	7·6	9·3
1,8	8·8	19·4
2,3	7·3	6·7
2,6	14·1	10·6
2,7	4·8	7·2

9.4.3.3 *Other compounds*

The molecular geometry of non-rigid aromatic molecules (e.g. the polyphenyls, stilbenes and diphenylpolyenes) commonly differs in its ground and excited electronic states. Lim and Li[139] have studied the luminescence of biphenyl in relation to the geometry of its excited states. Hudson and Kohler[140] have observed a low-lying weak transition in diphenyloctatetraene (DPO). Schulten and Karplus[141] have discussed the origin of such transitions in the diphenylpolyenes and related molecules. Klueger *et al.*[142] have observed the fluorescence properties of the *cis* and *trans* forms of the phenyl, naphthylethylenes and the dinaphthylethylenes, which are the analogues of the stilbenes. The influence of the solvent environment on non-rigid aromatic molecules is discussed in §9.4.4.

Reiser *et al.*[143] have measured the fluorescence spectra and quantum yields and the absorption spectra of a group of benzoxazoyl-substituted aromatic hydrocarbons.

9.4.4 *Influence of environment on fluorescence and absorption spectra*

The influence of vibrational relaxation and thermal equilibrium on the fluorescence spectra of aromatic molecular vapours has been discussed by Stockburger (Chapter 2, Vol. 1). The influence of pressure on the absorption and fluorescence properties of aromatic molecules in solution has been considered by Offen (Chapter 3, Vol. 1).

9.4.4.1 *Solvent viscosity effects*

The solvent viscosity η influences the solution fluorescence behaviour of many non-rigid aromatic molecules, in which aromatic groups are linked by single bonds or polyene chains. A dependence of Φ_{FM} on η has been observed in *cis*-stilbene and some substituted *trans*-stilbenes,[144] diphenylmethane dyes,[145] triphenylmethane dyes[146] and tetraphenylbutadienes.[147] The effect is not observed with naphthalene, anthracene or other rigid planar molecules. In low-viscosity solution *trans*-stilbene, which is a planar molecule, is fluorescent, but *cis*-stilbene and sterically hindered substituted *trans*-stilbenes, which are non-planar, have a very low fluorescence quantum yield Φ_{FM}. Sharafy and Muszkat[144] observed that Φ_{FM} increases with increase in η for the latter compounds. The

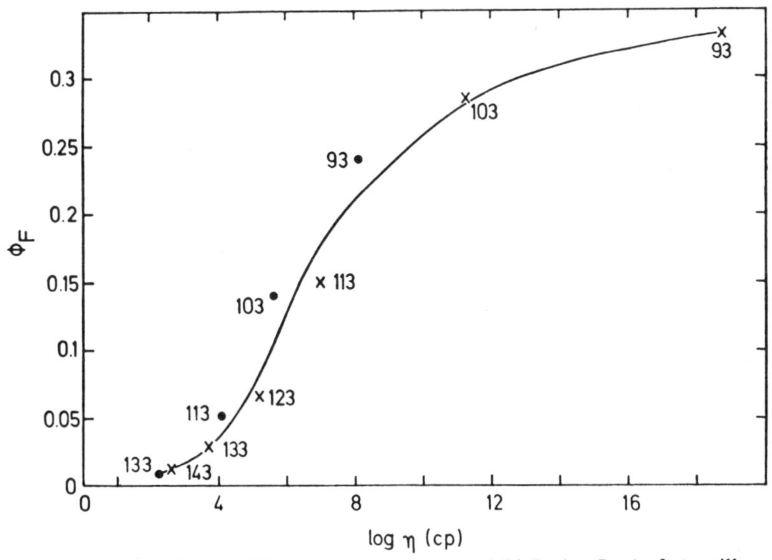

Figure 9.3 Dependence of fluorescence quantum yield Φ_F (= Φ_{FM}) of *cis*-stilbene on viscosity (η) and temperature. ● solutions in ethanol +6·6% water; X solutions in *n*-propanol. Temperatures (°K) indicated for each point (after Sharafy and Muszkat[144])

dependence of Φ_{FM} on η for *cis*-stilbene at various temperatures is plotted in Figure 9.3.

The phenomenon can be explained (p. *114*) in terms of

(a) initial excitation from the ground state S_0 to the Franck–Condon state S_1', which has the same nuclear configuration as S_0, and

(b) subsequent molecular relaxation, involving nuclear rearrangement, to the equilibrium excited state S_1.

The solvent viscosity η determines the rate of molecular relaxation (b). At high η process (b) is inhibited, and $S_1' - S_0$ fluorescence occurs with quantum yield Φ_{FM}' and radiative and radiationless rate parameters k_{FM}' and k_{IM}', respectively. At low η process (b) is rapid, and $S_1 - S_0$ fluorescence occurs with quantum yield Φ_{FM}^0 and radiative and radiationless rate parameters k_{FM}^0 and k_{IM}^0, respectively. At intermediate η the corresponding parameters are Φ_{FM}, k_{FM} and k_{IM}, where experimentally (Figure 9.3)

$$\Phi_{FM}' > \Phi_{FM} > \Phi_{FM}^0 \tag{9.9}$$

Relation (9.9) is consistent with either

$$k_{FM}' > k_{FM} > k_{FM}^0 \tag{9.10a}$$

or

$$k_{IM}' < k_{IM} < k_{IM}^0 \tag{9.10b}$$

or both.

Evidence for (9.10a) has been obtained by Birks and Dyson[104] who observed k_{FM}^0 for diphenylhexatriene (DPH) and diphenyloctatetraene (DPO) in benzene solution, and calculated $(k_{FM})_{theo}$ from the integrated absorption spectrum. The latter corresponds to the $S_0 - S_1'$ Franck–Condon transition, so that $(k_{FM})_{theo} = k_{FM}'$ refers to the $S_1' - S_0$ fluorescence. The $S_1' - S_1$ molecular relaxation (b) introduces a departure from mirror symmetry between the absorption and $S_1 - S_0$ fluorescence spectra,[104] corresponding to a red shift of the latter relative to the (unobserved) $S_1' - S_0$ fluorescence spectrum. For DPH $k_{FM}' = 5{\cdot}6\, k_{FM}^0$ and for DPO $k_{FM}' = 22\, k_{FM}^0$, consistent with (9.10a). Similar behaviour has been observed in several retinal polyenes by Thomson.[148]

From the isomerization quantum yields Sharafy and Muszkat[144] inferred that the triplet quantum yield Φ_{TM} of *cis*-stilbene and the sterically hindered *trans*-stilbenes in solution is independent of η and the temperature, and they evaluated

$$f = \frac{\Phi_{FM}}{\Phi_{FM}' - \Phi_{FM}} = \frac{\Phi_{FM}}{\Phi_{GM}} = \frac{k_{FM}}{k_{GM}} \tag{9.11}$$

where Φ_{GM} and k_{GM} are the quantum yield and rate of $S_1 - S_0$ internal conversion, respectively, assuming that $\Phi_{FM} + \Phi_{TM} + \Phi_{GM} = 1$ and that

$\Phi'_{GM} = 0$. They observed a linear relation between $\log f$ and $\log \eta$, which they interpreted in terms of a major increase of k_{GM} with decrease of η, i.e. in terms of (9.10b). They observed the absorption spectra to be independent of η, and they assumed (incorrectly) that this indicates that k_{FM} is independent of η (from the Franck–Condon principle the absorption is proportional to k'_{FM}, and not to k_{FM}). Parallel observations of Φ_{FM} and τ_M as a function of η are required to separate the contributions of k_{FM} and k_{GM} to the viscosity dependence of Φ_{FM}.

Hudson and Kohler[140] observed a weak low-lying transition in the absorption spectra of DPO in crystal bibenzyl and in polycrystalline *n*-paraffin solutions at 4°K which they attributed to a forbidden ${}^1A_g \leftarrow {}^1A_g$ transition of lower energy than the intense allowed ${}^1B_u \leftarrow {}^1A_g$ transition. Previous free-electron model calculations on the polyenes had placed the excited ${}^1A_g$ state above ${}^1B_u$, but calculations by Schulten and Karplus[141] reversed the state order. Moore and Song[305] made a high-resolution study of the polarization of the fluorescence excitation spectra and fluorescence spectra of DPO and all-*trans*-retinol in ether at 77°K. The main absorption and fluorescence bands are both long-axis polarized corresponding to ${}^1B_u \leftrightarrow {}^1A_g$ transitions, in contradiction to the hypothesis of Hudson and Kohler.[140] The mirror symmetry of the fluorescence and absorption spectra[305] shows them to be complementary transitions. Hudson and Kohler[612] have subsequently proposed that the 'excited ${}^1A_g$ state' is strongly mixed with the adjacent ${}^1B_u$ state, and thereby acquires its polarization characteristics. If so, the symmetry of the 'mixed state' is ${}^1B_u$, and not ${}^1A_g$.

Hudson and Kohler[140] proposed that the anomalous fluorescence properties (long fluorescence lifetime, spectral red shift and departure from mirror symmetry) of the diphenylpolyenes in fluid solution are due to the low-lying '${}^1A_g$ state', rather than to a change in molecular configuration. Their hypothesis does not explain why there is mirror symmetry between the fluorescence and absorption spectra of DPO in rigid solution,[305] but not in fluid solution.[104] Nor does it account for the viscosity effect on fluorescence observed in the stilbenes,[144] which are the simplest diphenylpolyenes, and in the tetraphenylbutadienes,[147] which are diphenylpolyene derivatives. The viscosity effect implies that the long-lived fluorescent state has a different molecular configuration to the initially excited Franck–Condon state. Related effects in di-aromatic molecules are considered in §9.5.3.2.

9.4.4.2 *Solvent dielectric relaxation effects*

The viscosity effects discussed above involve the relaxation of an excited solute molecule from its Franck–Condon state S'_1 into its equilibrium excited state S_1. The only role of the solvent is the influence of its viscosity on the $S_1' - S_1$ relaxation time. There are other solution systems in which the solvent dielectric relaxation is also important. Laser-spectroscopic studies of molecular reorientation and relaxation processes in solution have been described by Lippert (Chapter 1, Vol. 2).

Dielectric relaxation occurs in polar solvents when the excitation of a solute molecule modifies its dipole moment (pp. *113–6*). A solute molecule of dipole moment μ in a spherical cavity of radius a in a solvent of dielectric constant ε polarizes the solvent, thus producing a reaction field of

$$R(\mu, \varepsilon) = \frac{2\mu}{a^3} \frac{(\varepsilon - 1)}{(2\varepsilon + 1)} \tag{9.12}$$

$R(\mu, \varepsilon)$ reduces the energy of the solute molecule relative to that of an isolated molecule in the same electronic state (p. *113*). The suffix s refers to a solute molecule in equilibrium with the solvent.

The excitation of a solute molecule from its equilibrium ground state S_{0s} to its equilibrium first excited singlet state S_{1s} involves the following sequence of processes, in which the solute is characterized by its dipole moment, the solvent by its effective dielectric constant, and the system by the reaction field.

(i) Ground state S_{0s}. Solute μ_0; solvent ε_0 (= static dielectric constant); system $R(\mu_0, \varepsilon_0)$.

(ii) Franck–Condon excitation to S_1'. Solute μ_0; solvent n^2(= optical dielectric constant); system $R(\mu_0, n^2)$.

(iii) Solute relaxation to S_1. Solute μ_1; solvent n^2; system $R(\mu_1, n^2)$.

(iv) Solvent relaxation to S_{1s}. Solute μ_1; solvent ε_0; system $R(\mu_1, \varepsilon_0)$.

Unless process (iii) involves a viscosity-dependent change in molecular configuration (as in *cis*-stilbene), the intramolecular solute relaxation is normally rapid compared with the dipolar reorientation of the solvent molecules (iv). Apart from the above non-specific dielectric (universal) interaction, specific solvent–solute interactions may occur such as the formation of hydrogen bonds,[149,150] donor–acceptor complexes or exciplexes (p. *113*). The dipole moments of such complexes differ in the ground and excited states, and they differ from those of the solute molecules.

The solvent dielectric relaxation effect has been studied extensively by Bakhshiev[151–153] and others.[154–159] In many of these studies 3- and 4-aminophthalimide were used as solutes because of the relatively large changes in the magnitude and direction of their dipole moments on excitation. Large temperature-dependent shifts of the fluorescence spectra of the phthalimide derivatives in alcohol solvents are observed at low temperatures, where the solvent dielectric relaxation time is comparable with the fluorescence lifetime. Ware *et al.*[160] confirmed, by nanosecond time-resolved fluorescence spectrometry, that the spectral shift is time-dependent. In a more detailed study Ware *et al.*[75] investigated the fluorescence behaviour of 3- and 4-aminophthalimide in *n*-propanol, glycerol, propylene glycol and toluene/*n*-propanol mixtures. Typical time-resolved fluorescence spectra are shown in Figure 9.4. They concluded that the

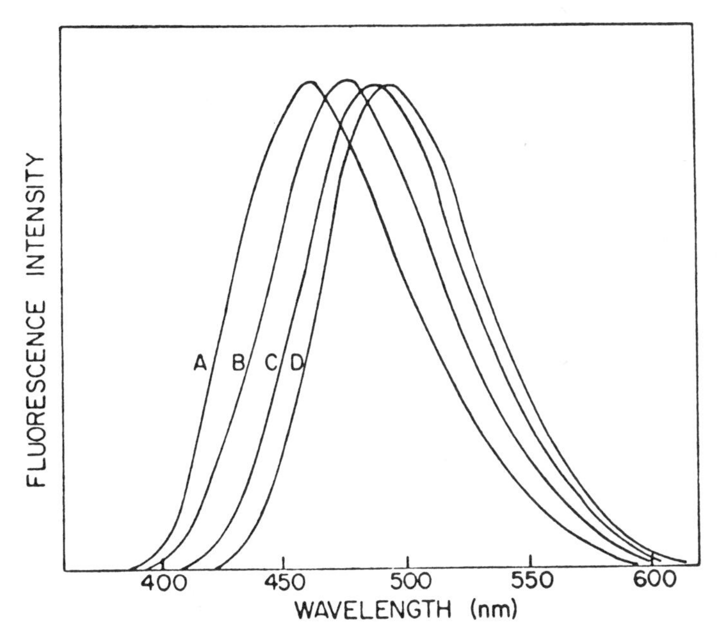

Figure 9.4 Time-resolved fluorescence spectrum of 4-amino phthalimide at −70°C in *n*-propanol. A, 4 ns; B, 8 ns; C, 15 ns; D, 23 ns (Ware, Lee, Brant and Chow;[75] reproduced by permission of the American Institute of Physics)

phenomenon is characterized by at least two relaxation times. The shorter subnanosecond relaxation time may be associated with exciplex formation between the excited solute and unexcited solvent molecules. The longer relaxation time is probably associated with the non-specific solvent dipolar reorientation round the excited solute molecule or exciplex, but its characteristics differ somewhat from the solvent dielectric relaxation time.

Chakrabarti and Ware[76] made similar studies of 1-anilino-8-naphthalene sulphonate (ANS) in alcohol solvents. ANS binds to specific sites in certain proteins, and it is used as a fluorescence probe of polarity. The observed time-dependent shifts in the fluorescence spectrum are interpreted in terms of solvent dipolar reorientation round the excited ANS molecule, but no subnanosecond relaxation component is observed. Table 9.10 lists the fluorescence maxima λ_F, quantum yields Φ_{FM} and

Table 9.10 Fluorescence parameters of 1-anilino-8-naphthalene sulphonate (ANS) in different solvents at 25°C (Chakrabarti and Ware[76])

Solvent	ε	λ_F(nm)	Φ_{FM} [a]	τ_M(ns)	k_{FM} (10^7 s^{-1})	k_{IM} (10^7 s^{-1})
n-Butanol	19·2	477	0·56	21·6	2·6	2·1
n-Propanol	22·0	479	0·46	21·0	2·2	2·5
Ethanol	25·8	485	0·37	18·0	2·1	3·5
Methanol	31·2	490	0·22	13·6	1·6	5·7
Ethylene glycol	37·7	495	0·15	11·8	1·3	7·2
Glycerol	42·5	501	—	9·0		

[a] Data on Φ_{FM} from Stryer.[162]

lifetimes τ_M, and the radiative and radiationless decay parameters k_{FM} and k_{IM} for ANS in various solvents at 25°C. An increase in the solvent dielectric constant ε causes a decrease in k_{FM} and an increase in k_{IM}, so that both factors contribute to the decrease in Φ_{FM}. Seliskar *et al.*[161] have proposed that k_{IM} is due to intersystem crossing, and that the solvent polarity influences the singlet–triplet energy gap, but experimental evidence for these hypotheses is lacking. In the 2-(*N*-arylamino)-6-naphthalene sulphonates, which are structurally related to ANS, Kosower and Tanizawa[163] concluded from the solvent polarity dependence of the fluorescence spectrum, which shows two different slopes, that emission occurs from two different excited states, a non-coplanar state and an intramolecular charge-transfer state. The plot of the intersystem crossing rate against the singlet–triplet energy gap also exhibits two slopes.

9.4.4.3 *The Shpol'skii effect*

The Shpol'skii effect (pp. *118–9*) is another environmental effect, which relates to solid solutions of aromatic molecules in *n*-alkane crystalline lattices. Appropriate matching of the lengths of the aromatic solute molecules and the *n*-alkane solvent molecules yields quasilinear absorption and fluorescence spectra.[164] Nakhimovskaya[165] has studied the conditions for the realization of the Shpol'skii effect. Kembrovskii *et al.*[166] have observed the quasilinear fluorescence and phosphorescence spectra of benzene in a cyclohexane matrix at 77°K. The vibronic spectral analyses show that the D_{6h} symmetry of the ground state S_0 ($^1A_{1g}$) is maintained in the S_1 ($^1B_{2u}$) and T_1 ($^3B_{1u}$) states. MacNab and Sauer[167] have made a systematic study of the quasilinear absorption and fluorescence spectra of anthracene, anthracene-d_{10} and acridine in *n*-hexane, *n*-heptane and *n*-octane matrices at 4°K. Morantz[168] has observed the quasilinear spectra of other aromatic compounds in aliphatic hydrocarbon solid solutions.

Richards and Rice[169,170] have studied solute–solvent interactions in Shpol'skii matrices. They observed the energy, intensity and linewidth of the vibronic absorption bands of solutions of naphthalene in *n*-pentane, anthracene and coronene in *n*-heptane, and 1:12-benzoperylene in *n*-hexane, at temperatures down to 4°K. The results are discussed in §9.5.2.

9.5 Radiationless transitions

The theory of radiationless transitions has been treated by Henry and Siebrand (Chapter 4, Vol. 1; Chapter 6, Vol. 2) and from a different standpoint by Voltz (Chapter 5, Vol. 2). The subject has also been reviewed by Henry and Kasha,[171] Jortner, Rice and Hochstrasser,[172] Schlag, Schneider and Fischer[173] and Freed.[174] Singlet–triplet intersystem crossing has been considered by Wilkinson (Chapter 3, Vol. 2), and data on these and other radiationless transitions have been discussed by Birks (Chapter 1, Vol. 1). Stockburger (Chapter 2, Vol. 1) has considered radiative and radiationless transitions in aromatic molecular vapours.

The present discussion is limited to a consideration of transitions from higher excited singlet states of the aromatic hydrocarbons. Data on the lifetime τ_H or decay rate k_H $(= 1/\tau_H)$ of a higher singlet excited π-electronic state S_p $(p > 1)$ of an aromatic molecule can be obtained from observations of

(a) $S_p - S_0$ fluorescence, or

(b) the linewidth $\Delta\bar{\nu}$ of a vibronic band, preferably the $0 - 0$ transition, of the $S_0 - S_p$ absorption spectrum.

9.5.1 $S_p - S_0$ *fluorescence* (pp. *162–71*)

The observation of $S_p - S_0$ fluorescence depends on some or all of the following parameters:

(i) the $S_p - S_0$ radiative decay rate k_{FH};

(ii) the S_p total decay rate k_H $(= 1/\tau_H)$;

(iii) the $S_p - S_1$ energy gap E_{HM};

(iv) the molecular density ρ;

(v) the temperature T;

(vi) the S_1 total decay rate k_M ($= 1/\tau_M$) and

(vii) the excitation wavelength λ_{ex}.

Three cases may be distinguished:

I. ρ high, E_{HM} large;

II. ρ high, E_{HM} small; and

III. ρ negligible.

E_{HM} is normally large for all S_p states with $p > 2$, and also for the S_2 states of some molecules, such as azulene and the alkyl benzenes. A list of aromatic molecules in which the $S_2 - S_1$ energy gap is small has been given previously (Table 1.6, Chapter 1, Vol. 1).

In Cases I and II the high molecular density ρ normally ensures that all radiationless transitions from S_p are instantaneously irreversible because of the subsequent rapid vibrational relaxation, although there is evidence[613] that in some cases the latter may be inhibited at reduced temperatures (see §9.5.1.2).

In Case I the magnitude of E_{HM} prevents the thermal repopulation of S_p from S_1 during the S_1 lifetime τ_M. This will be described as the *irreversible case*.

In Case II the small $S_2 - S_1$ energy gap E_{HM}, if associated with a long S_1 lifetime τ_M and a near-ambient temperature T, permits the repopulation of S_2 by thermal activation and internal conversion from S_1. This is the *equilibrated case*. It is analogous to E-type delayed fluorescence (pp. *373–5*) in which T_1, lying below but adjacent to S_1, acts as a reservoir for thermally activated $S_1 - S_0$ fluorescence. In the present case S_1, lying below but adjacent to S_2, acts as a reservoir for thermally activated $S_2 - S_0$ fluorescence.

In Case III the excited molecule is isolated from collisions with other molecules or the walls of the vessel by the low vapour density ρ, and vibrational relaxation is thus inhibited. Radiationless transitions may occur between S_p and isoenergetic vibronic levels of lower excited singlet and triplet states, but these are potentially reversible. This corresponds to the *isolated molecule case*.

9.5.1.1 *The irreversible case*

The S_2 state of azulene ($E_{MH} = 14{,}100\ \text{cm}^{-1}$) is one of the few higher excited states for which τ_H has been measured directly. Berlman[101] observed $\tau_H = 1{\cdot}4$ ns for azulene in cyclohexane solution at 25°C, which with his corrected value[175] of $q_{FH} = 0{\cdot}03$ gives $k_{FH} = 2{\cdot}1 \times 10^7\ s^{-1}$ and an S_2 radiationless decay rate of $k_{IH} = 7 \times 10^8\ s^{-1}$. Other photophysical properties of azulene and its derivatives are considered in §9.5.3.1.

Hirayama *et al.*[4,5] observed the $S_2 - S_0$ fluorescence of benzene, toluene, *p*-xylene and mesitylene in the liquid and/or vapour phases, the $S_3 - S_0$ fluorescence of toluene, *p*-xylene, mesitylene, naphthalene and pyrene in *iso*-octane solution and/or the vapour, and two higher transitions in pyrene in *iso*-octane solution, tentatively assigned to $S_4 - S_0$ and $S_5 - S_0$ fluorescence. In each case the $S_p - S_1$ energy gap and molecular density ρ are sufficiently large to satisfy the irreversibility condition I. The $S_3 - S_0$, $S_2 - S_0$ and $S_1 - S_0$ fluorescence spectra of *p*-xylene vapour[5] are shown in Figure 9.5. The low $S_1 - S_0$ fluorescence yield, which facili-

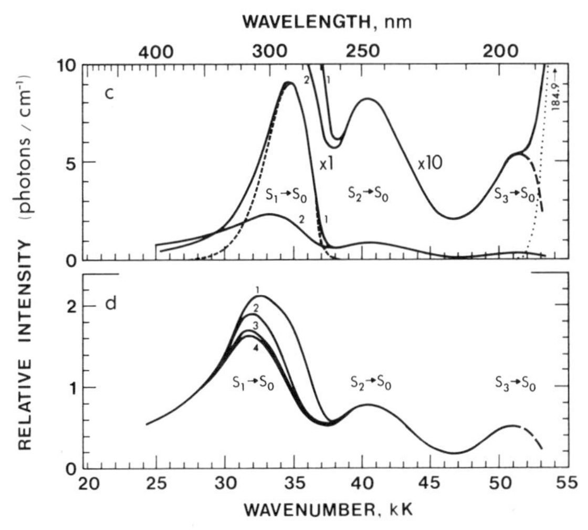

Figure 9.5 Corrected fluorescence spectra of *p*-xylene vapour. (c) Curve 1: 115 torr, 95°C, $\lambda_{ex} = 184{\cdot}9$ nm. Curve 2: 5 torr, 95°C, $\lambda_{ex} = 184{\cdot}9$ nm; ------: 5 torr, 95°C, $\lambda_{ex} = 253{\cdot}7$ nm, normalized to peak of Curve 1; ·······: scattered 184·9 nm light from empty cell; ——: S_3 emission corrected for 184·9 nm scattered light. (d) *p*-xylene (20 torr, 95°C, $\lambda_{ex} = 184{\cdot}9$ nm) at O_2 pressures of 0 (curve 1), 80 (Curve 2), 240 (Curve 3) and 760 torr (Curve 4). (After Gregory, Hirayama and Lipsky[5])

tates the observation of the weak $S_3 - S_0$ and $S_2 - S_0$ emissions, is due to the short excitation wavelength $\lambda_{ex} = 184{\cdot}9$ nm. The $S_3 - S_0$ fluorescence spectra of naphthalene and pyrene in deoxygenated *iso*-octane

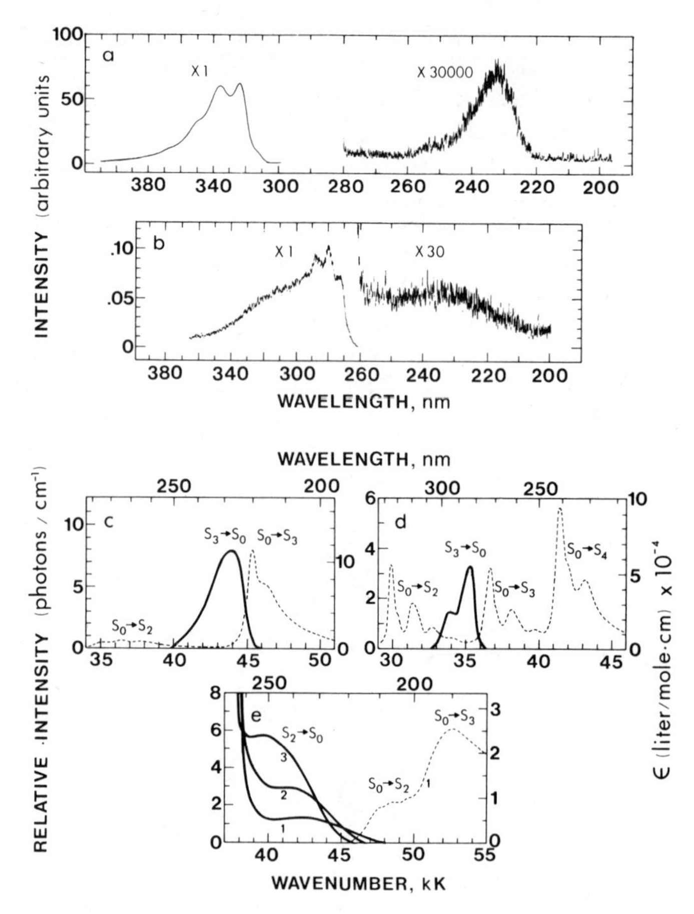

Figure 9.6 Recorded front-surface fluorescence spectra of (a) 0·1 M naphthalene in deoxygenated *iso*octane and (b) oxygenated neat benzene. A chlorine gas filter was used for recording spectrum (a) from 200 to 280 nm. Absorption and 'corrected' fluorescence spectra of (c) 0·1 M naphthalene in deoxygenated *iso*octane, (d) 0·01 M pyrene in deoxygenated *iso*octane and (e) oxygenated neat benzene (Curve 1), toluene (Curve 2) and *p*-xylene (Curve 3). Only neat benzene absorption is shown in (e). Excitation was at 184·9 nm for spectra (a), (b) and (e), at 206·2 nm for (c) and at 253·7 nm for (d). (After Hirayama, Gregory and Lipsky[4])

solution, and the $S_2 - S_0$ fluorescence spectra of oxygenated benzene, toluene and *p*-xylene liquids,[4] are shown in Figure 9.6 in relation to their absorption spectra. Data on these and other $S_p - S_0$ fluorescence transitions are listed in Table 9.11.

Table 9.11 $S_p - S_0$ fluorescence: excitation wavelength λ_{ex} (nm), fluorescence maximum λ_{max} (nm), fluorescence quantum yield Φ_{FH} and S_p decay rate k_H (10^{12} s^{-1})

Compound	Condition	S_p	λ_{ex}	λ_{max}	Φ_{FH}	k_H	Ref.
Benzene	Vapour (4 torr, 25°C)	S_2	184·9	233			5
	Liquid, O_2-saturated (25°C)	S_2	184·9	~235	8×10^{-6}	9·9	4, 12
Toluene	Vapour (4 torr, 25°C)	S_2	184·9	238			5
		S_3	184·9	~192			5
	Liquid, O_2-saturated (25°C)	S_2	184·9	~240	$1{\cdot}4 \times 10^{-5}$		4
p-Xylene	Vapour (5–115 torr, 95°C)	S_2	184·9	243	$1{\cdot}8 \times 10^{-5}$		5
		S_3	184·9	195	$1{\cdot}0 \times 10^{-5}$	150	5
	Liquid, O_2-saturated (25°C)	S_2	184·9	~245	$1{\cdot}9 \times 10^{-5}$		4
	Iso-octane solution (<0·1 M)	S_3	184·9	~205			4
Mesitylene	Vapour (4 torr, 25°C)	S_2	184·9	246			5
		S_3	184·9	~205	$\sim 10^{-5}$	~150	5
	Iso-octane solution (<0·1 M)	S_3	184·9	~210			4
Naphthalene	Vapour (0·2 torr, 35°C)	S_2	264·6	293	$6 \times 10^{-4}\ \Phi_{FM}$		188
	Vapour (23 torr, 120°C)	S_3	206·2	~220	2×10^{-5}	80	5
	Iso-octane solution (0·1 M)	S_3	206·2	227	$2{\cdot}2 \times 10^{-5}$	150	4
Azulene	Cyclohexane solution	S_2			0·03	$7{\cdot}1 \times 10^{-4}$	101
Pyrene	Vapour (low *p*, 145°C)	S_2	308·6	~326	$2 \times 10^{-3}\ \Phi_{FM}$		178
	Vapour (3 torr, 200°C)	S_2	236	~335			179
	Vapour (0·2 torr, 170°C)	S_2	266 (S_3^0)		$5{\cdot}4 \times 10^{-3}$		180
		S_2	233 (S_4^0)		$9{\cdot}8 \times 10^{-3}$		180
	Vapour (10^{-3} torr, 150°C)	S_2	264·6	329			183

Table 9.11 (*continued*)

Compound	Condition	S_p	λ_{ex}	λ_{max}	Φ_{FH}	k_H	Ref.
Pyrene	*Iso*-octane solution (0·01 M)	S_3	253·7	284	5×10^{-6}	100	4
		S_4(?)	184·9	255			4
		S_5(?)	184·9	~213			4
1:2-Benzanthracene	*n*-Heptane solution (5×10^{-4} M)	S_2	253·7	359·5 (0 – 0)			176
3:4-Benzopyrene	*n*-Heptane solution (5×10^{-6} M)	S_2	296·7	394	$\leqslant 0{\cdot}02$	0·01	177
1:12-Benzoperylene	*n*-Heptane solution (5×10^{-6} M)	S_2	296·7	396	$\leqslant 0{\cdot}006$	0·037	177

9.5.1.2 *The equilibrated case*

Easterly *et al.*[176,177,617] observed that the fluorescence spectrum of several aromatic compounds in solution near or above room temperature extends beyond the S_1 zero-point energy S_1^0 up to the S_2 zero-point energy S_2^0. This behaviour is observed in 1:12-benzoperylene (E_{HM} = 1420 cm^{-1}), 3:4-benzopyrene (E_{HM} = 1240 cm^{-1}), 1:2-benzanthracene (E_{HM} = 1900 cm^{-1}) and its 2′- and 7-methyl derivatives, 3:4-benzotetraphene (E_{HM} = 1360 cm^{-1}) and 3-methylpyrene in *n*-heptane solution. Thermal equilibration of S_2^0 and S_1^0 is favoured by the long S_1^0 (= ^1L_b) lifetime τ_M, the small $S_2^0 - S_1^0$ energy gap E_{MH}, the high molecular density ρ, and the relatively high temperature T.

Easterly *et al.*[177] assumed that only $S_1^0 - S_0$ and $S_2^0 - S_0$ fluorescence occurs, and that the two spectra have similar intensity distributions. By extrapolation of the spectrum between S_2^0 and S_1^0, they separated the overlapping $S_2^0 - S_0$ and $S_1^0 - S_0$ spectra and evaluated the respective fluorescence quantum yields Φ_{FH} and Φ_{FM} from the spectral areas. The data were analysed in terms of the two-level (S_1^0, S_X) fluorescence scheme of Figure 9.7(a), and the rate parameters were evaluated from the corresponding kinetic relations (Chapter 1, Vol. 1). A preliminary analysis, indicating that the activation energy W_{XM} of the hot fluorescence corresponded to E_{HM}, consistent with $S_2^0 = S_X$, proved incorrect. A fuller analysis[177] gave $W_{XM} = 690 \pm 60$ cm^{-1} for 1:12-benzoperylene (E_{HM} = 1420 cm^{-1}) and $W_{XM} = 629 \pm 33$ cm^{-1} for 3:4-benzopyrene (E_{HM} = 1240 cm^{-1}).

Although Easterly *et al.*[177] still considered the hot fluorescence to be primarily due to thermal population of S_2^0 from S_1^0, this is not consistent with the observation of $W_{XM} \simeq \frac{1}{2}E_{HM}$ for these and other compounds.[600,617] Under photostationary conditions at room temperature most singlet-excited molecules are in S_1^0, and the kinetic analysis[177] shows that most hot fluorescence occurs from a higher state S_X at an energy $E_{XM}(\simeq W_{XM})$ above S_1^0. The most intense peak in the hot fluorescence spectrum does indeed occur at $S_1^0 + E_{XM}$, although the spectrum extends at lower intensity to S_2^0.[177] The three-level (S_1^0, S_X, S_2^0) fluorescence scheme of Figure 9.7(b) is introduced to describe the system. Recent studies by Nakajima[594] on 1:12-benzoperylene in the vapour phase and in hexadecane and glycerol solutions, and by Van den Bogaardt *et al.*[613] on 3:4-benzopyrene solutions, discussed below, support the three-level model. S_X is interpreted as a vibrationally excited state S_1^*.

Hoytink[614] made a vibronic analysis of the fluorescence spectrum of 1:12-benzoperylene in *n*-heptane solution,[177] in terms of symmetric $\bar{\nu}_3$ (1400 cm^{-1}) vibrational progressions based on the electronic and two

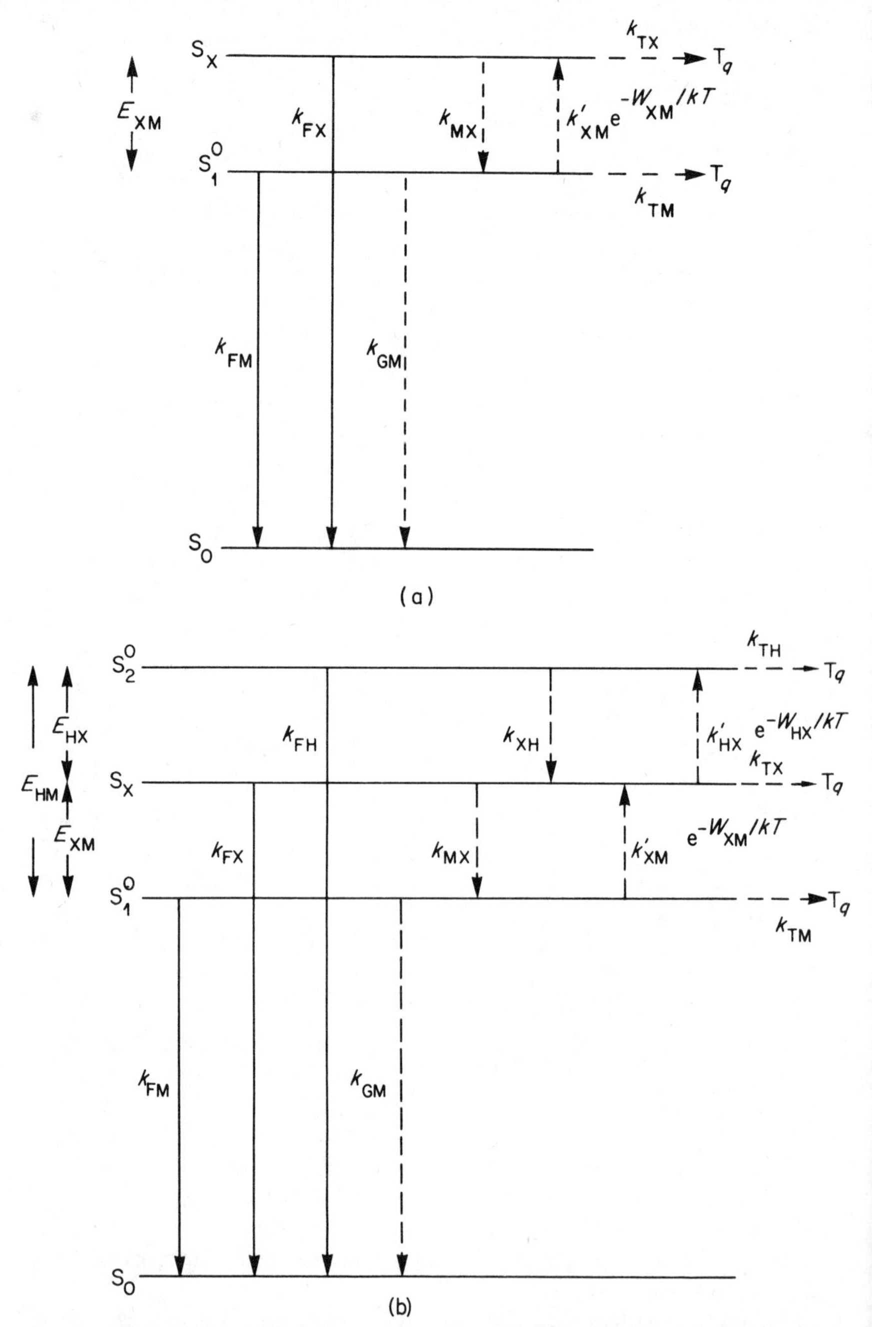

Figure 9.7 The equilibrated case. (a) Two-level (S_1^0, S_X) fluorescence scheme. (b) Three-level (S_1^0, S_X, S_2^0) fluorescence scheme. Solid horizontal lines represent energy levels, solid vertical lines radiative transitions, broken lines radiationless transitions

antisymmetric $\bar{\nu}_1$ (600 cm^{-1}) and $\bar{\nu}_2$ (800 cm^{-1}) vibrational origins. The normal S_1^0 fluorescence consists of three progressions originating at S_1^0, $S_1^0 - \bar{\nu}_1$ and $S_1^0 - \bar{\nu}_2$, in increasing order of intensity. The hot fluorescence was analysed in terms of three progressions originating at S_2^0, $S_1^0 + \bar{\nu}_2$ ($= S_2^0 - \bar{\nu}_1$) and $S_1^0 + \bar{\nu}_1$ ($= S_2^0 - \bar{\nu}_2$), respectively. Hoytink[614] attributed the S_2^0 progression to S_2^0 fluorescence, and the other two progressions to vibronically induced S_1^* fluorescence (in agreement with the W_{XM} observations[177,594]). The analysis[614] indicates that the intensity distributions of the $S_1^0 - S_0$ and hot fluorescence spectra differ, contrary to the assumption of Easterly *et al.*[177]

Van den Bogaardt *et al.*[613] made a detailed study of the fluorescence of 3:4-benzopyrene in 2-methylpentane solution from 293 to 77°K. The fluorescence spectrum of a 4×10^{-6} M solution at 293°K is shown in Figure 9.8. The two weak bands preceding the normal S_1^0 fluorescence are detailed in the inset. The 25,800 cm^{-1} band is assigned to S_2^0 fluorescence, and the 25,350 cm^{-1} band to S_1^* fluorescence. These assignments are confirmed by the temperature dependence of the band positions. The S_1^0 and S_1^* fluorescence and absorption bands have negligible temperature

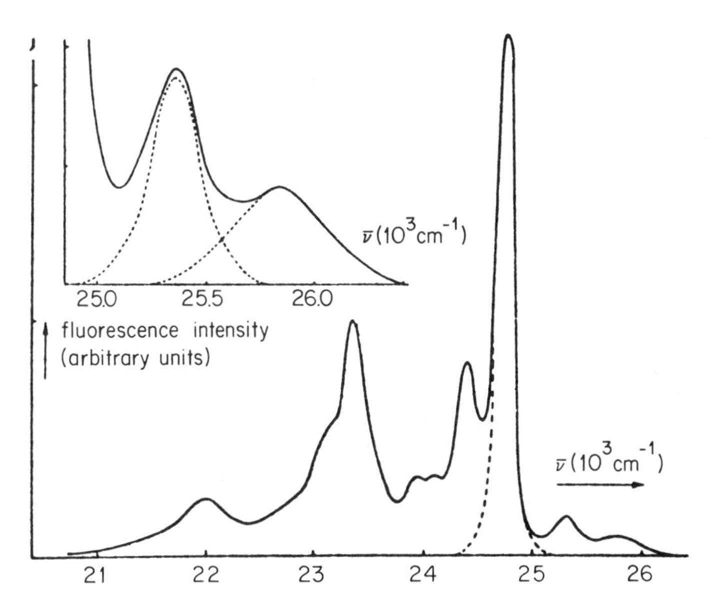

Figure 9.8 Fluorescence spectrum of 3:4-benzopyrene solution (4×10^{-6} M) in 2-methylpentane at 293°K. In the inset the fluorescence spectrum beyond S_1^0 is shown in more detail. (Van den Bogaardt, Rettschnick and van Voorst;[613] reproduced by permission of the North-Holland Publishing Company)

shifts. The S_2^0 fluorescence and absorption bands exhibit parallel red shifts on cooling.

The intensities I_H and I_X of the S_2^0 and S_1^* 0 – 0 fluorescence bands were evaluated from the band areas relative to the S_1^0 0 – 0 fluorescence band intensity.[613] Assuming mirror symmetry of the S_2^0 absorption and fluorescence spectra, I_X was reduced by $\frac{1}{3}I_H$ to allow for the S_2^0 fluorescence progression underlying the S_1^* band. This procedure circumvents the problems of separating the three overlapping fluorescence spectra.[177,614] With S_1^0 (= 24,850 cm^{-1}) excitation, the observed values of ln I_H and ln I_X against $1/T$ yield Arrhenius plots corresponding to activation energies of $W_{HM} = 950$ cm^{-1} and $W_{XM} = 520$ cm^{-1}, which are equal to the $S_2^0 - S_1^0$ and $S_1^* - S_1^0$ energy gaps, E_{HM} and E_{XM}, respectively. 3:4-Benzopyrene is a three-level fluorescent system (Figure 9.7b), and not a two-level system as originally proposed.[177]

With S_1^0 excitation S_1^* and S_2^0 can only be populated thermally from S_1^0. With S_2^*(= 28,700 cm^{-1}) excitation, S_2^0 and S_1^* can also be populated directly during the initial relaxation through the singlet vibronic manifold. At high temperatures ($T > 190$°K) the same values of I_H and I_X are observed[613] with S_2^* and S_1^0 excitation, i.e. $I_H(S_2^*) = I_H(S_1^0)$ and $I_X(S_2^*) = I_X(S_1^0)$, showing that the vibronic relaxation and thermal equilibration occur rapidly compared with the fluorescence emission. At $T < 190$°K, $I_H(S_2^*) > I_H(S_1^0)$, and the additional component of I_H has an activation energy $W_{HX} = 230$ cm^{-1}, which approximates to the $S_2^0 - S_1^*$ energy gap $E_{HX} \simeq 300$ cm^{-1} in this temperature range. At $T < 150$°K, $I_X(S_2^*) > I_X(S_1^0)$, and the additional component of I_X is temperature-independent. These results indicate that at low temperatures the vibrational relaxation of S_1^* is inhibited, and that thermal population of S_2^0 from S_1^* may occur.

Observations of the fluorescence lifetimes of $S_1^0(\tau_M)$, $S_1^*(\tau_X)$ and $S_2^0(\tau_H)$ with laser S_2^* excitation support these conclusions.[613] τ_M *decreases* from 55 ns to 40 ns on cooling from 293 to 77°K, due to the decrease in E_{HM} and the consequent increase of vibronic coupling between S_2 and S_1. This interesting effect does not appear to have been observed previously. The S_1^* and S_2^0 emissions show very rapid initial decays (direct fluorescence?) followed by slower decay times of τ_X and τ_H, respectively. At $T > 190$°K, $\tau_M = \tau_X = \tau_H$, corresponding to the equilibrated case with the complete system decaying with the S_1^0 lifetime τ_M. At T between 190 and 195°K, $\tau_M = \tau_X > \tau_H$, and at $T < 150$°K, $\tau_M > \tau_X > \tau_H$. At 90°K, where only direct S_1^* fluorescence is observed,[613] $\tau_X = 25$ ns corresponding to a S_1^* decay rate of $k_X = 4 \times 10^7$ s^{-1}. Comparison with the S_1^0 decay rate of $k_M = 2{\cdot}5 \times 10^7$ s^{-1} $\simeq k_{FM}$, which is mainly radiative (k_{FM}), suggests that k_X is also mainly radiative, i.e. $k_X \simeq k_{FX}$. Due to vibronic coupling of S_1^* and S_1^0 to S_2, it is to be expected that $k_{FX} > k_{FM}$, since $E_{HX} < E_{HM}$. Van den

Bogaardt *et al.*[613] were more conservative: they considered k_X simply as the upper limit of the vibrational relaxation rate.

Although the rate parameters derived by Easterly *et al.*[177] are only approximate,[614] they may be used to estimate k_{FX} for 1:12-benzoperylene in *n*-heptane, by applying the parameters derived for the two-level scheme (Figure 9.7a) to the lower two levels of the three-level scheme (Figure 9.7b). There is thermal equilibrium within the S_1^* vibrational manifold so that $k_{MX} = k'_{XM} \simeq 2{\cdot}8 \times 10^9\ s^{-1}$, and since $k_{FX}/k_{MX} \simeq 0{\cdot}006$[177] a value of $k_{FX} \simeq 1{\cdot}7 \times 10^7\ s^{-1}$ is obtained for the S_1^* fluorescence rate. k_{FX} is the geometric mean of the S_2^0 fluorescence rate k_{FH} $(= 2{\cdot}2 \times 10^8\ s^{-1})$ and the S_1^0 fluorescence rate k_{FM} $(= 1{\cdot}3 \times 10^6\ s^{-1})$, and the S_1^* energy is the mean of the S_2^0 and S_1^0 energies. S_1^* acquires its oscillator strength by vibronic coupling to the adjacent intense $S_0 - S_2^0$ transition ($\epsilon = 31{,}600$). Similar vibronically induced S_1^* hot band fluorescence is to be expected in other molecules with $S_1 = {}^1L_b$, $S_2 = {}^1L_b$, and a small energy gap E_{HM}.

Solutions of molecules such as 3:4-benzopyrene and 1:12-benzoperylene, in which $S_1 = {}^1L_b$, $S_2 = {}^1L_a$ and E_{HM} is small, thus exhibit several interesting effects.

(i) S_2^0 fluorescence. S_2^0 may be directly excited or produced by thermal population from S_1^* or S_1^0.

(ii) S_1^* fluorescence, vibronically induced by coupling to S_2. S_1^* may be directly excited or populated thermally from S_1^0.

(iii) S_1^0 fluorescence, mainly vibronically induced by coupling to S_2. k_{FM} and k_{FX} increase with decrease in *T*, due to the decrease in E_{HM} and E_{HX}.

(iv) At low temperatures vibrational relaxation of S_1^* is inhibited, probably due to the low density of vibrational states of the solvent isoenergetic with E_{XM}.[613] Under these circumstances the behaviour approximates to the isolated molecule case considered in §9.5.1.3.

9.5.1.3 *Isolated molecule case*

This relates to vapours at pressures sufficiently low to eliminate molecular collisions during the lifetime of the excited state. Stockburger (Chapter 2, Vol. 1) has reviewed the fluorescence of aromatic molecular vapours. The present discussion will be limited to the $S_2 - S_0$ fluorescence of isolated excited molecules and its relation to radiationless transitions.

Several authors[178–184] have studied the fluorescence of pyrene vapour at low pressures. Geldof *et al.*[178] first identified the $S_2 - S_0$ fluorescence of pyrene vapour from the mirror symmetry of the emission spectrum and the $S_0 - S_2$ absorption spectrum. The 0 – 0 and 0 – 1 fluorescence bands at 30,700 and 29,300 cm^{-1} correspond to the 0 – 0 and 0 – 1 absorption bands at 31,000 and 32,400 cm^{-1}. Nakajima and Baba[179] distinguished the $S_2 - S_0$ and $S_1 - S_0$ fluorescences by the difference in their excitation spectra. The $S_1 - S_0$ fluorescence excitation spectrum corresponds to the absorption spectrum, but the $S_2 - S_0$ fluorescence excitation spectrum increases further in intensity with increasing excitation energy.

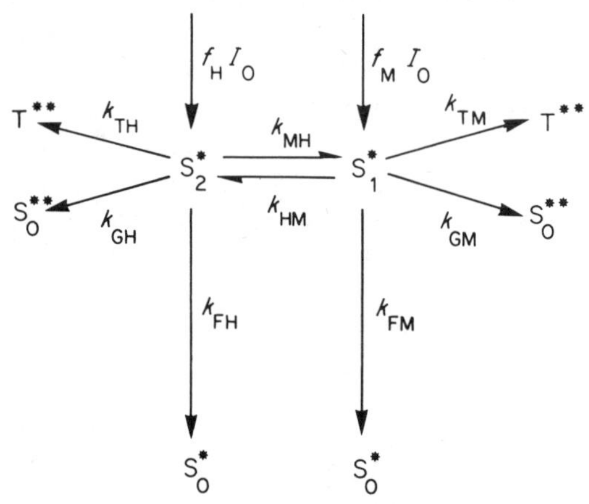

Figure 9.9 The isolated molecule case. S_2^*, S_1^*, T^{**} and S_0^{**} are isoenergetic states. f_H and f_M are the initial fractions of excited molecules in S_2^* and S_1^*, respectively

Figure 9.9 shows a simple kinetic model for the isolated molecule case. Asterisks denote vibrationally excited species, and the states S_2^*, S_1^*, S_0^{**} and T^{**} are isoenergetic. The S_1^*, S_0^{**} and T^{**} vibronic manifolds are considered quasi-continua, corresponding to the statistical limit of radiationless transition theory (Henry and Siebrand, Chapter 4, Vol. 1). The $S_2^* - S_1^*$ internal conversion is considered reversible, but the transitions to S_0^{**} which have much smaller Franck–Condon factors, and the transitions to T^{**} which are spin-forbidden, are considered irreversible within the excitation lifetime.

f_H and f_M $(= 1 - f_H)$ are the fractions of excited molecules initially in S_2^* and S_1^*, respectively. We define

$$k_H^0 = k_{FH} + k_{TH} + k_{GH} \tag{9.13}$$

$$k_H = k_H^0 + k_{MH} \tag{9.14}$$

$$k_M^0 = k_{FM} + k_{TM} + k_{GM} \tag{9.15}$$

$$k_M = k_M^0 + k_{MH} \tag{9.16}$$

Under photostationary conditions the $S_2^* - S_0^*$ and $S_1^* - S_0^*$ fluorescence quantum yields are, respectively,

$$\Phi_{FH} = \frac{q_{FH}(f_H + \phi_{HM})}{1 + \phi_{HM} + \phi_{MH}} \tag{9.17}$$

$$\Phi_{FM} = \frac{q_{FM}(f_M + \phi_{MH})}{1 + \phi_{HM} + \phi_{MH}} \tag{9.18}$$

where q_{FH} ($= k_{FH}/k_H^0$) and q_{FM} ($= k_{FM}/k_M^0$) are the $S_2^* - S_0^*$ and $S_1^* - S_0^*$ fluorescence quantum efficiencies in the absence of $S_2^* \leftrightarrow S_1^*$ internal conversion, and ϕ_{MH} ($= k_{MH}/k_H^0$) and ϕ_{HM} ($= k_{HM}/k_M^0$) relate the efficiencies of the latter processes. The first and second terms in brackets in (9.17) and (9.18) refer to the direct excited fluorescence and to that excited by $S_2^* \leftrightarrow S_1^*$ internal conversion, respectively, and the denominators indicate the degree of fluorescence quenching due to the latter process.

Under most experimental conditions $f_H \gg f_M$, so that (9.17) and (9.18) can be simplified by substituting $f_H = 1$, $f_M = 0$. Under these conditions, with δ-function excitation of S_2^* at time $t = 0$, the S_2^* and S_1^* fluorescence response functions are, respectively,

$$i_H(t) \propto e^{-\lambda_1 t} + A\,e^{-\lambda_2 t} \tag{9.19}$$

$$i_M(t) \propto e^{-\lambda_1 t} - e^{-\lambda_2 t} \tag{9.20}$$

where

$$\lambda_{1,2} = \tfrac{1}{2}[k_H + k_M \mp \{(k_H - k_M)^2 + 4k_{MH}k_{HM}\}^{\frac{1}{2}}] \tag{9.21}$$

$$A = \frac{k_H - \lambda_1}{\lambda_2 - k_H} \tag{9.22}$$

Since k_H and k_{MH} ($\gg k_H^0$) are much greater than the other rate parameters, (9.21) may be approximated by the binomial theorem to give

$$\lambda_1 = k_M - \frac{k_{MH}k_{HM}}{k_H - k_M} \simeq k_M^0 \tag{9.23}$$

$$\lambda_2 = k_H + \frac{k_{MH}k_{HM}}{k_H - k_M} \simeq k_H \tag{9.24}$$

$$A = 1 + \frac{(k_H - k_M)^2}{k_{MH}k_{HM}} \simeq \frac{k_{MH}}{k_{HM}} \tag{9.25}$$

The kinetic analysis predicts a rapid rise (k_H) for the $S_1^* - S_0^*$ fluorescence and a rapid initial decay (k_H) for the $S_2^* - S_0^*$ fluorescence, followed by a common slow decay (k_M^0) for the two fluorescences.

For pyrene vapour at 145°C, $\lambda_{ex} = 308{\cdot}6$ nm, Geldof *et al.*[178] observed $\Phi_{FH}/\Phi_{FM} = 2 \times 10^{-3}$; estimated $k_M \sim 10^6\ s^{-1}$ from oxygen quenching data; calculated $k_{FH} = 2 \times 10^8\ s^{-1}$ from the $S_0 - S_2$ absorption spectrum; assumed $\Phi_{FM} \sim 0{\cdot}7$, the value for pyrene in ethanol solution (later measurements[180] show $\Phi_{FM} \leqslant 0{\cdot}14$ for pyrene vapour); and estimated $k_{HM} \sim 5k_M \sim 5 \times 10^6\ s^{-1}$ and $k_{MH} \sim 10^{12}\ s^{-1}$ from the quenching of Φ_{FH} by the addition of 9 torr of *n*-hexane, which eliminates k_{HM} by introducing vibrational relaxation of S_1^*. From the temperature dependence of Φ_{FH} they concluded that in the pyrene/*n*-hexane mixture about half the S_2^* emission is directly excited and about half is due to repopulation of S_2^* from S_1^*. They modified their initial values of k_{HM} and k_{MH} by a factor of 2 to obtain estimates of $k_{HM} \sim 10^7\ s^{-1}$ and $k_{MH} \sim 2 \times 10^{12}\ s^{-1}$.

Baba *et al.*[180] observed the fluorescence characteristics of pyrene vapour (170°C, 0·21 torr) as a function of the excitation energy E_{ex}. The $S_0^0 \rightarrow S_1^0, S_2^0, S_3^0$ and S_4^0 transitions are at 27,100, 31,000, 37,600 and 42,900 cm^{-1}, respectively, where the superscript 0 refers to the zero-point vibrational level of each state. The observations of Φ_{FM} and Φ_{FH}/Φ_{FM} as a function of E_{ex} are plotted in Figure 9.10. Φ_{FM} decreases from 0·14 (S_2^0 excitation) to 0·08 (S_4^0 excitation) while Φ_{FH} increases from $\sim 1{\cdot}4 \times 10^{-4}$ (S_2^0 excitation) to $9{\cdot}8 \times 10^{-3}$ (S_4^0 excitation). The $S_1^* - S_0^*$ fluorescence lifetime τ_M decreases from 210 ns (S_2^0 excitation) to 62 ns (S_4^0 excitation). Increase in E_{ex} intensifies the $S_2^* - S_0^*$ fluorescence, while the $S_1^* - S_0^*$ fluorescence spectrum is shifted to the red and becomes more diffuse. With S_4^0 excitation, addition of cyclohexane vapour quenches the $S_2^* - S_0^*$ fluorescence, and the $S_1^* - S_0^*$ fluorescence spectrum tends towards that obtained with S_2^0 excitation in the absence of cyclohexane. Baba *et al.*[180] analysed their data for S_4^0 (42,900 cm^{-1}) excitation, using the observed values of Φ_{FM}, Φ_{FH} and τ_M and a calculated value of $k_{FH} = 1{\cdot}7 \times 10^8\ s^{-1}$. They obtained values of $k_{MH} = 5{\cdot}8 \times 10^{11}\ s^{-1}$, $k_{HM} = 3{\cdot}2 \times 10^8\ s^{-1}$ and $k_{HM}/k_{MH} = 5{\cdot}5 \times 10^{-4}$.

The general expression (p. *151*) for the radiationless transition probability from a state *u* to a state *l* is

$$k_{lu}^{nr} = \frac{4\pi^2 \rho_l}{h} |H_{lu}|^2 \tag{9.26}$$

where H_{lu} is the perturbation Hamiltonian and ρ_l is the density of vibrational states of *l* isoenergetic with the initial state *u*. Baba *et al.*[180] assumed that for the $S_2^* \rightleftharpoons S_1^*$ internal conversion $H_{MH} = H_{MH}$, so that

$$\frac{k_{HM}}{k_{MH}} = \frac{\rho_H}{\rho_M} \tag{9.27}$$

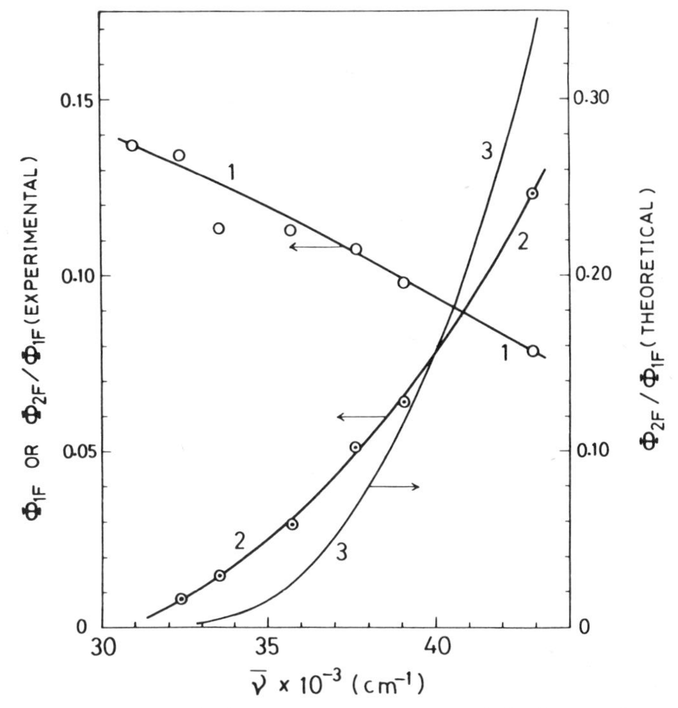

Figure 9.10 S_1^* and S_2^* fluorescence quantum yields Φ_{FM} $(=\Phi_{1F})$ and Φ_{FH} $(=\Phi_{2F})$ of pyrene vapour (170°C, 0·21 torr) as a function of excitation energy E_{ex}. (1) Φ_{FM} (experimental), (2) Φ_{FH}/Φ_{FM} (experimental), (3) Φ_{FH}/Φ_{FM} (theoretical from (9.27)). The relevant ordinate scale for each curve is indicated by an arrow. (Baba, Nakajima, Aoi and Chihara;[180] reproduced by permission of the American Institute of Physics)

where ρ_H and ρ_M are the densities of isoenergetic vibrational states of S_2^* and S_1^*, respectively. The ratio ρ_H/ρ_M increases with E_{ex}, so that the model accounts qualitatively for the observed dependence of Φ_{FH}/Φ_{FM} on E_{ex} (Figure 9.10).

Werkhoven *et al.*[181] observed the fluorescence decay of 0·1 torr pyrene vapour excited at $\lambda_{ex} = 347$ nm into S_1^* $(= S_1^0 + 1720\ \text{cm}^{-1})$ in the presence and absence of 50 torr *n*-hexane. With *n*-hexane present the system is thermally equilibrated, and it has an exponential fluorescence decay rate k_M^0. In the absence of *n*-hexane the initial decay is more rapid and it is non-exponential, tending towards a rate of k_M^0 at long t. This behaviour is attributed to an increase in k_M^0 with excess vibrational energy, and to incomplete vibrational relaxation of S_1^*. The fluorescence quantum yield Φ_{FM} and lifetime τ_M $(= 1/k_M^0)$ of the pyrene/*n*-hexane vapour system were measured from 159 to 346°C. Taking $k_{GM} \simeq 0$ from the data of Jones and Siegel[185] and the value of $k_{FM} \simeq 10^6\ s^{-1}$ estimated by Baba *et al.*,[180] the measurements show that k_{FM} increases from $10^6\ s^{-1}$ to $1{\cdot}4 \times 10^6\ s^{-1}$ and that k_{TM} increases from $1{\cdot}8 \times 10^6\ s^{-1}$ to $5.1 \times 10^6\ s^{-1}$. The latter increase is attributed to $S_1^* - T_2^*$ intersystem crossing (Chapter 1, Vol. 1).

For pyrene vapour excited by a frequency-doubled ruby laser at $\lambda_{ex} = 347$ nm into S_1^* ($= S_1^0 + 1720\ cm^{-1}$), but at a much lower pressure (10^{-3} torr) than previously,[181] Werkhoven *et al.*[182] observed a fluorescence decay of the form

$$i(t) \propto e^{-\lambda_1 t} + A\, e^{-\lambda_2 t} \tag{9.19a}$$

with $\lambda_1 = 3{\cdot}5 \times 10^6\ s^{-1}$ and $\lambda_2 = 1{\cdot}4 \times 10^7\ s^{-1}$. They interpreted the observations in terms of the initial excitation of a specific vibronic state $(S_1^*)_A$ and a subsequent vibrational redistribution into isoenergetic states $(S_1^*)_B$ of different radiative and radiationless decay rates from $(S_1^*)_A$. The situation is analogous to that of Figure 9.9 with $(S_1^*)_A \equiv S_2^*$ and $(S_1^*)_B \equiv S_1^*$, and a similar kinetic analysis is applicable. Werkhoven *et al.*[182] attributed λ_2 ($\equiv k_H$) mainly to the rate of vibrational redistribution k_{BA} ($\equiv k_{MH}$). In a subsequent analysis[184] they concluded that $0{\cdot}8 \times 10^7\ s^{-1} < k_{BA} < 1{\cdot}4 \times 10^7\ s^{-1}$. When the energy of the 2·5 ns laser pulse exceeds 1·5 mJ, a short ($\tau \simeq 4$ ns) fluorescence decay component appears due to two-photon excitation into a higher excited state.

The decay times of the $S_2^* - S_0^*$ fluorescence of low-pressure vapours of naphthalene, 3:4-benzopyrene and pyrene have been observed by Wannier *et al.*[6,186] and Deinum *et al.*[183] In each case the observed lifetime exceeds the $S_2 - S_0$ radiative lifetime τ_{FH} ($= 1/k_{FH}$) and it is equal to that of the $S_1^* - S_0^*$ fluorescence. Wannier *et al.*[6] proposed that this behaviour be interpreted in terms of the intermediate strong coupling model, and this description was also adopted by Deinum *et al.*[183]

In the intermediate case the density of S_1^* states approximately isoenergetic with S_2^* is relatively low because of the small $S_2 - S_1$ energy gap. Strong coupling is postulated between the zero-order S_2^* level ϕ_H and a sparse manifold $\{\phi_M\}$ of S_1^* vibronic levels. The model[6] assumes that $S_2^* \leftrightarrow S_1^*$ radiationless transitions do not occur in the isolated molecule, and that the quantum yield Φ_{FH} of resonance $S_2^* - S_0^*$ fluorescence is inversely proportional to the number, n, of effectively coupled levels. The $S_2^* - S_0^*$ and $S_1^* - S_0^*$ fluorescences have a common decay time[6]

$$\tau = \frac{\hbar}{\gamma_H/n + \langle \gamma_M \rangle} \tag{9.28}$$

where

$$\gamma_H = \Gamma_H + \Delta_H \quad (9.29)$$

$$\gamma_M = \Gamma_M + \Delta_M \quad (9.30)$$

are the total linewidths of the ϕ_H and ϕ_M states, respectively, Γ_H and Γ_M are the radiative contributions to the linewidths, Δ_H and Δ_M are the contributions due to radiationless transitions to T** or S_0^{**} ($S_2^* \leftrightarrow S_1^*$ transitions do not contribute to Δ_H or Δ_M) and $\langle\ \rangle$ denotes an average value over the $\{\phi_M\}$ manifold. Provided n is not too small and that $\Gamma_H \gg \langle\gamma_M\rangle$ then $\tau \gg \tau_{FH}\ (= \hbar/\Gamma_H)$.

The statistical limit model (Figure 9.9) also predicts a common lifetime τ for the $S_2^* - S_0^*$ and $S_1^* - S_0^*$ fluorescences, provided $k_{HM} > 0$. From (9.19) and (9.20), after a short time $\Delta t \simeq 1/\lambda_2 = 1/k_H \simeq 1$ ps, the two fluorescences decay at the same rate $\lambda_1 = k_M^0 = 1/\tau$ (9.23). The observation of a common τ has been previously interpreted as evidence of intermediate strong coupling (ISC) conditions,[6,183,186] but since it is also compatible with statistical limit (SL) conditions, each case needs to be considered individually. The excitation energy E_{ex} ($= S_2^* = S_1^*$) determines the energy gaps ΔE_{MH} ($= S_2^* - S_1^0$) and ΔE_{HM}($= S_1^* - S_2^0$), and these determine the corresponding densities of isoenergetic S_1^* and S_2^* states, ρ_M and ρ_H, respectively. When $\Delta E > \Delta E_0 = 5000\ cm^{-1}$, ρ is high and SL conditions are valid. When $\Delta E < \Delta E_0$, ρ is reduced, the SL model may be still applicable, but at low ρ the ISC conditions may be satisfied.

Pyrene vapour at $E_{ex} = 37{,}800\ cm^{-1}$ [183] has $\Delta E_{MH} = 10{,}900\ cm^{-1}$ and $\Delta E_{HM} = 6800\ cm^{-1}$. This corresponds to SL conditions, and not to ISC conditions as proposed by Deinum *et al.*[183] The Φ_{FH}/Φ_{FM} data for pyrene[180] (Figure 9.10) are consistent with the SL model, and they indicate that k_{HM} decreases monotonically with E_{ex}, remaining finite down to $\Delta E_{HM} = 1300\ cm^{-1}$. The values of $k_{MH} \simeq 2 \times 10^{12}\ s^{-1}$ and $k_{HM} \simeq 10^7\ s^{-1}$ obtained for pyrene vapour[178] at $E_{ex} = 32{,}400\ cm^{-1}$ ($\Delta E_{MH} = 5500\ cm^{-1}$, $\Delta E_{HM} = 1400\ cm^{-1}$) support these conclusions. Thus all the available data on pyrene vapour appear consistent with the SL model.

Naphthalene, with pulsed laser excitation at $E_{ex} = 37{,}730\ cm^{-1}$,[6] has $\Delta E_{MH} = 5710\ cm^{-1}$ and $\Delta E_{HM} = 1810\ cm^{-1}$. The values of k_{MH} and k_{HM} are expected to be similar to those of pyrene,[178] cited above, because of the similar values of ΔE_{MH} and ΔE_{HM}. Wannier *et al.*[6] proposed that the SL process k_{MH} can occur in parallel with the ISC process, which they considered responsible for the common τ of the S_2^* and S_1^* fluorescences, but they neglected the reverse SL process k_{HM}. If $k_{HM} > 0$, the SL model provides an alternative explanation of the common τ. Deinum *et al.*[188] found that the value of $\Phi_{FH}/\Phi_{FM} = 0{\cdot}03$, observed with pulsed laser excitation of naphthalene vapour,[6] exceeds by a factor of 50 that observed with less intense continuous excitation at the same E_{ex}. They proposed that the laser experiments on naphthalene[6] involved biphotonic excitation into higher excited singlet states, thereby increasing Φ_{FH}/Φ_{FM} and the spectral diffuseness. If this suggestion is correct, the values of ΔE_{MH} and ΔE_{HM} for naphthalene need to be increased by E_{ex}, corresponding to extreme SL conditions, and the common τ is then explained by the SL model.

3:4-Benzopyrene at $E_{ex} = 28{,}800\ \text{cm}^{-1}$ [186] has $\Delta E_{MH} = 4100\ \text{cm}^{-1}$, $\Delta E_{HM} = 2000\ \text{cm}^{-1}$, so that $\Delta E_0 > \Delta E_{MH}, \Delta E_{HM}$. These are ideal conditions for the observation of ISC phenomena. The interpretation of the common τ of the S_2^* and S_1^* fluorescences of 3:4-benzopyrene in terms of the ISC model by Wannier *et al.*[186] is probably correct, provided biphotonic excitation can be excluded. The oscillatory modulation of the fluorescence decay, observed in 3:4-benzopyrene,[186] is probably characteristic of intermediate strong coupling. The statistical limit model predicts a simple exponential decay, as observed in naphthalene[6] and pyrene.[183]

For isolated pyrene molecules k_{FM} and k_{IM} increase from $0{\cdot}7 \times 10^6\ \text{s}^{-1}$ and $4{\cdot}1 \times 10^6\ \text{s}^{-1}$ for S_2^0 excitation to $1{\cdot}3 \times 10^6\ \text{s}^{-1}$ and $14{\cdot}8 \times 10^6\ \text{s}^{-1}$ for S_4^0 excitation.[180] Similar increases of k_{IM} with excitation energy E_{ex} are observed in isolated molecules of naphthalene,[187,189–193] anthracene,[187] aniline[189,194] and 2-naphthylamine.[195–197] Lim and Huang[187] noted discontinuities in the plots of $\log k_{IM}$ against E_{ex} for naphthalene, 2-naphthylamine and quinoxaline, coincident with S_2^0 or other electronic energy levels.

In most aromatic hydrocarbons in solution (pp. *200–1*)

$$\Phi_{FM} + \Phi_{TM} \simeq 1{\cdot}0 \tag{9.31}$$

so that in the equilibrated case $k_{IM} \simeq k_{TM}$, the singlet–triplet intersystem crossing rate. If this remains valid in the isolated molecule, the increase of k_{IM} with E_{ex} is attributable to intersystem crossing to higher triplet states.[181,187] In isolated anthracene molecules Laor *et al.*[615] observed a stepwise increase of k_{IM} with E_{ex}, with mean k_{IM} values of $1{\cdot}59 \times 10^8\ \text{s}^{-1}$ at 27,000–27,900 cm^{-1}, $1{\cdot}70 \times 10^8\ \text{s}^{-1}$ at 28,300–36,300 cm^{-1}, $1{\cdot}83 \times 10^8\ \text{s}^{-1}$ at 37,700–38,500 cm^{-1}, $2{\cdot}27 \times 10^8\ \text{s}^{-1}$ at 39,200–42,900 cm^{-1} and $2{\cdot}54 \times 10^8\ \text{s}^{-1}$ at 44,200–46,200 cm^{-1}. They interpreted the behaviour as evidence for intersystem crossing to higher triplet states.

For low-pressure naphthalene vapour Ashpole *et al.*[198,199] observed $\Phi_{TM} = 0{\cdot}08$, $\Phi_{FM} = 0{\cdot}92$ consistent with (9.31). Beddard *et al.*[200] in the same laboratory concluded from the low value of Φ_{TM} that 'the decrease of Φ_{FM} with (increasing) excitation energy cannot be due to increasing triplet formation.' They calculated that the $S_1 - S_0$ internal conversion rate k_{GM} increases more rapidly with E_{ex} than the $S_1 - T_1$ intersystem crossing rate $(k^0_{TM})_1$, and they concluded that k_{GM} contributes significantly to the increase of k_{IM} with E_{ex}. However, the calculations neglected the $S_1 - T_2$ intersystem crossing rate $(k^0_{TM})_2$, the activated $S_1 - T_3$ intersystem crossing rate $(k'_{TM})_3$, and any higher S − T intersystem crossing processes. For naphthalene in solution $(k^0_{TM})_2 \sim 2 \times 10^6\ s^{-1}$ (p. *286*) and $(k'_{TM})_3 \sim 7 \times 10^7\ s^{-1}$ (p. *180*) and both parameters are $\gg (k_{TM})_1$. The neglect of two major components of k_{TM} may thus invalidate the conclusions of Beddard *et al.*[200] about the relative significance of k_{GM}.

9.5.2 *Absorption linewidths*

From the classical theory of radiation[201] the natural linewidth $\Delta\bar{\nu}_H$ (in cm^{-1}) of an optical transition from a ground state S_0 to an excited state S_p of lifetime τ_H is given by the relation,

$$k_H = 1/\tau_H = 2\pi \mathbf{c}\, \Delta\bar{\nu}_H \tag{9.32}$$

where **c** is the speed of light. $\Delta\bar{\nu}_H$ is defined as the full width of the $S_0 - S_p$ absorption line at half maximum intensity. The quantum theory of radiation[201] gives the same relation, and substitution of $\Delta E_H = \mathbf{hc}\, \Delta\bar{\nu}_H$, where **h** is Planck's constant, in (9.32) gives the uncertainty relation

$$\Delta E_H \,.\, \tau_H = \mathbf{h}/2\pi = \hbar \tag{9.33}$$

$\Delta\bar{\nu}_H$ is the natural linewidth of the $S_0 - S_p$ transition. In an atom or simple molecule $\Delta\bar{\nu}_H$ is determined by the radiative decay rate k_{FH}. In a complex molecule $\Delta\bar{\nu}_H$ is determined by $k_H = k_{FH} + k_{IH}$. For S_p with $p \geqslant 2$ the radiationless decay rate $k_{IH} \gg k_{FH}$, so that $k_H \simeq k_{IH}$. Apart from $\Delta\bar{\nu}_H$ and the optical resolution of the spectrometer, other factors can contribute to the total observed linewidth $\Delta\bar{\nu}$. These factors include temperature-dependent effects such as collisional and Doppler broadening, exciton–phonon interactions, and other effects associated with the interaction of the molecule with its environment. For vibronic transitions, other than 0 − 0 transitions, vibrational progression congestion may introduce overlap of two or more adjacent vibronic transitions and thus increase the apparent linewidth. Observations of $\Delta\bar{\nu}$ should preferably be made in the vapour phase at very low pressures or in the condensed phase in suitable media at very low temperatures, to minimize environmental effects, and they should be restricted to 0 − 0 transitions or to distinct vibronic transitions. Even with these precautions, the observation of $\Delta\bar{\nu}$ provides only an upper limit to the value of $\Delta\bar{\nu}_H$, unless proper account is taken of environmental and thermal effects.

Richards and Rice[170] measured $\Delta\bar{\nu}$ for $S_0 - S_p$ absorption transitions in anthracene, coronene and 1:12-benzoperylene in their appropriate

Shpol'skii matrices, which were chosen to minimize $\Delta\bar{\nu}$. The $S_0 - S_1$ transitions and the $S_0 - S_2$ transition in coronene were observed at 4°K, and the other transitions were observed at 77°K. The observations were of the $0 - 0$ transitions where these could be identified; otherwise $\Delta\bar{\nu}$ was determined for the most intense vibronic band, usually the $0 - 1$ transition. The results, including the corresponding values of k ($= 2\pi \mathbf{c}\, \Delta\bar{\nu}$), are summarized in Table 9.12, together with independent values of k_H obtained from fluorescence data for four of the transitions. For these transitions $k \sim 10^4$–$10^6\, k_H$, showing that even in a low-temperature Shpol'skii matrix, environmental effects provide the major contribution to $\Delta\bar{\nu}$ and k. Under these conditions the environmental contribution to k is $\sim 10^{12}\,\mathrm{s}^{-1}$ at 4°K and $\sim 10^{14}\,\mathrm{s}^{-1}$ at 77°K.

Callomon *et al.*[203] made a high-resolution study of the rotational fine structure of the $S_0 - S_1^*$ vibronic absorption bands of low-pressure benzene vapour. When the excitation energy $S_1^* \leqslant S_1^0 + 2900\,\mathrm{cm}^{-1}$, the fine structure is well resolved, and the S_1^* radiative and radiationless rate parameters k_{FM}^* and k_{IM}^* have been evaluated from observations of the lifetime and quantum yield of the $S_1^* - S_0^*$ resonance fluorescence[204–210] (see Stockburger, Chapter 2, Vol. 1). When $S_1^* \geqslant S_1^0 + 3300\,\mathrm{cm}^{-1}$, no fluorescence is observed, and the $S_0 - S_1^*$ absorption linewidth $\Delta\bar{\nu}$ is increased corresponding to an S_1^* decay rate k_M^* ($\simeq k_{IM}^*$) of 10^{11}–$10^{12}\,\mathrm{s}^{-1}$. The data on k_{FM}^*, k_{IM}^*, k_M^* and $\Delta\bar{\nu}$ as a function of the S_1^* vibrational energy E_v ($= S_1^* - S_1^0$) are listed in Table 9.13.

The dramatic increase in k_{IM}^* at $E_v \geqslant 3300\,\mathrm{cm}^{-1}$ is due to the onset of the 'channel 3' radiationless transition, which has been attributed to $S_1^* - S_X$ internal conversion, rate k'_{XM}, into the physical isomer state S_X (Birks, Chapter 1, Vol. 1). Similar values of k'_{XM} are obtained from observations of the temperature dependence of the fluorescence quantum yields and lifetimes of benzene solutions[119,122] and crystals.[211,212] These observations on benzene in the vapour, solution and crystal phases show that the process responsible for k'_{XM} and for the increase in $\Delta\bar{\nu}$ at $E_v \geqslant 3300\,\mathrm{cm}^{-1}$ is intramolecular, and that it is relatively insensitive to the molecular environment. There is, however, a solvent shift in the threshold vibrational energy W_{XM} of the channel 3 process from 2100–2200 cm^{-1} in the crystal,[211,212] to 2100–2800 cm^{-1} in solution[119,122] and 2900–3300 cm^{-1} in the vapour phase.[203,208]

The temperature dependence of the linewidth of the $S_0 - S_2$ $0 - 0$ transition in crystal C_6H_6, C_6D_6 and *sym*-$C_6H_3D_3$ has been measured from 4·2 to 135°K by Birks *et al.*[12] The transition energy S_2^0 shows a blue shift of 40 cm^{-1} per substituent deuteron, compared with 33 cm^{-1} for the $0 - 0$ transitions of the $S_0 \leftrightarrow S_1$ absorption and fluorescence and the $S_0 \leftrightarrow T_1$ absorption and phosphorescence in crystal benzene.[213,214] For

Table 9.12 $S_0 - S_p$ transitions in Shpol'skii matrices (Richards and Rice[170]). Energy E_p (cm^{-1}), linewidth $\Delta\bar{\nu}$ (cm^{-1}), $k = 2\pi c\Delta\bar{\nu}$ (s^{-1}) and S_p decay rate k_H (s^{-1}) from fluorescence measurements

Compound	Matrix	Temperature °K	S_p	E_p	$\Delta\bar{\nu}$	k	k_H	Ref.
Anthracene	*n*-Heptane	4	S_1 (^1L_a)	26234	6·2	$1·2 \times 10^{12}$	$6·3 \times 10^{7}$	p. *127*
		77	S_3 (^1B_b)	39140	280	$5·3 \times 10^{13}$		
			S_4	45080	400	$7·5 \times 10^{13}$		
			S_5	46660	400 (±200)	$7·5\,(\pm 3·8) \times 10^{13}$		
Coronene	*n*-Heptane	4	S_1	23450	2·1	$4·0 \times 10^{11}$	$3·1 \times 10^{5}$	202
			S_2	28440	4·1	$7·7 \times 10^{11}$		
		77	S_3	32870	300	$5·7 \times 10^{13}$		
			S_4	43840	670	$1·3 \times 10^{14}$		
			S_5	46600	350	$6·6 \times 10^{13}$		
			S_6	50000	4000 (±2000)	$7·5\,(\pm 3·8) \times 10^{14}$		
1:12-Benzoperylene	*n*-Hexane	4	S_1	24636	2·4	$4·5 \times 10^{11}$	$4·1 \times 10^{6}$	202
		77	S_2	25800	420	$7·9 \times 10^{13}$	$3·7 \times 10^{10}$	177
			S_3	32980	440	$8·3 \times 10^{13}$		
			S_4	34230	700 (±300)	$13·2\,(\pm 5·7) \times 10^{13}$		
			S_5	44580	1400 (±300)	$26·4\,(\pm 5·7) \times 10^{13}$		
			S_6	47100	800 (±200)	$15·1\,(\pm 3·8) \times 10^{13}$		
			S_7	49090	2500 (±200)	$47·1\,(\pm 3·8) \times 10^{13}$		

Table 9.13 Single vibronic levels of $^1B_{2u}(S_1)$ benzene of vibrational energy E_v. Lifetime (τ_M), fluorescence quantum yield (Φ^*_{FM}), absorption linewidth ($\Delta\bar{\nu}$) and total (k^*_M), radiative (k^*_{FM}) and radiationless (k^*_{IM}) decay rates. (τ_M and Φ_{FM} data from Spears and Rice;[208] $\Delta\bar{\nu}$ data from Callomon *et al.*[203])

Vibronic state	E_v(cm^{-1})	τ_M(ns)	Φ^*_{FM}	$\Delta\bar{\nu}$(cm^{-1})	k^*_M(10^6 s^{-1})	k^*_{FM}(10^6 s^{-1})	k^*_{IM}(10^6 s^{-1})
0	0	100	0·22	—	10·0	2·2	7·8
ν_6	521	79	0·27	—	12·7	3·4	9·3
$(\nu_6 + \nu_{16})$	764	77	0·26	—	13·0	3·4	9·6
ν_1	923	83	0·21	—	12·1	2·5	9·6
$(\nu_6 + 2\nu_{16})$	1007	66	0·23	—	15·2	3·5	11·7
$2\nu_6$	1042	72	0·29	—	13·9	4·0	9·9
$2\nu_{10}$	1170	52	0·16	—	19·2	3·1	16·1
$(\nu_1 + \nu_6)$	1444	71	0·21	—	14·1	2·9	11·2
?	1470	65	0·19	—	15·4	2·9	12·5
$(\nu_6 + 2\nu_{11})$	1547	51	0·18	—	19·6	3·5	16·1
$(\nu_1 + \nu_6 + \nu_{16})$	1687	61	0·16	—	16·4	2·6	13·8
$(\nu_6 + 2\nu_{10})$	1691	52	0·24	—	19·2	4·7	14·5
$2\nu_1$	1846	62	0·16	—	16·1	2·6	13·5
$(\nu_1 + \nu_6 + 2\nu_{16})$	1930	49	0·20	—	20·4	4·1	16·3
$(\nu_1 + 2\nu_6)$	1965	61	0·22	—	16·4	3·6	12·8
$(2\nu_5 + \nu_6)$	2070	55	0·06	—	18·2	1·1	17·1
$(2\nu_1 + \nu_6)$	2367	55	0·17	—	18·2	3·1	15·1
?	2393	39	0·26	—	25·6	6·7	18·9
$(2\nu_1 + \nu_6 + \nu_{16})$	2610	42	0·08	—	23·8	1·9	21·9
$(\nu_1 + \nu_6 + 2\nu_{10})$	2614	42	0·06	—	23·8	1·4	22·4
$3\nu_1$	2769	49	0·07	—	20·4	1·4	19·0
$(2\nu_1 + 2\nu_6)$	2888	47	0·03	—	21·3	0·7	20·6
$(3\nu_1 + \nu_6)$	3288	—	—	0·3	$5·6 \times 10^4$	—	—
$(2\nu_1 + \nu_6 + 4\nu_{16})$?	3314	—	—	0·7	$1·3 \times 10^5$	—	—

$(3v_1 + v_6 + v_{16})$	3527	—	—	1·0	$1{\cdot}9 \times 10^5$	—	—
$(3v_1 + 2v_6)$	3802	—	—	0·5	$9{\cdot}4 \times 10^4$	—	—
$(3v_1 + 2v_6 + v_{16})$	4045	—	—	5	$9{\cdot}5 \times 10^5$	—	—
$(4v_1 + v_6)$	4207	—	—	1·3	$2{\cdot}5 \times 10^5$	—	—
$(3v_1 + v_6 + 4v_{16})$?	4233	—	—	3	$5{\cdot}8 \times 10^5$	—	—
$(4v_1 + v_6 + v_{16})$	4442	—	—	>3	$>5{\cdot}8 \times 10^5$	—	—
$(4v_1 + 2v_6)$	4717	—	—	1·5	$2{\cdot}8 \times 10^5$	—	—
$(4v_1 + 2v_6 + v_{16})$	4960	—	—	>5	$>10^6$	—	—
$(5v_1 + v_6)$	5126	—	—	2·5	$4{\cdot}7 \times 10^5$	—	—

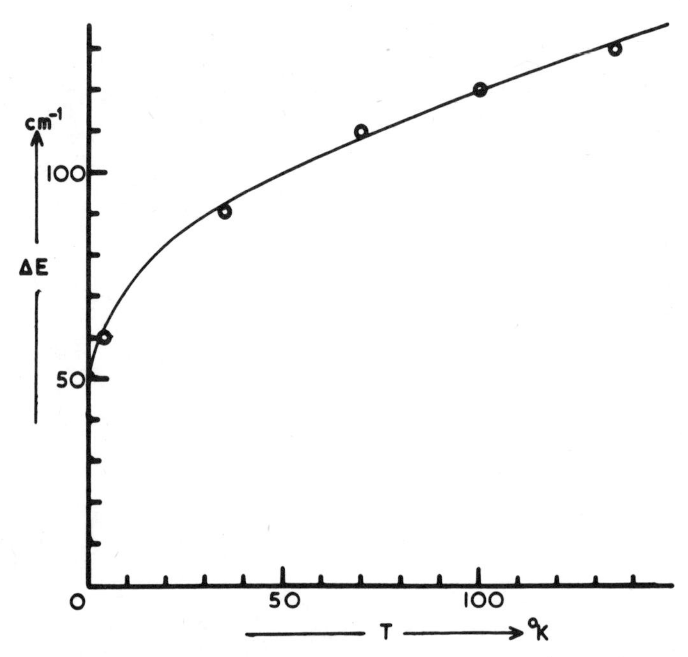

Figure 9.11 Linewidth $\Delta E (= \Delta\bar{\nu})$ of the $S_0 - S_2$ ($^1B_{1u} \leftarrow {}^1A_{1g}$) $0-0$ absorption transition in crystal C_6H_6 as a function of temperature T. (After Birks, Pantos and Hamilton[12])

the three benzenes $\Delta\bar{\nu}$ extrapolates to a common value of $\Delta\bar{\nu}_0 = 50\ \text{cm}^{-1}$ at 0°K. Figure 9.11 plots the observations of $\Delta\bar{\nu}$ for crystal C_6H_6 as a function of T. The curve in Figure 9.11 corresponds to the relation

$$\Delta\bar{\nu} = \Delta\bar{\nu}_0 + aT^{\frac{1}{2}} \tag{9.34}$$

with $\Delta\bar{\nu}_0 = 50\ \text{cm}^{-1}$ and $a = 7\ \text{cm}^{-1}\ \text{deg}^{-\frac{1}{2}}$, which is the form predicted for strong exciton–phonon coupling.[215] Substitution of $\Delta\bar{\nu}_0 = 50\ \text{cm}^{-1}$ in (9.33) gives $k_H(S_2) = 9{\cdot}4 \times 10^{12}\ \text{s}^{-1}$ for crystal benzene at 0°K.

Two main processes compete for the S_2 excitation energy: $S_2 - S_1$ internal conversion, rate k_{MH}; and $S_2 - S_X$ internal conversion (channel 3), rate k_{XH}. From an analysis[12] of the fluorescence excitation spectrum of crystal benzene[216] at 77°K, it is concluded that the $S_2 - S_1$ internal conversion quantum yield is

$$\Phi_{MH} = k_{MH}/k_H \simeq 0{\cdot}7$$

and that

$$k_{MH} \simeq 6{\cdot}6 \times 10^{12}\ \text{s}^{-1}; \quad k_{XH} \simeq 2{\cdot}8 \times 10^{12}\ \text{s}^{-1}$$

The value of k_{XH} may be compared with that of $k'_{XM} = 4{\cdot}5(\pm 1{\cdot}3) \times 10^{11}$ s^{-1} obtained[211,212] for the frequency factor of the temperature-dependent radiationless transition (channel 3) in crystal benzene from observations of q_{FM} and τ_M. k'_{XM} is the channel 3 rate at the threshold energy ($W_{XM} \sim$ 2200 cm^{-1}) of the process, while k_{XH} is the rate at ~6400 cm^{-1} above threshold. k'_{XM} shows a similar, but irregular, increase with excitation energy in the vapour phase (Table 9.13).

Lawson *et al.*[217] observed the $S_3 - S_1$ internal conversion quantum yield Φ_{ML} of benzene in different solvents at room temperature (pp. *174–5*, *189*). Φ_{ML} decreases from 0·45 in liquid benzene to 0·22–0·25 in aliphatic hydrocarbons and 0·04 in perfluoro-*n*-hexane (PFH) solution. Φ_{ML} is the product of the quantum yields of four rapid sequential processes: $S_3 - S_2^*$ internal conversion (Φ_{VL}), S_2^* vibrational relaxation (Φ_{HV}), $S_2 - S_1^*$ internal conversion (Φ_{VH}) and S_1^* vibrational relaxation (Φ_{MV}). The channel 3 radiationless transition competes with each of these processes and reduces their quantum yields below unity. Further studies are required to determine which of these parameters are sensitive to the solvent environment. The highly resolved absorption and fluorescence spectra of benzene in PFH solution (p. *117*) and the negligible solvent shift,[217] are indicative of minimal solvent–solute interactions, suggesting that the vibrational relaxation rate is reduced in this solvent, with a resultant decrease in Φ_{HV} and Φ_{MV}, and hence in Φ_{ML}. On the present evidence the solvent dependence of Φ_{ML} does not necessarily imply any solvent dependence of the individual internal conversion yields Φ_{VL} and Φ_{VH}.

A value of k_H may also be derived from the $S_2 - S_0$ fluorescence quantum yield Φ_{FH} of liquid benzene at 25°C observed by Hirayama *et al.*[4] The liquid was excited at $\lambda_{ex} = 184{\cdot}9$ nm (S_3) so that

$$\Phi_{FH} = \frac{k_{FH}}{k_H} \cdot \Phi_{HL} \tag{9.35}$$

where k_{FH} and k_H are the radiative and total S_2 decay rates, and Φ_{HL} ($= \Phi_{HV}\Phi_{VL}$) is the $S_3 - S_2$ internal conversion quantum yield. Substituting the experimental values of $\Phi_{FH} = 8 \times 10^{-6}$,[4] and $\Phi_{HL} = 0{\cdot}61$,[218] and $k_{FH} = 1{\cdot}3 \times 10^8\ s^{-1}$ calculated from the $S_0 - S_2$ absorption spectrum, we obtain $k_H = 9{\cdot}9 \times 10^{12}\ s^{-1}$. This value for liquid benzene at 25°C, obtained from fluorescence data, is in remarkable agreement with that of $9{\cdot}4 \times 10^{12}\ s^{-1}$ for crystal benzene at 0°K, obtained from linewidth data, considering the difference in the experimental conditions. It suggests that the $S_2 - S_1$ internal conversion and channel 3 processes which contribute to k_H are insensitive to the environment.

9.5.3 *Dual luminescences* (pp. *162–71*)

The multiple fluorescences of condensed aromatic hydrocarbons have been discussed in §9.5.1. Other molecules with dual luminescences will now be considered.

9.5.3.1 *Azulene and its derivatives* (pp. *166–8, 625–8*)

The photophysical properties of the pentaheptacyclic molecule, azulene, differ remarkably from those of its bihexacyclic isomer, naphthalene. The $S_2 - S_0$ fluorescence quantum yield Φ_{FH} greatly exceeds the $S_1 - S_0$ fluorescence quantum yield Φ_{FM}, and weak $S_2 - S_1$ fluorescence of quantum yield Φ_{FHM} has also been observed.[1–3] The rate parameters of Figure 9.12, which supercede previous incorrect values (p. *168*), have been

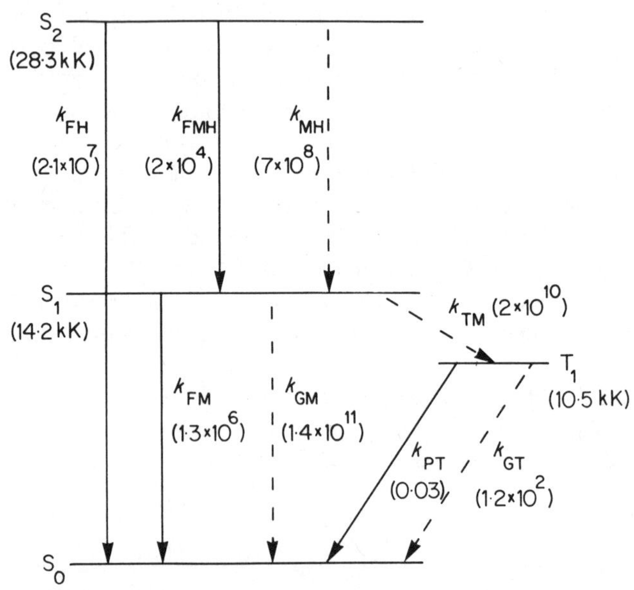

Figure 9.12 Rate parameters (in s^{-1}) of azulene estimated from the available experimental and theoretical data.[219] Radiative and radiationless transitions are indicated by solid and broken arrows respectively

derived from experimental and theoretical data on azulene.[219] They are based on a value[175] of $\Phi_{FH} = 0{\cdot}03$ for azulene in cyclohexane at 25°C.

No direct experimental data are available to separate k_{IH} into its components k_{MH} and k_{TH}, but Siebrand and Williams[222] have calculated that $S_2 - S_1$ internal conversion k_{MH} is the dominant S_2 decay mode. Perdeuteration of azulene should thus reduce k_{MH}, because of the large $S_2 - S_1$

energy gap of 14,000 cm^{-1}, and hence increase Φ_{FH}, Φ_{FMH} and τ_H. The perdeuteration increase in Φ_{FH} has been observed by Ross and co-workers,[223,224] that in Φ_{FMH} has been observed by Rentzepis *et al.*,[1] and that in τ_H by Knight and Selinger.[225] Table 9.14 lists the observations of τ_H for azulene-h_8 and azulene-d_8 in the vapour and solution phases.[225]

Table 9.14 τ_H(ns) of S_2 state of azulene-h_8 and azulene-d_8 (Knight and Selinger[225])

Phase	Temperature °K	λ_{ex}(nm)	$\tau_H(h_8)$	$\tau_H(d_8)$	$\tau_H(d_8)/\tau_H(h_8)$
Vapour, 0·02 torr	296	315·9	1·08	1·73	1·6
		337·1	2·09	2·94	1·4
Methylcyclohexane					
solution, 10^{-4} M	296	337·1	1·63	2·08	1·3
	77	337·1	1·82	2·67	1·5

Murata *et al.*[220,221] observed Φ_{FH}, and calculated k_{FH} from the $S_0 - S_2$ absorption spectra, for several azulene derivatives in cyclohexane solution. Table 9.15 lists their values of S_2^0, S_1^0, E_{HM} $(= S_2^0 - S_1^0)$, Φ_{FH}, k_{FH} and k_{IH}. k_{IH} increases as E_{HM} decreases, and there is an approximate linear correlation between $\log k_{IH}$ and E_{HM}, which is consistent with $S_2 - S_1$ internal conversion via CH vibrations (pp. *157–8*).

Table 9.15 Properties of azulene derivatives. Energies S_2^0, S_1^0 and E_{HM} $(= S_2^0 - S_1^0)$ in cm^{-1}. Rate parameters k_{FH} and k_{IH} in 10^8 s^{-1} (Murata *et al.*[220,221])

Substituents[a]	S_2^0	S_1^0	E_{HM}	Φ_{FH}	k_{FH}	k_{IH}
1,3-Dichloro	27·0	12·7	14·3	0·058	0·36	5·8
1-Chloro	27·7	13·4	14·3	0·036	0·23	6·1
5-Chloro	27·9	13·7	14·2	0·032	0·24	7·3
—	28·3	14·3	14·0	0·031	0·22	6·9
3,8-Dimethyl, 5-*iso*propyl	27·2	13·8	13·4	0·014	0·28	20
1,2,3-Trichloro	26·7	13·7	13·0	0·012	0·29	24
2-Methyl	27·8	14·9	12·9	0·0081	0·20	25
1,2-Dichloro	27·3	14·5	12·8	0·0062	0·24	39
2-Chloro	27·9	15·4	12·5	0·0066	0·27	41
2-Iodo	27·1	15·0	12·1	0·0034	0·41	119

[a] Cyclic numeration starting with 5-membered ring.

From the Siebrand radiationless transition model (pp. *152–61*), Siebrand and Williams[222] obtained $k_{GM} = 6 \times 10^9$ s^{-1} for azulene-h_8, and Birks (pp. *185*, *187*) estimated $k_{GM} = k_{MH} = 2{\cdot}5 \times 10^9\ s^{-1}$ for azulene·h_8 and $k_{GM} = k_{MH} = 1{\cdot}3 \times 10^9\ s^{-1}$ for azulene·d_8. The k_{MH} values are of the correct order of magnitude, but those of k_{GM} are up to two orders of magnitude less than the experimental values. The exceptionally high values of $k_{GM} \sim 1{\cdot}2 \times 10^{11}\ s^{-1}$ and of $k_{TM} \sim 2 \times 10^{10}\ s^{-1}$, which reduce Φ_{FM} to $\sim 10^{-5}$, are two major unexplained anomalies in the photophysics of azulene, probably associated with its peculiar molecular structure. The anomalous behaviour of azulene is associated with its S_1 state, rather than with its S_2 state, the fluorescence of which is typical of 'normal' $S_1 - S_0$ fluorescence.

9.5.3.2 *Di-aromatic molecules* (pp. *163–4*)

Heteropolar di-aromatic molecules X—$(CH_2)_n$—Y, where X and Y are different aryl groups joined by an alkane chain $(CH_2)_n$, exhibit two fluorescences and phosphorescences characteristic of X and Y, the relative yields depending on the excitation wavelength (pp. *163–4*). For $n = 3$ the homopolar and heteropolar di-aromatic molecules X—$(CH_2)_3$—X and X—$(CH_2)_3$—Y exhibit intramolecular excimer and exciplex fluorescence, characteristic of X_2^* and $(X.Y)^*$, respectively (p. *324*; Klöpffer, Chapter 7, Vol. 1).

The homopolar diaryl molecules X—X, such as 1,1′-binaphthyl,[226] 2,2′-binaphthyl[227] and 9,9′-dianthryl[228] also have unusual fluorescence properties. The fluorescence spectrum consists of two components,

(i) a structured emission F_1, the approximate mirror image of the $S_0 - S_1$ absorption spectrum, and similar to the fluorescence of X, and

(ii) a structureless emission F_2 at lower energies than F_1.

In the crystal phase, in solid solution and in highly viscous solvents only F_1 is observed.[226,228] As the solvent viscosity η is reduced, the ratio of the F_2 to F_1 intensities increases, and at low viscosities only F_2 is observed. The temperature dependence of the fluorescence spectrum F of 9,9′-dianthryl in glycerin[228] illustrates this behaviour (Figure 9.13). At $T \leqslant$ 183°K (high η) $F \to F_1$, and at $T \geqslant$ 344°K (low η) $F \to F_2$. The energy $(F_2)_{max}$ of the F_2 maximum remains constant up to a certain solvent polarity f_c, and it then shifts linearly to lower energies with increase in solvent polarity f[228] (see Lippert, Chapter 1, Vol. 2).

In 1,1′-binaphthyl Hochstrasser[226] interpreted the behaviour in terms of steric hindrance to twisting about the 1,1′-bond. Expressing the angle of twist relative to the *cis*-planar configuration ($\theta = 0°$), the ground-state S_0 equilibrium configuration is $\theta_0 = 78°$, while the *trans*-planar configuration corresponds to $\theta = 180°$. Due to steric hindrance the S_0

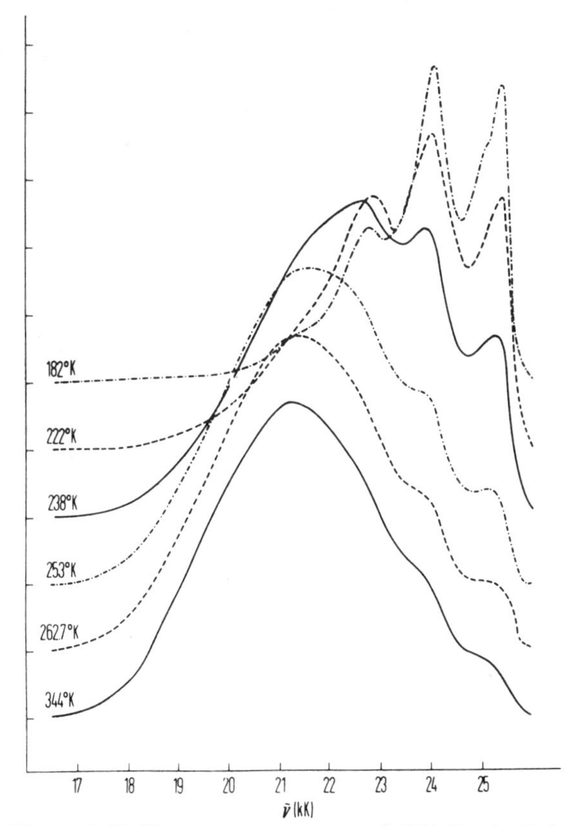

Figure 9.13 Fluorescence spectra of 9,9′-dianthryl in glycerin solutions at different temperatures. The ordinate (intensity) scale is shifted upwards by one unit for each incremental reduction in temperature. (Schneider and Lippert;[228] reproduced by permission of Verlag Chemie)

potential energy *vs.* θ curve is much steeper on the low-θ (*cis*) side of θ_0 than on the high-θ (*trans*) side, so that the S_0 potential energy curve is asymmetrical. Hochstrasser[226] proposed that the corresponding potential energy curve for the excited state has two minima, one at $\theta_1 \simeq \theta_0$ from which F_1 occurs and a deeper one at $\theta_2 \sim 120°$ from which F_2 occurs into high vibrational levels of S_0.

For 9,9′-dianthryl Schneider and Lippert[229] calculated the S_0 and S_1 potential energy curves as a function of θ. The S_0 curve is symmetrical with a single minimum at $\theta_0 = 90°$. The S_1 curve is symmetrical with a maximum at $\theta_1 = 90°$ and two minima at $\theta_2 = 78°$ and 102°. $S_1(\theta_1)$ is the Franck–Condon state which, subject to viscous restraint, relaxes to

$S_1(\theta_2)$, so that F_1 is the transition from $S_1(\theta_1)$ and F_2 is that from $S_1(\theta_2)$. The model[229] thus explains the viscosity dependence of the fluorescence, which is similar to that of *cis*-stilbene and other compounds in which molecular relaxation occurs subsequent to Franck–Condon excitation (§9.4.4).

The shift of $(F_2)_{max}$ with solvent polarity f for 9,9′-dianthryl is equivalent to an excited state dipole moment of $\mu_1 = 20{\cdot}6$ D for $f \geqslant f_c$, and of $\mu_1 = 0$ for $f < f_c$.[229] This behaviour is unusual, since the solute molecule is symmetrical and non-polar, and it might be expected to remain so in the excited state. An explanation of the phenomenon has been proposed by Beens and Weller.[230] The wavefunction of an excited species consisting of two identical aromatic moieties, such as an aromatic excimer or a diaryl molecule, can be expressed (p. *420*) as the linear combination of the wavefunctions of exciton resonance states

$$\Phi_E^{\pm} = \frac{1}{\sqrt{2}}[\psi(A^*D) \pm \psi(AD^*)] \tag{9.36}$$

and charge resonance states

$$\Phi_R^{\pm} = \frac{1}{\sqrt{2}}[\psi(A^-D^+) \pm \psi(A^+D^-)] \tag{9.37}$$

where A and D are identical. The excited state wavefunction, which is of the form

$$\Psi = c_1\Phi_R^+ + c_2\Phi_R^- + c_3\Phi_E^+ + c_4\Phi_E^- \tag{9.38}$$

has an electric dipole moment of

$$\mu_1 = 2c_1c_2\mu_0 \tag{9.39}$$

where μ_0 is the A^-D^+ dipole moment. In a highly polar solvent this dipole is solvated and the resultant reaction field R (p. *113*) corresponds to a perturbation which mixes Φ_R^+ and Φ_R^-. Beens and Weller[230] showed that in solution the non-polar ($\mu_1 = 0$) symmetrical state of the excited molecule or excimer corresponds to a (relative) energy minimum, but that for $f \geqslant f_c$ there is a solvated polar ($\mu_1 > 0$) asymmetrical species of lower energy. The critical polarity f_c depends on the matrix element

$$|\alpha| = |\langle\Phi_R^+|H^0|\Phi_E^+\rangle| \tag{9.40}$$

For $|\alpha|$ small, as in 9,9′-dianthryl, f_c lies within the normal range of solvent polarities, and a red shift of $(F_2)_{max}$ occurs at $f \geqslant f_c$.[229] For $|\alpha|$ large as in the pyrene excimer, f_c lies above the normal range of f, and the fluorescence maximum F_{max} is independent of solvent polarity[231] (p. *478*).

9.5.3.3 *Other molecules and ions*

The absorption spectra of the heterocyclic isomers, benzo[e]cyclohept[b]indole and benzo[g]cyclohept[b]indole, each consist of a weak, diffuse $S_0 - S_1$ band system and a more intense, structured $S_0 - S_2$ band system. In cyclohexane solution at room temperature the fluorescence spectra correspond to $S_2 - S_0$ transitions, as in azulene.[232]

The absorption spectrum of the decacylene anion[233] consists of an intense infrared band system at $\geqslant 4950\ cm^{-1}$ and a moderately intense visible band system at $\geqslant 16{,}500\ cm^{-1}$. These correspond respectively to the $D_0 - D_1$ and $D_0 - D_2$ transitions from the ground doublet state D_0 of the anion to the first and second excited doublet states D_1 and D_2. In dilute rigid solutions in 2-methyltetrahydrofuran two luminescences are observed: (a) phosphorescence[234] at $14{,}500\ cm^{-1}$, corresponding to the $Q_1 - D_0$ transition from the first excited quartet state Q_1 of the ion, and (b) fluorescence at $15{,}700\ cm^{-1}$, corresponding to the $D_2 - D_0$ transition. The phosphorescence lifetime of 2·6 ms is consistent with $Q_1 - D_1$ intersystem crossing through an energy gap of $9500\ cm^{-1}$. Brugman *et al.*[233] estimated a $D_2 - D_1$ internal conversion rate of $<6{\cdot}7 \times 10^{10}\ s^{-1}$ and a $D_2 - Q$ intersystem crossing rate of $6{\cdot}7 \times 10^{10}\ s^{-1}$. No data are available on any $D_1 - D_0$ fluorescence.

The phenylalkyl ketones, $C_6H_5COCH_2R$, and related compounds exhibit two or more phosphorescences in rigid organic glass matrices at 77°K.[235–241] It is generally agreed that the major, short-lived, emission is from an (n, π^*) triplet state, but the origin of the minor, long-lived, emission is uncertain. Wagner *et al.*[241] observed two components P_x and P_y of the short-lived (n, π^*) phosphorescence with $0 - 0$ bands at 384 nm and 397 nm, respectively. In a rigid methylcyclohexane glass at 77°K P_x is dominant, and it is attributed to (n, π^*) triplets held in the ground state molecular configuration. If the viscosity is reduced by the addition of 2-methyl butane to the glass, the relative intensity of P_y is increased, and in *iso*pentane solution at 77°K and in benzene solution at room temperature P_y is dominant. P_y is attributed to (n, π^*) triplets in which the molecular configuration has relaxed prior to emission.[240] These viscosity effects on phosphorescence are analogous to those on fluorescence discussed in §§9.4.4 and 9.5.3.2.

Other anomalous luminescence properties have been observed in Michler's ketone (MK)

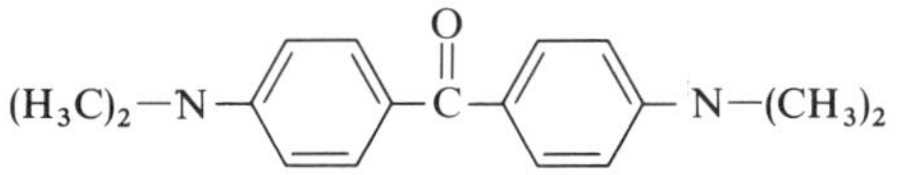

$$(H_3C)_2{-}N{-}C_6H_4{-}\overset{O}{\overset{\|}{C}}{-}C_6H_4{-}N{-}(CH_3)_2 \qquad \text{(MK)}$$

Klöpffer[242] reported that the fluorescence and phosphorescence spectra of

MK in rigid ethanol and methyltetrahydrofuran glasses at 77°K each consist of two components, the relative intensities of which depend on the excitation wavelength. He proposed that the MK phosphorescence occurs from two different triplet states T_1 and T_2, which are selectively populated from two different singlet states S_1 and S_2, respectively. Callis and Wilson[243] pointed out the very anomalous behaviour represented by this hypothesis. Since the $T_2 - T_1$ energy gap $\sim 1000\ \mathrm{cm}^{-1}$, the expected $T_2 - T_1$ internal conversion rate of $> 10^{11}\ \mathrm{s}^{-1}$ would need to be reduced to $< 10\ \mathrm{s}^{-1}$ for $T_2 - S_0$ emission to be observed.

Callis and Wilson[243] studied the 77°K absorption and emission spectra of MK in a variety of solvent mixtures, and they concluded that MK exists in at least three distinct ground state spectroscopic species, characterized by first absorption maxima at 360, 385 and 410 nm, respectively. Each species apparently represents a discrete solute–solvent complex or local environment, and the relative proportions of the three species depend on the nature of the solvent mixture. The changes in the luminescence spectra correlate with the changes in the absorption spectra, and it is not necessary to invoke changes in the excited state or inhibited internal conversion[242] to explain the unusual luminescence behaviour of Michler's ketone. The luminescence of mixed molecular species in solid solution has been discussed previously (pp. *375–8*).

p-N,N-Dimethylaminobenzonitrile (DMAB) exhibits two fluorescence bands *a* and *b*, whose positions and relative intensities I_a and I_b depend on the solvent polarity (pp. *164–6*, *188*). Lippert *et al.*[595] attributed the *a*- and *b*-bands to a strongly polar excited state ^1L_a and a less polar excited state ^1L_b, respectively. Khalil *et al.*[596] attributed the *a*- and *b*-bands to the excimer and molecular fluorescence of DMAB, respectively, and they also observed a weak concentration-dependent absorption and a corresponding fluorescence, which they attributed to a ground-state dimer.

Rotkiewicz *et al.*[597] have studied the spectral characteristics of the 2-methyl and 2-methoxy derivatives of DMAB. They observed no change in I_a/I_b on increasing the concentration by a factor of 10^3, and they concluded that excimers are not responsible for the *a*-band emission. I_a/I_b is also independent of λ_{ex}, so that both bands have the same excitation spectrum. The polarizations of the *a*- and *b*-bands are observed to be parallel, and not perpendicular, as required if they originated from ^1L_a and ^1L_b states. Rotkiewcz *et al.*[597] tentatively assign the emissions to two excited species differing in polarity and in the orientation of the $N(CH_3)_2$ group.

9.6 The triplet state

Many of the topics considered in Chapter 6 of *PAM* have been discussed by other contributors. These include

(a) triplet quantum yields and

(b) singlet–triplet intersystem crossing (Wilkinson, Chapter 3, Vol. 2);

(c) triplet–triplet absorption spectra (Labhart and Heinzelmann, Chapter 6, Vol. 1);

(d) assignment of the electronic states of the polyacenes,

(e) phosphorescence excitation spectrometry and

(f) electron impact spectrometry (Birks, Chapter 1, Vol 1);

(g) photo-ionization in condensed media (Lesclaux and Joussot-Dubien, Chapter 9, Vol. 1);

(h) vapours of benzene and its derivatives (Stockburger, Chapter 2, Vol. 1) and

(i) the benzene triplet state (Cundall, Chapter 2, Vol. 2).

Topics still to be discussed include spectrophosphorimetry, phosphorescence lifetimes and quantum efficiencies, and the heavy-atom effect.

9.6.1 *Spectrophosphorimetry* (pp. *201–6*)

Langelaar *et al.*[244] have designed a high sensitivity spectrophosphorimeter, which has been used to observe the phosphorescence spectra of the naphthalene triplet excimer in solution,[245] the pyrene crystal 'triplet excimer'[246] and aromatic hydrocarbon vapours.[247] Hamilton and Naqvi[357] have designed an instrument for time-resolved spectrophosphorimetry using an electronically gated photomultiplier.

Expressions relating the observed and instantaneous phosphorescence intensities to the phosphorimeter characteristics and the triplet lifetime have been derived by O'Haver and Winefordner,[248] and Naqvi[249] has extended their treatment to include P-type delayed fluorescence. His expression for the ratio of the delayed and prompt fluorescence quantum yields, Φ^{d}_{FM}/ϕ_{FM}, differs from that of Parker.[250,251] Naqvi's assertion,[249] that Parker's Φ^{d}_{FM} values were too low by a factor of 4, was shown by Parker[252] to be based on a misunderstanding of his experimental conditions. The light absorption rate I_a and the prompt and delayed fluorescence intensities were measured with the excitation chopper running, while Naqvi's expression[249] is for the hypothetical case of the chopper permanently open. Naqvi[253] has now accepted the validity of Parker's values of $\Phi^{d}_{FM}/\Phi_{FM}I_a$, although he has questioned other features of Parker's analysis.[250,251] As noted previously (p. *206*) the precise experimental determination of Φ^{d}_{FM}/Φ_{FM} is particularly difficult. It depends not only on the phosphorimeter factor, the chopper speeds and the triplet lifetime, but also on the light absorption rate I_a and on the complex first-order and second-order reaction kinetics which determine the time dependence of the triplet concentration and the delayed fluorescence intensity (pp. *378–84*).

9.6.2 *Phosphorescence lifetimes and quantum efficiencies* (pp. *207–8*)

9.6.2.1 *Higher aromatic hydrocarbons*

The first observations of the phosphorescence of aromatic hydrocarbon vapours have been made by van Leeuwen *et al.*[247] The phosphorescence spectra of naphthalene and pyrene vapours were identified by comparison with those of the compounds in rigid solution. For $p \leqslant 10$ torr, the phosphorescence parameters $\tau_T = 3$ ms, $\Phi_{PT} \sim 2 \times 10^{-6}$ for pyrene vapour and $\tau_T = 0{\cdot}5$ ms, $\Phi_{PT} \sim 10^{-5}$–10^{-6} (depending on λ_{ex}) for naphthalene vapour, are practically independent of pressure and temperature, but quenching due to triplet–triplet interaction occurs at higher p. The triplet radiative lifetimes of naphthalene and pyrene vapours are each estimated as $\tau_{PT} \sim 100$ s.

The determination of the triplet radiative and radiationless rate parameters k_{PT} $(= 1/\tau_{PT})$ and $k_{GT}(= 1/\tau_{GT})$ requires the observation of the phosphorescence lifetime τ_T $(= 1/k_T)$ and quantum efficiency q_{PT} $(= k_{PT}/k_T)$. Few direct observations have been made of q_{PT} (pp. *206–7*). It can alternatively be evaluated from measurements of the phosphorescence and triplet quantum yields, Φ_{PT} and Φ_{TM}, where $q_{PT} = \Phi_{PT}/\Phi_{TM}$, so that

$$k_{PT} = \frac{\Phi_{PT}}{\Phi_{TM}\tau_T} \tag{9.41}$$

$$k_{GT} = k_T - k_{PT} \tag{9.42}$$

The determination of Φ_{TM} presents difficulties, particularly in solid solutions, and many observers instead measure the fluorescence quantum yield Φ_{FM} and assume (p. *208–9*) that

$$\Phi_{TM} = 1 - \Phi_{FM} \tag{9.43}$$

This assumption is probably valid for most, but not all, compounds in rigid solution at 77°K or less (see Wilkinson, Chapter 3, Vol. 2).

Langelaar *et al.*[254] observed Φ_{PT} and τ_T for naphthalene, anthracene, phenanthrene and pyrene in dilute ethanol solution at 293°K (fluid) and 77°K (rigid glass). Table 9.16 summarizes their observations and the data on Φ_{FM} and Φ_{TM} used in the analysis. Within the experimental error the triplet radiative lifetime τ_{PT} is observed to be practically independent of the temperature and molecular environment. The values of $\tau_{PT} \sim 100$ s for naphthalene and pyrene vapours are consistent with this conclusion.

Table 9.16 Phosphorescence parameters of solutions in ethanol (Langelaar *et al.*[254])

Component	Temperature (°K)	Φ_{FM}	Φ_{TM}	Φ_{PT}	τ_T(s)	q_{PT}	τ_{PT}(s)
Naphthalene (5×10^{-4} M)	293	0·21[a]	0·71[a]	2×10^{-5}	0·003[a]	$2{\cdot}8 \times 10^{-5}$	100 (±50)
	77	0·40	0·60[b]	2×10^{-2}	2·5	$3{\cdot}3 \times 10^{-2}$	75 (±25)
Anthracene (10^{-5} M)	293	0·30[a]	0·70[b]	—	0·009	—	—
	77	0·27[a]	0·73[b]	2×10^{-4}	0·05	$2{\cdot}7 \times 10^{-4}$	180 (±70)
Phenanthrene (10^{-3} M)	293	0·13[a]	0·80[a]	2×10^{-5}	0·001	$2{\cdot}5 \times 10^{-5}$	40 (±20)
	77	0·14[a]	0·86[b]	$1{\cdot}2 \times 10^{-1}$	3·6	$1{\cdot}4 \times 10^{-1}$	25 (±8)
Pyrene (5×10^{-3} M)	293	0·72[a]	0·27[a]	3×10^{-5}	0·0094	$1{\cdot}1 \times 10^{-4}$	80 (±40)
	77	0·91	0·09[b]	$1{\cdot}7 \times 10^{-3}$	0·5	$1{\cdot}9 \times 10^{-2}$	30 (±15)

[a] Data from Parker.[251]
[b] Values estimated by assuming $\Phi_{FM} + \Phi_{TM} = 1$.

The values of τ_{PT} for anthracene and naphthalene exceed the value of $\tau_{PT} \simeq 30$ s for the unsubstituted aromatic hydrocarbons assumed by Siebrand,[253] although they are not necessarily as high as indicated by the data of Table 9.16. Other observers have measured $\tau_{PT} = 60$ s in anthracene single crystals,[256] $\tau_{PT} = 50$ s and 36·2 s for naphthalene in solid solution at 77°K,[257,258] and $\tau_{PT} = 30$ s for naphthalene in an argon matrix at 20°K.[259]

Li and Lim[258] observed Φ_{FM}, Φ_{PT} and τ_T for twenty aromatic and heteroaromatic molecules and their perdeuterated derivatives in EPA glass at 77°K. A solution of 9,10-diphenylanthracene, which has $q_{FM} = 1{\cdot}00$ at 77°K,[260] was used as a luminescence quantum yield standard, and Φ_{TM} was evaluated from (9.43). The experimental and derived parameters for perprotonated and perdeuterated hydrocarbons are listed in Table 9.17. For similar data on nitrogen-heterocyclics and aromatic carbonyl compounds, reference should be made to the original paper.[258]

Table 9.17 Phosphorescence parameters of aromatic hydrocarbons in EPA glass at 77°K (Li and Lim[258])

Compound	Φ_{FM}	Φ_{PT}	τ_T(s)	τ_{PT}(s)	τ_{GT}(s)
Benzene-h_6	0·19	0·18	6·3	28·4	8·1
Benzene-d_6	0·27	0·24	9·4	28·6	14·1
Naphthalene-h_8	0·37	0·04	2·3	36·2	2·5
Naphthalene-d_8	0·40	0·18	18·4	61·3	26·3
Phenanthrene-h_{10}	0·13	0·10	3·7	32·2	4·2
Phenanthrene-d_{10}	0·11	0·29	15·2	46·6	22·6
Chrysene-h_{12}	0·23	0·07	2·7	29·7	3·0
Chrysene-d_{12}	0·24	0·25	13·2	40·1	19·7
Triphenylene-h_{12}	0·07	0·41	14·0	32·0	24·9
Triphenylene-d_{12}	0·08	0·53	21·4	37·3	50·3
Biphenyl-h_{10}	0·13	0·24	4·6	16·7	6·4
Biphenyl-d_{10}	0·13	0·53	9·9	16·3	25·3

The influence of perdeuteration on Φ_{FM} and Φ_{TM} in benzene and other aromatic hydrocarbons is discussed in §9.6.2.2. Its influence on τ_{GT} (Table 9.17) is particularly marked, and it is consistent with the Siebrand model of $T_1 - S_0$ intersystem crossing. The data for naphthalene, phenanthrene and chrysene (but not benzene, triphenylene and biphenyl) also show a marked increase in the triplet radiative lifetime τ_{PT} on perdeuteration. Fischer and Lim[261] have proposed a model involving non-Born–Oppenheimer coupling, in association with spin–orbit coupling, between excited states to account for this effect.

Johnson and Studer[259] observed Φ_{FM}, Φ_{PT} and τ_T for perprotonated and perdeuterated toluene and naphthalene in an argon matrix at 20°K, and their results (Table 9.18) show a negligible effect of perdeuteration on k_{PT} ($= 1/\tau_{PT}$). Argon as a 'heavy-atom' solvent can potentially influence k_{TM}, k_{PT} and k_{GT} by external spin–orbit coupling. Comparison of the data of Tables 9.17 and 9.18 shows that the argon matrix increases k_{TM} (and hence decreases Φ_{FM}), but it has no systematic effect on k_{GT} or k_{PT}. Further experiments are required to establish whether the perdeuteration effect on k_{PT} in naphthalene exists[258] or does not exist.[259]

Table 9.18 Phosphorescence parameters of aromatic hydrocarbons in argon matrix at 20°K (Johnson and Studer[259])

Compound	Φ_{FM}	Φ_{PT}	τ_T(s)	k_{PT}(s^{-1})
Toluene-h_8	0·04 ± 0·01	0·47 ± 0·07	12·6 ± 0·1	0·040 ± 0·006
Toluene-d_8	0·05 ± 0·01	0·65 ± 0·07	16·8 ± 0·1	0·041 ± 0·006
Naphthalene-h_8	0·03 ± 0·01	0·05 ± 0·01	1·61 ± 0·02	0·033 ± 0·006
Naphthalene-d_8	0·03 ± 0·01	0·45 ± 0·08	15·0 ± 0·1	0·031 ± 0·006

Partial deuteration of benzene (pp. *249*, *629*) or naphthalene (p. *628*) increases τ_T, due to a reduction of k_{GT}, and the effect is position-dependent. Charlton and Henry[262] have observed and explained theoretically a similar position-dependent deuteration effect on τ_T in anthracene, 1,4,5,8-anthracene-d_4 and 2,3,6,7-anthracene-d_4.

Cundall and Pereira[263] have described a method of determining k_{PT} and k_{GT} which avoids the assumption

$$\Phi_{TM} = 1 - \Phi_{FM} \tag{9.43}$$

They determined k_{FM} and k_{TM} for the compound in liquid methylcyclohexane solution from 60 to −60°C, from observations of Φ_{FM}, Φ_{TM} and τ_M, and they extrapolated k_{FM} and k_{TM} to 77°K. They then observed Φ_{PT}/Φ_{FM} and τ_T in methylcyclohexane glass at 77°K, and evaluated

$$k_{PT} = \frac{\Phi_{PT}}{\Phi_{FM}} \frac{k_{FM}}{k_{TM}} \tau_T \tag{9.44}$$

and k_{GT} from (9.42). They obtained $k_{PT} = 0{\cdot}049\ \text{s}^{-1}$, $k_{GT} = 0{\cdot}090\ \text{s}^{-1}$ for toluene, and $k_{PT} = 0{\cdot}043\ \text{s}^{-1}$, $k_{GT} = 0{\cdot}140\ \text{s}^{-1}$ for tetralin (1,2,3,4-tetrahydronaphthalene). For toluene in methylcyclohexane below −60°C, relation (9.43) is found to be valid with $\Phi_{FM} = 0{\cdot}28$, $\Phi_{TM} = 0{\cdot}72$.

Graves *et al.*[264] measured the phosphorescence lifetime τ_T and intensity I_p for various aromatic hydrocarbons in polymethylmethacrylate

(PMMA) solutions as a function of temperature from 77 to 400°K. Like previous observers[265–267] they found two distinct temperature-dependent, triplet-depletive processes. The first process, characterized by a low activation energy W, is dominant at low temperatures, and the second process, which has a much larger activation energy ΔE, dominates at high temperatures. The T_1 and S_1 energies and the values of ΔE and W obtained by different observers are listed in Table 9.19.

The values of $T_X = T_1 + \Delta E$ (or $S_1 + \Delta E$, in the case of benzoperylene) are constant within the experimental error, and $T_X = 25{,}100\ (\pm 500)\ \text{cm}^{-1}$ is identified[264] as the energy of the first excited triplet state of PMMA. The main temperature-dependent triplet quenching process is thus attributed to $T_1 - T_X$ energy transfer. This process does not occur with benzene in PMMA since $T_1 > T_X$, and the temperature dependence of τ_T for benzene is due to the intramolecular process associated with W ($= W'_{GT}$) (pp. *248–9*). The values of W for the other compounds may also be intramolecular in origin, but the present evidence is inadequate. For naphthalene in EPA solution τ_T only decreases from 2·58 s at 4°K to 2·50 s at 77°K, and similar small decreases of τ_T by ~0·1–0·2 s are observed in the methyl, dimethyl and trimethyl naphthalenes.[271] This behaviour contrasts with that of benzene in which τ_T decreases from ~16 s at 4°K to ~7 s at 77°K (p. *248*).

9.6.2.2 *Benzene and derivatives*

Many of the photophysical properties of benzene differ from those of the higher aromatic hydrocarbons (see Cundall, Chapter 2, Vol. 2). Benzene is the only aromatic hydrocarbon studied by Li and Lim[258] in which Φ_{FM} increases significantly on perdeuteration (Table 9.12). If (9.43) is valid and if k_{FM} is unaffected by perdeuteration, then the effect of perdeuteration on $S_1 - T$ intersystem crossing corresponds to $(k_{TM})_H/(k_{TM})_D = 1{\cdot}6$. On the Siebrand model (p. *148*) this is consistent with the $S_1 - T_1$ energy gap of ~9000 cm^{-1}, but not with the $S_1 - T_2$ energy gap ~900 cm^{-1}. Li and Lim[258] concluded that $S_1 - T_1$ intersystem crossing is dominant in benzene, and that, despite the proximity of S_1 and T_2 and the consequent large Franck–Condon factor, $S_1 - T_2$ intersystem crossing is inhibited by the low density of vibrational states of T_2 in the vicinity of S_1. Breuer and Lee[272] observed that τ_M for benzene vapour is increased by perdeuteration, and they interpreted this in a similar manner.

In contrast to benzene, perdeuteration of phenanthrene (Table 9.17) and anthracene[273] decreases Φ_{FM} and τ_M, and this 'inverse deuterium effect' is consistent with dominant $S_1 - T_q$ $(q \geqslant 2)$ intersystem crossing.[173,274] The magnitude of k_{TM} (~$10^7\ \text{s}^{-1}$) for benzene is similar to the values of k_{TM} for these and other compounds in which $S_1 - T_2$ (or

Table 9.19 Energies and activation energies (cm^{-1}) for various solutes in PMMA (after Graves *et al.*[264])

Compound	T_1	S_1	ΔE	Ref.	$T_1 + \Delta E$	W	Ref.
Naphthalene	21300	31550	3000	264	24300	267	264
			(3320)		(24620)	(275)	
						480	268
Phenanthrene	21730	28630	3880	264	25610	159	264
Chrysene	19800	27600	5360	264	25160	512	264
			(5610)		(25410)		
Coronene-h_{12}	19700	23300	4900	265, 266	24600		
Coronene-d_{12}	19700	23300	5800	265, 266	25500		
Benzophenone	24500	28000	1050	269	25550		
			(1050)		(25550)		
Benzoperylene	16200	24675	800	270	25475		
					($S_1 + \Delta E$)		
Benzene	29500	37000				380	264
						(390)	
				Average	25100		
					± 500		

Bracketed values from I_p data, other values from τ_T data.

T_3) intersystem crossing is dominant, and it is much greater than k_{TM} ($\sim 3 \times 10^4\ s^{-1}$) for pyrene where only $S_1 - T_1$ intersystem crossing occurs. Thus the magnitude of k_{TM} indicates $S_1 - T_2$ intersystem crossing, while the perdeuteration effect indicates $S_1 - T_1$ intersystem crossing. The strong pseudo-Jahn–Teller coupling[275] between the T_2 ($^3E_{1u}$) and T_1 ($^3B_{1u}$) states of benzene may be associated with this apparent anomaly.

The triplet lifetime τ_T of benzene and its derivatives depends critically on the temperature and on the nature of the solvent environment (pp. *248–9*). Hatch *et al.*[276] studied the temperature dependence of τ_T for benzene and deuterated benzenes in various rigid glass solutions, and they analysed the data in terms of the relation

$$k_T = 1/\tau_T = 1/(\tau_T)_0 + k'_{GT} \exp(-W'_{GT}/\mathbf{k}T) \tag{9.45}$$

Tables 9.20 and 9.21 list the experimental values of $(\tau_T)_0$ for benzene-h_6 and benzene-d_6 and for benzene derivatives in different solvents. Kilmer and Spangler[267] made similar studies of benzene, benzene-d_6, toluene, toluene-α-d_3 and toluene-d_8 in various rigid solvents. Their observations of the temperature dependence of τ_T for C_6H_6 and C_6D_6 in cyclohexane are shown in Figure 9.14. Their data are consistent with (9.45), except for C_6H_6 and C_6D_6 in cyclopentane where a modified relation of the form

$$1/\tau_T = 1/(\tau_T)_0 + k'_{GT} \exp(-W'_{GT}/\mathbf{k}T) + k''_{GT} \exp(-W''_{GT}/\mathbf{k}T) \tag{9.46}$$

Table 9.20 Low-temperature (4·2°K) triplet lifetimes $(\tau_T)_0$ of benzene-h_6 and benzene-d_6

Solvent	$(\tau_T)_0$(s) C_6H_6	$(\tau_T)_0$(s) C_6D_6	Ref.
Cyclohexane	11·8	15·3	276
Cyclohexane	11·44	14·82	267
Cyclopentane	7·46	14·94	267
Borazine	10·2	12·5	276
Benzene-d_6	9·5	—	276
Methane	18	22	277
Methane	16	22	278
Argon	10	26	277
Argon	16	25·8	278
Argon (20°K) (high-energy site)	11·7	14·5	280
Argon (20°K) (low-energy site)	17·5	29·5	280
Krypton	1·8	1·9	277
Krypton	1	1·1	278
Xenon	0·04	0·08	277
Xenon	0·07	0·08	278

Table 9.21 Low-temperature (4·2°K) triplet lifetimes $(\tau_T)_0$ of benzene derivatives

Compound	Solvent	$(\tau_T)_0$(s)	Ref.
Toluene-h_8	Methylcyclohexane	10·59	267
	Perdeuteromethylcyclohexane	11·20	267
	Argon (20°K)	12·6	259
Toluene-α-d_3	Methylcyclohexane	10·73	267
Toluene-d_8	Methylcyclohexane	13·81	267
	Argon (20°K)	16·8	259
Mesitylene	B-Trimethylborazine (50–100°K)	7·2 (±0·2)	279

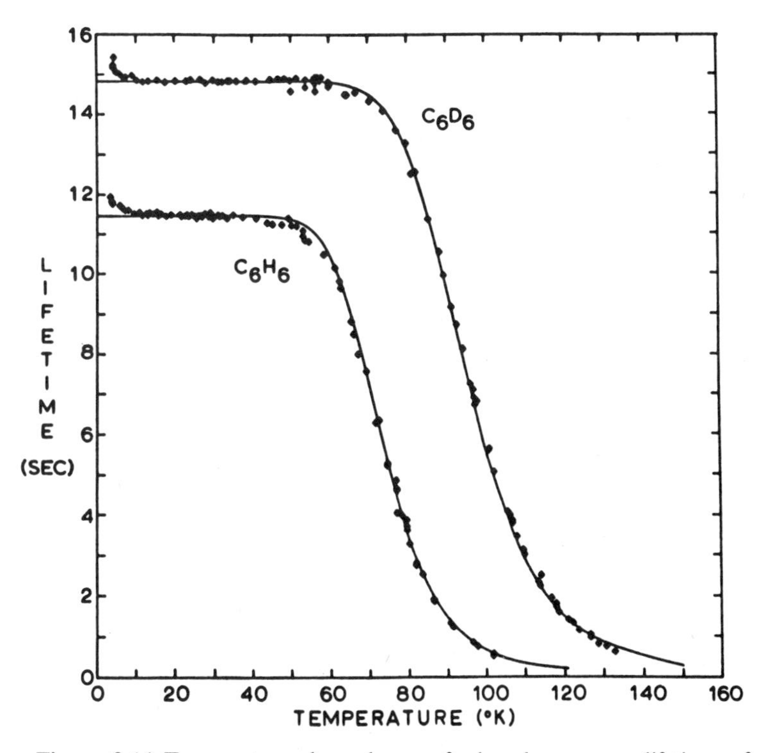

Figure 9.14 Temperature dependence of phosphorescence lifetime of C_6H_6 and C_6D_6 in cyclohexane. (Kilmer and Spangler;[267] reproduced by permission of the American Institute of Physics)

Table 9.22 Frequency factors (k'_{GT}, k''_{GT}) and activation energies (W'_{GT}, W''_{GT}) of triplet decay parameters of benzene derivatives

Compound	Solvent	$k'_{GT}(s^{-1})$	$W'_{GT}(cm^{-1})$	$k''_{GT}(s^{-1})$	$W''_{GT}(cm^{-1})$	Ref.
C_6H_6	Cyclohexane (S)	$4{\cdot}6 \times 10^3$	570	—	—	267
	Cyclohexane (M)	$6{\cdot}7 \times 10^3$	480	—	—	276
	Cyclohexane	$4{\cdot}2 \times 10^3$	554	—	—	267
	Borazine	$6{\cdot}8 \times 10^4$	1000	—	—	276
	Cyclopentane	$4{\cdot}1 \times 10^2$	347	0·15	65	267
	Polymethylmethacrylate	—	380	—	—	264
C_6H_5D	Cyclohexane (S)	$6{\cdot}3 \times 10^3$	560	—	—	276
	Borazine	$2{\cdot}4 \times 10^4$	970	—	—	276
C_6HD_5	Cyclohexane (S)	$1{\cdot}8 \times 10^4$	720	—	—	276
	Borazine	$1{\cdot}4 \times 10^5$	1280	—	—	276
C_6D_6	Cyclohexane (S)	$9{\cdot}8 \times 10^3$	790	—	—	276
	Cyclohexane (M)	$1{\cdot}0 \times 10^4$	670	—	—	276
	Cyclohexane	$2{\cdot}8 \times 10^3$	705	—	—	267
	Perdeuterocyclohexane (S)	$1{\cdot}6 \times 10^4$	860	—	—	276
	Perdeuterocyclohexane (M)	$5{\cdot}6 \times 10^3$	640	—	—	276
	Borazine	$2{\cdot}0 \times 10^5$	1400	—	—	276
	Cyclopentane	$2{\cdot}9 \times 10^3$	527	34	278	267
Toluene	Perdeuteromethylcyclohexane	$2{\cdot}6 \times 10^2$	505	—	—	267
Toluene-α-d_3	Methylcyclohexane	$1{\cdot}4 \times 10^2$	425	—	—	267
Toluene-d_8	Methylcyclohexane	$3{\cdot}1 \times 10^2$	673	—	—	267
Mesitylene	B-Trimethylborazine	—	1800	—	—	279

S. Stable (monoclinic) phase. M. Metastable (cubic) phase.

is required to describe the behaviour. The parameters of (9.45) and (9.46) are listed in Table 9.22.

Johnson and Ziegler[280] analysed the phosphorescence spectra of C_6H_6 and C_6D_6 in an argon matrix at 20°K, and they concluded that the emissions occurred from several crystalline sites. At the two major sites $\tau_T(C_6H_6) = 11{\cdot}7$ and $17{\cdot}5$ s, and τ_T $(C_6D_6) = 14{\cdot}5$ and $19{\cdot}5$ s. From the lower emission intensity in certain C_6D_6 vibrational progressions relative to those in C_6H_6, they concluded that perdeuteration reduces the triplet radiative decay rate k_{PT} of C_6H_6 and that $(k_{PT})_D/(k_{PT})_H = 0{\cdot}82 \pm 0{\cdot}04$. This result contrasts with those of Li and Lim[258] and Johnson and Studer,[259] who observed k_{PT} to be independent of perdeuteration for benzene and toluene (Tables 9.17 and 9.18).

There have been many observations[122,258,263,264,271,281–291] of τ_T for benzene and its deutero, fluoro and alkyl derivatives in various organic glasses at 77°K, and attempts have been made to correlate τ_T with the molecular structure and the molecular rate parameters. This is an unproductive exercise, since τ_T depends on the temperature and on at least four parameters k_{PT}, k^0_{GT}, k'_{GT} and W_{GT}, the values of which depend not only on the solute molecule, but also on the nature, form and thermal history of the solvent environment. Except for the few cases where k_{PT} has been evaluated from Φ_{FM}, Φ_{PT} and τ_T, the observations of τ_T at 77°K for the benzene derivatives provide little meaningful data. Martin and Kalantar[281–287] observed a systematic increase in τ_T with increasing deuteration in the deuterobenzenes (pp. *249*, *295–6*) in three different solvents at 77°K, but the results provide no indication as to which of the parameters in (9.45) is responsible for the change. Similar comments apply to the comparative studies of τ_T (77°K) for the alkyl benzenes[122,271,288,291] and the fluorobenzenes.[291] Although broad conclusions have sometimes been drawn from the relative order of the τ_T (77°K) values, this order may be reversed by a small temperature change, depending on the glass transition temperature and the solute mobility.[292,293]

Fischer and Schneider[294] have analysed the accepting vibrational modes in $T_1 - S_0$ intersystem crossing in benzene and benzene-d_6. They calculated that only 66% (55%) of the energy goes into the C–H (C–D) breathing modes in C_6H_6 (C_6D_6), but this is adequate to account for the perdeuteration effect on k_{GT}. The temperature dependence of τ_T for benzene has been interpreted in terms of (a) a pseudo-Jahn–Teller distortion of triplet benzene, leading to a low-frequency mode, corresponding to an interchange between two different distorted hexagons,[295] (b) an active low-lying vibrational state associated with such pseudorotational motion[296] and (c) intersystem crossing to a non-planar physical isomer of benzene, analogous to the $S_1 - S_X$ channel 3 process (Birks, Chapter 1, Vol. 1). Van Egmond *et al.*[275] have obtained further evidence for pseudo-Jahn–Teller coupling between the T_1 ($^3B_{1u}$) and T_2 ($^3E_{1u}$) states of benzene, by observations of the Zeeman effect on the intense vibronic doublet in the phosphorescence excitation spectrum of benzene.

The data of Table 9.22 show that

$$W'_{GT}(C_6D_6) \simeq 1{\cdot}4W'_{GT}(C_6H_6) \tag{9.47}$$

irrespective of the solvent, and that W'_{GT} increases with the polarity of the solvent. Fischer[296] has proposed a model to account for these and other features of the temperature dependence of τ_T. He postulated a low-lying vibrational state of T_1 of vibrational energy Δ, which has a strong Franck–Condon overlap with the corresponding vibrational state of S_0, and has a distorted molecular configuration.[295] The observed activation energy W'_{GT} ($\gg \Delta$) is then considered as part of a multiphonon activation process in which Δ is multiplied by the numbers of vibrational quanta released in the radiationless transition. The factor of 1·4 in (9.47) is shown to be equivalent to the isotope factor $(\mu_{CD}/\mu_{CH})^{\frac{1}{2}} = \omega_{CH}/\omega_{CD} = 1{\cdot}36$ (pp. *155–6*) for CH vibrations. The model is in good quantative agreement with the experimental data,[276] and it gives values of $\Delta = 56\ \text{cm}^{-1}$, $68\ \text{cm}^{-1}$ and $118\ \text{cm}^{-1}$ for benzene in cyclohexane (M), cyclohexane (S) and borazine, respectively. The influence of the solvent on $(\tau_T)_0$ (Table 9.20) has yet to be explained.

9.6.3 *The heavy-atom effect*

The internal heavy-atom effect in phenanthrene has been studied by Masetti *et al.*,[297] who observed Φ_{FM}, Φ_{PT} and τ_T for 3-substituted phenanthrenes (3-Ph) in EPA at 77°K, and evaluated k_{TM}, q_{PT}, k_{PT} and k_{GT} (Table 9.23), assuming (9.43) to be valid. There are some interesting differences between 3-Ph and the 1-substituted naphthalenes (1-N) (p. *264*). With heavy atom substitution k_{TM} increases less rapidly in 3-Ph than in 1-N, k_{TM} is approximately proportional to k_{PT} in 3-Ph, while in 1-N it is proportional to k_{PT}^2 (p. *210*); and k_{PT} increases more rapidly than k_{GT} in 3-Ph, causing an increase in q_{PT}, while in 1-N q_{PT} remains approximately constant.

The external heavy-atom effect on the phosphorescence of several perprotonated and perdeuterated aromatic hydrocarbons has been studied by Webber.[298] The hydrocarbon phosphorescence was sensitized by triplet–triplet energy transfer from benzophenone. This method, which is due to Ermolaev,[299] eliminates $S_1 - T_q$ intersystem crossing in the hydrocarbon, the process most strongly influenced by the heavy-atom effect (pp. *210–1*). If Φ_{DO} and Φ_D are the phosphorescence quantum yields

Table 9.23 Internal heavy-atom effect. Luminescence properties of phenanthrene derivatives in EPA at 77°K (Masetti *et al.*[297])

	Experimental parameters			Derived parameters			
Compound	Φ_{FM}	Φ_{PT}	τ_T(s)	$k_{TM}(10^6\ s^{-1})$	q_{PT}	$k_{PT}(s^{-1})$	$k_{GT}(s^{-1})$
Phenanthrene (Ph)	0·14	0·11	3·90	12	0·13	0·033	0·22
3-Methyl Ph	0·26	0·09	4·03	9	0·12	0·03	0·22
3-Chloro Ph	0·026	0·19	1·03	112	0·19	0·19	0·78
3-Bromo Ph	0·001	0·29	0·05	>3000	0·29	5·8	14·2
9-Bromo Ph	0·001	0·22	0·013	$\gg 3000$	0·22	17·3	59·7

of the benzophenone donor in the absence and presence of the hydrocarbon acceptor, respectively, and Φ_A is the sensitized phosphorescence quantum yield of the latter, then the phosphorescence quantum efficiency q_{PT} of the hydrocarbon is given by

$$q_{PT} = \frac{\Phi_A \Phi_{DO}}{\Phi_{DO} - \Phi_D} = \gamma \Phi_{DO} \tag{9.48}$$

a relation which is independent of the energy transfer rate parameter and the acceptor concentration. The relative values of Φ_{DO}, Φ_D and Φ_A, and hence γ, can be determined from the areas under the corrected phosphorescence spectra. Webber[298] evaluated a related quantity $\bar{\gamma}$ from his uncorrected spectra.

Observations were made at 77°K in ethanol, ethanol/*n*-propylbromide (1:1) and ethanol/*n*-propyl iodide (1:1). In ethanol the phosphorescence decay is exponential, but in the other solvents the decay is non-exponential, and the average triplet lifetime $\langle \tau_T \rangle$ was determined. The results (Table 9.24) show that the heavy-atom solvents (a) reduce $\langle \tau_T \rangle$, (b) usually increase, but sometimes decrease, $\bar{\gamma}$ which is roughly proportional to q_{PT} and (c) increase $\bar{\gamma}/\langle \tau_T \rangle$ which is roughly proportional to k_{PT}. In ethanol solution there are major differences between the values of $\bar{\gamma}$ and τ_T for the perprotonated and perdeuterated compounds, but in the iodide solvent these differences are eliminated, within the experimental error. Webber[298] proposed that the $T_1 - S_0$ intersystem crossing rates, k_{GT}, in these isotopic pairs become essentially equal when there is strong external spin–orbit coupling.

The quenching of triplet naphthalene-d_8 by alkali halides in ethanol solution at 77°K has been studied by Hofeldt *et al.*[300] They observed a non-exponential phosphorescence decay, which they analysed as the sum of two exponentials

$$i_P(t) \propto A \exp(-k_T t) + B \exp(-k'_T t) \tag{9.49}$$

with $k_T < k'_T$. $k_T (= 0{\cdot}055\ \mathrm{s}^{-1})$ corresponds to the decay of triplet naphthalene-d_8 in the absence of the heavy-atom quencher. k'_T depends on the quencher: LiBr ($0{\cdot}16\ \mathrm{s}^{-1}$), LiI ($0{\cdot}32\ \mathrm{s}^{-1}$), NaI ($0{\cdot}30\ \mathrm{s}^{-1}$), KI ($0{\cdot}41\ \mathrm{s}^{-1}$), RbI ($0{\cdot}38\ \mathrm{s}^{-1}$). Hofeldt *et al.*[300] attributed k'_T to the decay of a 1:1 naphthalene/alkali halide triplet exciplex formed in the rigid glass solution subsequent to excitation. The dissociation of the quencher molecules into ions, and the possible formation of ground-state donor–acceptor complexes between the naphthalene and quencher molecules or ions in room-temperature solution prior to freezing were not apparently considered, although these processes do not appear to be excluded by the

Table 9.24 External heavy-atom effect. Sensitized phosphorescence in different solvents at 77°K (Webber[298])

Solvent	Ethanol			Ethanol/*n*-propyl bromide (1 : 1)			Ethanol/n-propyl iodide (1 : 1)		
Solute	$\bar{\gamma}$	τ_T(s)	$\bar{\gamma}/\tau_T(s^{-1})$	$\bar{\gamma}$	$\langle\tau_T\rangle$(s)	$\bar{\gamma}/\langle\tau_T\rangle(s^{-1})$	$\bar{\gamma}$	$\langle\tau_T\rangle$(s)	$\bar{\gamma}/\langle\tau_T\rangle(s^{-1})$
Naphthalene-h_8	0·039 (±0·004)	2·3	0·017 (±0·002)	0·20 (±0·03)	0·56	0·35 (±0·05)	0·50 (±0·10)	0·09	5·5 (±1·1)
Naphthalene-d_8	0·30 (±0·02)	16·1	0·019 (±0·001)	0·40 (±0·04)	0·98	0·41 (±0·04)	0·49 (±0·05)	0·11	4·5 (±0·5)
Phenanthrene-h_{10}	0·16 (±0·02)	3·4	0·047 (±0·006)	0·28 (±0·07)	0·86	0·33 (±0·08)	0·41 (±0·08)	0·13	3·2 (±0·6)
Phenanthrene-d_{10}	0·50 (±0·012)	13·7	0·037 (±0·009)	0·42 (±0·07)	0·99	0·42 (±0·07)	0·36 (±0·05)	0·21	1·7 (±0·3)
Biphenyl-h_{10}	0·35 (±0·05)	4·0	0·088 (±0·013)	0·41 (±0·06)	0·92	0·44 (±0·06)	0·50 (±0·06)	0·17	2·9 (±0·4)
Biphenyl-d_{10}	0.54 (±0·14)	9·6	0·056 (±0·014)	0·45 (±0·05)	1·23	0·37 (±0·04)	0·57 (±0·05)	0·16	3·6 (±0·3)

experimental conditions. The general conclusion[300] that the external heavy-atom effect involves complex formation agrees with previous work on excited donor–acceptor complexes and exciplexes (pp. *415–20*, *425–39*).

Dimethylmercury (DMM) has been used as a heavy-atom solvent by Vander Donckt and co-workers.[301,304] In solution in neat DMM or ethanol-diluted DMM a singlet-excited aromatic hydrocarbon ($^1M^*$) exhibits a series of related phenomena:

(a) strong quenching of the $^1M^*$ fluorescence, due to

(b) the diffusion-controlled formation of singlet exciplexes, $^1(M.DMM)^*$, which yield

(c) exciplex fluorescence and/or

(d) enhanced intersystem crossing to triplet exciplexes, $^3(M.DMM)^*$, which dissociate yielding

(e) enhanced $^3M^*$ phosphorescence.

In the presence of DMM the $S_0 - T_1$ absorption of the aromatic hydrocarbon is increased, but its $S_0 - S_p$ absorption is unchanged.[303] The ground-state complex $^1(M.DMM)$ between 1-chloronaphthalene and DMM has an observed binding energy $-\Delta H = 0{\cdot}019$ eV, indicating that it is dissociated at room temperature ($\mathbf{k}T = 0{\cdot}025$ eV), and that the enhanced $S_0 - T_1$ absorption is due to contact complexes.[304]

The phosphorescence of aromatic hydrocarbons in fluid DMM solution at room temperature is of interest.[301] It has been observed in 1:2-benzanthracene, 3:4-benzophenanthrene, 1:2:5:6-dibenzanthracene and octahelicene. The phosphorescence quantum yield Φ_{PT} is about 100 times that in ethanol solution at room temperature,[254] and it is comparable in magnitude with the (quenched) fluorescence quantum yield Φ_{FM}. For 3:4-benzophenanthrene in neat DMM solution at 25°C, $\Phi_{PT} \sim 10^{-3}$. Room temperature DMM solutions provide a more homogeneous and convenient environment than low-temperature rigid glasses for the study of the external heavy-atom effect. With Φ_{PT} observable, the measurement of Φ_{FM}, Φ_{TM}, τ_M and τ_T, the other quantities required to evaluate all the S_1 and T_1 rate parameters, presents little difficulty. Moreover the radiative rate parameters k_{PT} and k_{FM} can be obtained independently from the $S_0 - T_1$ and $S_0 - S_1$ absorption, respectively.

9.7 Excimers

The following topics have already been discussed:

(a) diffusion-controlled excimer formation in fluid solutions (Alwattar *et al.*, Chapter 8, Vol. 1);

(b) the influence of pressure on excimer formation in fluid solutions, rigid solutions and aromatic crystals (Offen, Chapter 3, Vol. 1);

(c) the benzene excimer (Cundall, Chapter 2, Vol. 2);

(d) intramolecular excimers and excimers in polymers (Klöpffer, Chapter 7, Vol. 1); and

(e) the theory of excited molecular complexes, including excimers (Beens and Weller, Chapter 4, Vol. 2).

9.7.1 *Excimers, photodimers and dimers* (pp. *316–23*)

The original definition of an excimer as an excited dimer which is *dissociated* in the ground state (p. *301*) is valid for intermolecular excimers in fluid solutions, but it requires slight modification for intramolecular excimers and for excimers in rigid media, where there are restraints on molecular motion. To include these cases, an excimer is redefined as an excited dimer which is *dissociative* (i.e. would dissociate in the absence of external restraints) in the ground state.

The types of dimer formed from a pair of identical aromatic molecules can be classified as follows.

(i) A *dimer complex* ^{1}(M.M) may be formed by interaction between two unexcited molecules,

$$^1M + {}^1M \rightleftharpoons {}^1(M.M) \qquad (9.50)$$

The complex is normally also associated in its excited triplet ^{3}(M.M)* and singlet ^{1}(M.M)* states. Its absorption, phosphorescence and fluorescence spectra are more diffuse than those of the parent molecule, and they are usually shifted to lower energies.

(ii) A *triplet excimer* $^3D^*$ may be formed by interaction between a triplet-excited molecule $^3M^*$ and an unexcited molecule,

$$^3M^* + {}^1M \rightleftharpoons {}^3D^* \qquad (9.51)$$

or otherwise (e.g. intersystem crossing from $^1D^*$, or $^2M^+ - {}^2M^-$ association). $^3D^*$ has a dissociative ground state and an associated excited singlet state $^1D^*$. Its phosphorescence spectrum is more diffuse and at lower energies than that of the parent molecule.

(iii) A *singlet excimer* $^1D^*$ may be formed by interaction between a singlet-excited molecule $^1M^*$ and an unexcited molecule,

$$^1M^* + {}^1M \rightleftharpoons {}^1D^* \qquad (9.52)$$

or otherwise (pp. *339–42*). $^1D^*$ has a dissociative ground state, and it may undergo intersystem crossing to $^3D^*$, which is usually dissociative. The excimer fluorescence spectrum is structureless and at lower energies than that of the parent molecule.

(iv) A *photodimer* $^1M_2$ may be formed by the chemical interaction of a singlet-excited molecule $^1M^*$ and an unexcited molecule,

$$^1M^* + {}^1M \rightarrow {}^1M_2 \tag{9.53}$$

$^1M_2$ is stable in the ground state, but it usually dissociates in the excited state $^1M_2^*$ (see below).

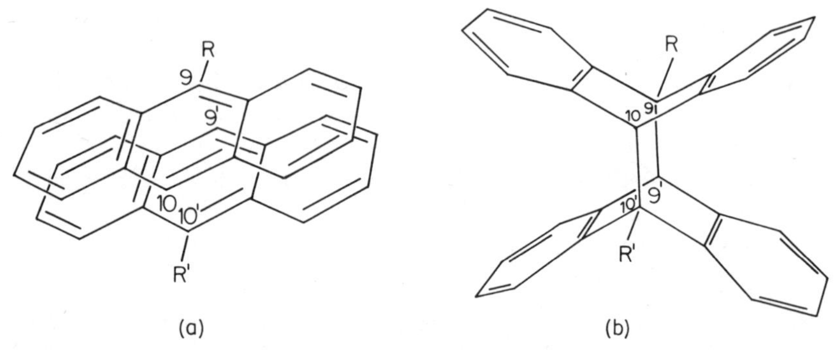

Figure 9.15 Structural diagrams of (a) the anthracene sandwich dimer (left) and (b) the anthracene photodimer, dianthracene (right). (Tomlinson, Chandross, Fork, Pryde and Lamola, *Applied Optics*, **11**, 533 (1972))

Figure 9.15 shows the structure of

(a) a sandwich pair of anthracene molecules, and

(b) the photodimer, dianthracene, formed therefrom.[306]

The molecular structure of the photodimer (b) indicates that the sandwich configuration (a) is that of the interacting molecules immediately prior to the reaction (9.53). The $^1M_2$ absorption spectrum lies at much shorter wavelengths than that of ^1M and it is unrelated to it.

(v) A *sandwich dimer* $^1(M\|M)$ may be formed by the photolysis of a photodimer $^1M_2$ in a rigid matrix, reversing process (9.53)

$$^1M_2 + h\nu \rightarrow {}^1M_2^* \rightarrow {}^1(M\|M)^* \rightarrow {}^1(M\|M) \tag{9.54}$$

and yielding a pair of adjacent molecules in a sandwich configuration (Figure 9.15a). $^1(M\|M)$ also occurs naturally in type B (e.g. pyrene) crystals (pp. *517–9*).

(vi) A *non-sandwich dimer* $^1(M \times M)$ corresponds to a pair of identical molecules which are not in a sandwich or displaced sandwich configuration, but which are sufficiently close and suitably oriented to interact in their ground and/or excited states.

Excimer ($^1D^*$) fluorescence may occur from various $^1(M\|M)$ and $^1(M \times M)$ dimer configurations, depending on the conditions.

(a) In fluid solutions singlet excimers ($^1D^*$) are in thermodynamic equilibrium with the dissociated excited molecular system ($^1M^* + {}^1M$) (9.52). The excimer energies lie between the zero-point energy of $^1D^*$ and the zero-point energy of $^1M^*$, covering an energy range equal to the excimer stabilization or binding energy B. This range includes all possible configurations of the excimer for which the excimer interaction potential exceeds the intermolecular repulsive potential. In the absence of steric hindrance the equilibrium $^1D^*$ configuration, corresponding to the excimer fluorescence maximum D_m (p. *326*), is a sandwich structure $^1(M\|M)^*$. The more energetic excimer configurations include more widely spaced and vibrationally excited $^1(M\|M)^*$ sandwich structures and a variety of $^1(M \times M)^*$ non-sandwich structures.

(b) Intramolecular excimers (Klöpffer, Chapter 7, Vol. 1) in fluid solution behave in a similar manner, except that the molecular chain linking the two interacting chromophores imposes a steric restraint on their possible relative positions. Depending on this restraint the equilibrium $^1D^*$ configuration may be $^1(M \times M)^*$ as in 1,3-diphenylpropane (p. *324*) or $^1(M\|M)^*$ as in the 1,3-dinaphthylpropanes. Chandross and Dempster[307] studied the fluorescence of various dinaphthylalkanes, but they only observed strong intramolecular excimer fluorescence in the two 1,3-dinaphthylpropanes, from which they concluded that the stable excimer configuration is $^1(M\|M)^*$.

(c) In the pyrene crystal, which has a type **B** lattice (pp. *317–9*), the molecules are arranged in $^1(M\|M)$ pairs with an intermolecular spacing of $r_0 = 3{\cdot}53$ Å. Following $^1M^*$ excitation excimer interaction reduces the spacing to $r_m = 3{\cdot}37$ Å to form the equilibrium $^1(M\|M)^*$ excimer configuration (pp. *331–5*). The intermediate excimer states with energies between $^1D^*$ and $^1M^*$ correspond to vibrationally excited states of the $^1(M\|M)^*$ configuration.

(d) In the anthracene crystal, which has a type A lattice (pp. *317–9*). there are two types of nearest-neighbour dimers: $^1(M\|M)$, corresponding to translationally equivalent molecules; and $^1(M \times M)$, corresponding to translationally inequivalent molecules. The former are responsible for the exciton shift, and the latter for the Davydov splitting, of the crystal absorption (pp. *523–8*).

In neither dimer configuration in the normal lattice are the molecules sufficiently close or suitably oriented for the excimer interaction to exceed the lattice forces, so that in a perfect crystal the fluorescence is molecular and no excimers are formed. If the crystal contains defects or vacancies, these provide sites where adjacent molecules have different orientations and/or more freedom of movement than in the perfect lattice. $^1(M\|M)^*$ and $^1(M \times M)^*$ excimers may be formed at these sites, which function as traps for $^1M^*$ excitons, so that an excimer emission band appears in the crystal fluorescence spectrum (p. *319*). Photodimers ($^1M_2$) are also formed at such sites. Thomas and Williams[308–310] have shown by optical and electron microscopy that anthracene crystals photodimerize preferentially at dislocations.

(e) The excimer fluorescence of aromatic polymers (Klöpffer, Chapter 7, Vol. 1) is due to intermolecular and intramolecular $^1(M \times M)$ and $^1(M\|M)$ dimers, suitably oriented and with sufficient freedom of motion to form excimers. These sites act as traps for $^1M^*$ excitons which migrate along and between the polymer chains.

9.7.2 *Dimer formation*

Several dye molecules are considered to form dimer complexes $^1(M.M)$ in fluid solution, but their molecular and dimer spectra possess little structure and they yield little information about the dimer geometry. Ferguson and Mau[503] have studied the absorption spectra of neutral molecules of acridine orange, 3,6-diaminoacridine and Rhodamine B in ethanol solutions. Contrary to previous observers,[504–507] they conclude that the neutral dye molecules do not dimerize in ethanol solution. They attribute the concentration and temperature dependent spectral changes, previously thought to involve monomer–dimer systems, to acid–base equilibria.

Hoytink and co-workers[311] have studied 9,10-diazaphenanthrene, which forms dimer complexes in fluid solution, but is subject to photodecomposition. To avoid the latter, the monomer–dimer equilibrium (9.50) was established at a given temperature in a liquid solution, which was then rapidly frozen to 77°K, where the fluorescence and fluorescence excitation spectra were observed. The dimer stabilization enthalpy $-\Delta H = 1260\ \text{cm}^{-1}$. Comparison of the spectra with configuration interaction calculations indicates that the dimer has a head-to-tail displaced sandwich configuration, with the nitrogen atoms adjacent to each other and an interplanar separation of about 3·5 Å.

Several studies have been made of the spectroscopic properties of aromatic hydrocarbons in rapidly frozen solutions. In such systems dimerization is the initial stage of crystallization and it is difficult to distinguish dimers from larger molecular aggregates or microcrystals, particularly of molecules like pyrene or perylene which crystallize in type B lattices (p. *318*), formed from $^1(M\|M)$ dimers.

McDonald and Selinger[312] observed the fluorescence and fluorescence excitation spectra of pyrene in frozen cyclohexane solutions at 77°K at concentrations from 10^{-5} to 10^{-1} M. The fluorescence spectra consist of molecular and structureless 'excimer' emission components, and the relative intensity of the latter increases with concentration. Previous observers[313–316] have attributed the 'excimer' fluorescence to the formation or presence of excimers in rigid solution. In fluid solution the excitation spectra of the molecular and excimer fluorescences are identical. In frozen solutions the two excitation spectra differ, showing that the emissions are from two distinct absorbing species, and that the 'excimers' are not formed subsequent to excitation.[312] In the 10^{-2} M solution where the 'excimer' emission dominates, its excitation spectrum corresponds to the pyrene crystal absorption spectrum. At lower concentrations competition from the molecular absorption modifies the 'excimer' excitation spectrum. McDonald and Selinger[312] conclude that the 'excimer' emission is due to pyrene microcrystals, as suggested earlier[317] (p. *315*). $^1(M\|M)$ dimers will be formed on freezing very dilute solutions of pyrene, but it is doubtful whether these can be distinguished spectroscopically from the microcrystalline aggregates of these $^1(M\|M)$ dimers formed in more concentrated solutions.

Microcrystallization and the associated appearance of 'excimer' fluorescence depends on the solvent. A rapidly frozen solution of pyrene in cyclohexane has equal intensities of molecular and 'excimer' emission at about 5×10^{-4} M; in *p*-xylene solution the required concentration is about 2×10^{-2} M; and in methylcyclohexane it is $> 10^{-1}$ M.[312]

A closely related phenomenon, the 'excimer' fluorescence of perylene in frozen and compressed solutions in cyclohexane, has been studied by Johnson and Offen.[318] In ethanol solution no 'excimer' emission is observed, and the blue molecular perylene fluorescence responds normally to changes in pressure (p) and temperature (T) (see Offen, Chapter 3, Vol. 1). In cyclohexane solution the molecular fluorescence converts to orange 'excimer' fluorescence on freezing at 0·38 kbar and 296°K. The 'excimer' fluorescence spectrum and lifetime are sensitive to p and T, but they become constant at $T \leqslant 160$°K. The fluorescence lifetime τ_D is given by

$$\tau_D^{-1} = (\tau_D)_0^{-1} + a \exp(-\Delta E/\mathbf{k}T) \tag{9.55}$$

with $(\tau_D)_0 = 80$ ns, $a = 1{\cdot}7 \times 10^9$ s^{-1} and $\Delta E = 670$ cm^{-1}. These results probably refer to microcrystals of perylene, which have a $^1(M\|M)$ structure and exhibit excimer fluorescence.

Dimers of molecules which form type A crystals, which do not exhibit excimer fluorescence, can more readily be identified spectroscopically. Two types of anthracene dimer have been observed: (a) the symmetrical sandwich dimer $^1(M\|M)$ formed by the photolysis of dianthracene (9.54) in solid solution which exhibits $^1D^*$ fluorescence (this is more closely spaced and symmetrical than the corresponding entity formed from two adjacent translationally equivalent molecules in the crystal lattice) and (b) the non-sandwich dimer $^1(M \times M)$ formed by successive freezing and thawing of a solution of anthracene. It has been proposed that the structure of the latter entity, described as a '55° dimer', approximates to that of two adjacent translationally inequivalent molecules in the crystal lattice (pp. *332–3*). Dupuy *et al.*[319] have studied the temperature and concentration conditions for the existence in frozen solution of the various associated species of anthracene (i.e. dimers, molecular aggregates, microcrystals). They have shown that the non-sandwich dimer $^1(M \times M)$ is formed abundantly in very dilute solution, but that larger aggregates and microcrystals are formed at higher concentrations.

From observations of the absorption and fluorescence spectra of frozen solutions of tetracene, Katul and Zahlan[320] identified an entity with different spectroscopic properties from the monomer, which they attributed to the tetracene dimer. They noted, however, that 'the dimer spectroscopic properties resemble markedly those of crystalline tetracene.' In a later study Fournie *et al.*[321] identified two auto-associated species in frozen solutions of tetracene: that observed by Katul and Zahlan[320] which is attributed to tetracene microcrystals, and a further species which appears to be the true dimer. It is proposed[321] that the tetracene dimer has a $^1(M \times M)$ configuration, corresponding to two adjacent translationally inequivalent molecules in the crystal and similar to the anthracene '55° dimer' (p. *332–3*).

9.7.3 *Excimer fluorescence in fluid solution* (pp. *316–7*)

The time-resolved fluorescence spectra of concentrated solutions of pyrene in cyclohexane, excited by a 3 ns pulsed nitrogen laser, have been observed from 3 ns to 150 ns by Yoshihara *et al.*[322] The results agree with those obtained using pulse-sampling fluorometry.[323]

Excimer fluorescence has been observed in fluorene and dibenzofuran in toluene solutions[324,501,502] and in quinoline and isoquinoline in ethanol solutions.[618] The excimer fluorescence of the 9-substituted anthracenes in solution is discussed in §9.7.5.

9.7.4 *Excimer fluorescence of crystals* (pp. *317–9*)

Kawakubo[325] has observed the fluorescence spectra of crystals of 1-methylnaphthalene and 1,6-dimethylnaphthalene at 77°K. Slow freezing of the compounds, which are liquid at room temperature, yields the molecular fluorescence spectrum, but rapid freezing yields a dominant excimer fluorescence spectrum. The former corresponds to the true crystal spectrum. The latter is attributed to the generation of a large number of defects, which function as excimer sites and exciton traps. Related effects are observed in crystals of naphthalene and 2,6-dimethylnaphthalene, in which continued optical irradiation induces changes in the fluorescence spectra attributable to excimer site formation.

Cohen *et al.*[329] have observed excimer fluorescence and photodimerization in 2,4-dichloro-*trans*-stilbene crystals. In solution the compound emits molecular ($^1M^*$) fluorescence, but not excimer ($^1D^*$) fluorescence, and it undergoes photodimerization in the $^1M^*$ state and *cis*-isomerization in the $^3M^*$ state. The crystal excimer is probably the photodimer intermediate.

Inoue *et al.*[326] have used time-resolved spectroscopy to study the excimer and monomer defect fluorescence[327,328] of pyrene and perylene crystals. In crystal pyrene at 77°K the lifetimes of the excimer ($\bar{\nu}_D = 21{,}500\ \text{cm}^{-1}$) fluorescence and monomer defect ($\bar{\nu}_M = 25{,}300\ \text{cm}^{-1}$) fluorescence are $\tau_D = 180$ ns and $\tau_M = 8$ ns, respectively. In crystal perylene at 77°K $\bar{\nu}_D = 17{,}000\ \text{cm}^{-1}$, $\tau_D = 91$ ns and $\bar{\nu}_M = 19{,}000\ \text{cm}^{-1}$, $\tau_M = 8$ ns. It is concluded that the excited states from which the two emissions originate are populated independently. Chu and Kearns[367] have also studied the excimer and monomer defect fluorescence of pyrene and perylene crystals. Other studies of pyrene crystal luminescence are considered in §9.7.7, and exciton migration in pyrene crystals is discussed in §9.11.3.

Chaudhuri and Ganguly[346] observed the polarized absorption and fluorescence spectra of crystal pyrene at 90°K. They found that the components of the excimer transition moment along the long molecular *a*-axis and normal to the molecular planes contribute 62% and 38%, respectively, to the fluorescence intensity. In contrast, Hochstrasser and Malliaris[359] found the pyrene crystal excimer fluorescence to be predominantly polarized along the molecular *a*-axis. The latter result is consistent with the experimental excimer interaction potential and its assignment to the ^1B_a exciton resonance state (pp. *331–5*). While the latter may be partially stabilized by configuration interaction with a charge resonance state (p. *335*), the 38% contribution indicated by the polarization data[346] appears excessive.

9.7.5 *Photodimers and sandwich dimers* (pp. *322–3*)

Tomlinson *et al.*[306] have studied the reversible photodimerization (9.53), (9.54) of a variety of aromatic compounds in rigid matrices, with a view to their utilization for reversible phase holographic recording with non-destructive readout. The absorption spectra of the photodimer $^1M_2$ and sandwich dimer $^1(M \parallel M)$ differ appreciably, and associated with this is a difference in refractive index. The photodimerizing compounds studied included 1,3-dimethylthymine, 9-methylanthracene, amyl 9-anthroate, 2-aminopyridinium methylsulphonate, 2-aminopyridinium dihydrogen phosphate, acridizium toluenesulphonate, acridizium camphorsulphonate, benzacridizium toluenesulphonate and benzacridizium camphorsulphonate. The most promising results for the holographic application were obtained with the two acridizium compounds.

Chandross and Dempster[330] discovered that irradiation of 1,3-*bis*(α-naphthyl)-propane results in *intramolecular* photodimerization. 1,3-*bis*-(α-naphthyl)-propane forms intramolecular excimers, and the photodimers are formed therefrom. Photolysis of the intramolecular photodimer in rigid solution regenerates the original compound with the two naphthalene nuclei held in a sandwich $^1(M \parallel M)$ configuration.[331] The intramolecular sandwich dimer has a structureless excimer fluorescence spectrum, similar to the naphthalene excimer fluorescence in fluid solution. The excimer interaction is much weaker in the triplet state. The monomer and dimer phosphorescence spectra have similar vibrational structure, but the latter is red-shifted by 250 cm^{-1} and it is more diffuse and relatively more intense at lower energies (see §9.7.7). The monomer and dimer phosphorescence lifetimes are 2·4 s and 1·5 s, respectively.

Ferguson and Mau[332] have made an elegant study of the photophysics of the anthracene symmetrical sandwich dimers $^1(M \parallel M)$ produced by the photolysis of dianthracene molecules $^1M_2$ in a single crystal of dianthracene. Excitation of $^1(M \parallel M)$ leads to the following processes:

$$\begin{array}{c} {}^1(M \parallel M) + h\nu \rightarrow {}^1(M \parallel M)^* \rightarrow {}^1D^* \xrightarrow{k_{AD}} {}^1M_2 \\ \downarrow k_{FD} \\ {}^1(M \parallel M) + h\nu_D \end{array} \tag{9.56}$$

The quantum yields of $^1D^*$ fluorescence Φ_{FD} ($= k_{FD}/k_D$) and of $^1M_2$ formation Φ_{AD} ($= k_{AD}/k_D$) are observed to sum to unity at all temperatures from 4 to 300°K. At low temperatures $\Phi_{FD} = 1{\cdot}0$ and $\Phi_{AD} = 0$, and at room temperature Φ_{FD} tends to zero and Φ_{AD} tends to unity. From the temperature dependence of Φ_{FD} and Φ_{AD} the activation energy W_{AD} of the $^1D^* \rightarrow {}^1M_2$ transition is found to be ~600 cm^{-1}. The results confirm that the excimer is the intermediate state in photodimer formation (p. *320*), and they show that excimer fluorescence and photodimerization are the only significant processes competing for the excimer excitation energy. In the relation

$$k_{AD} = k'_{AD} \exp(-W_{AD}/\mathbf{k}T) \tag{9.56a}$$

values of $k'_{AD} = 7{\cdot}6 \times 10^9\ s^{-1}$, $W_{AD} = 590\,(\pm 25)\ cm^{-1}$ and of $k'_{AD} = 4{\cdot}2 \times 10^9\ s^{-1}$, $W_{AD} = 530\,(\pm 40)\ cm^{-1}$, are obtained for the photodimerization of anthracene-h_{10} and anthracene-d_{10}, respectively.[332]

Aladekomo[510] observed the photodimerization quantum yield Φ_{AD} and excimer fluorescence quantum yield Φ_{FD} of 9-methylanthracene in toluene solution as a function of concentration and temperature. The results were analysed in terms of a reaction scheme (p. *320*) similar to (9.56), yielding

$$\frac{\Phi_{AD}}{\Phi_{FD}} = \frac{k'_{AD}}{k_{FD}} \exp(-W_{AD}/\mathbf{k}T) \tag{9.56b}$$

This relation was verified experimentally, and a mean value of $W_{AD} = 0{\cdot}23$ eV $= 1850\ cm^{-1}$ obtained.

Ferguson *et al.*[508] have studied the symmetrical sandwich dimers of anthracene and its 9-methyl, 9-phenyl, 9-chloro, 9-bromo and 9-cyano derivatives, and the asymmetrical sandwich (mixed) dimers of anthracene with its 9-methyl, 9-phenyl, 9-cyano, 9,10-dimethyl and 9,10-dichloro derivatives. The sandwich dimers were formed by photolysis in a rigid glass matrix at 77°K, and observations were made of their absorption spectra and fluorescence spectra and lifetimes. The results are summarized in Table 9.25. The glass was then allowed to soften to study the dissociation of the sandwich dimer. Initially the fluorescence spectrum shifts slightly to higher energies and the lifetime decreases from ~200 ns to ~100 ns. There is then an abrupt spectral broadening and appearance of vibrational structure, and the lifetime drops to the molecular value of ~5 ns. Ferguson *et al.*[509] have also made spectroscopic studies of the '55° dimers' of anthracene and 9,10-dichloroanthracene.

Meso-substitution in anthracene introduces steric hindrance to photodimer and excimer formation (pp. *320–2*). The photodimers of all the 9-substituted anthracenes studied in room-temperature solution have a head-to-tail *trans* molecular structure.[333–337] In topochemically controlled reactions in crystals the lattice structure usually determines the geometric configuration of the photodimer.[338] 9-cyanoanthracene (CNA) and other 9-substituted anthracenes are striking exceptions to the general

Table 9.25 Sandwich dimers of anthracene (A) and derivatives in methylcyclohexane glass at 77°K. Monomer fluorescence quantum yield (Φ_{FM}) and lifetime (τ_M). Dimer fluorescence quantum yield (Φ_{FD}), lifetime (τ_D) and spectral maximum (D_m)[508]

Dimer	Φ_{FM}	τ_M(ns)	Φ_{FD}	τ_D(ns)	D_m(cm^{-1})
A–A	0·3	6	0·27	200	20400
9 Me A–9 Me A	1·0	13	0·33	220	20000
9 Ph A–9 Ph A	1·0	12	0·38	260	18200
9 Cl A–9 Cl A	1·0	10	0·17	180	19700
9 Br A–9 Br A	1·0	10	0·014	30	19600
9 CN A–9 CN A	1·0	11	0·12	170	19100
A–9 Ph A	—	—	—	115	19100
A–9 Me A	—	—	—	130	19600
A–9 CNA	—	—	—	105	20160
A–9,10 diCl A	—	—	—	100	18740
A–9,10 diMe A	—	—	—	80	18860

pattern. They crystallize in a head-to-head *cis* $^1(M \parallel M)$ structure, but only *trans*-photodimers are formed within the crystal.[339–343]

Craig and Sarti-Fantoni[339] proposed that in CNA crystals photodimerization only occurs at defects or surfaces or in zones disordered by photodimerization. Cohen *et al.*[340–342] showed that the CNA crystal photodimerization occurs preferentially at dislocations emerging at the *bc* and *ac* crystal faces. Kawada and Labes[343] studied the photodimerization of single CNA crystals by observing the growth of 4-sided photo-etched pits as a function of time, light intensity and temperature. Growth along the (001) diagonal direction of the pits on the (010) plane is linear in time and light intensity, and it has an activation energy of 0·75 eV. They proposed that the *cis*-excimer $^1D_c^*$ is formed initially, followed by a rate-determining thermally activated, reorientation to the *trans*-excimer $^1D_t^*$ at the perimeter of the etch pit, followed by the rapid formation of the *trans*-photodimer $^1M_2$,

$$^1(M \parallel M)_c + h\nu \rightarrow {}^1D_c^* \rightarrow {}^1D_t^* \rightarrow {}^1M_2 \tag{9.57}$$

Because of exciton ($^1M^*$ and/or $^1D_c^*$) migration in the crystal, the $^1D_c^* \rightarrow {}^1D_t^*$ reorientation does not require the 180° internal rotation of a given dimer pair. The process will occur most readily when an exciton encounters a molecule which has a relative *trans*-orientation at the boundary of a dislocation.

In solution the 9-substituted anthracenes exhibit both excimer fluorescence and photodimerization (p. *321*). In 1963 Birks and Aladekomo[344] proposed that the excimer fluorescence of 9-methylanthracene occurs from the *cis*-excimer $^1D_c^*$ and that the photodimer $^1M_2$ is formed from the *trans*-excimer $^1D_t^*$. This hypothesis is consistent with the subsequent photodimer and crystal data, summarized above, and with (9.57). The *trans*-sandwich dimer of CNA, formed by the photolysis of the *trans*-photodimer in rigid solution, exhibits only weak *trans*-excimer ($^1D_t^*$) fluorescence, and it reverts rapidly to the *trans*-photodimer on exposure to long-wavelength light (> 300 nm).[345] This contrasts with the strong *cis*-excimer ($^1D_c^*$) fluorescence of CNA in concentrated solutions or crystals.

9.7.6 *Photophysical processes in excimers* (pp. *335–8*)

Fischer *et al.*[347] have studied the absorption edge spectra of pyrene single crystals from room temperature to liquid helium temperatures. Below the phase transition temperature of 123°K the first sharp absorption peak at 376 nm vanishes, but the fluorescence spectrum is unchanged. A new absorption band with a maximum at 410 nm appears below 50°K, and its intensity increases with decreasing temperature. This band is assigned to the $D_0 - D_1$ absorption of the pyrene crystal dimer $^1(M \parallel M)$ (p. *336*). The relation of the absorption to exciton migration in pyrene crystals is discussed in §9.11.3.

The $D_1 - D_p$ absorption spectrum of the pyrene excimer in solution has been studied between 400 and 600 nm using laser flash photolysis[348] and modulation excitation spectrometry.[349] A single structureless peak is observed with a maximum at 20,400 cm^{-1} and a bandwidth of 2800 cm^{-1}. Goldschmidt and Ottolenghi[348] compared their observations with the pyrene excimer configuration interaction calculations of Azumi *et al.*[350] For a sandwich dimer configuration with an intramolecular separation of $r_m = 3{\cdot}8$ Å, computational method **B** predicts values of 2·3 eV for the $^1B_{3g}^- \rightarrow {}^1A_g$ excimer fluorescence transition and 2·1 eV for the allowed $^1B_{2u}^+ \leftarrow {}^1B_{3g}^-$ excimer absorption transition, in agreement with experiment.

The influence of the solvent and temperature on the radiative (k_{FD}) and radiationless (k_{ID}) transition rates of the pyrene excimer has been studied by Birks *et al.*,[351] by observations of the excimer decay rate ($k_D = k_{FD} + k_{ID}$) and related parameters. k_{FD} depends on the solvent refractive index n

$$k_{FD} = n^2(k_{FD})_0 \tag{9.58}$$

but it is otherwise independent of temperature T. k_{ID} ($= k_{TD} + k_{GD}$) is the sum of the intersystem crossing (k_{TD}) and internal conversion (k_{GD}) rates, and it consists of a temperature-independent (k_{ID}^0) and a temperature-dependent (k_{ID}^T) component, where

$$k_{ID} = k_{ID}^0 + k_{ID}^T = k_{ID}^0 + k'_{ID} \exp(-W_{ID}/\mathbf{k}T) \tag{9.59}$$

Table 9.26 Pyrene excimer and solvent parameters. k-values in $10^7\ s^{-1}$; W-values in eV

Solvent	q^0_{FD}	k^0_D	k_{FD}	$(k_{FD})_0$	k^0_{ID}	k'_{ID}	W_{ID}	W_η	n	Refs.
Propylene glycol	0·29	3·4	1·0	0·49	2·4	1500	0·23	0·22	1·43	351
iso-Propanol	0·66	1·43	0·95	0·50	0·48	2700	0·22	0·23	1·38	351
n-Nonane	~1·0	~1·1	1·1	0·56	~0	220	0·15	0·10	1·40	352
Acetone	(0·78)	1·3	(1·01)	(0·55)	(0·3)	50	0·1	0·10	1·36	353
Ethanol	0·82	1·3	1·07	0·58	0·23	40	0·1	0·14	1·36	353–356
Cyclohexane	0·93	1·25	1·16	0·57	0·09	20	0·1	0·13	1·43	323, 353
n-Hexane	(0·88)	1·20	(1·05)	(0·55)	(0·15)	—	—	—	1·38	92, 358
Crystal	1·0	0·55	0·55	—	~0	4·7	0·07	—	—	328

Values in brackets obtained from (9.58) assuming $(k_{FD})_0 = 0{\cdot}55 \times 10^7\ s^{-1}$.

The temperature dependence of k_{ID}^{T} is related to that of the solvent viscosity η, given by

$$\eta = \eta_{\infty} \exp(W_{\eta}/\mathbf{k}T) \tag{9.60}$$

At low temperatures the decay rate k_D becomes

$$k_D^0 = k_{FD} + k_{ID}^0 \tag{9.61}$$

and the excimer fluorescence quantum efficiency q_{FD} becomes

$$q_{FD}^0 = k_{FD}/k_D^0 \tag{9.62}$$

The experimental values of the pyrene excimer parameters (q_{FD}^0, k_D^0, k_{FD}, $(k_{FD})_0$, k_{ID}^0, k'_{ID}, W_{ID}) in various solvents, and the solvent parameters (n, W_η) are listed in Table 9.26. The corresponding values of the pyrene excimer parameters (q_{FD}, k_D, k_{ID}, k_{ID}^0, k_{ID}^T, k_{TD}, k_{GD}) at 20°C in various solvents are listed in Table 9.27.

Table 9.27 Pyrene excimer parameters in various solvents at 20°C. k-values in 10^7 s^{-1}

Solvent	q_{FD}	k_D	k_{ID}	k_{ID}^0	k_{ID}^T	k_{TD}	k_{GD}	Refs.
Propylene glycol	0·29	3·4	2·4	2·4	0	—	—	351
iso-Propanol	0·52	1·82	0·87	0·48	0·39	—	—	351
n-Nonane	0·65	1·7	0·6	~0	~0·6	—	—	352
Acetone	(0·44)	2·3	(1·3)	(0·3)	(1·0)	—	—	353
Ethanol	0·55	1·95	0·88	0·23	0·65	0·25	0·63	353–356
Cyclohexane	0·75	1·55	0·39	0·09	0·30	—	—	323, 353
n-Hexane	0·55	(1·90)	(0·85)	(0·15)	(0·70)	0·15	0·70	92, 358
Crystal	0·64	0·88	0·33	0	0·33	—	—	328

Values in brackets obtained from (9.58) assuming $(k_{FD})_0 = 0{\cdot}55 \times 10^7$ s^{-1}.

Several conclusions may be drawn from the data.[351]

(a) The values of k_{FD} are consistent with (9.58), with a mean value of $(k_{FD})_0 = 5{\cdot}5 \times 10^6$ s^{-1}.

(b) For the pyrene crystal excimer $k_{FD} = (k_{FD})_0$, corresponding to an apparent refractive index $n = 1$. The crystal excimer is limited to $^1(M\|M)^*$ configurations, while the solution excimer includes $^1(M \times M)^*$ and $^1(M\|M)^*$ configurations, which will tend to reduce its mean radiative lifetime.

(c) For the two solvents (ethanol, n-hexane) for which data are available, the temperature-independent radiationless transition is due to intersystem crossing ($k_{ID}^0 = k_{TD}$) and the temperature-dependent radiationless transition is due to internal conversion ($k_{ID}^T = k_{GD}$).

(d) The thermally activated internal conversion is attributed to molecular motion which yields non-sandwich ^{1}(M × M)* excimer configurations and leads to increased intermolecular vibrational interaction.[351] This motion is subject to viscous constraint, and there is satisfactory agreement between the values of W_{ID} and W_η (Table 9.26). When $\eta \geqslant 4{\cdot}7$ cP the molecular motion and the thermally activated internal conversion are inhibited.[351]

(e) In the crystal pyrene excimer the thermally activated radiationless transition k_{ID}^T is attributed to intersystem crossing to ^{3}D* and/or ^{3}M* + ^{1}M. The observation of pyrene crystal phosphorescence[601] and delayed fluorescence due to ^{3}M* – ^{3}M* interaction[360] support this conclusion. Internal conversion via a ^{1}(M × M)* configuration is inhibited in the crystal lattice. On the experimentally determined potential energy diagram of the pyrene crystal excimer (p. *331*) the (^{3}M* + ^{1}M) potential curve 'crosses' the ^{1}D* curve at ~0·1 eV (~W_{ID}) above its minimum, which is consistent with thermally activated intersystem crossing.

Cundall and Pereira[361] determined the excimer fluorescence quantum efficiency q_{FD} and lifetime τ_D (= $1/k_D$) for 1-methylnaphthalene in ethanol from −30 to 60°C. They found that k_{FD} is independent of temperature, and that $k_{ID}^0 = 5 \times 10^6\ s^{-1}$, $k'_{ID} = 3{\cdot}2 \times 10^{11}\ s^{-1}$ and $W_{ID} = 0{\cdot}29$ eV. From the magnitudes of k_{ID}^0 and k'_{ID} they concluded that the former corresponds to intersystem crossing and the latter to internal conversion, as in the pyrene excimer in solution.[351]

9.7.7 *Triplet excimers* (pp. *343–8, 630–1*)

The triplet (^{3}M*) energy $E_T = 16{,}900\ cm^{-1}$ of crystal pyrene is known from the delayed fluorescence excitation spectrum[360] (p. *215*). Gijzeman *et al.*[246,362–365] have reported three long-lived emissions from pyrene crystals.

(a) A broad structureless emission with a maximum at 21,800 cm^{-1}, a lifetime of $\frac{1}{2}\tau_T$, and an intensity proportional to I_0^2, where I_0 is the incident light intensity, is assigned to the delayed ^{1}D* fluorescence resulting from ^{3}M* – ^{3}M* interaction (homofusion).

(b) A structured emission with a 0 – 0 band at 15,200 cm^{-1} (= E_T – 1700 cm^{-1}), a lifetime of τ_T and an intensity proportional to I_0, occurs at temperatures below 200°K. Gijzeman *et al.*[246,362] attribute it to the phosphorescence of 'ordered ^{3}(M ∥ M)* excimers' in the crystal lattice. From the temperature dependence of τ_T, they concluded that the 'regular triplet excimer' has a stabilization enthalpy $B = 800\ cm^{-1}$ and a ground-state repulsion energy $R = 900\ cm^{-1}$.[364]

(c) A structureless emission band with a maximum at 13,800 cm^{-1} ($= E_T - 3100$ cm^{-1}), a lifetime of τ_T and an intensity proportional to I_0, was observed at temperatures from 77 to 300°K. Gijzeman *et al.*[364] attributed it to 'triplet excimers' formed at lattice defects, and they concluded that the 'defect triplet excimer' has $B = 2000$ cm^{-1} and a mean $R = 1100$ cm^{-1}.

Application of high pressures to pyrene crystal powder increases the defect concentration and modifies the emission characteristics.[362,363] Emission (c) increases in intensity and two new luminescences appear.

(d) A weak (quantum yield ~10^{-6}) luminescence is observed, originating at $E_T = 16{,}900$ cm^{-1} and having a structured spectrum similar to the $^3M^*$ phosphorescence in solid and liquid solutions. Gijzeman *et al.*[362,363] assign it to 'defect $^3M^*$ phosphorescence' from pyrene molecules which lack a dimer partner, analogous to the defect $^1M^*$ fluorescence of pyrene crystals (p. *319*, §9.7.4). They consider that its observation supports the assignment of emission (b), which has a similar spectrum red-shifted by 1700 cm^{-1}, to 'regular $^3D^*$ phosphorescence'.

(e) At reduced temperatures the structureless delayed $^1D^*$ fluorescence (a) at 21,800 cm^{-1} is replaced by another structureless delayed fluorescence with a maximum at 19,500 cm^{-1}. This is attributed[362] to 'defect delayed $^1D^*$ fluorescence' produced by the interaction of free and trapped triplet excitons.

The various luminescences reported from crystal pyrene are listed in Table 9.28. Prompt and delayed $^1D^*$ fluorescence occurs from pure

Table 9.28 Luminescence of pyrene crystals

Luminescence assignment	Nature	Energy (cm^{-1})	Refs.
Defect monomer fluorescence (DMF)	S	~25300	326–328, 367
Delayed DMF	S	~25600	367
Excimer fluorescence (EF)	SX	21500	326–328, etc.
Delayed EF	SX	21800	246, 360, 367, 601
Delayed defect EF[a]	SX	19500	362
Monomer phosphorescence (MP)	S	16900	601
Defect MP[a]	S	16900	363
Excimer phosphorescence (EP)[a]	S	15200	246
Defect EP[a]	SX	13800	246

S = structured spectrum; SX = structureless spectrum.
[a] The data of Peter and Vaubel[366,601] and Chu and Kearns[367] indicate that these assignments are incorrect.

single crystals, and crystals containing a sufficient defect concentration also emit prompt and delayed defect $^1M^*$ fluorescence. Peter and Vaubel[601] have recently observed molecular ($^3M^*$) phosphorescence and prompt and delayed $^1D^*$ fluorescence from a high-purity pyrene crystal ($^1D^*$ lifetime $\tau_D = 135$ ns; $^3M^*$ lifetime $\tau_T = 120$ ms), but they were not accompanied by any of the other long-lived emissions reported by Gijzeman *et al.*[246,362–365]

Peter and Vaubel[366] purified pyrene by chromatography and successive zone-refining in the absence of oxygen, and they prepared pyrene crystals in the dark by (i) zone-refining, (ii) slow growth from the melt under vacuum by the Bridgman method, (iii) sublimation in a nitrogen atmosphere and (iv) crystallization from solution. The triplet lifetime τ_T, determined from the delayed excimer fluorescence lifetime $\frac{1}{2}\tau_T$, provides a criterion of crystal quality. The zone-refined crystal (i) has the highest value of $\tau_T = 140 \pm 10$ ms from 300 to 160°K. At 300°K $\tau_T = 10$–40 ms for the melt-grown crystals (ii) and $\tau_T \leqslant 1{\cdot}5$ ms for the vapour-grown crystals (iii) and solution-grown crystals (iv). Crystal irradiation in the 380–420 nm region in the presence of air causes a major reduction in τ_T, which correlates with the penetration depth of the incident light. Similar effects on the excimer fluorescence lifetime τ_D of pyrene crystals have been observed by Birks *et al.*[328] Irradiation in the absence of oxygen does not affect τ_T or τ_D. It appears that photo-oxidation of pyrene crystals due to irradiation in the presence of air produces impurity centres which function as singlet and triplet exciton traps.

The prompt and delayed fluorescence spectra of fresh, unoxidized, pyrene crystals prepared by the different methods are identical, as are the delayed excimer fluorescence excitation spectra, which agree with that of Avakian and Abramson.[360] Peter and Vaubel[366,601] and Chu and Kearns[367] did not observe the other long-lived emissions reported by Gijzeman *et al.*[246,362–364] The latter used vapour-grown crystal flakes and powdered specimens with τ_T (300°K) = 0·25 ms, which is 560 times less than τ_T (300°K) = 140 ms, obtained for the zone-refined crystal.[366] It is concluded that the emissions attributed by Gijzeman *et al.*[246,362–364] to (b) $^3D^*$ phosphorescence, (c) defect $^3D^*$ phosphorescence and possibly (e) delayed defect $^1D^*$ fluorescence, are associated with impurities, probably photo-oxides, and that the emission attributed to (d) defect $^3M^*$ phosphorescence corresponds to the normal $^3M^*$ phosphorescence of the pure crystal[601] (Table 9.28).

For the zone-refined pyrene crystal τ_T and k_{TT}, the $^3M^* - {}^3M^*$ interaction rate parameter (p. *560*), are independent of temperature from 300 to 160°K.[366] This is consistent with $^3M^*$ exciton migration with an activation energy $\ll \mathbf{k}T$, but it cannot be reconciled with the 'regular triplet excimers' $^3D^*$ with $B = 800\ cm^{-1}$, $R = 900\ cm^{-1}$, proposed by Gijzeman *et al.*[364] At $T < 160°K$ τ_T decreases rapidly from 140 ms at 160°K to 6 ms at 80°K, probably due to the presence of shallow traps of unknown origin in the crystal.

The concentration quenching of the triplet lifetime of naphthalene, phenanthrene, anthracene, pyrene[368] and *o*-xylene[369] in fluid solutions has been attributed to triplet excimer formation (pp. *347–8*, *630–1*). Triplet excimer phosphorescence of naphthalene in ethanol solution at 160°K has been reported by Langelaar,[245,370] who observed a structureless band with a maximum about 2600 cm^{-1} below the molecular phosphorescence 0 – 0 band (p. *347*). In contrast Chandross and Dempster[330] observed a structured phosphorescence spectrum from the 1,3-*bis* (α-naphthyl) propane sandwich dimer in rigid solution, which includes two naphthalene groups in a closely spaced, symmetrical $^1(M \parallel M)$ configuration (§9.7.5). The dimer phosphorescence is shifted only 250 cm^{-1} below the molecular phosphorescence.

Avouris and El-Bayoumi[371] have presented evidence for intramolecular triplet excimer formation in 1,3-diphenylpropane (DPP). A 10^{-3} M solution of DPP in 3-methylpentane at 77°K exhibits toluene-type emissions with a $^1M^*$ fluorescence maximum at 285 nm and a structured $^3M^*$ phosphorescence with a maximum at 390 nm. A 10^{-3} M solution of DPP in *iso*-pentane at 77°K has a more diffuse phosphorescence spectrum, which on warming to 115°K shifts to a maximum at 420 nm, which is attributed to excimer phosphorescence.

Most of the spectroscopic evidence for triplet excimer formation by aromatic hydrocarbons lacks confirmation. In contrast there is definitive evidence for triplet excimer emission in the halobenzenes (pp. *343–4*). George and Morris[372] studied the absorption, fluorescence and phosphorescence spectra of single crystals of 1,4-dichlorobenzene (DCB) and 1,2,4,5-tetrachlorobenzene (TCB) at low temperatures. At 4·2°K the phosphorescence of TCB is predominantly from an X-trap (pp. *557–8*) lying 48 cm^{-1} below the triplet exciton band, and the excimer phosphorescence is very weak. At 12°K the intrinsic triplet exciton phosphorescence of TCB is dominant, and strong excimer phosphorescence is observed. In DCB at 4·2°K trapped triplet exciton phosphorescence does not occur, and intense triplet excimer emission is observed.

The luminescence of tetraphenyl compounds of Group IV and V elements has been considered by Gouterman and Sayer,[373] following earlier work by LaPaglia.[374,375] The crystal fluorescence (XF) and phosphorescence (XP) spectra of the tetraphenyls are red-shifted by about 4000 to 7000 cm^{-1} relative to the fluorescence (F) and phosphorescence (P) spectra of the compounds in EPA solution at 77°K. The crystal spectra are more diffuse than the solution spectra, but they have similar Franck–Condon envelopes.

Gouterman and Sayer[373] consider that the crystal luminescences XF and XP are due to singlet and triplet excimers, respectively (or to a common biphenyl-type impurity). The molecular and crystal geometry of the tetraphenyls inhibit $^1(M \parallel M)$ sandwich dimer structures, and they propose that the crystal 'excimers' are σ-bonded structures consisting of coplanar phenyl groups. They express the belief that 'generally a σ-bonded triplet excimer is responsible for the XP luminescence, although in some cases the crystal luminescence may arise from impurities'.[373]

Several data appear inconsistent with the excimer hypothesis.

(a) The similarity of the Franck–Condon envelopes of the crystal and solution luminescence spectra suggests that XF and XP occur to similar ground states to F and P.

(b) The vibrational structure of the XF and XP spectra indicates that the transitions occur to an associated ground state.

(c) At 4°K the fluorescence spectra of ϕ_4Ge and ϕ_4Sn crystals are normal (F) and similar to the solution spectra.[376] For other Group IV tetraphenyls the crystal fluorescence spectra change from XF at 300°K to F at 77°K.[373] Thus the effect attributed to the crystal 'excimers' disappears on cooling, unlike conventional excimers.

(d) Dilute EPA solutions of ϕ_4Pb at 77°K exhibit intense fluorescence and both P and XP phosphorescences. The internal heavy-atom effect in ϕ_4Pb strongly quenches the normal fluorescence, and Gouterman and Sayer[373] therefore attribute the intense fluorescence to a photoproduct impurity, since ϕ_4Pb readily photodecomposes. They assign the XP emission to intramolecular σ-bonded 'triplet excimers,' a questionable assignment in a system containing significant impurities.

The present evidence for σ-bonded singlet or triplet excimers in the crystal tetraphenyls appears inadequate. Photoproducts, impurities and/or defects offer a more plausible explanation of the unusual crystal emissions.

9.8 Delayed fluorescence

9.8.1 E-*type delayed fluorescence* (pp. *373–5*)

Benzophenone has a very low fluorescence quantum yield and a high phosphorescence quantum yield. The phosphorescence spectrum $P(\bar{\nu})$ in solid methylcylohexane solution at 77°K is structured, with the $0-0$

transition maximum at 23,900 cm^{-1}, and it lies entirely at energies $\leqslant$ 24,700 cm^{-1}.[377] At room temperature and above, a weak higher energy (>25,000 cm^{-1}) long-lived emission is observed from benzophenone in the vapour phase,[378] in fluid solution[379] and in polymer matrices,[381] and this has been tentatively attributed to E-type delayed fluorescence. A weak short-lived emission with a decay time of 6–20 ps has also been observed in this spectral region, and attributed to prompt fluorescence.[382]

Brown *et al.*[377] have observed the prompt fluorescence spectrum $F(\bar{\nu})$ of benzophenone in benzene solution at room temperature, using a 6 ns, 50 kW pulsed nitrogen laser for excitation. The fluorescence spectrum is structureless: it originates at about 27,800 cm^{-1}, reaches a maximum at about 24,400 cm^{-1}, and then decreases monotonically over the whole region of the phosphorescence spectrum. The delayed luminescence spectra, recorded at different intervals ($\gg \tau_M$) after the excitation pulse, are composites of $P(\bar{\nu})$ and $F(\bar{\nu})$. The phosphorescence and the delayed fluorescence decay with a common lifetime τ_T, consistent with E-type delayed fluorescence originating from thermally activated $T_1 - S_1$ intersystem crossing.

9.8.2 P-*type delayed fluorescence in fluid solutions* (pp. *385–9*)

In aromatic molecular crystals the application of a magnetic field modifies the rate parameter of triplet–triplet interaction (homofusion) and the resultant delayed fluorescence intensity (pp. *563–4*; Swenberg and Geacintov, Chapter 10, Vol. 1; Chapter 8, Vol. 2). Faulkner and Bard[383] found that a magnetic field H modifies the intensity I_{FM}^{d} of the P-type delayed fluorescence of anthracene in N,N-dimethylformamide solution at room temperature. I_{FM}^{d} decreases monotonically by about 4·5% on increasing H from 0 to 8 kG, tending to a limiting value at higher H. The magnetic field effect is less than in crystal anthracene, and no low-field enhancement of I_{FM}^{d} is observed (Chapter 10, Vol. 1). The phenanthrene-sensitized P-type delayed fluorescence (p. *396*) of anthracene in solution has an identical magnetic field dependence to the normal P-type delayed fluorescence, and it is concluded that the anthracene triplet quantum yield Φ_{TM} is independent of H.[383] In dilute solution I_{FM}^{d} is given by (p. *384*)

$$I_{FM}^{d} = \tfrac{1}{2}k_{MTT}q_{FM}\left(\frac{\Phi_{TM}I_0}{k_T}\right)^2 \tag{9.63}$$

The magnetic field dependence of I_{FM}^{d} is attributed entirely to k_{MTT}, the rate parameter of the $^3M^* - {}^3M^*$ interaction process yielding $^1M^*$, since the other parameters in (9.63) are known, or are expected, to be independent of H. These results have been confirmed by Avakian *et al.*,[384] who observed a monotonic decrease of I^{d}_{FM} with increasing H to a limiting value of 0·92 of the zero-field intensity, for anthracene in ethanol solution at room temperature. They proposed a model to explain the monotonic decrease of k_{MTT} with increase in H, which indicates that the triplet fusion process in solution is diffusion-controlled.

Wyrsch and Labhart[385] have studied the influence of a magnetic field H on the delayed emission of 1:2-benzanthracene in fluid ethanol solutions from -70 to -160°C. The solutions have three long-lived emissions, phosphorescence (intensity I_P), delayed molecular fluorescence (intensity I^{d}_{FM}) and delayed excimer fluorescence (intensity I^{d}_{FM}). Using relatively low magnetic fields ($H \leqslant 1·65$ kG) I_P is observed to be independent of H. At -107°C I^{d}_{FM} increases by about 2·2% to a maximum at $H = 0·4$ kG and then decreases steadily to about its zero-field value, while I^{d}_{FD} increases by about 4% to a maximum at $H = 0·4$ kG and then remains constant up to $H = 1·65$ kG (Figure 9.16). I^{d}_{FM} and I^{d}_{FD} behave in a similar manner at

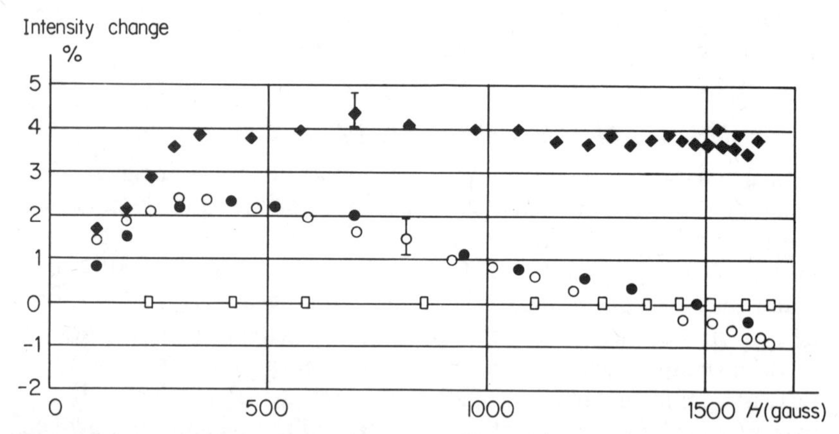

Figure 9.16 1:2-Benzanthracene in ethanolic solution. Relative changes of luminescence intensities as a function of magnetic field strength H. □ Phosphorescence (596 nm) at -133°C; ○ Delayed monomer fluorescence (410 nm) at -133°C; ● Delayed monomer fluorescence (410 nm) at -107°C; ◆ Delayed excimer fluorescence (550 nm) at -107°C. (Wyrsch and Labhart, *Chem. Phys. Letters*, **8**, 217 (1971))

other temperatures. The increase in I^{d}_{FM} is between 0·5 and 2·2% at $H =$ 0·4 kG, and between 0 and $-1·0$% at $H = 1·5$ kG, at -80 to -150°C. The increase in I^{d}_{FD} is between 1·6 and 4% and it is constant from $H = 0·4$ kG to 1·5 kG, at -70 to -110°C.

The reaction kinetics of a solution system exhibiting molecular and excimer delayed fluorescence are complicated (pp. *378–84*). I^{d}_{FM} and I^{d}_{FD} both depend on the rate parameters k_{MTT} and k_{DTT} of the $^3M^* - {}^3M^*$

interaction yielding $^1M^*$ and $^1D^*$, respectively, and also on the other $^1M^*$, $^1D^*$ and $^3M^*$ rate parameters. Knowledge of these rate parameters and their temperature dependence would be required for a complete analysis of the 1:2-benzanthracene solution data,[385] but a simplified analysis may be applied. For a specific solution system (given solvent, solute, concentration and temperature) we may write

$$I_{FM}^{d} = a\{k_D k_{MTT} + k_{MD}(k_{MTT} + k_{DTT})\} \tag{9.64}$$

$$I_{FD}^{d} = b\{k_M k_{DTT} + k_{DM}[^1M](k_{MTT} + k_{DTT})\} \tag{9.65}$$

where a and b depend on the other rate parameters, the concentration and the excitation light intensity, and are independent of H. For dilute ethanolic solutions of 1:2-benzanthracene[386] below $-70°C$,

$$k_M, k_D \gg k_{DM}[^1M], k_{MD}$$

so that (9.64) and (9.65) simplify to

$$I_{FM}^{d} = a k_D k_{MTT} \tag{9.64a}$$

$$I_{FD}^{d} = b k_M k_{DTT} \tag{9.65a}$$

The magnetic field dependence of I_{FM}^{d} and I_{FD}^{d} (Figure 9.16) is that of k_{MTT} and k_{DTT}, respectively, and it confirms that these two processes are distinct[385] and do not occur via a common excimer intermediate (pp. *391–4, 631–2*).

Wyrsch and Labhart[386] have also studied the P-type delayed fluorescence of 1:2-benzanthracene in solution in ethanol and 1:1 ethanol/ethylene glycol as a function of temperature in the absence of a magnetic field. Their observations of the triplet quantum yield Φ_{TM}, the $^3M^*$ first-order decay rate k_T, and the total $^3M^* - {}^3M^*$ interaction rate parameter k_{TT}, are listed in Tables 9.29 and 9.30. Comparison of the values of k_{TT} with $2k_{diff}$ (two $^3M^*$ may annihilate per encounter), where k_{diff} is the diffusion-controlled rate parameter calculated from the Stokes–Einstein relation (Chapter 8, Vol. 1) and the viscosity, shows that k_{TT} is two or three times larger than $2k_{diff}$, indicating that the $^3M^* - {}^3M^*$ interaction radius exceeds the Stokes molecular radius.[386] No evidence is obtained from the experiments for a viscosity-independent long-range $^3M^* - {}^3M^*$ interaction process,[386] such as was found from previous studies of pyrene solutions in propylene glycol (pp. *386–9, 393*). Such a process is, however, necessary to account for P-type delayed fluorescence in rigid solutions (pp. *389–90*).

Labhart and Wyrsch[387] have discussed the temperature dependence of $V\ (= I_{FM}^{d}/I_{FD}^{d})$ for solutions of 1:2-benzanthracene in ethanol (E) and 1:1

Table 9.29 1:2-Benzanthracene in ethanol. Triplet parameters[386]

		$[^1M] = 2{\cdot}35 \times 10^{-5}$ M		$[^1M] = 2{\cdot}85 \times 10^{-4}$ M	
T(°C)	Φ_{TM} (± 10%)	k_T (s^{-1}) (± 5%)	k_{TT} (M^{-1} s^{-1}) (± 30%)	k_T (s^{-1}) (± 5%)	k_{TT} (M^{-1} s^{-1}) (± 30%)
−140	0·60	—	—	3·8	401 × 10^5
−130	0·61	3·8	3·0 × 10^6	3·8	2·9 × 10^6
−120	0·62	4·6	1·3 × 10^7	5·5	1·7 × 10^7
−110	0·62	6·6	4·5 × 10^7	6·9	4·1 × 10^7
−100	0·63	7·5	1·8 × 10^8	10·5	1·3 × 10^8
−90	0·64	9·7	2·3 × 10^8	17·5	2·2 × 10^8
−80	0·64	16·4	5·0 × 10^8	42·9	5·8 × 10^8
−60	0·67	26·0	1·4 × 10^9	—	—

Table 9.30 1:2-Benzanthracene in 1:1 ethanol/ethylene glycol. Triplet parameters[386]

		$[^1M] = 2{\cdot}65 \times 10^{-5}$ M		$[^1M] = 3{\cdot}1 \times 10^{-4}$ M	
T(°C)	Φ_{TM} (± 10%)	k_T (s^{-1}) (± 5%)	k_{TT} (M^{-1} s^{-1}) (± 30%)	k_T (s^{-1}) (± 5%)	k_{TT} (M^{-1} s^{-1}) (± 30%)
−90	0·62	—	—	4·8	1·1 × 10^6
−80	0·63	5·0	5·6 × 10^6	—	—
−70	0·64	—	—	5·5	2·5 × 10^7
−60	0·65	5·0	1·0 × 10^8	—	—
−50	0·66	—	—	9·5	1·3 × 10^8
−40	0·66	10	5·2 × 10^8	—	—
−30	0·68	—	—	28	7·8 × 10^8
−20	0·68	21	1·2 × 10^9	—	—
−10	0·69	—	—	67	2·4 × 10^9
0	0·70	49	4·2 × 10^9	—	—
+10	0·71	—	—	—	—
+20	0·72	99	1·0 × 10^{10}	260	9·1 × 10^9

ethanol/ethylene glycol (EE) in terms of the temperature dependence of α (= k_{DTT}/k_{MTT}) and k_{MD}, the $^1D^*$ dissociation rate. At $T > T_c$ (T_c = −70°C in E, and −30°C in EE) α is constant, and the temperature dependence of V is due to k_{MD}. At $T < T_c$, k_{MD} is negligible and the temperature dependence of V is due to α. To account for the magnitude of α they propose a model, similar to that of Stevens[388] (pp. *631–2*), in which the $^3M^* - {}^3M^*$ interaction process leading to $^1M^* + {}^1M$ (k_{MTT}) occurs at a greater distance than that leading to $^1D^*$ (k_{DTT}).

Dupuy[389,390] has proposed that the only significant $^3M^* - {}^3M^*$ process in solution is that yielding $^1M^* + {}^1M$ (k_{MTT}), that the rate (k_{DTT}) of the parallel process yielding $^1D^*$ is negligible, and that any delayed $^1D^*$ fluorescence is due to subsequent $^1M^* - {}^1M$ interaction. This hypothesis is valid for phenanthrene and chrysene, two of the compounds studied, since these compounds do not yield any prompt or delayed $^1D^*$ fluorescence. It is not valid for pyrene and naphthalene, two of his other compounds, nor for other compounds exhibiting $^1D^*$ fluorescence in solution. Numerous observations on several compounds in a variety of solvents have shown consistently (p. *381*) that

$$\frac{I^{\mathrm{d}}_{\mathrm{FD}}}{I^{\mathrm{d}}_{\mathrm{FM}}} > \frac{\Phi_{\mathrm{FD}}}{\Phi_{\mathrm{FM}}} \tag{9.66}$$

This demonstrates clearly that α $(= k_{\mathrm{DTT}}/k_{\mathrm{MTT}}) > 0$. Several observers have determined α and studied its temperature dependence. Against this weight of experimental evidence, Dupuy's hypothesis[389,390] that $\alpha = 0$ is untenable.

9.8.3 *Recombination luminescence* (pp. *397–9*)

The photoionization of aromatic molecules in condensed media and the delayed fluorescence and phosphorescence resulting from the subsequent recombination of the ionic species have been discussed by Lesclaux and Joussot-Dubien (Chapter 9, Vol. 1).

The corresponding processes in aromatic molecular crystals (pp. *238–9*) have been considered by Swenberg and Geacintov (Chapter 10, Vol. 1). Ewald and Durocher[391] have studied the photo-ionization and recombination delayed fluorescence in pure anthracene single crystals and Schwob *et al.*[392] have investigated the recombination luminescence and normal fluorescence of doped anthracene crystals.

9.9 Molecular complexes and exciplexes

9.9.1 CT *absorption by* DA *complexes* (pp. *401–10*).

There have been two recent books on the subject. *Molecular Complexes* by Mulliken and Person[393] provides a comprehensive theoretical treatment. *Organic Charge-transfer Complexes* by Foster,[394] which is more experimental in approach, includes extensive data on the DA complexes of organic molecules.

Blaunstein and Christophorou[395] have made a useful compilation of the experimental and calculated ionization potentials of 262 molecules, including 75 aromatic molecules, and experimental and calculated electron affinities of 198 molecules. A critical comparison is made of the different experimental and theoretical methods.

Anderson[396] has considered the symmetry classification and optical selection rules of crystalline DA complexes. Systems of the benzidine-chloranil type, where the component molecules D and A each possess high symmetry (e.g. D_{2h}), are discussed in detail. The DA complex is stabilized in a sandwich configuration of relatively large intermolecular π–π overlap (compare the $^1(\mathrm{M} \parallel \mathrm{M})$ dimer and excimer configurations, §9.7.1) and the CT transitions are predicted to have unique polarization characteristics. The theory accounts for the orthogonal polarizations of the first and second CT absorption bands of the benzidine-chloranil and benzidine-TCNQ crystal DA complexes, observed by Amano *et al.*[397] The results are also applied to the following DA complex systems: TMPD-chloranil; anthracene-TCNQ; TMPD-TCNQ; naphthalene-TCNB; perylene-fluoranil; mesitylene-TCNB; naphthalene-TCNE; acenaphthene-TCNE; pyrene-TCNE; and phenanthrene-TCNE.

Infrared absorption transitions in DA complexes have been used to study vibrational and rotational motion. Larkindale and Simkin[398] observed a new band at 115 cm^{-1} in the absorption spectrum of tetracyanoethylene (TCNE) in mesitylene, which they assigned to the intermolecular stretching frequency of the mesitylene-TCNE DA complex. Intermolecular vibrations of this type have been previously observed in strong (e.g. pyridine-iodine) complexes, but not in weak π–π complexes. Rothschild[399] calculated the rotational motion of the permanent dipole moment of $CDCl_3$ (chloroform) in 1:4 $CDCl_3/C_6H_6$ mixtures from the infrared band contour at 362 cm^{-1}. The results show that the intermolecular forces in the chloroform-benzene complex do not prevent a considerable amount of practically frictionless motion of $CDCl_3$ away from the direction of the intermolecular bond.

The CT absorption spectrum is influenced by an external electric field, and this electrochromic effect may be used to determine the dipole moments, μ_E and μ_N, of a DA complex in its excited and ground states, respectively. In electrochromic studies the light is linearly polarized, and the electric field F is applied normal to the light propagation direction at an angle X to its electric vector. The parameter

$$L = \frac{\varepsilon_F - \varepsilon_0}{F^2 \varepsilon_0} \tag{9.67}$$

is observed as a function of $\bar{\nu}$ and X, where ε_F and ε_0 are the molar extinction coefficients in the presence and absence of F.

Liptay[400–402] has developed the theory relating $L(\bar{\nu}, X)$ to the DA complex parameters, which has been used by Eckhardt[403] to determine μ_E and μ_N for the *trans*-stilbene–TCNE DA complex. Varma and Oosterhoff[404] have reformulated the theory in a more convenient form for comparison with experiment. The values of μ_N and $(\mu_E - \mu_N)$ obtained by these and other workers for various DA complexes are listed in Table 9.31. Values of μ_N for other complexes have been given previously (p. *453*).

Table 9.31 Dipole moments (in Debyes) of DA complexes in ground state (μ_N) and excited state (μ_E)

D	A	μ_N	$\mu_E - \mu_N$	Ref.
Benzene	TCNE	0·8	−9·8	404
		1·0		405
		0·75		406
Mesitylene	TCNE	1·1	−9·4	404
		1·1		405
		1·12		406
Hexamethylbenzene	TCNE	1·6	4·5	404
			5·0	407
		1·5		405
		1·64		406
		1·35		408
	p-Chloranil	<0·6		404
		1·0		408
Naphthalene	TCNE	0·6	−6·4	404
		1·28		408
Anthracene	TNB	1·1	−10·9	404
PPDA	TNB	1·2	11·2	404
trans-Stilbene	TCNE	1·1	10·5	403

TCNE, tetracyanoethylene; TNB, trinitrobenzene; PPDA, *p*-phenylanediamine.

Ishida and Tsubomura[409] have measured the absorption spectra of aromatic hydrocarbons and aliphatic amines adsorbed on silica gel at 77°K. New absorption bands are observed which are attributed to the CT absorption of hydrocarbon-amine DA complexes. The wavelengths of the CT absorption maxima with tri-*n*-propylamine as acceptor are as follows: perylene (300 nm), anthracene (273 nm), 1:2-benzanthracene (270 nm) and pyrene (250 nm). Aromatic hydrocarbons and aliphatic amines exhibit exciplex fluorescence in room-temperature solution.[410]

Among other studies of DA complexes, Schug and co-workers[411,412] have calculated the interaction potentials of weak benzene-halogen DA complexes in axial and other configurations, Armitage *et al.*[413] have studied the thermodynamics of DA complexes with aromatic amines, and Davis[414] has reviewed spectrometric methods for the study of DA complexes.

9.9.2 *Contact* CT *absorption* (pp. *412–5*)

The CT absorption spectrum of the benzene-oxygen contact complex $^3(Bz{\cdot}O_2)$ in the vapour phase has been determined from the difference in the absorption spectra of 70 torr benzene vapour in the presence and absence of 760 torr oxygen.[415] The $^3(Bz{\cdot}O_2)$ CT absorption spectrum is a smooth continuum with maximum intensity at $E_{CT} = 45{,}650\ cm^{-1} = 5{\cdot}66$ eV. This agrees closely with the value of $E_{CT} = 45{,}200\ cm^{-1}$ for the $^3(Bz{\cdot}O_2)$ complex in carbon tetrachloride solution observed by Lim and Kowalski.[416] We may write (p. *407*)

$$E_{CT} = I_D - A_A - (\Delta E_E - \Delta E_N) \tag{9.68}$$

where I_D is the ionization potential of the donor (Bz), A_A is the electron affinity of the acceptor (O_2), ΔE_E is the energy of formation of the excited complex $^3(Bz{\cdot}O_2)^*$ from Bz^+ and O_2^-, and ΔE_N is the energy of formation of the unexcited complex $^3(Bz{\cdot}O_2)$ from Bz and O_2. For the $^3(Bz{\cdot}O_2)$ complex in solution Dijkgraaf and Hoytink[417,418] observed that $\Delta E_N < 0{\cdot}02$ eV, confirming its contact nature. From the mean of thirteen independent determinations[395] $I_D\,(Bz) = 9{\cdot}23\,(\pm 0{\cdot}02)$ eV, and from the mean of the five most consistent measurements[395] $A_A\,(O_2) = 0{\cdot}67\,(\pm 0{\cdot}2)$ eV, so that substitution in (9.68) gives $\Delta E_E = 2{\cdot}90\ (\pm 0{\cdot}22$ eV).

The energy of formation of an electric dipole from two elementary charges $+e$ and $-e$ is

$$\Delta E_E = \frac{e^2}{R} = \frac{14{\cdot}41}{R\ (\text{in Å})}\ (\text{in eV}) \tag{9.69}$$

where R is the electrical length of the dipole. Substituting ΔE_E in (9.69) we obtain $R = 5{\cdot}0\ (\pm 0{\cdot}4)$ Å for the dipole length of the excited $^3(Bz{\cdot}O_2)^*$ complex. These and corresponding data for benzene complexes with other electron acceptors are listed in Table 9.32.

Yamamoto *et al.*[426] observed the absorption spectrum of a thin polystyrene (PS) film in high pressure O_2. Subtraction of the PS and O_2 absorption yields a spectrum, attributed to the $(PS{\cdot}O_2)$ contact CT absorption, which originates at 330 nm, rises to a maximum at 273 nm, has a minimum at 260 nm, and increases steeply at 250 nm, the shortest wavelength of observation. Polarization measurements on drawn PS films show that the contact CT absorption transition moment is approximately normal to the plane of the benzene ring.

Immersion of a PS film in iodine (I_2) solution yields an absorption band attributed to the CT absorption of the $(PS{\cdot}I_2)$ DA complex.[426] Comparison

Table 9.32 Benzene complexes with different electron acceptors (all energies in eV).[415] (I_D = 9·23 eV)

Acceptor	E_{CT}	Ref.	A_A	Ref.	ΔE_N	Ref.	ΔE_E	R(Å)
Oxygen	5·66	415	0·67 (±0·2)	395	<0·02	417	2·90 (±0·22)	5·0 (±0·4)
Iodine	4·24	419	1·56	420	0·06	421	3·49	4·13
			1·7	422			3·35	4·30
1,3,5-Trinitrobenzene	4·37	423	1·8	424	0·07	423	3·13	4·60
p-Chloranil	3·56	408	2·45	425	0·07	408	3·29	4·38
			2·5	422			3·24	4·44

of the observed value of $E_{CT} = 31{,}550\ \text{cm}^{-1}$ with that of $E_{CT} = 33{,}670\ \text{cm}^{-1}$ for the $(Bz{\cdot}I_2)$ DA complex gives a value of $\Delta E_{CT} = -2120\ \text{cm}^{-1}$ for the substitution shift in going from Bz to PS. Applying the same value of ΔE_{CT} to the $(Bz{\cdot}O_2)$ complex ($E_{CT} = 45{,}650\ \text{cm}^{-1}$), the expected value for the $(PS{\cdot}O_2)$ contact complex is $E_{CT} = 43{,}530\ \text{cm}^{-1}$, corresponding to a wavelength of 230 nm. It is suggested that the assignment of the $(PS{\cdot}O_2)$ contact CT absorption is correct, but that the minimum at 260 nm and the consequent maximum at 273 nm are experimental artefacts, and that the true intensity maximum lies near 230 nm beyond the limit of the observations.

9.9.3 *Luminescence of* DA *complexes* (pp. *415–20*)

Kobayashi *et al.*[427] have measured the fluorescence spectra, lifetimes and quantum yields of DA complexes of tetracyanobenzene (TCNB) with benzene and its methyl derivatives. Observations were made in liquid solutions at room temperature, in the same solutions frozen to 77°K, and in polymethylmethacrylate (PMMA) solutions at room temperature and at 77°K. Table 9.33 lists the donor ionization potential I_D, and the DA complex fluorescence lifetime τ_E, fluorescence quantum yield Φ_{FE}, fluorescence decay rate k_{FE}, radiationless decay rate k_{IE}, fluorescence maximum $\bar{\nu}_f$, fluorescence $0-0$ transition $\bar{\nu}_0$, absorption maximum $\bar{\nu}_a$, Franck–Condon fluorescence shift $(\bar{\nu}_0 - \bar{\nu}_f)$ and Stokes shift $(\bar{\nu}_a - \bar{\nu}_f)$ in liquid solutions at room temperature. Table 9.34 lists the corresponding parameters in PMMA solution at 77°K.

The fluorescence parameters and spectra of the DA complexes in PMMA solutions are practically independent of temperature from 77 to 300°K, and the frozen solutions at 77°K have similar values. In contrast the fluorescence parameters and spectra in liquid solutions show a major temperature dependence, changing from the values of Table 9.33 to those of Table 9.34 in cooling from 300 to 77°K. The TCNB solution in toluene has been studied over the full temperature range.[427]

(i) $\bar{\nu}_f$ decreases from 20 kK (300°K) to 19 kK (150°K), increases abruptly to 24 kK (130°K) and remains almost constant to 23·8 kK (77°K) ($1\ \text{kK} = 10^3\ \text{cm}^{-1}$)

(ii) k_{FE} decreases from $0{\cdot}65 \times 10^6\ \text{s}^{-1}$ (300°K) to $\sim 0{\cdot}2 \times 10^6\ \text{s}^{-1}$ (200°K), increases to $\sim 1{\cdot}2 \times 10^6\ \text{s}^{-1}$ (140°K), jumps sharply to $12 \times 10^6\ \text{s}^{-1}$ (130°K) and decreases slightly to $11 \times 10^6\ \text{s}^{-1}$ (77°K).

(iii) k_{IE} is almost constant from $7{\cdot}5 \times 10^6\ \text{s}^{-1}$ (300°K) to $\sim 9 \times 10^6\ \text{s}^{-1}$ (160°K), passes through a maximum of $\sim 25 \times 10^6\ \text{s}^{-1}$ (140°K) and decreases to $9 \times 10^6\ \text{s}^{-1}$ (77°K).

Table 9.33 Fluorescence of DA complexes of benzene and methyl benzenes with TCNB in liquid solution at room temperature[427]

Donor	I_D (eV)	Solvent	τ_E (ns)	Φ_{FE}	k_{FE} (10^6 s^{-1})	k_{IE}	$\bar{\nu}_f$ (10^3 cm^{-1})	$\bar{\nu}_0$	$\bar{\nu}_a$	$\bar{\nu}_0 - \bar{\nu}_f$	$\bar{\nu}_a - \bar{\nu}_f$
Benzene	9·24	Benzene	147	0·09	0·61	6·2	20·5	23·5	32·5	3·0	12·0
Toluene	8·82	Toluene	122	0·08	0·65	7·5	20·0	23·0	31·8	3·0	11·8
m-Xylene	8·45	*m*-Xylene	53	0·02	0·38	15	19·6	22·0	30·3	2·4	10·7
p-Xylene	8·56	*p*-Xylene	52	0·02	0·39	15	19·6	21·8	30·5	2·2	10·9
Mesitylene	8·39	Mesitylene	43	0·01	0·23	23	19·2	21·5	28·3	2·3	9·1
Durene	8·03	MMA[a]	8	0·0003	0·04	120	17·2	19·8	26·4	2·6	9·2
Hexamethylbenzene	7·85	MMA[a]	5	~0·0001	0·02	~200	16·7	19·0	24·6	2·3	7·9

[a] Methyl methacrylate.

Table 9.34 Fluorescence of DA complexes of benzene and methylbenzenes with TCNB in PMMA solution at 77°K[427]

Donor	τ_E (ns)	Φ_{FE}	k_{FE} (10^6 s^{-1})	k_{IE}	$\bar{\nu}_f$ (10^3 cm^{-1})	$\bar{\nu}_0$	$\bar{\nu}_a$	$\bar{\nu}_0 - \bar{\nu}_f$	$\bar{\nu}_a - \bar{\nu}_f$
Benzene	38	0·63	17	9·5	25·0	27·5	32·5	2·5	7·5
Toluene	50	0·55	11	9·0	23·8	26·5	31·8	2·7	8·0
m-Xylene	74	0·51	6·9	6·6	22·2	25·0	30·3	2·8	8·1
p-Xylene	72	0·46	6·4	7·5	22·2	25·0	30·5	2·8	8·3
Mesitylene	85	0·47	5·5	6·5	22·0	24·7	28·3	2·7	6·3
Durene	58	0·25	4·3	13	20·4	22·8	26·4	2·4	6·0
Hexamethylbenzene	39	0·16	4·1	22	20·1	21·8	24·6	1·7	4·5

The three parameters each change dramatically in the region of 140°K, where the supercooled liquid fuses to the glassy or crystalline state.[429] At lower temperatures the DA complex is in a rigid environment and the fluorescence occurs from the Franck–Condon excited state E_1', which has the same configuration as the ground state E_0 of the DA complex. At higher temperatures the DA complex is in a fluid environment, and it is free to relax from E_1' to its equilibrium excited state E_1. This intermolecular relaxation and fluorescence behaviour is analogous to the intramolecular relaxation and fluorescence behaviour of *cis*-stilbene and the diphenyl-polyenes, discussed in §9.4.4. The increase of Φ_{FE} from $\Phi_{FE}^0 \sim 0{\cdot}04$ at 160°K (E_1 fluorescence) to $\Phi_{FE}' \simeq 0{\cdot}4$ at 130°K (E_1' fluorescence) in the toluene-TCNB complex is a viscosity effect similar to that observed in *cis*-stilbene.[144] It is of interest to note that over this temperature range

$$k_{FE}'(12 \times 10^6\ \mathrm{s}^{-1}) > k_{FE} > k_{FE}^0\ (0{\cdot}5 \times 10^6\ \mathrm{s}^{-1}) \tag{9.70}$$

analogous to (9.10a), but that

$$k_{IE}'\ (25 \times 10^6\ \mathrm{s}^{-1}) > k_{IE} > k_{IE}^0\ (9 \times 10^6\ \mathrm{s}^{-1}) \tag{9.71}$$

which is the opposite of (9.10b). In the toluene-TCNB complex the viscosity effect on Φ_{FE} is primarily due to the change in k_{FE} (9.70).

Egawa *et al.*[428,429] observed the time-resolved fluorescence spectra of the toluene-TCNB complex in the supercooled liquid state at 147°K. At this temperature the $E_1' \rightarrow E_1$ relaxation time is comparable with the fluorescence lifetime τ_E, and the time-resolved fluorescence spectrum is observed to shift progressively to lower energies with increasing time after the laser excitation pulse. No time dependence of the fluorescence is observed at room temperature, where the relaxation is much more rapid than the instrumental time. The fluorescence spectrum is also time-independent in the glassy state at low temperatures (137 to 77°K), where the relaxation is inhibited by the rigidity of the system. Somewhat surprisingly, the fluorescence spectrum of the toluene-TCNB system in the crystalline state, which is produced by a phase transition from the glassy state on warming from 77 to 137°K, is observed to be time-dependent, indicative of molecular relaxation.[429] In contrast in the benzene-TCNB, *o*-xylene-TCNB and *m*-xylene-TCNB systems, where the crystalline state is formed simply by cooling from the liquid state, the crystal fluorescence spectra are time-independent. It is proposed that the toluene-TCNB crystal system, being formed from the glassy state, contains many defects which allow molecular motion during the excited state lifetime.[429]

Masuhara and Mataga[430,431] have shown that giant pulse laser excitation of benzene-TCNB and toluene-TCNB generates a high concentration of excited DA complexes, which partially absorb the DA complex fluorescence. In the toluene-TCNB DA complex at 300°K, the $E_1 - E_p$ absorption has a maximum at 470 nm, in the region of the structureless $E_1 - E_0$ fluorescence band. Comparison of the laser-excited and normal fluorescence spectra enables the $E_1 - E_p$ absorption spectrum to be determined, without the external spectroscopic source used in conventional flash photolysis. The $E_1 - E_p$ absorption spectrum is similar to that of the TCNB anion. Similar results are obtained for excited DA complexes of other alkyl benzenes with TCNB.[511,512]

Masuhara *et al.*[432] observed the absorption and emission spectra of the fluorescent (E'_1) and phosphorescent (X'_1) states of the toluene-TCNB DA complex in rigid propanol and PMMA solutions at 77°K. They measured the rise and decay curves of the two states using a Q-switched frequency-doubled ruby laser for excitation. Similar rapid risetimes were observed for the $E'_1 - E_0$ fluorescence, the $E'_1 - E'_p$ absorption, the $X'_1 - E_0$ phosphorescence and the $X'_1 - X'_q$ absorption. Similar fast risetimes were observed for the $X'_1 - X'_q$ absorption[433] of the DA complexes of mesitylene, hexamethylbenzene and naphthalene with TCNB.[434]

If the triplet state X'_1 were formed only by intersystem crossing from E'_1, its risetime τ_R would correspond to the E'_1 decay time τ_E (p. *195*). To account for their observations of $\tau_R \ll \tau_E$, Masuhara *et al.*[432] postulated an initial short-lived Franck–Condon state E'_{FC} from which E'_1 and X'_1 are formed rapidly and independently. This hypothesis presents several difficulties.

(i) It contradicts the earlier model in which the fluorescent state E'_1 in rigid systems is equated to the Franck–Condon state.

(ii) The time-independent fluorescence spectra observed in rigid systems[428,429] show that E'_1 remains unchanged during its lifetime τ_E, and that it does not relax to the equilibrium state E_1 observed in fluid solutions. A rapid preceding relaxation from E'_{FC} to E'_1 is hardly consistent with this behaviour.

(iii) CT interaction facilitates intersystem crossing (p. *437*). The E'_1 radiationless transition rate k_{IE} ($= 9 \times 10^6$ s^{-1} for toluene-TCNB at 77°K) is probably mainly due to $E'_1 - X'_1$ intersystem crossing, a process neglected in the Masuhara hypothesis.[432] The latter only considers $E'_{FC} - X'_1$ intersystem crossing, which needs to occur at a rate $\gg k_{IE}$ to compete with the rapid $E'_{FC} - E'_1$ transition.

To circumvent these difficulties, it is suggested that the observed behaviour ($\tau_R \ll \tau_E$) could be due to direct excitation of X'_1 by the laser pulse. Observations of the phosphorescence risetime at normal light intensities would provide a simple test and establish the validity or otherwise of the postulated E'_{FC} state.

Potashnik and Ottolenghi[435] observed the excited singlet–singlet ($E_1 - E_p$) absorption spectra of DA complexes at room temperature using a pulsed nitrogen (337·1 nm) laser for excitation. In the systems studied, toluene-TCNB, benzene-TCNB and mesitylene-PMDA (pyromellitic anhydride), the excitation is absorbed only by the DA complexes, and not by D or A. The lifetime of the transient absorption is the same as that of the CT fluorescence τ_E. In each case the $E_1 - E_p$ absorption spectrum corresponds to that of the acceptor anion, $(TCNB)^-$ or $(PMDA)^-$, showing that the $E_1 - E_p$ transition is dissociative.

The ionic dissociation of a DA complex depends on the nature of the complex. DA complexes of aromatic hydrocarbons with TCNB dissociate into ions in the excited singlet (E_1) state.[436,437] DA complexes of mesitylene and hexamethylbenzene with PMDA, and of several donors with TCNE dissociate into ions in the excited triplet (X_1) state.[435,437–440,442] The ionic dissociation also depends on the solvent. For the mesitylene-PMDA complex in nonpolar solvents ionic dissociation occurs in the triplet (X_1) state, while in polar solvents it occurs in the singlet (E_1) state.[435,437,441] From observations of the rise and decay of the laser-excited fluorescence and photocurrent of the benzene-TCNB DA complex in 1,2-dichloroethane Shimada *et al.*[441] concluded that the ionic dissociation occurs from the excited Franck–Condon (E'_1) state. Achiba *et al.*[442] have made flash spectroscopic and kinetic studies of the tetrahydrofuran-TCNE DA complex in liquid paraffin at room temperature, and concluded that the ionic dissociation of the complex occurs from its excited triplet (X_1) state.

Observations of the triplet–triplet absorption spectra of DA complexes of toluene, durene and hexamethylbenzene with TCNB in rigid solution at 77°K have been made by Matsumoto *et al.*[443] Two $X'_1 - X'_q$ transitions are observed, one in the visible, which shifts to higher frequencies with decreasing donor ionization potential I_D, and one in the infra-red, which is attributed to the $X'_1 - X'_2$ transition. Krebs *et al.*[444] have observed the ESR spectra of the triplet states of DA complexes of naphthalene, anthracene, phenanthrene, pyrene and triphenylene with TCNB in oriented liquid crystal glasses, and of a single crystal of naphthalene-TCNB. Torihashi *et al.*[445] have studied the influence of external pressure on the absorption and fluorescence of toluene-TCNB and mesitylene-TCNB DA complexes at room temperature. The external pressure causes a decrease in the fluorescence quantum yield (Φ_{FE}) and lifetime (τ_E) and a red-shift of the absorption and fluorescence spectra.

9.9.4 *Exciplexes*

Beens and Weller (Chapter 4, Vol. 2) have considered theoretical and experimental aspects of exciplexes and other excited molecular π-complexes in solution. Other work on exciplexes will be discussed here.

A comprehensive review (with 484 references) of the photochemical reactions of aromatic molecules which proceed via an exciplex intermediate has been presented by Lablanche-Combier.[446] Tavares[447] has observed the fluorescence of AD exciplexes of anthracene, pyrene, 1:2-benzanthracene, chrysene, triphenylene and perylene with *N,N*-dimethylaniline (DMA) in cyclohexane and *n*-hexane solutions.

An aromatic hydrocarbon in its singlet excited state $^1M^*$ can form three types of bimolecular exciplex (pp. *420–1*).

(i) With an electron donor (D), such as an amine, $^1M^*$ acts as an electron acceptor (A) to form an AD exciplex.

(ii) With an electron acceptor, such as a cyano-substituted aromatic hydrocarbon, $^1M^*$ acts as an electron donor to form a DA exciplex.

(iii) With a 'neutral' molecule of the same or closely related (e.g. alkyl-substituted) species, $^1M^*$ forms excimers or mixed excimers (p. *421*), which are intermediate between (i) and (ii).

Knight, McDonald and Selinger[448,513] have studied exciplex formation in non-polar cyclohexane solutions, with initial excitation of either the acceptor

$$A^* + D \rightarrow (AD)^* \tag{9.72a}$$

or the donor

$$A + D^* \rightarrow (AD)^* \tag{9.72b}$$

Two of the systems studied, A = naphthalene or biphenyl, D = diethylaniline (DEA), form AD exciplexes. The third system, D = naphthalene, A = 1-cyanonaphthalene, forms DA exciplexes. Knight and Selinger[448] compared the exciplex fluorescence response functions for the two modes of excitation, and the results are listed in Table 9.35. As expected, the time t_d for the (AD)* fluorescence intensity to reach a maximum depends on the lifetime τ of the initially excited state.

McDonald and Selinger[513] studied the fluorescence quenching, spectra and excitation spectra of the three systems as a function of concentration and temperature. They showed that (9.72b) is reversible, so that each system can be described by the equilibrium scheme

$$A + D^* \rightleftharpoons (AD)^* \rightleftharpoons A^* + D \tag{9.73}$$

Table 9.35 Exciplexes in cyclohexane solution. Excitation wavelength (λ_{ex}), fluorescence lifetime (τ) of initially excited species, and time (t_d) to reach maximum exciplex fluorescence intensity[448]

Acceptor	Donor	λ_{ex} (nm)	τ(ns)	t_d(ns)
Naphthalene[a] (2×10^{-2} M)	DEA (8×10^{-4} M)	285	119	36
Naphthalene (2×10^{-2} M)	DEA[a] (4×10^{-2} M)	316	2·8	~1
Biphenyl[a] (0·15 M)	DEA ($3·7 \times 10^{-3}$ M)	240	15	8
Biphenyl ($5·1 \times 10^{-2}$ M)	DEA[a] ($1·4 \times 10^{-2}$ M)	316	2·8	~1
1-Cyanonaphthalene[a] (10^{-3} M)	Naphthalene ($2·5 \times 10^{-3}$ M)	316	17	12
1-Cyanonaphthalene ($2·3 \times 10^{-4}$ M)	Naphthalene[a] (5×10^{-2} M)	285	119	58

[a] Initially excited species.
DEA = diethylaniline.

but they were unable to demonstrate the 'broken-line' step in (9.73), corresponding to the reversal of (9.72a).

In polar solvents, increase of the solvent polarity reduces the exciplex fluorescence quantum yield Φ_{FE} and lifetime τ_E, and it shifts the fluorescence maximum E_m to lower energies (pp. *428–30*). Similar changes are observed in the fluorescence properties of the anthracene-DEA exciplex in cyclohexane solution when excess DEA is added. Selinger and McDonald[514] have shown that this is not due to any concentration quenching effect, but to the increase in solvent polarity. In ethyl acetate, which has a similar dielectric constant to DEA, the exciplex fluorescence is unaffected by excess DEA since the solution polarity remains constant.

To account for the dependence of the exciplex fluorescence parameters on solvent polarity, Selinger and McDonald[514] proposed the initial formation of a 'quench complex', which either dissociates into solvated ions (or otherwise) or forms an exciplex, which fluoresces, dissociates, or dissipates its energy radiationlessly. Taniguchi *et al.*[449,450] concluded that ion formation in anthracene-DMA and pyrene-DMA exciplexes occurs mainly from the non-relaxed (E_1') exciplex state (corresponding to the quench complex) before relaxation to the equilibrium (E_1) exciplex state. Related models of exciplexes in polar solvents by Knibbe *et al.* (pp. *429–30*) and Mataga *et al.* (pp. *438–9*) have been described previously. McDonald and Selinger[515] have pointed out errors in the published enthalpy values of exciplexes in polar solvents due to the absence of unperturbed equilibrium conditions. Prochorow[599] has studied the effect of solvent polarity on the fluorescence spectra, quantum yields and lifetimes of DA complexes of PMDA with toluene, *o*-xylene, *p*-xylene, mesitylene and 1,2,4-trimethylbenzene. The behaviour is similar to that of exciplexes in polar solvents, and it is interpreted in a similar manner.

Goldschmidt, Ottolenghi and co-workers[451,452] have observed the absorption spectra of the exciplexes of naphthalene, anthracene, pyrene and biphenyl with DEA as a common donor in toluene solution at room temperature, using pulsed laser photolysis. The numerous bands appearing in the spectra are analysed and attributed to transitions to higher excited CT states, $(A^{-*}D^{+})$ and $(A^{-}D^{+*})$ and to locally excited states (A^*D) and (AD^*). Unlike the absorption spectra of excited DA complexes, the absorption spectra of AD exciplexes are not simply related to those of the component molecular ions.

Masuhara and Mataga[453] observed a reduction in the fluorescence lifetime τ_E of the anthracene-DMA exciplex in room-temperature solution, when a concentrated solution is excited with a high-intensity laser pulse. A kinetic analysis shows that the quenching is due to a bimolecular interaction between the exciplexes. A similar self-quenching effect at high light intensities is observed with pyrene excimers.[453,454]

9.9.5 *Intramolecular exciplexes*

Intramolecular exciplex formation in the naphthyalkylamines[455] has been described by Klöpffer (Chapter 7, Vol. 1). Related studies on the (9-anthryl)—$(CH_2)_n$—(*p*-N,N-dimethylaminophenyl) molecular system, with $n = 0, 1, 2$ and 3, have been made by Okada *et al.*[456] The absorption spectra of the $n = 1, 2$ and 3 compounds are almost identical to that of an equimolar mixture of 9-methyl anthracene and *N*,*N*-dimethyl-*p*-toluidine, showing that there is negligible ground state interaction. The absorption spectrum of the $n = 0$ compound is more diffuse and extends to longer wavelengths, indicative of weak CT interaction. The fluorescence spectra of all the compounds in polar solvents include broad structureless bands, attributed to intramolecular AD exciplexes, which are shifted to lower energies on increase of the solvent polarity. Excited state dipole moments of 18, 27, 25 and 15 D are estimated from the solvent shifts for the $n = 0, 1$, 2 and 3 compounds, respectively. The $S_1 - S_p$ absorption spectra of the $n = 1$, 2 and 3 compounds in moderately polar solvents are similar to those of the anthracene anion, confirming the polar nature of the excited state.

The fluorescence spectra of the $n = 1, 2$ and 3 compounds also include a structured anthracene-like emission, which is absent from the $n = 0$ compound. The structureless exciplex component is absent from the fluorescence spectra of the $n = 1$ and 2 compounds in non-polar solvents. Table 9.36 presents data on the lifetimes (τ_A, τ_E) and relative quantum yields (Q_A, Q_E) of the structured and structureless fluorescence emissions, respectively, for the four compounds in different solvents at room temperature.[456]

It is interesting to compare the results with the corresponding data on the intermolecular exciplexes of anthracene with DEA[457] (pp. *429–30*) and with DMA[458,459] (pp. *430*, *479*). For the anthracene-DEA exciplex Q_E decreases monotonically by a factor of ~ 100, while τ_E decreases by a factor of ~ 5, on increasing ε from 2 to 15. This corresponds to a steady decrease in k_{FE}, the exciplex radiative rate parameter, and a steady increase in k_{IE}, its radiationless rate parameter, with increase in ε. This behaviour is attributed to the formation of solvated ion pairs ($A_s^- \ldots D_s^+$) in polar solvents.[457] The increase in k_{IE} is attributed to spontaneous reversal of the electron transfer,[457] and the decrease in k_{FE} to the increased $A^- - D^+$ separation (pp. *438–9*).

The behaviour of the anthracene-DMA intramolecular exciplexes with $n = 1$, 2 and 3 (Table 9.36) is closely related to that of the inter-

Table 9.36 (Anthracene)-$(CH_2)_n$-(DMA). Lifetimes (ns) and relative quantum yields of intramolecular exciplex (τ_E, Q_E) and anthracene-like (τ_A, Q_A) fluorescence in solvents of dielectric constant (ε) at room temperature[456]

Solvent	ε	$n = 0$		$n = 1$				$n = 2$				$n = 3$			
		τ_E	Q_E	τ_A	Q_A	τ_E	Q_E	τ_A	Q_A	τ_E	Q_E	τ_A	Q_A	τ_E	Q_E
n-Hexane	1·9	6	1·00	8	1·00	—	—	10	1·00	—	—	5·5	1·00	140	2·09
Decalin	2·3	6	1·69	9	0·77	—	—	12	1·2	—	—	6	1·91	120	1·63
Butyl ether	3·1	7	1·85	10	0·06	11	0·09	50	0·8	75	0·8	6	1·06	130	1·50
Ethyl ether	4·3	9	1·93	20	0·01	17	0·03	25	0·2	130	0·29	6	0·24	150	1·42
*iso*Butyl acetate	5·3	12	2·44	11	0·001	12	0·01	12	0·03	90	0·15	6	0·12	170	1·07
THF	7·4	8	—	15	0·002	15	0·004	14	0·03	35	0·16	10	0·03	120	0·73
*iso*Butanol	15·8	18	2·33	10	0·002	10	0·002	8	0·04	14	0·009	5·5	0·11	55	0·10
Acetone	20·7	25	1·89	10	0·002	10	0·003	10	0·02	8	0·007	10	0·13	10	0·02
Acetonitrile	37·5	32	1·24	12	0·001	8	0·001	12	0·015	—	0·001	6	0·05	10	~0

molecular exciplexes, but it is modified by the steric restraint of the $(CH_2)_n$ alkane chain. The properties of the $n = 3$ compound are most similar to those of the intermolecular exciplex, consistent with the $n = 3$ rule found for intramolecular excimer formation in the diphenylalkanes (p. *324*). The $n = 1$ and 2 compounds do not form intramolecular exciplexes in non-polar solvents due to steric restraint. This steric restraint is still present in polar solvents, but the solvent molecules provide a polarizable 'link' between A and D, leading to the formation of intramolecular solvated ion pairs ($A_s^- \ldots D_s^+$). In the $n = 0$ compound A and D are adjacent, and due to the strong CT interaction no anthracene-like fluorescence is observed. As expected, τ_E and Q_E are relatively insensitive to the solvent polarity. The solvated ion pair in the $n = 2$ compound appears to be similar to the exciplex in the $n = 3$ compound, and the solvated ion pair in the $n = 1$ compound appears to be similar to the CT state of the $n = 0$ compound.

Intramolecular exciplex formation has also been observed in the following compounds.

(a) $(CH_3)_2$—N—C_6H_4—$(CH_2)_n$—Ar

with Ar = 9-anthryl, $n = 0, 1, 2, 3$[456,516]
Ar = 1-pyrenyl, $n = 1, 2, 3$[516]
Ar = 2-anthryl, $n = 3$[516]

(b) $(CH_2)_nNR_2$ (1-naphthyl) and (2-naphthyl)—$(CH_2)_nNR_2$

Exciplexes are formed only if $n \leq 3$[455,517,518]

(c) $CH_2CO_2CH_2CH_2N(R)CH_3$ (1-naphthyl)

with R = CH_3 or C_6H_5.[518]

(d) $(CH_2)_3NEt_2$ (9-anthryl)

which is related to the type (a) compounds.[517,518]

9.9.6 *Chemiluminescence* (p. *433*)

Weller and Zachariasse[460–465] have made an extensive study of the chemiluminescence resulting from the interaction of solvated aromatic molecular anions (A_s^-) and solvated donor cations (D_s^+). The free enthalpy of the solvated ion pair ($A_s^- \ldots D_s^+$) is

$$\Delta G = E_{\frac{1}{2}}(D/D^+) + E_{\frac{1}{2}}(A^-/A) + \Delta\Delta G_s \qquad (9.74)$$

$E_{\frac{1}{2}}(D/D^+)$ and $E_{\frac{1}{2}}(A^-/A)$ are the polarographic oxidation and reduction potentials of the donor D and acceptor A, respectively, as measured against a saturated calomel electrode in acetonitrile or dimethylformamide ($\varepsilon = 37$) solution. $\Delta\Delta G_s$ is the difference in solvation energy in another solvent. For the ether-type solvents ($\varepsilon = 7$) used in these studies, $\Delta\Delta G_s = 0{\cdot}20$ eV.

In the initial studies the radical cation of *N*,*N*,*N*′,*N*′-tetramethyl-*p*-phenylenediamine (TMPD), known as Wurster's Blue ($TMPD^+$), was used as the donor cation D^+. Electron-transfer reactions between Wurster's Blue perchlorate ($D^+ClO_4^-$) and aromatic radical anions A^- were observed in tetrahydrofuran (THF) or 1,2-dimethoxyethane (DME) solution. For TMPD $E_{\frac{1}{2}}(D/D^+) = 0{\cdot}16$ eV, and the energies of its first excited triplet and singlet states are $E(^3D^*) = 2{\cdot}83$ eV and $E(^1D^*) = 3{\cdot}52$ eV, respectively.

The chemiluminescence (CL) spectra[465] show four distinct types of behaviour.

(I) The CL emission consists of acceptor $^1A^*$ fluorescence, sometimes accompanied by acceptor excimer $^1(A.A)^*$ fluorescence, but with no exciplex $^1(A^-D^+)$ fluorescence (Table 9.37).

(II) No emission is observed (Table 9.38).

(III) The CL emission consists of $^1(A^-D^+)$, $^1A^*$ and $^1D^*$ fluorescence (Table 9.39).

(IV) The CL spectrum includes emissions not originating from $^1A^*$, $^1(A.A)^*$, $^1D^*$ or $^1(A^-D^+)$.

To obtain chemiluminescence one or more excited states of the system must be energetically accessible from the solvated ion-pair ($A_s^- \ldots D_s^+$) state, i.e. they must have an energy $E \leqslant \Delta G$.

For Type I behaviour (Table 9.37)

$$E(^3A^*) \leqslant \Delta G < E(^3D^*), E(^1A^*), E(^1D^*) \qquad (9.75)$$

Table 9.37 Chemiluminescent (CL) electron-transfer reactions between Wurster's Blue perchlorate ($D^+ ClO_4^-$) and radical anions (A^-) in tetrahydrofuran (THF) or 1,2-dimethoxyethane (DME) solution[465]

No.	A	ΔG (eV)	$E(^3A^*)$ (eV)	$E(^1A^*)$ (eV)	CL emitter
1.	1-Phenylnaphthalene	2·76	2·55	3·85	$^1A^*$
2.	8,8′-Dimethyl-1,1′-binaphthyl	>2·69	2·48		$^1A^*$
3.	1,1′-Binaphthyl	2·69	2·53	3·69	$^1A^*$, $^1Pe^*$
4.	*p*-Terphenyl	2·67	2·55	3·97	$^1A^*$
5.	1,2-Dimethylchrysene	>2·66	2·39	3·39	$^1A^*$ ($^3A^*$?)
6.	Chrysene	2·66	2·49	3·43	$^1A^*$ ($^3A^*$?)
7.	Picene	2·64	2·49	3·30	$^1A^*$
8.	3:4-Benzophenanthrene	2·63	2·47	3·35	$^1A^*$ ($^3A^*$?)
9.	3:4:5:6-Dibenzo-phenanthrene	2·61	2·45	3·14	$^1A^*$ ($^3A^*$?)
10.	2,2′-Binaphthyl	2·53	2·42	3·72	$^1A^*$
11.	*p*-Quaterphenyl	2·53		3·72	$^1A^*$
12.	Pyrene	2·46	2·08	3·33	$^1A^*$ $^1(A.A)^*$
13.	1:2:3:4:5:6:7:8-Tetra-benzonaphthalene	2·45	2·14	3·16	$^1A^*$ ($^3A^*$?)
14.	1:2-Benzopyrene	2·44	2·31	3·37	$^1A^*$ $^1(A.A)^*$ ($^3A^*$?)
15.	1:2:3:4-Dibenzanthracene	2·40	2·21	3·31	$^1A^*$ $^1(A.A)^*$
16.	Coronene	2·40	2·37	2·95	$^1A^*$
17.	9,10-Dimethylanthracene	2·39	1·80	3·10	$^1A^*$ $^1(A.A)^*$
18.	1:2:5:6-Dibenzanthracene	2·39	2·27	3·14	$^1A^*$ ($^3A^*$?)
19.	1:2-Benzanthracene	2·38	2·05	3·17	$^1A^*$ $^1(A.A)^*$ ($^3A^*$?)
20.	1,4,5,8-Tetraphenyl-naphthalene	2·34	2·19		$^1A^*$
21.	9-Methylanthracene	2·34	1·81	3·19	$^1A^*$ $^1(A.A)^*$
22.	9,9′-Bianthryl	2·33	1·81		$^1A^*$
23.	Anthracene	2·32	1·82	3·28	$^1A^*$
24.	3:4-Benzotetraphene	2·30		3·17	$^1A^*$ ($^3A^*$?)
25.	9-Phenylanthracene	2·30	1·81	3·13	$^1A^*$ $^1(A.A)^*$
26.	9,10-Diphenylanthracene	2·28	1·77	3·16	$^1A^*$
27.	1:12-Benzoperylene	2·28	2·00	3·10	$^1A^*$ $^1(A.A)^*$
28.	9,10-Di-*o*-naphthyl-anthracene	2·26	1·77	3·16	$^1A^*$
29.	1,3,4,7-Tetraphenyl-*iso*benzofuran	2·26		2·70	$^1A^*$
30.	3:4-Benzopyrene	2·21	1·82	3·07	$^1A^*$ $^1(A.A)^*$
31.	9-Cyanoanthracene	2·06	1·82		$^1A^*$
32.	Perylene	2·01	1·56	2·85	$^1A^*$
33.	Tetracene	1·94	1·3	2·60	$^1A^*$
34.	Rubrene	1·83	1·2	2·39	$^1A^*$

Table 9.38 Non-chemiluminescent systems with Wurster's Blue perchlorate in THF or DME solution[465]

No.	A	ΔG (eV)	$E(^3A^*)$ (eV)	$E(^1A^*)$ (eV)	Reason for no CL
35.	Tetramethylpyrazine	2·82	3·23		(i) $\Delta G < E(^3A^*)$
36.	*p*-Dimethylamino-benzonitrile	2·76	2·96		(i)
37.	Benzonitrile	2·70	3·33	4·53	(i)
38.	*p,p'*-Dimethyldiphenyl-acetylene	2·58	2·71		(i)
39.	*iso*-Quinoline	2·56	2·63	3·97	(i)
40.	Quinoline	2·50	2·71	3·95	(i)
41.	Pyrazine	2·48	3·30	3·79	(i)
42.	Diphenylacetylene	2·47	2·71		(i)
43.	Phthalazine	2·33	2·85	3·79	(i)
44.	1-Cyanonaphthalene	2·21	2·49		(i)
45.	Fluoranthene	2·13	2·29	3·14	(i)
46.	Benzophenone	2·07	3·01	3·27	(i)
47.	2,3,6-Trimethyl*para*-dicyanobenzene	2·06	2·92		(i)
48.	3:4-Benzacridine	2·05	2·16	3·20	(i)
49.	Quinoxaline	2·00	2·63	3·20	(i)
50.	Acridine	1·94	1·96	3·19	(i)
51.	*p*-Dicyanobenzene	1·91	3·16	4·28	(i)
52.	Decacyclene	1·85	2·00	2·64	(i)
53.	Phenazine	1·51	1·94	2·78	(i)
54.	Pentacene	1·72	$\leqslant$0·99	2·12	(ii) $2E(^3A^*) < E(^1A^*)$
55.	*trans*-Stilbene	2·55	2·20		(iii) $^3A^*$ lifetime too short
56.	*cis*-Stilbene	2·55	2·48		(iii)
57.	Tetraphenylethylene	2·41	>2·05		(iii)
58.	Biphenylene	2·59	2·36	3·16	(iv) $^1A^*$ fluorescence efficiency low
59.	Azulene	1·98	1·04	1·76(S_1) 3·51(S_2)	(iv) (ii)

The observed $^1A^*$ and $^1(A.A)^*$ fluorescences are due to the triplet homo-fusion processes[460,462,465]

$$^3A^* + {}^3A^* \rightarrow {}^1A^* + {}^1A \tag{9.76}$$

$$^3A^* + {}^3A^* \rightarrow {}^1(A.A)^* \tag{9.77}$$

For type II behaviour (Table 9.38), there are various reasons why no chemiluminescence is observed.

(i) $\Delta G < E(^3A^*)$, $E(^3D^*)$, so that no $^3A^*$ or $^3D^*$ states are generated, e.g. compounds 35–53.

Table 9.39 Chemiluminescent (CL) systems with Wurster's Blue perchlorate in THF or DME solution in which exciplex $^{1}(A^{-}D^{+})$ fluorescence occurs. $\bar{\nu}_C$ and $\bar{\nu}_E$ are the $^{1}(A^{-}D^{+})$ emission maxima observed in CL and fluorescence quenching experiments, respectively in THF, DME and hexane (H). All energies in eV[465]

No.	A	ΔG	$E(^{3}A^{*})$	CL emitters	$\bar{\nu}_C$(THF)	$\bar{\nu}_C$(DME)	$\bar{\nu}_E$(THF)	$\bar{\nu}_E$(DME)	$\bar{\nu}_E$(H)
60.	*N*-Ethylcarbazole	3·16	3·04	$^{1}(A^{-}D^{+})$, $^{1}A^{*}$, $^{1}D^{*}$	2·31	2·25		2·24	2,68
61.	2-Methylbiphenyl	3·13	3·01	$^{1}(A^{-}D^{+})$, $^{1}A^{*}$, $^{1}D^{*}$		2·21			2·68
62.	4.4′-Dimethylbiphenyl	3·13	2·95	$^{1}(A^{-}D^{+})$, $^{1}A^{*}$, $^{1}D^{*}$, $(^{3}D^{*}?)$	2·28	2·23	2·28	2·21	2·68
63.	4-Methylbiphenyl	3·07	2·98	$^{1}(A^{-}D^{+})$, $^{1}A^{*}$, $^{1}D^{*}$		2·18	2·23	2·18	2·63
64.	1,4,5,8-Tetramethylnaphthalene	3·06	2·42	$^{1}(A^{-}D^{+})$, $^{1}A^{*}$, $^{1}D^{*}$		2·20		2·20	2·66
65.	*N*-Phenylcarbazole (CL from reaction product)	3·06	3·05	$^{1}(A^{-}D^{+})$, $^{1}A^{*}$, $^{1}D^{*}$		2·23	2·26	2·21	2·65
66.	1,8-Dimethylnaphthalene	3·02	2·56	$^{1}(A^{-}D^{+})$, $^{1}A^{*}$, $^{1}D^{*}$	2·21	2·18	2·21		2·64
67.	Biphenyl	3·01	3·01	$^{1}(A^{-}D^{+})$, $^{1}A^{*}$, $^{1}D^{*}$	2·17	2·14	2·18	2·14	2·57
68.	2,6-Dimethylnaphthalene	2·98	2·56	$^{1}(A^{-}D^{+})$, $^{1}A^{*}$, $^{1}D^{*}$		2·15		2·15	2·63
69.	1,3-*bis*(1-Naphthyl)propane	2·96	2·57	$^{1}(A^{-}D^{+})$, $^{1}A^{*}$, $^{1}(A.A)^{*}$, $^{1}D^{*}$ (with A linked below A^{-} and below $^{1}A^{*}$)		2·15			
70.	*o*-Terphenyl	2·95	2·67	$^{1}(A^{-}D^{+})$, $^{1}A^{*}$, $^{1}D^{*}$		2·18		2·18	2·45
71.	Naphthalene	2·94	2·64	$^{1}(A^{-}D^{+})$, $^{1}A^{*}$, $^{1}D^{*}$	2·15	2·11	2·16	2·12	2·59
72.	3.6-Dimethylphenanthrene	2·93	2·64	$^{1}(A^{-}D^{+})$, $^{1}A^{*}$, $^{1}D^{*}$	2·20	2·15	2·21	2·14	2·53
73.	*m*-Terphenyl	2·87	2·79	$^{1}(A^{-}D^{+})$, $^{1}A^{*}$, $^{1}D^{*}$	2·08		2·08		2·38
74.	Triphenylene	2·87	2·89	$^{1}(A^{-}D^{+})$, $^{1}A^{*}$, $^{1}D^{*}$		2·08	2·12	2·06	2·45
75.	1-Methylphenanthrene	2·85	2·67	$^{1}(A^{-}D^{+})$, $^{1}A^{*}$, $^{1}D^{*}$			2·14		2·45
76.	Phenanthrene	2·84	2·69	$^{1}(A^{-}D^{+})$, $^{1}A^{*}$, $^{1}D^{*}$	2·14	2·06	2·14	2·08	2·44

(ii) $\Delta G > E(^3A^*) < \frac{1}{2}E(^1A^*)$, so that although $^3A^*$ is formed, it lacks sufficient energy to generate $^1A^*$ by triplet homofusion (9.76), e.g. compound 54.

(iii) The $^3A^*$ lifetime is very short, and its unimolecular decay competes strongly with triplet homofusion (9.76), e.g. compounds 55–57.

(iv) The $^1A^*$ fluorescence quantum efficiency is very low, e.g. compounds 58–59.

(v) The instability of A^- is considered to account for the absence of CL with $TMPD^+$ in the following compounds: 1-fluoronaphthalene, methoxynaphthalenes, pyridine, 2,5-diphenyloxazole, trifluoromethylbenzene and biacetyl.[465]

For type III behaviour (Table 9.39) the experimental condition is that $E_{\frac{1}{2}}(A^-/A) \leqslant 2{\cdot}46$ eV, i.e. that $\Delta G \geqslant 2{\cdot}82$ eV $= E(^3D^*)$, so that

$$E(^3A^*),\ E(^3D^*) \leqslant \Delta G < E(^1A^*),\ E(^1D^*) \qquad (9.78)$$

The observed $^1A^*$ and $^1D^*$ fluorescences are attributed to $^3A^* - {}^3A^*$ homofusion (9.76) and to

$$^3D^* + {}^3D^* \rightarrow {}^1D^* + {}^1D \qquad (9.79)$$

The $^1(A^-D^+)$ fluorescence is attributed[461] to the triplet heterofusion process

$$^3A^* + {}^3D^* \rightarrow {}^1(A^-D^+) \qquad (9.80)$$

Formation of $^1(A^-D^+)$ directly from radical ions

$$^2A^- + {}^2D^+ \rightarrow {}^1(A^-D^+) \qquad (9.81)$$

is relatively inefficient in the Wurster's Blue systems.[465] The quantum yield of the $^1(A^-D^+)$ fluorescence, and its intensity relative to those of the $^1A^*$ and $^1D^*$ fluorescences, increases with ΔG.

Type IV behaviour, the appearance of emissions not due to $^1A^*$, $^1(A\,.\,A)^*$, $^1D^*$ or $^1(A^-D^+)$, has been observed in several systems.

(i) An emission in the region of the $^3A^*$ phosphorescence has been observed in chrysene, 1,2-dimethylchrysene, 1:2-benzanthracene, 3:4-benzophenanthrene, 1:2-benzopyrene, 1:2:5:6-dibenzanthracene, 3:4-benzotetraphene, 3:4:5:6-dibenzophenanthrene and 1:2:3:4:5:6:7:8-tetrabenzonaphthalene. It is uncertain whether the emission is due to $^3A^*$ phosphorescence or to an impurity.[465]

(ii) In the type III naphthalene$^-$/TMPD$^+$ and biphenyl$^-$/TMPB$^+$ systems, there is a structureless emission which is attributed to the $^3D^*$ phosphorescence of TMPD.[465]

(iii) In the Type I 1,1′-binaphthyl$^-$/TMPD$^+$ system there is a further emission attributed to perylene, formed by the ring closure of 1,1′-binaphthyl during the chemiluminescence experiment.

The CL quantum yield Φ_{CL} of aromatic anions with Wurster's Blue perchlorate ($TMPD^+ClO_4^-$) is relatively small ($\sim 10^{-6}$–10^{-4}). For aromatic anions with the radical cation of tri-*p*-tolylamine (TPTA) $\Phi_{CL} \simeq 10^{-2}$–10^{-1}.[463] Weller and Zachariasse[464] have studied the CL temperature dependence of anthracene, 9-methylanthracene and 9,10-dimethylanthracene anions with TPTA$^+$ in THF and MTHF solutions. Two CL components are observed, due to $^1A^*$ and $^1(A^-D^+)$ fluorescence, respectively. Unlike the TMPD$^+$ systems (Table 9.37), no $^1(A{\cdot}A)^*$ fluorescence is observed from 9-methyl or 9,10-dimethyl anthracene, indicating an alternative mechanism to triplet homofusion (9.76), (9.77). Weller and Zachariasse[464] propose the following mechanism

$$A^- + D^+ \rightarrow {}^1(A^-D^+) \rightleftharpoons {}^1A^* + D \qquad (9.82)$$
$$\begin{array}{cc} \downarrow \Phi_{FE} & \downarrow \Phi_{FA} \\ h\nu_E & h\nu_A \end{array}$$

as the main reaction pathway, with triplet homofusion (9.76) making a smaller contribution. From observations of Φ_{FA}/Φ_{FE} as a function of temperature, they determined values of $\Delta H_d = 9{\cdot}5$, $12{\cdot}5$ and 15 kcal/mole for the dissociation enthalpies of the $^1(A^-D^+)$ exciplexes formed by 9,10-dimethylanthracene, 9-methylanthracene and anthracene anions, respectively, with TPTA$^+$ in THF and MTHF.

The results are consistent with mechanism (9.82). Even in the anthracene$^-$/TPTA$^+$ system where a considerable part of the $^1A^*$ fluorescence is generated by triplet homofusion (9.76), over 95% of Φ_{CL} is due to $^1(A^-D^+)$ emission.[464] An external magnetic field modifies the relative intensities of the $^1A^*$, $^1D^*$ and $^1(A^-D^+)$ CL emissions of the 4,4′-dimethylbiphenyl$^-$/TMPD$^+$ system in DME, while it has no effect on the $^1A^*$ and $^1(A^-D^+)$ CL emissions of 9-methylanthracene$^-$/TPTA$^+$ in MTHF.[465] This is consistent with triplet fusion being dominant in the former case, and negligible in the latter case. Zachariasse[465] has also studied the CL emission of other aromatic radical anions with the radical cations of TPTA, tri-(*p*-dimethylaminophenyl) amine (TPDA), and tri-*p*-anisylamine (TPAA), and his thesis includes data similar to Tables 9.37–9.39 for the interactions with these donors.

9.9.7 *Impurity quenching of fluorescence* (pp. *433–47*)

The efficiency of the quenching of the fluorescence of anthracene and other aromatic molecules in solution by chlorinated alkanes (CCl_4, $CHCl_3$, etc.) depends on the excitation wavelength.[466–472] For the chlorinated alkanes the Stern–Volmer impurity quenching coefficient K_H with excitation of the aromatic molecule ^1M into a higher excited singlet state $^1M^{**}$ exceeds the corresponding coefficient K_M with excitation into the first excited singlet state $^1M^*$. In contrast, for the brominated alkanes (CBr_4, $CHBr_3$, C_2H_5Br, etc.) and the iodinated alkanes (C_2H_5I) it is observed that $K_H = K_M$.

The observations of Leite and Naqvi[468] of K_M and K_H for various systems are listed in Table 9.40. These authors adopted the exciplex model of

Table 9.40 Stern–Volmer coefficients of impurity quenching of fluorescence for excitation into $^1M^*$ (K_M) and $^1M^{**}$ (K_H)[468]

Hydrocarbon	Quencher	Solvent	K_M (M^{-1})	K_M (M^{-1})
Anthracene (4×10^{-4} M)	$CHCl_3$	Cyclohexane	0·07	0·26
	CCl_4	Cyclohexane	0·92	1·45
	CCl_4	Ethanol	14·3	15·9
	$CHBr_3$	Cyclohexane	45	45
	CBr_4	Cyclohexane	71	71
	C_2H_5Br	Cyclohexane	1·08	1·08
	C_2H_5I	Cyclohexane	18·8	18·8
	C_2H_5I	Liquid paraffin	2·98	2·98
Tetracene (2×10^{-4} M)	$CHCl_3$	Cyclohexane	0	+
	CCl_4	Cyclohexane	0·09	0·65
9,10-Diphenylanthracene (5×10^{-3} M)	$CHCl_3$	Cyclohexane	0	+
	CCl_4	Cyclohexane	2·7	2·95

dynamic quenching (p. *433*), which attributes the quenching to interaction between $^1M^{**}$ (or $^1M^*$) and the quencher molecule ^1Q to form an exciplex $^1E^*$. Assuming $^1E^*$ to be a DA exciplex in which the aromatic hydrocarbon acts as the electron donor, they estimated the $^1E^*$ energies from the ^1M ionization potentials and the ^1Q electron affinities. For the brominated and iodinated alkane exciplexes, they found the $^1E^*$ energy to lie below the $^1M^*$ energy, consistent with $K_H = K_M$. For the chlorinated alkane exciplexes they found the $^1E^*$ energy to lie between the $^1M^*$ and $^1M^{**}$ energies, and they proposed that $^1E^*$ formation from $^1M^*$ requires thermal activation energy, while that from $^1M^{**}$ does not, to account for $K_H > K_M$. Using an excitation bandwidth of 2–5 nm,

Leite and Naqvi[468] observed no wavelength dependence of fluorescence quenching within a given electronic absorption band system. They concluded that the vibrational relaxation rate in room-temperature solution greatly exceeds the rates of intramolecular radiationless transitions (internal conversion and/or intersystem crossing) and of impurity quenching, so that competition between these processes occurs at the lowest vibrational levels of $^1M^{**}$ or $^1M^*$.

If the model of Leite and Naqvi[468] were correct, it would provide a useful method of estimating the lifetimes of higher excited states of aromatic molecules. To first order, the Stern–Volmer coefficients are given by

$$K_M = k_{QM}\tau_M \tag{9.83}$$

$$K_H - K_M = k_{QH}\tau_H \tag{9.84}$$

where k_{QM} and k_{QH} are the impurity quenching rate parameters, and τ_M and τ_H are the lifetimes in the absence of the quencher, of $^1M^*$ and $^1M^{**}$, respectively. Applying these relations to the data of Table 9.40 for solutions of anthracene in cyclohexane for which $\tau_M = 5$ ns (p. *127*), we obtain $k_{QM}(CCl_4) = 1{\cdot}84 \times 10^8\ \text{M}^{-1}\ \text{s}^{-1}$ and $k_{QM}(CBr_4) = 1{\cdot}42 \times 10^{10}\ \text{M}^{-1}\ \text{s}^{-1}$. The latter value, which corresponds to the diffusion-controlled rate parameter, represents the upper limit for the quenching rate parameter in non-polar cyclohexane. Assuming therefore that $k_{QH}(CCl_4) \leqslant k_{QM}(CBr_4)$, we obtain from (9.84) $\tau_H \geqslant 38$ ps for the $^1M^{**}$ ($S_3 = {}^1B_b$) lifetime of anthracene. This lower limit value of τ_H is several orders of magnitude greater than the values of τ_H obtained from fluorescence and linewidth measurements for anthracene and similar molecules (Tables 9.11 and 9.12), indicating that this model of impurity quenching of $^1M^{**}$ is invalid.

An alternative model of chlorinated alkane quenching has been proposed by Lewis and Ware,[469,470] who studied the fluorescence quenching of cyclohexane solutions of anthracene by CCl_4 as a function of excitation wavelength, using an excitation bandwidth of 0·4 nm. Typical Stern–Volmer plots obtained at different excitation wavelengths are shown in Figure 9.17. Excitation at other wavelengths within the anthracene $S_0 - S_1$ absorption band ($^1M^*$ excitation) yielded plots identical with that at 365 nm. The quenching efficiency increases at 260 nm, near the onset of the $S_0 - S_3$ absorption band ($^1M^{**}$ excitation), and it then *decreases* with decrease of excitation wavelength to 258 and 254 nm (Figure 9.17). This behaviour is not compatible with the model of Leite and Naqvi,[468] nor with their observations made using a broader excitation bandwidth of 2–5 nm.

Lewis and Ware[469,470] proposed that the wavelength dependence of the fluorescence quenching is due to competitive absorption between the

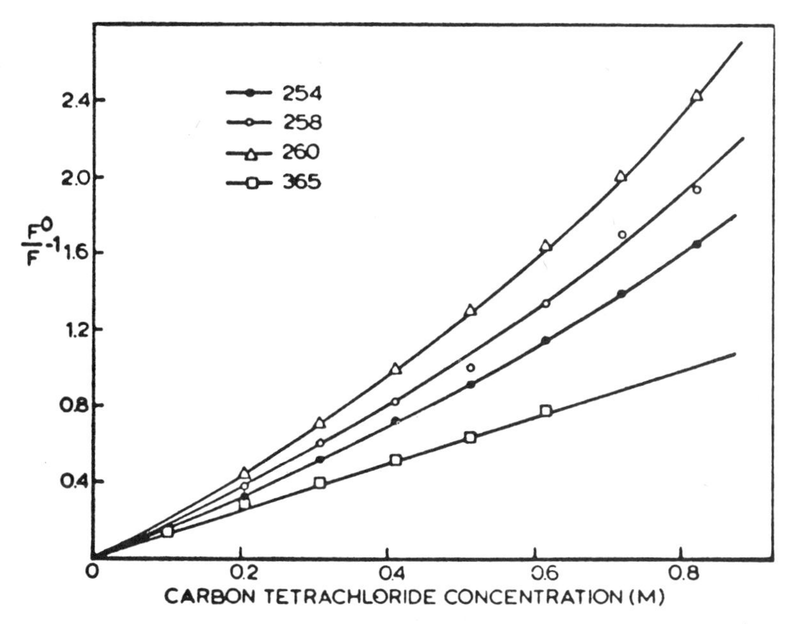

Figure 9.17 Stern–Volmer plots of fluorescence quenching of 8×10^{-4} M solution of anthracene in cyclohexane by carbon tetrachloride at different excitation wavelengths (nm). $(F^0/F - 1)$ against CCl_4 concentration [Q], where F is fluorescence intensity and $F = F^0$ at [Q] = 0. (Lewis and Ware, *Chem. Phys. Letters*, **15**, 290 (1972))

anthracene molecules and anthracene-CCl_4 contact DA complexes. From a comparison of the absorption spectra of cyclohexane solutions of anthracene in the absence and presence of CCl_4, they showed that the CT absorption spectrum of the DA complex originates at about 280 nm and increases in the vicinity of the anthracene $S_0 - S_3$ absorption edge. After testing various alternative reaction schemes, Ware and Lewis[470] concluded that their results are only consistent with the simultaneous static quenching (DA complex formation and excitation) and dynamic quenching (exciplex formation) model, described in *PAM* (pp. *442–3*). The wavelength dependence enters through the parameter K, which is proportional to ε'/ε, where ε' and ε are the molar extinction coefficients of the DA complex and ^1M, respectively.

The fluorescence response function of an excited molecular system $^1M^*$ subject to impurity quenching is

$$i_M(t) = k_{FM} \exp\{-(k_M + k_{QM}[^1Q])t\} \tag{9.85}$$

where k_{FM} is the $^1M^*$ radiative decay rate, k_M is the $^1M^*$ decay rate in the absence of ^1Q, and $[^1Q]$ is the molar concentration of ^1Q. From the Smoluchowski relation for a diffusion-controlled process (p. *445*; Chapter 8, Vol. 1), the rate parameter k_{QM} of diffusion-controlled impurity quenching is time-dependent and is given by

$$k_{QM} = 4\pi \mathrm{N}' D R' \left(1 + \frac{R'}{\sqrt{(\pi D t)}}\right) \tag{9.86}$$

where N′ is Avogadro's number per millimole, D is the sum of the diffusion coefficients of $^1M^*$ and ^1Q, and R' (= pR) is their effective interaction distance. Combining (9.85) and (9.86) we obtain

$$i_M(t) = k_{FM} \exp(-At - Bt^{\frac{1}{2}}) \tag{9.87}$$

with

$$A = k_M + 4\pi \mathrm{N}' D R' [^1Q] \tag{9.88}$$

$$B = 4\pi^{\frac{1}{2}} \mathrm{N}' D^{\frac{1}{2}} R'^2 [^1Q] \tag{9.89}$$

Ware and Nemzek[471] have made the first direct observations of the non-exponential decay of a quenched fluorescent solution, and they have verified relation (9.87). Solutions of 1:2-benzanthracene in propane-1,2-diol, quenched by CBr_4, were chosen for study between 10 and 35°C, because CBr_4 is a very strong ($p = 1$) quencher, accurate diffusion coefficients for propane-1,2-diol are available[472] and estimates of A and B indicated that non-exponential decay should be observable. No reasonable fit to the observed decay curves $f_M(t)$ was obtained by convolution (p. *95*) of the instrumental lamp function $I(t)$ with simple exponential decay functions, but convolution with fluorescence response functions $i_M(t)$ of the form of (9.87) yielded excellent fits with $f_M(t)$ over several decades. Such fits were obtained at all temperatures from 10 to 35°C, taking different values for the parameters A and B. A new deconvolution procedure[473] was used to check the results.

The values of R' and D, determined from the experimental values of A and B for a CBr_4 concentration of 0·18 M using (9.88) and (9.89), are listed in Table 9.41. These values do not fit the corresponding steady-state concentration quenching curves. To do so it is necessary to postulate

Table 9.41 1:2-Benzanthracene solutions in propane-1,2-diol quenched by CBr_4. Values of effective interaction distance (R') and total diffusion coefficient (D) obtained from fluorescence decay curves[471]

Temperature (°C)	15	20	25	30	35
R' (Å)	9·0	9·0	9·0	8·5	8·0
D (10^{-5} cm^2 s^{-1})	2·4	3·2	4·4	6·5	9·5

static quenching via a ground-state DA complex (p. *442*) with $K = 3{\cdot}6\ \text{M}^{-1}$ at 15°C. As discussed in *PAM* (pp. *443–6*), the CBr_4 quenching of anthracene fluorescence was described in terms of static quenching and time-independent dynamic quenching by Bowen and Metcalf,[474] and in terms of transient (time-dependent) and time-independent dynamic quenching by Ware and Novros.[475] The results of Ware and Nemzek[471] indicate that all three quenching processes, static, transient dynamic and time-independent dynamic, are operative.

Selinger and McDonald[514] observed the rate parameter k_{QM} for the fluorescence quenching of anthracene by *N*,*N*-diethylaniline (DEA) in seven solvents of different dielectric constant ε and viscosity η. They compared k_{QM} with the Stokes–Einstein diffusion-controlled rate parameter k_{SE} ($= 8\mathbf{R}T/2000\eta$). The values of $k_{QM}/k_{SE} = 0{\cdot}9 \pm 0{\cdot}2$ indicate the quenching is diffusion-controlled and independent of ε.

Okada *et al.*[476] have studied the fluorescence quenching of pyrene by *N*,*N*-dimethylaniline (DMA), tetracyanoethylene (TCNE), 1,2,4,5-tetracyanobenzene (TCNB), and the dicyanobenzenes (*o*-DCNB and *p*-DCNB). In agreement with previous studies (pp. *433–9*) they concluded that in non-polar and slightly polar solvents, the quenching is due to the formation of exciplexes, which may be fluorescent or non-fluorescent. In slightly polar solvents a solvated ion-pair (p. *429*) is formed via the exciplex. In more polar solvents they considered that direct electron transfer can occur between the fluorescent molecule and the quencher without exciplex formation, in agreement with Leonhardt and Weller,[477] who proposed that a solvated ion pair is the primary intermediate in the quenching process in polar solvents (p. *438*). The quenching of the fluorescence of pyrene in acetonitrile by TCNE is exceptionally high, and this and the high free energy $\Delta F_0 = -2{\cdot}5$ eV for the electron transfer process has led Mataga[478] to propose the possibility of four electron transfer channels in the pyrene-TCNE system corresponding to the ground and lower excited levels of the pyrene cation and anion radicals.

Davis and co-workers[479,480] have studied the quenching of anthracene fluorescence by various inorganic anions in acetonitrile and ethanol-water solutions. The anions exhibit CT absorption bands in solution due to DA complex formation with the solvent molecules. Correlations are observed between $\log_{10} k_{QM}$, where k_{QM} is the quenching rate parameter, and the anion redox potential and the anion-solvent CT absorption maximum. The data are interpreted in terms of exciplex formation. Brimage and Davidson[481] have studied the fluorescence quenching of naphthalene, anthracene and perylene by amino-alcohols and amines, in an attempt to assess the importance of hydrogen bond formation in electron transfer processes.

Vander Donckt and van Bellinghen[482] have studied the mechanism of the fluorescence quenching of anthracene solutions by dimethyl mercury, tetramethyl lead and tetramethyl tin, using fluorescence quantum yield and flash photolysis methods (pp. *197–8*). The results show that the fluorescence quenching by these organometallic compounds is entirely due to enhanced $S_1 - T_1$ intersystem crossing, corresponding to the sequence

$$^1M^* + {}^1Q \rightarrow {}^1(M{\cdot}Q)^* \rightarrow {}^3(M{\cdot}Q)^* \rightarrow {}^3M^* + {}^1Q \qquad (9.90)$$

as in the case of fluorescence quenching by xenon and by bromine and iodine compounds (pp. *436–7*) (see Wilkinson, Chapter 3, Vol. 2).

9.9.8 *Impurity quenching of triplet states* (pp. *447–51*)

The organometallic and other heavy-atom quenchers also enhance $T_1 - S_0$ intersystem crossing. Vander Donckt and van Bellinghen[482] observed that the S_1 quenching rate parameter k_{QM} is about 10^5–10^7 times the T_1 quenching rate parameter k_{QT} and that

$$k_{QM}/k_{QT} \simeq k_{TM}/k_{GT} \qquad (9.91)$$

where k_{TM} and k_{GT} are the rates of intersystem crossing from S_1 and T_1, respectively, in the absence of the heavy-atom quencher. In terms of the reaction scheme of Figure 9.18

$$k_{QM} = \frac{k_{EM}}{1 + k_{ME}/k_{XE}} \qquad (9.92)$$

$$k_{QT} = \frac{k_{XT}}{1 + k_{TX}/k_{GX}} \qquad (9.93)$$

where k_{EM} and k_{XT} are diffusion-controlled rate parameters of similar magnitude. The relative magnitudes of k_{QM} and k_{QT} are determined by k_{ME}/k_{XE} and k_{TX}/k_{GX}, the ratio of the dissociation and intersystem crossing rates of $^1(M{\cdot}Q)^*$ and $^3(M{\cdot}Q)^*$, respectively. $^1(M{\cdot}Q)^*$ is relatively stable with $k_{ME}/k_{XE} \simeq 0{\cdot}1$–$10$, giving $k_{QM} \simeq 0{\cdot}09$–$0{\cdot}9\ k_{EM}$. $^3(M{\cdot}Q)^*$ is very weakly bound with $k_{TX}/k_{GX} \simeq 10^5$–$10^7$, giving $k_{QT} \simeq 10^{-7}$–$10^{-5}\ k_{XT}$. The exciplex binding energy, which is unrelated to intersystem crossing, has such a strong influence on the quenching behaviour in fluid solutions, that there appears to be no theoretical basis for the empirical relation (9.91).

The external heavy-atom effect can be studied in the absence of diffusion and exciplex dissociation effects in rigid solutions (§9.6.3). Such studies on naphthalene (pp. *210–1, 266, 473*) show that an increase in the atomic number of the external heavy-atom causes only a small increase in k_{GT}, a larger increase in k_{PT} and a much larger increase in k_{TM} ($\propto k_{PT}^2$). Thus

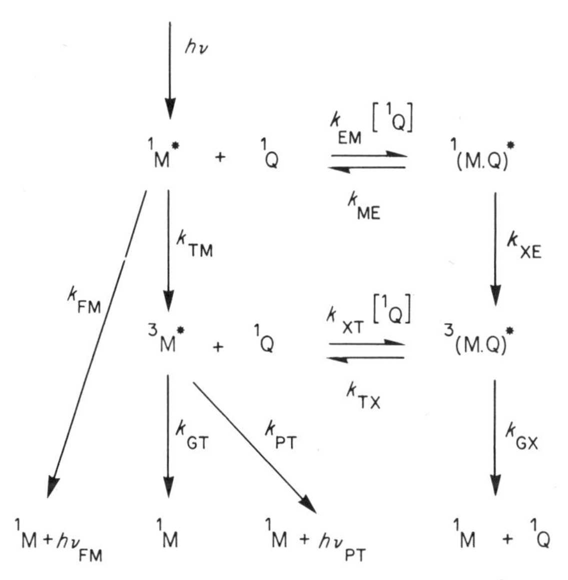

Figure 9.18 Reaction scheme for quenching of $^1M^*$ and $^3M^*$ by ^1Q

the heavy-atom 'quencher' reduces the triplet lifetime, but it *increases* the phosphorescence quantum efficiency and quantum yield.

Azumi[519] has considered the effect of external heavy-atom perturbation on each of the three spin subcomponents of the (π, π^*) triplet state. It is found that the radiative decay rate from the subcomponent, whose spin function is within the molecular plane, is most effectively enhanced. The main perturbing singlet state is the CT state arising from the excitation of an electron from a π orbital of the molecule to the σ^* orbital of the perturber.

9.10 Interactions with oxygen and nitric oxide

Reviews of the properties and interactions of singlet-excited molecular oxygen have been published.[483,484] The contact CT absorption of complexes of oxygen with benzene[415,416] and polystyrene[426] has been discussed in §9.9.2.

9.10.1 *Quenching by oxygen* (pp. *496–502*)

The oxygen quenching of the excited singlet state $^1M^*$ of an aromatic molecule is considered to involve the following sequence of spin-allowed processes.

$$^1M^* + {}^3O_2 \underset{k_{ME}}{\overset{k_{EM}(^3O_2)}{\rightleftharpoons}} {}^3(M{\cdot}O_2)^*_5 \tag{9.95a}$$

$$\downarrow k_{XE}$$

$$^3(M{\cdot}O_2)^*_4 \dashrightarrow {}^3M^{**} + {}^3O_2 \tag{9.95b}$$

$$\downarrow$$

$$^3(M{\cdot}O_2)^*_3 \dashrightarrow {}^3M^* + {}^1O_2^* \tag{9.95c}$$

$$\downarrow$$

$$^3(M{\cdot}O_2)^*_2 \dashrightarrow {}^2M^+ + {}^{2,4}O_2^- \tag{9.95d}$$

$$\downarrow$$

$$^3M^* + {}^3O_2 \xleftarrow{k_{TX}} {}^3(M{\cdot}O_2)^*_1 \tag{9.95e}$$

$$\downarrow k_{GX}$$

$$^1M + {}^3O_2 \tag{9.95f}$$

The more probable processes are indicated by solid arrows, and the less probable processes by broken arrows. The energies and sequence of the intermediate $^3(M{\cdot}O_2)^*_{2,3,4}$ states depend on the specific aromatic hydrocarbon ^1M. The CT state $^3(M{\cdot}O_2)^*_2$ and higher states of the complex are associated, so that internal conversion to a lower state of the complex is usually more probable than dissociation. In fluid solution $k_{XE} \gg k_{ME}$, so that the $^1M^*$ oxygen quenching rate parameter $k_{QM} \simeq k_{EM}$, the diffusion-controlled collisional rate parameter. The $^3(M{\cdot}O_2)^*_1$ state is only weakly associated, and in low-viscosity solutions $k_{TX} \gg k_{GX}$, so that the overall $^1M^*$ quenching process is

$$^1M^* + {}^3O_2 \rightarrow {}^3M^* + {}^3O_2 \tag{9.96a}$$

In very viscous or rigid solutions the $^3(M{\cdot}O_2)^*_1$ dissociation is inhibited, $k_{GX} \gg k_{TX}$ and the overall $^1M^*$ quenching process becomes

$$^1M^* + {}^3O_2 \rightarrow {}^1M + {}^3O_2 \tag{9.96b}$$

Potashnik *et al.*[485] studied the influence of oxygen on the fluorescence and triplet quantum yields of anthracene, phenanthrene, pyrene, chrysene, 1:2-benzanthracene, 3:4-benzopyrene, coronene and 1:2:5:6-dibenzanthracene in toluene and in acetonitrile solutions at room temperature using a pulsed nitrogen laser. They observed the decrease $\Delta\Phi_{FM}$ in the fluorescence quantum yield, and the increase $\Delta\Phi_{TM}$ in the triplet quantum yield, due to the presence of oxygen. For all the toluene solutions they observed that $\Delta\Phi_{FM} = \Delta\Phi_{TM}$, consistent with the overall $^1M^*$ quenching process (9.96a) (see Wilkinson, Chapter 3, Vol. 2, Table 3.2).

For the acetonitrile solutions the values of $\Delta\Phi_{FM}$ were similar to those for the toluene solutions, corresponding to similar values of k_{QM}, but the values of $\Delta\Phi_{TM}$ were generally lower, notably for phenanthrene, pyrene and 1:2-benzanthracene, so that $\Delta\Phi_{TM} \leqslant \Delta\Phi_{FM}$. The results indicate that an additional process, competitive with (9.96a), occurs in polar solvents.

In pyrene the reduction in $\Delta\Phi_{TM}$ is accompanied by the observation[485] of absorption due to the pyrene positive ion $^2M^+$. The behaviour can be explained in terms of the reaction scheme (9.95). In polar solvents solvation of the CT state $^3(M{\cdot}O_2)_2^*$ may occur at or above the energy of the CT state, leading to the formation of a solvent-shared ion pair ($^2M_s^+ \ldots {}^{2,4}O_{2s}^-$) (pp. *429–30*) and its dissociation into solvated ions $^2M_s^+$ and $^{2,4}O_{2s}^-$. The process

$$^1M^* + {}^3O_2 \rightarrow {}^3(M{\cdot}O_2)_2^* \rightarrow ({}^2M_s^+ \ldots {}^{2,4}O_{2s}^-) \rightarrow {}^2M_s^+ + {}^{2,4}O_{2s}^- \quad (9.97)$$

is competitive with (9.96a), but it does not increase the $^1M^*$ oxygen quenching rate parameter k_{QM}, which is diffusion-controlled.

The oxygen quenching of the excited triplet state $^3M^*$ of an aromatic molecule involves the following spin-allowed processes (pp. *500–1*),

$$^3M^* + {}^3O_2 \underset{k_{T1X}}{\overset{k_{1XT}[^3O_2]}{\rightleftharpoons}} {}^1(M{\cdot}O_2)^* \quad (9.98a)$$

$$^1(M{\cdot}O_2)^* \xrightarrow{k_{\Sigma 1X}} {}^1M + {}^1O_2^*({}^1\Sigma_g^+) \quad (9.98b)$$

$$^1(M{\cdot}O_2)^* \xrightarrow{k_{\Delta 1X}} {}^1M + {}^1O_2^*({}^1\Delta_g) \quad (9.98c)$$

$$^3M^* + {}^3O_2 \underset{k_{T3X}}{\overset{k_{3XT}[^3O_2]}{\rightleftharpoons}} {}^3(M{\cdot}O_2)^* \quad (9.98d)$$

$$^3(M{\cdot}O_2)^* \xrightarrow{k_{G3X}} {}^1M + {}^3O_2({}^3\Sigma_g^-) \quad (9.98e)$$

$$^3M^* + {}^3O_2 \underset{k_{T5X}}{\overset{k_{5XT}[^3O_2]}{\rightleftharpoons}} {}^5(M{\cdot}O_2)^* \quad (9.98f)$$

If there is no mixing of the $^{1,3,5}(M{\cdot}O_2)^*$ states, then

$$k_{1XT} = \tfrac{1}{9}k_{XT}; \quad k_{3XT} = \tfrac{1}{3}k_{XT}; \quad k_{5XT} = \tfrac{5}{9}k_{XT} \quad (9.99)$$

where k_{XT} is the total diffusion-controlled collisional rate parameter. The total $^3M^*$ oxygen quenching rate parameter is

$$k_{QT} = k_{XT}\left[\frac{k_{\Sigma 1X} + k_{\Delta 1X}}{9(k_{\Sigma 1X} + k_{\Delta 1X} + k_{T1X})} + \frac{k_{G3X}}{3(k_{G3X} + k_{T3X})}\right] \quad (9.100)$$

In low-viscosity solutions $k_{T3X} \gg k_{G3X}$ and this is confirmed by the $^1M^*$ oxygen quenching results[485] demonstrating the validity of (9.96a) for toluene solutions. Under these conditions (9.100) reduces to

$$k_{QT} = \frac{k_{XT}}{9(1 + \alpha)} \quad (9.101)$$

where

$$\alpha = \frac{k_{T1X}}{k_{\Sigma 1X} + k_{\Delta 1X}} \quad (9.102)$$

If the interaction distances for the diffusion-controlled interactions of $^3M^*$ and $^1M^*$ with $^3O_2$ are assumed to be the same, then we may put $k_{XT} = k_{EM}\,(= k_{QM})$.

Patterson *et al.*[486] determined k_{QM} and k_{QT} for the $^3O_2$ quenching of the $^1M^*$ and $^3M^*$ states, respectively, of nine aromatic hydrocarbons in cyclohexane solution at 25°C ($[^3O_2] = 2 \times 10^{-3}$ M[487]). Pulse-sampling fluorometry and laser flash photolysis were used to determine τ_M, $\tau_M(O_2)$, τ_T and $\tau_T(O_2)$, the $^1M^*$ and $^3M^*$ lifetimes in the absence and presence of oxygen, respectively.

The results, listed in Table 9.42, show that

(i) the oxygen quenching of $^1M^*$ is diffusion-controlled and has a rate parameter of $k_{QM} = 28\,(\pm 3) \times 10^9\ \text{M}^{-1}\,\text{s}^{-1}$;

(ii) the $^3M^*$ oxygen quenching rate parameter is

$$k_{QT} \leqslant k_{QM}/9 \tag{9.103}$$

(iii) k_{QT} decreases approximately linearly with increase in the $^3M^*$ energy E_T.

The experimental relation (9.103) agrees with the theoretical relation (9.101) and supports the assumptions on which it is based. Patterson *et al.*[486] attribute the observed decrease in k_{QT} (Table 9.42) with increase in E_T to a decrease in the Franck–Condon factors for processes (9.98b) and (9.98c) and a consequent decrease in $k_{\Sigma 1X}$ and $k_{\Delta 1X}$. The energies of the $^1\Sigma_g^+$ and $^1\Delta_g$ states of $^1O_2^*$ are 13,121 cm^{-1} and 7882 cm^{-1}. For anthracene and chrysene, the $^3M^* - {}^1\Sigma_g^+$ energy gaps $\Delta E_{T\Sigma}$ are 1580 cm^{-1} and 6680 cm^{-1}, respectively, and the $^3M^* - {}^1\Delta_g$ energy gaps $\Delta E_{T\Delta}$ are 6820 cm^{-1} and 11,920 cm^{-1}. From the Siebrand model (p. *148*; Henry and Siebrand, Chapter 4, Vol. 1) the increase in $\Delta E_{T\Sigma}$ reduces $k_{\Sigma 1X}$ by a factor of ~3, similar to the observed decrease[486] in k_{QT}, while the increase in $\Delta E_{T\Delta}$ reduces $k_{\Delta 1X}$ by a factor of > 10. These results support previous evidence that (9.98b) is the dominant $^3M^*$ oxygen quenching process.

Patterson *et al.*[486] assumed that k_{T1X} is independent of the aromatic hydrocarbon. k_{T1X} and k_{T3X} are determined by the $^1(M{\cdot}O_2)^*$ and $^3(M{\cdot}O_2)^*$ binding energies, which depend on the magnitude of the coupling to the corresponding higher CT states of energy E_{CT}, which is proportional to the inverse square of the energy gap $(E_{CT} - E_T)$. From (9.68), assuming $\Delta E_N = 0$, $\Delta E_E = 2{\cdot}9$ eV, $A_A = 0{\cdot}67$ eV, as for the benzene-O_2 complex, we obtain

$$E_{CT} = I_D - 3{\cdot}57\ \text{eV}$$

where I_D is the ionization potential of the aromatic hydrocarbon (pp. *457–61*). Values of $(E_{CT} - E_T)^{-2}$ between 0·214 eV^{-2} (1:2-benzanthracene)

Table 9.42 Oxygen quenching of excited singlet and triplet states of aromatic molecules in cyclohexane solution at 25°C [485]

Compound	E_T (cm^{-1})	$\tau_T(O_2)$ (ns)	k_{QT} (10^9 M^{-1} s^{-1})	τ_M (ns)	$\tau_M(O_2)$ (ns)	k_{QM} (10^9 M^{-1} s^{-1})	k_{QM}/k_{QT}
3:4:8:9-Dibenzopyrene	14100	170	2·94	143	16·5	27	9·2
Anthracene	14700	170	2·94	4·9	3·8	31	10·1
3:4-Benzopyrene	14800	190	2·63	57·5	13·0	29	11·0
1:2-Benzanthracene	16500	260	1·93	49·4	12·5	30	15·5
Pyrene	16800	310	1·61	450	20	25	15·5
1:2:3:4-Dibenzanthracene	17800	340	1·47	53·5	14·3	26	17·7
1:2:5:6-Dibenzanthracene	18300	390	1·28	37·5	12·0	29	22·7
Chrysene	19800	480	1·04	44·7	12·5	29	27·9
3:4-Benzophenanthrene	20600	490	1·02	70	—	—	—

and 0·32 eV^{-2} (chrysene) are obtained, corresponding to a 50% variation in the magnitude of the coupling. Such variations are expected to influence the magnitude of k_{T1X}.

Adamczyk and Wilkinson[488] used flash and laser photolysis to determine k_{QT} for the $^3O_2$ quenching of the $^3M^*$ states of perprotonated and perdeuterated naphthalene, anthracene and phenanthrene and of 9,10-dibromoanthracene in 95% ethanol solution at 23°C. The magnitude of k_{QT} is not affected by perdeuteration, and it is concluded that the radiationless transition responsible for the $^3M^*$ oxygen quenching involves a relatively small energy gap. The experimental values of k_{QT}, and the corresponding energy gaps $\Delta E_{T\Sigma}$, $\Delta E_{T\Delta}$ and E_T for processes (9.98b), (9.98c) and (9.98e), respectively, are listed in Table 9.43. Only $\Delta E_{T\Sigma}$ is

Table 9.43 Oxygen quenching of excited triplet states of aromatic hydrocarbons in 95% ethanol solution at 23°C [488] and energy gaps

Compound	k_{QT} (10^9 M^{-1} s^{-1})	$\Delta E_{T\Sigma}$ (cm^{-1})	$\Delta E_{T\Delta}$ (cm^{-1})	E_T (cm^{-1})
Naphthalene (h_8 and d_8)	2·5	8180	13420	21300
Anthracene (h_{10} and d_{10})	3·3	1580	6820	14700
Phenanthrene (h_{10} and d_{10})	4·2	8480	13720	21600
9,10-Dibromoanthracene	3·5	—	—	—

sufficiently small for the associated Franck–Condon factor to be insensitive to perdeuteration on the Siebrand model (p. *148*; Henry and Siebrand, Chapter 4, Vol. 1). The results thus indicate that (9.98b) is the dominant process in the compounds studied. The k_{QT} values of Table 9.43 are consistent with (9.101). The similar k_{QT} values for anthracene and 9,10-dibromoanthracene indicate that the $^3M^*$ oxygen quenching is spin-allowed, and uninfluenced by the internal heavy-atom effect.

In most aromatic hydrocarbons the $^3M^*$ energy exceeds the $^1O_2(^1\Sigma_g^+)$ energy, i.e. $\Delta E_{T\Sigma} > 0$, and process (9.98b) is observed to dominate. When $\Delta E_{T\Sigma} < 0$, $\Delta E_{T\Delta} > 0$, as for tetracene ($E_T = 10{,}300\ cm^{-1}$), process (9.98b) is energetically forbidden other than by thermal activation, and it is to be expected that process (9.98c) will dominate. When $\Delta E_{T\Sigma} < 0$, $\Delta E_{T\Delta} < 0$, as for pentacene ($E_T \simeq 6500\ cm^{-1}$), then both (9.98b) and (9.98c) are inhibited, leaving process (9.98e) as the sole quenching mode. Experimental data, currently lacking, on processes (9.98c) and (9.98e) could be obtained from studies of the oxygen quenching of the $^3M^*$ states of tetracene, pentacene and similar molecules.

Merkel and Kearns[489] observed the phosphorescence lifetimes of various carbonyl compounds in the absence (τ_T) and presence of dissolved oxygen in carbon tetrachloride solutions at 25°C ($[^3O_2] = 2{\cdot}5 \times 10^{-3}$ M), and they determined the oxygen $^3M^*$ quenching rate parameters k_{QT} (Table 9.44). The k_{QT} values lie between $\sim\frac{1}{10}$ and $\frac{1}{30}$ of k_{QM}, which is

Table 9.44 Carbonyl compounds in carbon tetrachloride solution at 25°C.[489] Triplet lifetime (τ_T) and oxygen triplet quenching rate parameter k_{QT}

Compound	$\tau_T(\mu s)$	$k_{QT}(10^9\ M^{-1}\ s^{-1})$
Acetophenone	24	2·8
Anthrone	260	1·2
Benzaldehyde	25	3·1
Benzil	320	1·4
Benzophenone	94	1·4
4-Bromobenzophenone	5	1·1
4,4′-Dibromobenzophenone	5	1·1
1-Indanone	8	3·8

estimated[489] as $\sim 30 \times 10^9$ M^{-1} s^{-1}. Although these k_{QT} values cover a similar range to those of the aromatic hydrocarbons (Table 9.42) and might be explained in a similar manner, the authors[489] propose that oxygen is less effective at quenching $^3(n, \pi^*)$ states than $^3(\pi, \pi^*)$ states. They note that k_{QT} is smallest for the carbonyl molecules which have the largest separation between their $^3(n, \pi^*)$ and $^3(\pi, \pi^*)$ states.

Kearns and Stone[490] have considered the effect of spin–spin interactions on k_{QT} for the quenching of $^3M^*$ by $^3O_2$. Such interactions could couple the $(M{\cdot}O_2)^*$ states of different multiplicity in (9.98). For the $(M{\cdot}O_2)^*$ complex the spin–spin interaction is of the form

$$\mathscr{H}_{ss} = D'[3S_{z'}^2 - S'(S'+1)] + D[3S_z^2 - S(S+1)] + E[S_x^2 - S_y^2] \quad (9.104)$$

where the unprimed quantities refer to $^3M^*$ (the x-, y- and z-axes are the principal axes of the $^3M^*$ zero-field splitting tensor) and the primed quantities refer to $^3O_2$. The matrix elements coupling $^1(M{\cdot}O_2)^*$ and $^3(M{\cdot}O_2)^*$ are found to be identically zero, but there are non-zero elements coupling $^5(M{\cdot}O_2)^*$ to the other states of the complex. The magnitude of such coupling depends on the zero-field splitting parameters D' (= 4 cm^{-1} for $^3O_2$[491]) and D ($\simeq$ 0·1–0·2 cm^{-1} for aromatic hydrocarbons[492]) and on the energy gap ΔE between $^5(M{\cdot}O_2)^*$ and the $^1(M{\cdot}O_2)^*$ and $^3(M{\cdot}O_2)^*$ states. The latter are lowered, relative to $^5(M{\cdot}O_2)^*$, by ΔE, previously estimated[493] as $\geqslant 10$ cm^{-1}, through interaction with higher energy states of the complex.

If $\Delta E \gg D, D'$ (Case I), the $^{1,3,5}(M{\cdot}O_2)^*$ states are uncoupled, and the reaction scheme (9.98) leads to relation (9.100) for k_{QT}. If $\Delta E \sim D, D'$

(Case II), there is strong mixing of the $^{1,3,5}(M{\cdot}O_2)^*$ states. Kearns and Stone[490] estimate an increased value of $k_{QT} \simeq 0{\cdot}55\, k_{XT}$, compared with $k_{QT} \simeq 0{\cdot}1\, k_{XT}$ for Case I. They implicitly assume a common $(M{\cdot}O_2)^*$ dissociation rate (k_{-D} in their notation), independent of the multiplicity and degree of mixing. Since k_{-D} ($\equiv k_{T1X}, k_{T3X}, k_{T5X}$) depends on the $(M{\cdot}O_2)^*$ binding energy, which is influenced by the multiplicity and the degree of mixing, the assumption of constant k_{-D} may invalidate their conclusions. All the experimental and theoretical evidence to date indicates that the $^3O_2$ quenching of $^3M^*$ corresponds to Case I, and that mixing of the $^{1,3,5}(M{\cdot}O_2)^*$ states is insignificant.

The $^3O_2$ quenching of $^3M^*$ (9.98) is a heterofusion process (Swenberg and Geacintov, Chapter 10, Vol. 1). Kearns and Stone[490] and Geacintov and Swenberg[494] have considered the possible influence of an external magnetic field on the $^3O_2$ quenching rate parameter k_{QT}. Because the oxygen zero-field splitting parameter D' ($= 4\,\text{cm}^{-1}$) is relatively high, very large magnetic fields must be used to quantize the spins along the external field. Geacintov and Swenberg[494] used fields of up to 145 kG, but they were unable to detect any influence on k_{QT}. To explain the null result they proposed that the $^{1,3}(M{\cdot}O_2)^*$ states (9.98) include significant components of the $^{1,3}(M^+O_2^-)$ states, which split the $^1(M{\cdot}O_2)^*$ and $^3(M{\cdot}O_2)^*$ states by an amount $\Delta E_{TS} > 30\,\text{cm}^{-1} \gg D' = 4\,\text{cm}^{-1}$.

9.10.2 *Quenching by nitric oxide* (pp. *504–8*)

The nitric oxide quenching of $^1M^*$ is considered to involve the following sequence of spin-allowed processes (pp. *504–5*),

$$^1M^* + {}^2NO \underset{k_{ME}}{\overset{k_{EM}[^2NO]}{\rightleftharpoons}} {}^2(M{\cdot}NO)_3^* \tag{9.105a}$$

$$\downarrow k_{XE}$$

$$^2(M{\cdot}NO)_2 \dashrightarrow {}^2M^+ + {}^{1,3}NO^- \tag{9.105b}$$

$$\downarrow$$

$$^3M^* + {}^2NO \xleftarrow{k_{TX}} {}^2(M{\cdot}NO)_1 \tag{9.105c}$$

$$\downarrow k_{GX}$$

$$^1M + {}^2NO \tag{9.105d}$$

Solid arrows indicate more probable processes, and broken arrows less probable processes. In normal fluid solutions $k_{XE} \gg k_{ME}$, so that the $^1M^*$ quenching rate parameter $k_{QM} = k_{EM}$. $^2(M{\cdot}NO)_1$ is only weakly associated and in low-viscosity solutions at room temperature it is expected that $k_{TX} \gg k_{GX}$, so that the overall process (9.105) is

$$^1M^* + {}^2NO \rightarrow {}^3M^* + {}^2NO \tag{9.106a}$$

analogous to (9.96a). In very viscous or rigid solutions $k_{GX} \gg k_{TX}$, and the overall $^1M^*$ quenching process becomes

$$^1M^* + ^2NO \rightarrow ^1M + ^2NO \tag{9.106b}$$

analogous to (9.96b). In polar solvents solvation of the $^2(M{\cdot}NO)_2^*$ CT state may occur, leading to dissociation into solvated ions, analogous to (9.97).

The NO quenching of $^3M^*$ involves the following spin-allowed processes (pp. *504–5*)

$$^3M^* + ^2NO \underset{k_{T2X}}{\overset{k_{2XT}[^2NO]}{\rightleftharpoons}} {}^2(M{\cdot}NO)^* \tag{9.107a}$$

$$^2(M{\cdot}NO)^* \xrightarrow{k_{G2X}} {}^1M + ^2NO \tag{9.107b}$$

$$^3M^* + ^2NO \underset{k_{T4X}}{\overset{k_{4XT}[^2NO]}{\rightleftharpoons}} {}^4(M{\cdot}NO)^* \tag{9.107c}$$

If there is no mixing of the $^{2,4}(M{\cdot}NO)^*$ states, then

$$k_{2XT} = \tfrac{1}{3}k_{XT}; \qquad k_{4XT} = \tfrac{2}{3}k_{XT} \tag{9.108}$$

where k_{XT} is the total diffusion-controlled rate parameter. Under these conditions the $^3M^*$ NO quenching rate parameter is

$$k_{QT} = \frac{k_{XT} \, . \, k_{G2X}}{3(k_{G2X} + k_{T2X})} \tag{9.109}$$

For $k_{T2X} \gg k_{G2X}$ (low-viscosity solutions)

$$k_{QT} = \frac{k_{XT}k_{G2X}}{3k_{T2X}} \tag{9.109a}$$

For $k_{G2X} \gg k_{T2X}$ (very viscous or rigid solutions)

$$k_{QT} = k_{XT}/3 \tag{9.109b}$$

k_{QT} is best determined from observations of the $^3M^*$ lifetimes τ_T $(= 1/k_T)$ and $(\tau_T)_Q$ in the absence and presence of [Q], respectively, where

$$k_{QT}[Q] = 1/(\tau_T)_Q - 1/\tau_T \tag{9.110}$$

Several observers[495–497] have also studied $^3M^*$ quenching by measurements of the phosphorescence quantum yields Φ_{PT} and $(\Phi_{PT})_Q$ in the absence and presence of [Q], respectively, but the results are more difficult to analyse. With initial $^1M^*$ excitation, $(\Phi_{PT})_Q$ depends not only on k_{QT}, but also on the $^1M^*$ quenching rate parameter k_{QM} and on its component

k'_{QM} ($\leqslant k_{QM}$) which yields $^3M^*$, so that

$$\frac{\Phi_{PT}}{(\Phi_{PT})_Q} = \left(1 + \frac{k_{QT}[Q]}{k_T}\right)\left(1 + \frac{k_{QM}[Q]}{k_M}\right)\left(1 + \frac{k'_{QM}[Q]}{k_{TM}}\right)^{-1} \quad (9.111)$$

$$\simeq 1 + \left(\frac{k_{QT}}{k_T} + \frac{k_{QM}}{k_M} - \frac{k'_{QM}}{k_{TM}}\right)[Q] = 1 + K[Q] \quad (9.112)$$

k_M is the $^1M^*$ decay rate and k_{TM} the $^1M^* - {}^3M^*$ intersystem crossing rate, in the absence of [Q].

Geacintov *et al.*[497] used an unconventional fluff-like polystyrene matrix with an aromatic hydrocarbon adsorbed on its surface to study the relative $^3M^*$ quenching probabilities by gaseous oxygen and nitric oxide at 25°C. Their observations of k_{QT}, obtained from τ_T, $(\tau_T)_Q$ and (9.110), and of K, obtained from Φ_{PT}, $(\Phi_{PT})_Q$ and (9.112), are listed in Table 9.45.

Table 9.45 Triplet energies (E_T) and quenching parameters (k_{QT}, K)[a] of aromatic hydrocarbons with oxygen and nitric oxide[497]

Quencher		O_2		NO	
Compound	E_T(cm^{-1})	k_{QT}	K	k_{QT}	K
Coronene	19400	0·056	0·99	—	~0·0032
Chrysene	19800	0·12	0·23	0·0058	0·011
1:2:5:6-Dibenzanthracene	18300	0·17	0·20	0·00089	0·0015
1:2-Benzanthracene	16500	0·34	—	—	—
Eight monomethyl 1:2-benzanthracenes	~16500	0·21–0·27 (depending on methyl position)	—	—	—

[a] k_{QT} in units of $\mu m^{-1}\,s^{-1}$ (1 μm = 10^{-3} torr); K in units of μm^{-1}.

It is found that $k_{QT}(O_2) \gg k_{QT}(NO)$, confirming previous observations[498] in fluid solutions, and providing further evidence for the importance of processes (9.98b) and (9.98c) in $^3M^*$ oxygen quenching. $k_{QT}(O_2)$ tends to decrease with increase in E_T in a similar manner to the data of Table 9.42.[485] The inverse behaviour of coronene cannot be explained in terms of Franck–Condon factors,[497] but k_{T1X} in (9.102) also depends on the aromatic molecule, as noted in §9.10.1. The increase of $k_{QT}(NO)$ with increase in E_T shows that parameters other than the Franck–Condon factors influence the behaviour.

Kearns and Stone[490] have considered the effect of spin–spin interaction on $k_{QT}(NO)$ in terms of the relative magnitude of the

^{4}(M·NO)* – ^{2}(M·NO)* splitting factor ΔE and the zero-field splitting parameter D. If $\Delta E \gg D$ (Case I) there is no spin–spin interaction, but Kearns and Stone[490] find that only two of the ^{4}(M·NO)* states have pure quartet character, the other two being strongly mixed with the ^{2}(M·NO)* states. They assume four equivalent (M·NO)* states, each with one-half doublet character (quenching rate $\frac{1}{2}k_{G2X}$) and a common dissociation rate k_{T2X} (= k_{-D} in their notation), and hence obtain

$$k_{QT} = \frac{2k_{XT} \cdot k_{G2X}}{3(k_{G2X} + 2k_{T2X})} \tag{9.113}$$

For $k_{T2X} \gg k_{G2X}$ (low-viscosity solutions) (9.113) reduces to (9.109a). For $k_{G2X} \gg k_{T2X}$ (very viscous or rigid solutions) (9.113) reduces to

$$k_{QT} = 2k_{XT}/3 \tag{9.113b}$$

which differs from (9.109b) by a factor of 2.

If $\Delta E \sim D$ (Case II) all six (M·NO)* states are mixed, and they are each assumed[490] to have one-third doublet character (quenching rate $\frac{1}{3}k_{G2X}$) and a common dissociation rate k_{T2X}, giving

$$k_{QT} = \frac{k_{XT} \cdot k_{G2X}}{k_{G2X} + 3k_{T2X}} \tag{9.114}$$

For $k_{T2X} \gg k_{G2X}$, (9.114) again reduces to (9.109a). For $k_{G2X} \gg k_{T2X}$, it reduces to

$$k_{QT} = k_{XT} \tag{9.114b}$$

It is not possible to distinguish experimentally the three models, corresponding to relations (9.109), (9.113) and (9.114) for k_{QT} (NO), from observations in low-viscosity solutions since they reduce to the common relation (9.109a). Studies of k_{QT} (NO) in very viscous solutions are required, to determine whether the behaviour is consistent with (9.109b), (9.113b) or (9.114b).

9.10.3 *Static quenching* (pp. *506–8*)

The experiments of Siegel and Judeikis[495] on the O_2 and NO luminescence quenching of naphthalene-d_8 in 3-methylpentane (3-MP) glass solutions at 77°K have been discussed previously (pp. *506–8*). Jones and Siegel[496] have extended these experiments to naphthalene-h_8 and to 67°K and they have modified the interpretation of the results.

Molecular diffusion occurs at 77°K in 3-MP glass[496] and in *iso*pentane (IP) glass.[499] Froehlich and Morrison[499] observed that the quenching of toluene phosphorescence by *cis*-piperylene in IP glass solution at 77°K obeys Stern–Volmer kinetics corresponding to dynamic quenching. This contrasts with the static quenching observed in rigid (e.g. ethanol or ethanol/ether) glass solutions at 77°K. The original interpretation[495] of the 77°K 3-MP glass solution data in terms of static quenching is incorrect.

To eliminate diffusion Jones and Siegel[496] reduced the temperature to 67°K. They analysed their data on the Perrin 'active sphere' static quenching model (p. *441*), in which any $^1M^*$ molecule lying within a distance R_{QM} of a quencher is quenched instantaneously, and any lying beyond this quenching radius is unquenched. Similarly any $^3M^*$ molecule lying within a distance R_{QT} of a quencher is quenched instantaneously, and any lying beyond this is unquenched. The experimental values of R_{QM} and R_{QT} are listed in Table 9.46.

Table 9.46 Static quenching radii R_{QM} and R_{QT} of $^1M^*$ and $^3M^*$ states by O_2 and NO in 3-methylpentane glass at 67°K [496]

Compound	Quencher	R_{QM}(Å)	R_{QT}(Å)
Naphthalene-h_8	Oxygen	$\leqslant 7{\cdot}1$	$9{\cdot}9 \pm 0{\cdot}3$
	Nitric oxide	$12{\cdot}4 \pm 0{\cdot}3$	$\leqslant 12{\cdot}4$
Naphthalene-d_8	Oxygen	$\leqslant 7{\cdot}1$	$9{\cdot}8 \pm 0{\cdot}3$
	Nitric oxide	$13{\cdot}0 \pm 0{\cdot}3$	$\leqslant 13{\cdot}0$

The initially excited $^1M^*$ molecules can be divided into groups depending on their distance R from a quencher. With NO static quenching ($R_{QM} \geqslant R_{QT}$) there are three groups of molecules with (i) $R \leqslant R_{QT}$, (ii) $R_{QT} \leqslant R \leqslant R_{QM}$, and (iii) $R > R_{QM}$. Group (i) is quenched through the complete $^2(M{\cdot}NO)^*$ manifold (9.105) to ($^1M + {}^2NO$); group (ii) is quenched to ($^3M^* + {}^2NO$) (9.105c); and group (iii) is unquenched. Increase of [^2NO] should (a) reduce the fluorescence quantum yield Φ_{FM}, due to quenching of groups (i) and (ii), (b) reduce the phosphorescence quantum yield Φ_{PT}, due to quenching of group (i), (c) not affect the fluorescence lifetime τ_M and (d) not affect the phosphorescence lifetime τ_T. Jones and Siegel[496] observed all these features in the static NO quenching of naphthalene. They also observed that Φ_{FM} and Φ_{PT} decrease in an identical manner with increase in [^2NO]. This result indicates that group (ii) is negligible, i.e. that $R_{QM}(NO) \simeq R_{QT}(NO)$.

With O_2 static quenching ($R_{QM} < R_{QT}$) there are three groups of molecules with (i) $R \leqslant R_{QM}$, (ii) $R_{QM} < R \leqslant R_{QT}$ and (iii) $R > R_{QT}$. Group (i) is quenched through the complete $^3(M{\cdot}O_2)^*$ manifold (9.95) to ($^1M + {}^3O_2$); group (ii), which is unquenched in the $^1M^*$ state, emits fluorescence or undergoes intersystem crossing to the $^3M^*$ state, where it forms $^{1,3,5}(M{\cdot}O_2)^*$ complexes (9.98) and is subject to $^3M^*$ static quenching;

and group (iii) is unquenched in either the $^1M^*$ or $^3M^*$ states. An increase of $[^3O_2]$ should (a) reduce Φ_{FM} due to quenching of group (i), (b) reduce Φ_{PT} due to quenching of groups (i) and (ii), (c) not affect τ_M and (d) not affect τ_T. All these features are observed in the static $^3O_2$ quenching of naphthalene.[496] It may be noted that the oxygen quenching of $^3M^*$ molecules in Group (i) occurs only by process (9.98e) via $^3(M{\cdot}O_2)^*$, while for group (ii) it also occurs by processes (9.98b) and (9.98c) via $^1(M{\cdot}O_2)^*$.

Jones and Siegel[496] observed that with pulsed (33 ms) $^1M^*$ excitation the initial phosphorescence decay of naphthalene-d_8 in 3-MP solution containing 0·26 M oxygen at 67 K was less rapid than that of a similar solution of naphthalene-h_8. They interpreted this as evidence of a significant perdeuteration decrease in the oxygen $^3M^*$ static quenching rate, indicative of a large energy gap for the quenching transition, and they proposed that the static $^3M^*$ oxygen quenching occurs by process (9.98e), rather than by processes (9.98b) or (9.98c). A contrary conclusion is drawn from their values of R_{QT} (O_2) for naphthalene-h_8 and naphthalene-d_8 (Table 9.46), which show the absence of any perdeuteration effect in static $^3M^*$ oxygen quenching. These results, which appear more reliable than the pulsed $^1M^*$ excitation data, indicate that (9.98b) and (9.98c) are the dominant static $^3M^*$ oxygen quenching processes, just as they are in dynamic quenching.

9.11 Energy migration and transfer

Exciton interactions in organic crystals have been reviewed by Swenberg and Geacintov (Chapter 10, Vol. 1; Chapter 8, Vol. 2). The theory of charge transport in organic molecular solids has been considered by Druger (Chapter 7, Vol. 2).

9.11.1 *Radiative migration and transfer* (pp. *521–3*)

Reabsorption commonly influences the fluorescence properties of pure and mixed organic crystals and of other organic molecular systems.

9.11.1.1 *Unitary systems*

A unitary system is defined as one containing a single fluorescent and absorbing species ^1M. Pure aromatic vapours, neat pure liquids and perfect pure crystals are examples of unitary systems. A real imperfect crystal is not a unitary system, since defects and residual impurities are present and function as guest fluorescent species ^1X_n. A fluorescent solution may be considered as a unitary system, provided the solvent does not participate in absorption, fluorescence, energy migration or transfer, and only one solute ^1M is present. The *molecular fluorescence parameters* of the excited species $^1M^*$, its fluorescence lifetime τ_M, quantum yield Φ_{FM} and spectrum $F_M(\bar{\nu})$, are defined in the absence of reabsorption.[102] The corresponding *technical fluorescence parameters* τ^t_M, Φ^t_{FM} and $F^t_M(\bar{\nu})$ are those observed in the presence of reabsorption, caused by the overlap of $F_M(\bar{\nu})$ and the absorption spectrum $\kappa_M(\bar{\nu})$.

$\kappa_M(\bar{\nu}) = c_M\sigma_M(\bar{\nu})$ is the absorption coefficient (cm^{-1}), where $\sigma_M(\bar{\nu})$ is the molecular absorption cross-section (cm^2— and c_M is the number of molecules/cm^3 (pp. *46–7*). For a guest species ^1Y the absorption coefficient is $\kappa_Y(\bar{\nu})$ $[^1Y]'$, where $[^1Y]'$ is the *mole fraction*, and $c_Y = c_M\,[^1Y]'$ is the number of ^1Y molecules/cm^3. In solutions, where $[^1Y]$ is the molar concentration, $c_Y = N'\,[^1Y]$, where N′ is Avogadro's number per millimole.

The fluorescence reabsorption depends on the mode of excitation which determines the distribution of the primary excited molecules $[^1M^*]'_1$. With optical excitation at wavenumber $\bar{\nu}_{ex}$, the $[^1M^*]'_1$ distribution decreases exponentially below the surface to 1/e of its surface value at the optical penetration depth $(\kappa_M(\bar{\nu}_{ex}))^{-1}$. With surface excitation by ionizing particles (α-particles, protons, etc.) the $[^1M^*]'_1$ distribution is nearly uniform along the particle track up to its range r within the specimen. With excitation by penetrating radiations (X-rays, γ-rays, ruby laser photons, etc.) the $[^1M^*]'_1$ distribution may be uniform throughout the whole of the specimen volume or, if collimation is used, through part of it.

The probability a of reabsorption of a fluorescence photon depends on its point of origin within the specimen, its direction of propagation, its wavenumber $\bar{\nu}_f$, its reabsorption coefficient $\kappa_M(\bar{\nu}_f)$, the specimen dimensions and the refractive index n, which determines the critical angle. The photon escape probability $e = (1 - a)$. For a particular specimen under given excitation conditions a_1 is defined as the mean value of a for the fluorescence photons originating from the primary excited molecules, mole fraction $[^1M^*]'_1$. Reabsorption produces a second generation of excited molecules $[^1M^*]'_2$, yielding secondary fluorescence photons which have a mean reabsorption probability a_2. The pth generation of excited molecules $[^1M^*]'_p$ yields fluorescence photons which have a reabsorption probability a_p.

k_{FM} and $k_M (= 1/\tau_M)$ are the radiative and total $^1M^*$ decay rates, respectively, and the decay is described by a series of rate equations,

$$-\frac{d[^1M^*]'_1}{dt} = k_M\,[^1M^*]'_1 \qquad (9.115.1)$$

$$-\frac{d[^1M^*]'_p}{dt} = k_M\,[^1M^*]'_p - a_{p-1}\,k_{FM}\,[^1M^*]'_{p-1} \qquad (9.115.p)$$

Summation over all values of p gives

$$-\frac{\mathrm{d}[^1\mathrm{M}^*]'}{\mathrm{d}t} = (k_\mathrm{M} - a_\mathrm{MM}k_\mathrm{FM})\,[^1\mathrm{M}^*]' \tag{9.116}$$

where

$$[^1\mathrm{M}^*]' = \sum_p [^1\mathrm{M}^*]'_p$$

and

$$a_\mathrm{MM} = \frac{\sum_p a_p[^1\mathrm{M}^*]'_p}{[^1\mathrm{M}^*]'}$$

is the overall mean reabsorption probability.

If n_p is the number of *escaping* fluorescence photons which originate from $[^1\mathrm{M}^*]'_p$, the photon escape rate is given by a similar series of equations

$$\frac{\mathrm{d}n_p}{\mathrm{d}t} = (1 - a_p)\,k_\mathrm{FM}\,[^1\mathrm{M}^*]'_p \tag{9.117p}$$

which sum to give

$$\frac{\mathrm{d}n}{\mathrm{d}t} = (1 - a_\mathrm{MM})\,k_\mathrm{FM}\,[^1\mathrm{M}^*]' \tag{9.117}$$

where

$$n = \sum_p n_p$$

The technical fluorescence response function,[520] obtained from (9.116) and (9.117) is

$$i^\mathrm{t}_\mathrm{M}(t) = (1 - a_\mathrm{MM})\,k_\mathrm{FM}\exp(-k^\mathrm{t}_\mathrm{M}t) \tag{9.118}$$

where

$$k^\mathrm{t}_\mathrm{M} = k_\mathrm{M} - a_\mathrm{MM}\,k_\mathrm{FM} \tag{9.119}$$

This is equivalent to

$$\tau^\mathrm{t}_\mathrm{M} = \frac{\tau_\mathrm{M}}{1 - a_\mathrm{MM}\,\Phi_\mathrm{FM}} \tag{9.120}$$

Integration of (9.118) yields the corresponding expression

$$\Phi^\mathrm{t}_\mathrm{FM} = \frac{(1 - a_\mathrm{MM})\,\Phi_\mathrm{FM}}{1 - a_\mathrm{MM}\,\Phi_\mathrm{FM}} \tag{9.121}$$

The mean fluorescence photon absorption probability a_{MM} is determined experimental[500] from

$$A_M^t = (1 - a_{MM})\, A_M \tag{9.122}$$

where A_M^t and A_M are the integrated areas of $F_M^t(\bar{\nu})$ and $F_M(\bar{\nu})$, respectively, the spectra being normalized in the low-wavenumber region where the profile is not affected by reabsorption.

Relations (9.120) and (9.121) were originally derived[102,500] by assuming $a_{MM} = a_1 = a_p$, an invalid assumption avoided in the above derivation.[545] The molecular parameters τ_M and Φ_{FM} can be evaluated from the technical parameters τ_M^t, Φ_{FM}^t and $F_M^t(\bar{\nu})$ and the molecular spectrum $F_M(\bar{\nu})$, using relations (9.120)–(9.122), provided the technical parameters are determined under *identical* excitation and observation conditions, since the latter influence the magnitude of a_{MM}.

9.11.1.2 *Multiple systems*

A real crystal is an example of a multiple fluorescent system consisting of a host species ^1M and n guest species ^1X_n (molecular parameters τ_n, Φ_{Fn}, $F_n(\bar{\nu})$ and $\kappa_n(\bar{\nu})$ with $n = 1, 2, 3 \ldots$). The exciton trap depth of ^1X_n is $\Delta E_n = \bar{\nu}_{0M} - \bar{\nu}_{0n}$, where $\bar{\nu}_{0M}$ and $\bar{\nu}_{0n}$ are the zero-point energies of the singlet-excited species $^1M^*$ and $^1X_n^*$, respectively. Radiative and radiationless (exciton) transfer occurs from $^1M^*$ to ^1X_n, but $[^1X_n]'$ is considered sufficiently small that energy migration or transfer between the guest species can be neglected. k_{FM} and k_{Fn} are the radiative decay rates of $^1M^*$ and $^1X_n^*$, respectively; k_M $(= 1/\tau_M)$ and k_n $(= 1/\tau_n)$ are the total decay rates of $^1M^*$ and $^1X_n^*$; $k_{nM}\,[^1X_n]'$ is the *net* rate of $^1M^*$ exciton transfer to ^1X_n; and a_{MM} and $a_{nM}\,[^1X_n]'$ are the probabilities of reabsorption of $^1M^*$ fluorescence by ^1M and ^1X_n, respectively.

The rate equations are as follows

$$-\frac{\mathrm{d}[^1M^*]'}{\mathrm{d}t} = \left(k_M - a_{MM}\,k_{FM} + \sum_n k_{nM}\,[^1X_n]'\right)[^1M^*]' \tag{9.123}$$

$$-\frac{\mathrm{d}[^1X_n^*]'}{\mathrm{d}t} = k_n[^1X_n^*]' - (a_{nM}\,k_{FM} + k_{nM})[^1X_n]'\,[^1M^*]' \tag{9.124.n}$$

The $^1M^*$ fluorescence response function is

$$i_M^t(t) = \left(1 - a_{MM} - \sum_n a_{nM}\,[^1X_n]'\right) k_{FM}\exp(-k_M^t t) \tag{9.125}$$

where

$$k_M^t = k_M - a_{MM}\,k_{FM} + \sum_n k_{nM}\,[^1X_n]' \tag{9.126}$$

The technical $^1M^*$ fluorescence parameters are

$$\tau_M^t = 1/k_M^t = \frac{\tau_M}{1 - a_{MM}\,\Phi_{FM} + \sum_n \sigma_{nM}\,[^1X_n]'} \tag{9.127}$$

$$\Phi_{FM}^t = \frac{\left(1 - a_{MM} - \sum_n a_{nM}\,[^1X_n]\right)\Phi_{FM}}{1 - a_{MM}\,\Phi_{FM} + \sum_n \sigma_{nM}\,[^1X_n]'} \tag{9.128}$$

$$A_M^t = \left(1 - a_{MM} - \sum_n a_{nM}\,[^1X_n]'\right) A_M \tag{9.129}$$

where

$$\sigma_{nM} - k_{nM}/k_M \tag{9.130}$$

Comparison with relations (9.120)–(9.122) for a unitary system shows that in the multiple system τ_M^t and Φ_{FM}^t are reduced by $^1M^*$ exciton transfer,

$$\sum_n \sigma_{nM}[^1X_n]'$$

and that Φ_{FM}^t and A_M^t are reduced by $^1M^*$ fluorescence reabsorption,

$$\sum_n a_{nM}[^1X_n]'$$

Relations (9.120)–(9.122) are not valid for a multiple system.

The $^1X_n^*$ fluorescence response function is, from (9.123) and (9.124.n),

$$i_n^t(t) = \frac{(a_{nM}k_{FM} + k_{nM})\,[^1X_n]'k_{Fn}}{(k_M^t - k_n)}\{\exp(-k_n t) - \exp(-k_M^t t)\} \tag{9.131.n}$$

The technical $^1X_n^*$ fluorescence parameters are $\tau_n^t = \tau_n$, $A_n^t = A_n$ and

$$\Phi_{F_n}^t = \frac{(a_{nM}\,\Phi_{FM} + \sigma_{nM})\,[^1X_n]'\,\Phi_{Fn}}{1 - a_{MM}\,\Phi_{FM} + \sum_n \sigma_{nM}\,[^1X_n]'} \tag{9.132.n}$$

The parameters k_{nM} and σ_{nM} decrease with increase in temperature T. At low temperatures $k_{nM} = k_{nM}^0$ and $\sigma_{nM} = \sigma_{nM}^0$. At higher temperatures thermally activated reverse transfer from $^1X_n^*$ to ^1M at a rate

$$k_{Mn} = k'_{Mn}\exp(-\Delta E_n/\mathbf{k}T) \tag{9.133.n}$$

reduces the net rate $k_{nM}[{}^1X_n]'$ of ${}^1M^*$ transfer to ${}^1X_n$, so that

$$k_{nM} = \frac{k_n k^0_{nM}}{k_n + k_{Mn}} = \frac{k^0_{nM}}{1 + \sigma_{Mn}}$$

$$= \frac{k^0_{nM}}{1 + \sigma'_{Mn} \exp(-\Delta E/\mathbf{k}T)} \qquad (9.134.n)$$

where $\sigma_{Mn} = k_{Mn}/k_n$ and $\sigma'_{Mn} = k'_{Mn}/k_n$. Equation (9.127) thus becomes

$$\tau^t_M = \frac{\tau_M}{1 - a_{MM}\Phi_{FM} + \sum_n \{\sigma^0_{nM}[{}^1X_n]'/[1 + \sigma'_{Mn}\exp(-\Delta E_n/\mathbf{k}T)]\}} \qquad (9.135)$$

describing the influence of reabsorption and exciton trapping on the fluorescence lifetime of the host crystal.

9.11.2 *Fluorescence of anthracene crystals*

9.11.2.1 *Guest species*

All anthracene crystals contain fluorescent defects[521,522] and a fluorescent impurity, 2-methylanthracene.[525,528] The principal guest species ${}^1X_n$ identified in 'pure' anthracene crystals may be categorized as follows.

(${}^1X_1$) 2-Methylanthracene (2-MA) is present in zone-refined anthracene. Its fluorescence bands (the 0_n bands) are observed in the low-temperature crystal fluorescence spectra with $\Delta E_1 = \bar{\nu}_{OM} - \bar{\nu}_{01} = 191\ \text{cm}^{-1}$.[525,528] The identification of the 2-MA impurity as the source of the 0_n bands resolves several outstanding questions. It accounts for most of the 'anti-reflection lines' observed by Morris *et al.*,[529] for the 24,908 cm^{-1} band attributed by Sidman[530] to a 'trapped exciton', and for the A, B and C series of Glockner and Wolf.[524] The triplet lifetime τ_T ($\sim$20 ms) is insensitive to the presence of 2-MA.

Helfrich and Lipsett[521] studied the defect fluorescence spectrum $F_n(\bar{\nu})$ of vapour-grown (*vg*) and melt-grown (*mg*) anthracene crystals at 4·2°K, using either ${}^1M^*$ excitation ($\bar{\nu}_{ex} > \bar{\nu}_{OM}$) or ${}^1X^*_n$ excitation ($\bar{\nu}_{0n} < \bar{\nu}_{ex} < \bar{\nu}_{OM} - 100\ \text{cm}^{-1}$), where $\bar{\nu}_{OM} = 25{,}103\ \text{cm}^{-1}$ is the crystal anthracene $0 - 0$ fluorescence transition.[524,525] Three distinct types of defect fluorescence are observed.

(${}^1X_2$) In *mg* anthracene crystals at 4·2°K at least 70% of the total fluorescence yield is a structureless emission $F_2(\bar{\nu})$, attributed to dislocations.[521] A similar, but less intense, emission is observed in *vg* crystals.[525] $F_2(\bar{\nu})$ originates near $\bar{\nu}_{OM}$, it has a maximum at about $\bar{\nu}_{OM} - 1500\ cm^{-1}$, and it extends beyond $F_M(\bar{\nu})$ to about $\bar{\nu}_{OM} - 7400\ cm^{-1}$.[523] The spectrum is similar to that of an excimer or sandwich dimer. $F_2(\bar{\nu})$ is excited at $\bar{\nu} \geqslant \bar{\nu}_{OM} - 100\ cm^{-1}$,[521] so that the $^1X_2$ trap depth $\Delta E_2 \leqslant 100\ cm^{-1}$. The $^1X_2^*$ fluorescence lifetime $\tau_2 \simeq 15$ ns.[532]

($^1X_3$) A structured emission $F_3(\bar{\nu})$, attributed to point defects, occurs in *mg* and *vg* anthracene crystals at 4·2°K with either $^1M^*$ or $^1X_n^*$ excitation.[521] $F_3(\bar{\nu})$ has rather broad maxima at 24,830, 24,460, 23,900, 23,000, 21,500 and 19,100 cm^{-1}. The emission may be excited at $\bar{\nu} \geqslant \bar{\nu}_{OM} - 2300\ cm^{-1}$,[521] so that $\Delta E_3 \leqslant 2300\ cm^{-1}$.

($^1X_4$) A structureless emission $F_4(\bar{\nu})$, attributed to disordered crystal regions, occurs in *mg* and *vg* anthracene crystals at 4·2°K with $^1X_n^*$ excitation.[521] With excitation at $\bar{\nu} = \bar{\nu}_{OM} - 1100\ cm^{-1}$ $F_4(\bar{\nu})$ decreases monotonically with $\bar{\nu}$,[521] so that $\Delta E_4 \leqslant 1100\ cm^{-1}$.

Other defect species have been observed, but not categorized.[533] The guest species $^1X_1 - {}^1X_4$ in a 'pure' anthracene crystal may be divided into two groups:

(a) shallow traps ^1X_S of depth $\Delta E_S \sim 100$–$300\ cm^{-1}$, including $^1X_1$ and $^1X_2$; and

(b) deep traps ^1X_D of depth $\Delta E_D \sim 1000$–$3000\ cm^{-1}$, including $^1X_3$ and $^1X_4$.

The nature and concentration of the defects depend on the method of crystal growth. Thus $[^1X_S]'$ is expected to be lower for a *vg* crystal than for a *mg* crystal of the same purity, because of the much lower value of $[^1X_2]'$. Lupien *et al.*[532] used a simple Bridgman crystal-growing furnace to study the effect of the heat exchange gas (He, N_2, Ar or Xe) within the crystal growing tube on the anthracene crystal properties. The triplet lifetime τ_T (= 20–25 ms), which is commonly used as a criterion of anthracene crystal purity and quality,[531] is insensitive to the nature of the gas. The space-charge-limited current characteristics show an increase in the concentration $[^1X_n]'$ and depth ΔE_n of traps with increase in the molecular weight of the gas. The electroluminescence (EL) spectra include a structureless background defect emission, similar to $F_2(\bar{\nu})$, the relative intensity of which increases with the molecular weight of the gas or with reduction in temperature. The fluorescence spectra exhibit similar behaviour, but the background emission is less intense than in the EL spectra. The crystals contained concentrations of $\sim 10^{-8}$ mole/mole Xe and $\sim 10^{-7}$ mole/mole He. The larger atom produces a much greater disturbance of the anthracene crystal lattice, despite its lower solubility. The defects, which appear to be mainly of the $^1X_2$ (dislocation) type, may be minimized in *mg* anthracene crystals by using low-pressure ($\leqslant 1$ torr) helium as the heat exchange medium.[532]

9.11.2.2 *Absorption*

Fluorescence reabsorption is due to the overlap of $F_M(\bar{\nu})$ with the absorption spectrum at $\bar{\nu} < \bar{\nu}_{0M}$. The total crystal absorption spectrum at $\bar{\nu} < \bar{\nu}_{0M}$ is

$$\kappa(\bar{\nu}) = \kappa_M(\bar{\nu}) + \sum_n \kappa_n(\bar{\nu})\,[^1X_n]' \qquad (9.136)$$

where

$$\kappa_M(\bar{\nu}) = \kappa_{0M}(\bar{\nu}) + \kappa_{1M}(\bar{\nu}) + \kappa_{2M}(\bar{\nu}) \qquad (9.137)$$

has three components $\kappa_{0M}(\bar{\nu})$, $\kappa_{1M}(\bar{\nu})$ and $\kappa_{2M}(\bar{\nu})$ due to the $0-0$ absorption band edge and the $1-0$ and $2-0$ absorption hot bands, respectively.

Nakada[526] observed $\kappa(\bar{\nu})$ at $\bar{\nu} < \bar{\nu}_{0M}$ for crystal anthracene with *a*- and *b*-polarized light normal to the (*ab*) crystal plane at temperatures from 93 to 353°K. $\kappa_{0M}(\bar{\nu})$ obeys the Urbach rule.[527]

$$\kappa_{0M}(\bar{\nu}) = \kappa_{0M}(\bar{\nu}_0)\exp\{-\alpha\mathbf{hc}(\bar{\nu}_0 - \bar{\nu})/\mathbf{k}T\} \qquad (9.138)$$

where $\mathbf{hc}\bar{\nu}$ is the photon energy, $\mathbf{hc}\bar{\nu}_0$ is the extrapolated ($0-0$ absorption transition) photon energy at 0°K, and $\kappa_{0M}(\bar{\nu}_0)$ and α are constants. For *a*- and *b*-polarized light $\kappa_{0M}(\bar{\nu}_0) = 6 \times 10^4$ and $6 \times 10^4\ \mathrm{cm}^{-1}$, $\bar{\nu}_0 = 25{,}602$ and $25{,}126\ \mathrm{cm}^{-1}$, and $\alpha = 1{\cdot}68$ and $0{\cdot}88$, respectively.[526]

The $1-0$ hot-band absorption $\kappa_{1M}(\bar{\nu})$, obtained[526] from $(\kappa(\bar{\nu}) - \kappa_{0M}(\bar{\nu}))$, is a diffuse band with a maximum at $\bar{\nu}_1$ given by the Boltzmann relation,

$$\kappa_{1M}(\bar{\nu}_1) = \kappa'_{1M}\exp\{-\mathbf{hc}(\bar{\nu}_0 - \bar{\nu}_1)/\mathbf{k}T\} \qquad (9.139)$$

For *a*- and *b*-polarized light $\bar{\nu}_1 = 24{,}240$ and $24{,}125\ \mathrm{cm}^{-1}$, $\kappa'_{1M} = 4{\cdot}9 \times 10^4$ and $1{\cdot}2 \times 10^5\ \mathrm{cm}^{-1}$, and $(\bar{\nu}_0 - \bar{\nu}_1) = 1360$ and $1300\ \mathrm{cm}^{-1}$, respectively.[526] The $2-0$ hot-band absorption $\kappa_{2M}(\bar{\nu})$, which lies beyond the limit of Nakada's observations,[526] has a maximum at $\bar{\nu}_2$ given by

$$\kappa_{2M}(\bar{\nu}_2) = \kappa'_{2M}\exp\{-\mathbf{hc}(\bar{\nu}_0 - \bar{\nu}_2)/\mathbf{k}T\} \qquad (9.140)$$

with $\kappa'_{2M} \simeq \kappa'_{1M}$ and $(\bar{\nu}_0 - \bar{\nu}_2) \simeq 2(\bar{\nu}_0 - \bar{\nu}_1)$.

$\kappa_M(\bar{\nu})$ at $\bar{\nu} < \bar{\nu}_{0M}$ is strongly temperature-dependent. $\kappa_{2M}(\bar{\nu})$ is negligible ($<1\ \mathrm{cm}^{-1}$) at $T \leqslant 353$°K and $\kappa_{1M}(\bar{\nu})$ becomes negligible below about 100°K. The magnitude of $\kappa_{0M}(\bar{\nu})$ and its overlap with the $0-0$ fluorescence band also decrease with T, and at low temperatures a_{MM} tends to zero.

The absorption spectrum $\kappa_n(\bar{\nu})$ of each guest species ^1X_n is red-shifted relative to $\kappa_M(\bar{\nu})$ by the trap depth ΔE_n. If it is assumed that the absorption spectra of 2-MA and the anthracene defects are similar to that of anthracene, then

$$\kappa_n(\bar{\nu}) \simeq \kappa_M(\bar{\nu} + \Delta E_n) \tag{9.141}$$

and the total guest absorption is

$$\sum_n \kappa_n(\bar{\nu})\,[^1X_n]' \simeq \sum_n \kappa_M(\bar{\nu} + \Delta E_n)\,[^1X_n]' \tag{9.142}$$

Unlike $\kappa_M(\bar{\nu})$, the guest absorption is practically independent of temperature, since the guest hot band absorption is negligible.

In the technical fluorescence spectrum of a 1 cm thick *mg* anthracene crystal at room temperature, the 0 – 0, 0 – 1 and 0 – 2 fluorescence bands are eliminated or strongly attenuated by the reabsorption which extends to about 3000 cm^{-1} below $\bar{\nu}_{0M}$.[96] $\kappa_{1M}(\bar{\nu})$ and $\kappa_{2M}(\bar{\nu})$ are inadequate to account for the magnitude and extent of the reabsorption, which is largely due to the absorption $\kappa_D(\bar{\nu})$ $[^1X_D]'$ of the deep traps.[520] The reabsorption data indicate that ΔE_D extends down to 3000 cm^{-1}.

Direct evidence for $\kappa_D(\bar{\nu})$ comes from the $^1X^*$ excitation ($\bar{\nu} < \bar{\nu}_{0M}$) of $F_3(\bar{\nu})$ and $F_4(\bar{\nu})$ in anthracene crystals at 4·2°K.[521] The magnitude of $\kappa_D(\bar{\nu})\,[^1X_D]'$ can be estimated from Nakada's data.[526] With *b*-polarized light $\kappa(\bar{\nu}) \simeq 2$ cm^{-1} at $\bar{\nu}_{0M} - 2200$ cm^{-1} is practically independent of T from 93 to 263°K, and it is therefore equated to $\kappa_D(\bar{\nu})\,[^1X_D]'$. If $\kappa_{0M}(\bar{\nu}_0) = 6 \times 10^4$ cm^{-1} for crystal anthracene[526] is taken as the upper limit of $\kappa_D(\bar{\nu})$ at $\bar{\nu} < \bar{\nu}_{0M}$, then $[^1X_D]' \geqslant 3{\cdot}3 \times 10^{-5}$ mole/mole. A similar value of $[^1X_D]' \geqslant 4 \times 10^{-5}$ mole/mole has been estimated from fluorescence reabsorption data.[520] Low-temperature absorption spectroscopy of anthracene crystals at $\bar{\nu} < \bar{\nu}_{0M}$ is required to obtain fuller data on the defect absorption spectra.

9.11.2.3 *Fluorescence*

Three effects influence the host crystal fluorescence,

(a) host crystal reabsorption (a_{MM})

(b) host–guest exciton transfer

$$\left(\sum_n \sigma_{nM}\,[^1X_n]'\right)$$

and

(c) guest reabsorption

$$\left(\sum_n a_{nM}\,[^1X_n]'\right)$$

The lifetime τ^{t}_{M} (9.127) is affected by (a) and (b). The spectrum $F^{t}_{M}(\bar{\nu})$ and its area A^{t}_{M} (9.129) are affected by (a) and (c). The quantum yield Φ^{t}_{FM} (9.128) is affected by all three factors. Thus τ^{t}_{M} and $F_{M}(\bar{\nu})$ appear the most suitable parameters to study to elucidate these effects.

The host and guest fluorescence occur in the same spectral region. The total fluorescence spectrum

$$F^{t}(\bar{\nu}) = \Phi^{t}_{FM}F^{t}_{M}(\bar{\nu}) + \sum_{n} \Phi^{t}_{Fn}F_{n}(\bar{\nu}) \tag{9.143}$$

is difficult to resolve into its components, since each component spectrum is either a complex sequence of bands or a broad continuum. Although the low-temperature anthracene crystal fluorescence spectrum has been studied for many years, only recently has 2-MA ($^{1}X_{1}$) been identified as an inherent component.[525,528] It is thus difficult to determine $F^{t}_{M}(\bar{\nu})$ from observations of $F^{t}(\bar{\nu})$.

The total fluorescence response function is, from (9.125) and (9.131.n),

$$i^{t}(t) = i^{t}_{M}(t) + \sum_{n} i^{t}_{n}(t) \tag{9.144}$$

$$= A \exp(-k^{t}_{M}t) + \sum_{n} B_{n} \exp(-k_{n}t) \tag{9.145}$$

Each component is represented by a single decay parameter, and there is little difficulty in identifying the major component k^{t}_{M} $(= 1/\tau^{t}_{M})$. Measurements of τ^{t}_{M} provide the simplest means of studying the influence of (a) host crystal reabsorption and (b) host–guest exciton transfer, on the host crystal fluorescence.

9.11.2.4 *Reabsorption*

The host–guest exciton transfer is described by the parameter

$$z = \sum_{n} \sigma_{nM} [^{1}X_{n}]' \tag{9.146}$$

Equation (9.127) may be rewritten as

$$\tau^{t}_{M} = \frac{\tau'_{M}}{1 - a'_{MM}\,\Phi_{FM}} \tag{9.147}$$

analogous to (9.120), where $\tau'_{M} = \tau_{M}/(1 + z)$ and $a'_{MM} = a_{MM}/(1 + z)$. τ'_{M} is the fluorescence lifetime corrected for reabsorption, but not for exciton trapping. τ'_{M} and $a'_{MM}\,\Phi_{FM}$ may be determined from observations of τ^{t}_{M} for crystals of different thickness d of the same purity and growth method, i.e. with similar values of z. Such measurements have been made by Powell[547] for anthracene crystals of d = 56 to 470 μm at 4, 100 and 300°K, and his

Table 9.47 Fluorescence lifetime τ_M^t of anthracene crystals as a function of thickness d and temperature T [547]

T (°K)	d (μm)	0	56	160	355	470
4		6·5	6·7	7·0	7·0	7·5
100	τ_M^t (ns)	10·3	10·5	10·6	10·8	11·0
300		11·0	11·5	13·2	18·5	22·0

results are listed in Table 9.47. Analysis of the data in terms of (9.147) leads to the following conclusions.

(i) The host crystal reabsorption of the thinnest ($d = 56\ \mu$m) crystal is small ($a'_{MM}\,\Phi_{FM} = 0{\cdot}02$–$0{\cdot}04$) from 4 to 300°K.

(ii) The host crystal reabsorption of the thickest ($d = 470\ \mu$m) crystal is small ($a'_{MM}\,\Phi_{FM} = 0{\cdot}065$) up to 100°K, but it increases to $a'_{MM}\,\Phi_{FM} = 0{\cdot}50$ at 300°K.

(iii) The 'zero-thickness' lifetime $\tau'_M = 6{\cdot}5$, 10·3 and 11·0 ns at 4, 100 and 300°K, respectively. This temperature dependence of τ'_M is due to the temperature-dependent exciton transfer z and to any intrinsic temperature dependence of τ_M.

9.11.2.5 *Host–guest exciton transfer*

The influence of this process on the anthracene crystal fluorescence can be studied by observation of τ_M^t, or preferably τ'_M, as a function of T. Tomura *et al.*[604] attributed an observed decrease of τ'_M from 10 ns at room temperature to about 6 ns at 100°K to the presence of two Davydov states with different fluorescence decay times in thermal equilibrium. The present interpretation in terms of host–guest exciton transfer is preferred.

Powell[547] observed the temperature dependence of τ_M^t ($\simeq \tau'_M$) for a 56 μm thick anthracene crystal from 4 to 100°K. The data (Figure 9.19) were fitted to a single shallow trap relation, equivalent to (9.135), with $\Delta E_S = 250\ \text{cm}^{-1}$ from observations of Avakian and Wolf,[548] $[^1X_S]' = 1{\cdot}8 \times 10^{-4}$ mole/mole, $\tau_M = 10{\cdot}5$ ns and $\tau_M^t = (\tau_M^t)_0 = 6{\cdot}6$ ns at low temperatures.

Munro *et al.*[523] observed a more complex temperature dependence of τ_M^t ($\simeq \tau'_M$) for a thin anthracene sublimation flake. τ_M^t rises sigmoidally from $(\tau_M^t)_0 \sim 5$ ns at 4°K to a plateau of $(\tau_M^t)_1 \sim 6$ ns at 75 to 140°K, and it then rises to a second plateau of $(\tau_M^t)_2 \sim 10{\cdot}5$ ns at 300°K. The data are consistent with a multi-component system containing shallow and deep traps $[^1X_S]$ and $[^1X_D]$, respectively. From (9.135), neglecting $a_{MM}\,\Phi_{FM}$ since

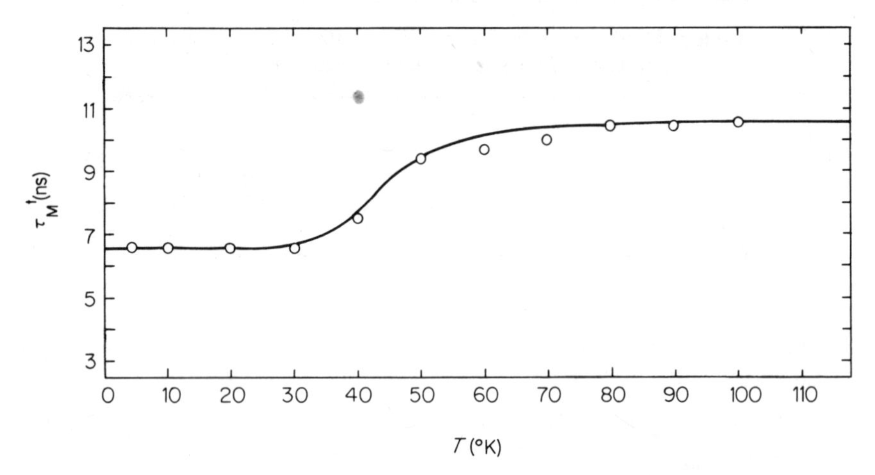

Figure 9.19 Temperature dependence of fluorescence lifetime τ_M^t of a 56 μm anthracene crystal. ○ Experimental data. Curve from single shallow trap relation, equivalent to (9.135). (Powell;[547] reproduced by permission of the American Institute of Physics)

the crystal is thin, we obtain

$$(\tau_M^t)_0 = \frac{\tau_M}{1 + \sigma_{SM}^0 [^1X_S] + \sigma_{DM}^0 [^1X_D]'} \tag{9.148}$$

$$(\tau_M^t)_1 = \frac{\tau_M}{1 + \sigma_{DM}^0 [^1X_D]'} \tag{9.149}$$

$$(\tau_M^t)_2 = \tau_M \tag{9.150}$$

Comparison with the data[523] gives $\sigma_{SM}^0 [^1X_S]' = 0{\cdot}35$, $\sigma_{DM}^0 [^1X_D]' = 0{\cdot}75$ and $\tau_M = 10{\cdot}5$ ns. The $^1M^*$ exciton capture parameter of various guest species ^1Y in crystal anthracene is $\sigma_{YM} \simeq 4 \times 10^4$ (p. *603*). If it is assumed that $\sigma_{DM}^0 = \sigma_{SM}^0 = \sigma_{YM}$, then $[^1X_D]' \simeq 2 \times 10^{-5}$ mole/mole and $[^1X_S]' \simeq 9 \times 10^{-6}$ mole/mole. $[^1X_D]'$ is identified with $^1X_3$ and $^1X_4$ defects, and $[^1X_S]'$ with $^1X_1$ (2-MA) and possibly some $^1X_2$ defects.

Munro *et al.*[523] also observed the temperature dependence of τ_M^t for a melt-grown anthracene crystal. (The same data were earlier attributed[549] to 'a defect-free (*sic*) anthracene crystal flake', but the later assignment is probably the correct one). In this case τ_M^t rises linearly from $(\tau_M^t)_0 \sim 5$ ns at 4°K to a plateau of $(\tau_M^t)_1 \sim 8$ ns from 75 to 140°K, and it then rises to a maximum of $(\tau_M^t)_2 \sim 17$ ns at 300°K. The higher value of $(\tau_M^t)_2$ is due to host reabsorption, but since the latter is small at $T \leqslant 100$°K, its effect on $(\tau_M^t)_0$ and $(\tau_M^t)_1$ may be neglected. The higher value of $(\tau_M^t)_1$ is attributed[545] to an increased value of $[^1X_S]' \simeq 3 \times 10^{-5}$ mole/mole, due to a higher concentration of $^1X_2$ defects. The linear non-sigmoid dependence of τ_M^t in the 4–75°K region is consistent with the presence of at least two guest species ($^1X_1$ and $^1X_2$) of different ΔE_S.

A similar anthracene crystal defect concentration of $\sim 4 \times 10^{-5}$ mole/mole has been estimated by Benz and Wolf[546] from fluorescence intensity data.

9.11.2.6 *Molecular fluorescence lifetime*

The above results indicate that the net host–guest exciton transfer probability is small in anthracene crystals at room temperature, so that the unitary relations (9.120)–(9.122) may be used to correct for reabsorption. Such an analysis[97] of previous anthracene crystal data gave a value of $\tau_M = 11 \pm 1$ ns, and all subsequent data, including those cited above,[523,547,604] appear consistent with this value. It corresponds to a radiative lifetime $\tau_{FM} = 1/k_{FM} = 12 \pm 1$ ns for crystal anthracene. This agrees with the experimental values of $\tau_{FM} = 15$–20 ns for anthracene in fluid solutions (p. *127*) and plastic solutions,[523] when allowance is made for the higher refractive index of the crystal,[104] and it also agrees with the theoretical value obtained from the integrated absorption spectrum.[104]

In contrast, Munro *et al.*[523] concluded from an analysis of their data described above (§9.11.2.5), using the unitary relations (9.120) and (9.122), that $\tau_M = 6 \pm 1$ ns for crystal anthracene, irrespective of the crystal growth method or the temperature. Apart from the incorrect procedure of applying (9.120) and (9.122) to a multiple system at reduced temperatures or to observations of τ_M^t and $F_M^t(\bar{\nu})$ made under different excitation conditions, there are errors in the published fluorescence spectra,[523] which have been described elsewhere.[545] These probably account for the incorrect value of τ_M.

9.11.3 *Exciton migration in pyrene crystals*

Singlet exciton ($^1M^*$) migration in anthracene and other type A molecular crystals occurs by a hopping mechanism at normal temperatures (pp. *528–37*). In pyrene and other type **B** molecular crystals, which crystallize in $^1(M \parallel M)$ pairs, the singlet excimer ($^1D^*$) state lies $\sim 0{\cdot}4$ eV below that $^1M^*$ state (pp. *331–5*), so that $^1M^*$ migration is inhibited. It might be expected that the singlet excitation energy would be trapped in the $^1D^*$ states, but several observers[534–541] have found that singlet–singlet energy transfer occurs from pyrene host crystals to various guest molecules. To account for the observed decrease in the fluorescence lifetime of pyrene crystals with crystal size, associated with the appearance of monomer defect bands in the fluorescence spectrum, Birks *et al.*[328] proposed that $^1D^*$ exciton migration can occur to surface monomer defects.

In general, transfer of singlet excitation energy from a host crystal to a guest species ^1Y may involve

(i) $^1D^*$ exciton migration (k_{DD}) and/or

(ii) $^1M^*$ exciton migration (k_{MM})

Energy transfer from $^1D^*$ to ^1Y may occur

(iii) radiationlessly by dipole–dipole interaction (k_{YD}) or

(iv) radiatively by $^1D^*$ emission and ^1Y absorption (a_{YD}).

Energy transfer from $^1M^*$ to ^1Y may occur

(v) radiationlessly (k_{YM}), or

(vi) radiatively (a_{YM}).

Klöpffer *et al.*[541] observed that energy transfer does not occur from crystal pyrene to any guest molecule whose singlet excitation energy $^1Y^*$ lies between 26,700 cm^{-1} and 25,400 cm^{-1} (anthracene, picene, tetraphenylethylene, 1:2-benzanthracene, 9,10-diphenylanthracene and 1:2:5:6-dibenzanthracene). Energy transfer does occur to guest molecules for which $^1Y^* \leqslant$ 23,350 cm^{-1} (coronene, perylene, anthanthrene, ovalene, tetracene, violanthrene, dinaphthopyrene and *iso*violanthrene). For pyrene $^1M^*$ = 26,600 cm^{-1} and $^1D^*$ = 23,460 cm^{-1} at 0°K (p. *333*). The results show that the energy transfer only occurs if $^1D^* > {}^1Y^*$, and that it does not occur if $^1M^* \geqslant {}^1Y^* > {}^1D^*$. These observations indicate that the $^1M^*$ processes k_{MM}, k_{YM} and a_{YM} do not occur in pyrene crystals, and that the host–guest energy transfer involves some or all of the $^1D^*$ processes k_{DD}, k_{YD} and a_{YD}.

Klöpffer *et al.*[541] used thin crystal specimens and chose guest molecules with weak absorption in the pyrene fluorescence region, in order to minimize the radiative (a_{YD}) and radiationless (k_{YD}) transfer probabilities. With coronene as guest they observed energy transfer at 298°K, but not at 77°K, showing that the $^1D^*$ exciton migration (k_{DD}) requires thermal activation. The pyrene and coronene fluorescence spectra overlap and are difficult to distinguish, and two non-fluorescent guests, anthraquinone and acenaphthylene, were used as quenchers in further studies. The hopping time τ_h (= 5 × 10^{-11} s at 298°K) and activation energy W_{DD} (= 0·055 eV) of $^1D^*$ exciton migration were determined from the temperature dependence of the fluorescence quantum yields of pure and doped pyrene crystals. The results show that $^1D^*$ exciton migration (k_{DD}) occurs by a thermally activated hopping process, and that in the mixed crystals studied it terminates in collisional transfer to ^{1}Y.

Chu *et al.*[539] studied the fluorescence parameters of pure and perylene-doped pyrene crystals down to 1·7°K. Perylene has a strong absorption in the pyrene fluorescence region, so that processes k_{YD} and a_{YD} are not excluded. They concluded that

(a) ^{1}D* exciton migration (k_{DD}) is only important above 130°K, and it has an activation energy of $W_{DD} = 0{\cdot}014$ eV;

(b) long-range dipole–dipole transfer (k_{YD}) from ^{1}D* to ^{1}Y occurs at perylene concentrations $> 10^{-3}$ mole/mole; and

(c) radiative transfer (a_{YD}) from ^{1}D* to ^{1}Y occurs at high perylene concentrations or with thick crystals.

Chu *et al.*[539] observed a short-lived (13 ns) decay component in the fluorescence of a pyrene crystal containing 10^{-3} mole/mole perylene at 77°K, which they attributed to perylene excited by ^{1}M* exciton migration (k_{MM}). Other explanations, such as direct excitation of the perylene by the flash lamp, do not appear to be excluded by the experimental conditions. The 0·4 eV binding energy of the pyrene excimer and the data of Klöpffer *et al.*[541] do not support the hypothesis of ^{1}M* exciton migration (k_{MM}) in crystal pyrene at 77°K.

The ^{1}D* exciton migration (k_{DD}) and the dipole–dipole transfer (k_{YD}) are consecutive processes, which can only be distinguished experimentally when either is negligible. This occurs under two limiting conditions:

(a) At low temperatures where k_{DD} is absent, k_{YD} occurs between static molecules, and Förster kinetics (pp. *570–1*) are applicable.

(b) With a weakly absorbing guest, such as coronene, the k_{YD} critical transfer distance R_0 is small and the ^{1}D* exciton migration (k_{DD}) culminates in collisional transfer (pp. *518–20*). Stern–Volmer kinetics (p. *570*) are then applicable.

Perylene-doped pyrene crystals at normal temperatures are an intermediate case, with k_{DD} and k_{YD}, with large R_0, occurring consecutively. The intermediate kinetics (p. *577*) of Yokota and Tanimoto,[542] who considered the influence of diffusion (≡ exciton migration) on dipole–dipole transfer, may be applied. Tomura and Takahashi[540] studied the temperature dependence of the fluorescence response functions of pure and perylene-doped pyrene crystals. They analysed their data using an approximate form of the Yokota–Tanimoto relation, and they concluded that the ^{1}D* exciton migration (k_{DD}) coefficient of crystal pyrene is of the form

$$D = D_0 \exp(-W_{DD}/\mathbf{k}T) \tag{9.151}$$

with $D_0 = 10^{-2}\ \mathrm{cm^2\ s^{-1}}$ and $W_{DD} = 0{\cdot}071$ eV.

Almost all experimental data on singlet excitons in organic crystals have been derived from mixed crystal studies (pp. *528–37*). The inclusion of a guest species in a host crystal perturbs the system, and it introduces processes absent from the host crystal, whose properties are to be determined. These aspects of mixed crystals are discussed in §9.11.4. The initial data on triplet ($^3M^*$) excitons were also obtained from mixed crystal studies (pp. *544–59*), but the discovery of $^3M^* - {}^3M^*$ interaction (homofusion) has enabled $^3M^*$ excitons to be studied in pure crystals (pp. *559–67*).

The corresponding $^1M^* - {}^1M^*$ interaction process in a pure anthracene crystal was demonstrated by Bergman *et al.*,[543] who observed an increase in the $^1M^*$ fluorescence decay rate with increase in laser excitation intensity. Goldschmidt *et al.*[454] observed a decrease in the $^1D^*$ fluorescence quantum yield of pyrene crystals at high excitation intensities, which they attributed to $^1D^* - {}^1D^*$ exciton interaction. Similar processes occur in solution. Babenko *et al.*[598] observed a reduction in Φ_{FM} of anthracene solutions at high light intensities, which they attributed to $^1M^* - {}^1M^*$ and $^1M^* - {}^3M^*$ interactions. Masuhara and Mataga[453] observed a reduction in the fluorescence lifetimes of the pyrene excimer ($^1D^*$) and the anthracene-DEA exciplex ($^1E^*$) in solution, consistent with $^1D^* - {}^1D^*$ and $^1E^* - {}^1E^*$ interactions, respectively.

Inoue *et al.*[544] have studied the temperature dependence of the rate parameter k_{AA} of singlet–singlet exciton annihilation in anthracene, pyrene and perylene crystals. Using a 4 ns pulsed nitrogen laser for excitation, they observed the fluorescence response functions at different excitation intensities. In general

$$\frac{d[^1A^*]'}{dt} = -k_A[^1A^*]' - k_{AA}[^1A^*]'^2 \tag{9.152}$$

where A = M for molecular excitons (e.g. anthracene) and A = D for excimer excitons (e.g. pyrene, perylene). Integration of (9.152) gives

$$1/[^1A^*]' = \{1/[^1A^*]'_0 + k_{AA}/k_A\} \exp(k_A t) - k_{AA}/k_A \tag{9.153}$$

where $[^1A^*]'_0$ is the initial exciton concentration. As predicted by (9.153), a linear relation is observed between $1/[^1A^*]'$ and $\exp(k_A t)$. k_{AA}/k_A is evaluated from its intercept and k_A is determined from the fluorescence decay at low excitation intensity. For anthracene k_{MM} is found to be independent of temperature from 77 to 293°K, in agreement with previous mixed crystal studies (p. *535*). For pyrene and perylene at 293°K values of $k_{DD} = 9 \times 10^{-15}$ and 8×10^{-14} cm^3 s^{-1} are estimated, respectively. From the temperature dependence of k_{DD}, values of $W_{DD} = 0{\cdot}041$ and 0·036 eV are obtained for the activation energies of $^1D^*$ exciton migration in pyrene and perylene crystals, respectively.

The values of W_{DD} obtained for pyrene crystals with different guest species (shown in brackets) are as follows: 0·014 eV (perylene);[539] 0·041 eV (none);[544] 0·055 eV (anthraquinone);[541] 0·055 eV (acenaphthylene;[541] 0·071 eV (perylene).[544] Disregarding the first value, a mean of $W_{DD} = 0{\cdot}056\ (\pm 0{\cdot}015)$ eV is obtained, in excellent agreement with the two identical values of Klöpffer *et al.*[541] The latter authors considered three possible mechanisms of $^1D^*$ exciton migration in pyrene crystals, only one of which is consistent with the experimental potential diagram of the pyrene crystal dimer and excimer (p. *331*). This mechanism involves thermal activation of the excimer to its ground-state equilibrium intermolecular separation ($r_0 = 3{\cdot}53$ Å) and energy migration therefrom to a neighbouring ground-state dimer. The 410 nm absorption band, observed by Fischer *et al.*[347] in pyrene crystals at low temperatures and assigned by them to $D_0 - D_1$ absorption (§9.7.6), is consistent with the pyrene crystal potential diagram and with this model of $^1D^*$ exciton migration.

9.11.4 *Mixed crystals* (pp. *528–37*)

9.11.4.1 *Singlet–singlet energy transfer*

Host–guest energy transfer in a mixed crystal may be described by the time-dependent (TD) exciton diffusion model in which the energy transfer rate parameter is given by the Smoluchowski relation (p. *445*)

$$k_{YM}(t) = 4\pi DR(1 + R(\pi Dt)^{-\frac{1}{2}})c_M \qquad (9.154)$$

D is the $^1M^*$ exciton diffusion coefficient, R is the $^1M^* - {}^1Y$ interaction distance and c_M is the host crystal concentration in molecules/cm^3.

In the time-independent (TI) exciton diffusion model, the transient term in (9.154) is neglected, so that

$$k_{YM} = 4\pi DRc_M \qquad (9.155)$$

The time independence of k_{YM}, corresponding to Stern–Volmer kinetics (p. *36*), was assumed in the derivation of the multi-component system relations of §9.11.1.2. Most studies of mixed crystal systems have been concerned with observations of the host and guest fluorescence quantum yields Φ^t_{FM} and Φ^t_{FY} or the host fluorescence lifetime τ^t_M as a function of the guest mole fraction $[^1Y]'$. From the multi-component system relations (§9.11.1.2), neglecting radiative migration and transfer, and putting Y = X, we obtain

$$\Phi_{FM}^{t} = \frac{\Phi_{FM}}{1 + \sigma_{YM}[^1Y]'} \tag{9.156}$$

$$\frac{\Phi_{FY}^{t}}{\Phi_{FM}^{t}} = \frac{\Phi_{FY}}{\Phi_{FM}} \sigma_{YM}[^1Y]' \tag{9.157}$$

$$\tau_{M}^{t} = \frac{\tau_M}{1 + \sigma_{YM}[^1Y]'} \tag{9.158}$$

where Φ_{FM} and τ_M are the $^1M^*$ parameters in the absence of $[^1Y]'$. These expressions have been confirmed for many mixed crystal systems, and reasonably consistent values of σ_{YM} and k_{YM} ($= \sigma_{YM}k_M$) have been derived therefrom (p. *603*). These results agree with the TI exciton diffusion model.

In 1969 Powell and Kepler[550–553] initiated a series of studies of the fluorescence response functions of mixed crystals. The experimental studies have been continued by Powell[547,554–557] and analysed and interpreted by Soos and Powell.[558,559] On the TI exciton diffusion model, the $^1M^*$ and $^1Y^*$ fluorescence response functions are

$$i_M^t(t) \propto \exp(-k_M^t t) \tag{9.125a}$$

$$i_Y^t(t) \propto \exp(-k_Y^t t) - \exp(-k_M^u t) \tag{9.131a}$$

with $k_M^t = k_M^u$. For thick tetracene-doped anthracene crystals Powell and Kepler[550–553] observed that the $^1Y^*$ fluorescence rise time is less than the $^1M^*$ fluorescence decay time. This behaviour, known as the PK effect, also occurs in other mixed crystal systems.[547,554–557] In terms of (9.125a) and (9.131a), the PK effect corresponds to $k_M^u > k_M^t$, although $i_M^t(t)$ and $i_Y^t(t)$ cannot properly be expressed by simple exponentials.

For three years Powell[555,558] was the sole arbiter of the several models proposed to account for the PK effect. Independent quantitative comparisons with his data were inhibited, because his observed fluorescence response functions $f(t)$ were uncorrected for the unpublished instrumental response function $I(t)$ (p. *95*). The ultimate publication by Soos and Powell[557,559] of $k_{YM}(t)$ values derived from the original data ended a frustrating triennium.

The following models have been considered in attempts to explain the PK effect.

(i) Powell and Kepler[550–553] showed that the observed fluorescence response functions do not agree with relations (9.125a) and (9.131a) of the TI exciton diffusion model.

(ii) They also showed that their results are not consistent with long-range dipole–dipole (LRDD) transfer between static molecules with a critical transfer radius R_0.

(iii) Powell and Kepler[550–553] found that the only reasonable fit to their data was obtained using a model combining TI exciton diffusion and LRDD transfer.[542,560] The fit involved apparently improbable values of D and R_0. For tetracene-doped anthracene crystals $D \simeq 4 \times 10^{-6}\ \text{cm}^2\ \text{s}^{-1}$ is two orders of magnitude less than the normally accepted value, and $R_0 \simeq 100$ Å is effectively three orders of magnitude greater than the calculated value of $R_0 \simeq 30$ Å, since LRDD transfer varies as r^{-6} (p. *569*).

(iv) Rosenstock[561] formulated a nearest-neighbour random walk (NNRW) model for exciton propagation and guest trapping, and he showed that it agreed satisfactorily with the fluorescence *decay* data for tetracene-doped anthracene crystals.[550]

Powell[555] pointed out that the NNRW and the TI exciton diffusion models give identical results, provided the number of steps is large, as in the present case. He reported that his experimental k_M^t values[550] were consistent with the TI exciton diffusion relation (9.158) and with corresponding fluorescence quantum yield data (9.156) and (9.157), taking $D = 5 \times 10^{-4}\ \text{cm}^2\ \text{s}^{-1}$ and $R = 10$ Å in (9.155). Since (9.155) does not distinguish D and R, it is preferable to write $DR = 5 \times 10^{-11}\ \text{cm}^3\ \text{s}^{-1}$. These results indicate that the PK effect is transient and limited to a short period after excitation, and that it does not affect the subsequent fluorescence decay or the photostationary behaviour.

(v) Birks[520] proposed a TI exciton diffusion model involving defect traps ^{1}X, populated by ^{1}M* exciton capture or by ^{1}M* radiative transfer, to account for the PK effect in thick mixed crystals. ^{1}X* is assumed to have the same fluorescence parameters as ^{1}M*, but to be unable to transfer energy to ^{1}Y. On this model k_M^u is the decay rate of the ^{1}M* excitons which populate ^{1}Y*, while k_M^t is the decay rate of the combined ^{1}M* and ^{1}X* fluorescence. For thick mixed crystals, in which radiative transfer to ^{1}X* occurs, the model predicts $k_M^u > k_M^t$ in qualitative agreement with the data of Powell and Kepler.[550–553] Pearlstein *et al.*[562] extended the model to include partially reversible trapping, mainly by host reabsorption of the ^{1}X* fluorescence, and they concluded that the extended model gives a good quantitative fit to the data. Powell and Soos[558] drastically modified the Birks[520] model by the omission of radiative migration and transfer, by the inclusion of thermal detrapping of $^1X^*$, and by the inclusion of energy transfer from $^1X^*$ to ^1Y, and they found that the modified model does not fit the data.

(vi) Powell[547] observed that the PK effect is not limited to thick mixed crystals. It also occurs in a 267 μm anthracene crystal containing 10^{-6} mole/mole tetracene at 300 and 100°K. Since the Birks[520] model does not account for this thin crystal behaviour, an alternative explanation of the PK effect must be sought. Radiative transfer to defects may, however, contribute to the thick crystal behaviour, and defects may influence the overall energy transfer process in other ways.

(vii) Powell and Soos[558] attempted to detect $^1X^*$ emission from a thick pure anthracene crystal at 300°K by observing the response function of the fluorescence at $\lambda > 600$ nm. The 'equipment-limited' rise time of 5 ns and the decay time of 27 ns did not differ from those of the host crystal, and they concluded that most defect traps in pure anthracene crystals are thermally depopulated at room temperature, in agreement with the results of §9.11.2.6. This does not necessarily exclude them from consideration (see (xiii) below)).

(viii) The significance of guest-induced defects in a mixed crystal has been suggested previously (p. *534*). Lyons and Warren[528] observed the defect bands introduced into the fluorescence spectra of anthracene crystals at 4·2°K by various guest species. The defect bands, which are distinct from the host and guest fluorescence bands, are attributed to host molecules displaced by the guest from their normal lattice positions. In anthracene the effects increase in the following order of guest species: carbazole, biphenyl, *p*-terphenyl, anthraquinone, naphthalene, tetracene, phenanthrene, pyrene and chrysene. Such guest-induced defects in the vicinity of a guest molecule ^1Y can increase its effective $^1M^*$ exciton interaction radius R (p. *534*).

(ix) Powell[556] observed the fluorescence response functions of thin naphthalene crystals containing (a) anthracene, (b) tetracene and (c) anthracene and tetracene. For (a) and (b) good fits to the data were obtained with $D = 0$, $R_0 = 60$ Å, compared with theoretical static values of $R_0 = 23$ Å and 12 Å for anthracene and tetracene, respectively. For energy transfer from anthracene to tetracene in crystal (c), the data were fitted with $D = 0$, $R_0 = 28$ Å, in agreement with the theoretical values. The results show that guest–guest LRDD transfer is not affected by the PK effect, which is restricted to host–guest transfer. The high value of $R_0 = 60$ Å in (a) and (b) is independent of the guest, suggesting that it is primarily due to the host.

(x) Soos and Powell[559] have formulated a generalized random walk (GRW) model of exciton motion and transfer in mixed crystals. The model includes

(I) extended trapping regions about each guest molecule, corresponding to guest-induced host traps;

(II) anisotropy of the exciton motion;

(III) anisotropy of the extended trapping regions; and

(IV) exciton jumps to molecules which are not nearest neighbours.

There is experimental evidence for I (see (viii) above)) and for II (p. *532–3*). The guest-induced extended trapping regions are probably irregular in shape and size, like dislocations and disordered regions in pure crystals, but it is questionable if they have any quantifiable anisotropy III. Process IV is improbable for singlet excitons in organic molecular crystals where nearest-neighbour interactions dominate (pp. *523–8*), and it is ignored in the NNRW singlet exciton hopping model (pp. *528–9*).

As Powell[557] has recently noted, 'the time dependence of the energy transfer rate is of central importance in interpreting time-resolved spectroscopy results' in mixed crystal systems. The GRW model[559] predicts a time-dependent energy transfer rate parameter

$$k_{YM}(t) = k_{MM}C\{1 + (C/\pi\sigma^2)(2\pi\sigma^2 k_{MM}t)^{-\frac{1}{2}}\} \tag{9.159}$$

k_{MM} $(= 1/\tau_{MM})$ is the $^1M^*$ exciton hopping frequency, C is the capacity of the random walk, defined, in the limit of many hops, as the number of new sites sampled by the exciton on each hop. C depends on the size (I) and shape (III) of the extended trapping region, and on the anisotropy (II) and the distribution of step lengths (IV) in the random walk. For a spherical extended trapping region of radius R, C is proportional to R. σ is a parameter which depends on the exciton hopping probabilities in different directions (II) and of different lengths (IV). For an isotropic random walk $2\sigma^2 \simeq 1$; for an anisotropic random walk $2\sigma^2 \simeq 0{\cdot}5$.

Soos and Powell[559] analysed the previous $f_M^t(t)$ and $f_Y^t(t)$ data,[547,550–556] and they showed that the energy transfer rate parameter $k_{YM}(t)$ is time-dependent. Although there is considerable scatter in the data from different mixed crystal specimens, it is clear that $k_{YM}(t)$ decreases with increase of t towards an asymptotic value

k_{YM}. In doped anthracene crystals $k_{YM}(t)$ decreases from $\sim 3\, k_{YM}$ at $t = 2{\cdot}5$ ns to $\sim k_{YM}$ at $t > \sim 20$ ns. In doped naphthalene crystals $k_{YM}(t)$ decreases from $\sim 3\, k_{YM}$ at $t = 2{\cdot}5$ ns to $\sim k_{YM}$ at $t > \sim 40$ ns.

A comparison of the $k_{YM}(t)$ data with the GRW relation (9.159) gives values of $C \sim 6$–60 and $\tau_{MM} \sim 4$–40×10^{-11} s for doped naphthalene crystals, and values of $C \sim 30$–80 and $\tau_{MM} \sim 1$–6×10^{-11} s for doped anthracene crystals.[559] A re-analysis of the data previously fitted[550–553] to the combined TI exciton diffusion and LRDD transfer model (see (iii) above)) yields values of $C \sim 19$–38 and $\tau_{MM} \sim 1$–3×10^{-10} s for doped naphthalene crystals, and $C \sim 35$–70 and $\tau_{MM} \sim 3$–6×10^{-11} s for doped anthracene crystals.[559] In a subsequent study[557] of pyrene-doped naphthalene crystals, the relation

$$k_{YM}(t) = a + bt^{-\frac{1}{2}} \tag{9.160}$$

was observed with $b/a = 4{\cdot}5\,(\text{ns})^{\frac{1}{2}}$. Assuming an anisotropic random walk, corresponding to $2\sigma^2 \simeq 0{\cdot}5$, values of $C \sim 12$ and $\tau_{MM} \sim 1{\cdot}5 \times 10^{-10}$ s are derived.

(xi) The substitution of

$$C = 2\pi\sigma^2 c_M^{\frac{1}{3}} R \tag{9.161}$$

$$k_{MM} = 2c_M^{\frac{2}{3}} D/\sigma^2 \tag{9.162}$$

in the GRW relation (9.159) gives the TD exciton diffusion relation

$$k_{YM}(t) = 4\pi DR(1 + R(\pi Dt)^{-\frac{1}{2}})c_M \tag{9.154}$$

This relation assumes isotopic exciton diffusion, so that $2\sigma^2 = 1$.

For crystal anthracene ($c_M = 4{\cdot}2 \times 10^{21}$ molecules/cm^3) the experimental $k_{YM}(t)$ data can be fitted to (9.154) with $R = 1{\cdot}97C$ Å and $D = 9{\cdot}6 \times 10^{-16} k_{MM}$ cm^2 s^{-1}. The C and τ_{MM} values[559] for doped anthracene crystals thus correspond to $R \sim 60$–160 Å and $D \sim 1{\cdot}6$–$9{\cdot}6 \times 10^{-5}$ cm^2 s^{-1}. For crystal naphthalene ($c_M = 5{\cdot}8 \times 10^{21}$ molecules/cm^3) $R = 1{\cdot}77C$ Å and $D = 7{\cdot}7 \times 10^{-16} k_{MM}$ cm^2 s^{-1}. The C and τ_{MM} values[559] for doped naphthalene crystals correspond to $R \sim 11$–106 Å and $D \sim 1{\cdot}9$–19×10^{-6} cm^2 s^{-1}.

(xii) The values of k_{MM} and D are one or two orders of magnitude less than the previously accepted values, obtained from (9.155) assuming collisional energy transfer (p. *604*).

On the TD exciton diffusion model the radii R of the spherical extended trapping regions in doped naphthalene and anthracene crystals correspond to ~5–20 lattice spacings. On the GRW model[559] it is proposed that anisotropy of exciton motion (II), anisotropy of the trapping regions (III) and exciton jumps to non-nearest neighbours (IV), can each increase C beyond the value given by (9.161), and thus reduce the size of the trapping regions required to fit the data. Soos and Powell[559] made GRW calculations for simple cubic crystals. By a qualitative extrapolation they concluded that trapping regions extending over ~5 lattice spacings around the guest molecules are sufficient to account for the behaviour of doped monoclinic naphthalene and anthracene crystals, provided factors II, III and IV are present.

Powell[557] considers that the GRW model has distinct advantages over the TD exciton diffusion model in that 'parameters such as C and σ have well-defined physical meanings, whereas R is an artificial parameter whose exact interpretation is difficult.' The author takes a contrary view. In the TD exciton diffusion model the parameters D and R also have 'well-defined physical meanings.' D is a property of the host crystal, which can be determined independently from singlet exciton–exciton annihilation studies[543,544] in pure crystals (§9.11.3). The size and shape of the extended trapping regions could be determined using the various techniques used for the study of defects and dislocations in organic crystals.[564] Until such measurements have been shown to conflict with the TD exciton diffusion model, it would be premature to discard it.

In contrast, none of the three parameters k_{MM}, C and σ of the GRW model are amenable to independent observation. The factors II and IV of this model can be incorporated in the TD exciton diffusion model, if required. The trapping region anisotropy III needs parameterization before it can be included quantitatively in either model. To date, the GRW model has only been formulated for a simple cubic lattice, and its quantitative application to monoclinic crystals is a major extrapolation. As Powell[557] states, 'the only problem with using the GRW model is that its quantitative application to lattices with low symmetry and more than one molecule per unit cell is very complicated.' Until these complications are resolved, the GRW model supplements, but does not replace, the TD exciton diffusion model.

(xiii) Hemenger *et al.*[565] have used a master equation approach, equivalent to the random walk method, to examine various aspects of exciton transfer and quenching on cubic lattices. They conclude that the host and guest fluorescence response functions are considerably modified by

(V) slow back-transfer from the guest to the host and

(VI) the presence of host traps (defects),

but they are insensitive to

(II) anisotropy of the exciton motion.

The GRW model of Soos and Powell[557] ignores V and VI, while it considers II to be important in determining σ^2 and influencing C.

In mixed crystals slow back-transfer V may occur from defects near the boundaries of the guest-induced extended trapping regions. The inherent assumption that exciton capture by such defects automatically results in $^1Y^*$ fluorescence is questionable. The concentration of host defects VI, other than guest-induced defects, commonly exceeds the guest concentration.[520] At room temperature these defects will be transiently populated by capture of $^1M^*$ excitons and subsequently depopulated thermally. This process, which takes a finite time, brings the random walk to a temporary halt, thus reducing the mean exciton hopping frequency k_{MM} and the mean $^1M^*$ exciton diffusion coefficient D. It thus appears that both host defects and guest-induced defects play important roles in mixed crystals, the former by decreasing the effective value of D, and the latter by increasing the effective value of R.

(xiv) Selsby and Swenberg[566] proposed that the Smoluchowski relation (9.154) should be modified to include $^1M^*$ exciton decay, and they obtained the relation

$$k_{YM}(t) = 4\pi DR \exp(-k_M t)(1 + R(\pi Dt)^{-\frac{1}{2}})c_M \tag{9.163}$$

If the $R(\pi Dt)^{-\frac{1}{2}}$ factor is neglected, this reduces to

$$k_{YM}(t) = 4\pi DR \exp(-k_M t)c_M \tag{9.164}$$

Using an analytic expression for the excitation function $I(t)$ ($= \exp[-(t - 2{\cdot}9)^2/2{\cdot}8]$ with t in ns) and experimental values of the rate parameters, Selsby and Swenberg[566] showed that (9.164) gives a good fit to the observed fluorescence response functions of tetracene-doped anthracene crystals,[550–553] using the usually accepted value of $D = 10^{-3}$ cm^2 s^{-1} and an unspecified value of R. Similar agreement was obtained for anthracene-doped naphthalene crystals.

Despite the empirical success of this model, it is open to various objections.

(1) The $^1M^*$ exciton decay rate

$$-\frac{d[^1M^*]}{dt} = (k_M + k_{YM}[^1Y])[^1M^*] \qquad (9.165)$$

is already included in the energy transfer rate $k_{YM}[^1Y][^1M^*]$ through the time-dependent $[^1M^*]$, and its further inclusion in k_{YM} does not appear to be justified.

(2) In a mixed crystal the $^1M^*$ exciton decay rate is $(k_M + k_{YM}[^1Y])$ and not k_M, as assumed in (9.163) and (9.164).

(3) The observation[559] that $k_{YM}(t)$ tends to an asymptotic value k_{YM}, and the experimental relation (9.160), are not consistent with (9.163) unless $k_M = 0$.

(xv) The experimental and theoretical studies of the PK effect have already radically changed ideas about exciton migration, trapping and transfer processes in mixed crystals. A more quantitative understanding of these processes could be obtained from the following types of experiments.

(a) The observation of the fluorescence response function (9.152) of a pure crystal with intense pulsed laser excitation, and the evaluation of the singlet exciton–exciton annihilation rate parameter k_{AA} from (9.153), provides a method for the determination of the $^1M^*$ diffusion coefficient D in the absence of a guest. Inoue *et al.*[544] have determined absolute values of k_{AA} for pyrene and perylene crystals, and relative values of k_{AA} for anthracene crystals from 77 to 293°K.

(b) 'Pure' anthracene crystals contain 2-MA impurity molecules and certain defect species, the nature and concentration of the defect species being determined by the method of crystal growth (§9.11.2). Time-resolved spectroscopy of 'pure' crystals at low temperatures, where the fluorescence of the different species can be resolved, offers a mean of studying energy migration and transfer in the absence of guest perturbation. Such measurements have already been made,[523,549] but they have not been analysed from this viewpoint.

(c) Guest-induced defects can be eliminated by the choice of guest molecules which are isomorphic with the host crystal. Isomorphic mixed crystals can be studied at normal temperatures, provided the host and guest fluorescences are spectroscopically resolved.

(d) Various physical methods, such as microscopy, electron microscopy, X-ray crystallography and radiography, are available to determine the structure and size of the guest-induced extended defect regions.

(e) The time resolution of the mixed crystal fluorescence studies can be greatly improved by using single-photon fluorometry with either short duration light flashes (e.g. from a mode-locked laser) or individual ionizing radiations for excitation (§9.4.1.2). The 'equipment-limited' fluorescence rise time of 5 ns, observed by Powell and Soos,[558] is an order of magnitude longer than that obtained using the single-photon technique.[69,70,91–94] Improved time resolution is essential for the more precise determination of $k_{YM}(t)$.

Studies of this type should show whether the TD exciton diffusion model (9.154) is adequate, or whether a more complicated random-walk model is required, to describe singlet host ($^1M^*$)–singlet guest ($^1Y^*$) energy transfer in organic mixed crystals.

9.11.4.2 *Singlet–triplet energy transfer*

The much less probable process of singlet guest ($^1Y^*$)–triplet host ($^3M^*$) energy transfer has been observed in tetracene-doped anthracene crystals by Vaubel.[567] The tetracene ($^1Y^*$) was directly excited by a krypton ion laser of variable intensity, and the intensity of the delayed anthracene ($^1M^*$) fluorescence produced by $^3M^* - {}^3M^*$ interaction was observed. A small correction was made for direct population of $^3M^*$. The observed $^1Y^* - {}^3M^*$ energy transfer rate is $\sim 4 \times 10^{-4} k_Y = 3 \times 10^4\ s^{-1}$, taking the experimental value of $k_Y = 7{\cdot}7 \times 10^7\ s^{-1}$ for tetracene in crystal anthracene,[550–553] rather than the value of $k_Y = 10^7\ s^{-1}$ assumed by Vaubel.[567]

All four possible energy transfer processes from a singlet ($^1A^*$) or triplet ($^3A^*$) excited molecule to an unexcited molecule in the singlet state (^1B) have now been observed.

$$^3A^* + {}^1B \rightarrow {}^1A + {}^3B^* \tag{9.166}$$

$$^3A^* + {}^1B \rightarrow {}^1A + {}^1B^* \tag{9.167}$$

$$^1A^* + {}^1B \rightarrow {}^1A + {}^1B^* \tag{9.168}$$

$$^1A^* + {}^1B \rightarrow {}^1A + {}^3B^* \tag{9.169}$$

Processes (9.166)–(9.168) have been considered previously (pp. *528–59, 567–80, 590–4*).

The corresponding energy transfer processes in which the acceptor is in a triplet state ($^3B^*$ or $^3O_2$) are as follows:

$$^3A^* + {}^3B^* \rightarrow {}^1A + {}^3B^{**} \tag{9.170}$$

$$^3A^* + {}^3B^* \rightarrow {}^1A + {}^1B^* \quad (9.171a)$$

$$^3A^* + {}^3O_2 \rightarrow {}^1A + {}^1O_2^* \quad (9.171b)$$

$$^1A^* + {}^3B^* \rightarrow {}^1A + {}^3B^{**} \quad (9.172)$$

$$^1A^* + {}^3B^* \rightarrow {}^1A + {}^1B^* \quad (9.173)$$

All of these processes, except (9.173), have been observed.

9.11.5 *Aromatic liquid solutions* (pp. *580–90*)

The rate parameter k_{YM} of energy transfer from an excited aromatic liquid solvent $^1M^*$ to a solute ^1Y exceeds the rate parameter k_{diff} of a diffusion-controlled collisional process. Similarly the rate parameter k_{QM} of quenching of $^1M^*$ by an efficient ($p = 1$) quencher ^1Q exceeds the rate parameter k_{QY} ($= k_{diff}$) of quenching of $^1Y^*$ by ^1Q. These effects have been interpreted as evidence for $^1M^*$ migration in aromatic liquids at a rate exceeding that of molecular diffusion, and several models of $^1M^*$ migration have been proposed.[568–576]

Birks *et al.*[568] have determined k_{YM}, k_{QM} and k_{QY} for the solution system: deoxygenated toluene = ^1M, 2,5-diphenyloxazole (PPO) = ^1Y, and carbon tetrabromide = ^1Q, from -20 to $+50$°C. The results, which are plotted in Figure 9.20, have been analysed using the Smoluchowski–Sveshnikov relations,

$$k_{QY} = 4\pi N' p_{QY} R_{QY} (D_Q + D_Y)[1 + p_{QY} R_{QY} \{(D_Q + D_Y)\tau_Y\}^{-\frac{1}{2}}] \quad (9.174)$$

$$k_{QM} = 4\pi N' p_{QM} R_{QM} (D_Q + Z_M)[1 + p_{QM} R_{QM} \{(D_Q + Z_M)\tau_M\}^{-\frac{1}{2}}] \quad (9.175)$$

$$k_{YM} = 4\pi N' p_{YM} R_{YM} (D_Y + Z_M)[1 + p_{YM} R_{YM} \{(D_Y + Z_M)\tau_M\}^{-\frac{1}{2}}] \quad (9.176)$$

D_Q, D_Y and D_M are the diffusion coefficients; $Z_M = D_M + \Lambda_M$, where Λ_M is the $^1M^*$ migration coefficient; τ_M and τ_Y are the $^1M^*$ and $^1Y^*$ lifetimes; p_{QY}, p_{QM} and p_{YM} are the interaction probabilties, and R_{QY}, R_{QM} and R_{YM} are the interaction distances, of the three processes, respectively.

Experimental values of τ_M [121] and τ_Y (p. *137*) are available. The molecular radii r_M ($= 3{\cdot}17$ Å), r_Y ($= 3{\cdot}94$ Å) and r_Q ($= 3{\cdot}23$ Å) were evaluated from

$$r = 0{\cdot}91 \left(\frac{3M}{4\pi N \rho} \right)^{\frac{1}{3}} \quad (9.177)$$

where M is the molecular weight, ρ is the density at 20°C and **N** is Avogadro's number. Values of D_Q ($= D_M r_M / r_Q$) and D_Y ($= D_M r_M / r_Y$) were obtained from experimental data on the self-diffusion coefficient D_M of toluene,[577,578] assuming Stokes' law.

The solute excitation quenching k_{QY} is consistent with (9.174) with $p_{QY} = 1{\cdot}0$, $R_{QY} = 5{\cdot}35$ Å over the complete temperature (t) range. Over

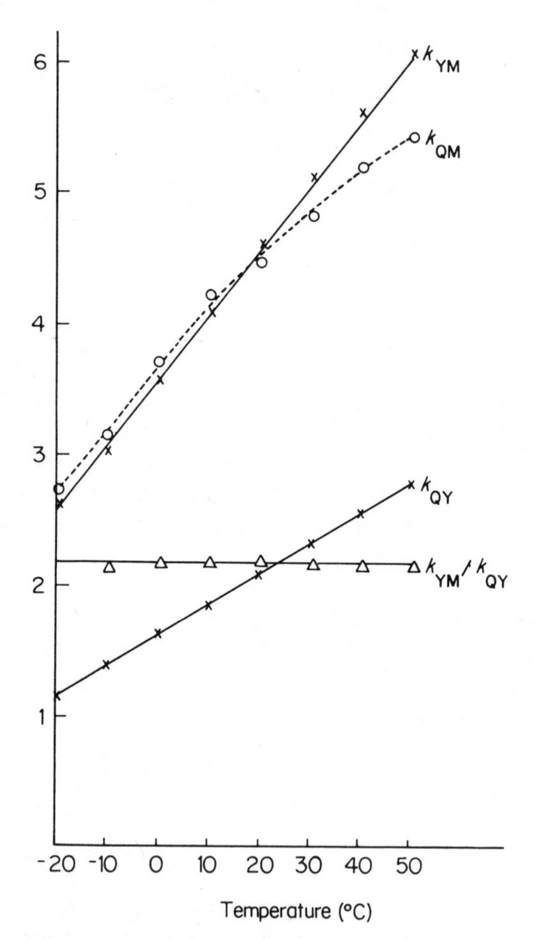

Figure 9.20 Deoxygenated toluene (^{1}M) solutions of PPO (^{1}Y) and carbon tetrachloride (^{1}Q). Temperature dependence of rate parameters (in 10^{10} M^{-1} s^{-1}) of solvent–solute energy transfer (k_{YM}), solvent fluorescence quenching (k_{QM}), solute fluorescence quenching (k_{QY}) and the ratio k_{YM}/k_{QY}. (After Birks, Najjar and Lumb[568])

the same range k_{YM} is proportional to k_{QY} (Figure 9.20) with $k_{YM}/k_{QY} = 2{\cdot}18\ (\pm 0{\cdot}02)$. Since the ratio of the transient terms (in square brackets) in (9.174) and (9.176) is practically independent of t, this result shows that $Z_M \propto D_M$ over the full temperature range. Two alternative models may be considered to account for the behaviour:

(a) $Z_M = D_M$ (no ^{1}M* migration), and $R_{YM} \simeq R_{QM} > R_{QY}$; or

(b) $Z_M = D_M + \Lambda_M$ (with $\Lambda_M \propto D_M$) and $R_{YM} \simeq R_{QM} \simeq R_{QY}$.

On model (a) the solvent–solute energy transfer is consistent with (9.176) with $Z_M = D_M$, $p_{YM} = 1{\cdot}0$, $R_{YM} = 12{\cdot}7$ Å over the full temperature range, and the solvent excitation quenching is consistent with (9.175) with $Z_M = D_M$, $R_{QM} = 12{\cdot}1$ Å and $p_{QM} = 1{\cdot}0$ at $t \leqslant 10°C$ and $p_{QM} < 1{\cdot}0$ at $t > 10°C$ ($p_{QM} = 0{\cdot}87$ at 50°C). The values of R_{QY} $(= \bar{r}_Q + \bar{r}_Y)$, R_{YM} $(= \bar{r}_Y + \bar{r}_M)$ and R_{QM} $(= \bar{r}_Q + \bar{r}_M)$ correspond to molecular interaction radii of $\bar{r}_Q = 2{\cdot}38$ Å $(= 0{\cdot}74 r_Q)$, $\bar{r}_Y = 2{\cdot}98$ Å $(= 0{\cdot}76 r_Y)$ and $\bar{r}_M = 9{\cdot}73$ Å $(= 3{\cdot}07 r_M)$. On this model the $^1M^*$ interaction radius, either for transfer to ^1Y or for quenching by ^1Q, exceeds the ^1M molecular radius by a factor of ~3. The magnitude of $\bar{r}_M$ corresponds to delocalization of the solvent excitation over three adjacent molecules, possibly due to excimer interaction of $^1M^*$ with its nearest neighbours.

On model (b) $Z_M \simeq 3D_M$ offers an alternative description of the data, provided that Λ_M $(\simeq 2D_M)$ is proportional to D_M over the full temperature range. None of the existing theories of $^1M^*$ migration in aromatic liquids predicts this type of behaviour.[569–576] It thus remains open to question whether

(a) $\bar{r}_M \simeq 3r_M$, $Z_M = D_M$, or

(b) $\bar{r}_M \simeq r_M$, $Z_M \simeq 3D_M$,

is the appropriate method of describing energy transfer and quenching of an aromatic liquid solvent. Further experimental and theoretical work are required to resolve this question.

9.12 Saturated compounds

9.12.1 *Alkanes*

The discovery of the fluorescence of saturated hydrocarbons by Hirayama and Lipsky[579] was described in *PAM* (pp. *636*, *638*). Subsequent studies of the fluorescence spectra,[580,581] quantum yields,[580,581] lifetimes,[582–584] energy transfer[585–588] or quenching[579,583] of 128 alkanes, including linear, branched, cyclic and polycyclic compounds, have established the generality of the phenomenon, and they have shown several interesting correlations with molecular structure.

Hirayama *et al.*[579,580] observed the fluorescence spectrum F_M and quantum yield $\Phi_{FM}(147)$ of 43 liquid alkanes at 25°C with an excitation wavelength $\lambda_{ex} = 147$ nm. In each case F_M is structureless with an intensity maximum at wavelength λ_F and a bandwidth of σ_F. In a subsequent study,[581] which included 85 other alkanes, they used $\lambda_{ex} = 165$ nm, which is within 5–25 nm of the absorption onset, defined as the wavelength λ_A

Table 9.48 Fluorescence and absorption characteristics of *n*-alkane liquids at 25°C [581]

Compound[a]	λ_F (nm)	σ_F (10^3 cm^{-1})	$\Phi_{FM}(165)$ ($\times 10^3$)	$\Phi_{FM}(147)$ ($\times 10^3$)	β	λ_A (nm)	Δ (10^3 cm^{-1})
n-5	206	8·4	0·35	0·063	0·18	169·0	10·6
n-6	206	8·2	0·59	0·17	0·29	170·7	10·0
n-7	206	7·9	1·6	0·69	0·43	171·5	9·8
n-8	207	7·7	2·5	1·1	0·44	171·8	9·9
n-9	207	7·7	3·3	1·6	0·48	173·0	9·5
n-10	207	7·7	4·2	2·0	0·48	172·8	9·6
n-11	207	7·6	5·0	2·8	0·56	173·3	9·4
n-12	207	7·6	5·5	3·2	0·58	173·8	9·2
n-13	207	7·6	6·7	3·5	0·52	174·0	9·2
n-14	207	7·6	6·8	3·9	0·57	174·0	9·2
n-15	208	7·6	7·3	4·0	0·55	175·0	9·1
n-16	207	7·7	7·6	4·4	0·58	174·2	9·1
n-17	209	7·7	7·4	4·6	0·62	176·0	9·0

[a] The number in the compound designation refers to the number of carbon atoms in the *n*-alkane.

at which the molar extinction coefficient $\varepsilon = 5\ \mathrm{M}^{-1}\ \mathrm{cm}^{-1}$. The increase in λ_{ex} does not change F_M, but it increases the fluorescence quantum yield from $\Phi_{FM}(147)$ to $\Phi_{FM}(165)$. The parameter $\beta = \Phi_{FM}(147)/\Phi_{FM}(165)$ is the quantum yield of internal conversion between the two corresponding excitation levels. The parameter Δ ($= \lambda_A^{-1} - \lambda_F^{-1}$) relates to the Stokes shift between the absorption and fluorescence spectra. The data[581] are presented in Tables 9.48–9.56.

Table 9.49 Fluorescence characteristics of *n*-alkane liquids at −78°C [581]

Compound	λ_F (nm)	σ_F (10^3 cm^{-1})	$\Phi_{FM}(165)$ ($\times 10^3$)	$\Phi_{FM}(147)$ ($\times 10^3$)	β
n-4	215	9·0	2·5	0·6	0·24
n-5	208	8·4	5·5	1·9	0·35
n-6	209	8·2	8·5	4·2	0·49
n-7	206	7·9	10·5	7·0	0·67

Tables 9.48 and 9.49 list data on liquid *n*-alkanes at 25°C and −78°C, respectively. Corresponding data on liquid monosubstituted *n*-alkanes at 25°C are given in Table 9.50. Dialkyl substitution on the same carbon atom is described as *geminal*, and that on adjacent carbon atoms as *vicinal*. No geminal dialkyl-substituted alkanes have been observed to fluoresce.[581] Tables 9.51 and 9.52 list data at 25°C for liquid alkanes with (a) two or more non-geminal, non-vicinal alkyl branches and (b) two or more vicinal alkyl branches, respectively.

Of the liquid unsubstituted monocycloalkanes, only cyclohexane is observed to fluoresce at 25°C,[581] although solid cyclododecane is also fluorescent.[580] Energy transfer has been observed[579,585–588] from cyclohexane to benzene and other aromatic molecules. Tables 9.53, 9.54 and 9.55 present data at 25°C for liquid monocyclohexanes with (a) one or no substituent alkyl branch, (b) two or more non-vicinal branches and (c) two or more vicinal branches, respectively. Table 9.56 gives data for several liquid polycycloalkanes at 25°C. Cyclohexane, decalin and the other polycycloalkanes are the perhydrogenated derivatives of benzene, naphthalene and other polycyclic aromatic hydrocarbons.

The fluorescence lifetimes τ_M of several liquid alkanes at 25°C have been measured with pulsed X-ray excitation by Henry and Helman[582] and with pulsed vacuum ultraviolet excitation by Ware and Lyke.[584] Table 9.57 lists the values of τ_M, $\Phi_{FM}(165)$, $k_{FM}(165)$ ($= \Phi_{FM}(165)/\tau_M$) and k_{IM} ($= \tau_M^{-1} - k_{FM}(165) \simeq \tau_M^{-1}$). Since $\Phi_{FM}(165) \leqslant \Phi_{FM}$, the fluorescence quantum yield of the equilibrated system, $k_{FM}(165) \leqslant k_{FM}$, the corresponding radiative transition rate.

Table 9.50 Fluorescence and absorption characteristics of monosubstituted alkane liquids at 25°C [581]

Compound[a]	λ_F (nm)	σ_F (10^3 cm^{-1})	$\Phi_{FM}(165)$ ($\times 10^3$)	$\Phi_{FM}(147)$ ($\times 10^3$)	β	λ_A (nm)	Δ (10^3 cm^{-1})
2-M4	233	9·0	0·10	<0·01	<0·1	171·0	15·6
2-M5	230	9·7	0·14	0·01	0·07	173·0	14·3
3-M5	231	10·0	0·16	0·03	0·2	173·0	14·5
2-M6	223	9·5	0·13	0·03	0·2	173·0	13·0
3-M6	225	9·2	0·20	0·07	0·35	174·5	12·9
3-E5	227	8·7	0·44				
2-M7	220	9·4	0·24	0·07	0·29	173·8	12·1
3-M7	223	9·5	0·25	0·11	0·44		
4-M7	218	9·4	0·30	0·12	0·40		
3-E6	227	9·7	0·58				
2-M8	217	9·5	0·31	0·11	0·35	173·8	11·5
3-M8	220	8·6	0·33	0·16	0·48	174·5	11·9
4-M8	217	8·8	0·44	0·22	0·50	174·8	11·1
3-E7	222	9·0	0·67			175·4	12·0
4-E7	222	9·2	0·74			175·5	12·0
2-M8	215	10·2	0·36	0·17	0·47	174·2	10·9
3-M9	220	9·6	0·36	0·22	0·61	174·8	11·8
4-M9	216	9·3	0·55	0·28	0·51	175·0	10·8
5-M9	216	8·7	0·53	0·35	0·66	175·0	10·8
3-E8	220	9·0	0·79			175·5	11·5
4-E8	220	10·0	0·85			176·0	11·4
4-Pr7	215	9·1	0·76			176·2	10·2
2-M10	212	9·3	0·49	0·25	0·51	174·2	10·2
3-M10	215	9·6	0·45	0·29	0·64		
4-M10	212	8·9	0·67	0·39	0·58		
3-E9	212	9·4	0·90				
2-M11	212	8·3	0·60	0·36	0·60	174·5	10·1
2-M19	208	7·7	2·0			175·5	8·9

[a] The first number gives the position of the branch, the letter the nature of the branch (M = methyl, E = ethyl, Pr = propyl), and the last number

Table 9.51 Fluorescence and absorption characteristics at 25°C of liquid alkanes with multiple and non-vicinal and non-germinal branches[581]

Compound[a]	λ_F (nm)	σ_F (10^3 cm^{-1})	$\Phi_F(165)$ ($\times 10^3$)	λ_A (nm)	Δ (10^3 cm^{-1})
2,4-M5	216	9·2	0·43	175·2	10·8
2,4-M6	215	9·5	0·48		
2,5-M6	222	9·5	0·13		
2,4-M7	210	9·5	0·64		
2,5-M7	218	9·5	0·26	175·2	11·2
2,6-M7	215	9·9	0·29	175·5	10·5
3,5-M7	215	9·5	0·55	176·0	10·3
3,5-M8	212	9·5	0·66		
3,6-M8	217	9·3	0·34	176·0	10·7
2,6-M8	218	9·7	0·37	177·0	10·6
2,4,6-M7	206	8·8	1·6	177·7	7·7
2,6,10,14-M15	210	8·3	1·6	177·5	8·7

[a] The first numbers give the positions of the branches, the letters the nature of the branch (M = methyl), the last numbers give the number of main chain carbon atoms.

The behaviour of the *n*-alkanes depends on N_c, the number of carbon atoms.

(i) *Fluorescence spectra.* For $N_c \geqslant 8$, F_M is practically independent of N_c with mean values of $\lambda_F = 207$ nm, $\sigma_F = 7650$ cm^{-1} (Table 9.48). For $N_c < 8$, F_M is practically independent of temperature, but σ_F increases and λ_F tends to increase with reduction in N_c (Tables 9.48 and 9.49). Fluorescence is also observed from *n*-butane, *n*-pentane and *n*-hexane ($N_c = 4$–6) vapours, but only with $\lambda_{ex} \geqslant 160$ nm, implying a predissociation threshold at this wavelength.[581] The fluorescence spectra of the vapour and liquid are similar with the former blue-shifted about 1 nm, indicating that the liquid fluorescence is molecular in origin, and not due to excimers or other excited aggregates.

(ii) *Absorption spectra.* λ_A increases and Δ decreases with increase in N_c, even in compounds for which λ_F remains constant (Table 9.48). Thus the energy S_1 of the first excited Franck–Condon singlet state observed in absorption decreases with increase in N_c.

(iii) *Internal conversion.* β increases with N_c (Table 9.48), and it also increases by a factor of 1·5–2 on decreasing the temperature from 25 to -78°C (Table 9.49). As in benzene (pp. *171–7*), the deviation of β from unity might be due to (a) a radiationless transition in competition with the internal conversion and vibrational relaxation to the equilibrated fluorescent state S_1, and/or (b) external quenching by

Table 9.52 Fluorescence and absorption characteristics at 25°C of liquid alkanes with two or more vicinal branches[581]

Compound[a]	λ_F (nm)	σ_F (10^3 cm^{-1})	$\Phi_{FM}(165)$ ($\times 10^3$)	$\Phi_{FM}(147)$ ($\times 10^3$)	β	λ_A (nm)	Δ (10^3 cm^{-1})
2,3-M4	242	8·7	6·1	0·9	0·15	175·5	15·7
2,3-M5	242	8·7	4·0	0·7	0·18		
2,3-M6	241	8·7	4·8	1·7	0·35	177·0	15·0
3,4-M6	242	8·7	2·2				
2-M, 3-E5	242	8·7	2·2				
2,3,4-M5	244	8·9	1·6				
2,3-M7	241	8·7	5·6	2·5	0·45	177·4	14·9
3,4-M7	241	8·8	2·9			177·0	15·0
2,3,4-M6	244	9·0	1·1				
2,3-M8	241	8·7	5·8	2·8	0·48	177·4	14·9
3,4-E6	243	8·8	0·30			178·2	15·0
2-M, 3-E7	242	9·0	2·0				
3,4,5-M7	244	9·1	0·88			179·7	14·7

[a] The first numbers give the positions of the branches, the letter gives the nature of the branch (M = ethyl, E = ethyl) and the last number gives the number of main chain carbon atoms.

Table 9.53 Fluorescence and absorption characteristics at 25°C of liquid cycloalkanes with a single branch[581]

Compound[a]	λ_F (nm)	σ_F (10^3 cm^{-1})	$\Phi_{FM}(165)$ ($\times 10^3$)	$\Phi_{FM}(147)$ ($\times 10^3$)	β	λ_A (nm)	Δ (10^3 cm^{-1})
*c*6	201	7·7	8·8	3·5	0·40	177·2	6·7
M*c*6	213	8·9	11·2	5·5	0·49	179·6	8·7
E*c*6	210	8·7	11·6	7·0	0·60		
n-Pr*c*6	212	8·4	14·4	9·8	0·68	180·0	8·4
n-Bu*c*6	212	8·4	14·5			180·2	8·3
n-Pe*c*6	212	8·2	15·9				
n-Hex*c*6	212	8·1	16·1	12·0	0·75		
n-Dod*c*6	211	8·2	15·6	11·0	0·71	180·2	8·1
M*c*7	225		0·18	0·04	0·22	177·5	11·9

[a] The substituent on the ring is indicated first (M = methyl, E = ethyl, *n* = normal, Pr = propyl, Bu = butyl, Pe = pentyl, Hex = hexyl, Dod = dodecyl) followed by the number of carbon atoms in the ring (*c*6 = six membered ring, *c*7 = seven membered ring).

Table 9.54 Fluorescence and absorption characteristics at 25°C of liquid cycloalkanes with two or more non-vicinal branches[581]

Compound[a]	λ_F (nm)	σ_F (10^3 cm^{-1})	$\Phi_{FM}(165)$ ($\times 10^3$)	λ_A (nm)	Δ (10^3 cm^{-1})
cis-1,3Mc6	210	8·8	18·6	180·7	7·7
trans-1,3Mc6	212	8·6	2·0	182·3	7·7
cis-1,4Mc6	213	8·9	7·5	182·3	7·9
trans-1,4Mc6	211	9·3	18·6	180·0	8·2
1,3,5-Mc6	210	9·0	5·4	182·7	7·1
*iso*Buc6	212	9·1	12·2	181·0	8·1
1,1-Mc6	225	9·0	1·2		
1,1,3-Mc6	221	8·9	2·2		
1,1,4-Mc6	224	8·9	2·7		

[a] Unless otherwise indicated, compounds are mixtures of stereoisomers. Nomenclature follows convention of Table 9.53.

Table 9.55 Fluorescence and absorption characteristics at 25°C of liquid cycloalkanes with two or more vicinal branches[581]

Compound[a]	λ_F (nm)	σ_F (10^3 cm^{-1})	$\Phi_{FM}(165)$ ($\times 10^3$)	λ_A (nm)	Δ (10^3 cm^{-1})
cis-1,2Mc6	235	8·2	13·3	184·0	11·8
trans-1,2Mc6	231	8·7	3·1	179·8	12·3
*iso*Prc6	232	8·8	11·9	181·6	12·0
sec-Buc6	232	9·0	11·4	181·8	11·9
1,2,3-Mc6	232	8·3	20·9		
1,2,4-Mc6	232	8·5	14·8		
1,3-*iso*Prc6	229	8·8	10·8		
1,4-*iso*Prc6	232	9·0	9·3	183·5	11·4
tert-Buc6	252	8·5	0·56	183·7	14·8
tert-Pec6	250	8·5	0·16		
1,1,2-Mc6	248	7·9	1·2	183·0	14·3

[a] Unless otherwise indicated, compounds are mixtures of stereoisomers. Nomenclature follows convention of Table 9.53.

photolysis products, e.g. olefins. Addition of either 1 atm. O_2 or 10^{-2} M CCl_4 to the liquid alkane is observed[561] to change β by less than 10%, showing that $\beta < 1$ is primarily due to (a) rather than (b). Predissociation, which occurs in the vapour phase at $\lambda_{ex} < 160$ nm, is the most probable process competing with internal conversion.

(iv) *The radiative transition rate* $k_{FM}(165)$ is practically independent of N_c for the *n*-alkanes, with a mean value of $1{\cdot}64 \times 10^6$ s^{-1} (Table 9.57).

Table 9.56 Fluorescence and absorption characteristics of liquid polycycloalkanes at 25°C[581]

Compound	λ_F (nm)	σ_F (10^3 cm^{-1})	$\Phi_{FM}(165)$ ($\times 10^3$)	λ_A (nm)	Δ (10^3 cm^{-1})
1,2-Dicyclohexylethane	211	8·0	11·3		
Bicyclohexyl	226	8·5	20·1		
Cyclopentylcyclohexane	228	9·5	1·0		
Bicyclopentyl	225		0·14	185·5	9·5
cis-Decalin	231	8·1	21·0	187·9	9·9
trans-Decalin	217	8·9	23·0	186·2	7·6
cis-Hydroindane	235	7·7	5·2	186·0	11·2
trans-Hydroindane	231	7·2	9·5	183·5	11·2
Perhydrophenanthrene	236	8·0	35·8	192·8	7·6
Perhydrofluorene	243	8·0	0·94		

Table 9.57 Fluorescence lifetimes of saturated hydrocarbon liquids at 25°C

Compound	$\Phi_{FM}(165)$ [581] ($\times 10^3$)	τ_M [584] (ns)	$k_{FM}(165)$ (10^6 s^{-1})	k_{IM} (10^8 s^{-1})	τ_M [582] (ns)	$k_{FM}(165)$ (10^6 s^{-1})
n-Heptane	1·6	0·90	1·77	11·09	0·73	2·19
n-Octane	2·5	1·47	1·70	6·98		
n-Nonane	3·3	2·05	1·61	4·86		
n-Decane	4·2	2·5	1·68	3·98	2·6	1·62
n-Dodecane	5·5	2·85	1·45	2·58	3·1	1·77
n-Tridecane	6·7	4·2	1·63	2·35	3·7	1·81
n-Pentadecane	7·3	4·5	1·62	2·20	3·9	1·87
n-Heptadecane	7·4				4·5	1·64
Cyclohexane	8·8	0·68	12·9	14·58	0·3	29·3
Bicyclohexyl	20·1	1·58	12·6	6·17	1·6	12·6
cis-Decalin	21·0	2·18	9·63	4·49		
trans-Decalin	23·0	2·82	8·15	3·47		
Decalin	(22·0)				2·3	9·57
Methylcyclohexane	11·2	0·65	17·2	15·21		

This is much less than the theoretical value of k_{FM} calculated from the absorption spectrum, and it corresponds to a forbidden transition. No experimental evidence[581] has been found for a forbidden transition lying below the first absorption band system, and molecular orbital calculations[592] do not indicate such a transition.

The behaviour is therefore attributed to a major difference between the equilibrium molecular configurations in the ground (S_0) and excited fluorescent (S_1) states.[581] The $S_0 - S_1'$ absorption transition to the Franck–Condon excited state S_1' is followed by $S_1' - S_1$ molecular relaxation, leading to the $S_1 - S_0'$ fluorescence transition to the Franck–Condon ground state S_0'. The analogous behaviour of the diphenylpolyenes, leading to $k_{FM}' > k_{FM}$ (9.10a), has been discussed in §9.4.4. The large Stokes shift factor Δ (= 7700 – 9000 cm^{-1} for the *n*-alkanes, compared with 500–1500 cm^{-1} for condensed aromatic hydrocarbons), and the lack of mirror symmetry between the fluorescence and absorption spectra, are also consistent with a large difference between the molecular configurations in S_1' and S_1. In the *n*-alkanes, λ_A increases and the S_1' energy decreases, with increase in N_c. On the other hand, the common values of λ_F, σ_F and k_{FM} for the *n*-alkanes with $N_c \geqslant 8$ indicates that they relax to fluorescent S_1 states of similar energies and molecular configurations.

(v) *The radiationless transition rate* k_{IM} decreases with increase in N_c (Table 9.57). The increase of $\Phi_{FM}(165)$ by an order of magnitude in cooling from 25 to −78°C (Table 9.49) is probably due to a corresponding decrease in k_{IM}. The alternative possibility, an increase in k_{FM} associated with hindered molecular relaxation due to the increased viscosity, is discounted because there is no corresponding change in F_M. Extrapolation of the $\Phi_{FM}(165)$ vs. N_c data of Table 9.49 to $N_c = 3$ would indicate a value of $\Phi_{FM}(165) = 10^{-3}$ for *n*-propane, but no fluorescence ($\Phi_{FM}(165) < 10^{-5}$) was observed from *n*-propane at −78 or −90°C, nor from ethane ($N_c = 2$) at −95°C.[581]

Perdeuteration of *n*-octane and cyclohexane has little effect on their absorption or fluorescence spectra, but it increases $\Phi_{FM}(165)$ by factors of 4 and 2·5, respectively.[581] The increases are probably due to corresponding reductions in k_{IM}, and they indicate that CH vibrational modes are responsible for the radiationless transitions.

(vi) *Energy transfer and quenching.* The experimental rate parameters of energy transfer and impurity quenching of liquid alkanes exceed those calculated from the Stokes–Einstein relation by an order of magnitude.[585–587] Rothman *et al.*[581] considered the possibility of excitation migration in a liquid alkane, analogous to that in a liquid aromatic hydrocarbon (§9.11.5). They chose *cis*-decalin ($\lambda_F = 231$ nm) as the fluor, CCl_4 as the quencher and *n*-pentadecane ($\lambda_A = 175$ nm) as the diluent. With $\lambda_{ex} = 180$ nm, they observed no difference in the quenching efficiency of neat decalin and of decalin in a 2% solution. Rothman *et al.*[581] concluded that, although this negative result casts doubts on the excitation migration, the latter might still occur via intermediate states of the pentadecane solvent. The use of a non-fluorescent, preferably non-hydrocarbon solvent, such as a fluoro-carbon, is required to eliminate this possibility. Meanwhile the anomalously high energy transfer and quenching rates of the saturated hydrocarbons remain unexplained.

(vii) *Monoalkyl substitution.* Comparison of Tables 9.48 and 9.50 shows that the addition of a branch to an *n*-alkane increases the spectral parameters λ_F, σ_F and Δ and decreases the quantum yield parameters β, $\Phi_{FM}(147)$ and $\Phi_{FM}(165)$. The spectral parameters of any given branched alkane series (e.g. the 2-methylalkanes) tend to approach the *n*-alkane values with increase of N_c. The quantum yield parameters also increase with N_c, but at a slower rate than in the *n*-alkanes. The position and nature of the branch also influence the spectral and quantum yields parameters in a systematic way.

For a discussion of the other compounds in Tables 9.51–9.56 the reader is referred to the original paper of Rothman *et al.*[581]

9.12.2 *p-Dioxan*

The saturated cyclic ether, *p*-dioxan, is used as a liquid scintillator solvent for aqueous solution specimens because of its miscibility with water.[96] The fluorescence of liquid *p*-dioxan was first observed by Baxendale *et al.*[589] using pulsed electron beam excitation, and confirmed by Hirayama *et al.*[590] using optical excitation at $\lambda_{ex} = 184{\cdot}9$ nm, and by Halpern and Ware[591] who measured τ_M using a nanosecond flashlamp. The results of the three studies are listed in Table 9.58.

The fluorescence spectrum of liquid *p*-dioxan at 25°C is structureless with $\lambda_F = 247$ nm.[590] On dilution with *iso*-octane (chosen as a diluent because of its negligible fluorescence[580]), Φ_{FM} is strongly reduced and F_M is blue-shifted. No fluorescence ($\Phi_{FM} < 10^{-5}$) is observed from *p*-dioxan vapour at 25 torr with $\lambda_{ex} = 184{\cdot}9$ nm (the $S_0 - S_1$ $0 - 0$ absorption transition is at 189·6 nm) nor from solid *p*-dioxan at -78°C, and it is concluded that the *p*-dioxan molecule does not fluoresce.[590] The liquid fluorescence is attributed to some form of excited aggregate whose composition varies with dilution. The liquid absorption spectrum gives no evidence of extensive ground-state aggregation. On the other hand, increase of the diluent viscosity by a factor of 4 reduces Φ_{FM} by a factor of 2, indicating that the *p*-dioxan aggregation occurs in the excited state and is diffusion-controlled.[590] Further evidence for aggregation comes from the lifetime measurements[591] (Table 9.58), where *iso*-octane dilution is found to decrease k_{FM} and increase k_{IM}, and from the oxygen quenching data[590–591] where the measured diffusion coefficients yield an effective encounter radius for quenching of $\geqslant 13$ Å, which is much larger than the sum of the molecular radii of *p*-dioxan and oxygen.

Table 9.58 Fluorescence parameters of *p*-dioxan

System	Temperature (°C)	Excitation	λ_F (nm)	Φ_{FM}	τ_M (ns)	k_{FM} (10^7 s^{-1})	k_{IM} (10^8 s^{-1})	Ref.
Neat liquid (deaerated)	Ambient	12 MeV e's	245	—	2–3	—	—	589
	25	184·9 nm	247	0·029	—	—	—	590
		D_2 flash	—	—	2·15	1·35	4·5	591
	12	184·9 nm	250	0·040	—	—	—	590
		D_2 flash	—	—	3·2	1·25	3·0	591
Neat liquid (O_2 saturated)	25	184·9 nm	247	0·020	—	—	—	590
		D_2 flash	—	—	1·70	—	—	591
	12	184·9 nm	250	0·025	—	—	—	590
		D_2 flash	—	—	2·3	—	—	591
10 M in *iso*-octane (deaerated)	25	184·9 nm	242	0·020	—	—	—	590
		D_2 flash	—	—	1·75	1·14	5·6	591
8 M in *iso*-octane (deaerated)	25	184·9 nm	241	0·016	—	—	—	590
		D_2 flash	—	—	1·5	1·07	6·6	591

The fluorescence of liquid *p*-dioxan is extremely sensitive to the presence of water or methanol.[590] Traces of water noticeably reduce Φ_{FM} and red-shift F_M. With addition of water λ_F increases from 247 nm (neat *p*-dioxan) to 256 nm (0·1 vol. % water), 276 nm (0·5 %), 290 nm (1 %), 324 nm (5 %), 336 nm (10 %) and 350 nm (20 %). This behaviour is tentatively attributed to the formation of fluorescent *p*-dioxan–water complexes.[590] It is clearly relevant to the performance of *p*-dioxan liquid scintillator solutions for the internal counting of aqueous solution specimens.

No fluorescence has been observed from *m*-dioxan[590] or tetrahydrofuran,[589–590] but weak fluorescence has been detected from tetrahydropyran and oxepane.[590]

9.13 Conclusion

To return to the musical analogy of the Preface, the previous chapters have provided comprehensive modern renditions of many variations on the theme of 'Organic Molecular Photophysics'. This final chapter was intended as a scavenging exercise 'picking up the crumbs from the rich men's tables', and hopefully finding, like Mr. Durante, a few lost chords. That it should have proved so lengthy is an indication of the remarkable number of variations that the composer, Nature, and the different instrumentalists have introduced into the work. For many variations a reasonably complete score has been deciphered; for others only a few isolated bars or fragmentary chords have been struck; and there are some variations as yet unknown. For those who enjoy the 'Enigma Variations' of organic molecular photophysics, there is much new music still to be heard.

9.14 References

1. P. M. Rentzepis, J. Jortner and R. P. Jones, *Chem. Phys. Letters*, **4**, 599 (1970).
2. D. Huppert, J. Jortner and P. M. Rentzepis, *Chem. Phys. Letters*, **13**, 225 (1972).
3. D. Huppert, J. Jortner and P. M. Rentzepis, *J. Chem. Phys.*, **56**, 5826 (1972).
4. F. Hirayama, T. A. Gregory and S. Lipsky, *J. Chem. Phys.*, **58**, 4696 (1973).
5. T. A. Gregory, F. Hirayama and S. Lipsky, *J. Chem. Phys.*, **58**, 4697 (1973).
6. P. Wannier, P. M. Rentzepis and J. Jortner, *Chem. Phys. Letters*, **10**, 193 (1971).
7. E. Pantos and T. D. S. Hamilton, *Chem. Phys. Letters*, **17**, 588 (1972).
8. E. Pantos, Ph.D. thesis, University of Manchester (1973).
9. B. Katz, M. Brith, B. Sharf and J. Jortner, *J. Chem. Phys.*, **52**, 88 (1970).
10. M. Brith, R. Lubart and I. T. Steinberger, *J. Chem. Phys.*, **54**, 5104 (1971).
11. B. Katz, M. Brith, B. Sharf and J. Jortner, *J. Chem. Phys.*, **54**, 3924 (1971).
12. J. B. Birks, E. Pantos and T. D. S. Hamilton, *Chem. Phys. Letters*, **21**, 426 (1973).
13. G. C. Morris, and J. G. Angus, *J. Molec. Spectrosc.*, **45**, 271 (1973).
14. A. M. Taleb, I. H. Munro and J. B. Birks, *Chem. Phys. Letters*, **21**, 454 (1973).

15. E. E. Koch and A. Otto, *Chem. Phys. Letters*, **12**, 476 (1972).
16. T. Inagaki, *J. Chem. Phys.*, **57**, 2526 (1972).
17. B. L. Sowers, E. T. Arakawa and R. D. Birkhoff, *J. Chem. Phys.*, **54**, 2319 (1971).
18. M. W. Williams, R. A. MacRae, R. N. Hamm and E. T. Arakawa, *Phys. Rev. Letters*, **22**, 1088 (1969).
19. D. W. Turner, C. Baker, A. D. Baker and C. R. Brundle, *Molecular photoelectron Spectroscopy: A Handbook of the 584 Å Spectra*, John Wiley and Sons, Inc., New York, 1970.
20. E. Lindholm and B. Ö. Jonsson, *Chem. Phys. Letters*, **1**, 501 (1967).
21. B. Ö. Jonsson and E. Lindholm, *Arkiv Fysik*, **39**, 65 (1969).
22. L. Åsbrink, O. Edqvist, E. Lindholm and L. E. Selin, *Chem. Phys. Letters*, **5**, 192 (1970).
23. E. N. Lassettre, A. Skerbele, M. A. Dillon and K. J. Ross, *J. Chem. Phys.*, **48**, 5066 (1968).
24. E. E. Koch, A. Otto and K. Radler, *Chem. Phys. Letters*, **16**, 131 (1972).
25. J. H. D. Eland and C. J. Danby, *Z. Naturforsch.*, **239**, 355 (1968).
26. J. G. Angus and G. C. Morris, *Molec. Cryst. Liq. Cryst.*, **11**, 257 (1970).
27. R. Pariser, *J. Chem. Phys.*, **24**, 250 (1956).
28. R. L. Hummel and K. Ruedenberg, *J. Phys. Chem.*, **66**, 2335 (1962).
29. D. P. Craig, J. M. Hollas, M. F. Redies and S. C. Wait, *Phil. Trans. Roy. Soc.*, **A 253**, 543 (1961).
30. H. Sponer and C. D. Cooper, *J. Chem. Phys.*, **23**, 646 (1955).
31. G. A. George and G. C. Morris, *J. Molec. Spectrosc.*, **26**, 67 (1968).
32. J. G. Angus and G. C. Morris, *Aust. J. Chem.* (see reference 26).
33. J. Jortner and G. C. Morris, *J. Chem. Phys.*, **51**, 3689 (1969).
34. B. Sharf, *Chem. Phys. Letters*, **5**, 456 (1970).
35. B. Sharf, *Chem. Phys. Letters*, **5**, 459 (1970).
36. J. G. Angus, B. J. Christ and G. C. Morris, *Aust. J. Chem.*, **21**, 2153 (1968).
37. R. Scheps, D. Florida and S. A. Rice, *J. Chem. Phys.*, **56**, 295 (1972).
38. A. D. Buckingham and P. J. Stephens, *Ann. Rev. Phys. Chem.*, **17**, 399 (1966).
39. P. N. Schatz and A. J. McCaffery, *Quart. Rev.*, **23**, 552 (1969).
40. J. G. Foss and M. E. McCarville, *J. Chem. Phys.*, **44**, 4350 (1966).
41. D. A. Schooley, E. Bunnenberg and C. Djerassi, *Proc. Natl. Acad. Sci.*, **56**, 1377 (1966).
42. P. J. Stephens, P. N. Schatz, A. B. Ritchie and A. J. McCaffery, *J. Chem. Phys.*, **48**, 132 (1968).
43. J. P. Larkindale and D. J. Simkin, *J. Chem. Phys.*, **55**, 5688 (1971).
44. P. N. Skancke, *Acta Chem. Scand.*, **18**, 1671 (1964).
45. J. E. Bloor and N. Blearley, *Canad. J. Chem.*, **43**, 1761 (1965).
46. A. Bergman and J. Jortner, *Chem. Phys. Letters*, **15**, 309 (1972).
47. J. J. Hopfield, J. M. Worlock and K. Park, *Phys. Rev. Letters*, **11**, 414 (1963).
48. J. J. Hopfield and J. M. Worlock, *Phys. Rev.*, **137**, A 1455 (1965).
49. D. Frohlich and H. Mahr, *Phys. Rev. Letters*, **18**, 593 (1966).
50. K. B. Eisenthal, M. W. Dowley and W. L. Peticolas, *Phys. Rev. Letters*, **20**, 93 (1968).
51. M. W. Dowley and W. L. Peticolas, *IBM J. Res. Develop.*, **12**, 188 (1968).
52. P. R. Monson and W. M. McClain, *J. Chem. Phys.*, **53**, 29 (1970).
53. P. R. Monson and W. M. McClain, *J. Chem. Phys.*, **56**, 4817 (1972).
54. F. C. Strome, Jr., *Phys. Rev. Letters*, **20**, 3, (1968).

55. I. Webman and J. Jortner, *J. Chem. Phys.*, **20**, 2706 (1969).
56. J. T. Richards and J. K. Thomas, *Chem. Phys. Letters*, **5**, 527 (1970).
57. B. Honig, J. Jortner and A. Szöke, *J. Chem. Phys.*, **46**, 2714 (1967).
58. T. M. Dunn and C. K. Ingold, *Nature*, **176**, 65 (1955).
59. W. M. McClain, *J. Chem. Phys.*, **55**, 2789 (1971).
60. W. M. McClain, *J. Chem. Phys.*, **57**, 2264 (1972).
61. W. M. McClain, *J. Chem. Phys.*, **58**, 324 (1973).
62. J. B. Birks, L. G. Christophorou and R. H. Huebner, *Nature*, **217**, 809 (1968).
63. T. Kobayashi and S. Nagakura, *Chem. Phys. Letters*, **13**, 217 (1972).
64. *The Single-photon Technique for Measuring Light Intensity and Decay Characteristics*, Ortec Appln. Note AN 35 (1971).
65. *Fluorescence Lifetime Apparatus*, Applied Photophysics Ltd., The Royal Institution, London (1972).
66. R. D. Spencer, Ph.D. thesis, University of Illinois at Urbana Champaign (1970).
67. L. M. Bollinger and G. E. Thomas, *Rev. Sci. Instr.*, **32**, 1044 (1961).
68. W. R. Ware. *Fluorescence Lifetime Measurements by Time-correlated single-photon Counting*. ONR Tech. Rpt. No. 3, March 1 (1969).
69. W. R. Ware. *Creation and Detection of the Excited State*, Vol. 1, p. 213 (Ed. A. A. Lamola), Marcel Dekker, New York, 1971.
70. A. E. W. Knight and B. K. Selinger, *Aust. J. Chem.*, **26**, 1 (1973).
71. G. A. Morton, *Appl. Opt.*, **7**, 1 (1968).
72. R. G. Tull, *Appl. Opt.*, **7**, 2023 (1968).
73. C. C. Davis and T. A. King, *Rev. Sci. Instr.*, **41**, 407 (1970).
74. C. C. Davis and T. A. King, *J. Phys. A.*, **3**, 101 (1970).
75. W. R. Ware, S. K. Lee, G. J. Brant and P. P. Chow, *J. Chem. Phys.*, **54**, 4729 (1971).
76. S. K. Chakrabarti and W. R. Ware, *J. Chem. Phys.*, **55**, 5494 (1971).
77. K. A. Bridgett and T. A. King, *Proc. Phys. Soc.*, **92**, 75 (1967).
78. K. A. Bridgett, T. A. King and R. J. Smith-Saville, *J. Phys. E* (*Sci. Instr.*), **3**, 767 (1970).
79. R. E. Imhof and F. H. Read, *Chem. Phys. Letters*, **3**, 652 (1969).
80. R. E. Imhof and F. H. Read, *J. Phys. B.* (*Atom. Molec. Phys.*), **4**, 450 (1971).
81. R. E. Imhof and F. H. Read, *J. Phys. B.* (*Atom. Molec. Phys.*), **4**, 1063 (1971).
82. R. E. Imhof and F. H. Read, *Chem. Phys. Letters*, **11**, 326 (1971).
83. R. E. Imhof, F. H. Read and S. T. Beckett, *J. Phys. B.* (*Atom. Molec. Phys.*), **5**, 896 (1972).
84. G. C. M. King, A. Adams and F. H. Read, *J. Phys. B.* (*Atom. Molec. Phys.*), **5**, L254 (1972).
85. Y. Koechlin, Doctoral thesis, University of Paris (1961).
86. G. Pfeffer, H. Lami, G. Laustriat and A. Coche, *C.R. Acad. Sci. Paris*, **257**, 434 (1963).
87. Y. Koechlin and A. Raviart, *Nucl. Instr. Meth.*, **29**, 45 (1964).
88. P. G. Sjölin, *Nucl. Instr. Meth.*, **37**, 45 (1965).
89. J. Kirkbride, E. C. Yates and D. G. Crandall, *Nucl. Instr. Meth.*, **52**, 293 (1967).
90. L. L. Nicholls and W. E. Wilson, *Appl. Opt.*, **7**, 167 (1968).
91. S. Ness, Ph.D. thesis, University of Manchester (1967).
92. R. J. Smith-Saville, Ph.D. thesis, University of Manchester (1968).
93. R. J. Smith-Saville and S. Ness, *Colloque d'Electronique Nucléaire et Radioprotection*, Vol. 2, Toulouse, 1968.

94. J. B. Birks and R. W. Pringle, *Proc. Roy. Soc. Edinburgh*, **A 70**, 233 (1972).
95. J. B. Birks, *J. Phys. B* (*Atom. Molec. Phys.*), **1**, 946 (1968).
96. J. B. Birks, *Theory and Practice of Scintillation Counting*, Pergamon Press, Oxford, 1964.
97. J. B. Birks, *Proc. Phys. Soc.*, **79**, 494 (1962).
98. M. Shinitzky, *J. Chem. Phys.*, **56**, 5979 (1972).
99. W. H. Melhuish, *J. Phys. Chem.*, **65**, 229 (1961).
100. W. H. Melhuish, *J. Opt. Soc. Am.*, **54**, 183 (1964).
101. I. B. Berlman, *Handbook of Fluorescence Spectra of Aromatic Molecules*, Academic Press, New York; (a) 1st edn., 1965, (b) 2nd edn., 1971.
102. J. B. Birks, *Phys. Rev.*, **94**, 1567 (1954).
103. A. Budó, J. Dombi and L. Szöllösy, *Acta Phys. et Chem. Szged* (*NS*), **2**, 18 (1956).
104. J. B. Birks and D. J. Dyson, *Proc. Roy. Soc.*, **A 275**, 135 (1963).
105. W. R. Ware and B. A. Baldwin, *J. Chem. Phys.*, **40**, 1703 (1964).
106. T. Medinger and F. Wilkinson, *Trans. Faraday Soc.*, **61**, 620 (1965).
107. B. Stevens and L. E. Mills, *Chem. Phys. Letters*, **15**, 381 (1972).
108. A. S. Cherkasov, V. A. Molchanov, T. M. Vember and K. G. Voldaikina, *Soviet Phys. Doklady*, **1**, 427 (1956).
109. S. J. Strickler and R. A. Berg, *J. Chem. Phys.*, **37**, 814 (1962).
110. J. B. Birks, S. Georghiou and I. H. Munro, *J. Phys. B.* (*Atom. Molec. Phys.*), **1**, 266 (1968).
111. W. S. Metcalf, private communication in reference 104.
112. R. B. Cundall and G. B. Evans, *J. Phys. E* (*Sci. Instr.*), **1**, 305 (1968).
113. F. R. Lipsett, G. Bechthold, F. D. Blair, F. V. Cairns and D. H. O'Hara, *Appl. Opt.*, **9**, 312 (1970).
114. J. W. Longworth, *Photochem. Photobiol.*, **8**, 589 (1968).
115. P. M. Froehlich and H. A. Morrison, *J. Phys. Chem.*, **76**, 3566 (1972).
116. A. Reiser and L. J. Leyshon, *J. Chem. Phys.*, **56**, 1011 (1972).
117. M. D. Lumb and D. A. Weyl, *J. Molec. Spectrosc.*, **23**, 365 (1967).
118. T. V. Ivanova, P. I. Kudryashov and B. Yu. Sveshnikov, *Soviet Phys. Doklady*, **6**, 407 (1961).
119. J. W. Eastman, *J. Chem. Phys.*, **49**, 4617 (1968).
120. W. P. Helman, *J. Chem. Phys.*, **51**, 354 (1969).
121. J. R. Greenleaf, M. D. Lumb and J. B. Birks, *J. Phys. B* (*Atom. Molec. Phys.*), **1**, 1157 (1968).
122. J. R. Greenleaf and T. A. King, *Proc. Intern. Conf. Luminescence*, *Budapest*, p. 212, Akadémiai Kiadó, Budapest, 1968.
123. V. M. Berenfel'd and V. A. Krongauz, *Theor. Exp. Chem. USSR*, **3**, 63 (1967).
124. E. J. Bowen and A. H. Williams, *Trans. Faraday Soc.*, **35**, 44 (1939).
125. J. Klein, F. Heisel, H. Lami and G. Laustriat, *Proc. Intern. Conf. Luminescence*, *Budapest*, p. 205, Akadémiai Kiadó, Budapest, 205 (1968).
126. J. Petruska, *J. Chem. Phys.*, **34**, 1120 (1961).
127. J. N. Murrell. *The Theory of the Electronic Spectra of Organic Molecules*, 199, Methuen, London, 1963.
128. J. W. Rabalais, H. J. Maria and S. P. McGlynn, *J. Chem. Phys.*, **51**, 2259 (1969).
129. D. Phillips, *J. Photochem.*, **1**, 97 (1972).
130. W. A. Noyes, Jr., D. A. Harter and W. A. Mulac, *J. Chem. Phys.*, **44**, 2100 (1966).
131. W. A. Noyes, Jr. and D. A. Harter, *J. Chem. Phys.*, **46**, 674 (1967).
132. M. Nishikawa and P. K. Ludwig, *J. Chem. Phys.*, **52**, 107 (1970).

133. C. S. Burton and W. A. Noyes, Jr., *J. Chem. Phys.*, **49**, 1705 (1968).
134. G. M. Breuer and E. K. C. Lee, *Chem. Phys. Letters*, **14**, 404 (1972).
135. W. A. Noyes, Jr. and D. A. Harter, *J. Amer. Chem. Soc.*, **91**, 7585 (1969).
136. W. A. Noyes, Jr. and D. A. Harter, *J. Phys. Chem.*, **75**, 2741 (1971).
137. T. R. Wright and A. Reiser, *IV IUPAC Symp. Photochemistry, Baden-Baden*, Paper 59 (1972).
138. J. N. Murrell and H. C. Longuet-Higgins, *Proc. Phys. Soc.*, **A 68**, 321 (1955).
139. E. C. Lim and Y. H. Li, *J. Chem. Phys.*, **52**, 6416 (1970).
140. B. S. Hudson and B. E. Kohler, *Chem. Phys. Letters*, **14**, 299 (1972).
141. K. Schulten and M. Karplus, *Chem. Phys. Letters*, **14**, 305 (1972).
142. J. Klueger, G. Fischer, E. Fischer, C. Goedicke and H. Stegenmeyer, *Chem. Phys. Letters*, **8**, 279 (1971).
143. A. Reiser, L. J. Leyshon, D. Saunders, M. V. Mijovic, A. Bright and J. Bogie, *J. Amer. Chem. Soc.*, **94**, 2414 (1972).
144. S. Sharafy and K. A. Muszkat, *J. Amer. Chem. Soc.*, **93**, 4119 (1971).
145. G. Oster and Y. Nishijima, *J. Amer. Chem. Soc.*, **78**, 1581 (1956).
146. G. Hoffman, Dissertation, University of Stuttgart (1969).
147. M. A. El-Bayoumi and F. M. Abdul-Halim, *J. Chem. Phys.*, **48**, 2536 (1968).
148. A. J. Thomson, *J. Chem. Phys.*, **51**, 4106 (1969).
149. A. Matsuyama and H. Baba, *Bull. Chem. Soc. Japan*, **44**, 854 (1971).
150. A. Matsuyama and H. Baba, *Bull. Chem. Soc. Japan*, **44**, 1162 (1971).
151. N. G. Bakhshiev, *Opt. Spectrosc.*; (a) **10**, 379 (1961); (b) **12**, 261 (1962); (c) **12**, 293 (1962); (d) **12**, 309 (1962); (e) **13**, 24 (1962); (f) **13**, 104 (1962); (g) **16**, 446 (1964); (h) **19**, 196 (1965).
152. N. G. Bakhshiev and I. V. Piterskaya, *Opt. Spectrosc.*, **20**, 437 (1966).
153. N. G. Bakhshiev and I. V. Piterskaya, *Proc. Intern. Conf. Luminescence, Budapest*, p. 305, Akadémiai Kiadó, Budapest, 1968.
154. A. S. Cherkasov and G. I. Dragneva, *Opt. Spectrosc.*, **10**, 238 (1961).
155. A. S. Cherkasov, *Opt. Spectrosc.*, **12**, 35 (1962).
156. T. V. Veselova, L. A. Limereva, A. S. Cherkasov and V. I. Shirokov, *Opt. Spectrosc.*, **19**, 39 (1965).
157. N. G. Bakhshiev, Yu. T. Maruzenko and I. V. Piterskaya, *Opt. Spectrosc.*, **21**, 307 (1966).
158. L. G. Pikulik and M. A. Solomakko, *Opt. Spectrosc.*, **8**, 176 (1960).
159. Th. Förster and K. Rokos, *Z. Phys. Chem. (NF)*, **63**, 208 (1969).
160. W. R. Ware, S. K. Lee and P. P. Chow, *Chem. Phys. Letters*, **2**, 356 (1968).
161. C. J. Seliskar, D. Turner, J. R. Gohlke and L. Brand, *Molecular Luminescence*, p. 677 (Ed. E. C. Lim), W. A. Benjamin, Inc., New York, 1969.
162. L. Stryer, *J. Molec. Biol.*, **13**, 482 (1965).
163. E. M. Kosower and K. Tanizawa, *Chem. Phys. Letters*, **16**, 419 (1972).
164. E. V. Shpol'skii, *Soviet Phys. Uspekhi.*; (a) **2**, 378 (1959); (b) **3**, 372 (1960); (c) **6**, 411 (1963).
165. L. A. Nakhimovskaya, *Opt. Spectrosc.*, **24**, 105 (1968).
166. G. S. Kembrovskii, V. P. Bobrovich and A. V. Sevchenko, *Opt. Spectrosc.*, **24**, 213 (1968).
167. R. M. MacNab and K. Sauer, *J. Chem. Phys.*, **53**, 2805 (1970).
168. D. J. Morantz, *Chem. Phys. Letters*, **5**, 20 (1970).
169. J. L. Richards and S. A. Rice, *J. Chem. Phys.*, **54**, 2014 (1971).
170. J. L. Richards and S. A. Rice, *Chem. Phys. Letters*, **9**, 444 (1971).
171. B. R. Henry and M. Kasha, *Ann. Rev. Phys. Chem.*, **19**, 161 (1968).

172. J. Jortner, S. A. Rice and R. M. Hochstrasser, *Adv. Photochem.*, **7**, 149 (1969).
173. E. W. Schlag, S. Schneider and S. F. Fischer, *Ann. Rev. Phys. Chem.*, **22**, 465 (1971).
174. K. F. Freed, *Topics Current Chem.*, **31**, 105 (1971).
175. The earlier value (reference 101a) of $\Phi_{FH} = 0{\cdot}24$ for azulene has since been corrected (reference 101b, 221, 605) to $\Phi_{FH} = 0{\cdot}03$.
176. C. E. Easterly, L. G. Christophorou, R. P. Blaunstein and J. G. Carter, *Chem. Phys. Letters*, **6**, 579 (1970).
177. C. E. Easterly, L. G. Christophorou and J. G. Carter, *J. Chem. Soc.* (*Faraday Trans. II*), **69**, 471 (1973).
178. P. A. Geldof, R. P. H. Rettschnick and G. J. Hoytink, *Chem. Phys. Letters*, **4**, 59 (1969).
179. A. Nakajima and H. Baba, *Bull. Chem. Soc. Japan*, **43**, 969 (1970).
180. H. Baba, A. Nakajima, M. Aoi and K. Chihara, *J. Chem. Phys.*, **55**, 2433 (1971).
181. C. J. Werkhoven, P. A. Geldof, M. F. M. Post, J. Langelaar, R. P. H. Rettschnick and J. D. W. van Voorst, *Chem. Phys. Letters*, **9**, 6 (1971).
182. C. J. Werkhoven, T. Deinum, J. Langelaar, R. P. H. Rettschnick and J. D. W. van Voorst, *Chem. Phys. Letters*, **11**, 478 (1971).
183. T. Deinum, C. J. Werkhoven, J. Langelaar, R. P. H. Rettschnick and J. D. W. van Voorst, *Chem. Phys. Letters*, **12**, 189 (1971).
184. C. J. Werkhoven, T. Deinum, J. Langelaar, R. P. H. Rettschnick and J. D. W. van Voorst, *Chem. Phys. Letters*, **18**, 171 (1973).
185. P. F. Jones and S. Siegel, *Chem. Phys. Letters*, **2**, 486 (1968).
186. P. Wannier, P. M. Rentzepis and J. Jortner, *Chem. Phys. Letters*, **10**, 102 (1971).
187. E. C. Lim and C.-S. Huang, *J. Chem. Phys.*, **58**, 1247 (1973).
188. T. Deinum, C. J. Werkhoven, J. Langelaar, R. P. H. Rettschnick and J. D. W. van Voorst, *Chem. Phys. Letters*, **19**, 29 (1973).
189. J. C. Hsieh, U. Laor and P. K. Ludwig, *Chem. Phys. Letters*, **10**, 412 (1971).
190. U. Laor and P. K. Ludwig, *J. Chem. Phys.*, **54**, 1054 (1971).
191. E. W. Schlag, S. Schneider and D. W. Chandler, *Chem. Phys. Letters*, **11**, 474 (1971).
192. J. O. Uy and E. C. Lim, *Chem. Phys. Letters*, **7**, 306 (1970).
193. E. C. Lim and J. O. Uy, *J. Chem. Phys.*, **56**, 3374 (1972).
194. H. von Weyssenhoff and F. Kraus, *J. Chem. Phys.*, **54**, 2387 (1971).
195. B. S. Neporent, *Zh. Fiz. Khim.*, **21**, 1111 (1947).
196. B. S. Neporent and S. V. Mirumyants, *Opt. Spectrosc.*, **8**, 635 (1960).
197. E. W. Schlag and H. von Weyssenhoff, *J. Chem. Phys.*, **51**, 2508 (1969).
198. C. W. Ashpole, S. J. Formosinho and G. Porter, *Proc. Roy. Soc.*, **A 323**, 11 (1971).
199. S. J. Formosinho and G. Porter, *Proc. Roy. Soc.*, **A**, in press (1973).
200. G. S. Beddard, G. R. Fleming, O. L. J. Gijzeman and G. Porter, *Chem. Phys. Letters*, **18**, 481 (1973).
201. W. Heitler, *The Quantum Theory of Radiation*, Oxford University Press, 1944.
202. J. L. Kropp and W. R. Dawson, *Molecular Luminescence*, p. 39 (Ed. E. C. Lim), W. A. Benjamin, Inc., New York, 1969.
203. J. H. Callomon, R. Lopez-Delgado and J. E. Parkin, *Chem. Phys. Letters*, **13**, 125 (1972).
244. C. S. Parmenter and M. W. Schuyler, *Chem. Phys. Letters*, **6**, 339 (1970).
205. W. R. Ware, B. K. Selinger, C. S. Parmenter and M. W. Schuyler, *Chem. Phys. Letters*, **6**, 342 (1970).

206. B. K. Selinger and W. R. Ware, *J. Chem. Phys.*, **53**, 3160 (1970).
207. W. M. Gelbart, K. G. Spears, K. F. Freed, J. Jortner and S. A. Rice, *Chem. Phys. Letters*, **6**, 345 (1970).
208. K. G. Spears and S. A. Rice, *J. Chem. Phys.*, **55**, 5561 (1971).
209. A. S. Abramson, K. G. Spears and S. A. Rice, *J. Chem. Phys.*, **56**, 2291 (1972).
210. C. S. Parmenter, *Adv. Chem. Phys.*, **22**, 365 (1972).
211. M. D. Lumb and L. C. Pereira, *Organic scintillators and liquid scintillation counting*, p. 239 (Eds. D. L. Horrocks and C.-T. Peng), Academic Press, New York, 1971.
212. M. D. Lumb and A. O. Al-Chalabi, *Chem. Phys. Letters*, **11**, 365 (1971).
213. S. D. Colson and E. R. Bernstein, *J. Chem. Phys.*, **43**, 2661 (1965).
214. S. D. Colson and G. W. Robinson, *J. Chem. Phys.*, **48**, 2550 (1968).
215. Y. Toyozawa, *Prog. Theor. Phys.*, **20**, 53 (1958).
216. M. D. Lumb, C. Lloyd Braga and L. C. Pereira, *Trans. Faraday. Soc.*, **65**, 1992 (1969).
217. C. W. Lawson, F. Hirayama and S. Lipsky, *J. Chem. Phys.*, **51**, 1590 (1969).
218. C. Fuchs, F. Heisel and R. Voltz, *J. Phys. Chem.*, **76**, 3867 (1972).
219. J. B. Birks, *Chem. Phys. Letters*, **17**, 370 (1972).
220. S. Murata, C. Iwanaga, T. Toda and H. Kokobun, *Chem. Phys. Letters*, **13**, 101 (1972).
221. S. Murata, C. Iwanaga, T. Toda and H. Kokobun, *Chem. Phys. Letters*, **15**, 152 (1972).
222. W. Siebrand and D. F. Williams, *J. Chem. Phys.*, **49**, 1860 (1968).
223. G. D. Johnson, L. M. Logan and I. G. Ross, *J. Molec. Spectrosc.*, **14**, 198 (1964).
224. J. P. Byrne, E. F. McCoy and I. G. Ross, *Aust. J. Chem.*, **18**, 1589 (1965).
225. A. E. W. Knight and B. K. Selinger, *Chem. Phys. Letters*, **12**, 419 (1971).
226. R. M. Hochstrasser, *Canad. J. Chem.*, **39**, 459 (1961).
227. D. L. Horrocks, *Molecular Luminescence*, p. 63 (Ed. E. C. Lim), W. A. Benjamin, Inc., New York, 1969.
228. F. Schneider and E. Lippert, *Ber. Bunsenges. Phys. Chem.*, **72**, 1155 (1968).
229. F. Schneider and E. Lippert, *Ber. Bunsenges. Phys. Chem.*, **74**, 624 (1970).
230. H. Beens and A. Weller, *Chem. Phys. Letters*, **3**, 666 (1969).
231. H. Beens, H. Knibbe and A. Weller, *J. Chem. Phys.*, **47**, 1183 (1967).
232. J. A. Poole, R. C. Dhingra and B. Gelernt, *J. Chem. Phys.*, **52**, 464 (1970).
233. C. J. M. Brugman, R. P. H. Rettschnick and G. J. Hoytink, *Chem. Phys. Letters*, **8**, 574 (1971).
234. C. J. M. Brugman, R. P. H. Rettschnick and G. J. Hoytink, *Chem. Phys. Letters*, **8**, 263 (1971).
235. N. C. Yang and S. L. Murov, *J. Chem. Phys.*, **45**, 4358 (1966).
236. A. A. Lamola, *J. Chem. Phys.*, **47**, 4810 (1967).
237. R. D. Rauh and P. A. Leermakers, *J. Amer. Chem. Soc.*, **90**, 2246 (1968).
238. Y. Kanda, J. Stanislaus and E. C. Lim, *J. Amer. Chem. Soc.*, **91**, 5085 (1969).
239. N. C. Yang and R. Dusenbury, *Molec. Photochem.*, **1**, 159 (1969).
240. P. J. Wagner, M. J. May, A. Haug and D. R. Graber, *J. Amer. Chem. Soc.*, **92**, 5269 (1970).
241. P. J. Wagner, M. J. May and A. Haug, *Chem. Phys. Letters*, **13**, 545 (1972).
242. W. Klöpffer, *Chem. Phys. Letters*, **13**, 545 (1972).
243. P. R. Callis and R. W. Wilson, *Chem. Phys. Letters*, **13**, 417 (1972).
244. J. Langelaar, G. A. de Vries and D. Bebelaar, *J. Sci. Instr.*, **46**, 149 (1969).

245. J. Langelaar, R. P. H. Rettschnick, A. M. F. Lambooy and G. J. Hoytink, *Chem. Phys. Letters*, **1**, 609 (1968).
246. O. L. J. Gijzeman, J. Langelaar and J. D. W. van Voorst, *Chem. Phys. Letters*, **5**, 269 (1970).
247. W. H. van Leeuwen, J. Langelaar and J. D. W. van Voorst, *Chem. Phys. Letters*, **13**, 622 (1972).
248. T. C. O'Haver and J. D. Winefordner, *Anal. Chem.*, **38**, 602 (1966).
249. K. Razi Naqvi, *Chem. Phys. Letters*, **5**, 171 (1970).
250. C. A. Parker, *Adv. Photochem.*, **2**, 321 (1964).
251. C. A. Parker, *Photoluminescence of Solutions*, Elsevier, Amsterdam, 1968.
252. C. A. Parker, *Chem. Phys. Letters*, **6**, 516 (1970).
253. K. Razi Naqvi, *Chem. Phys. Letters*, **6**, 518 (1970).
254. J. Langelaar, R. P. H. Rettschnick and G. J. Hoytink, *J. Chem. Phys.*, **54**, 1 (1971).
255. W. Siebrand, *J. Chem. Phys.*, **47**, 2411 (1967).
256. G. D. Smith, *Phys. Rev.*, **166**, 839 (1968).
257. R. E. Kellogg and R. G. Bennett, *J. Chem. Phys.*, **41**, 3042 (1964).
258. R. Li and E. C. Lim, *J. Chem. Phys.*, **57**, 605 (1972).
259. P. M. Johnson and M. C. Studer, *Chem. Phys. Letters*, **18**, 341 (1973).
260. E. C. Lim, J. D. Laposa and J. M. H. Yu, *J. Molec. Spectrosc.*, **19**, 412 (1966).
261. S. F. Fischer and E. C. Lim, *Chem. Phys. Letters*, **14**, 40 (1972).
262. J. L. Charlton and B. R. Henry, *Chem. Phys. Letters*, **15**, 369 (1972).
263. R. B. Cundall and L. C. Pereira, *Chem. Phys. Letters*, **16**, 371 (1972).
264. W. E. Graves, R. H. Hofeldt and S. P. McGlynn, *J. Chem. Phys.*, **53**, 1309 (1972).
265. J. L. Kropp and W. R. Dawson, *J. Phys. Chem.*, **71**, 4499 (1967).
266. J. L. Kropp and W. R. Dawson, *J. Phys. Chem.*, **73**, 693 (1969).
267. N. G. Kilmer and J. D. Spangler, *J. Chem. Phys.*, **54**, 604 (1971).
268. P. F. Jones and S. Siegel, *J. Chem. Phys.*, **50**, 1134 (1969).
269. W. H. Melhuish, *Trans. Faraday Soc.*, **62**, 3384 (1966).
270. W. R. Dawson and J. L. Kropp, *J. Phys. Chem.*, **73**, 1752 (1969).
271. J. W. Rabalais, A. Wahlborg and S. P. McGlynn, *J. Molec. Spectrosc.*, **36**, 155 (1970).
272. G. M. Breuer and E. K. C. Lee, *J. Chem. Phys.*, **51**, 3615 (1969).
273. E. C. Lim and H. Bhattacharjee, *Chem. Phys. Letters*, **9**, 249 (1971).
274. B. Sharf and R. Silbey, *Chem. Phys. Letters*, **4**, 314 (1970).
275. J. van Egmond, D. M. Burland and J. H. van der Waals, *Chem. Phys. Letters*, **12**, 206 (1971).
276. G. F. Hatch, M. D. Erlitz and G. C. Nieman, *Molecular Luminescence*, p. 21 (Ed. E. C. Lim), W. A. Benjamin, Inc., New York, 1969.
277. C. E. Thompson, *J. Opt. Soc. Am.*, **55**, 1184 (1965).
278. G. W. Robinson and R. P. Frosch, *J. Chem. Phys.*, **38**, 1187 (1963).
279. M. D. Erlitz and G. C. Nieman, *J. Chem. Phys.*, **50**, 1479 (1969).
280. P. M. Johnson and L. Ziegler, *J. Chem. Phys.*, **56**, 2169 (1972).
281. T. E. Martin and A. H. Kalantar, *J. Chem. Phys.*, **48**, 4996 (1968).
282. T. E. Martin and A. H. Kalantar, *J. Chem. Phys.*, **49**, 235 (1968).
283. T. E. Martin and A. H. Kalantar, *J. Chem. Phys.*, **49**, 244 (1968).
284. T. E. Martin and A. H. Kalantar, *J. Phys. Chem.*, **72**, 2265 (1968).
285. T. E. Martin and A. H. Kalantar, *J. Chem. Phys.*, **50**, 1486 (1969).
286. T. E. Martin and A. H. Kalantar, *J. Chim. Phys.*, **67**, 101 (1970).
287. T. E. Martin and A. H. Kalantar, *J. Phys. Chem.*, **74**, 2030 (1970).

288. J. W. Rabalais, H. J. Maria and S. P. McGlynn, *Chem. Phys. Letters*, **3**, 59 (1969).
289. A. H. Kalantar, *Chem. Phys. Letters*, **16**, 198 (1972).
290. J. D. Spangler and N. G. Kilmer, *J. Chem. Phys.*, **48**, 698 (1968).
291. J. E. Kent, M. F. O'Dwyer and R. J. Stephens, *Chem. Phys. Letters*, **10**, 201 (1971).
292. I. H. Leubner, *J. Phys. Chem.*, **73**, 2088 (1969).
293. I. H. Leubner and J. E. Hodgkins, *J. Phys. Chem.*, **73**, 2545 (1969).
294. S. Fischer and S. Schneider, *Chem. Phys. Letters*, **10**, 392 (1971).
295. M. S. de Groot and J. H. van der Waals, *Molec. Phys.*, **6**, 545 (1963).
296. S. Fischer, *Chem. Phys. Letters*, **10**, 397 (1971).
297. F. Masetti, U. Mazzucato and G. Galiazzo, *J. Luminescence*, **4**, 8 (1971).
298. S. E. Webber, *Chem. Phys. Letters*, **5**, 466 (1970).
299. V. L. Ermolaev, *Soviet Phys. Uspekhi*, **6**, 333 (1963).
300. R. H. Hofeldt, R. Sahai and S. H. Lin, *J. Chem. Phys.*, **53**, 4512 (1970).
301. E. Vander Donckt, M. Matagne and M. Sapir, *Chem. Phys. Letters*, **20**, 81 (1973).
302. E. Vander Donckt and D. Liétaer, *J. Chem. Soc.* (*Faraday Trans. I*), **68**, 112 (1972).
303. E. Vander Donckt and C. Vogels, *Spectrochim. Acta*, **27A**, 2157 (1971).
304. E. Vander Donckt and C. Vogels, *Bull. Soc. Chim. Belg.*, **82**, 151 (1973).
305. T. A. Moore and P.-S. Song, *Chem. Phys. Letters*, **19**, 128 (1973).
306. W. J. Tomlinson, E. A. Chandross, R. L. Fork, C. A. Pryde and A. A. Lamola, *App. Opt.*, **11**, 533 (1972).
307. E. A. Chandross and C. J. Dempster, *J. Amer. Chem. Soc.*, **92**, 3586 (1970).
308. J. M. Thomas and J. O. Williams, *Molec. Cryst. Liq. Cryst.*, **4**, 59 (1969).
309. J. M. Thomas, *Adv. in Catalysis*, **19**, 293 (1969).
310. J. O. Williams and J. M. Thomas, *Molec. Cryst. Liq. Cryst.*, **16**, 371 (1972).
311. D. N. de Vries Reilingh, R. P. H. Rettschnick and G. J. Hoytink, *J. Chem. Phys.*, **54**, 2722 (1971).
312. R. J. McDonald and B. K. Selinger, *Aust. J. Chem.*, **24**, 249 (1971).
313. G. Oster, N. Geacintov and A. U. Khan, *Nature*, **196**, 1089 (1962).
314. G. Oster, *Acta Phys. Polon.*, **26**, 873 (1964).
315. N. Mataga, Y. Torihashi and Y. Ota, *Chem. Phys. Letters*, **1**, 385 (1967).
316. E. Loewenthal, Y. Tomkiewicz and A. Weinreb, *Spectrochim. Acta*, **A 25**, 1501 (1969).
317. J. B. Birks, *Acta Phys. Polon.*, **26**, 879 (1964).
318. P. C. Johnson and H. W. Offen, *Chem. Phys. Letters*, **18**, 258 (1973).
319. F. Dupuy, M. Martinand, G. Nouchi and J. M. Turlet, *J. Chim. Phys.*, **69**, 614, 619 (1972).
320. J. A. Katul and A. B. Zahlan, *J. Chem. Phys.*, **47**, 1012 (1967).
321. G. Fournie, F. Dupuy, M. Martinand, G. Nouchi and J. M. Turlet, *Chem. Phys. Letters*, **16**, 332 (1972).
322. K. Yoshihara, T. Kasaya, A. Inoue and S. Nagakura, *Chem. Phys. Letters*, **9**, 469 (1971).
323. J. B. Birks, D. J. Dyson and I. H. Munro, *Proc. Roy. Soc.*, **A 275**, 575 (1963).
324. D. L. Horrocks and W. G. Brown, *Chem. Phys. Letters*, **5**, 117 (1970).
325. T. Kawakubo, *Molec. Cryst. Liq. Cryst.*, **16**, 333 (1972).
326. A. Inoue, K. Yoshihara, T. Kasaya and S. Nagakura, *Bull. Chem. Soc. Japan*, **45**, 720 (1972).
327. J. B. Birks and A. J. W. Cameron, *Proc. Roy. Soc.*, **A 249**, 297 (1959).

328. J. B. Birks, A. A. Kazzaz and T. A. King, *Proc. Roy. Soc.*, **A 291**, 556 (1966).
329. M. D. Cohen, B. S. Green, Z. Ludmer and G. M. J. Schmidt, *Chem. Phys. Letters*, **7**, 486 (1970).
330. E. A. Chandross and C. J. Dempster, *J. Amer. Chem. Soc.*, **92**, 703 (1970).
331. E. A. Chandross and C. J. Dempster, *J. Amer. Chem. Soc.*, **92**, 704 (1970).
332. J. Ferguson and A. W. H. Mau, *Molec. Phys.*, **27**, 377 (1974).
333. D. E. Applequist, E. C. Friedrich and M. T. Rogers, *J. Amer. Chem. Soc.*, **81**, 457 (1959).
334. R. Calas, R. Lalande and P. Mauret, *Bull. Soc. Chim. France*, **1960**, 148 (1960).
335. F. D. Greene, *Bull. Soc. Chim. France*, **1960**, 1356 (1960).
336. R. Calas, R. Lalande, J. Faugere and M. Moulines, *Bull. Soc. Chim. France*, **1960**, 119 (1965).
337. O. L. Chapman and K. Lee, *J. Org. Chem.*, **34**, 4166 (1969).
338. G. M. J. Schmidt, *Reactivity of the Photoexcited Organic Molecule*, p. 227, Interscience, London, 1967.
339. D. P. Craig and P. Sarti-Fantoni, *Chem. Comm.*, **1966**, 742 (1966).
340. M. D. Cohen, *Molec. Cryst. Liq. Cryst.*, **9**, 287 (1969).
341. M. D. Cohen, Z. Ludmer, J. M. Thomas and J. O. Williams, *Chem. Comm.*, **1969**, 1172 (1969).
342. M. D. Cohen, Z. Ludmer, J. M. Thomas and J. O. Williams, *Proc. Roy. Soc.*, **A 324**, 459 (1971).
343. A. Kawada and M. M. Labes, *Molec. Cryst. Liq. Cryst.*, **11**, 133 (1970).
344. J. B. Birks and J. B. Aladekomo, *Photochem. Photobiol.*, **2**, 415 (1963).
345. E. A. Chandross and J. Ferguson, *J. Chem. Phys.*, **45**, 3554 (1966).
346. M. K. Chaudhuri and S. C. Ganguly, *J. Phys. C.* (*Solid St. Phys.*), **3**, 1791 (1970).
347. D. Fischer, G. Naundorf and W. Klöpffer, *Z. Naturforsch.*, **28a**, 973 (1973).
348. C. R. Goldschmidt and M. Ottolenghi, *J. Phys. Chem.*, **74**, 2041 (1970).
349. M. A. Slifkin and A. O. Al-Chalabi, *Chem. Phys. Letters*, **20**, 211 (1973).
350. T. Azumi, A. T. Armstrong and S. P. McGlynn, *J. Chem. Phys.*, **41**, 3839 (1964).
351. J. B. Birks, A. J. H. Alwattar and M. D. Lumb, *Chem. Phys. Letters*, **11**, 89 (1971).
352. M. Hauser and G. Heidt, *Z. Phys. Chem.* (*NF*), **69**, 201 (1970).
353. J. B. Birks, M. D. Lumb and I. H. Munro, *Proc. Roy. Soc.*, **A 280**, 289 (1964).
354. C. A. Parker and C. G. Hatchard, *Trans. Faraday Soc.*, **59**, 284 (1963).
355. J. B. Birks, *J. Phys. Chem.*, **67**, 2199 (1963); **68**, 439 (1964).
356. T. Medinger and F. Wilkinson, *Trans. Faraday Soc.*, **62**, 1785 (1966).
357. T. D. S. Hamilton and K. Razi Naqvi, *Anal. Chem.*, **45**, 1581 (1973).
358. W. Heinzelmann and H. Labhart, *Chem. Phys. Letters*, **4**, 20 (1969).
359. R. M. Hochstrasser and A. J. Malliaris, *J. Chem. Phys.*, **42**, 2243 (1965).
360. P. Avakian and E. Abramson, *J. Chem. Phys.*, **43**, 821 (1965).
361. R. B. Cundall and L. C. Pereira, *Chem. Phys. Letters*, **15**, 383 (1972).
362. O. L. J. Gijzeman, Doctoral thesis, University of Amsterdam (1972).
363. O. L. J. Gijzeman, J. Langelaar and J. D. W. van Voorst, *Chem. Phys. Letters*, **11**, 526 (1971).
364. O. L. J. Gijzeman, W. H. van Leeuwen, J. Langelaar and J. D. W. van Voorst, *Chem. Phys. Letters*, **11**, 528 (1971).
365. O. L. J. Gijzeman, W. H. van Leeuwen, J. Langelaar and J. D. W. van Voorst, *Chem. Phys. Letters*, **11**, 532 (1971).
366. L. Peter and G. Vaubel, *Chem. Phys. Letters*, **18**, 531 (1973).

367. N. Y. C. Chu and D. R. Kearns, *Molec. Cryst. Liq. Cryst.*, **16**, 61 (1972).
368. J. Langelaar, G. Jansen, R. P. H. Rettschnick and G. J. Hoytink, *Chem. Phys. Letters*, **12**, 86 (1971).
369. R. B. Cundall and A. J. R. Voss, *Chem. Comm.*, **1969**, 116 (1969).
370. J. Langelaar, Doctoral thesis, University of Amsterdam (1969).
371. P. Avouris and M. A. El-Bayoumi, *Chem. Phys. Letters*, **20**, 59 (1973).
372. G. A. George and G. C. Morris, *Molec. Cryst. Liq. Cryst.*, **11**, 61 (1970).
373. M. Gouterman and P. Sayer, *Chem. Phys. Letters*, **8**, 126 (1971).
374. S. R. LaPaglia, *J. Molec. Spectr.*, **7**, 427 (1961).
375. S. R. LaPaglia, *Spectrochim. Acta*, **18**, 1295 (1962).
376. R. M. Hochstrasser and A. P. Marchetti, *J. Chem. Phys.*, **52**, 1360 (1970).
377. R. E. Brown, L. A. Singer and J. H. Parks, *Chem. Phys. Letters*, **14**, 193 (1972).
378. N. A. Borisevich and V. V. Gruzinskii, *Proc. Acad. Sci. USSR*, **175**, 578 (1967).
379. J. Saltiel, H. C. Curtis, L. Metts, J. M. Miley, J. Winterle and M. Wrighton, *J. Amer. Chem. Soc.*, **92**, 410 (1970).
380. P. F. Jones and A. R. Calloway, *J. Amer. Chem. Soc.*, **92**, 4997 (1970).
381. P. F. Jones and A. R. Calloway, *Chem. Phys. Letters*, **10**, 438 (1971).
382. P. M. Rentzepis, *Science*, **169**, 239 (1970).
383. L. R. Faulkner and A. J. Bard, *J. Amer. Chem. Soc.*, **91**, 6495 (1969).
384. P. Avakian, R. P. Groff, R. E. Kellogg, R. E. Merrifield and A. Suna, *Organic Scintillators and Liquid Scintillation Counting* (Eds. D. L. Horrocks and C.-T. Peng), Academic Press, New York, 1971.
385. D. Wyrsch and H. Labhart, *Chem. Phys. Letters*, **8**, 217 (1971).
386. D. Wyrsch and H. Labhart, *Chem. Phys. Letters*, **12**, 373 (1971).
387. H. Labhart and D. Wyrsch, *Chem. Phys. Letters*, **12**, 378 (1971).
388. B. Stevens, *Chem. Phys. Letters*, **3**, 233 (1969).
389. F. Dupuy, Doctoral thesis, University of Bordeaux (1970).
390. F. Dupuy, *Chem. Phys. Letters*, **13**, 537 (1972).
391. M. Ewald and G. Durocher, *Chem. Phys. Letters*, **12**, 385 (1971).
392. H. P. Schwob, J. Fünfschilling and I. Zschokke-Gränacher, *Molec. Cryst. Liq. Cryst.*, **10**, 39 (1970).
393. R. S. Mulliken and W. B. Person, *Molecular Complexes*, Wiley–Interscience, New York, 1969.
394. R. Foster, *Organic Charge-transfer Complexes*, Academic Press, London, New York, 1969.
395. R. P. Blaunstein and L. G. Christophorou, *Radiation Res. Rev.*, **3**, 69 (1971).
396. G. R. Anderson, *J. Amer. Chem. Soc.*, **92**, 3552 (1970).
397. T. Amano, H. Kuroda and H. Akamatu, *Bull. Chem. Soc. Japan*, **42**, 671 (1969).
398. J. P. Larkindale and D. J. Simkin, *J. Chem. Phys.*, **56**, 3730 (1972).
399. W. G. Rothschild, *Chem. Phys. Letters*, **9**, 149 (1971).
400. W. Liptay, *Z. Naturforsch.*, **20a**, 272 (1965).
401. W. Liptay, B. Dumbacher and H. Weisenberger, *Z. Naturforsch.*, **23a**, 1601 (1968).
402. W. Liptay, *Angew. Chem. Internal. Ed.*, **8**, 177 (1968).
403. C. J. Eckhardt, *J. Chem. Phys.*, **56**, 3947 (1972).
404. C. A. G. O. Varma and L. J. Oosterhoff, *Chem. Phys. Letters*, **9**, 406 (1971).
405. K. Bergmann, M. Eigen and L. de Maeyer, *Ber. Bunsenges. Physik. Chem.*, **67**, 819 (1963).
406. C. A. G. O. Varma and L. J. Oosterhoff, to be published (see reference 404).

407. R. K. Chan and S. C. Liao, *Canad. J. Chem.*, **48**, 299 (1970).
408. G. Briegleb, *Elektronen-Donator-Acceptor-Komplexe*, Springer-Verlag, Berlin, 1961.
409. H. Ishida and H. Tsubomura, *Chem. Phys. Letters*, **9**, 296 (1971).
410. M. G. Kuzmin and L. N. Guseva, *Chem. Phys. Letters*, **3**, 71 (1969).
411. E. G. Cook, Jr. and J. C. Schug, *J. Chem. Phys.*, **53**, 723 (1970).
412. J. C. Schug and M. C. Dyson, *J. Chem. Phys.*, **58**, 297 (1973).
413. D. A. Armitage, T. G. Beaumont, K. M. C. Davis, D. J. Hall and K. W. Morcom, *Trans. Faraday Soc.*, **67**, 2548 (1971).
414. K. M. C. Davis, *Photoelectric Spectr. Gp. Bull.*, No. 20, 603 (1972).
415. J. B. Birks, E. Pantos and T. D. S. Hamilton, *Chem. Phys. Letters*, **20**, 544 (1973).
416. E. C. Lim and V. L. Kowalski, *J. Chem. Phys.*, **36**, 1729 (1962).
417. C. Dijkgraaf and G. J. Hoytink, *Tetrahedron*, **19**, 179 (1963).
418. G. J. Hoytink, *Accts. Chem. Res.*, **2**, 114 (1969).
419. L. J. Andrews and R. M. Keefer, *J. Amer. Chem. Soc.*, **74**, 4500 (1952).
420. J. Jortner and U. Sukulov, *Nature*, **190**, 1003 (1961).
421. R. M. Keefer and L. J. Andrews, *J. Amer. Chem. Soc.*, **77**, 2164 (1955).
422. W. B. Person, *J. Chem. Phys.*, **38**, 109 (1963).
423. G. Briegleb and J. Czekalla, *Z. Elektrochem.*, **59**, 184 (1955).
424. M. Batley and L. E. Lyons, *Nature*, **196**, 573 (1962).
425. A. L. Farragher and F. M. Page, *Trans. Faraday Soc.*, **62**, 3072 (1966).
426. N. Yamamoto, S. Akaishi and H. Tsubomura, *Chem. Phys. Letters*, **15**, 458 (1972).
427. T. Kobayashi, K. Yoshihara and S. Nagakura, *Bull. Chem. Soc. Japan*, **44**, 2603 (1971).
428. K. Egawa, N. Nakashima, N. Mataga and Ch. Yamanaka, *Chem. Phys. Letters*, **8**, 108 (1971).
429. K. Egawa, N. Nakashima, N. Mataga and Ch. Yamanaka, *Bull. Chem. Soc. Japan*, **44**, 3287 (1971).
430. H. Masuhara and N. Mataga, *Chem. Phys. Letters*, **6**, 608 (1970).
431. H. Masuhara and N. Mataga, *Bull. Chem. Soc. Japan*, **44**, 43 (1972).
432. H. Masuhara, N. Tsujino and N. Mataga, *Chem. Phys. Letters*, **12**, 481 (1972).
433. N. Tsujino, H. Masuhara and N. Mataga, *Chem. Phys. Letters*, **15**, 360 (1972).
434. N. Tsujino, H. Masuhara and N. Mataga, *Chem. Phys. Letters*, **15**, 357 (1972).
435. R. Potashnik and M. Ottolenghi, *Chem. Phys. Letters*, **6**, 525 (1970).
436. H. Masuhara, M. Shimada and N. Mataga, *Bull. Chem. Soc. Japan*, **43**, 3316 (1970).
437. H. Masuhara, M. Shimada, N. Tsujino and N. Mataga, *Bull. Chem. Soc. Japan*, **44**, 3310 (1971).
438. R. Potashnik, C. R. Goldschmidt and M. Ottolenghi, *J. Phys. Chem.*, **73**, 3170 (1969).
439. D. F. Ilten and M. Calvin, *J. Chem. Phys.*, **42**, 3760 (1965).
440. F. E. Stewart, M. Eisner and W. R. Carper, *J. Chem. Phys.*, **44**, 2866 (1966).
441. M. Shimada, H. Masuhara and N. Mataga, *Chem. Phys. Letters*, **15**, 364 (1972).
442. Y. Achiba, S. Katsumata and K. Kimura, *Chem. Phys. Letters*, **13**, 213 (1972).
443. S. Matsumoto, S. Nagakura, S. Iwata and J. Nakamura, *Chem. Phys. Letters*, **13**, 463 (1972).
444. P. Krebs, E. Sackmann and J. Schwarz, *Chem. Phys. Letters*, **8**, 417 (1971).

445. Y. Torihashi, Y. Furatani, K. Yagii, N. Mataga and A. Sawaoko, *Bull. Chem. Soc. Japan*, **44**, 2985 (1971).
446. A. Lablanche-Combier, *Bull. Soc. Chim. France*, No. 12, 4791 (1972).
447. M. A. F. Tavares, *Trans. Faraday Soc.*, **66**, 2431 (1970).
448. A. E. W. Knight and B. K. Selinger, *Chem. Phys. Letters*, **10**, 43 (1971).
449. Y. Taniguchi, Y. Nishina and N. Mataga, *Bull. Chem. Soc. Japan*, **45**, 764 (1972).
450. Y. Taniguchi and N. Mataga, *Chem. Phys. Letters*, **13**, 596 (1972).
451. C. R. Goldschmidt and M. Ottolenghi, *Chem. Phys. Letters*, **4**, 570 (1970).
452. R. Potashnik, C. R. Goldschmidt, M. Ottolenghi and A. Weller, *J. Chem. Phys.*, **55**, 5344 (1971).
453. H. Masuhara and N. Mataga, *Chem. Phys. Letters*, **7**, 417 (1970).
454. C. R. Goldschmidt, Y. T. Tomkiewicz and I. B. Berlman, *Chem. Phys. Letters*, **2**, 520 (1968).
455. E. A. Chandross and H. T. Thomas, *Chem. Phys. Letters*, **9**, 393 (1971).
456. T. Okada, T. Fujita, M. Kubota, S. Masaki, N. Mataga, R. Ide, Y. Sakata and S. Misumi, *Chem. Phys. Letters*, **14**, 563 (1972).
457. H. Knibbe, K. Röllig, F. P. Schafer and A. Weller, *J. Chem. Phys.*, **47**, 1184 (1967).
458. N. Mataga, K. Ezumi and T. Okada, *Molec. Phys.*, **10**, 201 (1966).
459. N. Mataga, T. Okada and N. Yamamoto, *Bull. Chem. Soc. Japan*, **39**, 2562 (1966).
460. A. Weller and K. Zachariasse, *J. Chem. Phys.*, **46**, 4984 (1967).
461. A. Weller and K. Zachariasse, *Molecular Luminescence*, p. 895 (Ed. E. C. Lim), W. A. Benjamin, Inc., New York, 1969.
462. A. Weller and K. Zachariasse, *Chem. Phys. Letters*, **10**, 197 (1971).
463. A. Weller and K. Zachariasse, *Chem. Phys. Letters*, **10**, 424 (1971).
464. A. Weller and K. Zachariasse, *Chem. Phys. Letters*, **10**, 590 (1971).
465. K. A. Zachariasse, Doctoral thesis, University of Amsterdam (1972).
466. G. Kallmann-Oster, *Acta Phys. Polon.*, **26**, 435 (1964).
467. G. Kallmann-Oster and H. Kallmann, *International Symposium on Luminescence. The Physics and Chemistry of Scintillators*, p. 31 (Eds. N. Riehl and H. Kallmann), Verlag Karl Thiemig, Munich, 1966.
468. M. S. S. C. Leite and K. Razi Naqvi, *Chem. Phys. Letters*, **4**, 35 (1969).
469. C. Lewis and W. R. Ware, *Chem. Phys. Letters*, **15**, 290 (1972).
470. W. R. Ware and C. Lewis, *J. Chem. Phys.*, **57**, 3546 (1972).
471. W. R. Ware and T. L. Nemzek, private communication (1974).
472. M. Mitchell and H. J. V. Tyrell, *J. Chem. Soc. (Faraday Trans. II)*, **68**, 385 (1972).
473. W. R. Ware, L. J. Doemeny and T. L. Nemzek, *J. Phys. Chem.*, **77**, 2038 (1973).
474. E. J. Bowen and W. S. Metcalf, *Proc. Roy. Soc.*, **A 206**, 937 (1951).
475. W. R. Ware and J. S. Novros, *J. Phys. Chem.*, **70**, 3246 (1966).
476. T. Okada, H. Oohari and N. Mataga, *Bull. Chem. Soc. Japan*, **43**, 2750 (1970).
477. H. Leonhardt and A. Weller, *Z. Elektrochem.*, **67**, 791 (1963).
478. N. Mataga, *Bull. Chem. Soc. Japan*, **43**, 3623 (1970).
479. R. Beer, K. M. C. Davis and R. Hodgson, *Chem. Comm.*, **1970**, 840 (1970).
480. C. A. G. Brooks and K. M. C. Davis, *J. Chem. Soc. (Perkin Trans. II)*, **1972**, 1649 (1972).
481. D. R. G. Brimage and R. S. Davidson, *J. Photochem.*, **1**, 79 (1972).
482. E. Vander Donckt and J. P. van Bellinghen, *Chem. Phys. Letters*, **7**, 630 (1970).
483. R. P. Wayne, *Adv. Photochem.*, **7**, 311 (1969).

484. *Singlet Molecular Oxygen*, Bull. No. 38, Institute of Molecular Biophysics, Florida State University, 1970.
485. R. Potashnik, C. R. Goldschmidt and M. Ottolenghi, *Chem. Phys. Letters*, **9**, 424 (1971).
486. L. K. Patterson, G. Porter and M. Topp, *Chem. Phys. Letters*, **7**, 612 (1970).
487. C. S. Parmenter and J. D. Rau, *J. Chem. Phys.*, **51**, 2242 (1967).
488. A. Adamczyk and F. Wilkinson, *Organic Scintillators and Liquid Scintillation Counting*, p. 223, (Eds. D. L. Horrocks and C.-T. Peng), Academic Press, New York, 1971.
489. P. B. Merkel and D. R. Kearns, *J. Chem. Phys.*, **58**, 398 (1973).
490. D. R. Kearns and A. J. Stone, *J. Chem. Phys.*, **55**, 3383 (1971).
491. M. Tinkham and M. W. P. Strandberg, *Phys. Rev.*, **97**, 951 (1955).
492. C. A. Hutchison, Jr. and B. W. Mangam, *J. Chem. Phys.*, **34**, 908 (1961).
493. A. U. Khan and D. R. Kearns, *J. Chem. Phys.*, **48**, 3272 (1968).
494. N. E. Geacintov and C. E. Swenberg, *J. Chem. Phys.*, **57**, 378 (1972).
495. S. Siegel and H. S. Judeikis, *J. Chem. Phys.*, **48**, 1613 (1968).
496. P. F. Jones and S. Siegel, *J. Chem. Phys.*, **54**, 3360 (1971).
497. N. E. Geacintov, R. Benson and S. B. Pomeranz, *Chem. Phys. Letters*, **17**, 284 (1972).
498. S. Kusuhara and R. Hardwick, *J. Chem. Phys.*, **41**, 2386 (1964).
499. P. Froehlich and H. Morrison, *J. Chem. Soc. Chem. Comm.*, **1972**, 184 (1972).
500. J. B. Birks and W. A. Little, *Proc. Phys. Soc.*, **A 66**, 921 (1953).
501. J. P. Pinion, F. L. Minn and N. Filipescu, *J. Luminescence*, **3**, 245 (1971).
502. P. R. Nott and B. K. Selinger, *J. Luminescence*, **5**, 138 (1972).
503. J. Ferguson and A. W. H. Mau, *Chem. Phys. Letters*, **17**, 543 (1972).
504. V. L. Levshin, *Zh. Eksperim. i. Teor. Fiz.*, **28**, 213 (1955).
505. N. Mataga, *Bull. Chem. Soc. Japan*, **30**, 375 (1957).
506. V. I. Permogorov, L. A. Serdynkova and M. D. Frank-Kamenetskii, *Opt. Spectrosc.*, **25**, 38 (1968).
507. M. Hida and T. Sanuki, *Bull. Chem. Soc. Japan*, **43**, 2291 (1970).
508. J. Ferguson, A. W. H. Mau and J. M. Morris, *Aust. J. Chem.*, **26**, 91 (1973).
509. J. Ferguson, A. W. H. Mau and J. M. Morris, *Aust. J. Chem.*, **26**, 103 (1973).
510. J. B. Aladekomo, *J. Luminescence*, **6**, 83 (1973).
511. H. Masuhara and N. Mataga, *Z. Phys. Chem. N.F.*, **80**, 113 (1972).
512. H. Masuhara, N. Tsujino and H. Mataga, *Bull. Chem. Soc. Japan*, **46**, 1088 (1973).
513. R. J. McDonald and B. K. Selinger, *Aust. J. Chem.*, **24**, 1797 (1971).
514. R. J. McDonald and B. K. Selinger, *Aust. J. Chem.*, **25**, 897 (1972).
515. R. J. McDonald and B. K. Selinger, *Molec. Photochem.*, **3**, 99 (1971).
516. R. Ide, Y. Sakata, S. Misumi, T. Okada and N. Mataga, *J. Chem. Soc. Chem. Comm.*, **1972**, 1009 (1972).
517. D. R. G. Brimage and R. S. Davidson, *Chem. Comm.*, 1385 (1971).
518. G. S. Beddard, R. S. Davidson and A. Lewis, *J. Photochem.*, **1**, 491 (1972/3).
519. T. Azumi, *Chem. Phys. Letters*, **19**, 580 (1973).
520. J. B. Birks, *J. Phys. B.* (*At. Molec. Phys.*), **3**, 1704 (1970).
521. W. Helfrich and F. R. Lipsett, *J. Chem. Phys.*, **43**, 4368 (1965).
522. P. E. Fielding and R. C. Jarnagin, *J. Chem. Phys.*, **47**, 247 (1967).
523. I. H. Munro, L. M. Logan, F. D. Blair, F. R. Lipsett and D. F. Williams, *Molec. Cryst. Liq. Cryst.*, **15**, 297 (1972).
524. E. Glockner and H. C. Wolf, *Z. Naturforsch.*, **24a**, 943 (1969).

525. L. E. Lyons and L. J. Warren, *Aust. J. Chem.*, **25**, 1411 (1973).
526. I. Nakada, *J. Phys. Soc. Japan*, **20**, 346 (1965).
527. F. Urbach, *Phys. Rev.*, **92**, 1324 (1953).
528. L. E. Lyons and L. J. Warren, *Aust. J. Chem.*, **25**, 1427 (1973).
529. G. C. Morris, S. A. Rice and A. E. Martin, *J. Chem. Phys.*, **52**, 5149 (1970).
530. J. W. Sidman, *Phys. Rev.*, **102**, 96 (1956).
531. Y. Lupien and D. F. Williams, *Molec. Cryst.*, **5**, 1 (1968).
532. Y. Lupien, J. O. Williams and D. F. Williams, *Molec. Cryst. Liq. Cryst.*, **18**, 129 (1972).
533. D. F. Williams, private communication (1973).
534. D. C. Northrop and O. Simpson, *Proc. Roy. Soc.*, **A 234**, 136 (1956).
535. R. M. Hochstrasser, *J. Chem. Phys.*, **36**, 1099 (1962).
536. K. Kawaoka and D. R. Kearns, *J. Chem. Phys.*, **41**, 2095 (1964).
537. Y. Tomkiewicz and E. Loewenthal, *Molec. Cryst. Liq. Cryst.*, **6**, 211 (1969).
538. W. Klöpffer and H. Bauser, *Chem. Phys. Letters*, **6**, 279 (1970).
539. N. Y. C. Chu, K. Kawaoka and D. R. Kearns, *J. Chem. Phys.*, **55**, 3059 (1971).
540. M. Tomura and Y. Takahashi, *J. Phys. Soc. Japan*, **31**, 797 (1971).
541. W. Klöpffer, H. Bauser, F. Dolezalek and G. Naundorf, *Molec. Cryst. Liq. Cryst.*, **16**, 229 (1972).
542. M. Yokota and O. Tanimoto, *J. Phys. Soc. Japan*, **22**, 779 (1967).
543. A. Bergman, M. Levine and J. Jortner, *Phys. Rev. Letters*, **18**, 593 (1967).
544. A. Inoue, K. Yoshihara and S. Nagakura, *Bull. Chem. Soc. Japan*, **45**, 1973 (1972).
545. J. B. Birks, *Molec. Cryst. Liq. Cryst.*, in press.
546. K. W. Benz and H. C. Wolf, *Z. Naturforsch.*, **19a**, 177 (1964).
547. R. C. Powell, *Phys. Rev.*, **B 2**, 2090 (1970.
548. P. Avakian and H. C. Wolf, *Z. Phys.*, **165**, 439 (1961).
549. L. M. Logan, I. H. Munro, D. F. Williams and F. R. Lipsett, *Molecular Luminescence*, p. 773 (Ed. E. C. Lim), W. A. Benjamin, Inc., New York, 1969.
550. R. C. Powell and R. G. Kepler, *Phys. Rev. Letters*, **22**, 636 (1969).
551. R. C. Powell and R. G. Kepler, *Phys. Rev. Letters*, **22**, 1232 (1969).
552. R. C. Powell and R. G. Kepler, *J. Luminescence*, **1**, 254 (1969).
553. R. C. Powell and R. G. Kepler, *Molec. Cryst. Liq. Cryst.*, **11**, 349 (1970).
554. R. C. Powell, *Phys. Rev.*, **B 2**, 1159 (1970).
555. R. C. Powell, *Phys. Rev.*, **B 2**, 1207 (1970).
556. R. C. Powell, *Phys. Rev.*, **B 4**, 628 (1971).
557. R. C. Powell, *J. Chem. Phys.*, **58**, 920 (1973).
558. R. C. Powell and Z. G. Soos, *Phys. Rev.*, **B 5**, 1547 (1972).
559. Z. G. Soos and R. C. Powell, *Phys. Rev.*, **B 6**, 4035 (1972).
560. C. E. Swenberg and W. T. Stacy, *Phys. Stat. Sol.*, **36**, 717 (1969).
561. H. B. Rosenstock, *Phys. Rev.*, **187**, 1166 (1969).
562. R. M. Pearlstein, R. P. Hemenger and K. Lakatos-Lindenberg, *Phys. Rev. Letters*, **27**, 1509 (1971).
563. A. Hammer and H. C. Wolf, *Molec. Cryst.*, **4**, 191 (1968).
564. A. R. McGhie, A. M. Voshchenkov, P. J. Reucroft and M. M. Labes, *J. Chem. Phys.*, **48**, 186 (1968).
565. R. P. Hemenger, R. M. Pearlstein and K. Lakatos-Lindenberg, *J. Math. Phys.*, **13**, 1056 (1972).
566. R. G. Selsby and C. E. Swenberg, *Phys. Stat. Sol.*, **50**, 235 (1972).
567. G. Vaubel, *Chem. Phys. Letters*, **9**, 51 (1971).

568. J. B. Birks, H. Y. Najjar and M. D. Lumb, *J. Phys. B.* (*Atom. Molec. Phys.*), **4**, 1516 (1971).
569. R. Voltz, G. Laustriat and A. Coche, *C.R. Acad. Sci. Paris*, **257**, 1473 (1963).
570. R. Voltz, G. Laustriat and A. Coche, *J. Chim. Phys.*, **63**, 1253 (1966).
571. J. Klein, R. Voltz and G. Laustriat, *J. Chim. Phys.*, **67**, 704 (1970).
572. P. K. Ludwig and C. D. Amata, *J. Chem. Phys.*, **49**, 326; **49**, 333 (1968).
573. J. Yguerabide, *J. Chem. Phys.*, **49**, 1018; **49**, 1026 (1968).
574. J. B. Birks and J. C. Conte, *Proc. Roy. Soc.*, **A 303**, 85 (1968).
575. J. B. Birks, *The Current Status of Liquid Scintillation Counting*, p. 1 (Ed. E. D. Bransome, Jr.), Grune and Statton, New York, 1970.
576. J. B. Birks, *Organic Scintillators and Liquid Scintillation Counting*, p. 3 (Eds. D. L. Horrocks and C.-T. Peng), Academic Press, New York, 1971.
577. K. Graupner and E. R. S. Winter, *J. Chem. Soc.*, 1145 (1952).
578. G. M. Panchenkov, V. V. Erchenkov and N. N. Borisenko, *Sovrem. Probl. Fiz. Khim.*, **14**, 316 (1970).
579. F. Hirayama and S. Lipsky, *J. Chem. Phys.*, **51**, 3616 (1969).
580. F. Hirayama, W. Rothman and S. Lipsky, *Chem. Phys. Letters*, **5**, 296 (1970).
581. W. Rothman, F. Hirayama and S. Lipsky, *J. Chem. Phys.*, **58**, 1300 (1973).
582. M. S. Henry and W. P. Helman, *J. Chem. Phys.*, **56**, 5734 (1972).
583. W. P. Helman, *Chem. Phys. Letters*, **17**, 306 (1973).
584. W. R. Ware and R. L. Lyke, *Chem. Phys. Letters*, **24**, 195 (1974).
585. F. Hirayama and S. Lipsky, *Organic Scintillators and Liquid Scintillation Counting*, p. 205 (Eds. D. L. Horrocks and C.-L. Peng), Academic Press, New York, 1971.
586. G. Beck and J. K. Thomas, *Chem. Phys. Letters*, **16**, 318 (1972).
587. J. H. Baxendale and J. Mayer, *Chem. Phys. Letters*, **17**, 458 (1972).
588. D. Paligorič and J. Klein, *Intern. J. Radn. Phys. Chem.*, **4**, 359 (1972).
589. J. H. Baxendale, D. Beaumond and M. A. J. Rodgers, *Chem. Phys. Letters*, **4**, 3 (1969).
590. F. Hirayama, C. W. Lawson and S. Lipsky, *J. Phys. Chem.*, **74**, 2411 (1970).
591. A. M. Halpern and W. R. Ware, *J. Phys. Chem.*, **74**, 2413 (1970).
592. D. R. Salahub and C. Sandorfy, *Theor. Chem. Acta.* **20**, 227 (1971).
593. A. Brillante, C. Tarliani and C. Zauli, *Molec. Phys.*, **25**, 1263 (1973).
594. A. Nakajima, *Chem. Phys. Letters*, **21**, 200 (1973).
595. E. Lippert, W. Lüder and H. Boos, *Advances in Molecular Spectroscopy*, p. 443, Pergamon Press, Oxford, 1962.
596. O. S. Khalil, R. H. Hofeldt and S. P. McGlynn, *Chem. Phys. Letters*, **17**, 479 (1972).
597. K. Rotkiewicz, K. H. Grellmann and Z. R. Grabowski, *Chem. Phys. Letters*, **19**, 315 (1973).
598. S. D. Babenko, V. A. Benderskii, V. I. Gol'danskii, A. G. Lavrushko and V. P. Tychinskii, *Chem. Phys. Letters*, **8**, 598 (1971).
599. J. Prochorow, *Chem. Phys. Letters*, **19**, 596 (1973).
600. L. G. Christophorou, private communication (1973).
601. L. Peter and G. Vaubel, *Chem. Phys. Letters*, **21**, 158 (1973).
602. S. Lipsky, private communication (1973).
603. C. Lewis, W. R. Ware, L. J. Doemeny and T. L. Nemzek, *Rev. Sci. Instr.*, **44**, 107 (1973).
604. M. Tomura, Y. Takahashi, A. Matsui and Y. Ishii, *J. Phys. Soc. Japan*, **25**, 647 (1968).

605. I. B. Berlman, *Chem. Phys. Letters*, **21**, 344 (1973).
606. D. A. Bird and J. H. Callomon, *Molec. Phys.*, **26**, 1317 (1973).
607. A. Davidson and B. Norden, *Tetrahedron Letters*, 3093 (1972).
608. S. F. Mason and R. D. Peacock, *Chem. Phys. Letters*, **21**, 406 (1973).
609. E. E. Koch, A. Otto and K. Radler, *Chem. Phys. Letters*, **21**, 501 (1973).
610. E. E. Koch and A. Otto, *Phys. Stat. Sol.*, **51**b, 69 (1972).
611. H. B. Klevens and J. R. Platt, *J. Chem. Phys.*, **17**, 470 (1949).
612. B. S. Hudson and B. E. Kohler, *Chem. Phys. Letters*, **23**, 139 (1973).
613. P. A. M. van den Bogaardt, R. P. H. Rettschnick and J. D. W. van Voorst, *Chem. Phys. Letters*, **18**, 351 (1973).
614. G. J. Hoytink, *Chem. Phys. Letters*, **22**, 10 (1973).
615. U. Laor, J. C. Hsieh and P. K. Ludwig, *Chem. Phys. Letters*, **22**, 150 (1973).
616. E. Pantos, A. M. Taleb, T. D. S. Hamilton and I. H. Munro, in press.
617. C. E. Easterly and L. G. Christophorou, *J. Chem. Soc. Faraday Trans. II*, **70**, 267 (1974).
618. R. P. Blaunstein and K. S. Grant, *Photochem. Photobiol.*, **18**, 347 (1973).

Author Index

Subject Index